W9-CUZ-566

ELECTRIC MACHINES
Theory, Operation, Applications, Adjustment, and Control

Second Edition

Charles I. Hubert

Professor of Electrical Engineering
United States Merchant Marine Academy

Upper Saddle River, New Jersey
Columbus, Ohio

Library of Congress Cataloging-in-Publication Data

Hubert, Charles I.
Electric machines: theory, operation, applications, adjustment, and control/Charles I. Hubert.—2nd ed.
p. cm.
Includes bibliographical references and index.
ISBN 0-13-061210-3
1. Electric machinery. I. Title.

TK2182 .H83 2002
621.31'042—dc21

2001036240

Editor in Chief: Stephen Helba
Assistant Vice President and Publisher: Charles E. Stewart, Jr.
Production Editor: Alexandrina Benedicto Wolf
Production Coordination: Clarinda Publication Services
Design Coordinator: Diane Ernsberger
Cover Designer: Linda Sorrells-Smith
Production Manager: Matthew Ottenweller

This book was set in Times and Univers by The Clarinda Company. It was printed and bound by R. R. Donnelley & Sons Company. The cover was printed by The Lehigh Press, Inc.

Pearson Education Ltd., *London*
Pearson Education Australia Pty. Limited, *Sydney*
Pearson Education Singapore, Pte. Ltd.
Pearson Education North Asia Ltd., *Hong Kong*
Pearson Education Canada, Ltd., *Toronto*
Pearson Educación de Mexico, S.A. de C.V.
Pearson Education—Japan, *Tokyo*
Pearson Education Malaysia, Pte. Ltd.
Pearson Education, *Upper Saddle River, New Jersey*

10 9 8 7 6 5 4 3 2 1
ISBN 0-13-061210-3

Dedicated to my lovely wife Josephine for her understanding and encouragement.

Preface

This second edition retains the easy-to-understand student-oriented approach that was the hallmark of the first edition. Additional steps were added to some derivations, and some example problems were expanded for even greater clarity and ease of understanding. New example problems and new homework problems were also added to further enhance student learning. Also added is a section on the high-efficiency NEMA design E AC motors.

The text is designed to be used for a one- or two-semester course in electrical machinery. The minimum prerequisite for effective use of the text is a circuits course and a working knowledge of complex algebra and phasor diagrams. A review of current, voltage, and power relationships in the three-phase system, including applications of complex power, is provided in Appendix A for students who need extra help.

The text is unique in that it responds to the needs of faculty in many colleges who have expressed the desire that more attention be given to current industrial requirements. The most frequent request was for a text that allows faculty to devote more time to motors than to generators, and more time to machine characteristics than to different types of armature windings. To accomplish this, motors are presented before generators, and just enough material on armature windings is provided to acquaint the students with basic armature construction and associated technical terms. NEMA standards and tables are introduced in the solution of application-type problems similar to those found on professional engineering license examinations.

To make more efficient use of student time, transformers and AC machines are presented before DC machines; this sequence was developed and used at the United States Merchant Marine Academy for more than 30 years, where a one-quarter course in AC machines was followed by a one-quarter course in DC machines. Teaching transformers and AC machines when knowledge of AC circuits is still fresh is simple and straightforward. The application of phasor diagrams and complex algebra to equivalent series and equivalent series–parallel circuits of AC machines and transformers provides immediate reinforcement of AC circuit theory.

Alternating current machines and transformers are the building blocks of most present-day power and industrial systems and, as such, require greater emphasis than do DC machines. Furthermore, because conventional DC machines are in effect AC machines whose commutators provide the necessary AC/DC and DC/AC conversion, some additional efficiency may be obtained by presenting AC machines before DC machines.

Presenting motors before generators in both AC and DC machines and presenting DC machines as a stand-alone block of three chapters provide significant freedom in course development. Some easy choices include a one-semester course in only AC machines; a one-semester course in both AC and DC machines (de-emphasizing generators); and a two-semester course that includes all topics on motors and generators for both AC and DC machines. Suggested course outlines are included in the Instructor's Manual.

ORGANIZATION OF THE TEXT

The first chapter provides the basic background common to all machines and transformers. It includes such topics as the development of mechanical force by the interaction of magnetic fields, electromagnetically induced voltages, space angles, electrical degrees, magnetic circuits, and magnetization curves.

The substance of the machinery course begins with transformers in Chapters 2 and 3. Transformers are relatively easy to visualize, and tie in nicely with the ideal transformer covered in a prerequisite circuits course.

The study of induction machines in Chapters 4 and 5 follows naturally from transformers, where the stator is the primary and the rotor is the secondary. Furthermore, introducing induction machines immediately after transformers permits the newly developed equivalent-circuit model and associated phasor diagrams of the transformer to be easily applied to induction-motor theory, illustrating the common relationship they share. Where feasible, approximations are made that allow simplified and practical calculations.

Single-phase induction motors, discussed in Chapter 6, are a natural continuation of three-phase induction motors. Included are capacitor motors, and resistance split-phase motors. Special-purpose motors, such as shaded-pole motors, reluctance motors, hysteresis motors, stepper motors, universal motors, and linear-induction motors are covered in Chapter 7.

Synchronous motors are developed in Chapter 8, and tie in nicely with the rotating field theory of induction motors. The transition from synchronous motor action to synchronous generator operation is presented in Chapter 9. Changes in power angle, as the shaft load is gradually removed and a driving torque applied, are shown on a common phasor diagram. Also included is the parallel operation of synchronous machines, division of load, and power factor correction.

The material on DC machines is designed as a stand-alone block of three chapters (Chapters 10, 11, and 12), so that if desirable it may be taught effectively prior to AC machines and transformers. Thus, courses with special objectives, curriculum requirements, or laboratory constraints that require the early introduction of DC machines may be easily accommodated. Faculty teaching one-quarter or one-semester courses that emphasize AC machines, but still include a very brief introduction to DC machines, will find Chapter 10 (Principles of Direct-Current Machines) more than adequate for the purpose.

Chapter 13 provides a brief introduction to electronic and magnetic control of motors. Typical examples of reversing, speed control, braking, and ladder-type circuits are included. Programmable logic controllers (PLCs) are touched on briefly to provide an insight into this expanding field.

Common Core: The text provides a common core of minimum essentials, supplemented with optional material selected (by the instructor) from a wide range of topics in supplemental chapters. The common core, outlined in the Instructor's Manual, assures a basic understanding of electrical machines, while preparing the student for this millennium. This is accomplished by devoting more time to AC and special-purpose machines than to DC machines, devoting more time to motors than to generators, devoting more time to machine characteristics than to armature windings, and making extensive use of NEMA standards and tables in discussions, examples, and problems.

The common core requires approximately 27 periods and is recommended for all electrical machinery courses regardless of length (one quarter, one semester, two quarters, or two semesters). The limited time available in one-quarter machines courses (approximately 30 periods) will, in most cases, limit course content to the common core. However, if magnetic circuits and transformers are covered in previous courses, these common-core topics should be replaced with optional topics selected to meet regional industrial requirements.

One-semester, two-semester, and two-quarter courses provide ample opportunity for more extensive use of optional topics, enabling the instructor to tailor the course to meet specific objectives. A listing of optional topics available in supplemental chapters is given in the Instructor's Manual, along with a universal one-semester outline that is easily adaptable to different course requirements.

Boldface Letters in Equations: Boldface letters in equations and circuit diagrams are used throughout the text to designate the following as *complex numbers:* current phasors **I,** voltage phasors **V** and **E,** admittance **Y,** impedance **Z,** and phasor power (complex power) **S.** The corresponding magnitudes are printed as |**I**|, |**V**|, |**E**|, |**Y**|, |**Z**|, |**S**|, or *I, V, E, Y, Z, S.*

Boldface Numbers in Rectangular Brackets: Boldface numbers in rectangular brackets, e.g., **[5],** direct the student to specific end-of-chapter references. This encourages further investigation; students will not have to search a collection of general references for additional information on a specific topic.

Significant Figures: If the answer to one part of a problem is required data for the solution of another part, the unrounded answer is used to minimize continuing errors. Thus, where appropriate, the answers to multipart problems are given in both unrounded form and rounded form. For example, if the answer to part (a) of a problem calls for three significant figures, the text may show it as 127.1648 ⇒ $\underline{127}$. Although $\underline{127}$ is the answer, 127.1648 is substituted in parts (b), (c), etc., as appropriate.

Summary of Equations: A summary of equations at the end of each chapter helps guide the students in solving chapter problems, and is a handy reference for the electrical power portions of professional engineering exams. Furthermore, since the equations are keyed to the text, it is easy for the reader to find the associated application and derivation.

Problem Numbers: Problem numbers are keyed to chapter sections by a triple-number system. For example, Problem 5–9/12 indicates that Chapter 5, Problem 9, requires Section 5–12. This makes it easier for faculty to assign homework problems, and easier for a student to pick additional problems for review. Problems recommended for computer solution (using commercially available software) are indicated with an asterisk.

ACKNOWLEDGMENTS

The author takes this opportunity to acknowledge, with gratitude, the following reviewers, whose valued suggestions and constructive criticism helped shape the text during its formative stages.

Robert L. Anderson, Purdue University, Calumet, Indiana
Charles L. Bachman, Southern College of Technology, Georgia
Thomas J. Bingham, St. Louis Community College, Missouri
Luces M. Faulkenberry, University of Houston, Texas
Ahmet Fer, Purdue University, Indianapolis, Indiana
Brendan Gallagher, Middlesex County College, New Jersey
James L. Hales, University of Pittsburgh, Johnstown, Pennsylvania
Warren R. Hill, University of Southern Colorado
Gerald Jensen, Western Iowa, Iowa
Joseph Pawelczyk, Erie Community College, New York
John Stratton, Rochester Institute of Technology, New York
Conrad Youngren, SUNY-Maritime, New York
Kurdet Yurtseven, Penn State University, Harrisburg, Pennsylvania

A special thanks to Dr. George J. Billy (Chief Librarian), Mr. Donald Gill, Ms. Marilyn Stern, Ms. Laura Cody, and Ms. Barbara Adesso, of the United States Merchant Marine Academy Library, for their kind assistance in obtaining needed references.

An affectionate thanks to my wonderful wife Josephine for her encouragement, faith, counsel, patience, and companionship during the many years of preparing this and other manuscripts. Her early reviews of the manuscript, while in its formative stages, assisted in clarity of expression and avoidance of ambiguity. Her apparently endless years of pounding typewriters and word processors, for this and other texts, were truly a work of love.

Charles I. Hubert

Contents

1 MAGNETICS, ELECTROMAGNETIC FORCES, GENERATED VOLTAGE, AND ENERGY CONVERSION 1

1.1 Introduction 1
1.2 Magnetic Field 1
1.3 Magnetic Circuit Defined 2
1.4 Reluctance and the Magnetic Circuit Equation 4
1.5 Relative Permeability and Magnetization Curves 5
1.6 Analogies Between Electric and Magnetic Circuits 12
1.7 Magnetic Hysteresis and Hysteresis Loss 15
1.8 Interaction of Magnetic Fields (Motor Action) 17
1.9 Elementary Two-Pole Motor 18
1.10 Magnitude of the Mechanical Force Exerted on a Current-Carrying Conductor Situated in a Magnetic Field (BLI Rule) 19
1.11 Electromechanically Induced Voltages (Generator Action) 21
1.12 Elementary Two-Pole Generator 25
1.13 Energy Conversion in Rotating Electrical Machines 27
1.14 Eddy Currents and Eddy-Current Losses 28
1.15 Multipolar Machines, Frequency, and Electrical Degrees 29
Summary of Equations for Problem Solving 32
Specific References Keyed to Text 33
Review Questions 33
Problems 34

2 TRANSFORMER PRINCIPLES 37

2.1 Introduction 37
2.2 Construction of Power and Distribution Transformers 37
2.3 Principle of Transformer Action 40
2.4 Transformers with Sinusoidal Voltages 41
2.5 No-Load Conditions 43

2.6 Transient Behavior When Loading and Unloading 46
2.7 Effect of Leakage Flux on the Output Voltage of a Real Transformer 48
2.8 Ideal Transformer 49
2.9 Leakage Reactance and the Equivalent Circuit of a Real Transformer 51
2.10 Equivalent Impedance of a Transformer 55
2.11 Voltage Regulation 62
2.12 Per-Unit Impedance and Percent Impedance of Transformer Windings 64
2.13 Transformer Losses and Efficiency 71
2.14 Determination of Transformer Parameters 75
Summary of Equations for Problem Solving 79
Specific References Keyed to Text 81
General References 82
Review Questions 82
Problems 83

3 TRANSFORMER CONNECTIONS, OPERATION, AND SPECIALTY TRANSFORMERS 91

3.1 Introduction 91
3.2 Transformer Polarity and Standard Terminal Markings 92
3.3 Transformer Nameplates 94
3.4 Autotransformers 95
3.5 Buck-Boost Transformers 101
3.6 Parallel Operation of Transformers 104
3.7 Load Division Between Transformers in Parallel 106
3.8 Transformer In-Rush Current 109
3.9 Harmonics in Transformer Exciting Current 110
3.10 Three-Phase Connections of Single-Phase Transformers 113
3.11 Three-Phase Transformers 118
3.12 Beware the 30° Phase Shift When Paralleling Three-Phase Transformer Banks 119
3.13 Harmonic Suppression in Three-Phase Connections 121
3.14 Instrument Transformers 125
Summary of Equations for Problem Solving 126
Specific References Keyed to Text 127

General References 127
Review Questions 128
Problems 128

4 PRINCIPLES OF THREE-PHASE INDUCTION MOTORS 133

4.1 Introduction 133
4.2 Induction-Motor Action 133
4.3 Reversal of Rotation 135
4.4 Induction-Motor Construction 136
4.5 Synchronous Speed 137
4.6 Multispeed Fixed-Frequency Pole-Changing Motors 141
4.7 Slip and Its Effect on Rotor Frequency and Voltage 141
4.8 Equivalent Circuit of an Induction-Motor Rotor 143
4.9 Locus of the Rotor Current 146
4.10 Air-Gap Power 148
4.11 Mechanical Power and Developed Torque 150
4.12 Torque-Speed Characteristic 153
4.13 Parasitic Torques 156
4.14 Pull-Up Torque 157
4.15 Losses, Efficiency, and Power Factor 157
Summary of Equations for Problem Solving 161
Specific References Keyed to Text 162
Review Questions 162
Problems 163

5 CLASSIFICATION, PERFORMANCE, APPLICATIONS, AND OPERATION OF THREE-PHASE INDUCTION MACHINES 167

5.1 Introduction 167
5.2 Classification and Performance Characteristics of NEMA-Design Squirrel-Cage Induction Motors 168
5.3 NEMA Tables 170
5.4 Motor Performance as a Function of Machine Parameters, Slip, and Stator Voltage 178
5.5 Shaping the Torque-Speed Characteristic 182
5.6 Some Useful Approximations for Normal-Running and Overload Conditions of Squirrel-Cage Motors 186

5.7 NEMA Constraints on Voltage and Frequency 189
5.8 Effect of Off-Rated Voltage and Off-Rated Frequency on Induction Motor Performance 189
5.9 Wound-Rotor Induction Motor 195
5.10 Normal Running and Overload Conditions for Wound-Rotor Induction Motors 200
5.11 Motor Nameplate Data 202
5.12 Locked-Rotor In-Rush Current 205
5.13 Effect of Number of Starts on Motor Life 208
5.14 Reclosing Out-of-Phase Scenario 209
5.15 Effect of Unbalanced Line Voltages on Induction Motor Performance 209
5.16 Per-Unit Values of Induction-Motor Parameters 212
5.17 Determination of Induction-Motor Parameters 213
5.18 Induction Generators 219
5.19 Dynamic Braking of Induction Motors 227
5.20 Induction-Motor Starting 229
5.21 Motor Branch Circuits 238
Summary of Equations for Problem Solving 239
Specific References Keyed to Text 242
General References 243
Review Questions 243
Problems 245

6 SINGLE-PHASE INDUCTION MOTORS 253

6.1 Introduction 253
6.2 Quadrature Field Theory and Induction-Motor Action 253
6.3 Induction-Motor Action Through Phase Splitting 256
6.4 Locked-Rotor Torque 256
6.5 Practical Resistance-Start Split-Phase Motors 260
6.6 Capacitor-Start Split-Phase Motors 262
6.7 Reversing Single-Phase Induction Motors 269
6.8 Shaded-Pole Motors 269
6.9 NEMA-Standard Ratings for Single-Phase Induction Motors 270
6.10 Operation of Three-Phase Motors From Single-Phase Lines 270
6.11 Single Phasing (A Fault Condition) 272
Summary of Equations for Problem Solving 274

Specific References Keyed to Text 275
Review Questions 275
Problems 276

7 SPECIALTY MACHINES 279

7.1 Introduction 279
7.2 Reluctance Motors 279
7.3 Hysteresis Motors 282
7.4 Stepper Motors 286
7.5 Variable-Reluctance Stepper Motors 287
7.6 Permanent-Magnet Stepper Motors 291
7.7 Stepper-Motor Drive Circuits 292
7.8 Linear Induction Motor 295
7.9 Universal Motor 299
Summary of Equations for Problem Solving 301
Specific References Keyed to Text 302
Review Questions 303
Problems 303

8 SYNCHRONOUS MOTORS 305

8.1 Introduction 305
8.2 Construction 305
8.3 Synchronous Motor Starting 309
8.4 Shaft Load, Power Angle, and Developed Torque 311
8.5 Counter-EMF and Armature-Reaction Voltage 312
8.6 Equivalent-Circuit Model and Phasor Diagram of a Synchronous-Motor Armature 315
8.7 Synchronous-Motor Power Equation (Magnet Power) 316
8.8 Effect of Changes in Shaft Load on Armature Current, Power Angle, and Power Factor 318
8.9 Effect of Changes in Field Excitation on Synchronous-Motor Performance 320
8.10 V Curves 321
8.11 Synchronous-Motor Losses and Efficiency 323
8.12 Using Synchronous Motors to Improve the System Power Factor 324

8.13 Salient-Pole Motor 326
8.14 Pull-In Torque and Moment of Inertia 329
8.15 Speed Control of Synchronous Motors 330
8.16 Dynamic Braking 331
Summary of Equations for Problem Solving 331
Specific References Keyed to Text 332
General Reference 332
Review Questions 332
Problems 333

9 SYNCHRONOUS GENERATORS (ALTERNATORS) 337
9.1 Introduction 337
9.2 Motor-to-Generator Transition 338
9.3 Synchronous-Generator Power Equation 342
9.4 Generator Loading and Countertorque 344
9.5 Load, Power Factor, and the Prime Mover 344
9.6 Paralleling Synchronous Generators 345
9.7 Prime-Mover Governor Characteristics 350
9.8 Division of Active Power Between Alternators in Parallel 351
9.9 Motoring of Alternators 356
9.10 General Procedure for Safe Shutdown of AC Generators in Parallel With Other Machines 356
9.11 Characteristic Triangle as a Tool for Solving Load Distribution Problems Between Alternators in Parallel 357
9.12 Division of Reactive Power Between Alternators in Parallel 363
9.13 Accidental Loss of Field Excitation 367
9.14 Per-Unit Values of Synchronous Machine Parameters 367
9.15 Voltage Regulation 368
9.16 Determination of Synchronous Machine Parameters 373
9.17 Losses, Efficiency, and Cooling of AC Generators 377
Summary of Equations for Problem Solving 379
Specific References Keyed to Text 381
Review Questions 381
Problems 383

10 PRINCIPLES OF DIRECT-CURRENT MACHINES 389

10.1 Introduction 389
10.2 Flux Distribution and Generated Voltage in an Elementary DC Machine 389
10.3 Commutation 394
10.4 Construction 394
10.5 Layout of a Simple Armature Winding 396
10.6 Brush Position 398
10.7 Basic DC Generator 398
10.8 Voltage Regulation 400
10.9 Generator-to-Motor Transition and Vice Versa 401
10.10 Reversing the Direction of Rotation of a DC motor 403
10.11 Developed Torque 403
10.12 Basic DC Motor 403
10.13 Dynamic Behavior When Loading and Unloading a DC Motor 406
10.14 Speed Regulation 406
10.15 Effect of Armature Inductance on Commutation When a DC Machine Is Supplying a Load 408
10.16 Interpoles 410
10.17 Armature Reaction 412
10.18 Brush Shifting as an Emergency Measure 414
10.19 Compensating Windings 415
10.20 Complete Equivalent Circuit of a Separately Excited Shunt Generator 416
10.21 Complete Equivalent Circuit of a Shunt Motor 418
10.22 General Speed Equation for a DC Motor 419
10.23 Dynamic Behavior During Speed Adjustment 422
10.24 Precautions When Increasing Speed Through Field Weakening 424
10.25 Mechanical Power and Developed Torque 426
10.26 Losses and Efficiency 428
10.27 Starting a DC Motor 431
Summary of Equations for Problem Solving 435
Specific References Keyed to Text 436
General References 436
Review Questions 436
Problems 438

11 DIRECT-CURRENT MOTOR CHARACTERISTICS AND APPLICATIONS 443

11.1 Introduction 443
11.2 Straight Shunt Motors 443
11.3 Compound Motors 443
11.4 Beware the Differential Connection 445
11.5 Reversing the Direction of Rotation of Compound Motors 445
11.6 Series Motor 446
11.7 Effect of Magnetic Saturation on DC Motor Performance 447
11.8 Linear Approximations 455
11.9 Comparison of Steady-State Operating Characteristics of DC Motors 458
11.10 Adjustable-Voltage Drive Systems 459
11.11 Dynamic Braking, Plugging, and Jogging 461
11.12 Standard Terminal Markings and Connections of DC Motors 465
Summary of Equations for Problem Solving 466
Specific References Keyed to Text 467
General References 467
Review Questions 467
Problems 468

12 DIRECT-CURRENT GENERATOR CHARACTERISTICS AND OPERATION 475

12.1 Introduction 475
12.2 Self-Excited Shunt Generators 475
12.3 Effect of Speed on Voltage Buildup of a Self-Excited Generator 479
12.4 Other Factors Affecting Voltage Buildup 482
12.5 Effect of a Short Circuit on the Polarity of a Self-Excited Shunt Generator 482
12.6 Load-Voltage Characteristics of Self-Excited Shunt Generators 485
12.7 Graphical Approximation of the No-Load Voltage 486
12.8 Compound Generators 490
12.9 Series-Field Diverter 493
12.10 Compounding Effect of Speed 495
12.11 Paralleling Direct-Current Generators 495

12.12 Effect of Field-Rheostat Adjustment on the Load-Voltage Characteristics of DC Generators 496
12.13 Division of Oncoming Bus-Load Between DC Generators in Parallel 497
12.14 Characteristic Triangle as a Tool for Solving Load-Distribution Problems Between Paralleled DC Generators 499
12.15 Theory of Load Transfer Between DC Generators in Parallel 502
12.16 Compound Generators in Parallel 504
12.17 Reverse-Current Trip 505
Summary of Equations for Problem Solving 506
Specific References Keyed to Text 506
General Reference 507
Review Questions 507
Problems 508

13 CONTROL OF ELECTRIC MOTORS 513

13.1 Introduction 513
13.2 Controller Components 513
13.3 Motor-Overload Protection 515
13.4 Controller Diagrams 519
13.5 Automatic Shutdown on Power Failure 519
13.6 Reversing Starters for AC Motors 523
13.7 Two-Speed Starters for AC Motors 523
13.8 Reduced-Voltage Starters for AC Motors 524
13.9 Controllers for DC Motors 525
13.10 Definite-Time Starters for DC Motors 526
13.11 Counter-emf Starter for DC Motors 528
13.12 Reversing Starter with Dynamic Braking and Shunt Field Control for DC Motors 528
13.13 Solid-State Controllers 531
13.14 Thyristor Control of Motors 531
13.15 Solid-State Adjustable-Speed Drives 532
13.16 Cycloconverter Drives 534
13.17 Programmable Controllers 535
Specific References Keyed to Text 537
General References 537
Review Questions 538

APPENDIXES

A Balanced Three-Phase System 539

B Three-Phase Stator Windings 561

C Constant-Horsepower, Constant-Torque, and Variable-Torque Induction Motors 573

D Selected Graphic Symbols Used in Controller Diagrams 577

E Full-Load Current in Amperes, Direct-Current Motors 579

F Full-Load Current in Amperes, Single-Phase Alternating-Current Motors 581

G Full-Load Current, Two-Phase Alternating-Current Motors (Four-Wire) 583

H Full-Load Current, Three-Phase Alternating-Current Motors 585

I Representative Transformer Impedances for Single-Phase 60-Hz Transformers 587

J Unit Conversion Factors 589

ANSWERS TO ODD-NUMBERED PROBLEMS 591

INDEX 597

Tables in Chapters

Available Buck-boost Voltage Ratios 103

Typical Induction Losses for Four-Pole Motors 158

Minimum Locked-Rotor Torque, in Percent of Full-Load Torque, of Single Speed 60–50-Hz, Polyphase Squirrel-Cage Continuous-Rated, Medium Motors with Rated Voltage and Frequency Applied for NEMA Designs *A, B, C,* and *D* 171

Minimum Locked-Rotor Torque, in Percent of Full-Load Torque, of Single-Speed, 60–50-Hz, Polyphase, Squirrel-Cage, Continuous-Rated, Medium Motors with Rated Voltage and Frequency Applied for NEMA Design *E* 172

Minimum Breakdown Torque, in Percent of Full-Load Torque, of Single Speed 60–50-Hz, Polyphase, Squirrel-Cage, Continuous-Rated, Medium Motors with Rated Voltage and Frequency Applied for NEMA Designs *A, B,* and *C* 173

Minimum Breakdown Torque, in Percent of Full-Load Torque, of Single-Speed, 60–50-Hz, Polyphase, Squirrel-Cage, Continuous-Rated, Medium Motors with Rated Voltage and Frequency Applied for NEMA Design *E* 174

Minimum Pull-Up Torque, in Percent of Full-Load Torque of Single-Speed, 60–50-Hz, Polyphase, Squirrel-Cage, Continuous-Rated, Medium Motors with Rated Voltage and Frequency Applied for NEMA Designs *A* and *B* 175

Minimum Pull-Up Torque, in Percent of Full-Load Torque of Single-Speed, 60–50-Hz, Polyphase, Squirrel-Cage, Continuous-Rated, Medium Motors with Rated Voltage and Frequency Applied for NEMA Design *C* 176

Minimum Pull-Up Torque, in Percent of Full-Load Torque of Single-Speed, 60–50-Hz, Polyphase, Squirrel-Cage, Continuous-Rated, Medium Motors with Rated Voltage and Frequency Applied for NEMA Design *E* 177

Maximum Allowable Temperature Rise for Medium Single-Phase and Polyphase Induction Motors °C, Based on a Maximum Ambient Temperature of 40°C 205

NEMA Code Letters for Locked-Rotor kVA per Horsepower 205

Division of Blocked-Rotor Reactance for NEMA-Design Motors 216

Allowable Emergency Overspeed of Squirrel-Cage and Wound-Rotor Motors 223

Range of Standard Power Ratings, Single-Phase Motors 270

Synchronous-Motor Torques in Percent of Rated Torque (Minimum Values) 330

1

Magnetics, Electromagnetic Forces, Generated Voltage, and Energy Conversion

1.1 INTRODUCTION

This chapter starts with a brief review of electromagnetism and magnetic circuits, which are normally included in a basic circuits or physics course. This review is followed by a discussion of the development of the mechanical forces that are caused by the interaction of magnetic fields and that form the basis for all motor action. Faraday's law provides the basis from which all magnetically induced voltages are derived. The relationship between applied torque and countertorque is developed and visualized through the application of Lenz's law and the "flux bunching" rule.

1.2 MAGNETIC FIELD

A magnetic field is a condition resulting from electric charges in motion. The magnetic field of a permanent magnet is attributed to the uncompensated spinning of electrons about their own axis within the atomic structure of the material and to the parallel alignment of these electrons with similar uncompensated electron spins in the adjacent atoms. Groups of adjacent atoms with parallel magnetic spins are called domains. The magnetic field surrounding a current-carrying conductor is caused by the movement of electric charges in the form of an electric current.

For convenience in visualization and analysis, magnetic fields are represented on diagrams by closed loops. These loops, called magnetic flux lines, have been assigned a specific direction that is related to the polarity of a magnet, or the direction of current in a coil or a conductor.

The direction of the magnetic field around a current can be determined by the *right-hand rule:* Grasp the conductor with the right hand, with the thumb pointing in the direction of conventional current, and the fingers will curl in the direction of the magnetic field. This can be visualized in Figure 1.1(a).

FIGURE 1.1
Direction of magnetic flux: (a) around a current-carrying conductor; (b) in a coil; (c) about a magnet.

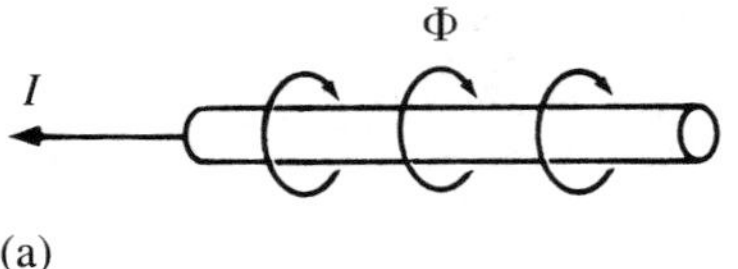

(a)

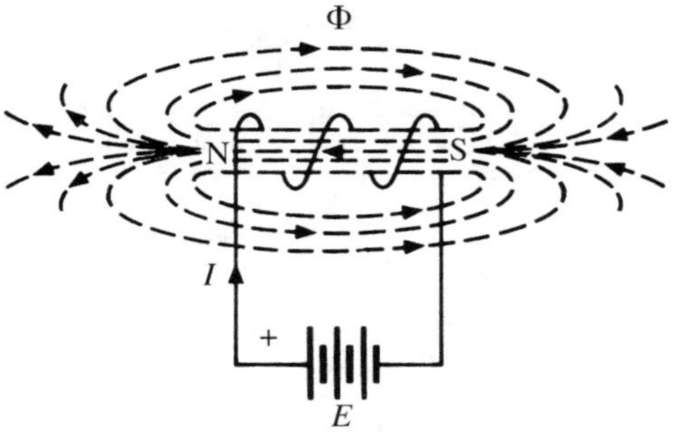

(b)

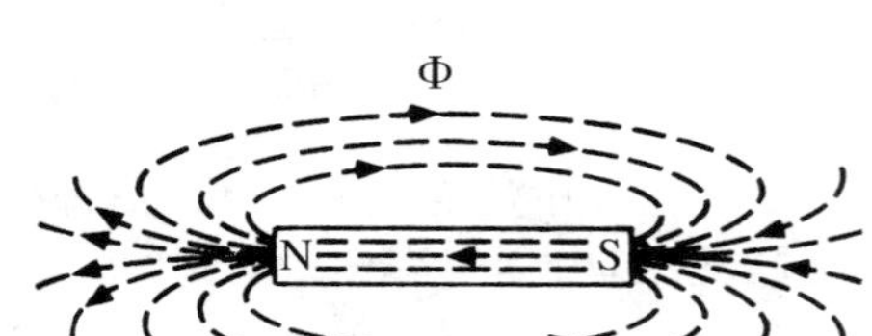

(c)

In a similar manner, to determine the direction of the magnetic field generated by a current through a coil of wire, grasp the coil with the right hand, with the fingers curled in the direction of the current, and the thumb will point in the direction of the magnetic field. This can be visualized in Figure 1.1(b).

The direction of the magnetic field supplied by a magnet is out from the north pole and into the south pole, but is south-to-north within the magnet, as shown in Figure 1.1(c).

1.3 MAGNETIC CIRCUIT DEFINED

Each magnetic circuit shown in Figure 1.2 is an arrangement of ferromagnetic materials called a *core* that forms a path to contain and guide the magnetic flux in a specific direction. The core shape shown in Figure 1.2(a) is used in transformers. Figure 1.2(b)

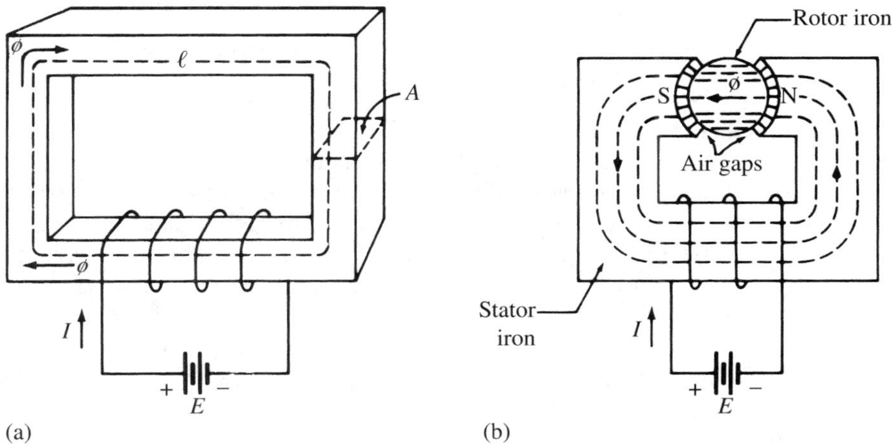

FIGURE 1.2
Magnetic circuit: (a) for a transformer; (b) for a simple two-pole motor.

shows the magnetic circuit of a simple two-pole motor; it includes a stator core, a rotor core, and two air gaps. Note that the flux always takes the shortest path across an air gap.

Magnetomotive Force

The ampere-turns (A-t) of the respective coils in Figure 1.2 represent the driving force, called *magnetomotive force* or *mmf,* that causes a magnetic field to appear in the corresponding magnetic circuits. Expressed in equation form,

$$\mathscr{F} = N \cdot I \tag{1–1}$$

where: $\mathscr{F}$ = magnetomotive force (mmf) in ampere-turns (A-t)
N = number of turns in coil
I = current in coil (A)

Magnetic Field Intensity

Magnetic field intensity, also called mmf gradient, is defined as the magnetomotive force per unit length of magnetic circuit, and it may vary from point to point throughout the magnetic circuit. The average magnitude of the field intensity in a *homogeneous* section of a magnetic circuit is numerically equal to the mmf across the section divided by the effective length of the magnetic section. That is,

$$H = \frac{\mathscr{F}}{\ell} = \frac{N \cdot I}{\ell} \tag{1–2}$$

where: H = magnetic field intensity (A-t/m)
ℓ = mean length of the magnetic circuit, or section (m)
$\mathcal{F}$ = mmf (A-t)

Note that in a homogeneous magnetic circuit of uniform cross section, the field intensity is the same at all points in the magnetic circuit. In composite magnetic circuits, consisting of sections of different materials and/or different cross-sectional areas, however, the magnetic field intensity differs from section to section.

Magnetic field intensity has many useful applications in magnetic circuit calculations. One specific application is calculating the *magnetic-potential difference,* also called *magnetic drop* or *mmf drop,* across a section of a magnetic circuit. The magnetic drop in ampere-turns per meter of magnetic core length in a magnetic circuit is analogous to the voltage drop in volts per meter of conductor length in an electric circuit.

Flux Density

The flux density is a measure of the concentration of lines of flux in a particular section of a magnetic circuit. Expressed mathematically, and referring to the homogeneous core in Figure 1.2(a),

$$B = \frac{\Phi}{A} \tag{1–3}$$

where: Φ = flux, webers (Wb)
A = cross-sectional area (m^2)
B = flux density (Wb/m^2), or teslas (T)

1.4 RELUCTANCE AND THE MAGNETIC CIRCUIT EQUATION

A very useful equation that expresses the relationship between magnetic flux, mmf, and the reluctance of the magnetic circuit is

$$\Phi = \frac{\mathcal{F}}{\mathcal{R}} = \frac{N \cdot I}{\mathcal{R}} \tag{1–4}$$

where: Φ = magnetic flux (Wb)
$\mathcal{F}$ = magnetomotive force (A-t)
$\mathcal{R}$ = reluctance of magnetic circuit (A-t/Wb)

Reluctance $\mathcal{R}$ is a measure of the opposition the magnetic circuit offers to the flux and is analogous to resistance in an electric circuit. The reluctance of a magnetic circuit, or section of a magnetic circuit, is related to its length, cross-sectional area, and permeability. Solving Eq. (1–4) for $\mathcal{R}$, dividing numerator and denominator by ℓ, and rearranging terms,

$$\mathcal{R} = \frac{N \cdot I}{\Phi} = \frac{N \cdot I/\ell}{\Phi/\ell} = \frac{H}{B \cdot A/\ell} = \frac{\ell}{(B/H) \cdot A}$$

Defining

$$\mu = \frac{B}{H} \tag{1–5}$$

$$\mathcal{R} = \frac{\ell}{\mu A} \tag{1–6}$$

where: B = flux density (Wb/m^2), or teslas (T)
H = magnetic field intensity (A-t/m)
ℓ = mean length of magnetic circuit (m)
A = cross-sectional area (m^2)
μ = permeability of material (Wb/A-t · m)

Equation (1–6) applies to a homogeneous section of a magnetic circuit of uniform cross section.

Magnetic Permeability

The ratio $\mu = B/H$ is called magnetic permeability and has different values for different degrees of magnetization of a specific magnetic core material.

1.5 RELATIVE PERMEABILITY AND MAGNETIZATION CURVES

Relative permeability is the ratio of the permeability of a material to the permeability of free space; it is, in effect, a figure of merit that is very useful for comparing the magnetizability of different magnetic materials whose relative permeabilities are known. Expressed in equation form,

$$\mu_r = \frac{\mu}{\mu_0} \tag{1–7}$$

where: μ_0 = permeability of free space = $4\pi 10^{-7}$ (Wb/A-t · m)
μ_r = relative permeability, a dimensionless constant
μ = permeability of material (Wb/A-t · m)

Representative graphs of Eq. (1–5) for some commonly used ferromagnetic materials are shown in Figure 1.3. The graphs, called *B–H curves, magnetization curves,* or *saturation curves,* are very useful in design, and in the analysis of machine and transformer behavior.

The four principal sections of a typical magnetization curve are illustrated in Figure 1.4. The curve is concave up for "low" values of magnetic field intensity, exhibits a somewhat (but not always) linear characteristic for "medium" field intensities, and then is concave down for "high" field intensities, eventually flattening to an almost horizontal line for "very high" intensities. The part of the curve that is concave down is known as the *knee* of the curve, and the "flattened" section is the *saturation region.*

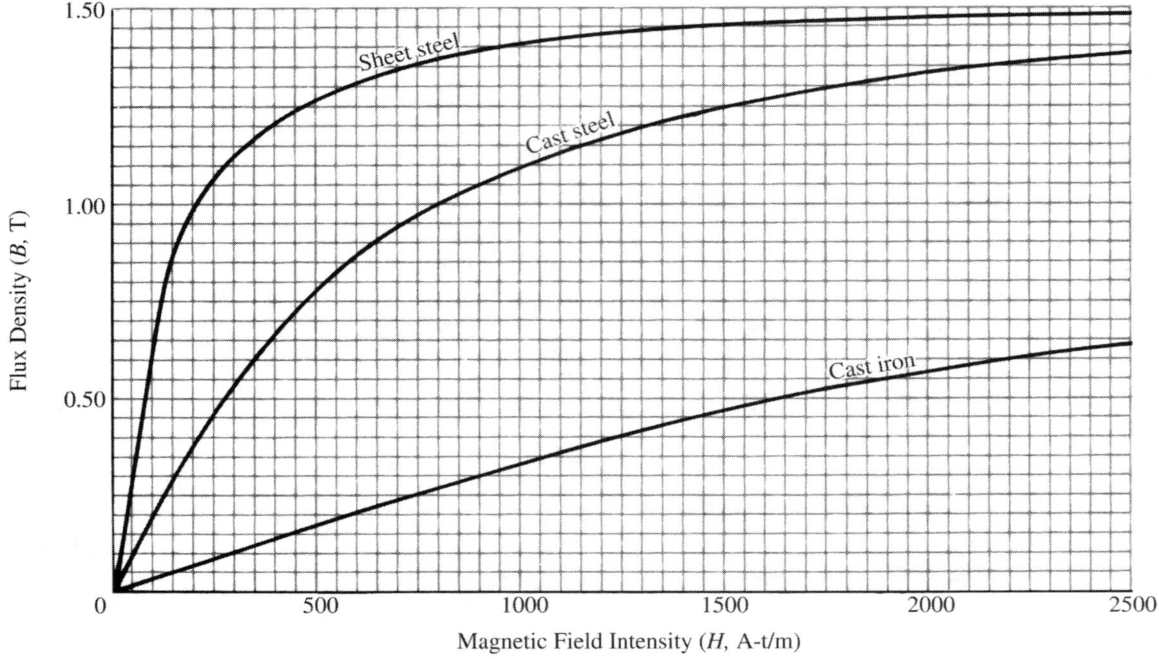

FIGURE 1.3
Representative B–H curves for some commonly used ferromagnetic materials.

Magnetic saturation is complete when all of the magnetic domains of the material are oriented in the direction of the applied magnetomotive force. Saturation begins at the start of the knee region and is essentially complete when the curve starts to flatten.

Depending on the specific application, the magnetic core of an apparatus may be operated in the linear region, and/or the saturation region. For example, transformers and AC machines are operated in the linear region and lower end of the knee; self-excited DC generators and DC motors are operated in the upper end of the knee region, extending into the saturation region; separately excited DC generators are operated in the linear and lower end of the knee region.

Magnetization curves supplied by manufacturers for specific electrical steel sheets or casting are usually plotted on semilog paper, and often include a curve of relative permeability vs. field intensity, as shown in Figure 1.5.[1]

---

[1] Figure 1.5, as furnished by the manufacturer, has the magnetic field intensity expressed in oersteds, a cgs unit. To convert to A-t/m multiply by 79.577. See Appendix K for other conversion factors. Although not shown, the minimum value of $\mu_r = 1.0$, and it occurs when saturation is complete, resulting in $\mu = \mu_0$.

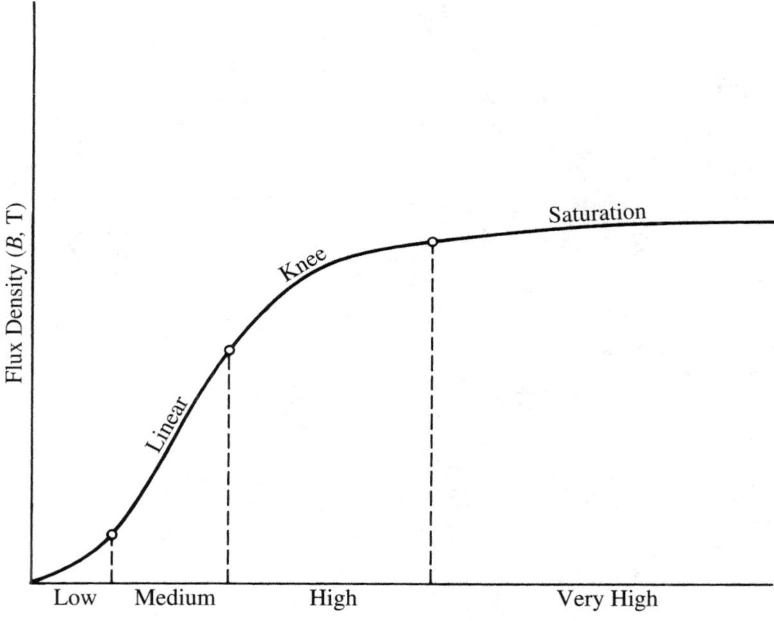

FIGURE 1.4
Exaggerated magnetization curve illustrating the four principal sections.

The relationship between the relative permeability and the reluctance of a magnetic core is obtained by solving Eq. (1–7) for μ, and then substituting into Eq. (1–6). The result is

$$\mathcal{R} = \frac{\ell}{\mu A} = \frac{\ell}{\mu_r \mu_0 A} \qquad \textbf{(1–8)}$$

Equation (1–8) indicates that the reluctance of a magnetic circuit is affected by the relative permeability of the material, which, as shown in Figure 1.5, is dependent on the magnetization, and hence is not constant.

EXAMPLE 1.1 **(a)** Determine the voltage that must be applied to the magnetizing coil in Figure 1.6(a) in order to produce a flux density of 0.200 T in the air gap. *Flux fringing,* which always occurs along the sides of an air gap, as shown in Figure 1.6(b), will be assumed negligible. Assume the magnetization curve for the core material (which is homogeneous) is that given in Figure 1.5. The coil has 80 turns and a resistance of 0.05 Ω. The cross section of the core material is 0.0400 m^2.

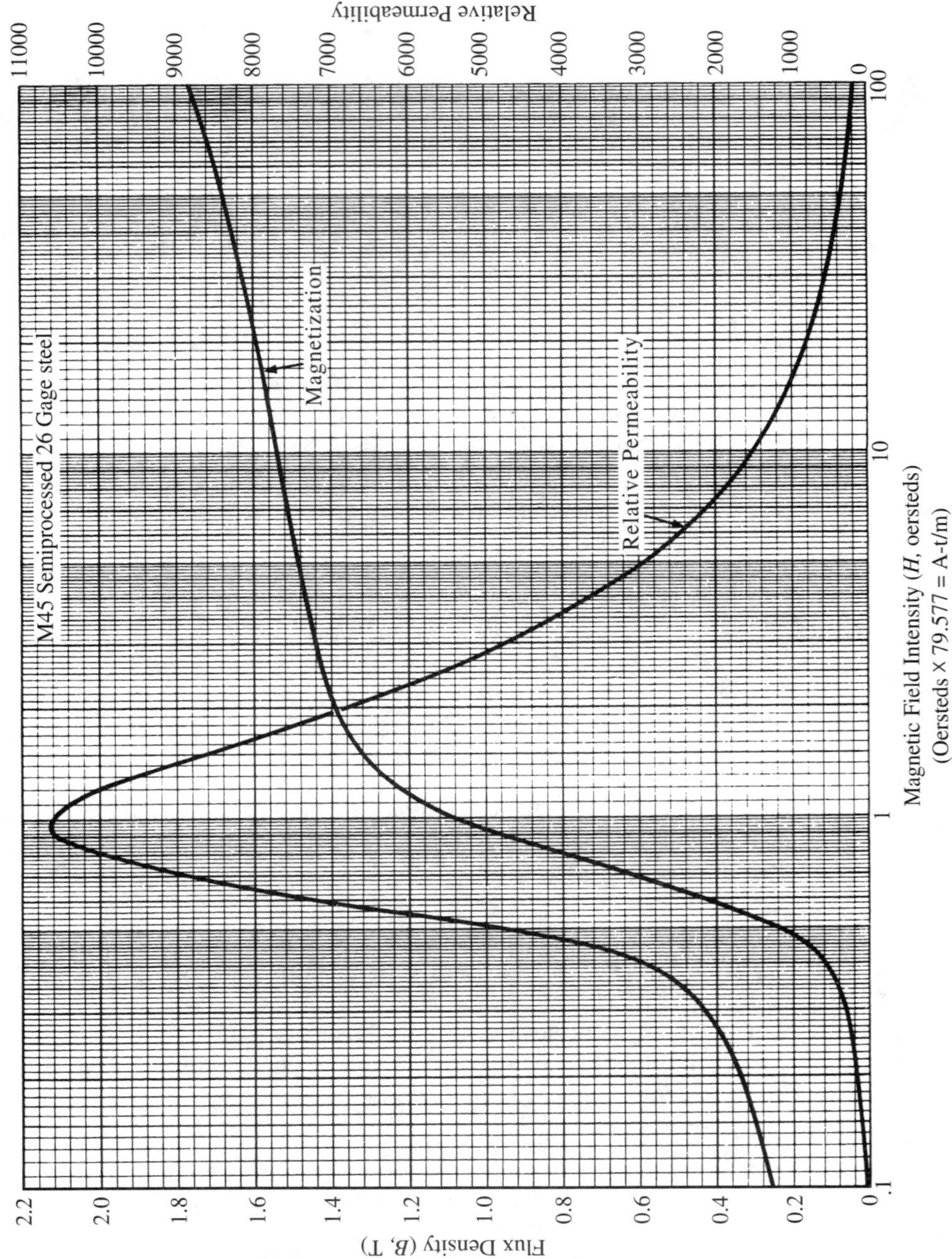

FIGURE 1.5
Magnetization and permeability curves for electrical sheet steel used in magnetic applications. (Courtesy USX Corp.)

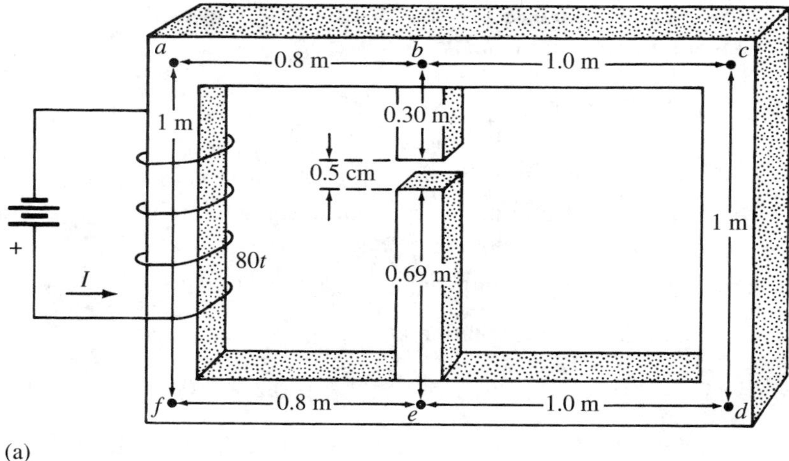

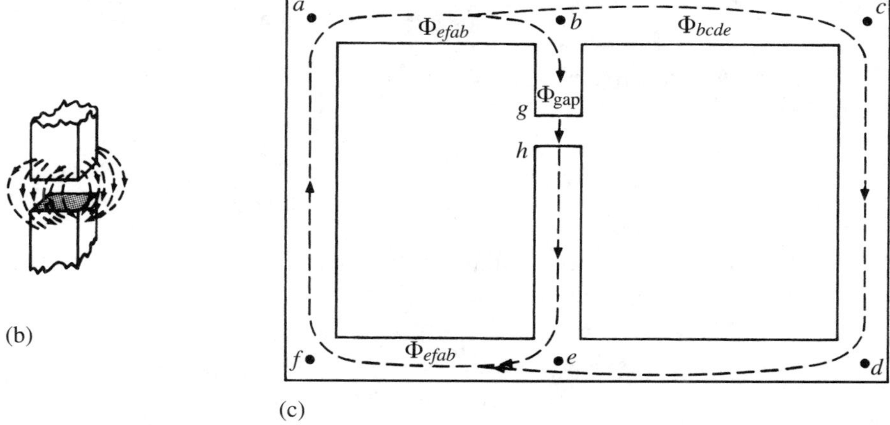

FIGURE 1.6
Magnetic circuit for Example 1.1: (a) physical layout and dimensions; (b) flux fringing; (c) flux distribution.

(b) Using Eqs. (1–5) and (1–7), determine the relative permeability of each of the three legs of the core, and compare the calculated values with the corresponding values obtained from the permeability curve in Figure 1.5.

Solution

(a) The physical layout and dimensions of the magnetic circuit shown in Figure 1.6(a) are used in conjunction with the *B–H* curve to determine the magnetic field intensity in the component parts. The flux distribution is shown in Figure 1.6(c). The procedure for solving the problem is as follows:

Step 1: Determine Φ_{gap}, and $\mathcal{F}_{bghe}$.

Step 2: Determine: H_{bcde}, B_{bcde}, and Φ_{bcde}.

Step 3: Determine Φ_{efab}, B_{efab}, H_{efab}, and $\mathcal{F}_{efab}$.

Step 4: Determine $\mathcal{F}_T$, and, knowing the number of turns in the coil, determine the required current.

Step 5: Using Ohm's law, determine the required voltage.

The flux in the center section is

$$\Phi_{\text{gap}} = B_{\text{gap}}A_{\text{gap}} = 0.2 \times 0.04 = 0.008 \text{ Wb}$$

The flux density throughout the two cores of the center leg is 0.2 T. The field intensity required to provide a flux density of 0.2 T in each of the two cores in the center leg is obtained from the magnetization curve in Figure 1.5. The corresponding field intensity, obtained from the curve is

$$H_{0.30} = H_{0.69} \approx 0.47 \times 79.577 = 37.4 \text{ A-t/m}$$

The resultant magnetic-potential difference across each core of the center leg is determined from Eq. (1–2):

$$\mathcal{F}_{0.30} = H \cdot \ell = 37.4 \times 0.30 = 11.22 \text{ A-t}$$

$$\mathcal{F}_{0.69} = H \cdot \ell = 37.4 \times 0.69 = 25.81 \text{ A-t}$$

The magnetic-potential difference required across the air gap to obtain a flux density of 0.20 T is obtained from Eq. (1–5), where $\mu_{\text{gap}} = \mu_0$.

$$\mu_{\text{gap}} = \frac{B_{\text{gap}}}{H_{\text{gap}}} \quad \Rightarrow \quad 4\pi 10^{-7} = \frac{0.2}{H_{\text{gap}}}$$

$$H_{\text{gap}} = 159{,}155 \text{ A-t/m}$$

The resultant magnetic-potential difference across the air gap is

$$\mathcal{F}_{\text{gap}} = H_{\text{gap}}\ell_{\text{gap}} = 159{,}155(0.005) = 795.77 \text{ A-t}$$

Thus, the total magnetic-potential difference across the center leg is

$$\mathcal{F}_{bghe} = \mathcal{F}_{0.30} + \mathcal{F}_{0.69} + \mathcal{F}_{\text{gap}} = 11.22 + 25.81 + 795.77 = 833 \text{ A-t}$$

Note that the magnetic-potential drop across the 0.005-m air gap is 795.77 A-t, whereas the combined magnetic drop across the 0.30-m and 0.69-m cores total only $11.22 + 25.81 = 37.03$ A-t. *The greatest magnetic-potential drop occurs across an air gap.* Thus, to reduce the amount of ampere-turns required to obtain a desired flux density, air gaps in electrical machinery are kept small.

Since $\mathscr{F}_{bghe}$ is also the magnetic-potential difference across section *bcde,* the magnetic field intensity in that region is

$$H_{bcde} = \frac{\mathscr{F}_{bcde}}{\ell_{bcde}} = \frac{833}{1 + 1 + 1} = 277.67 \text{ A-t/m}$$

Converting to oersteds,

$$277.67 \div 79.577 = 3.49 \text{ oersteds}$$

The corresponding flux density, as obtained from the magnetization curve in Figure 1.5 is

$$B_{bcde} \approx 1.45 \text{ T}$$

Thus, the flux in section *bcde* is

$$\Phi_{bcde} = BA = 1.45 \times 0.04 = 0.058 \text{ Wb}$$

The total magnetic flux supplied by the coil is

$$\Phi_{efab} = \Phi_{\text{gap}} + \Phi_{bcde} = 0.008 + 0.058 = 0.066 \text{ Wb}$$

$$B_{efab} = \frac{\Phi}{A} = \frac{0.066}{0.04} = 1.65 \text{ T}$$

The field intensity required to provide a flux density of 1.65 T in the left leg, as obtained from the magnetization curve in Figure 1.5, is $\approx$ 37 oersteds. Thus,

$$H_{efab} = 37 \times 79.577 = 2944.35 \text{ A-t/m}$$

The mmf drop in section *efab* is

$$\mathscr{F}_{efab} = H \cdot \ell = 2944.35(1 + 0.8 + 0.8) = 7655.31 \text{ A-t}$$

The total mmf that must be supplied by the magnetizing coil is

$$\begin{aligned} \mathscr{F}_T &= \mathscr{F}_{bghe} + \mathscr{F}_{efab} = 7655.31 + 833 = 8488.31 \text{ A-t} \\ \mathscr{F}_{\text{coil}} &= NI \quad \Rightarrow \quad 8488.31 = 80 \times I \\ I &= 106.1 \text{ A} \\ V &= IR = 106.1 \times 0.05 = \underline{5.30 \text{ V}} \end{aligned}$$

(b) Combining Eqs. (1–5) and (1–7),

$$\mu_r = \frac{\mu}{\mu_0} = \frac{B/H}{4\pi \times 10^{-7}} = \frac{B}{4\pi \times 10^{-7} \cdot H}$$

$$\mu_{\text{left}} = \frac{1.65}{4\pi \times 10^{-7} \times 2944} = 446$$

$$\mu_{\text{center}} = \frac{0.20}{4\pi \times 10^{-7} \times 37.4} = 4256$$

$$\mu_{\text{right}} = \frac{1.45}{4\pi \times 10^{-7} \times 277.67} = 4156.1$$

Note that even though the core is homogeneous throughout, the permeability is not the same in all parts of the core. The left leg, with the greater magnetization, is approaching saturation, and thus has a much lower permeability than the other legs.

The following table compares the relative permeability of the core legs, as obtained from the curve in Figure 1.5, with the calculated values.

Core	*H* (A-t/m)	*B* (T)	μ_r (calc)	μ_r (curve)
Left leg	2944	1.65	446	450
Center leg	37.4	0.20	4256	4000
Right leg	277.67	1.45	4156	4100

1.6 ANALOGIES BETWEEN ELECTRIC AND MAGNETIC CIRCUITS

The relationship between mmf, flux, and reluctance in a magnetic circuit is an analog of the relationship between emf, current, and resistance, respectively, in an electric circuit.

$$\Phi = \frac{\mathcal{F}}{\mathcal{R}} \qquad I = \frac{E}{R}$$

where: Φ corresponds to I
$\mathcal{F}$ corresponds to E
$\mathcal{R}$ corresponds to R

Continuing the analogy, the equivalent reluctance of n reluctances in series is

$$\mathcal{R}_{\text{ser}} = \mathcal{R}_1 + \mathcal{R}_2 + \mathcal{R}_3 + \cdots + \mathcal{R}_n \tag{1–9}$$

The equivalent reluctance of n reluctances in parallel is

$$\frac{1}{\mathcal{R}_{\text{par}}} = \frac{1}{\mathcal{R}_1} + \frac{1}{\mathcal{R}_2} + \frac{1}{\mathcal{R}_3} + \cdots + \frac{1}{\mathcal{R}_n}$$

or

$$\mathcal{R}_{\text{par}} = \frac{1}{1/\mathcal{R}_1 + 1/\mathcal{R}_2 + 1/\mathcal{R}_3 + \cdots + 1/\mathcal{R}_n} \tag{1–10}$$

An equivalent magnetic circuit that shows the analogous relationship to an electric circuit is often used to solve magnetic circuit problems that may otherwise be more difficult to visualize. For example, the components of the series–parallel circuit shown in Figure 1.7(a) are represented as lumped reluctances in the equivalent magnetic circuit shown in Figure 1.7(b). Using the methods developed for electric circuits, the total reluctance of the series–parallel magnetic circuit is

$$\mathcal{R}_T = \mathcal{R}_1 + \frac{\mathcal{R}_2 \cdot \mathcal{R}_3}{\mathcal{R}_2 + \mathcal{R}_3}$$

EXAMPLE 1.2 Assume that flux Φ_1 in Figure 1.7(a) is 0.250 Wb, and that the magnetic circuit parameters for this condition are

$$\mathcal{R}_1 = 10{,}500 \text{ A-t/Wb}$$
$$\mathcal{R}_2 = 40{,}000 \text{ A-t/Wb}$$
$$\mathcal{R}_3 = 30{,}000 \text{ A-t/Wb}$$

The magnetizing coil is wound with 140 turns of copper wire. Determine (a) the current in the coil; (b) the magnetic-potential difference across $\mathcal{R}_3$; (c) the flux in $\mathcal{R}_2$.

Solution

(a) Applying basic circuit concepts to the equivalent magnetic circuit in Figure 1.7(b),

$$\mathcal{R}_{\text{par}} = \frac{\mathcal{R}_2\mathcal{R}_3}{\mathcal{R}_2 + \mathcal{R}_3} = \frac{40{,}000 \times 30{,}000}{40{,}000 + 30{,}000} = 17{,}142.8571 \text{ A-t/Wb}$$

$$\mathcal{R}_{\text{circ}} = \mathcal{R}_1 + \mathcal{R}_{\text{par}} = 10{,}500 + 17{,}142.8571 = 27{,}642.8571 \text{ A-t/Wb}$$

$$\Phi = \frac{NI}{\mathcal{R}} \quad \Rightarrow \quad 0.250 = \frac{140 \times I}{27{,}642.8571}$$

$$I = 49.3622 \quad \Rightarrow \quad \underline{49.36 \text{ A}}$$

(b) The magnetic drop across $\mathcal{R}_1$ is

$$\mathcal{F}_1 = \Phi_T \cdot \mathcal{R}_1 = 0.25 \times 10{,}500 = 2625 \text{ A-t}$$

Referring to Figure 1.7(b),

$$\mathcal{F}_T = \mathcal{F}_1 + \mathcal{F}_{\text{par}} \quad \Rightarrow \quad 49.3622 \times 140 = 2625 + \mathcal{F}_{\text{par}}$$

$$\mathcal{F}_3 = \mathcal{F}_{\text{par}} = 4285.7143 \quad \Rightarrow \quad \underline{4285.71 \text{ A-t}}$$

(c)

$$\Phi_2 = \frac{\mathcal{F}_{\text{par}}}{\mathcal{R}_2} = \frac{4285.7143}{40{,}000} = \underline{0.1071 \text{ Wb}}$$

Or, using the magnetic analog of the current divider rule,

$$\Phi_2 = \Phi_T \times \frac{\mathcal{R}_3}{\mathcal{R}_2 + \mathcal{R}_3} = 0.25 \times \frac{30{,}000}{40{,}000 + 30{,}000} = \underline{0.1071 \text{ Wb}}$$

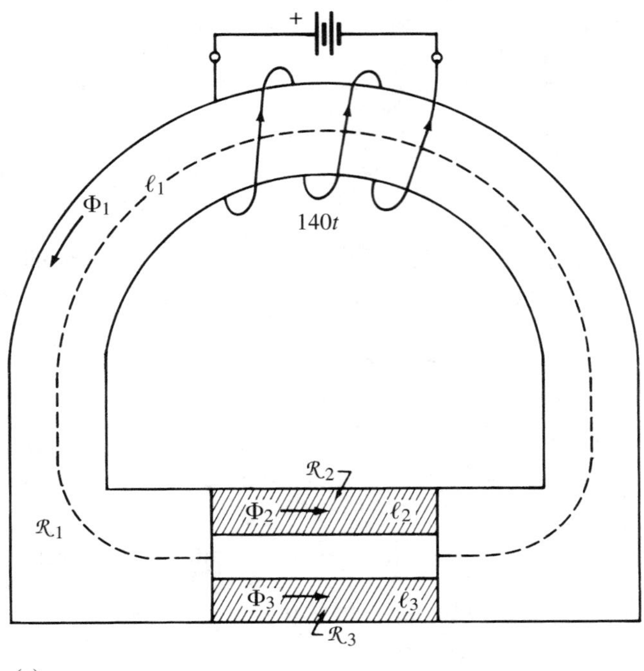

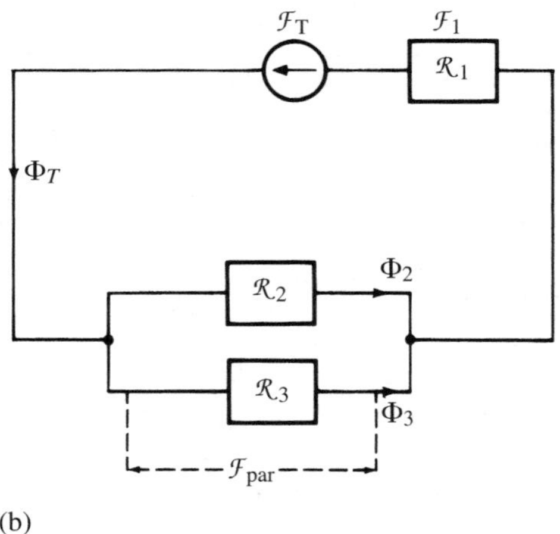

FIGURE 1.7
Magnetic circuit for Example 1.2: (a) physical layout; (b) equivalent magnetic circuit.

1.7 MAGNETIC HYSTERESIS AND HYSTERESIS LOSS

If an alternating magnetomotive force is applied to a magnetic material, as shown in Figure 1.8(a), and the flux density B plotted against the magnetic field intensity H, the resultant curve will indicate a lack of retraceability. This phenomenon, shown in Figure 1.8(b), is called *hysteresis,* and the resultant curve is called an *hysteresis loop.*

Starting with an unmagnetized ferromagnetic core, point O on the curve, $H = 0$ and $B = 0$. Increasing the coil current in the positive direction increases the ampere-turns, and hence the magnetic field intensity. From Eqs. (1–1) and (1–2),

$$H = \frac{NI}{\ell}$$

When the current reaches its maximum value, the flux density and magnetic field intensity have their respective maximum values, and the curve is at point a; this initial trace of the curve, drawn with a broken line, is called the *virgin section* of the curve. As the current decreases, the curve follows a different path, and when the current is reduced to zero, H is reduced to zero, but the flux density in the core lags behind, holding at point b on the curve. The flux density at point b is the *residual magnetism.* This lagging of flux behind the magnetizing force is the *hysteresis effect.*

As the alternating current and associated magnetic field intensity increase in the negative direction, the residual magnetism decreases but remains positive until point c is reached, at which time the flux density in the core is zero. The negative field intensity required to force the residual magnetism to zero is called the *coercive force,* and is

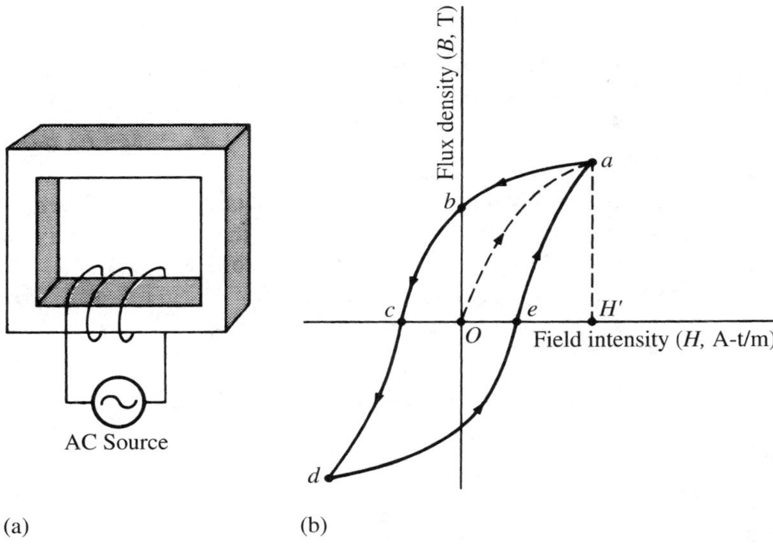

FIGURE 1.8
(a) Magnetic circuit with an alternating mmf; (b) representative hysteresis loop.

represented by line O–c on the H axis. As the current continues its alternations, the plot of B vs. H follows points c–d–e–a–b–c on the hysteresis loop.

Magnetic hysteresis affects the rate of response of magnetic flux to a magnetizing force. In electrical apparatus such as transformers, in which the desired characteristic necessitates a quick and proportional response of flux to a change in mmf, with little residual magnetism, a high-grade silicon steel is used. Machines such as self-excited generators require steel that retains sufficient residual magnetism to permit the buildup of voltage. Stepper motors and some DC motors require permanent magnets with a very high magnetic retentivity (high hysteresis). Thus, the choice of magnetic materials is dictated by the application.

Magnetic Hysteresis Loss

If an alternating voltage is connected to the magnetizing coil, as shown in Figure 1.8(a), the alternating magnetomotive force causes the magnetic domains to be constantly reoriented along the magnetizing axis. This molecular motion produces heat, and the harder the steel the greater the heat. The power loss due to hysteresis for a given type and volume of core material varies directly with the frequency and the nth power of the maximum value of the flux density wave. Expressed mathematically,

$$P_h = k_h \cdot f \cdot B_{\max}^n \tag{1–11}$$

where:
P_h = hysteresis loss (W/unit mass of core)
f = frequency of flux wave (Hz)
$B_{\max}$ = maximum value of flux density wave (T)
k_h = constant
n = Steinmetz exponent[2]

The constant k_h is dependent on the magnetic characteristics of the material, its density, and the units used. The area enclosed by the hysteresis loop is equal to the hysteresis energy in joules/cycle/cubic-meter of material.

EXAMPLE 1.3 The hysteresis loss in a certain electrical apparatus operating at its rated voltage and rated frequency of 240 V and 25 Hz is 846 W. Determine the hysteresis loss if the apparatus is connected to a 60-Hz source whose voltage is such as to cause the flux density to be 62 percent of its rated value. Assume the Steinmetz exponent is 1.4.

Solution

From Eq. (1–11),

$$\frac{P_{h1}}{P_{h2}} = \frac{[k_h \cdot f \cdot B_{max}^n]_1}{[k_h \cdot f \cdot B_{max}^n]_2} \quad \Rightarrow \quad P_{h2} = P_{h1} \times \frac{[k_h \cdot f \cdot B_{max}^n]_2}{[k_h \cdot f \cdot B_{max}^n]_1}$$

$$P_{h2} = 846 \times \frac{60}{25} \times \left[\frac{0.62}{1.0}\right]^{1.4} = \underline{1.04 \text{ kW}}$$

---

[2] The Steinmetz exponent varies with the core material and has an average value of 1.6 for silicon steel sheets.

1.8 INTERACTION OF MAGNETIC FIELDS (MOTOR ACTION)

When two or more sources of magnetic fields are arranged so that their fluxes, or a component of their fluxes, are parallel within a common region, a mechanical force will be produced that tends to either force the sources of flux together or force them apart. A force of repulsion will occur if the two magnetic sources have components of flux that are parallel and in the same direction; this will be indicated by a net increase in flux called "flux bunching" in the common region. A force of attraction will occur if the respective fluxes have components that are parallel and in opposite directions; this will be indicated by a net subtraction of flux in the common region.

Forces on Adjacent Conductors

The interaction of magnetic fields of adjacent current-carrying conductors produces mechanical forces that tend to bring together or separate the two conductors. If the currents in adjacent conductors are in opposite directions, as shown in Figure 1.9(a), the respective components of flux in the common region will be in the same direction,

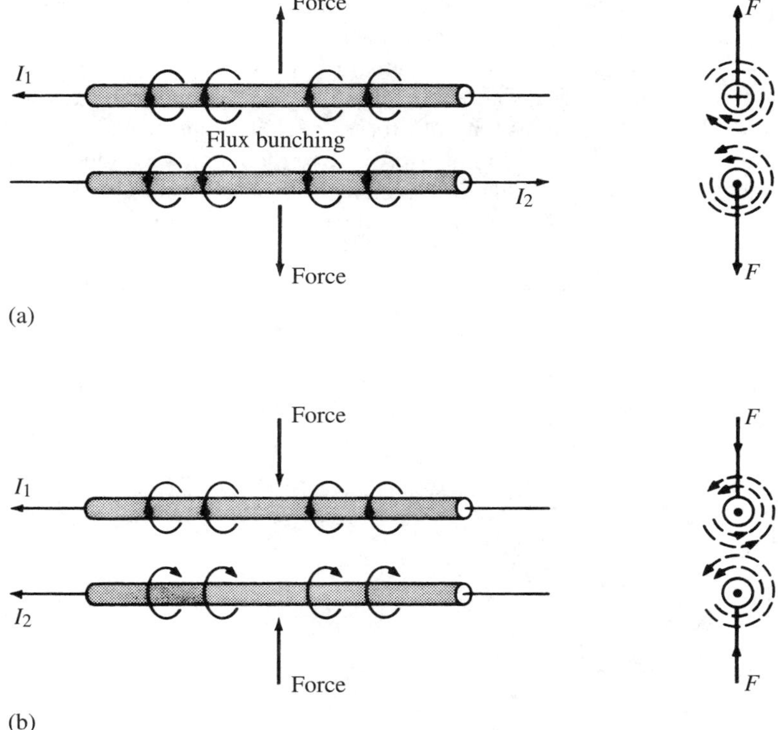

FIGURE 1.9
Interaction of magnetic fields of adjacent current-carrying conductors: (a) currents in opposite direction; (b) currents in same direction.

and as indicated by flux bunching, a separating force will be produced on the conductors. If the currents in adjacent conductors are in the same direction, as shown in Figure 1.9(b), the respective components of flux in the common region will be in opposite directions, and the net reduction in flux indicates a force of attraction.

Under severe short-circuit conditions, the forces between adjacent conductors can be high enough to physically crush the insulation of transformers, motors, and generators, bend bus bars, tear switchboards apart, and cause switches and circuit breakers to come apart with explosive violence. Thus, in those applications where the available short-circuit current is of a magnitude that would cause destruction of apparatus if a fault occurred, special current-limiting devices, as well as mechanical bracing and conductor support must be installed **[1], [2].**

1.9 ELEMENTARY TWO-POLE MOTOR

Figure 1.10 shows a rotor core, containing two insulated conductors in rotor slots, and the rotor centered between the poles of a stationary magnet (called the stator). The + mark on the end of conductor *A* is the tail end of an arrow that represents the direction of current in conductor *A*. The dot in the center of conductor *B* is the point of an arrow indicating the direction of current in conductor *B*. The direction of flux around each conductor is determined by the right-hand rule.

The broken lines show the paths of component fluxes, assuming the rotor and stator were energized at different times. The dotted line indicates the direction and path of the resultant flux with both rotor and stator energized at the same time. Note that the net flux on top of conductor *A*, due to the magnet and due to the current in the conductor, is additive (bunching), indicating a downward mechanical force *F*, as shown in Figure 1.10. A similar action occurs at the bottom of conductor *B*, causing an

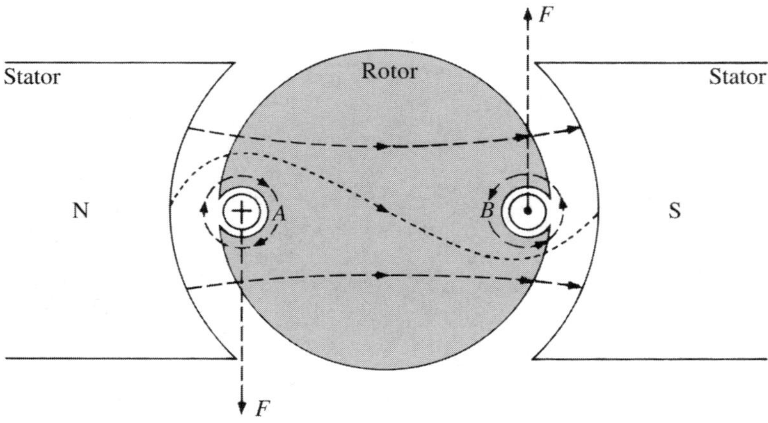

FIGURE 1.10
Motor action.

upward mechanical force. The net result is a counterclockwise (CCW)-turning moment or torque, called *motor action.*

1.10 MAGNITUDE OF THE MECHANICAL FORCE EXERTED ON A CURRENT-CARRYING CONDUCTOR SITUATED IN A MAGNETIC FIELD (BLI RULE)

The magnitude of the mechanical force exerted on a straight conductor that is carrying an electric current and situated within and perpendicular to a magnetic field, as shown in Figure 1.11(a), is expressed by

$$F = B \cdot \ell_{\text{eff}} \cdot I \tag{1–12}$$

where: F = mechanical force (N)
B = flux density of stator field (T)
I = current in rotor conductor (A)
ℓ_{eff} = effective length of rotor conductor (m)

The effective length of a conductor is that component of its length that is immersed in and normal to the magnetic field. Thus, if the conductor is not perpendicular to the magnetic field as shown in Figure 1.11(b), the effective length of the conductor is

$$\ell_{\text{eff}} = \ell \sin \alpha$$

Angle β is called the skewing angle, which may range from 0 to 30 degrees in electrical machines.

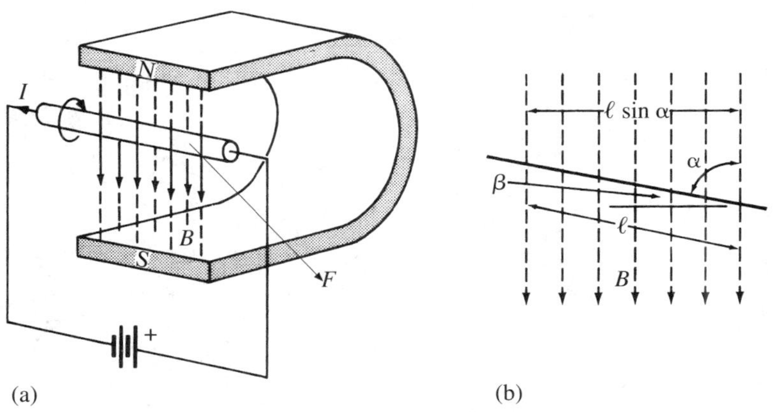

FIGURE 1.11
(a) Conductor carrying current, situated within and perpendicular to the B-field of a permanent magnet; (b) conductor skewed β°.

The direction of the mechanical force exerted on the conductor in Figure 1.11(a) is determined by flux bunching.

Developed Torque

Figure 1.12(a) shows a rotor coil made up of a single loop, situated in a two-pole stator field of uniform flux density. The effective length of each conductor (coil side) does not include the *end connections.* The end connections, also called *end turns,* are used to connect the conductors in series, but because they are not immersed in the field, they do not develop torque. The distance d between the center of the shaft and the center of a conductor is the moment arm.

The direction of developed torque may be determined from an end view of the conductors and magnet poles, as seen from the battery end in Figure 1.12(b). The direction of flux due to the known direction of current was determined by the right-hand

FIGURE 1.12
(a) Single-loop rotor coil carrying an electric current and situated in a two-pole field; (b) end view of coil, showing direction of developed force.

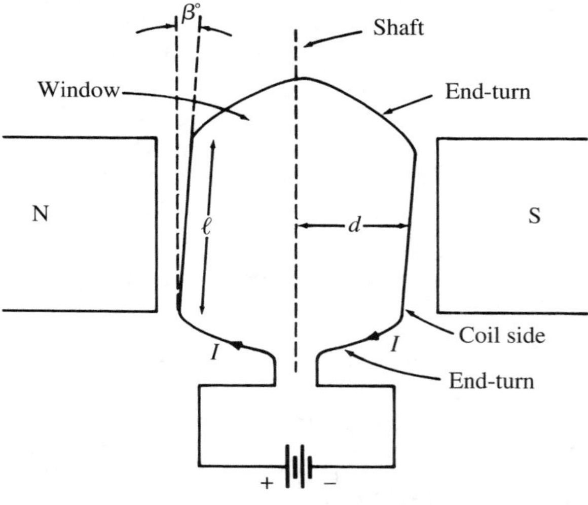

(a)

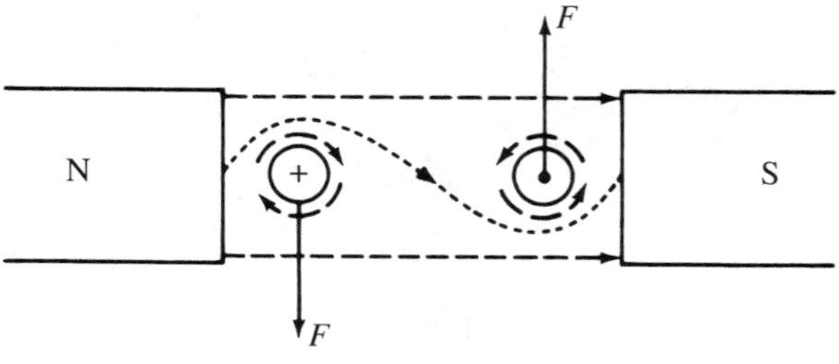

(b)

rule, and the direction of the mechanical force on each conductor, due to the interaction of the magnetic fields, was determined by the flux bunching effect. The resultant torque, produced by the two-conductor couple, is CCW and has a magnitude equal to

$$T_D = 2 \cdot F \cdot d \qquad \text{N} \cdot \text{m} \tag{1–13}$$

Substituting Eq. (1–12) into Eq. (1–13),

$$T_D = 2 \cdot B \cdot \ell_{eff} \cdot I \cdot d \qquad \text{N} \cdot \text{m} \tag{1–14}$$

EXAMPLE 1.4

Assume each coil side in Figure 1.12(a) has a length of 0.30 m and a skew angle of 15°. The distance between the center of each conductor and the center of the shaft is 0.60 m. The combined resistance of the coil and its connections to a 36-V battery is 4.0 Ω. If the stator field has a uniform flux density of 0.23 T between the poles, determine the magnitude and direction of the developed torque.

Solution

From Figure 1.11(b),

$$\alpha = 90° - \beta° = 90° - 15° = 75°$$

$$I = \frac{E_{\text{bat}}}{R} = \frac{36}{4.0} = 9.0 \text{ A}$$

$$T = 2 \cdot B \cdot I(\ell \sin \alpha) \cdot d = 2 \times 0.23 \times 9(0.3 \sin 75°) \times 0.60 = \underline{0.72 \text{ N} \cdot \text{m}}$$

The direction of the developed torque is counterclockwise, as indicated in Figure 1.12(b).

1.11 ELECTROMAGNETICALLY INDUCED VOLTAGES (GENERATOR ACTION)

The magnitude of the voltage induced in a coil by electromagnetic induction is directly proportional to the number of series-connected turns in the coil, and to the rate of change of flux through its window. This relationship, known as *Faraday's law,* is expressed mathematically as

$$e = N\frac{d\phi}{dt} \tag{1–15}$$

where: e = induced voltage (electromotive force, emf) (V)
N = number of series-connected turns
$d\phi/dt$ = rate of change of flux through window (Wb/s)

The basic Faraday relationship expressed in Eq. (1–15) is often converted by mathematical manipulation to other forms for solution of specific groups of problems.

Electromagnetically induced voltages are generated by relative motion or transformer action. Voltages generated by transformer action are due to flux varying with time through the window of a stationary coil. Voltages generated by relative motion involve a moving coil and a stationary magnet, or a moving magnet and a stationary coil. Voltages caused by relative motion are called *speed voltages* or "flux cutting" voltages.

In accordance with Lenz's law, the voltage, current, and associated flux, generated by transformer action, or relative motion between a conductor and a magnetic field, will always be induced in a direction to oppose the action that caused it.[3] In a transformer, the flux due to current generated in a transformer coil will be in a direction to oppose the change in flux that caused it.

In the case of a conductor driven by an applied force, the flux due to current generated in the conductor will set up a counterforce in opposition to the applied force. In a rotating machine, the flux due to generated current in the conductors will set up a countertorque (motor action) in opposition to the driving torque of the prime mover. In fact, as will be shown in subsequent chapters, all generators may be operated as motors and all motors may be operated as generators.

Speed Voltages and the BLV Rule

A closed loop consisting of two conductors X and Y, and a set of conducting rails, is situated within a uniform magnetic field, as shown in Figure 1.13(a); conductor Y is clamped and conductor X is moving to the right at velocity v meters per second. The window in Figure 1.13(a) is the area enclosed by conductor X, conductor Y, and the conducting rails. As conductor X moves to the right, the window area increases, causing the flux through the window to increase with time, inducing a voltage in the loop.

Expressing the flux in terms of the flux density and the area of the window,

$$\phi = B \cdot A$$

Taking the derivative with respect to time,

$$\frac{d\phi}{dt} = B \cdot \frac{dA}{dt}$$

Substituting into Eq. (1–15)

$$e = N \cdot B \frac{dA}{dt} \quad \textbf{(1–16)}$$

From Figure 1.13(a), the increment increase in window area, as conductor X moves to the right, may be expressed in terms of length ℓ and an increment increase in distance (ds) along the rails. That is,

$$dA = \ell ds \quad \textbf{(1–17)}$$

---

[3] When applying Lenz's law, it will be assumed that the circuit is complete, resulting in a current and its associated flux.

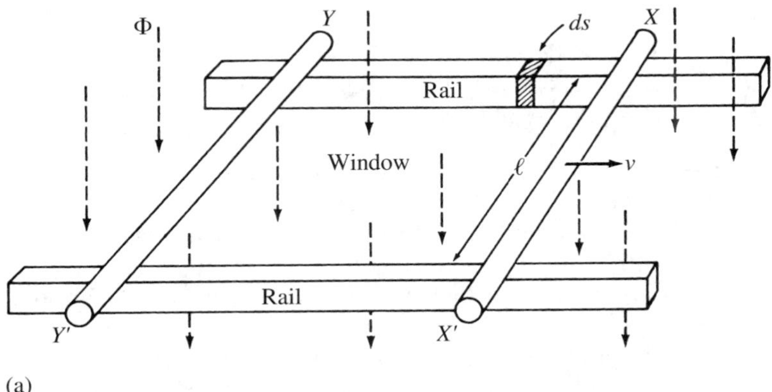

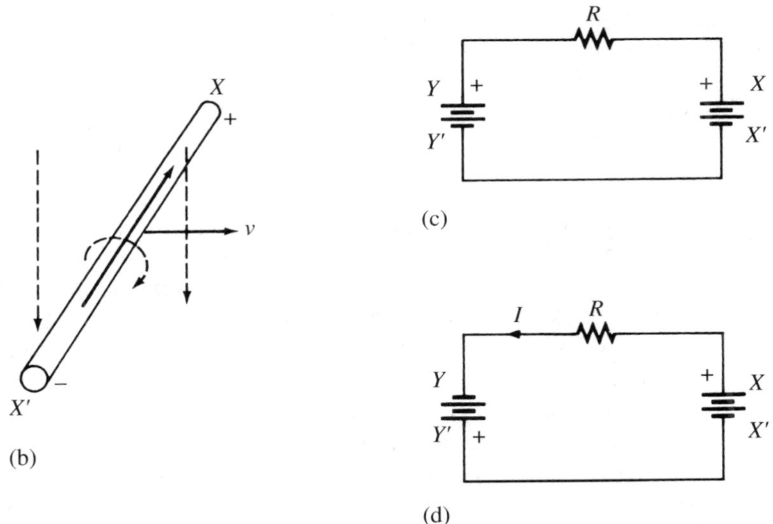

FIGURE 1.13
(a) Closed loop consisting of two conductors and a set of conducting rails; (b) direction of emf and current caused by conductor *X* moving to the right; (c) equivalent circuit, both conductors moving in the same direction; (d) equivalent circuit, conductors moving in opposite directions.

Substituting into Eq. (1–16), and noting that $N = 1$ for a single loop,

$$e = B \cdot \ell \cdot \frac{ds}{dt} \qquad \textbf{(1–18)}$$

Since *ds*/*dt* represents the velocity of the conductor, Eq. (1–18) may be rewritten as

$$e = B \cdot \ell \cdot v \qquad \textbf{(1–19)}$$

where: e = induced voltage (V)
B = flux density of field (T)
ℓ = effective length of conductor (m)
v = velocity of conductor (m/s)

Note: For the loop formed by the rod and rails in Figure 1.13(a), conductor X is the only moving conductor.

Since the emf was generated by an applied force driving conductor X to the right, the induced voltage and associated current will be in a direction to develop a counterforce. For this to happen, flux bunching must occur on the right side of conductor X, as shown in Figure 1.13(b). This establishes the direction of conductor flux, and the right-hand rule may then be used to determine the direction of the associated current, and hence the direction of the induced emf. Thus, the direction of induced emf *within the conductor* is away from the reader, as shown in Figure 1.13(b), causing terminal X to be positive with respect to terminal X'.

Equation (1–19) defines a speed-voltage generated by a conductor of length ℓ, *cutting flux lines* while moving at velocity v through (and normal to) a magnetic field of density B, and is called the *$B\ell v$ rule.*

The equivalence of the $B\ell v$ rule and the $d\phi/dt$ through the window method for determining the generation of an emf is further demonstrated in the following two examples.

1. If both conductor X and conductor Y in Figure 1.13(a) are moved to the right by an applied force and at the same speed, they would each cut the same number of flux lines, at the same speed and in the same direction, and thus generate the same voltage. The respective voltage directions *within the conductors* would be Y' to Y and X' to X. As a result, the net voltage around the loop (and thus the current in the loop) would be zero. The corresponding equivalent circuit is shown in Figure 1.13(c), resistor R is the equivalent total resistance of conductors and rails.

   Analyzing the same conditions, on the basis of $d\phi/dt$ through the window, indicates that with both coil sides moving at the same speed and in the same direction, $d\phi/dt$ through the window will be zero, resulting in zero voltage generated in the loop.
2. If conductor Y is moved to the left while conductor X is moved to the right, both at the same speed, they would each cut the same number of flux lines, at the same speed, *but in opposite directions.* Thus, the voltage *within* conductor Y would be from Y to Y', while the voltage *within* conductor X would be from X' to X. The net voltage in the loop formed by the conductors and the rails would be doubled. This is the case for almost all rotating machines that use coils[4]; the two coil sides always move in opposite directions with respect to the flux from the field poles. The corresponding equivalent circuit is shown in Figure 1.13(d).

---

[4]Acyclic machines, also called homopolar or unipolar machines, use conducting cylinders instead of coils **[3]**.

Analyzing the same conditions on the basis of $d\phi/dt$ through the window indicates that with the coil sides moving in opposite directions, the $d\phi/dt$ through the window will double, generating twice the voltage that would otherwise occur if only one coil side moved.

Thus, in the case of rotating machines, it is not necessary to look for a rate of change of flux through a window in order to determine whether or not a voltage is generated. *If a conductor "cuts flux," a voltage is generated.*

EXAMPLE 1.5 Determine the length of conductor required to generate 2.5 V when passing through and normal to a magnetic field of 1.2 T at a speed of 8.0 m/s.

Solution

$$e = B\ell v \qquad \Rightarrow \qquad 2.5 = 1.2 \times \ell \times 8.0$$

$$\ell = 0.26 \text{ m}$$

1.12 ELEMENTARY TWO-POLE GENERATOR

Figure 1.14(a) shows a closed coil situated within a magnetic field and driven in a clockwise direction by the prime mover. To satisfy Lenz's law, the induced voltage, current, and associated flux must be in a direction that will develop a *countertorque* to oppose the driving torque of the prime mover. For this to happen, flux bunching must occur on the top of coil side *B* and the bottom of coil side *A,* as shown in Figure 1.14(b). With the direction of conductor flux known, the direction of the respective emfs may be determined by applying the right-hand rule; the emf and current are toward the reader in *A,* and away from the reader in *B.* Thus, as viewed from the south pole in Figure 1.14(a) the current in the coil is in a CCW direction.

Sinusoidal Emfs

Referring to the elementary generator in Figure 1.14(a), if the coil rotates at a constant angular velocity in a uniform magnetic field, the variation of flux through the coil window will be sinusoidal.

$$\phi = \phi_{max} \sin(\omega t) \tag{1–20}$$

where: ωt = instantaneous angle that the plane of the coil makes with the flux lines (rad)

Φ_{max} = maximum flux through coil window (Wb)

Referring to Figure 1.14(a), the maximum flux through the coil window occurs when the window of the coil is parallel to the pole face.

The rate-of-change of flux through the window as the coil rotates within the magnetic field is

$$\frac{d\phi}{dt} = \omega\Phi_{max}\cos(\omega t) \tag{1–21}$$

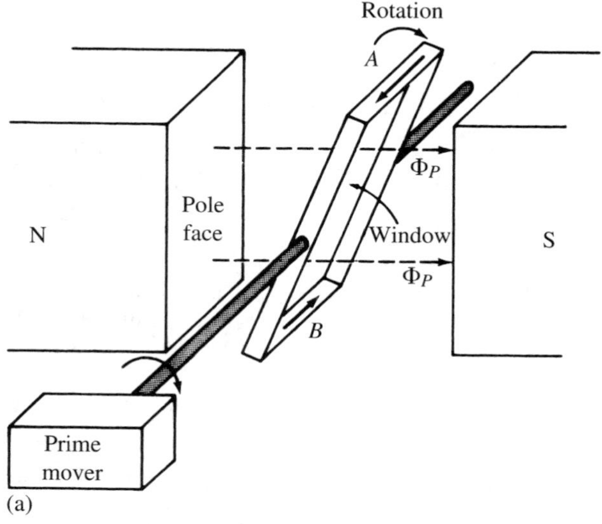

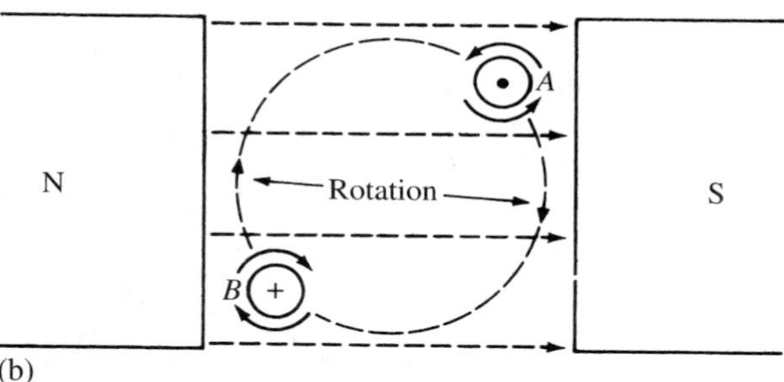

FIGURE 1.14
(a) Closed coil rotating CW within a magnetic field; (b) direction of emf and current for the instant shown in (a).

Substituting Eq. (1–21) into Eq. (1–15),

$$e = N\frac{d\phi}{dt} = N \cdot \omega\Phi_{max}\cos(\omega t) \tag{1–22}$$

The maximum value of the voltage wave in Eq. (1–22) is

$$E_{max} = \omega N\Phi_{max} = 2\pi f N\Phi_{max} \tag{1–23}$$

Dividing both sides by $\sqrt{2}$

$$\frac{E_{max}}{\sqrt{2}} = \frac{2\pi f N\Phi_{max}}{\sqrt{2}} \tag{1–24}$$

$$E_{rms} = 4.44\, fN\Phi_{max} \tag{1–25}$$

where: f = frequency of the sinusoidal flux through the window, and hence the frequency of the generated emf (Hz)

N = number of series-connected turns in coil

Note: Equation (1–25) may also be expressed in terms of rotational or angular velocity:

$$E_{rms} = n \cdot \Phi_{max} \cdot k_n \tag{1–26}$$

or

$$E_{rms} = \omega \cdot \Phi_{max} \cdot k_\omega \tag{1–27}$$

where: ω = angular velocity (rad/s)
n = rotational speed (r/s or r/min)
k_n, k_ω = constants[5]

Frequencies currently used in electrical power applications are 25 Hz, 50 Hz, 60 Hz, and 400 Hz. The 60-Hz system is used primarily in North America; the 50-Hz system is used throughout Europe and most other countries; the 400-Hz system is the preferred system for aircraft and spacecraft because of its light weight; and the 25-Hz system is used extensively for traction motors in railroads **[4], [5], [6].**

EXAMPLE 1.6 An elementary four-pole generator with a six-turn rotor coil generates the following voltage wave:

$$e = 24.2 \sin(36 \cdot t)$$

Determine (a) the frequency; (b) the pole flux.

Solution

(a) $\omega = 2\pi f \quad \Rightarrow \quad 36 = 2\pi f \quad \Rightarrow \quad f = 5.7296 \text{ Hz}$

(b) $E_{rms} = 4.44 f N \Phi_{max} \quad \Rightarrow \quad \dfrac{24.2}{\sqrt{2}} = 4.44 \times 5.7296 \times 6 \times \Phi_{max}$

$\Phi_{max} = \underline{0.112 \text{ Wb}}$

1.13 ENERGY CONVERSION IN ROTATING ELECTRICAL MACHINES

All rotating electrical machines may be operated as either motors or as generators. If mechanical energy is supplied to the shaft, the machine converts mechanical energy to electrical energy. If electrical energy is applied to the machine windings, the machine converts electrical energy to mechanical energy. *Regardless of the direction of energy flow, however, all electrical machines (when operating) generate voltage and develop torque at the same time.* If operating as a motor, it develops torque and a counter-emf; if operating as a generator, it develops an emf; and if supplying a load, it develops a countertorque.

---

[5] The constant depends on the units used and the number of series-connected turns in the coil.

1.14 EDDY CURRENTS AND EDDY-CURRENT LOSSES

Eddy currents are circulating currents produced by transformer action in the iron cores of electrical apparatus. Figure 1.15(a) shows a block of iron that may be viewed as an infinite number of concentric shells or loops. The eddy voltages generated in these shells by a changing magnetic field are proportional to the rate of change of flux through the window of the respective shells. Thus,

$$e_e \propto \frac{d\phi}{dt}$$

Expressed in terms of frequency and flux density, as obtained from Eq. (1–25),

$$E_e \propto f \cdot B_{\text{max}} \tag{1–28}$$

Slicing the core into many laminations and insulating one from the other will reduce the magnitude of the eddy currents by providing smaller paths, and hence lower eddy voltages. This is shown in Figure 1.15(b). Laminated cores are made by stacking insulated steel stampings to the desired thickness or depth. Each lamination is insulated by a coating of insulating varnish or oxide on one or both sides. Laminating the core results in much smaller shells, significantly reducing the heat losses in the iron.

The eddy-current loss, expended as heat power in the resistance of each shell, is proportional to the square of the eddy voltage.

$$P_e \propto E_e^2 \tag{1–29}$$

Substituting Eq. (1–28) into Eq. (1–29) and applying a proportionality factor results in

$$P_e = k_e f^2 B_{\text{max}}^2 \tag{1–30}$$

where: P_e = eddy-current loss (W/unit mass)
f = frequency of flux wave (Hz)

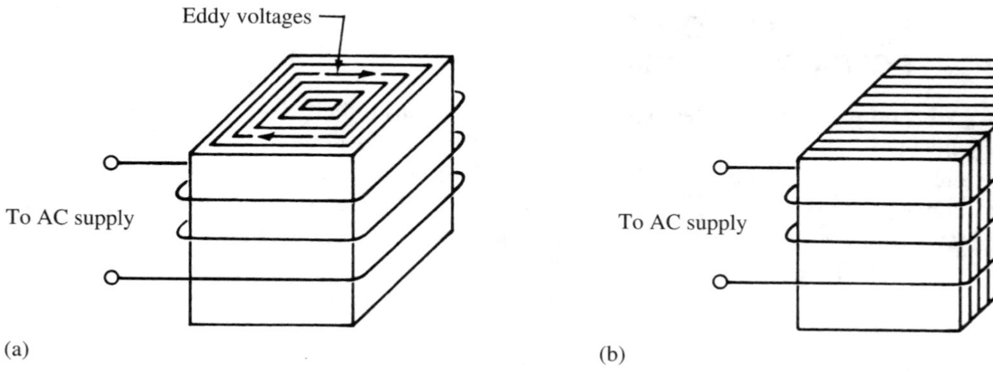

FIGURE 1.15
(a) Eddy currents in solid iron core; (b) laminated core.

B_{max} = maximum value of flux density wave (T)
k_e = constant

The constant k_e is dependent on the lamination thickness, electrical resistivity, density and mass of the core material, and the units used.

EXAMPLE 1–7 The eddy-current loss in a certain electrical apparatus operating at its rated voltage and rated frequency of 240 V and 25 Hz is 642 W. Determine the eddy-current loss if the apparatus is connected to a 60-Hz source whose voltage is such as to cause the flux density to be 62 percent of its rated value.

Solution
From Eq. (1–30),

$$\frac{P_{e1}}{P_{e2}} = \frac{[k_e f^2 B_{max}^2]_1}{[k_e f^2 B_{max}^2]_2} \quad \Rightarrow \quad P_{e2} = P_{e1} \times \left[\frac{f_2}{f_1}\right]^2 \times \left[\frac{B_{\max,2}}{B_{\max,1}}\right]^2$$

$$P_{e2} = 642 \times \left[\frac{60}{25}\right]^2 \times \left[\frac{0.62}{1.0}\right]^2 = \underline{1.42 \text{ kW}}$$

1.15 MULTIPOLAR MACHINES, FREQUENCY, AND ELECTRICAL DEGREES

The magnetic circuit for an elementary four-pole generator is shown in Figure 1.16(a). The four poles of the stator core are alternately north and south, and an *armature coil* wound on the rotor core spans one-quarter of the rotor circumference. The stator is marked off in *space degrees,* also called *mechanical degrees.* If the rotor coil is positioned at the 0° reference, as shown in Figure 1.16(a), maximum flux from the north pole will enter the outside face of the coil window. At the 45° position, shown in Figure 1.16(b), the net flux passing through the window is zero; the number of lines entering the window is equal to the number of lines leaving the same side of the window. At 90°, the flux through the window reaches its maximum value in the opposite direction, etc.

A plot of the variation of flux through the coil window for one revolution of the rotor is shown in Figure 1.16(c); the variation of flux is assumed to be essentially sinusoidal.

Note that for a four-pole machine, such as that shown in Figure 1.16, one revolution of the rotor causes two complete cycles of flux to pass through the coil window, *one cycle per pair of poles.* Similarly, a six-pole machine would produce three cycles per revolution, etc. Expressed as an equation,

$$f = \frac{Pn}{2} \qquad \textbf{(1–31)}$$

where: f = frequency (Hz)
P = number of poles
n = rotational speed (r/s)

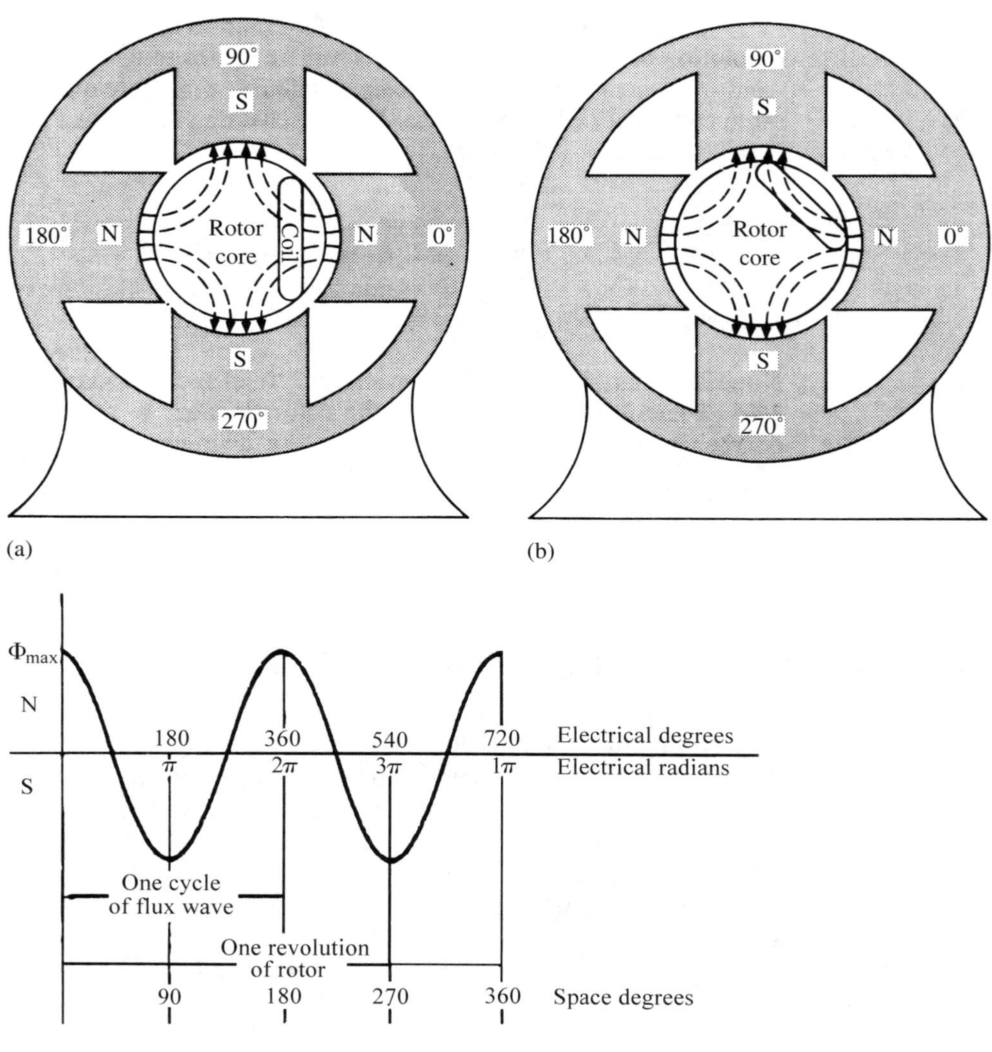

FIGURE 1.16
Four-pole generator: (a) flux through coil window is at maximum value; (b) net flux through coil window is zero; (c) variation of flux through coil window as rotor turns in CCW direction.

Note also that, for a four-pole machine, 720° of the periodic wave corresponds to 360° of angular displacement of the rotor. Hence, to differentiate between the degrees of an electrical quantity and the degrees of space displacement, the former are known as electrical degrees or time degrees, and the latter as space degrees. This distinction is also used in radian measure, namely, electrical radians and space radians.

As indicated in Figure 1.16(c), the relationship between electrical degrees and space degrees is

$$\text{Elec. deg.} = \text{space deg.} \times \frac{P}{2} \qquad \textbf{(1–32)}$$

where: P = the number of poles

Unless otherwise specified, angular measurements used in electrical transactions in this text, and in other electrical texts, are expressed in electrical degrees or electrical radians. Adjacent poles are always 180 electrical degrees (π electrical radians) apart.

EXAMPLE 1.8 A special-purpose 80-pole, 100-kVA generator is operating at 20 r/s. Determine (a) the number of cycles per revolution; (b) the number of electrical degrees per revolution; (c) the frequency in Hz.

Solution

(a) Two poles per cycle, or 40 cycles.

(b) $\text{Elec. deg.} = 360 \times \frac{80}{2} = 14{,}400$

(c) $f = \frac{Pn}{2} = \frac{80 \times 20}{2} = \underline{800\text{ Hz}}$

EXAMPLE 1.9 The voltage generated in a 15-turn armature coil by a four-pole rotating field is 100 V. If the flux per pole is 0.012 Wb, determine (a) frequency of the generated emf; (b) speed of the rotor.

Solution

(a) From Eq. (1–25),

$$f = \frac{E_{\text{rms}}}{4.44 \times N \times \phi_{\max}} = \frac{100}{4.44 \times 15 \times 0.012} = 125.13 \quad \Rightarrow \quad 125\text{ Hz}$$

(b) From Eq. (1–30),

$$n = \frac{2f}{P} = \frac{2 \times 125.13}{4} = 62.57\text{ r/s} \quad \text{or} \quad 60 \times 62.57 = 3754\text{ r/min}$$

SUMMARY OF EQUATIONS FOR PROBLEM SOLVING

$$\mathcal{F} = N \cdot I \qquad N \tag{1–1}$$

$$H = \frac{\mathcal{F}}{\ell} = \frac{N \cdot I}{\ell} \qquad \text{A-t/m} \tag{1–2}$$

$$B = \frac{\Phi}{A} \qquad \text{T} \tag{1–3}$$

$$\Phi = \frac{\mathcal{F}}{\mathcal{R}} = \frac{N \cdot I}{\mathcal{R}} \qquad \text{Wb} \tag{1–4}$$

$$\mu = \frac{B}{H} \qquad \text{Wb/A-t} \cdot \text{m} \tag{1–5}$$

$$\mathcal{R} = \frac{\ell}{\mu A} \qquad \text{A-t/Wb} \tag{1–6}$$

$$\mu_r = \frac{\mu}{\mu_0} \tag{1–7}$$

$$\mathcal{R} = \frac{\ell}{\mu_r \mu_0 A} \qquad \text{A-t/Wb} \tag{1–8}$$

$$\mathcal{R}_{\text{ser}} = \mathcal{R}_1 + \mathcal{R}_2 + \mathcal{R}_3 + \cdots + \mathcal{R}_n \tag{1–9}$$

$$\frac{1}{\mathcal{R}_{\text{par}}} = \frac{1}{\mathcal{R}_1} + \frac{1}{\mathcal{R}_2} + \frac{1}{\mathcal{R}_3} + \cdots + \frac{1}{\mathcal{R}_n}$$

or

$$\mathcal{R}_{\text{par}} = \frac{1}{1/\mathcal{R}_1 + 1/\mathcal{R}_2 + 1/\mathcal{R}_3 + \cdots + 1/\mathcal{R}_n} \tag{1–10}$$

$$P_h = k_h \cdot f \cdot B_{\max}^n \qquad \text{W} \tag{1–11}$$

$$F = B \cdot \ell_{eff} \cdot I \tag{1–12}$$

$$\alpha = 90° - \beta$$

$$T_D = 2 \cdot B \cdot \ell_{eff} \cdot I \cdot d \qquad \text{N} \cdot \text{m} \tag{1–14}$$

$$e = N\frac{d\phi}{dt} \qquad \text{V} \tag{1–15}$$

$$e = B \cdot \ell \cdot v \qquad \text{V} \tag{1–19}$$

$$E_{\text{rms}} = 4.44 fN\Phi_{\max} \tag{1–25}$$

$$E_{\text{rms}} = n \cdot \Phi_{\max} \cdot k_n \qquad \text{V} \tag{1–26}$$

$$E_{\text{rms}} = \omega \cdot \Phi_{\max} \cdot k_\omega \qquad \text{V} \tag{1–27}$$

$$P_e = k_e f^2 B_{\max}^2 \qquad \text{W} \tag{1–30}$$

$$f = \frac{Pn}{2} \qquad \text{Hz} \tag{1–31}$$

$$\text{Elec. deg.} = \text{space deg.} \times \frac{P}{2} \tag{1–32}$$

SPECIFIC REFERENCES KEYED TO TEXT

1. Barnett, R. D., "The frequency that wouldn't die." *IEEE Spectrum,* Nov. 1990, pp. 120–121.
2. Campbell, J. J., P. E. Clark, I. E. McShane, and K. Wakeley. Strains on motor end windings. *IEEE Trans. Industry Applications,* Vol. IA-20, No. 1, Jan./Feb. 1984.
3. Hubert, C. I. *Preventive Maintenance of Electrical Equipment.* Prentice Hall, Upper Saddle River, NJ, 2002.
4. Jones, Andrew J. Amtrack's Richmond static frequency converter project. *IEEE Vehicular Technology Society News,* May 2000, pp. 4–10.
5. Lamme, B. G. The technical story of the frequencies. Electrical Engineering Papers, Westinghouse Electric & Manufacturing Co., 1919, pp. 569–589.
6. Matsch, L. W., and J. D. Morgan. *Electromagnetic and Electromechanical Machines.* Harper & Row, New York, 1986.

REVIEW QUESTIONS

1. Sketch a coil connected to a DC source, and indicate the direction of current in the coil and the direction of magnetic flux around the connecting wires, coil, and battery.
2. Differentiate between magnetic field intensity and magnetomotive force, and state the units for each.
3. (a) How is the reluctance of a section of magnetic material related to the material and its dimensions? (b) Is the reluctance of a magnetic material dependent on the degree of magnetization? Explain.
4. Is the permeability of a given block of magnetic material constant? Explain.
5. Differentiate between permeability and relative permeability.
6. What is flux fringing and where does it occur?
7. Explain why, in any series magnetic circuit (or series branch of a magnetic circuit) containing an air gap, the greatest magnetic-potential drop occurs across the air gap.
8. List the analogous relationships that exist between an electric circuit and a magnetic circuit.
9. What is magnetic hysteresis, and what effect does it have on the rate of response of magnetic circuits to a magnetizing force?
10. Sketch an hysteresis loop and discuss the behavior of the loop as the magnetizing current goes through the first 1.5 cycles. Assume the magnetic core was initially in an unmagnetized state.
11. (a) What causes magnetic hysteresis loss, and how is it affected by the frequency and density of the flux wave? (b) How is the hysteresis loss related to the hysteresis loop?
12. Sketch two parallel conductors in the vertical plane with currents in opposite directions. Show the directions of current, component magnetic fields, resultant field, and direction of mechanical force exerted on each conductor.
13. Repeat Question 12, assuming the currents are in the same direction.

14. Sketch a conductor carrying direct current, situated in and normal to the magnetic field of a permanent magnet. Show the directions of current, component magnetic fields, resultant field, and direction of mechanical force exerted on the conductor and on the poles.
15. Make a sketch of a one-turn coil situated in the field of a permanent magnet, and explain how current in the coil produces torque. Indicate on the sketch the direction of current, respective directions of the component magnetic fields, and the direction of the resultant two-conductor couple.
16. Using appropriate sketches and Lenz's law, explain (a) how speed voltages are generated and indicate their direction; (b) how transformer voltages are generated and indicate their direction.
17. Explain why all electrical machines (when operating) develop torque and generate voltage at the same time.
18. Explain how eddy currents are generated in magnetic cores and how they can be minimized.
19. How are eddy-current losses affected by the frequency and density of the flux wave?

PROBLEMS

1–1/4 The magnetic circuit of an inductance coil has a reluctance of 1500 A-t/Wb. The coil is wound with 200 turns of aluminum wire, and draws 3 A when connected to a 24-V battery. Determine (a) the core flux; (b) the resistance of the coil.

1–2/4 A magnetic circuit constructed of sheet steel has an average length of 1.3 m and a cross-sectional area of 0.024 m^2. A 50-turn coil wound on the ring has a resistance of 0.82 Ω, and draws 2 A from a DC supply. The reluctance of the core for this condition is 7425 A-t/Wb. Determine (a) flux density; (b) voltage applied.

1–3/4 A magnetic circuit has an average length of 1.4 m and a cross-sectional area of 0.25 m^2. Excitation is provided by a 140-turn, 30-Ω coil. Determine the voltage required to establish a flux density of 1.56 T. The reluctance of the magnetic circuit, when operating at this flux density is 768 A-t/Wb.

1–4/5 A ferromagnetic core in the shape of a doughnut has a cross-sectional area of 0.11 m^2 and an average length of 1.4 m. The permeability of the core is 1.206×10^{-3} Wb/A-t · m. Determine the reluctance of the magnetic circuit.

1–5/5 A magnetic circuit has a mean length of 0.80 m, a cross-sectional area of 0.06 m^2, and a relative permeability of 2167. Connecting its 340-turn 64-Ω magnetizing coil to a DC circuit causes a 56-V drop across the coil. Determine the flux density in the core.

1–6/5 The magnetic circuit shown in Figure 1.17 has a mean core length of 52 cm and a cross-sectional area of 18 cm^2. The length of the air gap is 0.14 cm. Determine the battery voltage required to obtain a flux density of 1.2 T in the air gap. Use the magnetization curve shown in Figure 1.5.

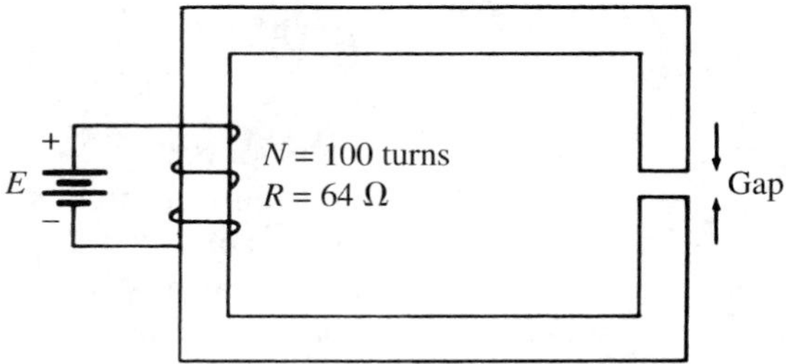

FIGURE 1.17
Magnetic circuit for Problem 1–6/5.

1–7/5 The mean length and cross-sectional area of the core shown in Figure 1.18 are 1.5 m and 0.08 m^2, respectively. The core is made of cast steel and the magnetization curve for the material is shown in Figure 1.3. The 260-turn magnetizing coil has a resistance of 27.75 Ω, and is connected to a 240-V DC supply. Determine (a) magnetic field intensity; (b) core flux density and core flux; (c) relative permeability of the core; (d) reluctance of the magnetic circuit.

1–8/5 Repeat Problem 1–7, assuming a sheet steel core.

1–9/5 Repeat Problem 1–7, assuming a cast-iron core.

1–10/6 A magnetic circuit, composed of two half-rings of different core materials, is joined at the ends to form a doughnut. The cross-sectional area of the core is 0.14 m^2, and the reluctances of the two halves are 650 A-t/Wb and 244 A-t/Wb, respectively. A coil of 268 turns and 5.2-Ω resistance is wound around the doughnut and connected to a 45-V battery. Determine (a) the core flux; (b) repeat (a), assuming the half-rings are separated 0.12 cm at each end (assume no fringing), and the reluctance of the half-rings does not change; (c) the magnetic drop across each air gap in (b).

1–11/7 A coil wound around a ferromagnetic core is supplied from a 25-Hz source. Determine the percent change in hysteresis loss if the coil is connected to a 60-Hz sources and the resultant flux density is reduced by 60%. Assume the Steinmetz coefficient is 1.65, and voltage is constant.

1–12/7 A certain electrical apparatus operating at rated voltage and rated frequency has an hysteresis loss of 250 W. Determine the hysteresis loss if the fre-

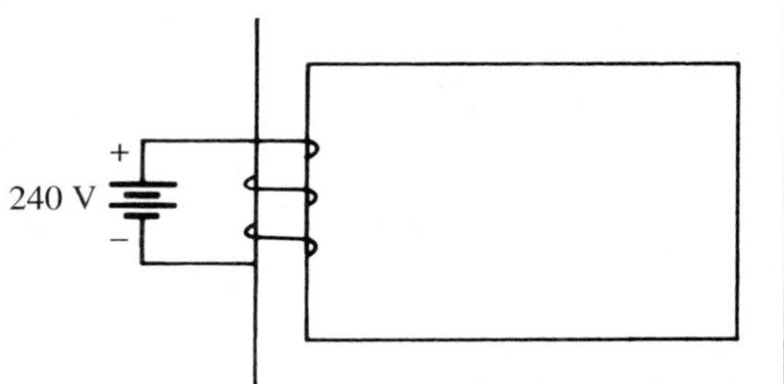

FIGURE 1.18
Magnetic circuit for Problem 1–7/5.

quency is reduced to 60.0 percent rated frequency, and the applied voltage is adjusted to provide 80.0 percent rated flux density. Assume the Steinmetz exponent is 1.6.

1–13/10 A conductor 0.32 m long with 0.25-Ω resistance is situated within and normal to a uniform magnetic field of 1.3 T. Determine (a) the voltage drop across the conductor that would cause a force of 120 N to be exerted on the conductor; (b) repeat part (a), assuming a skew angle of 25°.

1–14/10 A rotor coil consisting of 30 series-connected turns, with a total resistance of 1.56 Ω, is situated within a uniform magnetic field of 1.34 T. Each coil side has a length of 54 cm, is displaced 22 cm from the center of the rotor shaft, and has a skew angle of 8.0°. Sketch the system and determine the coil current required to obtain a shaft torque of 84 N · m.

1–15/11 Determine the required linear velocity of a 0.54-m conductor that will generate 30.6 V when cutting flux in a 0.86-T magnetic field.

1–16/11 A conductor 1.2 m long is moving at a constant velocity of 5.2 m/s through and normal to a uniform magnetic field of 0.18 T. Determine the generated voltage.

1–17/12 Determine the frequency and rms voltage generated by a three-turn coil rotating at 12 r/s within a four-pole field that has a pole flux of 0.28 Wb/pole.

1–18/12 Determine the rotational speed required to generate a sinusoidal voltage of 24 V in a 25-turn coil that rotates within a two-pole field of 0.012 Wb/pole.

1–19/12 The flux through the window of a 20-turn coil varies with time in the following manner:

$$\phi = 1.2 \sin(28 \cdot t) \qquad \text{Wb}$$

Determine (a) the frequency and rms value of voltage generated in the coil; (b) the equation representing the voltage wave.

1–20/14 A coil wound around a ferromagnetic core is supplied by a 120-V, 25-Hz source. Determine the percent change in eddy-current loss if the coil is connected to a 120-V, 60-Hz source.

1–21/14 A certain electrical apparatus operating at rated voltage and rated frequency has an eddy-current loss of 212.6 W. Determine the eddy-current loss if the frequency is reduced to 60.0 percent rated frequency and the applied voltage is adjusted to provide 80.0 percent rated flux density.

2

Transformer Principles

2.1 INTRODUCTION

The principle of transformer action is based on the work of Michael Faraday (1791–1867), whose discoveries in electromagnetic induction showed that, given two magnetically coupled coils, a changing current in one coil will induce an electromotive force in the other coil. Such electromagnetically induced emfs are called *transformer voltages,* and coils specifically arranged for such purposes are called *transformers.*

Transformers are very versatile. They are used to raise or lower voltage in AC distribution and transmission systems; to provide reduced-voltage starting of AC motors; to isolate one electric circuit from another; to superimpose an alternating voltage on a DC circuit; and to provide low voltage for solid-state control, for battery charging, door bells, etc.

The principle of transformer action is also applicable in many ways to motors, generators, and control apparatus. A specific example is the application of the equivalent-circuit model of the transformer, developed in this chapter, to the analysis of induction-motor performance in Chapter 4.

2.2 CONSTRUCTION OF POWER AND DISTRIBUTION TRANSFORMERS

The two basic types of transformer construction used for power and distribution applications are shown in Figure 2.1. Note that the high-voltage coils are wound with a greater number of turns of smaller cross-section conductor than the low-voltage coils. The core type, shown in Figure 2.1(a), has primary and secondary coils wound on different legs, and the shell type, shown in Figure 2.1(b), has both coils wound on the same leg. The wider spacing between primary and secondary in the core-type transformer gives it an advantage in high-voltage applications. The shell type, however, has the advantage of less leakage flux.

Transformer core material is made of nonaging, cold-rolled, high-permeability silicon steel laminations, and each lamination is insulated with a varnish or oxide coating

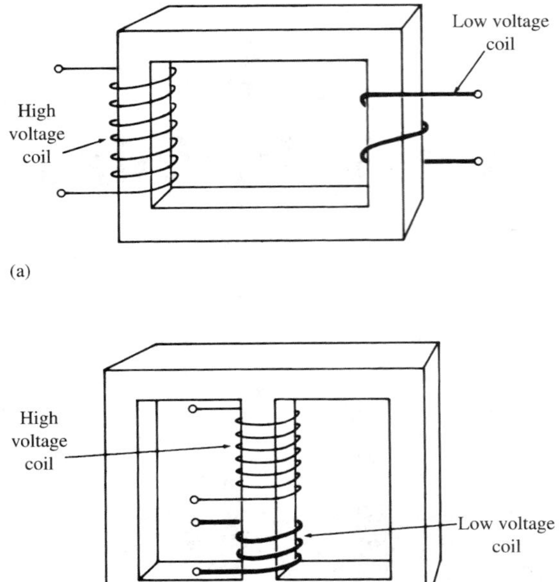

FIGURE 2.1
Transformer construction: (a) core type; (b) shell type.

to reduce eddy currents. The coils are wound with insulated aluminum conductor or insulated copper conductor, depending on design considerations. Cooling is provided by air convection, forced air, insulating liquids, or gas.

Ventilated Dry-Type Transformers

Ventilated dry-type transformers are cooled by natural air convection. The principal application for this type of transformer is in schools, hospitals, and shopping areas, where large groups of people are present and potential hazards to personnel from burning oil or toxic gases must be avoided. The ventilated dry-type transformer, however, requires periodic maintenance, such as removal of dust or dirt from the windings by light brushing, vacuuming, and/or blowing with dry air.

Gas-Filled Dry-Type Transformers

Gas-filled dry-type transformers are cooled with nitrogen or other dielectric gases, such as fluorocarbon C_2F_6 and sulfurhexafluoride SF_6. These transformers can be installed indoors, outdoors, or in underground environments. Gas-filled transformers are hermetically sealed and require only periodic checks of gas pressure and temperature.

Liquid-Immersed Transformers

Liquid-immersed transformers, such as that shown in Figure 2.2, have hermetically sealed tanks filled with insulating liquid to provide both insulation and cooling. Cooling fins on the tank provide for convection cooling of the insulating liquid. Forced cooling with pumps and/or fans is also provided on larger power transformers. The

FIGURE 2.2
Cutaway view of a large three-phase oil-cooled power transformer. (Courtesy, TECO Westinghouse)

insulating liquids used are mineral oil and silicone oil. Polychlorinated biphenyls (PCBs) called askarels[1] were used in earlier construction but are no longer permitted.

2.3 PRINCIPLE OF TRANSFORMER ACTION

The principle of transformer action is explained with the aid of Figure 2.3(a), which shows coil 1 connected to a battery through a switch, and coil 2 connected to a resistor. Closing the switch causes a clockwise (CW) buildup of flux in the iron core, generating

[1] The EPA has declared PCBs to be toxic liquids, and they are no longer permitted in new construction. Existing transformers and capacitors containing PCBs had to be replaced or detoxified and refilled with nontoxic liquids, and all work was to be completed by October 1, 1990.

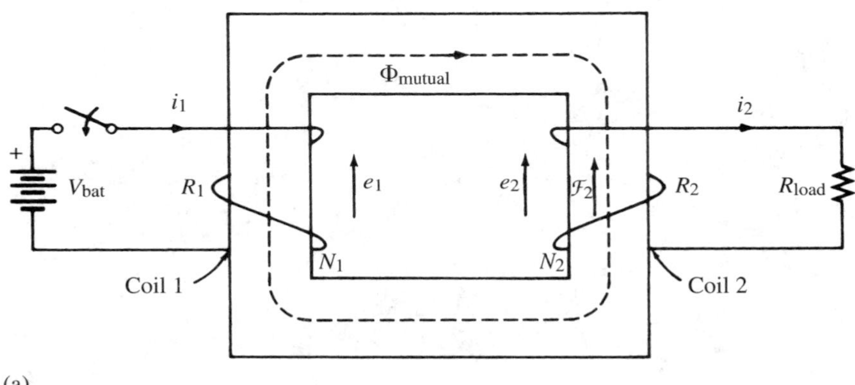

(a)

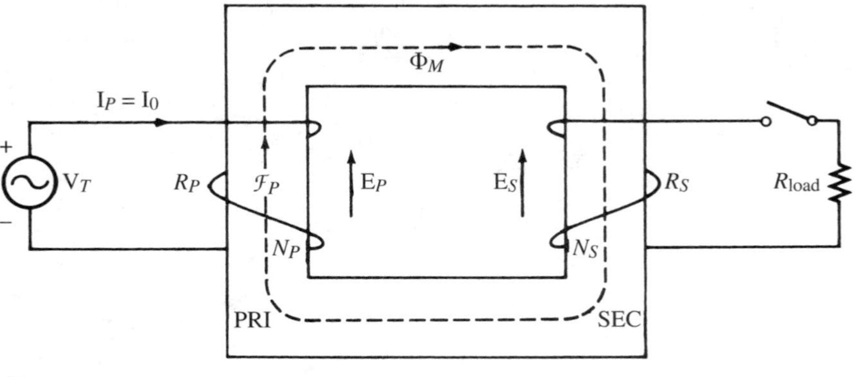

(b)

FIGURE 2.3
(a) Transformer with battery in primary circuit to aid in the explanation of transformer action; (b) transformer with sinusoidal source and no load on secondary.

a voltage in each coil that is proportional to the number of turns in the coil and the rate of change of flux through the respective coils. Assuming no leakage, the same flux (called the *mutual flux*) exists in both coils. Thus,

$$e_1 = N_1 \frac{d\phi}{dt} \qquad e_2 = N_2 \frac{d\phi}{dt}$$

where: N_1 = turns in coil 1
N_2 = turns in coil 2

In accordance with Lenz's law, the voltage generated in each coil will be induced in a direction that opposes the action that caused it. Thus, the induced emf in coil 1 must be opposite in direction to the battery voltage, as shown in Figure 2.3(a). This opposing voltage, shown as e_1 in Figure 2.3(a), is called a counter-emf (cemf).

In the case of coil 2, the induced emf and associated current must be in a direction that will develop a counterclockwise (CCW) mmf to oppose the buildup of flux in its window. Thus, with the direction of mmf known, the direction of induced emf and associated current may be determined by applying the right-hand rule to coil 2. *Note:* The induced emfs and secondary current in Figure 2.3(a) are transients. When Φ_{mutual} reaches steady state, $d\phi/dt = 0$, the induced emfs = 0, and $i_2 = 0$.

2.4 TRANSFORMERS WITH SINUSOIDAL VOLTAGES

Figure 2.3(b) shows a transformer with one winding (called the primary) connected to a sinusoidal source, and the other winding (called the secondary) connected to a switch and a resistor load. The currents and voltages are expressed as phasors. The directions of the induced voltages are the same as in Figure 2.3(a) and are determined in the same manner.

For this *preliminary discussion,* the following simplifying assumptions will be made: (1) The permeability of the core is constant over the range of transformer operation, and thus the reluctance of the core is constant; and (2) there is no leakage flux, hence the same flux links both primary and secondary windings.

The voltages induced in the primary and secondary windings by the sinusoidal variation of flux in the respective coil windows, expressed in terms of rms values, are[2]

$$E_P = 4.44 N_P f \Phi_{max} \tag{2–1}$$

$$E_S = 4.44 N_S f \Phi_{max} \tag{2–2}$$

Dividing Eq. (2–1) by Eq. (2–2),

$$\frac{E_P}{E_S} = \frac{N_P}{N_S} \tag{2–3}$$

---

[2] See Section 1.12, Chapter 1.

where: E_P = voltage induced in primary (V)
E_S = voltage induced in secondary (V)
N_P = turns in primary coil
N_S = turns in secondary coil

Thus, assuming no leakage flux, the ratio of induced voltages is equal to the ratio of turns.

EXAMPLE 2.1 Determine the peak value of sinusoidal flux in a transformer core that has a primary of 200 turns and is connected to a 240-V, 60-Hz, 50-kVA source.

Solution

$$E_P = 4.44 N_P f \Phi_{\max}$$

$$\Phi_{\max} = \frac{E_P}{4.44 N_P f} = \frac{240}{4.44 \times 200 \times 60} = 4.5 \times 10^{-3}\,\text{Wb}$$

EXAMPLE 2.2 A 15-kVA, 2400—240-V, 60-Hz transformer[3] has a magnetic core of 50-cm^2 cross section and a mean length of 66.7 cm. The application of 2400 V causes a magnetic field intensity of 450 A-t/m rms, and a maximum flux density of 1.5 T. Determine (a) the turns ratio; (b) the number of turns in each winding; (c) the magnetizing current.[4]

Solution

(a) The turns ratio is equal to the ratio of induced emfs, and is approximately equal to the nameplate voltage ratio. Thus,

$$\frac{N_P}{N_S} = \frac{E_P}{Es} \approx \frac{V_P}{V_S} = \frac{2400}{240} = 10$$

(b)

$$\Phi_{\max} = B_{\max} \times A = 1.5 \times \frac{50}{10^4} = 7.5 \times 10^{-3}\,\text{Wb}$$

$$E_P = 4.44 N_P f \Phi_{\max} \quad \Rightarrow \quad N_P = \frac{E_P}{4.44 f \Phi_{\max}}$$

$$N_P = \frac{2400}{4.44 \times 60 \times 7.5 \times 10^{-3}} = \underline{1201\text{ turns}}$$

$$\frac{N_P}{N_S} = 10 \quad \Rightarrow \quad \frac{1201}{N_S} = 10$$

$$N_S = \underline{120\text{ turns}}$$

---

[3] The long dash (called an em dash) indicates that the voltages are from different windings. See Section 3.3 on transformer nameplates.

[4] See Section 1.3, Chapter 1, for relationship between magnetic field intensity and magnetomotive force.

(c) From Eq. (1–2),

$$H = \frac{N_P I_M}{\ell}$$

where:
H = magnetic field intensity (A-t/m, rms)
N_P = turns in primary winding
I_M = magnetizing current (A, rms)
ℓ = mean length of core (m)

Solving for I_M and substituting known values,

$$I_M = \frac{H\ell}{N_P} = \frac{450 \times 0.667}{1201} = \underline{0.25 \text{ A}}$$

2.5 NO-LOAD CONDITIONS

With no load connected to the secondary, the current in the primary is just enough to establish the magnetic flux needed for transformer action, and to supply the hysteresis and eddy-current losses in the iron.[5] This no-load current, called the *exciting current,* varies between 1 and 2 percent of rated current in large power transformers, and may be as high as 6 percent of rated current in very small distribution transformers.

The exciting current can be divided into two right-angle components: a core-loss component that supplies the hysteresis and eddy-current losses in the iron, and a magnetizing component that establishes the *mutual flux* (Φ_M) that links both primary and secondary windings. These components are shown in Figure 2.4(a), and form the equivalent-circuit model of a transformer operating at *no load.* The corresponding phasor diagram for the exciting current and its right-angle components is shown in Figure 2.4(b). Note that the exciting current lags the applied voltage by a large angle;

[5] In effect, at no load, the transformer is nothing more than an impedance coil.

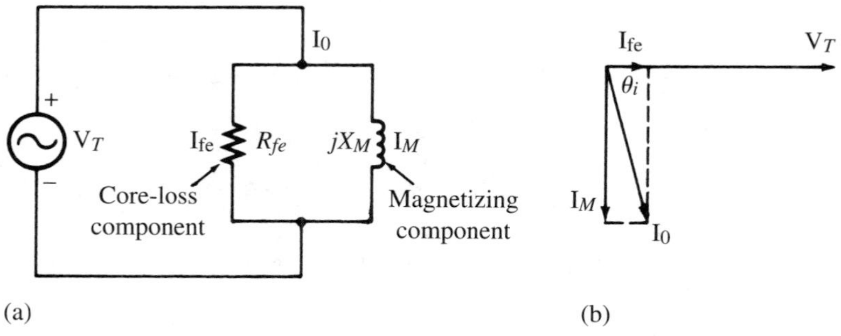

FIGURE 2.4
(a) Equivalent-circuit model of transformer with no load on secondary; (b) phasor diagram showing no-load conditions.

this may be as much as 85° in high-efficiency transformers. Expressing the exciting current in terms of its quadrature components,

$$\mathbf{I}_{\text{fe}} = \frac{\mathbf{V}_T}{R_{\text{fe}}}$$

$$\mathbf{I}_M = \frac{\mathbf{V}_T}{jX_M}$$

$$\mathbf{I}_0 = \mathbf{I}_{\text{fe}} + \mathbf{I}_M \qquad \textbf{(2–4)}$$

where:
$\mathbf{I}_0$ = exciting current
$\mathbf{I}_{\text{fe}}$ = core-loss component
$\mathbf{I}_M$ = magnetizing component
X_M = fictitious magnetizing reactance that accounts for the magnetizing current
R_{fe} = fictitious resistance that accounts for the core loss
$\mathbf{V}_T$ = voltage applied to primary

EXAMPLE 2.3 A 25-kVA, 2400—240-V, 60-Hz single-phase distribution transformer, operating at no load in the step-down mode, draws 138 W at a power factor (F_P) of 0.210 lagging. Using the equivalent circuit shown in Figure 2.4, determine (a) the exciting current and its quadrature components; (b) the equivalent magnetizing reactance and equivalent core-loss resistance; (c) and (d) repeat parts (a) and (b) for the transformer in the step-up mode.

Solution

(a) The phase angle of the exciting current is determined from the power-factor angle using[6]

$$\theta = (\theta_v - \theta_i)$$

where:
θ = power-factor angle
θ_i = angle of current phasor
θ_v = angle of applied voltage phasor

Unless otherwise specified, the phase angle of applied voltage is assumed to be zero.

$$\theta = \cos^{-1}(F_P) = \cos^{-1}0.210 = 77.8776 \quad \Rightarrow \quad 77.88^\circ$$

$$77.88^\circ = (0 - \theta_i)$$

$$\theta_i = -77.88^\circ$$

Referring to Figure 2.4(a), only the core-loss component draws active power. Hence,

$$P_{\text{core}} = V_T I_{\text{fe}} \quad \Rightarrow \quad 138 = 2400 \times I_{\text{fe}}$$

$$I_{\text{fe}} = \underline{0.0575\ \text{A}}$$

---

[6] See Appendix A.5, power relationships in a single-phase system.

The magnitudes $|\mathbf{I}_0|$ and $|\mathbf{I}_M|$ are determined from geometry of the corresponding phasor diagram in Figure 2.4(b),

$$\cos\theta_i = \frac{I_{\text{fe}}}{I_0} \quad \Rightarrow \quad 0.21 = \frac{0.0575}{I_0}$$

$$I_0 = \underline{0.2738 \text{ A}}$$

$$\tan\theta_i = \frac{I_M}{I_{\text{fe}}} \quad \Rightarrow \quad \tan(-77.8776°) = \frac{-I_M}{0.0575}$$

$$I_M = \underline{0.268 \text{ A}}$$

Or, applying the Pythagorean theorem to Figure 2.4(b),

$$I_0 = \sqrt{I_{\text{fe}}^2 + I_{\text{M}}^2} \quad \Rightarrow \quad I_M = \sqrt{I_0^2 - I_{\text{fe}}^2}$$

$$I_M = \sqrt{(0.2738)^2 + (0.0575)^2} = 0.268 \text{ A}$$

(b) Using high-side data,

$$I_M = \frac{V_T}{X_M} \quad \Rightarrow \quad 0.2677 = \frac{2400}{X_M}$$

$$X_M = 8965 \quad \Rightarrow \quad \underline{8.97 \text{ k}\Omega}$$

$$I_{\text{fe}} = \frac{V_T}{R_{\text{fe}}} \quad \Rightarrow \quad 0.0575 = \frac{2400}{R_{\text{fe}}}$$

$$R_{\text{fe}} = 41{,}739 \quad \Rightarrow \quad \underline{41.7 \text{ k}\Omega}$$

(c) The core loss and power factor are the same whether operating in the step-down or step-up mode. Hence, using the 240-V side as the primary,

$$P_{\text{core}} = V_T I_{\text{fe}} \quad \Rightarrow \quad 138 = 240 \times I_{\text{fe}}$$

$$I_{\text{fe}} = \underline{0.575 \text{ A}}$$

$$\cos\theta_i = \frac{I_{\text{fe}}}{I_0} \quad \Rightarrow \quad 0.21 = \frac{0.575}{I_0}$$

$$I_0 = 2.738 \quad \Rightarrow \quad \underline{2.74 \text{ A}}$$

$$\tan\theta_i = \frac{I_{\text{M}}}{I_{\text{fe}}} \quad \Rightarrow \quad \tan(-77.8776) = \frac{-I_M}{0.575}$$

$$I_M = 2.677 \quad \Rightarrow \quad \underline{2.68 \text{ A}}$$

(d) Using low-side data,

$$I_M = \frac{V_T}{X_M} \quad \Rightarrow \quad 2.677 = \frac{240}{X_M}$$

$$X_M = \underline{89.7 \ \Omega}$$

$$I_{\text{fe}} = \frac{V_T}{R_{\text{fe}}} \quad \Rightarrow \quad 0.575 = \frac{240}{R_{\text{fe}}}$$

$$R_{\text{fe}} = \underline{417.4 \ \Omega}$$

No-Load Ampere-Turns and Its Components

Multiplying Eq. (2–4) by the primary turns expresses the no-load mmf in terms of its quadrature components:

$$N_P\mathbf{I}_0 = N_P\mathbf{I}_{\text{fe}} + N_P\mathbf{I}_M \tag{2–5}$$

Component $N_P\mathbf{I}_{\text{fe}}$ does not contribute to the development of mutual flux, but serves only to oscillate the magnetic domains and to generate eddy currents in the core. If there were no core losses, component $N_P\mathbf{I}_{\text{fe}}$ would not exist, and the exciting ampere-turns would be reduced to only that required to establish the mutual flux.

Component $N_P\mathbf{I}_M$, called the magnetizing ampere-turns, produces the mutual flux and hence transformer action. The mutual flux expressed in terms of the rms magnetizing current is

$$\Phi_M = \frac{N_P I_M}{\mathcal{R}_{\text{core}}} \tag{2–6}$$

where: Φ_M = mutual flux produced by the magnetizing component of exciting current
I_M = magnetizing current
$\mathcal{R}_{\text{core}}$ = reluctance of transformer core

Applying Kirchhoff's voltage law to the primary circuit in Figure 2.3(b), and noting that $\mathbf{I}_P = \mathbf{I}_0$ at no load,

$$\mathbf{V}_T = \mathbf{I}_P R_P + \mathbf{E}_P \tag{2–7}$$

Solving for $\mathbf{I}_P$,

$$\mathbf{I}_P = \frac{\mathbf{V}_T - \mathbf{E}_P}{R_P} \tag{2–8}$$

where: $\mathbf{V}_T$ = applied voltage
$\mathbf{I}_P$ = primary current
$\mathbf{E}_P$ = voltage induced in the primary
R_P = resistance of primary winding

Voltage $\mathbf{E}_P$ is the cemf in the primary coil caused by the sinusoidal variation of flux in its window.

2.6 TRANSIENT BEHAVIOR WHEN LOADING AND UNLOADING

In accordance with Lenz's law, the emf induced in the secondary will be in a direction that opposes the change in flux that caused it. Hence, when a load is placed on the secondary winding, the instantaneous direction of the secondary current will set up an mmf of its own in opposition to the primary mmf. This is shown in Figure 2.5. Thus, for a very brief instant of time the core flux will decrease to

$$\phi_M = \frac{N_P i_M - N_S i_S}{\mathcal{R}_{\text{core}}} \tag{2–9}$$

The decrease in flux causes a decrease in cemf, which, in accordance with Eq. (2–8), causes an increase in primary current. The additional primary current ($\mathbf{I}_{P,\text{load}}$), called the *load component of primary current,* adds its mmf to the magnetizing component, causing the flux to increase. Thus,

$$\phi_M = \frac{N_P i_M + N_P i_{P,\text{load}} - N_S i_S}{\mathcal{R}_{\text{core}}} \tag{2–10}$$

The primary current increases until $N_P \mathbf{I}_{P,\text{load}} = N_S \mathbf{I}_S$, at which point both $\mathbf{\Phi}_M$ and $\mathbf{E}_P$ will have returned to essentially the same values they had before the switch was closed; any difference between $\mathbf{E}_P$ at no load and $\mathbf{E}_P$ under load conditions is due to the additional (but small) increase in voltage drop due to the resistance of the primary winding. Thus, the final steady-state primary current *under load conditions* will be

$$\mathbf{I}_P = \mathbf{I}_{\text{fe}} + \mathbf{I}_M + \mathbf{I}_{P,\text{load}}$$

$$\mathbf{I}_P = \mathbf{I}_0 + \mathbf{I}_{P,\text{load}} \tag{2–11}$$

Removing load from the secondary causes the opposite effect to take place. Opening the switch in Figure 2.5 causes $\mathbf{I}_S$ and hence $N_s\mathbf{I}_s$ to drop to zero. The resultant transient increase in mutual flux produces a transient increase in cemf, causing the primary current to drop back to its initial no-load value.

Although described as a step-by-step process, the actual behavior, when loading or unloading, is essentially simultaneous and takes place in a fraction of a second. *Note.* This entire discussion assumed constant permeability and no leakage flux.

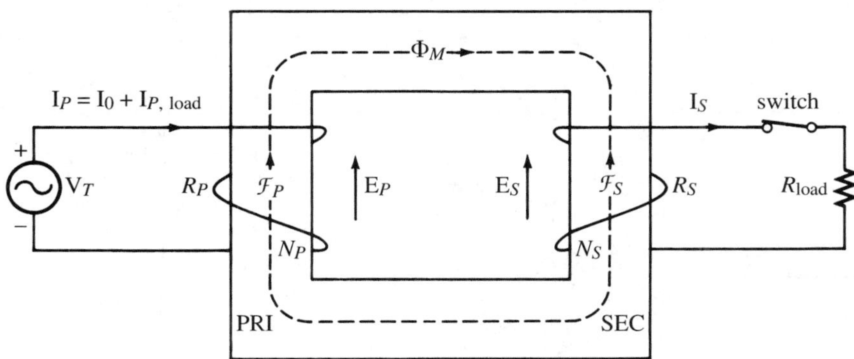

FIGURE 2.5
Relative directions of secondary current and secondary mmf for one-half cycle when load switch is closed.

2.7 EFFECT OF LEAKAGE FLUX ON THE OUTPUT VOLTAGE OF A REAL TRANSFORMER

All of the flux in a real transformer is not common to both primary and secondary windings. The flux in a real transformer has three components: mutual flux, primary leakage flux, and secondary leakage flux. This is shown in Figure 2.6, where, in order to simplify visualization and analysis, only a few representative leakage paths are shown. For the transformer shown in Figure 2.6, the primary leakage flux (caused by primary current) links only the primary turns, the secondary leakage flux (caused by secondary current) links only the secondary turns, and the mutual flux (due to the magnetizing component of the exciting current) links both windings.

The relationship between coil flux, leakage flux, and mutual flux, for the respective primary and secondary coils shown in Figure 2.6, are

$$\Phi_P = \Phi_M + \Phi_{\ell \mathrm{p}} \tag{2–12}$$

$$\Phi_S = \Phi_M - \Phi_{\ell \mathrm{s}} \tag{2–13}$$

where:
Φ_P = net flux in window of primary coil
Φ_S = net flux in window of secondary coil
Φ_M = mutual flux
$\Phi_{\ell \mathrm{p}}$ = leakage flux associated with the primary coil
$\Phi_{\ell \mathrm{s}}$ = leakage flux associated with the secondary coil

Equations (2–12) and (2–13) illustrate how the leakage flux in both windings serves to reduce the output voltage of the secondary; the mutual flux is less than the available primary flux because of primary leakage, and the net flux in the secondary is the mutual flux less the secondary leakage. Less flux in the secondary coil results in a lower secondary voltage than if no leakage were present.

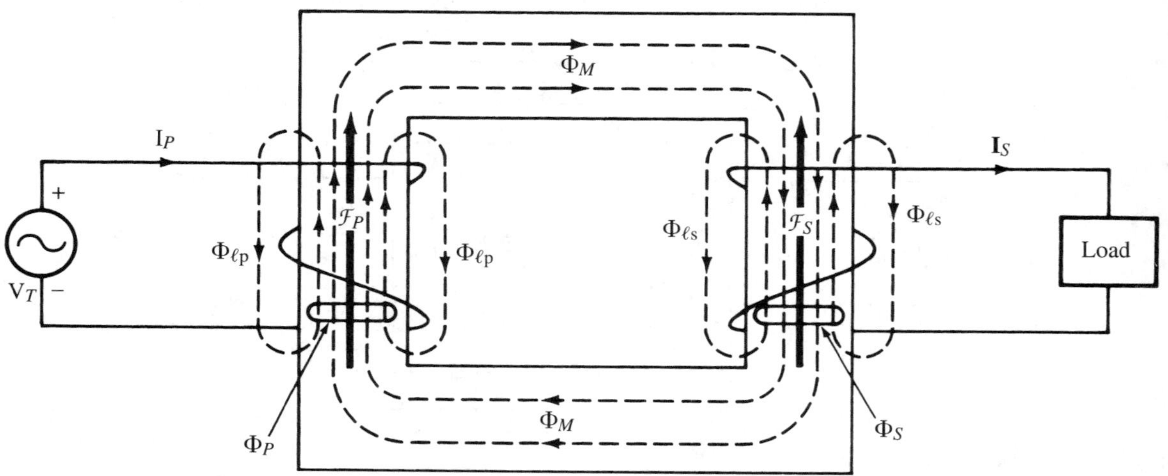

FIGURE 2.6
Component fluxes in the core of a loaded transformer.

The voltage drop caused by leakage flux is proportional to the load current. The greater the load current, the greater the magnitudes of both the primary and secondary ampere turns, and hence the greater the respective leakage fluxes in both primary and secondary windings. Although leakage flux has an adverse effect on the transformer output voltage, it proves an asset under severe short-circuit conditions; the large voltage drop caused by the intense leakage flux limits the current to a lower value than would otherwise occur if no leakage were present and thus helps to avoid damage to the transformer.

2.8 IDEAL TRANSFORMER

An ideal transformer is a hypothetical transformer that has no leakage flux and no core losses; the permeability of its core is infinite, it requires no exciting current to maintain the flux, and its windings have zero resistance. Although an ideal transformer does not exist, its mathematical relationships have practical applications in the development of equivalent circuits for real transformers, for the development of equivalent circuits for induction motors, and for impedance transformation applications.

The basic relationships for the ideal transformer are developed with the aid of Figure 2.7, which shows a load connected across the secondary terminals of an ideal transformer. The primed symbols are used to designate the induced voltages and input impedance of an ideal transformer.

Turns Ratio

The turns ratio a *is the ratio of the number of turns in the high-voltage winding to the number of turns in the low-voltage winding* **[3]**. It is equal to the ratio of voltages in the ideal transformer and is approximately equal to the voltage ratio of the real transformer (*high side to low side*), with no load connected to the secondary; the effects of leakage flux and winding resistance are insignificant at no load. Hence, when information

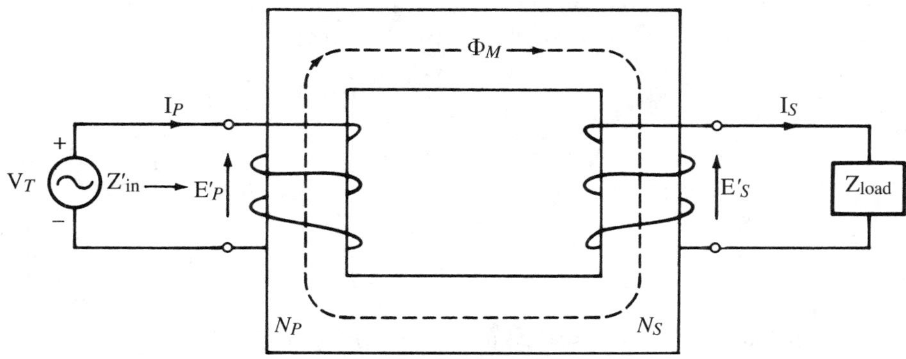

FIGURE 2.7
Ideal transformer.

about the turns ratio is not available, and voltage measurements at no load cannot be made, the nameplate voltage ratio can be used as an approximation of the turns ratio.

Thus, in terms of high-side (HS) and low-side (LS) values,

$$a = \frac{N_{\text{HS}}}{N_{\text{LS}}} = \frac{E'_{\text{HS}}}{E'_{\text{LS}}} \approx \frac{V_{\text{HS}}}{V_{\text{LS}}} \tag{2–14}$$

where: a = turns ratio
$V_{\text{HS}}/V_{\text{LS}}$ = nameplate voltage ratio
$E'_{\text{HS}}/E'_{\text{LS}}$ = ratio of induced voltages

Thus, referring to the ideal transformer in Figure 2.7, and *assuming* the primary is the high-voltage winding,

$$\frac{\mathbf{E}'_P}{\mathbf{E}'_S} = \frac{N_P}{N_S} = a$$

$$\mathbf{E}'_P = a\mathbf{E}'_S$$

Input Impedance of an Ideal Transformer

The input impedance looking into the primary terminals of the ideal transformer shown in Figure 2.7 is

$$\mathbf{Z}'_{\text{in}} = \frac{\mathbf{E}'_P}{\mathbf{I}_P} \tag{2–15}$$

Voltages $\mathbf{E}'_P$ and $\mathbf{E}'_S$ are induced by the same flux, and thus must have the same phase angle. Hence,

$$\frac{\mathbf{E}'_P}{\mathbf{E}'_S} = \frac{E'_P\angle\alpha}{E'_S\angle\alpha} = a$$

$$\mathbf{E}'_P = a\mathbf{E}'_S \tag{2–16a}$$

The apparent power input to the ideal transformer must equal the apparent power output. Expressed as phasor power (see Appendix A–5),

$$\mathbf{E}'_P\mathbf{I}^*_P = \mathbf{E}'_S\mathbf{I}^*_S$$

$$\mathbf{I}^*_P = \frac{\mathbf{E}'_S}{\mathbf{E}'_P}\cdot\mathbf{I}^*_S \qquad \Rightarrow \qquad \mathbf{I}^*_P = \frac{1}{a}\mathbf{I}^*_S$$

Therefore,

$$\mathbf{I}_P = \frac{1}{a}\mathbf{I}_S \tag{2–16b}$$

Substituting Eqs. (2–16a) and (2–16b) into Eq. (2–15),

$$\mathbf{Z}'_{\text{in}} = \frac{a\mathbf{E}'_S}{\mathbf{I}_S/a} = a^2\frac{\mathbf{E}'_S}{\mathbf{I}_S} \tag{2–17}$$

Applying Ohm's law to the secondary circuit in Figure 2.7,

$$\mathbf{Z}_{\text{load}} = \frac{\mathbf{E}'_S}{\mathbf{I}_S} \tag{2–18}$$

Substituting Eq. (2–18) into Eq. (2–17),

$$\mathbf{Z}'_{\text{in}} = a^2\mathbf{Z}_{\text{load}} \tag{2–19}$$

Equation (2–19) indicates that a well-designed transformer, with very low leakage flux, can be used as an *impedance multiplier.* The multiplication factor is equal to the square of the turns ratio. Transformers specifically designed for this purpose are called *impedance-matching transformers,* and have applications in audio systems **[2]**.

EXAMPLE 2.4 An ideal transformer with a primary of 200 turns and a secondary of 20 turns has its primary connected to a 120-V, 60-Hz supply, and its secondary connected to a 100 $\angle 30°$-Ω load. Determine (a) the secondary voltage; (b) the load current; (c) the input current to the primary; (d) the input impedance looking into the primary terminals. *Note:* For purposes of simplification in problem solving, unless otherwise specified it will be assumed (throughout the text) that the phase angle of the input voltage is zero degrees.

Solution

(a) Using Figure 2.7 as a guide,

$$a = \frac{N_{\text{HS}}}{N_{\text{LS}}} = \frac{200}{20} = 10$$

$$\frac{E'_P}{E'_S} = a \quad \Rightarrow \quad E'_{\text{LS}} = \frac{120}{10} = \underline{12 \text{ V}}$$

(b)
$$\mathbf{I}_S = \frac{\mathbf{E}'_S}{\mathbf{Z}_{\text{load}}} = \frac{12\angle 0°}{100\angle 30°} = \underline{0.12\angle -30° \text{ A}}$$

(c)
$$\mathbf{I}_P = \frac{1}{a} \cdot \mathbf{I}_S = \frac{0.12\angle -30°}{10} = \underline{0.012\angle -30 \text{ A}}$$

(d)
$$\mathbf{Z}'_{\text{in}} = a^2\mathbf{Z}_{\text{load}} = 10^2 \times 100\angle 30° = \underline{10\angle 30° \text{ k}\Omega}$$

2.9 LEAKAGE REACTANCE AND THE EQUIVALENT CIRCUIT OF A REAL TRANSFORMER

Calculations to determine the overall voltage drop in a transformer, for different magnitudes and different power factors of loads, must take into consideration the effect of leakage flux. To facilitate such calculations, voltage drops caused by leakage flux are expressed in terms of fictitious *leakage reactances;* these derived mathematical quantities, when multiplied by the current in them, will result in voltage drops equal to those brought about by the respective leakage fluxes.

Figure 2.8(a) represents a real transformer with all of its associated core losses and flux leakages. The induced emfs in the primary and secondary coils, due to the net flux through their respective windows, are

$$\mathbf{E}_P = 4.44 N_P f \mathbf{\Phi}_P \qquad \mathbf{E}_S = 4.44 N_S f \mathbf{\Phi}_S$$

From Eqs. (2–12) and (2–13), respectively,

$$\mathbf{\Phi}_P = \mathbf{\Phi}_M + \mathbf{\Phi}_{\ell p} \qquad \mathbf{\Phi}_S = \mathbf{\Phi}_M - \mathbf{\Phi}_{\ell s}$$

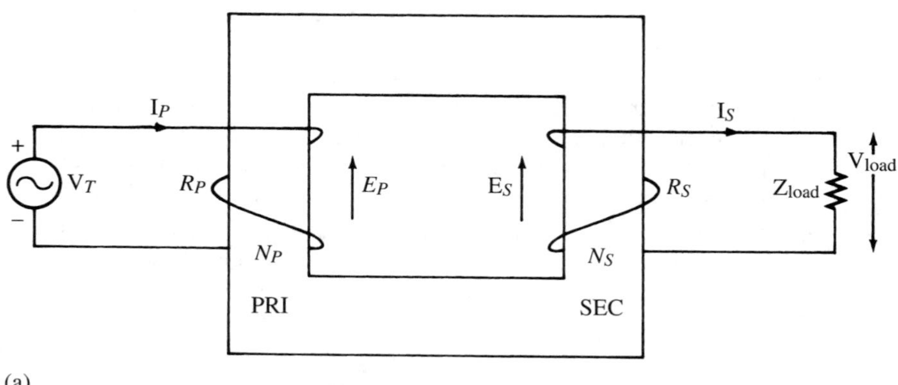

(a)

(b)

FIGURE 2.8
(a) Real transformer; (b) equivalent circuit of real transformer using an ideal transformer and external components.

Thus, expressed in terms of component fluxes,

$$\mathbf{E}_P = 4.44N_P f\mathbf{\Phi}_M + 4.44N_P f\mathbf{\Phi}_{\ell \mathrm{p}} \tag{2–20}$$

$$\mathbf{E}_S = 4.44N_S f\mathbf{\Phi}_M - 4.44N_S f\mathbf{\Phi}_{\ell \mathrm{s}} \tag{2–21}$$

Expressing Eqs. (2–20) and (2–21) in simplified form,

$$\mathbf{E}_P = \mathbf{E}'_P + \mathbf{E}_{\ell \mathrm{p}} \tag{2–22}$$

$$\mathbf{E}_S = \mathbf{E}'_S - \mathbf{E}_{\ell \mathrm{s}} \tag{2–23}$$

where: $\mathbf{E}_P$ = net voltage induced in primary
$\mathbf{E}'_P$ = voltage induced in primary due to mutual flux
$\mathbf{E}_{\ell \mathrm{p}}$ = voltage induced in primary due to primary leakage
$\mathbf{E}_S$ = net voltage induced in secondary
$\mathbf{E}'_S$ = voltage induced in secondary due to mutual flux
$\mathbf{E}_{\ell \mathrm{s}}$ = voltage induced in secondary due to secondary leakage

Applying Kirchhoff's voltage law to the primary circuit,

$$\mathbf{V}_T = \mathbf{E}_P + \mathbf{I}_P \boldsymbol{R}_P \tag{2–24}$$

Substituting Eq. (2–22) into Eq. (2–24),

$$\mathbf{V}_T = \mathbf{I}_P R_P + \mathbf{E}_{\ell \mathrm{p}} + \mathbf{E}'_P \tag{2–25}$$

where

$$\mathbf{I}_P = \mathbf{I}_{\mathrm{fe}} + \mathbf{I}_M + \mathbf{I}_{P,\mathrm{load}} \tag{2–26}$$

Applying Kirchhoff's law to the secondary in Figure 2.8(a),

$$\mathbf{E}_S = \mathbf{I}_S R_S + \mathbf{V}_{\mathrm{load}} \tag{2–27}$$

Substituting Eq. (2–23) into Eq. (2–27) and rearranging terms,

$$\mathbf{E}'_S = E_{\ell \mathrm{s}} + \mathbf{I}_S R_S + \mathbf{V}_{\mathrm{load}} \tag{2–28}$$

Induced voltages $\mathbf{E}'_P$ and $\mathbf{E}'_S$ are due to the mutual flux, and induced voltages $\mathbf{E}_{\ell \mathrm{p}}$ and $\mathbf{E}_{\ell \mathrm{s}}$ are due to their respective leakage fluxes.

Using Eqs. (2–25), (2–26), and (2–28) as a guide, the real transformer in Figure 2.8(a) can be redrawn as an equivalent circuit using an ideal transformer whose windings are in series with external components that account for the losses, the voltage drops, and the exciting current of the real transformer. This is shown in Figure 2.8(b). The leakage flux, shown in the equivalent "leakage coils" in Figure 2.8(b), may be expressed in terms of the respective coil currents, and the reluctances of the respective leakage paths of the real transformer. Thus,

$$\mathbf{\Phi}_{\ell \mathrm{p}} = \frac{N_P \mathbf{I}_P}{\mathcal{R}_{\ell \mathrm{p}}} \qquad \mathbf{\Phi}_{\ell \mathrm{s}} = \frac{N_S \mathbf{I}_S}{\mathcal{R}_{\ell \mathrm{s}}} \tag{2–29}$$

The voltage generated by the sinusoidal variation of flux through the window of any coil is expressed as[7]

$$e = 2\pi f N\Phi_{\max}\cos(2\pi f t)$$
$$E_{\max} = 2\pi f N\Phi_{\max} \tag{2–30}$$

Expressing equation set (2–29) in terms of maximum flux ($I_{\max}$ causes $\Phi_{\max}$), substituting each (in turn) in Eq. (2–30), and using appropriate subscripts,

$$E_{\ell p,\max} = 2\pi f N_P\left(\frac{N_P I_{P,\max}}{\mathcal{R}_{\ell p}}\right) \qquad E_{\ell s,\max} = 2\pi f N_S\left(\frac{N_S I_{S,\max}}{\mathcal{R}_{\ell s}}\right)$$
$$E_{\ell p,\max} = 2\pi f\left(\frac{N_P^2}{\mathcal{R}_{\ell p}}\right)I_{P,\max} \qquad E_{\ell s,\max} = 2\pi f\left(\frac{N_S^2}{\mathcal{R}_{\ell s}}\right)I_{S,\max}$$

Dividing both sides of each equation by $\sqrt{2}$ to obtain rms values,

$$E_{\ell p} = 2\pi f\left(\frac{N_P^2}{\mathcal{R}_{\ell p}}\right)I_P \qquad E_{\ell s} = 2\pi f\left(\frac{N_S^2}{\mathcal{R}_{\ell s}}\right)I_S \tag{2–31}$$

The inductance of a coil is related to the number of turns in the coil and the reluctance of its magnetic circuit in the following manner **[2]**:

$$L = \frac{N^2}{\mathcal{R}} \tag{2–32}$$

Substituting Eq. (2–32) into equation set (2–31),

$$E_{\ell p} = (2\pi f L_{\ell p})I_P \qquad E_{\ell s} = (2\pi f L_{\ell s})I_S \tag{2–33}$$

Thus,

$$E_{\ell p} = I_P X_{\ell p} \qquad E_{\ell s} = I_S X_{\ell s} \tag{2–34}$$

where:
$E_{\ell p}$ = leakage voltage of primary (rms)
$E_{\ell s}$ = leakage voltage of secondary (rms)
$X_{\ell p} = 2\pi f L_{\ell p}$ = leakage reactance of primary (Ω)
$X_{\ell s} = 2\pi f L_{\ell s}$ = leakage reactance of secondary (Ω)
$L_{\ell p}$ = leakage inductance of primary (H)
$L_{\ell s}$ = leakage inductance of secondary (H)

As indicated in equation set (2–34), the voltage drops due to leakage flux may be expressed in terms of the respective leakage reactances and the associated primary and secondary currents.

The final two-winding equivalent circuit of the transformer, expressing the voltage drops due to leakage flux in terms of leakage reactance drops, is shown in Figure 2.9. The parallel branch, representing the path for exciting current $\mathbf{I}_0$, contains

[7] See Section 1.12, Chapter 1.

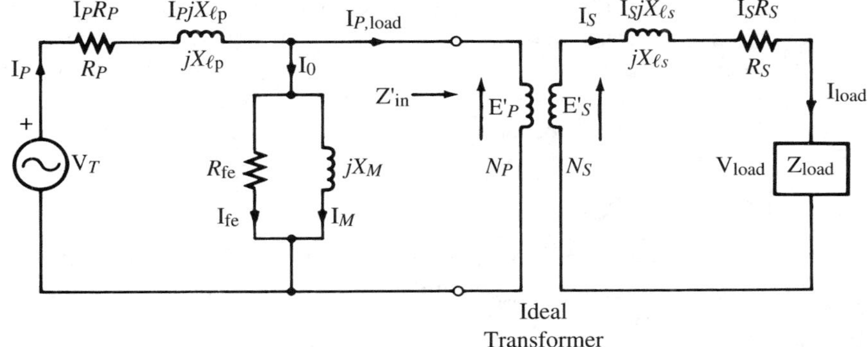

FIGURE 2.9
Leakage reactance in equivalent-circuit model.

a fictitious resistance R_{fe} that dissipates heat energy at the same rate as does the actual hysteresis and eddy-current losses in the core, and a fictitious magnetizing reactance X_M that draws the same magnetizing current as does the actual transformer.

2.10 EQUIVALENT IMPEDANCE OF A TRANSFORMER

The equivalent-circuit models of the transformer, shown in Figures 2.8(b) and 2.9, are very useful for analyzing the individual effects of winding resistance and leakage effects in primary and secondary windings. For purposes of simplified calculations of engineering problems, however, the actual transformer is replaced by an equivalent impedance in series with the source voltage and the load.

Applying the relationships developed for the ideal transformer in Section 2.8 to the ideal transformer in Figure 2.9,

$$\mathbf{Z}'_{\text{in}} = a^2 \frac{\mathbf{E}'_S}{\mathbf{I}_S} \tag{2–17}$$

Applying Ohm's law to the secondary circuit in Figure 2.9,

$$\mathbf{I}_S = \frac{\mathbf{E}'_S}{R_S + jX_{\ell s} + \mathbf{Z}_{\text{load}}} \tag{2–35}$$

$$\frac{\mathbf{E}'_S}{\mathbf{I}_S} = R_S + jX_{\ell s} + \mathbf{Z}_{\text{load}} \tag{2–36}$$

Substituting Eq. (2–36) into Eq. (2–17) and multiplying through,

$$\mathbf{Z}'_{\text{in}} = a^2(R_S + jX_{\ell s} + \mathbf{Z}_{\text{load}}) \tag{2–37}$$

$$\mathbf{Z}'_{\text{in}} = a^2 R_S + ja^2 X_{\ell s} + a^2 \mathbf{Z}_{\text{load}} \tag{2–38}$$

As indicated in Eqs. (2–37) and (2–38), impedance $\mathbf{Z}'_{\text{in}}$ is the impedance of the secondary and its connected load multiplied by the square of the turns ratio. Impedance $\mathbf{Z}'_{\text{in}}$ is called the *reflected impedance;* it is the impedance of the secondary and its connected load reflected (or *referred*) to the primary side. This is shown in Figure 2.10(a), where the reflected impedances are placed to the left of the ideal transformer.

Referring to Eq. (2–38),

$$a^2R_S = \text{resistance of secondary } \textit{referred} \text{ to primary}$$
$$a^2X_{\ell s} = \text{leakage reactance of secondary } \textit{referred} \text{ to primary}$$
$$a^2Z_{\text{load}} = Z_{\text{load},P} = \text{impedance of load } \textit{referred} \text{ to primary}$$

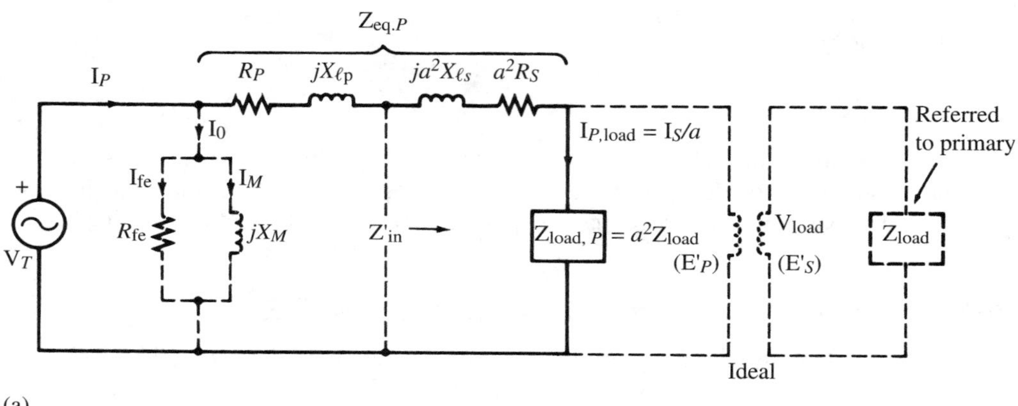

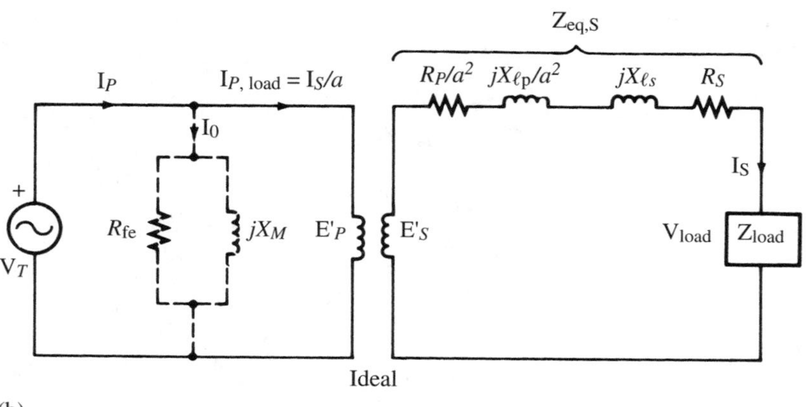

FIGURE 2.10
Equivalent circuits: (a) parameters referred to primary; (b) parameters referred to secondary.

Note that the parallel branch representing the exciting current in Figure 2.9 is shown shifted to the input terminals of the transformer in Figure 2.10. When operating at or near rated load, the load component of primary current is significantly greater than the exciting current ($\mathbf{I}_{P,\text{load}} >> \mathbf{I}_0$). Hence, shifting the exciting current component to the input terminals will not cause any appreciable error in calculations involving transformer behavior under rated or near-rated load conditions. Thus, neglecting the exciting current branch, the *equivalent impedance* of the transformer in Figure 2.10(a), with all parameters referred to the primary is

$$\mathbf{Z}_{\text{eq},P} = R_P + a^2R_S + j(X_{\ell\text{p}} + a^2X_{\ell s}) \quad \textbf{(2–39)}$$

$$\mathbf{Z}_{\text{eq},P} = R_{\text{eq},P} + jX_{\text{eq},P} \quad \textbf{(2–40)}$$

Figure 2.10(b) shows an equivalent circuit with all parameters *referred to the secondary,* where

R_P/a^2 = resistance of primary referred to secondary
$X_{\ell\text{p}}/a^2$ = leakage reactance of primary referred to secondary

Thus, the equivalent impedance of the transformer shown in Figure 2.10(b), with the primary parameters referred to the secondary, is

$$\mathbf{Z}_{\text{eq},S} = R_S + R_P/a^2 + j(X_{\ell s} + X_{\ell p}/a^2) \quad \textbf{(2–41)}$$

$$\mathbf{Z}_{\text{eq},S} = R_{\text{eq},S} + jX_{\text{eq},S} \quad \textbf{(2–42)}$$

Although the resistance and leakage parameters of a transformer, as expressed in Eqs. (2–39), (2–40), (2–41), and (2–42), are constant for a given frequency, the load connected to the secondary is adjustable. Hence, $\mathbf{Z}_{\text{load}}$ will be different for different loadings and different power factors. The equivalent-circuit parameters for a given transformer may be obtained from the transformer nameplate, from the manufacturer, or from a test procedure outlined in Section 2.14.

High-Side, Low-Side

Power and distribution transformers may be used to either step up or step down voltage. Hence, it is convenient to refer to the two windings as the high-voltage side (HS) and the low-voltage side (LS). This is shown in Figure 2.11 for *step-down* operation and is a modification of Figure 2.10.

The exciting current branch, shown with broken lines in Figure 2.11(a), may be omitted when making calculations involving operations at or near rated load; for such loadings, the load component of primary current is so much greater than the exciting current that the exciting current may be neglected. However, *when making calculations for loadings less than 25 percent rated load, the no-load components must be considered if significant errors in current calculations are to be avoided.*

The same circuits shown in Figure 2.11 for step-down operation may also be used for step-up operations by referring the transformer parameters to the low side, as

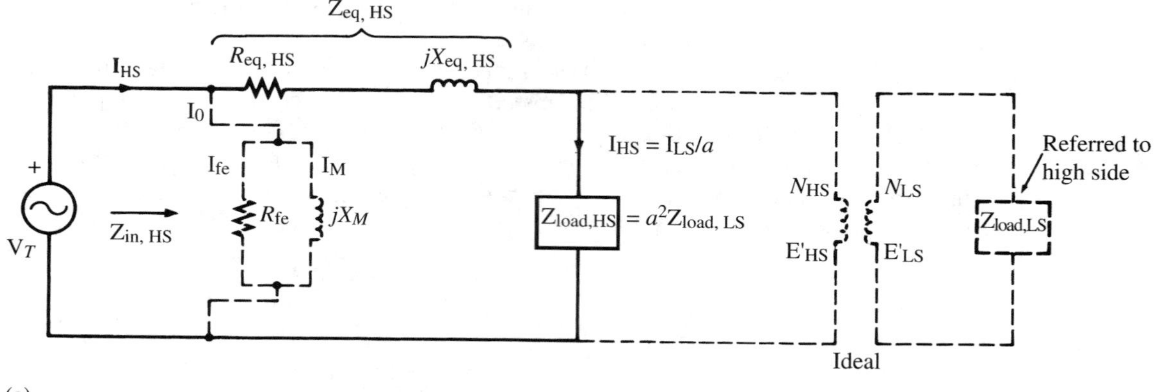

(a)

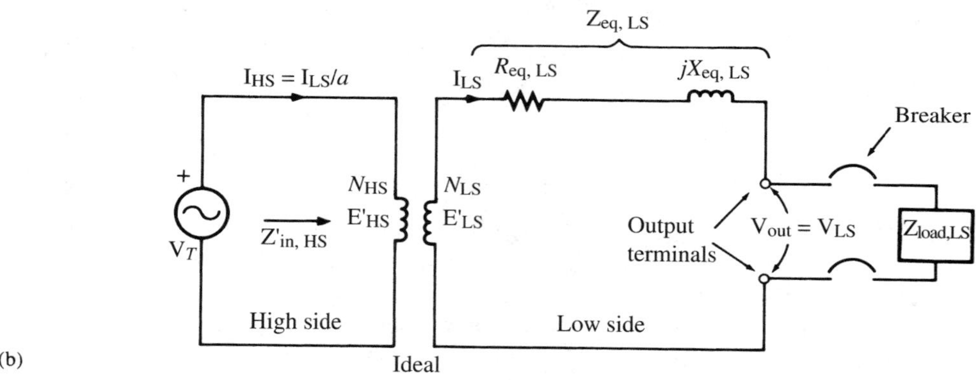

(b)

FIGURE 2.11
Equivalent circuits for step-down operation: (a) in terms of high-side values; (b) in terms of low-side values.

shown in Figure 2.12, and converting the high-side load impedance to the low side using

$$\mathbf{Z}_{\text{load,LS}} = \frac{1}{a^2}\mathbf{Z}_{\text{load,HS}} \qquad \textbf{(2–43)}$$

Note that the lower voltage rating and higher current rating of the low side (compared to the high side) requires the low side to have fewer turns of larger cross-sectional area conductor. Hence, the equivalent impedance of the transformer referred to the low side will always be less than the equivalent impedance referred to the high side.

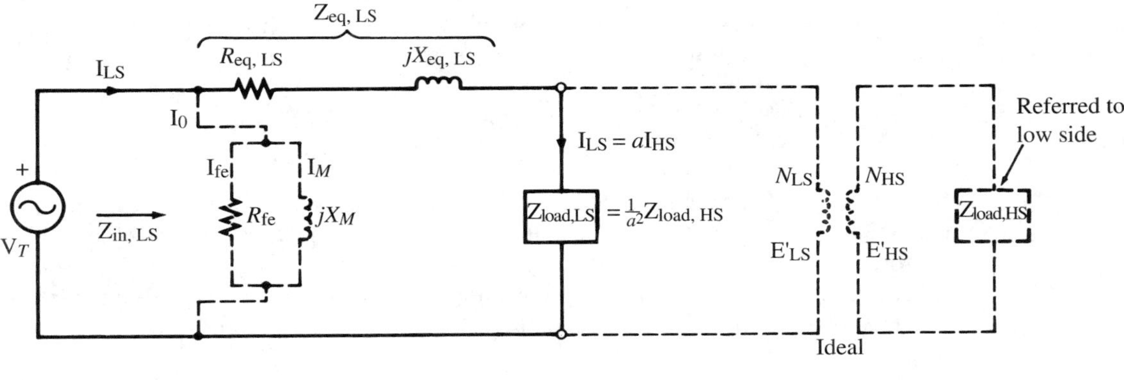

(a)

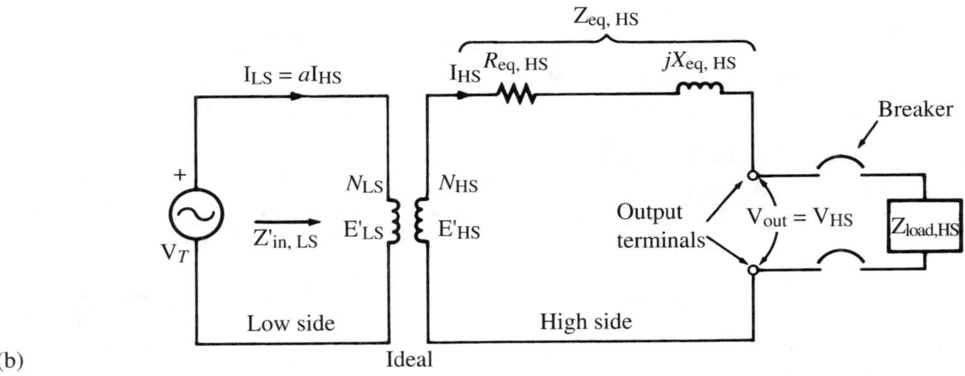

(b)

FIGURE 2.12
Equivalent circuits for step-up operation: (a) in terms of low-side values; (b) in terms of high-side values.

$$\mathbf{Z}_{eq,LS} = \frac{\mathbf{Z}_{eq,HS}}{a^2} \tag{2–43a}$$

The circuit models shown in Figures 2.11(a) and 2.12(a) are used to determine the input impedance of the combined transformer and load. The circuit models shown in Figures 2.11(b) and 2.12(b) are used to determine the no-load voltage and the voltage regulation.[8]

[8] See Section 2.11 for a discussion of voltage regulation.

EXAMPLE 2.5 A 75-kVA, 4800—240-V, 60-Hz, single-phase transformer has the following parameters expressed in ohms:

$$R_{LS} = 0.00600 \qquad R_{HS} = 2.488 \qquad R_{fe,HS} = 44{,}202$$
$$X_{LS} = 0.0121 \qquad X_{HS} = 4.8384 \qquad X_{M,HS} = 7798.6$$

The transformer is operating in the step-down mode, delivering one-half rated load at rated voltage and 0.96 F_p lagging. Determine (a) the equivalent impedance of the transformer referred to the high side; (b) the input impedance of the combined transformer and load; (c) the actual input voltage at the high side; (d) the input impedance if the load is disconnected; (e) the exciting current for the conditions in (d).

Solution

(a)

$$I_{LS} = \frac{S}{V_{LS}} = \frac{75{,}000 \times 1/2}{240} = \underline{156.25 \text{ A}}$$

For lagging power factor,

$$\theta = \cos^{-1} 0.96 = 16.26°$$
$$\theta = (\theta_v - \theta_i) \quad \Rightarrow \quad 16.26° = 0 - \theta_i \quad \Rightarrow \quad \theta_i = -16.26°$$
$$\mathbf{I}_{LS} = 156.25\angle{-16.26°} \text{ A}$$
$$\mathbf{Z}_{load,LS} = \frac{\mathbf{V}_{load}}{\mathbf{I}_{load}} = \frac{240\angle 0°}{156.25\angle{-16.26°}} = 1.536\angle 16.26° \ \Omega$$
$$a \approx \frac{V_{HS}}{V_{LS}} = \frac{4800}{240} = 20$$

Referring to Figure 2.11(a),

$$\mathbf{Z}_{eq,HS} = R_{eq,HS} + jX_{eq,HS} = R_{HS} + a^2R_{LS} + j(X_{HS} + a^2X_{LS})$$
$$\mathbf{Z}_{eq,HS} = 2.488 + 20^2(0.00600) + j(4.8384 + 20^2(0.0121))$$
$$\mathbf{Z}_{eq,HS} = 4.888 + j9.678 = \underline{10.84\angle 63.2° \ \Omega}$$

(b) Referring to Figure 2.11(a), and neglecting the exciting current branch,

$$\mathbf{Z}_{load,HS} = a^2\mathbf{Z}_{load,LS} = 20^2 \times 1.536\angle 16.26° = 614.40\angle 16.26° \ \Omega$$
$$\mathbf{Z}_{load,HS} = 589.82 + j172.03 \ \Omega$$
$$\mathbf{Z}_{in} = \mathbf{Z}_{load,HS} + \mathbf{Z}_{eq,HS} = (589.82 + j172.03) + (4.888 + j9.678)$$
$$\mathbf{Z}_{in} = 594.71 + j181.71 = \underline{621.85\angle 16.99° \ \Omega}$$

(c)

$$I_{HS} = \frac{I_{LS}}{a} = \frac{156.25}{20} = 7.81 \text{ A}$$
$$V_T = I_{HS}Z_{in} = 7.81 \times 621.85 = \underline{4857 \text{ V}}$$

(d) With the load disconnected, the output load and current are zero. Thus,

$$Z_{\text{load}} = \frac{V_{\text{load}}}{I_{\text{load}}} = \frac{V_{\text{load}}}{0} = \infty$$

Referring to Figure 2.11(a), with the load disconnected, the net impedance looking into the input terminals of the transformer is the impedance of the exciting branch. The rest of the model is an open circuit. Thus,

$$\mathbf{Z}_{\text{in}} = \frac{1}{(1/R_{\text{fe}}) + (1/jX_{\text{M}})} = \frac{1}{(1/44{,}202) + (1/j7798.6)}$$

$$\mathbf{Z}_{\text{in}} = \frac{1}{22.623 \times 10^{-6} - j128.228 \times 10^{-6}} = \frac{10^6}{130.208\angle{-79.99°}}$$

$$\mathbf{Z}_{\text{in}} = \underline{7680\angle{79.99°}\,\Omega}$$

(e)

$$\mathbf{I}_0 = \frac{\mathbf{V}_T}{\mathbf{Z}_{\text{in}}} = \frac{4857\angle{0°}}{7680\angle{79.99°}} = \underline{0.63\angle{-79.99°}\text{ A}}$$

EXAMPLE 2.6 The equivalent resistance and equivalent reactance (high side) for a 37.5-kVA, 2400—600-V, 60-Hz transformer are 2.80 Ω and 6.00 Ω, respectively. If a load impedance of $10.0\angle{20°}$ Ω is connected to the low side, determine (a) the equivalent input impedance of the transformer and load combination; (b) the primary current if 2400 V is supplied to the primary; (c) the voltage across the load.

Solution

The circuit used is that shown in Figure 2.11(a).

(a)

$$a \approx \frac{V_{\text{HS}}}{V_{\text{LS}}} = \frac{2400}{600} = 4.0$$

$$\mathbf{Z}_{\text{load,HS}} = a^2\mathbf{Z}_{\text{load,LS}} = 4^2 \times 10\angle{20°} = 160\angle{20°} = (150.351 + j54.723)\ \Omega$$

$$\mathbf{Z}_{\text{in}} = 2.8 + j6.0 + 150.351 + j54.723 = \underline{164.75\angle{21.63°}\ \Omega}$$

(b)

$$\mathbf{I}_{\text{HS}} = \frac{\mathbf{V}_T}{\mathbf{Z}_{\text{in}}} = \frac{2400\angle{0°}}{164.75\angle{21.63°}} = \underline{14.57\angle{-21.63°}\text{ A}}$$

(c) Referring to Figure 2.11(a), the voltage *across the reflected load* is

$$\mathbf{E}'_{\text{HS}} = \mathbf{I}_{\text{HS}}a^2\mathbf{Z}_{\text{load,LS}} = 14.57\angle{-21.63°} \times 4^2 \times 10\angle{20°} = \underline{2330.8\angle{-1.63°}\text{ V}}$$

The actual voltage across the *real load* is the voltage at the secondary of the ideal transformer in Figure 2.11(a). Thus,

$$\mathbf{E}'_{\text{LS}} = \frac{\mathbf{E}'_{\text{HS}}}{a} = \frac{2330.8\angle{-1.63°}}{4} = \underline{582.7\angle{-1.63°}\text{ V}}$$

2.11 VOLTAGE REGULATION

The effects of leakage flux and winding resistance in a transformer cause internal voltage drops that result in different output voltages for different loads. *The difference between the output voltage at no load and the output voltage at rated load, divided by the output voltage at rated load, is called the voltage regulation* of the transformer, and is commonly used as a figure of merit when comparing transformers **[2]**. Expressed mathematically,

$$\text{reg} = \frac{E_{\text{nl}} - V_{\text{rated}}}{V_{\text{rated}}} \tag{2–44}$$

where: E_{nl} = voltmeter reading at the output terminals when no load is connected to the transformer

V_{rated} = voltmeter reading at the output terminals when the transformer is supplying rated apparent power

Although Eq. (2–44) expresses the voltage regulation in decimal form, called *per-unit regulation,* it may also be expressed in percent.

The no-load and full-load voltages in Eq. (2–44) must be *all* high-side values or *all* low-side values. The voltage regulation will be the same, however, whether all high-side values or all low-side values are used.

The voltage regulation of a transformer, along with voltage, current, frequency, and apparent power ratings, are required data when specifying replacement transformers, when selecting transformers for parallel operation, when selecting transformers for polyphase arrangements, or when selecting transformers that will be used in distribution systems that feed large induction motors.

Although the regulation of a transformer may be determined from a set of no-load and full-load voltage measurements, as expressed in Eq. (2–44), this requires loading the transformer to its rated value at the desired power factor. Since this is seldom easy to accomplish, and in most cases is impractical, a mathematical determination using the equivalent circuit in Figure 2.11(b) or Figure 2.12(b) is preferred.

Referring to Figure 2.11(b), $\mathbf{E}'_{\text{LS}}$ is the no-load voltage. It is the voltage that appears across the output terminals when the load is removed (circuit breaker open); removing the load causes $\mathbf{I}_{\text{LS}} = 0$, which causes $\mathbf{I}_{\text{LS}}\mathbf{Z}_{\text{eq,LS}} = 0$, resulting in an output voltage equal to $\mathbf{E}'_{\text{LS}}$. Thus, the no-load voltage for *rated load conditions at a specified power factor* is determined by applying Kirchhoff's voltage law to the secondary and solving for $\mathbf{E}'_{\text{LS}}$. Referring to Figure 2.11(b), and assuming *rated load* on the secondary,

$$\mathbf{E}'_{\text{LS}} = \mathbf{I}_{\text{LS}}\mathbf{Z}_{\text{eq,LS}} + \mathbf{V}_{\text{LS}} \tag{2–45}$$

where: $\mathbf{I}_{\text{LS}}$ = rated low-side current at specified power factor

$\mathbf{V}_{\text{LS}}$ = rated low-side voltage (output V, breaker closed)

$\mathbf{E}'_{\text{LS}}$ = no-load low-side voltage (output V, breaker open)

$\mathbf{Z}_{\text{eq,LS}}$ = equivalent impedance of transformer referred to low side

EXAMPLE 2.7 The equivalent low-side parameters of a 250-kVA, 4160—480-V, 60-Hz transformer are $R_{eq,LS} = 0.00920\ \Omega$, and $X_{eq,LS} = 0.0433\ \Omega$. The transformer is operating in the step-down mode and is delivering rated current at rated voltage to a 0.840 power-factor lagging load. Determine (a) the no-load voltage; (b) the actual input voltage at the high side; (c) the high-side current; (d) the input impedance; (e) the voltage regulation; (f) the voltage regulation if the power factor of the load is 0.840 leading; (g) sketch the tip-to-tail phasor diagram of the secondary circuit for the 0.840 power-factor lagging load. Show all voltage drops.

Solution

(a)

$$I_{LS} = \frac{250{,}000}{480} = 520.83 \text{ A} \qquad \theta = \cos^{-1}0.840 = 32.86^\circ$$

For a lagging power-factor load, the load current lags the load voltage as shown in Figure 2.13(a). Thus, from Figure 2.13(a),

$$\mathbf{V}_{LS} = 480\angle 0^\circ \text{ V} \qquad \mathbf{I}_{LS} = 520.83\angle -32.86^\circ \text{ A}$$

Using Figure 2.11(b) as a guide,

$$\mathbf{E}'_{LS} = \mathbf{I}_{LS}R_{eq,LS} + \mathbf{I}_{LS}jX_{eq,LS} + \mathbf{V}_{LS}$$

$$\mathbf{E}'_{LS} = 520.83\angle -32.86^\circ \times 0.0092 + 520.83\angle -32.86^\circ \times j0.0433 + 480\angle 0^\circ$$

$$\mathbf{E}'_{LS} = 4.79\angle -32.86^\circ + 22.55\angle 57.14^\circ + 480\angle 0^\circ$$

$$\mathbf{E}'_{LS} = 4.024 - j2.599 + 12.235 + j18.94 + 480 + j0 = \underline{496.53\angle 1.886^\circ \text{ V}}$$

(b)

$$a = \frac{E'_{HS}}{E'_{LS}} \approx \frac{V_{HS}}{V_{LS}} = \frac{4160}{480} = 8.667$$

From Figure 2.11(b),

$$\mathbf{V}_T = \mathbf{E}'_{HS} = a\mathbf{E}'_{LS} = 8.667 \times 496.53\angle 1.886^\circ = \underline{4303.4\angle 1.886^\circ \text{ V}}$$

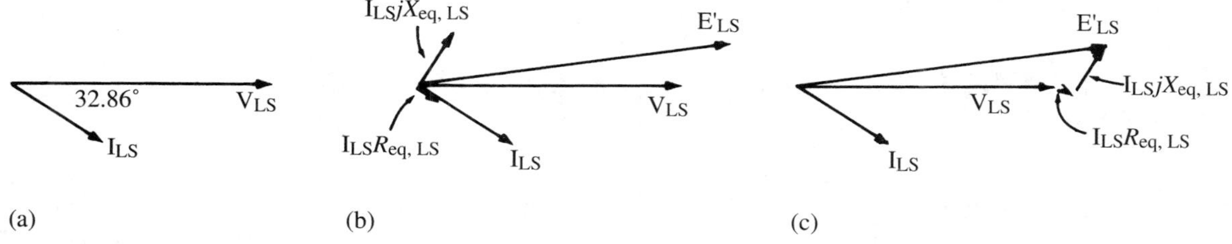

FIGURE 2.13
Phasor diagrams for Example 2.7: (a) low-side output; (b) component phasors; (c) tip-to-tail addition.

(c)
$$\mathbf{I}_{HS} = \frac{\mathbf{I}_{LS}}{a} = \frac{520.83\angle -32.86^\circ}{8.667} = \underline{60.09\angle -32.86^\circ \text{ A}}$$

(d) From Figure 2.11(b),

$$\mathbf{Z}'_{in} = \frac{\mathbf{V}_T}{\mathbf{I}_{HS}} = \frac{4303.4\angle 1.886^\circ}{60.09\angle -32.86^\circ} = \underline{71.62\angle 34.74^\circ\ \Omega}$$

(e)
$$\text{reg} = \frac{E_{nl} - V_{rated}}{V_{rated}} = \frac{496.53 - 480}{480} = \underline{0.0344 \quad \text{or} \quad 3.44\%}$$

(f) For a 0.840 leading power factor, $\mathbf{I}_{LS} = 520.83\angle +32.86^\circ$. Thus,

$$\mathbf{E}'_{LS} = 520.83\angle +32.86^\circ \times 0.0092 + 520.83\angle +32.86^\circ \times j0.0433 + 480\angle 0^\circ$$

$$\mathbf{E}'_{LS} = 4.79\angle 32.86^\circ + 22.55\angle 122.86^\circ + 480\angle 0^\circ$$

$$\mathbf{E}'_{LS} = 4.024 + j2.599 + (-12.235 + j18.94) + 480 + j0 = \underline{\underline{472.28\angle 2.61^\circ}}$$

$$\text{reg} = \frac{E_{nl} - V_{rated}}{V_{rated}} = \frac{472.28 - 480}{480} = \underline{-0.0161 \quad \text{or} \quad -1.61\%}$$

Note that the effect of a leading power-factor load is to cause a voltage rise in the transformer, resulting in a negative regulation. The voltage rise is a resonance effect caused by the combined leakage reactance of the transformer and the capacitance characteristic of the load.

(g) The voltage drops due to the equivalent resistance and equivalent reactance are

$$\mathbf{I}_{LS}R_{eq,LS} = 520.83\angle -32.86^\circ \times 0.0092 = 4.79\angle -32.86^\circ \text{ V}$$

$$\mathbf{I}_{LS}X_{eq,LS} = 520.83\angle -32.86^\circ \times j0.0433 = 22.6\angle 57.14^\circ \text{ V}$$

A phasor diagram showing the component voltages for the 0.840 power-factor lagging load is shown in Figure 2.13(b) and the corresponding tip-to-tail diagram is shown in Figure 2.13(c). Although the diagrams are not drawn to scale, because of the large difference in magnitudes between V_{LS} and the voltage drops, it does provide a perspective of how transformer resistance and leakage reactance affect the output voltage. The phasor diagrams in Figure 2.13 relate to the equivalent circuit in Figure 2.11(b).

2.12 PER-UNIT IMPEDANCE AND PERCENT IMPEDANCE OF TRANSFORMER WINDINGS

Information regarding the impedance of transformer windings is generally available from the manufacturer, or from the transformer nameplate as per-unit (PU) impedance or percent impedance.[9] Per-unit impedance (Z_{PU}) also called per-unit impedance

[9] Percent values are per-unit values times 100.

voltage, is the ratio of the voltage drop within the transformer caused by transformer impedance, to the rated voltage of the transformer, when operating at rated current. Thus,

$$\left.\begin{aligned} Z_{PU} &= \frac{I_{rated} Z_{eq}}{V_{rated}} \\ R_{PU} &= \frac{I_{rated} R_{eq}}{V_{rated}} \\ X_{PU} &= \frac{I_{rated} X_{eq}}{V_{rated}} \end{aligned}\right\} \tag{2–46}$$

where: Z_{PU} = per-unit impedance
R_{PU} = per-unit resistance
X_{PU} = per-unit reactance

Note: V_{rated} and I_{rated} are also called *base voltage* and *base current,* respectively.

Per-unit impedance of a transformer is often expressed in terms of a *base impedance* obtained from the transformer rating:

$$Z_{base} = \frac{V_{rated}}{I_{rated}} \tag{2–47}$$

To express Z_{base} in terms of transformer apparent power, multiply the numerator and denominator of Eq. (2–47) by V_{rated}. Thus,

$$Z_{base} = \frac{V_{rated}^2}{V_{rated}\, I_{rated}} = \frac{V_{rated}^2}{S_{rated}} \tag{2–48}$$

Solving Eq. (2–47) for V_{rated}, and substituting into equation set (2–46), expresses the per-unit values of *Z, R,* and *X* in terms of the base impedance:

$$\left.\begin{aligned} Z_{PU} &= \frac{I_{rated} Z_{eq}}{I_{rated} Z_{base}} = \frac{Z_{eq}}{Z_{base}} \\ R_{PU} &= \frac{I_{rated} R_{eq}}{I_{rated} Z_{base}} = \frac{R_{eq}}{Z_{base}} \\ X_{PU} &= \frac{I_{rated} X_{eq}}{I_{rated} Z_{base}} = \frac{X_{eq}}{Z_{base}} \end{aligned}\right\} \tag{2–49}$$

Note: I_{rated}, V_{rated}, R_{eq}, X_{eq}, and Z_{eq} must be all high-side values or all low-side values. The per-unit impedance (and percent impedance) has the same value whether calculated using all high-side values or all low-side values. This is a big advantage when making calculations involving systems that have more than one transformer, each at a different voltage level.

The per-unit system has its greatest application in the solution of network problems involving several voltage levels and it is used extensively in power system analysis **[4]**. All per-unit values are dimensionless.

The per-unit impedance, in terms of its components, is

$$\mathbf{Z}_{PU} = R_{PU} + jX_{PU} \tag{2–50}$$

$$Z_{PU} = \sqrt{R_{PU}^2 + X_{PU}^2} \tag{2–51}$$

$$\alpha = \tan^{-1}\left(\frac{X_{PU}}{R_{PU}}\right) \tag{2–52}$$

Angle α is the phase angle of the per-unit impedance, and is the same for percent impedance and equivalent impedance.[10]

Transformers rated above 100 kVA have conductors of such large cross-sectional area that $X_{PU} >> R_{PU}$. Thus, for very large transformers,

$$\underset{\text{kVA}>100}{Z_{PU}} \approx X_{PU} \tag{2–53}$$

Although Eq. (2–53) is a close approximation for large transformers, it is often applied to calculations involving smaller transformers if no other data are available.

EXAMPLE 2.8 The percent resistance and percent reactance of a 75-kVA, 2400—240-V, 60-Hz transformer are 0.90 and 1.30, respectively. Determine (a) percent impedance; (b) rated high-side current; (c) equivalent resistance and equivalent reactance referred to the high side; (d) high-side fault current if an accidental short circuit of 0.016 Ω (resistive) occurs at the secondary when 2300 V is impressed across the primary.

Solution

(a)

$$Z = \sqrt{(R)^2 + (X)^2} = \sqrt{0.90^2 + 1.30^2} = \underline{1.58\%}$$

(b)

$$I_{HS} = \frac{75{,}000}{2400} = \underline{31.25 \text{ A}}$$

(c)

$$R_{PU} = \frac{I_{HS}R_{eq,HS}}{V_{HS}} \qquad X_{PU} = \frac{I_{HS}X_{eq,HS}}{V_{HS}}$$

$$0.009 = \frac{31.25R_{eq,HS}}{2400} \qquad 0.013 = \frac{31.25X_{eq,HS}}{2400}$$

$$\underline{R_{eq,HS} = 0.691\ \Omega} \qquad \underline{X_{eq,HS} = 0.998\ \Omega}$$

---

[10] Impedance phase angle α may be approximated at 76° to 80° for single-phase transformers with apparent power ratings ≥500 kVA; 70° to 76° for apparent power ratings between 100 and 500 kVA. A table of representative impedances is given in Appendix J.

(d) The equivalent circuit is shown in Figure 2.14.

$$\mathbf{Z}_{\text{in}} = \mathbf{Z}_{\text{eq,HS}} + a^2\mathbf{Z}_{\text{short}} \qquad a = \frac{2400}{240} = 10$$

$$\mathbf{Z}_{\text{in}} = 0.691 + j0.998 + 10^2(0.016) = \underline{2.499\angle 23.54^\circ\ \Omega}$$

$$\mathbf{I}_{\text{HS}} = \frac{\mathbf{V}_{\text{HS}}}{\mathbf{Z}_{\text{HS}}} = \frac{2300\angle 0^\circ}{2.499\angle 23.54^\circ} = \underline{\underline{920\angle -23.54^\circ\ \text{A}}}$$

Calculating Voltage Regulation From Per-Unit Values

The voltage regulation of a transformer (operating at rated voltage and rated current) may be determined from the power factor of the load and the known per-unit values of transformer reactance and resistance, without having to calculate load currents and voltage drops. Referring to Figure 2.15(a),

$$\mathbf{E}'_{LS} = \mathbf{I}_{\text{LS}}R_{\text{eq,LS}} + \mathbf{I}_{\text{LS}}jX_{\text{eq,LS}} + \mathbf{V}_{\text{LS}} \qquad \textbf{(2–54)}$$

where: $\mathbf{V}_{\text{LS}}$ = output voltage, breaker closed
$\mathbf{E}'_{\text{LS}}$ = output voltage, breaker open

The component phasors in Eq. (2–54) are shown on the phasor diagram in Figure 2.15(b) for a lagging power-factor load, with the current phasor drawn as the reference phasor at 0°. The diagram is not drawn to scale.

The *magnitude* of the no-load low-side voltage is obtained by resolving $\mathbf{V}_{\text{LS}}$ into vertical and horizontal components, and applying the Pythagorean theorem. Thus, referring to Figure 2.15(b),

$$E'_{\text{LS}} = \sqrt{(I_{\text{LS}}R_{\text{eq,LS}} + V_{\text{LS}}\cos\theta)^2 + (I_{\text{LS}}X_{\text{eq,LS}} + V_{\text{LS}}\sin\theta)^2} \qquad \textbf{(2–55)}$$

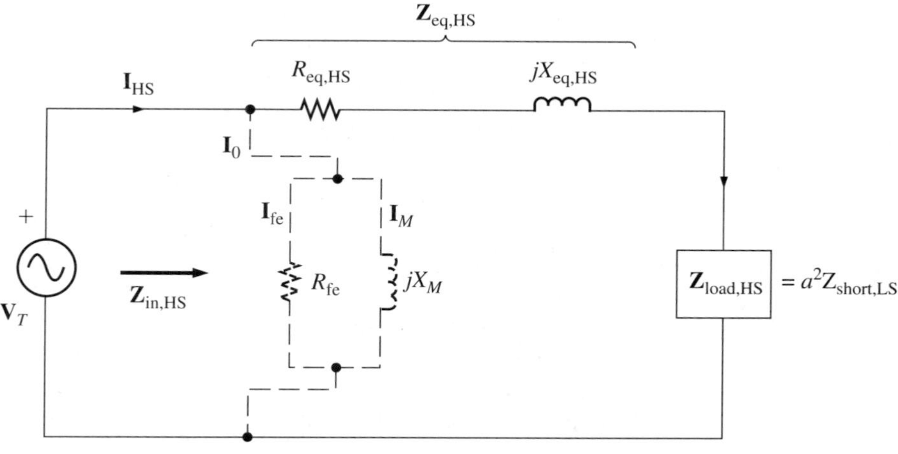

FIGURE 2.14
Equivalent circuit for Example 2.8.

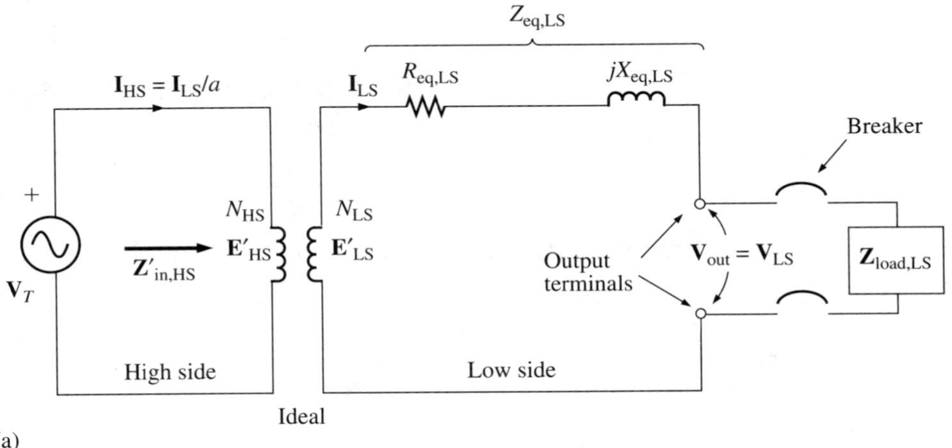

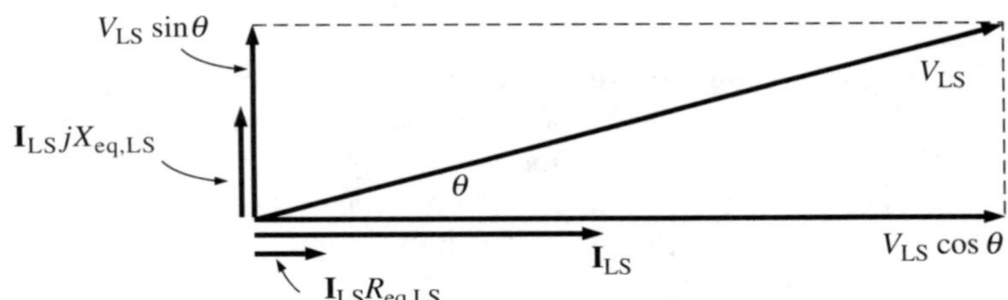

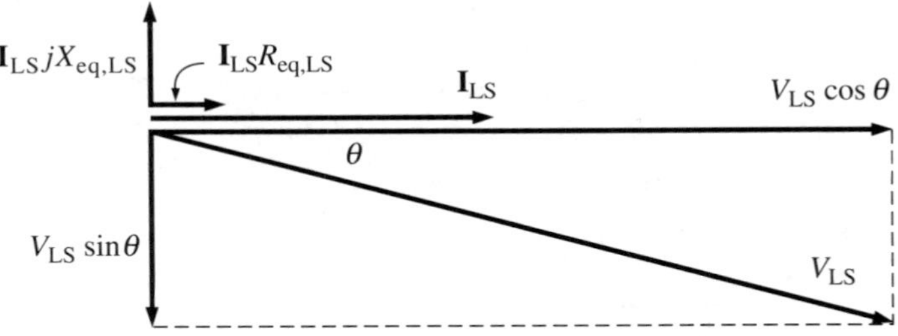

FIGURE 2.15
(a) Equivalent circuit; (b) phasor diagram, lagging power-factor load; (c) phasor diagram, leading power-factor load.

Substituting Eq. (2–55) into Eq. (2–44) and simplifying,

$$\text{reg}_{PU} = \frac{\sqrt{(I_{LS}R_{eq,LS} + V_{LS}\cos\theta)^2 + (I_{LS}X_{eq,LS} + V_{LS}\sin\theta)^2} - V_{LS}}{V_{LS}} \qquad \textbf{(2–56)}$$

Dividing numerator and denominator by V_{LS},

$$\text{reg}_{PU} = \sqrt{\left(\frac{I_{LS}R_{eq,LS}}{V_{LS}} + \cos\theta\right)^2 + \left(\frac{I_{LS}X_{eq,LS}}{V_{LS}} + \sin\theta\right)^2} - 1 \qquad \textbf{(2–57)}$$

Substituting appropriate equations from equation set (2–46) into Eq. (2–57),

$$\text{reg}_{PU} = \sqrt{(R_{PU} + \cos\theta)^2 + (X_{PU} + \sin\theta)^2} - 1 \qquad \textbf{(2–58)}$$

Note: Angle θ, shown in Figure 2.15, is called the *power-factor angle;* it is positive for lagging power-factor loads, and negative for leading power-factor loads.[11] The cosine of the power-factor angle is the power factor. Expressed mathematically,

$$\theta = \cos^{-1}F_P \qquad \text{for lagging power-factor loads}$$

$$\theta = -\cos^{-1}F_P \qquad \text{for leading power-factor loads}$$

[11] See Appendix A.5 for the relationship between power, power factor, and power-factor angle.

EXAMPLE 2.9 A single-phase distribution transformer, rated at 50 kVA, 7200—600 V, is supplying rated kVA at 600 V and 0.75 power-factor lagging. The percent resistance and percent reactance are 1.3 and 3.8, respectively. Determine (a) transformer regulation; (b) secondary voltage when the load is disconnected; (c) input voltage that must be applied to the primary in order to obtain rated secondary voltage when carrying rated load at 0.75 power-factor lagging.

Solution

(a)

$$\theta = \cos^{-1}F_P = \cos^{-1}0.750 = 41.41° \qquad \sin 41.41° = 0.661$$

$$\text{reg}_{PU} = \sqrt{(R_{PU} + \cos\theta)^2 + (X_{PU} + \sin\theta)^2} - 1$$

$$\text{reg}_{PU} = \sqrt{0.0130 + 0.750)^2 + (0.038 + 0.661)^2} - 1$$

$$\text{reg}_{PU} = 1.035 - 1 = \underline{0.035} \quad \text{or} \quad \underline{3.5\%}$$

(b)

$$\text{reg}_{PU} = \frac{V_{nl} - V_{rated}}{V_{rated}} \Rightarrow 0.035 = \frac{V_{nl} - 600}{600}$$

$$V_{nl} = \underline{621\ V}$$

(c) The voltage ratio obtained from the voltage ratings on the transformer nameplate is approximately equal to the turns ratio of the ideal transformer shown in Figure 2.15(a). Expressed mathematically,

$$\frac{E'_{HS}}{E'_{LS}} = \frac{7200}{600} \Rightarrow E'_{HS} = E'_{LS} \times \frac{7200}{600} \qquad \textbf{(2–56a)}$$

Referring to Figure 2.15(a), with no load connected to the secondary (load breaker open), $I_l = 0$. Thus, there are no voltage drops in the secondary circuit, and

$$E'_{LS} = V_{nl} = 621 \text{ V}$$

Substituting into Eq. (2–56a),

$$E'_{HS} = 621 \times \frac{7200}{600} = \underline{7452.5 \text{ V}}$$

EXAMPLE 2.10 Assume the transformer in Example 2.9 is operating at rated kVA and 600 V, but the power factor of the load is 0.75 *leading*. Determine (a) transformer regulation; (b) secondary voltage when the load is disconnected; (c) input voltage that must be applied to the primary in order to obtain rated secondary voltage when carrying rated load at 0.75 power factor leading.

Solution

(a) $\theta = -\cos^{-1} F_P = -\cos^{-1} 0.750 = -41.41°$ $\qquad \sin(-41.41°) = -0.661$

$$\text{reg}_{PU} = \sqrt{(R_{PU} + \cos\theta)^2 + (X_{PU} + \sin\theta)^2} - 1$$

$$\text{reg}_{PU} = \sqrt{(0.0130 + 0.750)^2 + (0.038 - 0.661)^2} - 1$$

$$\text{reg}_{PU} = 0.9853 - 1 = \underline{-0.0147 \text{ or } -1.5\%}$$

(b) $$\text{reg}_{PU} = \frac{V_{nl} - V_{rated}}{V_{rated}} \quad \Rightarrow \quad -0.0147 = \frac{V_{nl} - 600}{600}$$

$$V_{nl} = \underline{591.2 \text{ V}}$$

(c) $$E'_{HS} = 591.2 \times \frac{7200}{600} = \underline{7094 \text{ V}}$$

Note: As shown in part (b) of Example 2.9, for lagging power-factor loads the transformer regulation is positive; this is also true for unity power-factor loads. However, for loads that have sufficiently leading power factors, as shown in Example 2.10, the voltage regulation will be negative.

Voltage Regulation at Other Than Rated Load

Equation (2–58) is applicable to transformers operating at rated load. If operating at other than rated load, Eq. (2–58) must be modified to reflect the actual per-unit load connected to the secondary. Making the modification,

$$\text{reg}_{PU} = \sqrt{(S_{PU} \times R_{PU} + \cos\theta)^2 + (S_{PU} \times X_{PU} + \sin\theta)^2} - 1 \qquad \textbf{(2–58a)}$$

$$I_{PU} = \frac{I}{I_{rated}} = S_{PU} = \frac{S}{S_{rated}}$$

where: S_{PU} = per-unit apparent power of load (PU)
S = apparent power of load (VA)
S_{rated} = rated apparent power of transformer (VA)
I_{PU} = per-unit load current (PU)
I = load current (A)
I_{rated} = rated current of secondary (A)

EXAMPLE 2.11 A 25-kVA, 7620—480-V distribution transformer is supplying a 10-kVA load at 0.65 power-factor lagging. The percent *IR* drop and the percent *IX* drop are 1.2 and 1.4, respectively. Determine the transformer regulation for the specific load.

Solution

$$S_{PU} = \frac{S}{S_{rated}} = \frac{10}{25} = 0.0394 \qquad \theta = \cos^{-1} 0.65 = 49.49° \qquad \sin 49.49° = 0.76$$

Substituting into Eq. (2–58a),

$$\text{reg}_{PU} = \sqrt{(0.0394 \times 0.0124 + 0.65)^2 + (0.0394 \times 0.014 + 0.76)^2} - 1$$

$$\text{reg}_{PU} = \underline{0.738 \text{ or } 73.8\%}$$

2.13 TRANSFORMER LOSSES AND EFFICIENCY

Transformer losses include the I^2R losses in the primary and secondary windings, and the hysteresis and eddy-current losses (core losses) in the iron. These losses are the same whether operating in the step-up or step-down mode.

The efficiency of a transformer is the ratio of the power out to the power in, and may be expressed in decimal form, called *per-unit efficiency,* or expressed as *percent efficiency* by multiplying by 100.

$$\eta = \frac{P_{out}}{P_{in}} = \frac{P_{out}}{P_{out} + P_{core} + I_{HS}^2 R_{HS} + I_{LS}^2 R_{LS}} \qquad \textbf{(2–59)}$$

where: η = efficiency

$$P_{core} = P_h + P_e \qquad \textbf{(2–60)}$$

$$P_e = k_e f^2 B_{max}^2 \qquad \textbf{(2–61)}$$

$$P_h = k_h f B_{max}^{1.6} \qquad \textbf{(2–62)}$$

From Eq. (2–1),

$$\Phi_{max} \propto \frac{V_T}{f}$$

Thus,

$$B_{max} \propto \frac{V_T}{f}$$

Obtaining proportionalities by substituting into Eqs. (2–61) and (2–62), respectively,

$$P_e \propto f^2\left(\frac{V_T}{f}\right)^2 \propto V_T^2 \tag{2–63}$$

$$P_h \propto f\left(\frac{V_T}{f}\right)^{1.6} \tag{2–64}$$

The hysteresis component of the total core losses is generally greater than the eddy-current component, $P_h > P_e$.

As indicated in Eq. (2–63), the eddy-current losses are proportional to the square of the applied voltage. However, as shown in Eq. (2–64), the hysteresis losses are affected by both the frequency and the applied voltage. Hence, assuming the frequency and magnitude of applied voltage are constant, the core loss will be essentially constant for all load conditions up to the transformer rating; slight changes in leakage flux from no load to full load will have an insignificant effect on the core loss.

The combined conductor losses of both primary and secondary windings may be expressed in terms of the *equivalent resistance referred to the high side or referred to the low side.* That is,

$$(I_{\text{HS}}^2 R_{\text{HS}} + I_{\text{LS}}^2 R_{\text{LS}}) = I_{\text{HS}}^2 R_{\text{eq,HS}} = I_{\text{LS}}^2 R_{\text{eq,LS}} \tag{2–65}$$

Substituting Eq. (2–65) into Eq. (2–59),

$$\eta = \frac{P_{\text{out}}}{P_{\text{out}} + P_{\text{core}} + I^2 R_{\text{eq}}} \tag{2–66}$$

where I and R_{eq} *are both high-side values or are both low-side values.*

Depending on its apparent power rating, the efficiency of distribution transformers and power transformers varies from 96 to more than 99 percent; the larger transformers have the higher efficiencies.

EXAMPLE 2.12

A 50-kVA, 450—230-V, 60-Hz transformer has percent resistance and percent leakage reactance of 1.25 and 2.24, respectively. Its efficiency at rated voltage, rated frequency, and rated apparent power at 0.860 power-factor lagging is 96.5 percent. Determine (a) the core loss; (b) the core loss if operating at rated load current and 0.860 power factor from a 375-V, 50-Hz supply (assume the hysteresis loss is 71.0 percent of the total core loss); (c) the efficiency for the conditions in (b); (d) the efficiency if the load is disconnected.

Solution

(a)

$$I_{\text{HS}} = \frac{50{,}000}{450} = 111.11 \text{ A}$$

$$R_{\text{PU}} = \frac{I_{\text{rated}} R_{\text{eq}}}{V_{\text{rated}}} \quad \Rightarrow \quad R_{\text{eq}} = R_{\text{PU}} \cdot \frac{V_{\text{rated}}}{I_{\text{rated}}}$$

Using high-side values,

$$R_{eq,HS} = 0.0125 \times \frac{450}{111.11} = 0.0506\ \Omega$$

$$P_{out} = S_{rated} \cdot F_P = 50{,}000 \times 0.860 = 43{,}000\ \text{W}$$

$$P_{in} = \frac{P_{out}}{\eta} = \frac{43{,}000}{0.965} = 44{,}559.59\ \text{W}$$

$$P_{out} = P_{in} - P_{core} - I_{HS}^2 R_{eq,HS} \quad \Rightarrow \quad P_{core} = P_{in} - P_{out} - I_{HS}^2 R_{eq,HS}$$

$$P_{core} = 44{,}559.59 - 43{,}000 - (111.11)^2 \times 0.0506 = \underline{934.9\ \text{W}}$$

(b)

$$P_{h,60} = 0.71 \times 934.9 = 663.78\ \text{W}$$

$$P_{e,60} = 934.9 - 663.78 = 271.12\ \text{W}$$

From Eq. (2–63),

$$\frac{P_{e,60}}{P_{e,50}} = \frac{(V_T)_{60}^2}{(V_T)_{50}^2} \quad \Rightarrow \quad P_{e,50} = P_{e,60} \times \left[\frac{V_{T,50}}{V_{T,60}}\right]^2$$

$$P_{e,50} = 271.12 \times \left[\frac{375}{450}\right]^2 = 188.3\ \text{W}$$

From Eq. (2–64),

$$\frac{P_{h,60}}{P_{h,50}} = \frac{f_{60}(V_T/f)_{60}^{1.6}}{f_{50}(V_T/f)_{50}^{1.6}} \quad \Rightarrow \quad P_{h,50} = P_{h,60} \times \frac{50}{60} \times \left[\frac{V_{T,50}}{V_{T,60}} \times \frac{60}{50}\right]^{1.6}$$

$$P_{h,50} = 663.78 \times \frac{50}{60} \times \left[\frac{375}{450} \times \frac{60}{50}\right]^{1.6} = 553.15\ \text{W}$$

$$P_{core,50} = 188.3 + 553.15 = \underline{741.45\ \text{W}}$$

(c)

$$P_{out} = 375 \times 111.11 \times 0.860 = 35{,}832.98\ \text{W}$$

Discounting small changes in skin effect due to changes in frequency, the equivalent resistance is essentially the same. Thus,

$$I_{HS}^2 R_{eq,HS} = 111.11^2 \times 0.0506 = 624.68\ \text{W}$$

Substituting into Eq. (2–66),

$$\eta = \frac{35{,}832.98}{35{,}832.98 + 741.45 + 624.68} = \underline{0.963 \quad \text{or} \quad 96.3\%}$$

(d) With the load disconnected, $P_{out} = 0$. Thus, the efficiency is zero.

Calculating Efficiency From Per-Unit Values

Quick-and-easy calculations of efficiency may be accomplished if the transformer parameters and the core loss are given in per-unit values or percent values. The appropriate equation is derived by first expressing P_{out} in Eq. (2–66) in terms of apparent power. Thus,

$$\eta = \frac{S \times F_P}{S \times F_P + P_{\text{core}} + I^2 R_{\text{eq}}} \tag{2–67}$$

where: F_P = per-unit power factor
S = apparent power of connected load (VA)
I = load current

Dividing both numerator and denominator in Eq. (2–67) by the *rated apparent power* of the transformer,

$$\eta = \frac{(S/S_{\text{rated}}) \times F_P}{(S/S_{\text{rated}}) \times F_P + (P_{\text{core}}/S_{\text{rated}}) + (I^2 R_{\text{eq}}/S_{\text{rated}})} \tag{2–68}$$

where: η = per-unit efficiency
$S_{\text{rated}} = V_{\text{rated}} I_{\text{rated}}$ = rated apparent power, also called *base apparent power*

Defining:

$$\frac{S}{S_{\text{rated}}} = S_{\text{PU}} = \text{per-unit apparent power of load} \tag{2–69}$$

$$\frac{P_{\text{core}}}{S_{\text{rated}}} = P_{\text{core,PU}} = \text{per-unit core loss} \tag{2–70}$$

Multiplying the numerator and denominator of the $(I^2 R_{\text{eq}}/S_{\text{rated}})$ term in Eq. (2–68) by I_{rated}, and then rearranging its components,

$$\frac{I^2 R_{\text{eq}}}{S_{\text{rated}}} = \frac{I^2 R_{\text{eq}}}{V_{\text{rated}} I_{\text{rated}}} \cdot \frac{I_{\text{rated}}}{I_{\text{rated}}} = \left[\frac{I^2}{I_{\text{rated}}^2}\right]\left[\frac{I_{\text{rated}} R_{\text{eq}}}{V_{\text{rated}}}\right]$$

Defining $I_{\text{PU}} = (I/I_{\text{rated}})$

$$\frac{I^2 R_{\text{eq}}}{S_{\text{rated}}} = I_{\text{PU}}^2 \cdot R_{\text{PU}} \tag{2–71}$$

Substituting Eqs. (2–69), (2–70), and (2–71) into Eq. (2–68),

$$\eta = \frac{S_{\text{PU}} \times F_P}{S_{\text{PU}} \times F_P + P_{\text{core,PU}} + I_{\text{PU}}^2 \times R_{\text{PU}}} \tag{2–72}$$

Equation (2–72) is applicable to all loads. When operating at *rated conditions,* $S_{PU} = 1$, $I_{PU} = 1$, and Eq. (2–72) reduces to

$$\eta_{\text{rated}} = \frac{F_P}{F_P + P_{\text{core,PU}} + R_{\text{PU}}} \qquad \textbf{(2–73)}$$

The components in Eqs. (2–72) and (2–73) may be expressed *all in per-unit* or *all in percent.* However, the calculated efficiency will be in per-unit.

EXAMPLE 2.13 A 100-kVA, 4800—240-V, 60-Hz transformer is operating at *rated conditions* and 80.0 percent power factor. The core loss, resistance, and leakage reactance, *expressed in percent,* are 0.450, 1.46, and 3.38, respectively. Determine the efficiency at (a) rated load and 80% power factor; (b) 70% load and 80% power factor.

Solution

(a) Substituting into Eq. (2–73),

$$\eta_{\text{rated}} = \frac{0.800}{0.800 + 0.0045 + 0.0146} = \underline{0.977 \text{ or } 97.7\%}$$

(b)

$$S_{PU} = \frac{S_{\text{load}}}{S_{\text{rated}}} = \frac{70}{100} = 0.70$$

and since I_{load} is proportional to S_{load}

$$I_{PU} = S_{PU} = 0.70$$

Substituting into Eq. (2–72),

$$\eta = \frac{0.70 \times 0.80}{0.70 \times 0.80 + 0.0045 + 0.70^2 \times 0.0146} = \underline{0.979 \text{ or } 97.9\% \text{ efficient}}$$

Note: There is very little change in efficiency.

2.14 DETERMINATION OF TRANSFORMER PARAMETERS

If transformer parameters are not readily available from the nameplate or from the manufacturer, they can be approximated from an open-circuit test (also called a no-load test) and a short-circuit test.

Open-Circuit Test

The purpose of the open-circuit test is to determine the magnetizing reactance X_M and the equivalent core-loss resistance R_{fe}. The connections and instrumentation required for this test are shown in Figure 2.16(a).

For safety in testing and instrumentation, the open-circuit test is generally made on the low-voltage side. The test is performed at rated frequency and rated low-side

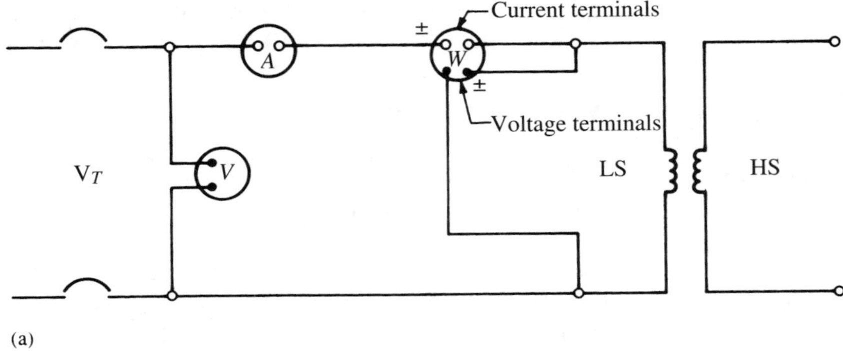

(a)

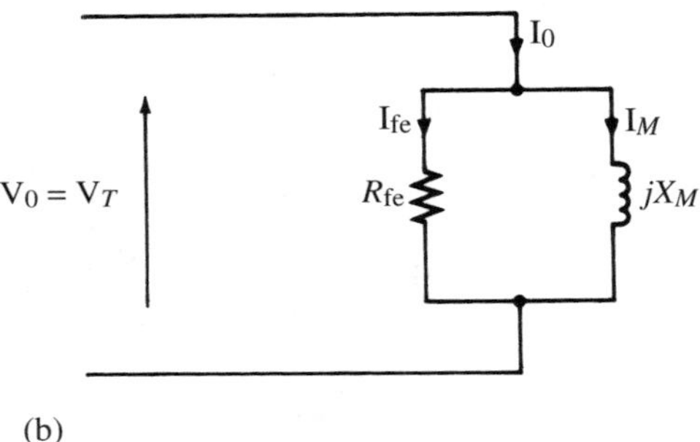

(b)

FIGURE 2.16
Open-circuit test: (a) connection diagram; (b) equivalent circuit.

voltage; the high-side terminals should be covered with insulating material to prevent accidental contact. Since no load is connected to the secondary, the copper losses in the secondary are zero, and the copper losses in the primary are negligible. Thus, the wattmeter reading (for the open-circuit test) is essentially core losses. The equivalent open-circuit model of the transformer is shown in Figure 2.16(b).

Assuming the wattmeter, voltmeter, and ammeter readings taken during the open-circuit test are P_{OC}, V_{OC}, and I_{OC}, respectively, and that the test was made on the low side, the open-circuit parameters referred to the low side may be obtained by substituting into the following equations:[12]

$$\left.\begin{aligned} P_{OC} &= V_{OC} I_{fe} & \qquad I_{OC} &= \sqrt{I_{fe}^2 + I_M^2} \\ R_{fe,LS} &= \frac{V_{OC}}{I_{fe}} & \qquad X_{M,\,LS} &= \frac{V_{OC}}{I_M} \end{aligned}\right\} \tag{2–74}$$

[12] If an analog wattmeter is used for the no-load test, wattmeter losses may be an appreciable part of the wattmeter reading, and must be subtracted before calculating the open-circuit parameters [**2**].

Short-Circuit Test

The purpose of the short-circuit test is to determine the equivalent resistance, equivalent leakage reactance, and equivalent impedance of the transformer windings. The connections and instrumentation required for the test are shown in Figure 2.17(a). The high side of the transformer under test is connected to the supply line through an adjustable-voltage autotransformer, and the low-voltage side is jumped by connecting a short piece of large cross-sectional area copper across its terminals; in effect, the secondary is shorted. The jumper represents a load impedance of almost zero ohms. That is, $Z_{\text{load}} \approx 0\ \Omega$. Thus, by jumping the secondary, the test provides data that include the effects of primary and secondary resistance, and primary and secondary leakage, but excludes the load impedance. Furthermore, short circuiting the secondary causes the flux density to be reduced to a very low value, making the core losses insignificant. Thus, the wattmeter reading (for the short-circuit test) is essentially copper losses. The equivalent series circuit model is shown in Figure 2.17(b).

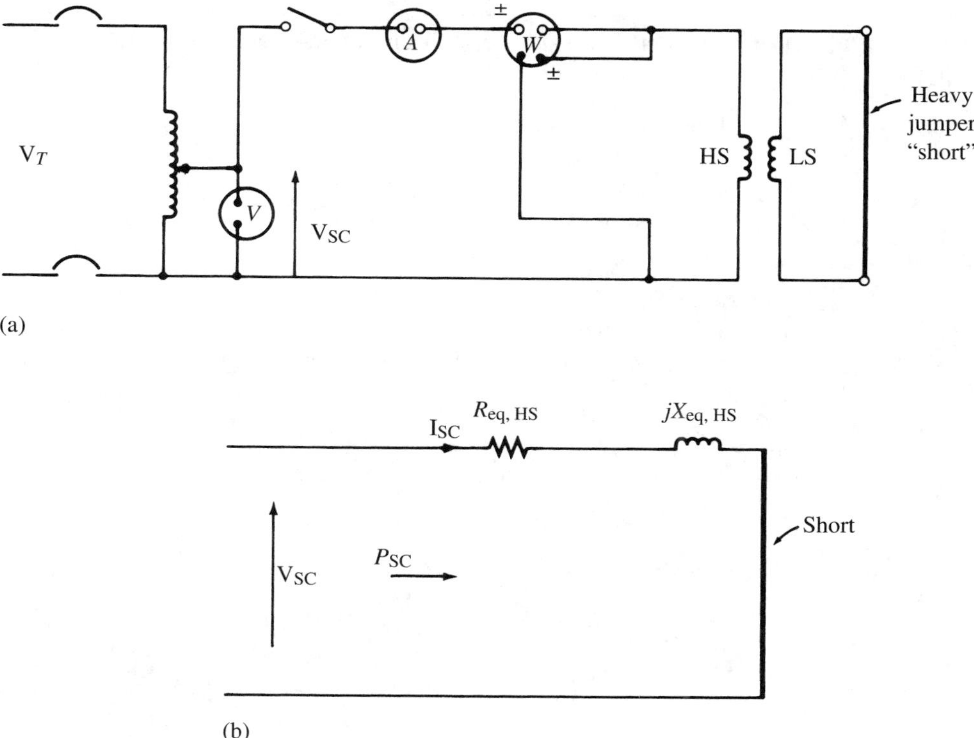

FIGURE 2.17
Short-circuit test: (a) connection diagram; (b) equivalent circuit.

Test Procedure With the variable voltage set to zero, the breaker is closed, the voltage is gradually raised until the ammeter indicates approximately rated high-side current, and the instruments are then read. Assuming the wattmeter, voltmeter, and ammeter readings taken during the short-circuit test are P_{SC}, V_{SC}, and I_{SC}, respectively, and the test is made on the high side, the equivalent resistance, equivalent reactance, and equivalent impedance may be obtained by substituting values into the following set of equations:

$$\left.\begin{aligned} I_{SC} &= \frac{V_{SC}}{Z_{eq,HS}} \qquad P_{SC} = I_{SC}^2 R_{eq,HS} \\ Z_{eq,HS} &= \sqrt{R_{eq,HS}^2 + X_{eq,HS}^2} \end{aligned}\right\} \tag{2–75}$$

The short-circuit test may be made using either winding. For reasons of lower current input and meter sizing, however, the high-voltage winding is preferred. If the measurements are made on the low side, the resultant values determined from the test would be the equivalent resistance, equivalent reactance, and equivalent impedance referred to the low side.

EXAMPLE 2.14 Data obtained from short-circuit and open-circuit tests of a 75-kVA, 4600—230-V, 60-Hz transformer are

Open-Circuit Test	**Short-Circuit Test**
(Low-Side Data)	(High-Side Data)
$V_{OC} = 230$ V	$V_{SC} = 160.8$ V
$I_{OC} = 13.04$ A	$I_{SC} = 16.3$ A
$P_{OC} = 521$ W	$P_{SC} = 1200$ W

Determine (a) the magnetizing reactance and equivalent core-loss resistance; (b) the per-unit resistance, per-unit reactance, and per-unit impedance of the transformer windings; (c) the voltage regulation when operating at rated load and 0.75 power-factor lagging.

Solution

(a) From the open-circuit test, shown in Figure 2.16,

$$P_{OC} = V_{OC} I_{fe} \qquad \Rightarrow \qquad 521 = 230 \times I_{fe}$$

$$I_{fe} = 2.265 \text{ A}$$

$$R_{fe,LS} = \frac{V_0}{I_{fe}} = \frac{230}{2.265} = \underline{101.5\ \Omega}$$

$$I_{OC} = \sqrt{I_{fe}^2 + I_M^2} \qquad \Rightarrow \qquad I_M = \sqrt{I_{OC}^2 - I_{fe}^2}$$

$$I_M = \sqrt{(13.04)^2 - (2.265)^2} = 12.842 \text{ A}$$

$$X_{M,LS} = \frac{V_{OC}}{I_M} = \frac{230}{12.842} = \underline{17.91\ \Omega}$$

(b) From the short-circuit test, shown in Figure 2.17,

$$I_{SC} = \frac{V_{SC}}{Z_{eq,HS}} \quad \Rightarrow \quad 16.3 = \frac{160.8}{Z_{eq,HS}}$$

$$Z_{eq,HS} = 9.865\ \Omega$$

$$P_{SC} = I_{SC}^2 R_{eq,HS} \quad \Rightarrow \quad 1200 = 16.3^2 R_{eq,HS}$$

$$R_{eq,HS} = 4.517\ \Omega$$

$$Z_{eq,HS} = \sqrt{R_{eq,HS}^2 + X_{eq,HS}^2} \quad \Rightarrow \quad X_{eq,HS} = \sqrt{Z_{eq,HS}^2 - R_{eq,HS}^2}$$

$$X_{eq,HS} = \sqrt{(9.865)^2 - (4.517)^2} = 8.77\ \Omega$$

$$I_{HS} = \frac{S}{V_{HS}} = \frac{75{,}000}{4600} = 16.3\ \text{A}$$

$$R_{PU} = \frac{I_{HS}R_{eq,HS}}{V_{HS}} = \frac{16.3 \times 4.517}{4600} = \underline{0.016}$$

$$X_{PU} = \frac{I_{HS}X_{eq,HS}}{V_{HS}} = \frac{16.3 \times 8.77}{4600} = \underline{0.031}$$

$$Z_{PU} = R_{PU} + jX_{PU} = 0.016 + j0.031 = \underline{\underline{0.035\angle 62.7^\circ}}$$

(c)

$$\text{reg}_{PU} = \sqrt{(R_{PU} + \cos\theta)^2 + (X_{PU} + \sin\theta)^2} - 1$$

$$\theta = \cos^{-1} 0.75 = 41.41^\circ \qquad \sin 41.41^\circ = 0.661$$

$$\text{reg}_{PU} = \sqrt{(0.016 + 0.75)^2 + (0.031 + 0.661)^2} - 1$$

$$\text{reg}_{PU} = \underline{0.0326 \quad \text{or} \quad 3.26\%}$$

SUMMARY OF EQUATIONS FOR PROBLEM SOLVING

$$E_P = 4.44 N_P f \Phi_{max} \tag{2–1}$$

$$E_S = 4.44 N_S f \Phi_{max} \tag{2–2}$$

$$\mathbf{I}_0 = \mathbf{I}_{fe} + \mathbf{I}_M \tag{2–4}$$

$$\Phi_M = \frac{N_P I_M}{\mathcal{R}_{core}} \tag{2–6}$$

$$\mathbf{I}_P = \mathbf{I}_0 + \mathbf{I}_{P,\,load} \tag{2–11}$$

Ideal Transformer

$$a = \frac{N_{HS}}{N_{LS}} \approx \frac{V_{HS}}{V_{LS}} \tag{2–14}$$

$$\mathbf{E}'_P = a\mathbf{E}'_S \tag{2–16a}$$

$$\mathbf{I}_P = \frac{1}{a}\mathbf{I}_S \tag{2–16b}$$

$$\mathbf{Z}'_{\text{in}} = a^2\mathbf{Z}_{\text{load}} \tag{2–19}$$

Real Transformer

$$\mathbf{Z}_{\text{eq,H}S} = R_{\text{HS}} + a^2R_{\text{LS}} + j(X_{\text{HS}} + a^2X_{\text{LS}}) \tag{2–39}$$

$$\mathbf{Z}_{\text{eq,HS}} = R_{\text{eq,HS}} + jX_{\text{eq,HS}} \tag{2–40}$$

$$\mathbf{Z}_{\text{eq,LS}} = R_{\text{LS}} + R_{\text{HS}}/a^2 + j(X_{\text{LS}} + X_{\text{HS}}/a^2) \tag{2–41}$$

$$\mathbf{Z}_{\text{eq,LS}} = R_{\text{eq,LS}} + jX_{\text{eq,LS}} \tag{2–42}$$

$$\mathbf{Z}_{\text{load,LS}} = \frac{1}{a^2}\mathbf{Z}_{\text{load,HS}} \tag{2–43}$$

$$\mathbf{Z}_{\text{eq,LS}} = \frac{\mathbf{Z}_{\text{eq,HS}}}{a^2} \tag{2–43a}$$

$$\text{reg} = \frac{E - V_{\text{rated}}}{V_{\text{rated}}} \tag{2–44}$$

$$\text{reg}_{\text{PU}} = \sqrt{(R_{\text{PU}} + \cos\theta)^2 + (X_{\text{PU}} + \sin\theta)^2} - 1 \tag{2–58}$$

$$\text{reg}_{\text{PU}} = \sqrt{(S_{\text{PU}} \times R_{\text{PU}} + \cos\theta)^2 + (S_{\text{PU}} \times X_{\text{PU}} + \sin\theta)^2} - 1 \tag{2–58a}$$

$$\left.\begin{aligned} Z_{\text{PU}} &= \frac{I_{\text{rated}}Z_{\text{eq}}}{\text{V}_{\text{rated}}} \\ R_{\text{PU}} &= \frac{I_{\text{rated}}R_{\text{eq}}}{\text{V}_{\text{rated}}} \\ X_{\text{PU}} &= \frac{I_{\text{rated}}X_{\text{eq}}}{\text{V}_{\text{rated}}} \end{aligned}\right\} \tag{2–46}$$

$$Z_{\text{base}} = \frac{V_{\text{rated}}}{I_{\text{rated}}} = \frac{V^2_{\text{rated}}}{S_{\text{rated}}} \tag{2–48}$$

$$\left.\begin{aligned} Z_{\text{PU}} &= \frac{Z_{\text{eq}}}{Z_{\text{base}}} \\ R_{\text{PU}} &= \frac{R_{\text{eq}}}{Z_{\text{base}}} \\ X_{\text{PU}} &= \frac{X_{\text{eq}}}{Z_{\text{base}}} \end{aligned}\right\} \tag{2–49}$$

$$\mathbf{Z}_{PU} = R_{PU} + jX_{PU} \tag{2–50}$$

$$\eta = \frac{P_{out}}{P_{in}} = \frac{P_{out}}{P_{out} + P_{core} + I_{HS}^2 R_{HS} + I_{LS}^2 R_{LS}} \tag{2–59}$$

$$P_{core} = P_h + P_e \tag{2–60}$$

$$P_e = k_e f^2 B_{max}^2 \tag{2–61}$$

$$P_h = k_h f B_{max}^{1.6} \tag{2–62}$$

$$P_e \propto f^2 \left(\frac{V_T}{f}\right)^2 \propto V_T^2 \tag{2–63}$$

$$P_h \propto f \left(\frac{V_T}{f}\right)^{1.6} \tag{2–64}$$

$$\eta = \frac{P_{out}}{P_{out} + P_{core} + I^2 R_{eq}} \tag{2–66}$$

$$\eta = \frac{S_{PU} \times F_P}{S_{PU} \times F_P + P_{core,PU} + I_{PU}^2 \times R_{PU}} \tag{2–72}$$

$$\eta_{rated} = \frac{F_P}{F_P + P_{core,PU} + R_{PU}} \tag{2–73}$$

Open-Circuit Test

$$\left.\begin{aligned} P_{OC} &= V_{OC} I_{fe} & I_{OC} &= \sqrt{I_{fe}^2 + I_M^2} \\ R_{fe,LS} &= \frac{V_{OC}}{I_{fe}} & X_{M,LS} &= \frac{V_{OC}}{I_M} \end{aligned}\right\} \tag{2–74}$$

Short-Circuit Test

$$\left.\begin{aligned} I_{SC} &= \frac{V_{SC}}{Z_{eq,HS}} \qquad P_{SC} = I_{SC}^2 R_{eq,HS} \\ Z_{eq,HS} &= \sqrt{R_{eq,HS}^2 + X_{eq,HS}^2} \end{aligned}\right\} \tag{2–75}$$

SPECIFIC REFERENCES KEYED TO TEXT

1. Hubert, C. I. *Electric Circuits AC/DC: An Integrated Approach.* McGraw-Hill, New York, 1982.
2. IEEE standard terminology for power and distribution transformers. ANSI/IEEE C57.12.80-1986, IEEE, New York.

3. Standards publication: Dry type transformers for general applications. NEMA Publication No. ST20-1972, National Electrical Manufacturers Association, Washington, DC.
4. Stevenson, W. D., Jr. *Elements of Power System Analysis.* McGraw-Hill, New York, 1982.

GENERAL REFERENCES

Heathcote, Martin J. *JSP Transformer Book: A Practical Technology of the Power Transformer,* 12th ed. Oxford, Boston, 1998.

Blume, L. F. *Transformer Engineering.* Wiley, New York, 1938.

MIT EE Staff. *Magnetic Circuits and Transformers.* Wiley, New York, 1943.

Lawrence, R. R. *Principles of Alternating Current Machinery.* McGraw-Hill, New York, 1940.

Westinghouse Staff. *Electrical Transmission and Distribution Reference Book.* Westinghouse Electric Corp., 1964.

REVIEW QUESTIONS

1. Describe the difference in construction between core-type and shell-type transformers and state the advantages of each.
2. Describe the different methods used for cooling power and distribution transformers.
3. Why is the iron core of a transformer laminated?
4. Explain why the core loss of a transformer does not change with changes in load.
5. Explain why the current in the primary of a transformer increases when a load is placed across the secondary. What causes the primary current to level off to a value that is just sufficient to carry the load plus losses?
6. What is leakage flux and how does it affect the output of a transformer?
7. Explain how leakage flux in the primary and leakage flux in the secondary affect the secondary voltage.
8. If it were possible to design a transformer with no leakage flux, would this be desirable? Explain.
9. What is meant by the voltage regulation of a transformer and how is this information useful to an applications engineer?
10. Explain why a leading power-factor load tends to cause a rise in voltage above the no-load value?
11. Distinguish between equivalent impedance, per-unit impedance, and percent impedance as they apply to a transformer, and indicate the unique advantages offered by the per-unit system.
12. (a) Explain the nature of hysteresis and eddy-current losses. (b) How are these losses affected by the magnitude and frequency of the applied voltage? (c) How are the core losses minimized during transformer design?

13. Explain why core loss remains essentially constant over the kilovoltampere load range of a transformer.
14. Explain the effect powering a transformer at higher than its rated primary voltage will have on (a) its output voltage; (b) its efficiency.
15. What transformer parameters are determined by the short-circuit test? Sketch the appropriate circuit and indicate how the parameters are determined.
16. What transformer parameters are determined by the open-circuit test? Sketch the appropriate circuit and indicate how the parameters are determined.
17. What precautions should be observed when making the open-circuit test?
18. Explain why the short-circuit test minimizes the core losses.
19. Explain why the open-circuit test minimizes the copper losses.

PROBLEMS

2–1/4 A 22,000-V, 60-Hz generator is connected to the high-voltage side of a 22,000—2200-V, 500-kVA step-down transformer. If the resultant core flux is 0.0683 Wb (max), determine (a) the number of turns of wire in the secondary coil; (b) the new core flux if the driving voltage is increased by 20 percent and the frequency is decreased by 5 percent.

2–2/4 A 2400—115-V transformer has a sinusoidal flux expressed by $\phi = 0.113 \sin 188.5t$. Determine the primary and secondary turns.

2–3/4 A core-type transformer rated at 37.5 kVA, 2400—480 V, and 60 Hz has a core whose mean length is 1.07 m and whose cross-sectional area is 95 cm^2. The application of rated voltage causes a magnetic field intensity of 352 A-t/m (rms), and a maximum flux density of 1.505 T. Determine (a) the number of turns in the primary and the secondary; (b) the magnetizing current when operating as a step-up transformer.

2–4/4 A 2000-kVA, 4800—600-V, 60-Hz, core-type transformer operating at no load in the step-down mode draws a magnetizing current equal to 2 percent rated current. The core has a mean length of 3.15 m, and is operated at a flux density of 1.55 T. The magnetic field intensity is 360 A-t/m. Determine (a) the magnetizing current; (b) the number of turns in the two coils; (c) the core flux; (d) the cross-sectional area of the core.

2–5/5 The exciting current for a certain 50-kVA, 480—240-V, 60-Hz transformer is 2.5 percent of rated current at a phase angle of 79.8°. Sketch the equivalent circuit and phasor diagram for the no-load conditions and, assuming operation is in the step-down mode, determine (a) the exciting current; (b) the core-loss component of the exciting current; (c) the magnetizing current; (d) the core loss.

2–6/5 A single-phase oil-cooled distribution transformer rated at 200 kVA, 7200—460 V, and 60 Hz has a core loss of 1100 W, of which 74 percent is due to hysteresis. The magnetizing current is 1.5 percent of rated current.

Sketch the appropriate equivalent circuit and phasor diagram and, assuming step-down operation, determine (a) the magnetizing current and the core-loss component of exciting current; (b) the exciting current; (c) the no-load power factor; (d) the eddy-current losses.

2–7/5 The hysteresis and eddy-current losses for a 75-kVA, 480—120 V, 60-Hz transformer are 215 W and 115 W, respectively. The magnetizing current is 2.5 percent rated current, and the transformer is operating in the step-up mode. Sketch the appropriate equivalent circuit and phasor diagram and determine (a) the exciting current; (b) the no-load power factor; (c) the reactive power input at no load.

2–8/8 A 480—120-V, 60-Hz transformer has its high-voltage winding connected to a 460-V system, and its low-voltage winding connected to a $24\angle 32.8^\circ$-Ω load. Assume the transformer is ideal. Determine (a) the secondary voltage; (b) secondary current; (c) primary current; (d) input impedance at the primary terminals; (e) active, reactive, and apparent power drawn by the load.

2–9/8 A 7200—240-V, 60-Hz transformer is connected for step-up operation, and a $144\angle 46^\circ$-Ω load is connected to the secondary. Assume the transformer is ideal and the input voltage is 220 V at 60 Hz. Determine (a) secondary voltage; (b) secondary current; (c) primary current; (d) input impedance at primary terminals of transformer; (e) active, reactive, and apparent power input to the transformer.

2–10/8 A 200-kVA, 2300—230-V, 60-Hz transformer operating at rated voltage in the step-down mode is supplying 150 kVA at 0.654 power-factor lagging. Assume the transformer is ideal. Determine (a) secondary current; (b) impedance of the load; (c) primary current.

2–11/8 A 50-Hz ideal transformer with a 5-to-1 turns ratio has a low-side current of $15.6\angle -32^\circ$ A when operating in the step-down mode and feeding a load impedance of $8\angle 32^\circ$ Ω. Sketch the circuit and determine (a) low-side voltage; (b) high-side voltage; (c) high-side current; (d) active, reactive, and apparent power input to the transformer.

2–12/10 A 100-kVA, 60-Hz, 7200—480-V, single-phase transformer has the following parameters.

$$R_{\text{HS}} = 2.98\ \Omega \qquad X_{\text{HS}} = 6.52\ \Omega$$

$$R_{\text{LS}} = 0.021\ \Omega \qquad X_{\text{LS}} = 0.031\ \Omega$$

Determine the equivalent impedance of the transformer (a) referred to the high side; (b) referred to the low side.

2–13/10 A 30-kVA, 60-Hz, 2400—600-V transformer has the following parameters in ohms:

$$R_{\text{HS}} = 1.86 \qquad X_{\text{HS}} = 3.41 \qquad X_{M,\text{HS}} = 4962$$

$$R_{\text{LS}} = 0.15 \qquad X_{\text{LS}} = 0.28 \qquad R_{\text{fe,HS}} = 19{,}501$$

Determine the equivalent impedance of the transformer (a) referred to the high side; (b) referred to the low side.

2–14/10 A single-phase, 25-kVA, 2200—600-V, 60-Hz transformer used for step-down operation has the following parameters expressed in ohms:

$$R_{\text{HS}} = 1.40 \qquad X_{\text{HS}} = 3.20 \qquad X_{M,\text{HS}} = 5011$$
$$R_{\text{LS}} = 0.11 \qquad X_{\text{LS}} = 0.25 \qquad R_{\text{fe,HS}} = 18{,}694$$

Sketch the appropriate equivalent circuit and determine (a) the input voltage required to obtain an output of 25 kVA at 600 V and 0.8 power-factor lagging; (b) the load component of primary current; (c) the exciting current.

2–15/10 A 100-kVA, 60-Hz, 7200—480-V, single-phase transformer has the following parameters expressed in ohms:

$$R_{\text{HS}} = 3.06 \qquad X_{\text{HS}} = 6.05 \qquad X_{M,\text{HS}} = 17{,}809$$
$$R_{\text{LS}} = 0.014 \qquad X_{\text{LS}} = 0.027 \qquad R_{\text{fe,HS}} = 71{,}400$$

The transformer is supplying a load that draws rated current at 480 V and 75 percent power-factor lagging. Sketch the appropriate equivalent circuit and determine (a) the equivalent resistance and equivalent reactance referred to the high side; (b) the input impedance of the combined transformer and load; (c) the load component of high-side current; (d) the input voltage to the transformer; (e) the exciting current and its components; (f) the input impedance at no load.

2–16/10 A 75-kVA, 60-Hz, 4160—240-V, single-phase transformer operating in the step-down mode is feeding a $1.45\angle{-38.74^\circ}$-Ω load at 270 V. The transformer parameters expressed in ohms are:

$$R_{\text{LS}} = 0.0072 \qquad X_{\text{LS}} = 0.0128$$
$$R_{\text{HS}} = 2.16 \qquad X_{\text{HS}} = 3.84$$

Sketch the appropriate equivalent circuit and determine (a) the equivalent impedance of the transformer referred to the high side; (b) the input impedance; (c) the voltage impressed at the high-side terminals that results in a load voltage of 270 V. (d) Sketch the phasor diagram for the low-side voltage and current, and determine the power factor at the high side of the transformer.

2–17/11 The parameters for a 250-kVA, 2400—480-V, single-phase transformer operating at rated voltage, rated kVA, and 0.82 power-factor lagging, are $X_{\text{eq,HS}} = 1.08\ \Omega$ and $R_{\text{eq,HS}} = 0.123\ \Omega$. The transformer is operating in the step-down mode. Sketch the appropriate equivalent circuit and determine (a) the equivalent low-side parameters; (b) the no-load voltage; (c) the voltage regulation at 0.82 power-factor lagging.

2–18/11 Re-solve Problem 2–17/11(b) and (c) assuming operation in the step-up mode and 0.70 power-factor leading.

2–19/11 A 333-kVA, 60-Hz, 4160—2400-V transformer operating in the step-down mode has an equivalent resistance and equivalent reactance referred to the high side of 0.5196 Ω and 2.65 Ω, respectively. Assume operation is at rated voltage, rated load, and 0.95 power-factor leading. Sketch the appropriate equivalent circuit and determine (a) the no-load voltage, (b) the voltage regulation; (c) the combined input impedance of the transformer and load.

2–20/11 A 100-kVA, 4800—480-V, 60-Hz, single-phase distribution transformer has 6 V/turn and an equivalent impedance referred to the high side of $8.48\angle 71^\circ$ Ω. The transformer is operating in the step-down mode supplying a 50-kVA, unity power-factor load at 480 V. Determine (a) the output voltage when the load is removed; (b) the inherent voltage regulation of the transformer when operating at 78 percent power-factor lagging. *Note:* By definition, inherent voltage regulation infers rated kVA.

2–21/11 A 37.5-kVA, 6900—230-V, 60-Hz, single-phase transformer is operating in the step-down mode at rated load, rated voltage, and 0.68 power-factor lagging. The equivalent resistance and reactance referred to the low side are 0.0224 Ω and 0.0876 Ω, respectively. The magnetizing reactance and equivalent core-loss resistance (high side) are 43,617 Ω and 174,864 Ω, respectively. Determine (a) the output voltage when the load is removed; (b) the voltage regulation: (c) the combined input impedance of transformer and load; (d) the exciting current and input impedance at no load.

2–22/11 A 500-kVA, 7200—600-V, 60-Hz transformer is operating in the step-down mode at rated kVA and 0.83 power-factor lagging. The output voltage when the load is removed is 625 V. Determine the equivalent impedance of the transformer referred to the high side (assume the equivalent resistance is negligible). *Hint:* Draw a phasor diagram showing **I**, **E**, **V**, and the impedance drop. Use trigonometry to solve for IX_{eq} and then determine X_{eq}.

2–23/12 A 25-kVA, 480—120-V, 60-Hz transformer has a 2.1 percent impedance. Determine (a) the equivalent impedance referred to the high side; (b) the equivalent impedance referred to the low side.

2–24/12 The percent impedance and the percent resistance of a 25-kVA, 7200—600-V, 60-Hz transformer are 2.3 and 1.6 percent, respectively. Determine (a) the percent reactance; (b) the equivalent resistance, equivalent reactance, and equivalent impedance referred to the high side; (c) repeat (b) for the equivalent low-side values.

2–25/12 A 500-kVA, 7200—240-V, 60-Hz transformer with a 2.2 percent impedance was severely damaged as a result of a dead short across the secondary terminals. Determine (a) the short-circuit current; (b) the required percent impedance of a replacement transformer that will limit the low-side short-circuit current to 60,000 A.

2–26/12 A 167-kVA, 60-Hz, 600—240-V, 60-Hz, 4.1 percent impedance distribution transformer with 46 turns on the high side is operating at rated load and 0.82 power-factor lagging. Determine (a) the voltage regulation; (b) the no-load voltage; (c) the core flux; (d) the cross-sectional area of the core if the transformer is operating at a maximum flux density of 1.4 T.

2–27/12 A 150-kVA, 2300—240-V, 60-Hz transformer is operating at rated load and 90 percent power-factor lagging. The resistance and reactance of the transformer, expressed in per-unit values, are 0.0127 and 0.0380, respectively. Determine the inherent voltage regulation.

2–28/12 A 75-kVA, 4160—460-V, 60-Hz transformer is operating at 76 percent rated load and 85 percent power-factor leading. The resistance and reactance of the transformer, expressed in per-unit values, are 0.0160 and 0.0311, respectively. Determine the inherent voltage regulation.

2–29/12 A 50-kVA, 4370—600-V, 60-Hz transformer is operating at 80 percent rated load and 75 percent power-factor lagging. The resistance and reactance of the transformer, expressed in per-unit values, are 0.0156 and 0.0316, respectively. Determine the inherent voltage regulation.

2–30/12 A single-phase distribution transformer, rated at 50 kVA, 450—120 volt, is supplying rated kVA at 120 V and 0.80 power-factor lagging. The percent resistance and percent reactance are 1.0 and 4.4, respectively. Determine (a) transformer regulation; (b) secondary voltage when the load is disconnected; (c) input voltage that must be applied to the primary in order to obtain rated secondary voltage when carrying rated load at 0.80 power-factor lagging.

2–31/12 A single-phase distribution transformer, rated at 75 kVA and 450—230 volt, is supplying rated kVA at 230 V and 0.9 power-factor lagging. The percent resistance and percent reactance are 1.8 and 3.7, respectively. Determine (a) transformer regulation; (b) secondary voltage when the load is disconnected; (c) input voltage that must be applied to the primary in order to obtain rated secondary voltage when carrying rated load at 0.9 power-factor lagging.

2–32/12 A single-phase distribution transformer, rated at 50 kVA and 480—240 volt, is supplying rated kVA at 240 V and 0.85 power-factor leading. The percent resistance and percent reactance are 1.1 and 4.6, respectively. Determine (a) transformer regulation; (b) secondary voltage when the load is disconnected; (c) input voltage that must be applied to the primary in order to obtain rated secondary voltage when carrying rated load at 0.85 power-factor leading.

2–33/12* A 200-kVA, 2300—230-V, 60-Hz transformer has percent resistance and percent leakage reactance of 1.24 and 4.03 respectively. For 10° increments of power-factor angle, tabulate and plot percent regulation vs. power factor for operation between 0.5 lagging and 0.5 leading.

2–34/13 A 150-kVA, 7200—600-V, 60-Hz, single-phase transformer operating at rated conditions has an hysteresis loss of 527 W, an eddy-current loss of 373 W, and a conductor loss of 2000 W. The transformer is to be used on a

---

*Solution by computer is recommended.

50-Hz system, with the restriction that it maintain the same maximum core flux and the same total losses. Determine (a) the new voltage rating; (b) the new kVA rating.

2–35/13 A 75-kVA, 450—120-V, 60-Hz, single-phase transformer has percent resistance and percent reactance of 1.75 and 3.92, respectively. Its efficiency when operating at rated voltage, rated frequency, and rated load at 0.74 power-factor lagging is 97.1 percent. Determine (a) the core loss; (b) the core loss and efficiency if the transformer is powered at the same voltage, load, and power factor, but at 50 Hz. Assume the core-loss ratio $P_h : P_e$ is 2.5.

2–36/13 A 200-kVA, 7200—600-V, 60-Hz transformer is operating at rated load and 90 percent power-factor lagging. The core loss, resistance, and reactance of the transformer, expressed in per-unit values, are 0.0056, 0.0133, and 0.0557, respectively. Determine (a) the efficiency; (b) the inherent voltage regulation; (c) the efficiency and regulation at 30 percent load and 80 percent power-factor lagging.

2–37/13* A 50-kVA, 2300—230-V, 60-Hz transformer supplies a 0.8 lagging power-factor load whose kVA is adjustable from no load to 120 percent rated kVA. The percent resistance, percent reactance, and percent core loss are 1.56, 3.16, and 0.42, respectively. For 2-kVA increments of load, tabulate and plot the efficiency of the transformer from no load to 120 percent rated load.

2–38/14 A short-circuit test performed on a 150-kVA, 4600—230-V, 60-Hz transformer provided the following data:

$$V_{SC} = 182 \text{ V} \qquad I_{SC} = 32.8 \text{ A} \qquad P_{SC} = 1902 \text{ W}$$

Determine (a) per-unit resistance and per-unit reactance; (b) regulation when operated at 0.6 power-factor lagging.

2–39/14 The following test data were obtained from short-circuit and open-circuit tests of a 50-kVA, 2400—600-V, 60-Hz transformer.

$$\begin{aligned} V_{OC} &= 600 \text{ V} & V_{SC} &= 76.4 \text{ V} \\ I_{OC} &= 3.34 \text{ A} & I_{SC} &= 20.8 \text{ A} \\ P_{OC} &= 484 \text{ W} & P_{SC} &= 754 \text{ W} \end{aligned}$$

Determine (a) the equivalent high-side parameters; (b) regulation; (c) efficiency at rated load and 0.92 power-factor lagging.

2–40/14 Data from short-circuit and open-circuit tests of a 25-kVA, 6900—230-V, 60-Hz transformer are:

$$\begin{aligned} V_{OC} &= 230 \text{ V} & V_{SC} &= 513 \text{ V} \\ I_{OC} &= 5.4 \text{ A} & I_{SC} &= 3.6 \text{ A} \\ P_{OC} &= 260 \text{ W} & P_{SC} &= 465 \text{ W} \end{aligned}$$

Determine (a) the magnetizing reactance referred to the high side; (b) the per-unit parameters; (c) efficiency; (d) voltage regulation at 0.65 per-unit load and 84 percent power-factor leading; (e) low-side voltage when the load is removed; (f) voltage that must be applied to the primary in order to obtain the low-side voltage in (e).

2–41/14 Data from short-circuit and open-circuit tests of a 60-Hz, 100-kVA, 4600—230-V transformer are:

$$V_{OC} = 230 \qquad V_{SC} = 172.3$$
$$I_{OC} = 14 \qquad I_{SC} = 20.2$$
$$P_{OC} = 60 \qquad P_{SC} = 1046$$

Determine (a) the magnetizing reactance referred to the high-side; (b) the per-unit parameters; (c) efficiency; (d) voltage regulation at 0.85 per-unit load and 89 percent power-factor lagging; (e) low-side voltage when the load is removed; (f) voltage that must be applied to the primary in order to obtain the low-side voltage in (e).

3

Transformer Connections, Operation, and Specialty Transformers

3.1 INTRODUCTION

The proper connections of transformers and the analysis of specific transformer behavior require the correct interpretation of transformer nameplate data and an understanding of transformer polarity and phase angle.

Transformers designed for special applications, such as autotransformers and instrument transformers, operate on the same principle as do all transformers, but have different circuit arrangements.

The autotransformer uses a single coil with one or more taps to provide transformer action. These transformers are used extensively in industry for reduced voltage starting of induction motors, for establishing a neutral on a three-phase system, for balance coils in connection with three-wire DC generators, for speed control of small motors, for voltage step up or step down at either end of high-voltage transmission lines, for buck and boost applications where the utilization voltage (voltage at the load) must be raised or lowered 5 or 10 percent, etc.

Instrument transformers are used to transform high currents and high voltages to low values for instrumentation and control. Instrument potential-transformers are used in voltage measurements, and instrument current-transformers are used in current measurements. Both types also serve to insulate the low-voltage instruments from the high-voltage system.

Three-phase power distribution and transmission systems require the use of three single-phase transformers, or one three-phase transformer, whose primaries and secondaries are connected in wye or delta. For certain three-phase connections, however, harmonics in the transformer exciting current can cause severe system overvoltages.

The in-rush current to a transformer during the first few cycles depends on the instantaneous value of the voltage wave at the moment the switch or breaker is closed. Since the in-rush current may exceed 25 times rated current, it is essential that this phenomenon be understood and taken into consideration when selecting fuses or circuit breakers.

Safe and efficient paralleling of transformers requires information on turns ratio, equivalent impedance, and phase angles of the corresponding secondaries. Failure to take these factors into consideration when paralleling transformers may overload and damage the transformers even though the load switch may be open.

3.2 TRANSFORMER POLARITY AND STANDARD TERMINAL MARKINGS

Transformer polarity refers to the relative phase relationship of transformer leads as brought outside the transformer tank. Knowledge of transformer polarity is a necessary consideration when paralleling distribution or power transformers, when connecting single-phase transformers in polyphase arrangements, or when connecting instrument transformers to synchroscopes, wattmeters, or power-factor meters. Failure to take relative polarity into consideration may cause severe short circuits, resulting in serious injury or death to operating personnel, as well as severe damage to electrical apparatus. Repaired or replaced transformers, including new transformers, should always have their terminals properly tested for accuracy of markings before being placed into service **[4],[7]**.

Standard Terminal Markings

The terminal markings of distribution and power transformers are stamped with a letter to indicate the relative voltage level and a numeral to indicate the relative phase relationships among the different windings. The *letter markings* of two-winding transformers are H for the high-voltage winding, and L for the low-voltage winding. Transformers with more than two windings have the highest voltage winding designated H; the other windings, in order of decreasing voltage, are designated X, Y, and Z, respectively. The numeral markings assigned to the terminals are such that:

1. Terminals with the same numeral markings have the same instantaneous polarity.
2. When current is entering a specific numbered terminal of one winding, current is leaving the corresponding numbered terminal(s) of the other winding(s), as shown in Figure 3.1(a).

In the case of transformers with tapped windings, such as that shown in Figure 3.1(b), the potential gradient follows the sequence of numerals. Thus, the relative potential difference between the tapped terminals in Figure 3.1(b) is

$$V_{X1\rightarrow X4} > V_{X1\rightarrow X3} > V_{X1\rightarrow X2}$$

Polarity markings such as the $\pm$ markings shown in Figure 3.1(c), or paint marks such as ●, ♦, etc., are also used to indicate the direction or sense of the coil windings with respect to one another. This type of marking is generally used in conjunction with current transformers and potential transformers, but may also be found in other appli-

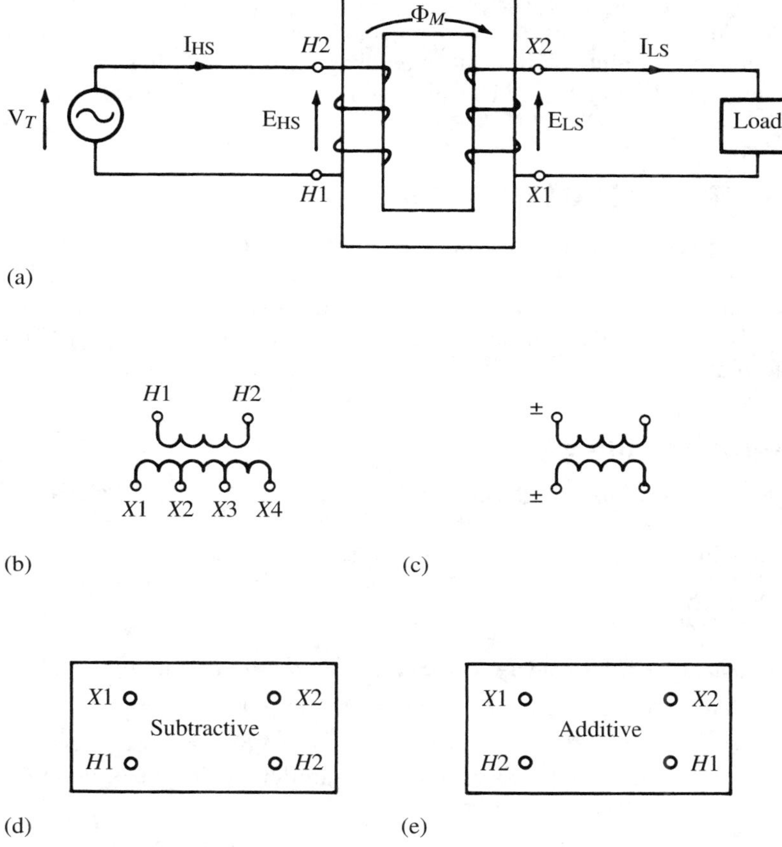

FIGURE 3.1
Transformer polarity marks.

cations. All terminals of a transformer with the same markings have the same instantaneous polarity.[1]

Additive and Subtractive Polarity

The position of the terminals with respect to the high- and low-voltage windings of the transformer affects the voltage stress on the external leads, especially in high-voltage transformers. In Figure 3.1(d), the terminals with the same instantaneous polarity are opposite each other, and if accidental contact between two adjacent terminals occurs—one from each winding—the voltage across the other ends will be the difference between the high and low voltages. This arrangement of terminals, called

[1] See References **[4]** and **[7]** for simple tests to check transformer polarity.

subtractive polarity, is the standard arrangement. If the terminals are arranged as shown in Figure 3.1(e), however, accidental contact between two adjacent terminals of opposite windings will result in a voltage across the other ends equal to the sum of the high and low voltages. This arrangement of terminals is called *additive polarity.*

3.3 TRANSFORMER NAMEPLATES

Transformer nameplate data include voltage rating, kilovoltampere rating, frequency, number of phases, temperature rise, cooling class, percent impedance, and name of manufacturer. The nameplates of large power transformers also include basic impulse level (BIL), phasor diagrams for three-phase operation, and tap-changing information **[7]**.

Voltage Ratings

The voltage ratings, high side and low side, are *no-load* values. Full-load values depend on the power factor of the connected load and hence are not given. Voltage ratings include a winding designator such as a long dash (—), slant (/), cross (×), or wye (Y) to indicate how the voltages are related to each other. These NEMA standard markings indicate the following.

Dash (—): Indicates voltages are from *different* windings.

Slant (/): Indicates voltages are from the *same* winding.

Cross (×): Indicates voltages that may be obtained by reconnecting a two-part winding in series or multiple (parallel). This type of winding is not suitable for three-wire operation.

Wye (Y): Indicates voltages in a wye-connected winding.

The following examples indicate how the winding designators are used in single-phase and three-phase applications.

Single Phase

240/120: 240-V winding with a center tap.

240 × 120: Two-part winding that may be connected in series for 240-V, or connected in parallel for 120-V.

240—120: A 240-V winding and a separate 120-V winding.

Three-Phase

4160—480Y/277: A 4160-V delta-connected winding, and a separate 480-V wye-connected winding with an available neutral connection. *Note:* The voltage of the delta winding is always given first.

Frequency: Rated frequency of transformer.

kVA: Rated apparent power of transformer.

Percent impedance: The percent impedance of the transformer measured at the indicated temperature. Temperature affects resistance, thus affecting the impedance.

Temperature rise: The maximum allowable temperature rise of the transformer based on an ambient temperature of 30°C.

Class: The insulating medium and the method of cooling.

BIL: The basic impulsive level of a transformer, or any other apparatus, is a measure of the *transient voltage stress* that the insulation can withstand without damage. The BIL rating of a transformer indicates that it was tested using an impulse voltage that rises to its peak value in 1.2 μs, and then decays to 50 percent peak voltage after a total of 50 μs has elapsed. The impulse test simulates a lightning surge induced in a transmission line, with the surge voltage modified by a lightning arrester **[7],[9]**.

3.4 AUTOTRANSFORMERS

An autotransformer, shown in Figure 3.2(a), uses a *single coil* with one or more taps to provide transformer action; the input/output connections for operation in the step-down mode are shown in Figure 3.2(b), where

N_{HS} = number of turns in the high side

N_{LS} = number of turns *embraced* by the low side

In those applications where continuous noninterruptible adjustment of voltage is required, slide-wire autotransformers are used. Voltage adjustment is accomplished by means of a carbon brush that slides along a sanded strip for the full length of the coil; the brush replaces tap T in Figure 3.2(b), and can slide the full length of the coil.

Due to their single-coil construction, autotransformers have less leakage flux, less copper, less iron, weigh less, take up less space, are more efficient, and cost less than their two-winding counterparts. Their major disadvantage is the lack of electrical isolation between the primary and secondary. Thus, *autotransformers should only be used in applications where lack of electrical isolation between the high-voltage side and the low-voltage side does not present a safety hazard.*

Another factor that must be considered when specifying autotransformers for large power applications is its equivalent impedance. Although the lower equivalent impedance of an autotransformer causes less of a voltage drop than does its two-winding counterpart, the lower impedance permits a higher short-circuit current should a major fault occur.

Theory of Load Transfer

The theory of load transfer in an autotransformer is explained by starting with the current and voltage relationships in a two-winding transformer, and then "merging" the windings to form a single winding that has the same input/output characteristics. Figure 3.3(a) shows a two-winding transformer whose 80-turn primary is connected to

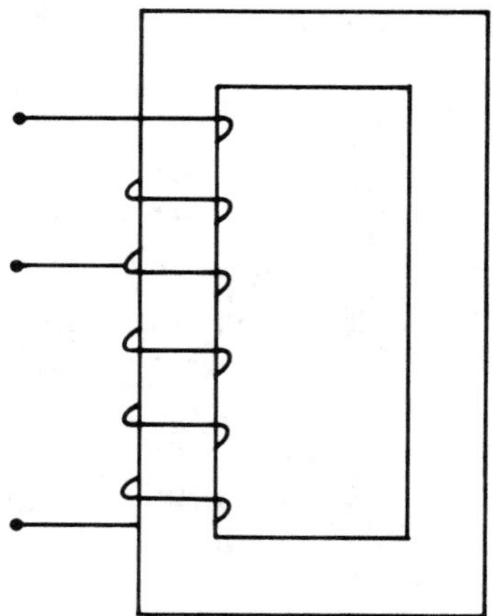

(a)

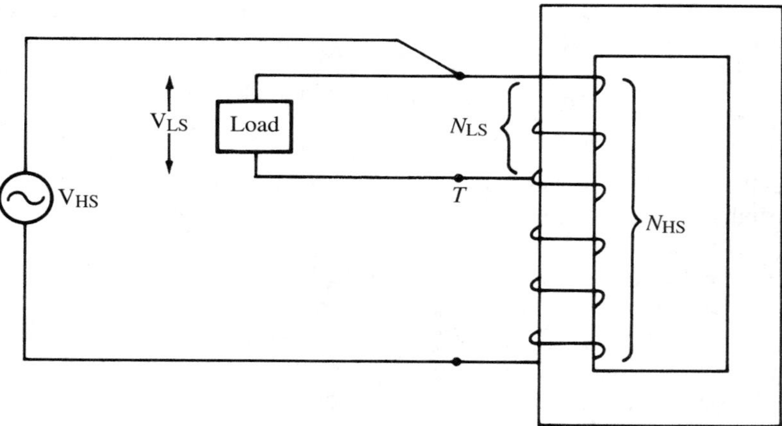

(b)

FIGURE 3.2
Autotransformer: (a) basic autotransformer circuit; (b) connections for step-down operation.

a 120-V system, and whose 20-turn secondary is connected to a 0.50-Ω resistor load. Tap *A*2 on the primary coil embraces 20 turns, which is equal to the 20 turns of the secondary coil. The polarity of the coils was determined from Lenz's law as outlined in Section 2.3. The secondary voltage, secondary current, and primary current shown in Figure 3.3(a) are determined from:

$$a = \frac{N_{HS}}{N_{LS}} = \frac{N_A}{N_B} = \frac{80}{20} = 4$$

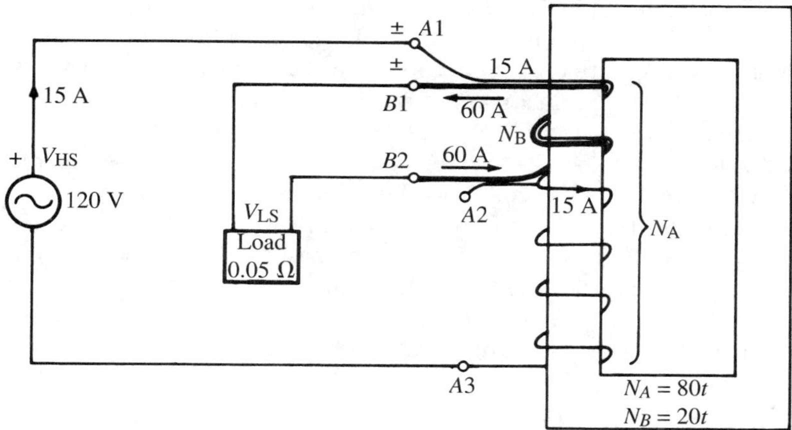

(a)

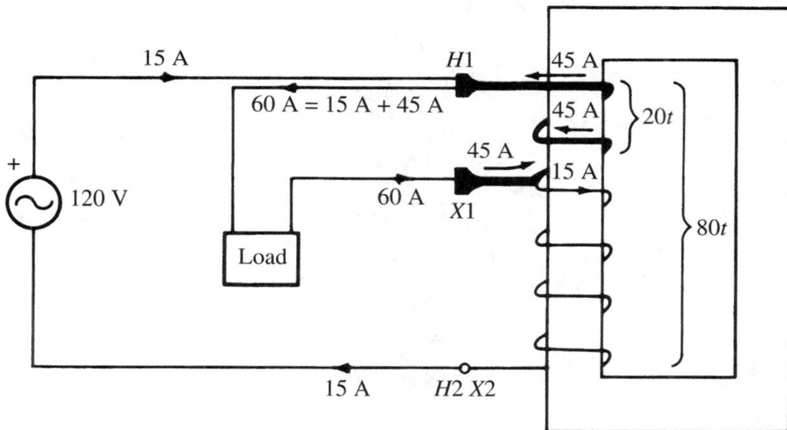

(b)

FIGURE 3.3
(a) Two-winding transformer connected for step-down operation; (b) "merging" primary and secondary turns.

$$V_{LS} = \frac{V_{HS}}{a} = \frac{120}{4} = 30 \text{ V}$$

$$\mathbf{I}_{LS} = \frac{\mathbf{V}_{LS}}{\mathbf{Z}_{load}} = \frac{30\angle 0^\circ}{0.50} = \underline{60 \text{ A}}$$

$$\mathbf{I}_{HS} = \frac{\mathbf{I}_{LS}}{a} = \frac{60}{4} = \underline{15 \text{ A}}$$

Assuming an ideal transformer, the same flux links both coils. Hence, the voltage induced in the 20 turns of coil B will be equal to the voltage induced in the top 20 turns of coil A. Furthermore, the polarity of terminal $A1$ is the same as the polarity of terminal $B1$, and the polarity of terminal $A2$ is identical to the polarity of terminal $B2$. Thus, connecting $A1$ to $B1$ and $A2$ to $B2$ will not change the input and output currents and voltages. This being the case, the 20 turns of coil B may be "merged" turn by turn with the top 20 turns of coil A, as shown in Figure 3.3(b). Note that the current in the "merged" coil is the phasor summation of the currents in the component coils of Figure 3.3(a). Inspection of Figure 3.3(b) shows that of the 60 A delivered to the load, 15 A is *conducted* directly to the load from the 120-V source and 45 A is delivered to the load by *transformer action.* The terminal markings for the autotransformer shown in Figure 3.3(b) are standard NEMA markings **[7]**.

EXAMPLE 3.1

A 400-turn autotransformer, operating in the step-down mode with a 25 percent tap, supplies a 4.8-kVA, 0.85 F_P lagging load. The input to the transformer is 2400-V, 60-Hz. Neglecting the small losses and leakage effects, determine (a) the load current; (b) the incoming line current; (c) the transformed current; (d) the apparent power conducted and the apparent power transformed.

Solution

The circuit diagram is shown in Figure 3.4.

(a)

$$a = \frac{N_{\text{HS}}}{N_{\text{LS}}} = \frac{400}{0.25 \times 400} = 4$$

$$V_{\text{LS}} = \frac{V_{\text{HS}}}{a} = \frac{2400}{4} = 600 \text{ V}$$

$$I_{\text{LS}} = \frac{4800}{600} = \underline{8 \text{ A}}$$

(b)

$$I_{\text{HS}} = \frac{I_{\text{LS}}}{a} = \frac{8}{4} = \underline{2 \text{ A}}$$

(c)

$$I_{\text{TR}} = I_{\text{LS}} - I_{\text{HS}} = 8 - 2 = \underline{6 \text{ A}}$$

(d)

$$S_{\text{cond}} = I_{\text{HS}} V_{\text{LS}} = 2 \times 600 = \underline{1200 \text{ VA}}$$

$$S_{\text{trans}} = I_{\text{TR}} V_{\text{LS}} = 6 \times 600 = \underline{3600 \text{ VA}}$$

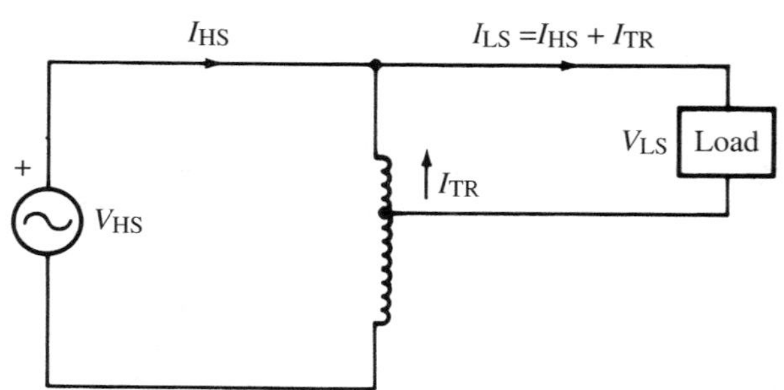

FIGURE 3.4
Circuit diagram for Example 3.1.

Relationship Between Turns Ratio and Power Rating of Two-Winding Transformers When Reconnected as Autotransformers

The relationship between the turns ratio of a two-winding transformer and its power rating when reconnected as an autotransformer is derived from Figure 3.5, where a two-winding transformer in Figure 3.5(a) is shown reconnected as an autotransformer in Figure 3.5(b).

Referring to Figure 3.5(b), the apparent power of the autotransformer connection (assuming rated coil currents) is

$$S_{at} = (V_1 + V_2) \cdot I_2 \tag{3–1}$$

From the turns ratio in Figure 3.5(a),

$$\frac{V_1}{V_2} = \frac{N_1}{N_2} \quad \Rightarrow \quad V_1 = V_2 \cdot \frac{N_1}{N_2} \tag{3–2}$$

Substituting Eq. (3–2) into Eq. (3–1),

$$S_{at} = \left(V_2 \cdot \frac{N_1}{N_2} + V_2\right) \cdot I_2 = \left(\frac{N_1}{N_2} + 1\right) \cdot V_2 I_2$$

$$S_{at} = (a + 1) \cdot S_{2w} \tag{3–3}$$

where: a = turns ratio, 2-winding transformer
S_{at} = apparent-power rating as an autotransformer
S_{2w} = apparent-power rating as a two-winding transformer

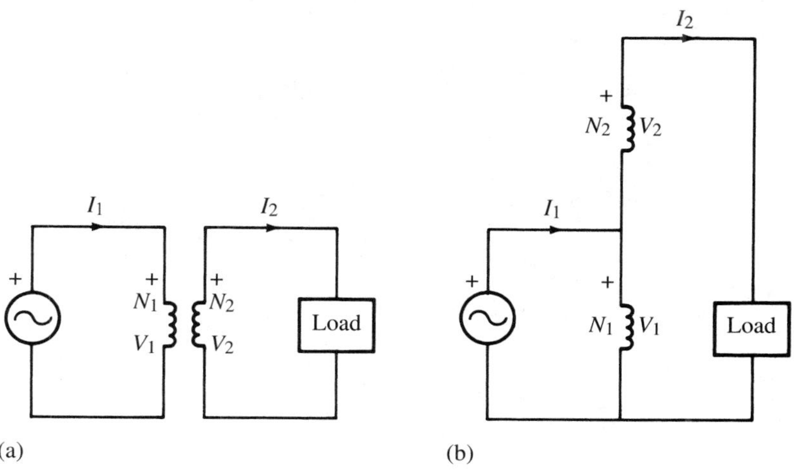

FIGURE 3.5
(a) Two-winding transformer; (b) reconnected as an autotransformer.

As indicated in Eq. (3–3), a two-winding transformer, reconnected as an autotransformer, has an apparent-power rating equal to the transformer rating multiplied by $(a + 1)$.

EXAMPLE 3.2 A 10-kVA, 60-Hz, 2400—240-V distribution transformer is reconnected for use as a step-up autotransformer with a 2640-V output and a 2400-V input. Determine (a) the rated primary and rated secondary currents when connected as an autotransformer; (b) the apparent-power rating when connected as an autotransformer.

Solution

(a) The circuit diagrams for connections as a two-winding transformer and as an autotransformer are shown in Figures 3.6(a) and (b), respectively. To prevent overheating, the rated current for the primary winding and for the secondary winding,

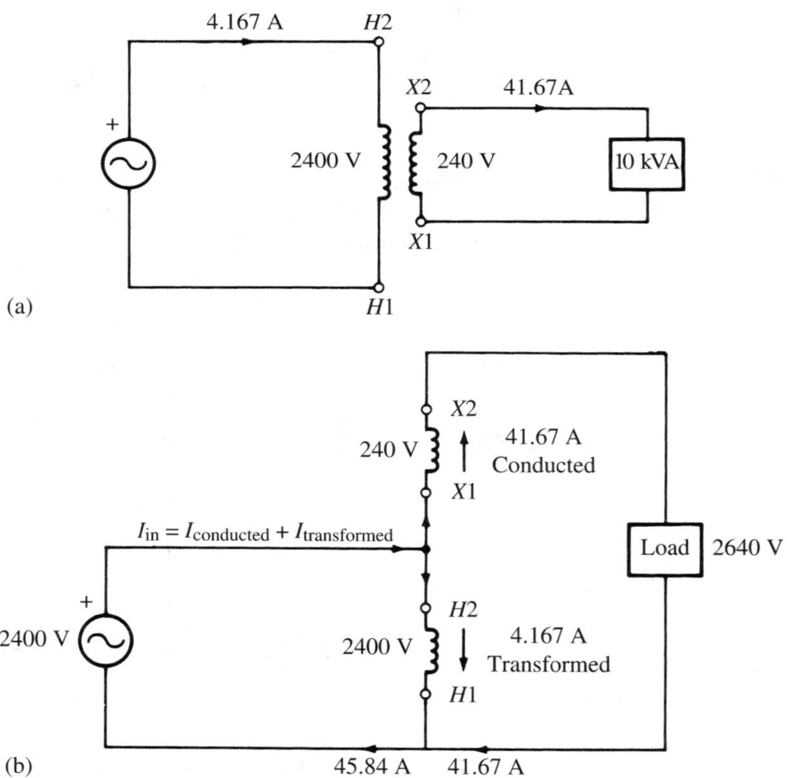

FIGURE 3.6
Circuit diagrams for Example 3.2.

when operated as an autotransformer, must be limited to the same ratings as when operated as a two-winding transformer. The rated currents, as determined for the two-winding transformer in Figure 3.6(a) are

$$I_{\text{LS, winding}} = \frac{10{,}000}{240} = \underline{41.67 \text{ A}} \qquad I_{\text{HS, winding}} = \frac{10{,}000}{2400} = \underline{4.167 \text{ A}}$$

(b) Referring to Figure 3.6(b),

$$S_{\text{at}} = (a + 1) \cdot S_{2w} = \left(\frac{2400}{240} + 1\right) \times 10 = \underline{110 \text{ kVA}}$$

3.5 BUCK-BOOST TRANSFORMERS

All electrical apparatus operate more effectively and more efficiently when the utilization voltage[2] corresponds to the nameplate voltage of the apparatus. A classic example of motor failures brought about by low voltage is air conditioners. The high starting torque required for air-conditioner motors varies as the square of the applied voltage. Thus, a 10 percent drop in utilization voltage will cause a 19 percent drop in starting torque; if the developed torque is not sufficient to start the motor, it will burn out.

Utilization voltages that are too high or too low may be corrected through the use of *buck-boost transformers.* These are special-purpose two-winding transformers whose windings are connected for use as autotransformers. The low-voltage output of the secondary is added to the line voltage (boost connection) or subtracted from the line voltage (buck connection) to obtain the desired utilization voltage. Buck-boost transformers are particularly useful in applications where the utilization voltage is 5 to 15 percent lower or 5 to 15 percent higher than the rated voltage of the apparatus it serves.

A representative buck-boost transformer, shown in Figure 3.7(a), uses a 120 × 240-V primary and a 12 × 24-V or 16 × 32-V secondary.[3] Primary connections for operation in the 120-V region (±15 percent) require paralleling the two primary windings by connecting *H*1 to *H*3 and *H*2 to *H*4; primary connections for operation in the 240-V region (±15 percent) require connecting the two primary windings in series by connecting *H*2 to *H*3. Similarly, a 12-V or 16-V buck or boost requires paralleling the secondaries, and a 24-V or 32-V buck or boost requires the series connection of the two secondaries.

The voltage ratios available from 120 × 240 buck-boost transformers with 12 × 24-V or 16 × 32-V secondaries are shown in Table 3.1.

---

[2] The utilization voltage is the voltage delivered to the load.

[3] Buck-boost transformers with higher nominal-voltage ratings are also available.

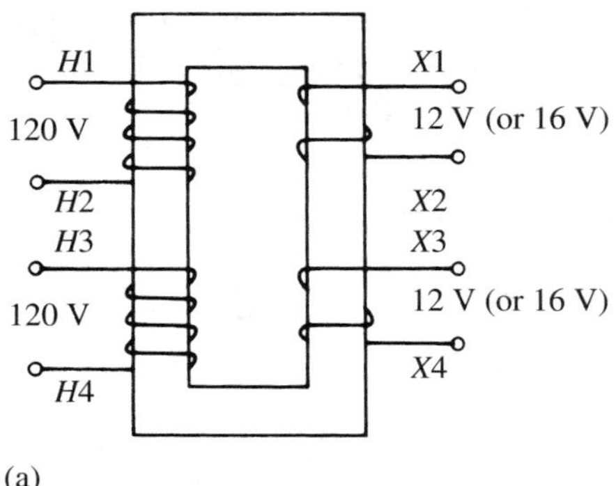

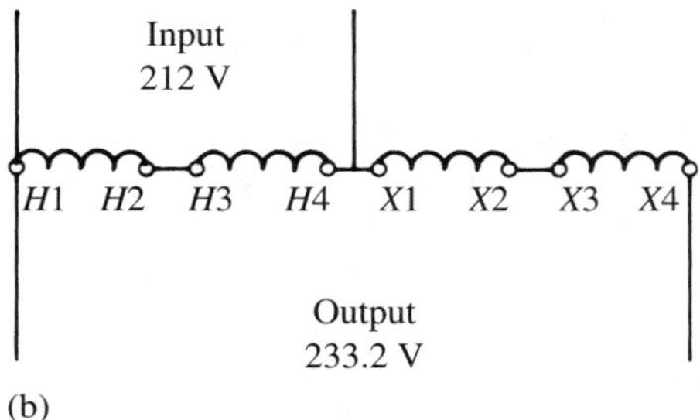

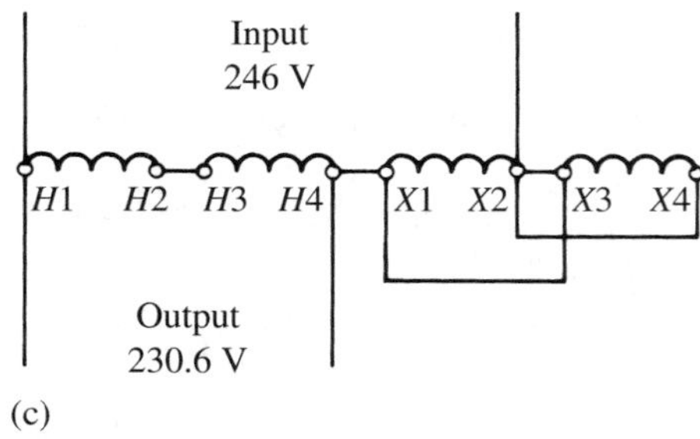

FIGURE 3.7
(a) Buck-boost transformer; (b) circuit for Example 3.3(a); (c) circuit for Example 3.3(b).

EXAMPLE 3.3 The rated voltage of an induction motor driving an air conditioner is 230-V. The utilization voltage is 212-V.

(a) Select a buck-boost transformer and indicate the appropriate connections that will closely approximate the required voltage.
(b) Repeat (a), assuming the utilization voltage is 246-V.

TABLE 3.1
Available buck-boost voltage ratios

A: High-side connected for 240 V:

$$a' = \frac{240 + 12}{240} = 1.050 \qquad a' = \frac{240 + 24}{240} = 1.100$$

$$a' = \frac{240 + 16}{240} = 1.0667 \qquad a' = \frac{240 + 32}{240} = 1.1333$$

B: High-side connected for 120 V:

$$a' = \frac{120 + 12}{120} = 1.100 \qquad a' = \frac{120 + 24}{120} = 1.200$$

$$a' = \frac{120 + 16}{120} = 1.1333 \qquad a' = \frac{120 + 32}{120} = 1.2667$$

where $a' = (V_{HS}/V_{LS})$ = voltage ratio of the autotransformer.

Solution

(a) The required step-up voltage ratio for satisfactory performance is

$$a' = \frac{V_{HS}}{V_{LS}} = \frac{230}{212} = 1.085$$

From Table 3.1A, the available voltage ratio that best suits the load requirement is 1.100. Thus, a buck-boost transformer with 12-V secondaries is required; the two 120 primary windings are connected in series, providing a *rated* 240-V primary; and the two 12-V secondaries are connected in series to provide a rated 24-V output. The correct connections are shown in Figure 3.7(b), and *the actual output voltage* supplied to the air conditioner is

$$V_{HS} = a' \cdot V_{LS} = 1.100 \times 212 = \underline{233.2 \text{ V}}$$

(b) The required step-down voltage ratio for satisfactory performance is

$$a' = \frac{V_{HS}}{V_{LS}} = \frac{246}{230} = 1.070$$

The available voltage ratio that best suits the load requirements is 1.0667. Thus, a buck-boost transformer with 16-V secondaries is required. The two 120-V windings are connected in series, the two 16-V windings are connected in parallel, and the circuit connections are as shown in Figure 3.7(c). Using this connection, the actual output voltage to the air conditioner is

$$V_{LS} = \frac{V_{HS}}{a'} = \frac{246}{1.0667} = \underline{230.6 \text{ V}}$$

3.6 PARALLEL OPERATION OF TRANSFORMERS

When increases in industrial or utility loads approach the full-load rating of a transformer, another transformer of similar rating is generally paralleled with the first, and the load shared between them. For optimum conditions when operating in parallel, however, transformers should have the same turns ratio, identical impedances, and identical ratios of resistance to reactance. Transformers with different turns ratios will have circulating currents in the paralleled loop formed by the transformer secondaries, and transformers with unlike impedances will divide the load in the inverse ratio of their impedances.

Effect of Different Turns Ratios on Parallel Operation

Figure 3.8(a) shows two transformers in parallel (called a bank), with no load connected to the secondaries. Assuming different turns ratios, the output voltages $\mathbf{E}_A$ and $\mathbf{E}_B$ will not be equal, and a current will circulate in the closed loop formed by the two secondaries. The circulating current is indicated by broken arrows in Figure 3.8(a). The phasor sum of the voltages around the loop is $\mathbf{E}_A - \mathbf{E}_B$, and the impedance of the loop is $\mathbf{Z}_A + \mathbf{Z}_B$. Thus, from Ohm's law,

$$\mathbf{I}_{\text{circulating}} = \frac{\mathbf{E}_A - \mathbf{E}_B}{\mathbf{Z}_A + \mathbf{Z}_B} \tag{3–4}$$

When the load switch is closed, as shown in Figure 3.8(b), the circulating current adds to the load current in one transformer and subtracts from the load current in the other transformer. Thus, if the transformer bank is operating at rated load, the transformer with the higher secondary voltage will be overloaded, and the other transformer will be underloaded.

EXAMPLE 3.4

Two 100-kVA single-phase, 60-Hz transformers *A* and *B* are to be operated in parallel. The respective no-load voltage ratios and respective impedances as obtained from the transformer nameplates are

Transformer	Voltage Ratio	%*R*	%*X*
A	2300—460	1.36	3.50
B	2300—450	1.40	3.32

Determine (a) the circulating current in the paralleled secondaries; (b) the circulating current as a percent of the rated current of transformer *A;* (c) the percent difference in secondary voltage that caused the circulating current.

Solution

(a) The rated low-side currents are

$$I_A = \frac{100 \times 1000}{460} = 217.39 \text{ A} \qquad I_B = \frac{100 \times 1000}{450} = 222.22 \text{ A}$$

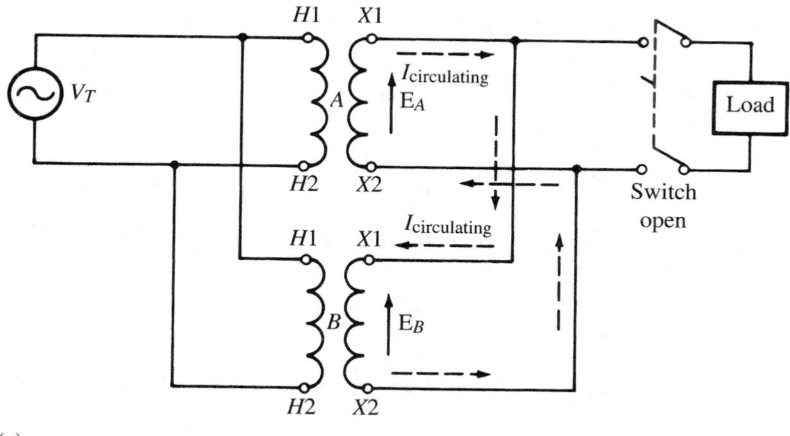

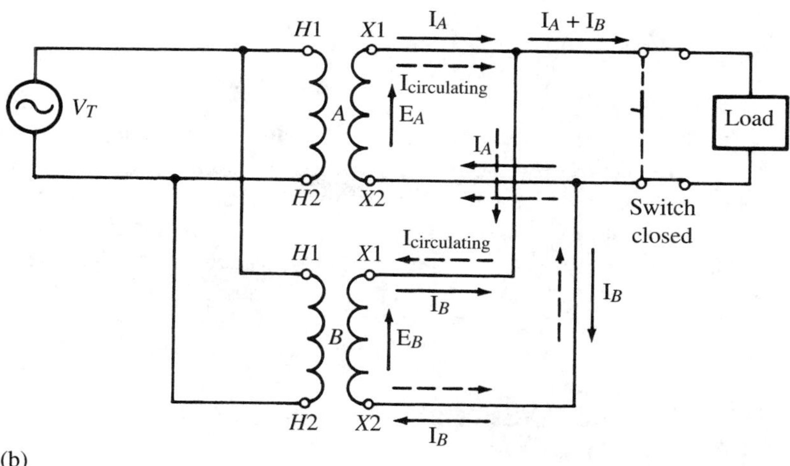

FIGURE 3.8
Circulating current in paralleled transformers: (a) load switch open; (b) load switch closed.

The equivalent resistance and equivalent reactance of each transformer referred to the low side is

$$R_{PU} = \frac{I_{rated} \cdot R_{eq}}{V_{rated}} \qquad X_{PU} = \frac{I_{rated} \cdot X_{eq}}{V_{rated}}$$

$$0.0136 = \frac{217.39 \cdot R_{A,eq}}{460} \qquad 0.0350 = \frac{217.39 \cdot X_{A,eq}}{460}$$

$$R_{A,\text{eq}} = 0.0288\ \Omega \qquad X_{A,\text{eq}} = 0.0741\ \Omega$$

$$0.0140 = \frac{222.22 \cdot R_{B,\text{eq}}}{450} \qquad 0.0332 = \frac{222.22 \cdot X_{B,\text{eq}}}{450}$$

$$R_{B,\text{eq}} = 0.0284\ \Omega \qquad X_{B,\text{eq}} = 0.0672\ \Omega$$

The impedance of the closed loop formed by the two secondaries is

$$\mathbf{Z}_{\text{loop}} = \mathbf{Z}_A + \mathbf{Z}_B = 0.0288 + j0.0741 + 0.0284 + j0.0672 = 0.0572 + j0.1413$$
$$\mathbf{Z}_{\text{loop}} = 0.1524\angle 67.97^\circ\ \Omega$$

From Eq. (3–4),

$$\mathbf{I}_{\text{circulating}} = \frac{460\angle 0^\circ - 450\angle 0^\circ}{0.1524\angle 67.97^\circ} = 65.62\angle -67.97^\circ\ \text{A} \quad \Rightarrow \quad \underline{65.6\angle -68.0^\circ \text{A}}$$

(b)
$$\frac{65.62}{217.39} \cdot 100 = \underline{30.2\%}$$

(c)
$$\frac{460 - 450}{450} \cdot 100 = \underline{2.2\%}$$

Note that a 2.2 percent difference in secondary voltages caused a circulating current equal to 30.2 percent of transformer *A* rated current. Although the circulating current is not apparent at the load terminals, loading the transformer bank until rated load is supplied will seriously overload transformer *A*, and the resultant overheating will damage the winding insulation. It is extremely important that the turns ratio of paralleled transformers be as close to identical as possible if circulating currents and their adverse effects are to be avoided.

3.7 LOAD DIVISION BETWEEN TRANSFORMERS IN PARALLEL

As was previously developed in Section 2.10, and illustrated by the solid lines in Figure 2.10(a), a transformer may be represented by an equivalent impedance in series with the supply voltage and the load impedance, with both the equivalent impedance and the load impedance referred to the primary. Thus, if the turns ratios of paralleled transformers are *alike*, they may be represented by paralleled impedances, as shown in Figure 3.9, for any number of paralleled transformers **[1]**.

The current in any one of the paralleled impedances may be determined by converting the impedances to admittances, and then using the current-divider rule.[4] Thus, for the transformers in Figure 3.9,

---

[4] See Appendix A.4.

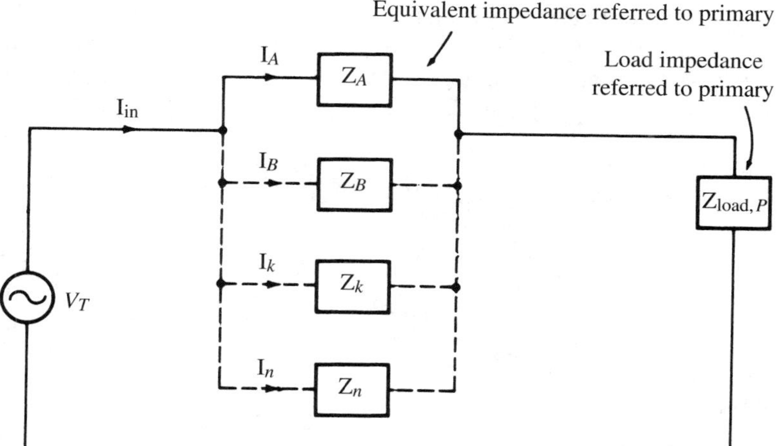

FIGURE 3.9
Equivalent circuit for paralleled transformers.

$$\mathbf{Y}_A = \frac{1}{\mathbf{Z}_A} \qquad \mathbf{Y}_B = \frac{1}{\mathbf{Z}_B} \qquad \mathbf{Y}_k = \frac{1}{\mathbf{Z}_k} \qquad \mathbf{Y}_n = \frac{1}{\mathbf{Z}_n}$$

Using the current-divider rule:

$$\mathbf{I}_k = \mathbf{I}_{\text{bank}} \cdot \frac{\mathbf{Y}_k}{\mathbf{Y}_P} \tag{3–5}$$

$$\mathbf{Y}_P = \mathbf{Y}_A + \mathbf{Y}_B + \cdots + \mathbf{Y}_k + \cdots + \mathbf{Y}_n \tag{3–6}$$

where: $\mathbf{I}_k$ = primary current in transformer k
$\mathbf{I}_{\text{bank}}$ = total input current to transformer bank
$\mathbf{Y}_P$ = admittance of paralleled transformers (siemens, S)
$\mathbf{Y}_k$ = equivalent admittance of transformer k (S)

Equation (3–5) is valid for paralleled transformers that have the same turns ratio. If the turns ratios are not alike, circulating currents will cause the calculated currents to differ from the actual values. *Note:* If the transformer parameters are given in percent impedance or per-unit impedance and they have the same base impedance, they may be used in place of the equivalent ohmic impedances to calculate the current drawn by each transformer.

EXAMPLE 3.5 A 75-kVA transformer (A) is to be paralleled with a 200-kVA transformer (B). Both transformers have a turns ratio equivalent to their 2400—240 voltage ratio, and are operated in the step-down mode. The percent impedance of transformers (A) and (B), based on individual transformer ratings, are 1.64 + j3.16 and 1.10 + j4.03, respectively. Determine (a) the rated high-side current of each transformer; (b) the percent of the total bank-current drawn by each transformer; (c) the maximum load that can be handled by the bank without overloading any one of the transformers.

Solution

(a) The high-side currents of the two transformers are

$$I_{A,\text{rated}} = \frac{75{,}000}{2400} = \underline{31.25 \text{ A}} \qquad I_{B,\text{rated}} = \frac{200{,}000}{2400} = \underline{83.333 \text{ A}}$$

(b)

$$\%\mathbf{Z}_A = 1.64 + j3.16 = 3.5602\underline{/62.571^\circ}$$
$$\%\mathbf{Z}_B = 1.10 + j4.03 = 4.1774\underline{/74.733^\circ}$$

From Eq. (2–47),

$$\mathbf{Z}_{\text{base},A} = \frac{2400}{31.25} = 76.80\ \Omega \qquad \mathbf{Z}_{\text{base},B} = \frac{2400}{83.333} = 28.80\ \Omega$$

$$\mathbf{Z}_{\text{eq},A} = \mathbf{Z}_{\text{base},A} \cdot \mathbf{Z}_{\text{PU},A} = 76.80 \times 0.035602\underline{/62.571^\circ} = 2.7342\underline{/62.571^\circ}\ \Omega$$
$$\mathbf{Z}_{\text{eq},B} = \mathbf{Z}_{\text{base},B} \cdot \mathbf{Z}_{\text{PU},B} = 28.80 \times 0.041774\underline{/74.733^\circ} = 1.2031\underline{/74.733^\circ}\ \Omega$$

$$\mathbf{Y}_{\text{eq},A} = \frac{1}{2.7342\underline{/62.571^\circ}} = 0.36573\underline{/-62.571^\circ} = (0.16847 - j0.32462)\ \text{S}$$

$$\mathbf{Y}_{\text{eq},B} = \frac{1}{1.2031\underline{/74.733^\circ}} = 0.83119\underline{/-74.733^\circ} = (0.21887 - j0.80185)\ \text{S}$$

$$\mathbf{Y}_P = (0.16847 - j0.32462) + (0.21887 - j0.80185) = 1.19121\underline{/-71.0245^\circ}\ \text{S}$$

From Eq. (3–5),

$$|\mathbf{I}_A| = |\mathbf{I}_{\text{bank}}| \cdot \left|\frac{\mathbf{Y}_A}{\mathbf{Y}_P}\right| = I_{\text{bank}} \cdot \frac{0.36573}{1.19121} = 0.307 \cdot I_{\text{bank}}$$

$$|\mathbf{I}_B| = |\mathbf{I}_{\text{bank}}| \cdot \left|\frac{\mathbf{Y}_B}{\mathbf{Y}_P}\right| = I_{\text{bank}} \cdot \frac{0.83119}{1.19121} = 0.6978 \cdot I_{\text{bank}}$$

Thus, the 75-kVA transformer (A) will carry ≈30 percent of the total load, and the 200-kVA transformer (B) will carry ≈70 percent of the total load.

(c) To prevent overloading of the smaller transformer, the bank rating must be based on the rated current of the smallest transformer. Thus,

$$I_A = 0.307 I_{\text{bank}} \quad \Rightarrow \quad I_{\text{bank}} = \frac{I_A}{0.307} = \frac{31.25}{0.307} = 101.79 \text{ A}$$

A bank rating of 101.79 A is selected to prevent overloading transformer A.

$$S_{\text{bank}} = V_{\text{bank}} \cdot I_{\text{bank}} = \frac{2400 \times 101.79}{1000} = \underline{244 \text{ kVA}}$$

Note: In this example, paralleling a 75-kVA transformer with a 200-kVA transformer results in a transformer bank that has less usable capacity than the 200-kVA transformer operating alone. If transformers with the same turns ratio, but different kVA ratings are paralleled, it is possible for the smaller transformer to be overloaded and the larger transformer to be lightly loaded.

EXAMPLE 3.6 Two 60-kVA, 2300—230-V, 60-Hz transformers *A* and *B* are to be operated in parallel. The percent impedances based on individual transformer ratings are:

$$\mathbf{Z}_A = (1.58 - j3.01)\% \qquad \mathbf{Z}_B = (1.09 + j3.98)\%$$

Determine the percent of total bank current drawn by each transformer.

Solution

Since the transformers have the same apparent power ratings and the same voltage ratings, they will also have the same base impedance, and the problem may be solved using per-unit values.

$$\mathbf{Z}_{A,\text{PU}} = 0.0158 + j0.0301 = \underline{0.033995\angle 62.3043^\circ}$$

$$\mathbf{Z}_{B,\text{PU}} = 0.0109 + j0.0398 = \underline{0.041266\angle 74.6840^\circ}$$

$$\mathbf{Y}_{A,\text{PU}} = \frac{1}{0.033995\angle 62.3043^\circ} = 29.416\angle -62.3043^\circ = 13.672 - j26.046$$

$$\mathbf{Y}_{B,\text{PU}} = \frac{1}{0.041266\angle 74.6840^\circ} = 24.233\angle -74.6840^\circ = 6.401 - j23.373$$

$$\mathbf{Y}_{P,\text{PU}} = \mathbf{Y}_{A,\text{PU}} + \mathbf{Y}_{B,\text{PU}}$$

$$\mathbf{Y}_{P,\text{PU}} = (13.672 - j26.046) + (6.401 - j23.373) = 53.340\angle -67.89^\circ$$

$$I_A = \frac{Y_{A,\text{PU}}}{Y_{P,\text{PU}}} \times 100 = \frac{29.416}{53.340} \times 100 = \underline{55.15\%}$$

$$I_B = 100 - 55.15 = \underline{44.85\%}$$

3.8 TRANSFORMER IN-RUSH CURRENT

When a switch is closed, connecting an AC source to an *R-L* series circuit (such as the equivalent series circuit of a transformer), the current will have a source-free response, called the *transient component* or *in-rush current,* and a forced response called the *steady-state component.* Although the in-rush component to a transformer decays rapidly, dropping to the normal no-load current within 5 to 10 cycles, it may exceed 25 times the full-load rating during the first half-cycle. This high in-rush must be taken into consideration when selecting fuses and/or circuit breakers **[2],[8]**.

The magnitude of the in-rush depends on the magnitude and phase angle of the voltage wave at the instant the switch is closed, and the magnitude and direction of the residual flux in the iron. If there is no residual magnetism, and the switch is closed at the instant the voltage wave has its maximum value, the current will be limited to the transformer no-load current, and there will be no in-rush.

Maximum in-rush will occur if the switch is closed at the instant the voltage wave is zero, and the buildup of flux due to the buildup of current is in a direction to reinforce the residual flux. If this occurs, saturation of the iron, caused by the resultant high flux density, will reduce $d\phi/dt$, and this will decrease the primary cemf, permitting a very high in-rush current.

Closing the switch to a transformer is a random event. Hence, the in-rush current may be zero, very large, or some value in between. The in-rush current is also affected by the type and magnitude of the load connected to the secondary. Inductive loads increase the in-rush, whereas resistive loads and capacitive loads decrease the in-rush.

3.9 HARMONICS IN TRANSFORMER EXCITING CURRENT

The nonlinear characteristics of ferromagnetic cores used in transformers cause the magnetizing current to be nonsinusoidal even though the mutual flux is sinusoidal. A Fourier series expansion of the nonsinusoidal magnetizing current shows it to be composed of many sine waves of different frequencies called *harmonics.* The third harmonic is particularly troublesome in certain three-phase connections of transformers, causing system overvoltages and telephone interference.

The Flux Wave

Figure 3.10(a) shows a sinusoidal voltage source connected to the primary winding of an unloaded transformer. Applying Kirchhoff's voltage law to the primary,

$$v_T = i_0 R_P + e_p \tag{3–7}$$

where: v_T = sinusoidal applied voltage
e_p = induced emf
i_0 = exciting current

Since the $i_0 R_P$ drop at no load is very small, Eq. (3–7) may be reduced to

$$v_T \approx e_p \tag{3–8}$$

Thus, the application of a sinusoidal driving voltage to the primary of a transformer will result in an essentially sinusoidal cemf. From Faraday's law,

$$e_p = N_P \frac{d\phi_M}{dt} \tag{3–9}$$

Solving for ϕ_M,

$$\phi_M = \frac{1}{N_P} \int e_p \, dt \tag{3–10}$$

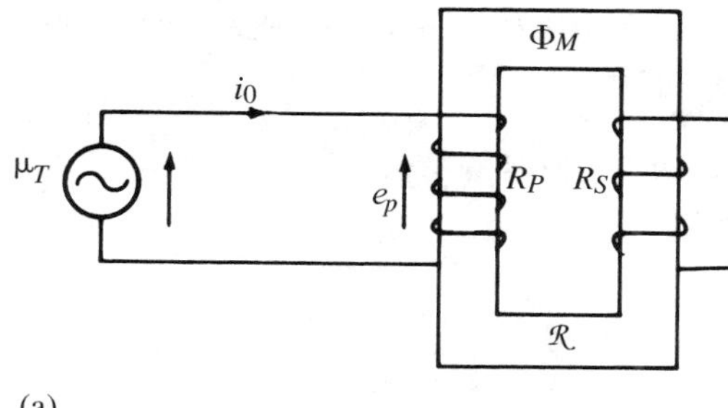

(a)

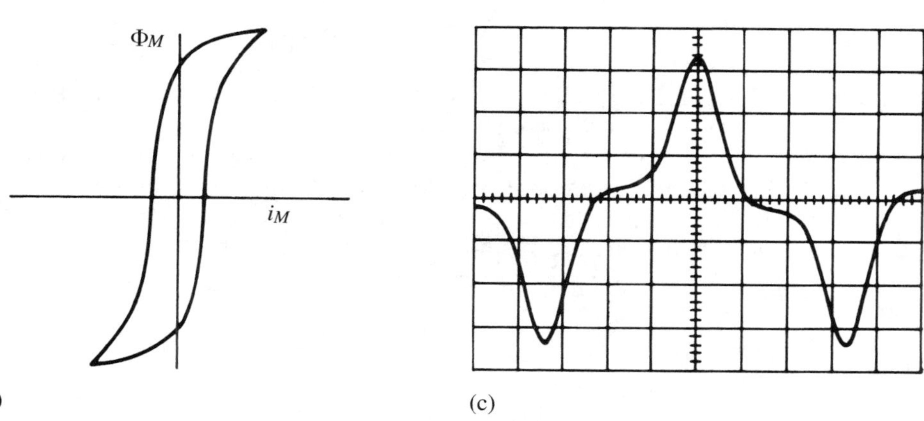

(b) (c)

FIGURE 3.10
Transformer harmonics: (a) circuit; (b) hysteresis loop; (c) nonsinusoidal magnetizing current.

As indicated in Eq. (3–10), the mutual flux is the integral of the sinusoidal cemf. Thus, *the application of a sinusoidal voltage to a transformer results in a sinusoidal cemf and a sinusoidal flux.*

The Current Wave

From Eq. (2–6) in Chapter 2, the magnetizing component of exciting current is,

$$i_M = \frac{\phi_M \mathcal{R}}{N_P} \tag{3–11}$$

A plot of Eq. (3–11) for ferromagnetic materials used in transformer cores results in the typical hysteresis loop shown in Figure 3.10(b). Since the characteristic is nonlinear and the flux is sinusoidal, as proved by Eq. (3–10), *the magnetizing current must be nonsinusoidal.* Figure 3.10(c) shows the *nonsinusoidal magnetizing current* for a representative 60-Hz transformer that produces a sinusoidal flux.

As previously developed and demonstrated in Example 2.2, Section 2.4, the exciting current is composed of a magnetizing component i_M and a core-loss component i_{fe}, where

$$i_0 = i_M + i_{fe}$$

However,

$$i_M \gg i_{fe}$$

Hence,

$$i_0 \approx i_M$$

Thus, for all practical considerations, the names exciting current and magnetizing current may be used interchangeably.

A Fourier series expansion of the 60-Hz exciting current wave in Figure 3.10(c) shows it to be composed of a pure 60-Hz sine wave plus odd multiples of the 60-Hz wave **[3],[6]**. The 60-Hz component is called the *first harmonic* or *fundamental.* The other harmonics are the third, fifth, seventh, ninth, etc., representing 180-Hz, 300-Hz, 420-Hz, 540-Hz, etc., respectively. Thus,

$$i_0 \approx i_M = i_{1h} + i_{3h} + i_{5h} + i_{7h} + \cdots + i_{kh} + \cdots + i_{nh}$$

where: i_0 = actual exciting current
i_{1h} = fundamental component
i_{kh} = component of magnetizing current whose frequency is $k \times$ frequency of the fundamental

The fundamental, third, and fifth harmonics are the dominant components of the magnetizing current and they are shown in Figure 3.11 for a specific distribution transformer.

Harmonic currents in power lines interfere with telephone communications by introducing an objectionable hum. Also, the power system is subject to overvoltages

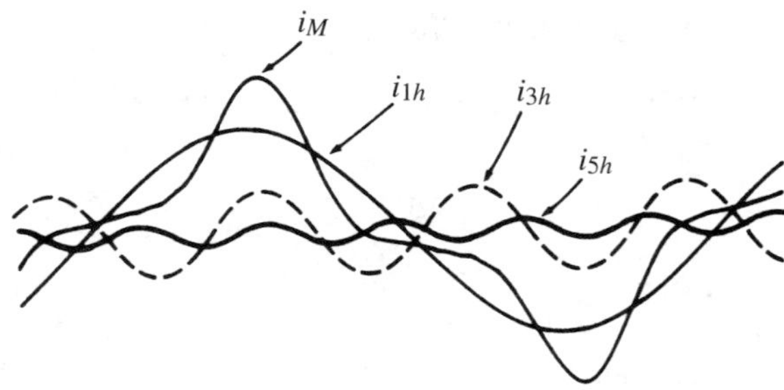

FIGURE 3.11
Magnetizing current and its first three harmonics.

caused by possible series resonance at a harmonic frequency; the third harmonic is the principal troublemaker and manifests itself when transformers are connected in certain three-phase arrangements.

3.10 THREE-PHASE CONNECTIONS OF SINGLE-PHASE TRANSFORMERS

Most AC power is generated and distributed as three phase. The voltage is raised or lowered with three-phase transformers, or with a *bank* of single-phase transformers connected in three-phase arrangements, as shown in Figure 3.12. The current and voltage relationships between phase and line values for a wye connection are[5]

$$V_{\text{line}} = \sqrt{3} \cdot V_{\text{phase}} \qquad I_{\text{line}} = I_{\text{phase}}$$

The current and voltage relationships between phase and line values for a delta connection are

$$I_{\text{line}} = \sqrt{3} \cdot I_{\text{phase}} \qquad V_{\text{line}} = V_{\text{phase}}$$

[5] See Appendix A for a review of current and voltage relationships in three-phase circuits.

EXAMPLE 3.7 A 150-kVA bank of wye–delta-connected step-down transformers has an input line voltage of 4160-V and an output line voltage of 240-V. Determine (a) bank ratio (b) transformer ratio; (c) rated line and phase currents for the high side; (d) rated line and phase currents for the low side.

Solution

(a) The wye–delta connection is shown in Figure 3.12(d). The bank ratio is the ratio of high-side to low-side *line voltages*:

$$\frac{V_{\text{line,HS}}}{V_{\text{line,LS}}} = \frac{4160}{240} = \underline{17.3}$$

(b) The transformer ratio is the ratio of phase voltages. For the wye primary,

$$V_{\text{line}} = \sqrt{3}\, V_{\text{phase}}$$

$$V_{\text{phase}} = \frac{4160}{\sqrt{3}} = 2402 \text{ V}$$

For the secondary,

$$V_{\text{phase}} = V_{\text{line}} = 240 \text{ V}$$

$$\frac{V_{\text{phase,HS}}}{V_{\text{phase,LS}}} = \frac{2402}{240} = 10.0$$

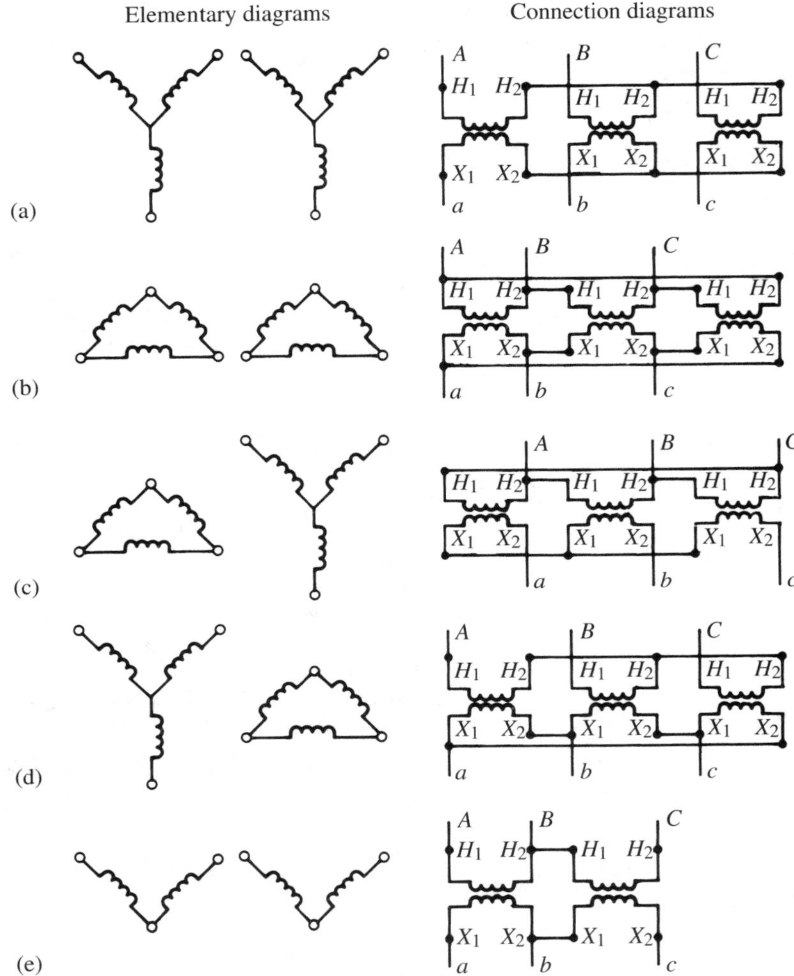

FIGURE 3.12
Three-phase connections of single-phase transformers.

(c)

$$S = \sqrt{3}\, V_{\text{line}} I_{\text{line}}$$

$$I_{\text{line}} = \frac{150{,}000}{\sqrt{3} \times 4160} = \underline{20.8 \text{ A}}$$

Since the high side is wye connected,

$$I_{\text{phase}} = I_{\text{line}} = \underline{20.8 \text{ A}}$$

(d)

$$S = \sqrt{3}\, V_{\text{line}} I_{\text{line}}$$

$$I_{\text{line}} = \frac{150{,}000}{\sqrt{3} \times 240} = \underline{360.8 \text{ A}}$$

For the low-side delta connection,

$$I_{\text{phase}} = \frac{360.8}{\sqrt{3}} = \underline{208.3 \text{ A}}$$

Delta–Delta and V–V Banks

The delta–delta bank, shown in Figure 3.12(b) and in Figure 3.13(a), has the advantage of being able to operate continuously with one of the three transformers disconnected from the circuit. This open-delta connection, also called a V–V connection, provides a convenient means for inspection, maintenance, testing, and replacing of transformers one at a time, with only a brief power interruption. The open-delta connection is also used to provide three-phase service in applications where a possible future increase in load is expected. The increase may be accommodated by adding the third transformer to the bank at a later date. Transformers selected for a delta–delta or open-delta connection must have the same turns ratio and the same percent impedances in order to share the load equally.

A phasor diagram illustrating the current and voltage relationships for a delta-connected secondary is given in Figure 3.13(b). The phase currents, also called *coil currents,* are $\mathbf{I}_{aa'}$, $\mathbf{I}_{bb'}$, and $\mathbf{I}_{cc'}$. The three line currents, determined by applying Kirchhoff's current law to the secondary junctions in Figure 3.13(a), are

$$\mathbf{I}_1 = \mathbf{I}_{aa'} + \mathbf{I}_{b'b}$$
$$\mathbf{I}_2 = \mathbf{I}_{bb'} + \mathbf{I}_{c'c}$$
$$\mathbf{I}_3 = \mathbf{I}_{cc'} + \mathbf{I}_{a'a}$$

Performing the indicated operations in Figure 3.13(b), the magnitudes of the three line currents, as determined by geometry, are shown to be equal to $\sqrt{3}$ or 1.73 times the phase currents.

Disconnecting one transformer, as shown in Figure 3.13(c), does not change the secondary line voltages; $\mathbf{V}_{1-2}$ and $\mathbf{V}_{2-3}$ are the same as before, and $\mathbf{V}_{3-1}$ as determined from phasor addition is

$$\mathbf{V}_{3-1} = \mathbf{V}_{c'c} + \mathbf{V}_{b'b}$$

Performing the indicated phasor additions in Figure 3.13(d) shows $\mathbf{V}_{3-1}$ to be the same whether connected delta–delta or open-delta.

Since the three secondary line voltages are the same whether operating delta–delta or open-delta, and the load impedance has not changed, line currents $\mathbf{I}_1$, $\mathbf{I}_2$, and $\mathbf{I}_3$ must also be the same when operating delta–delta or open-delta. As evidenced

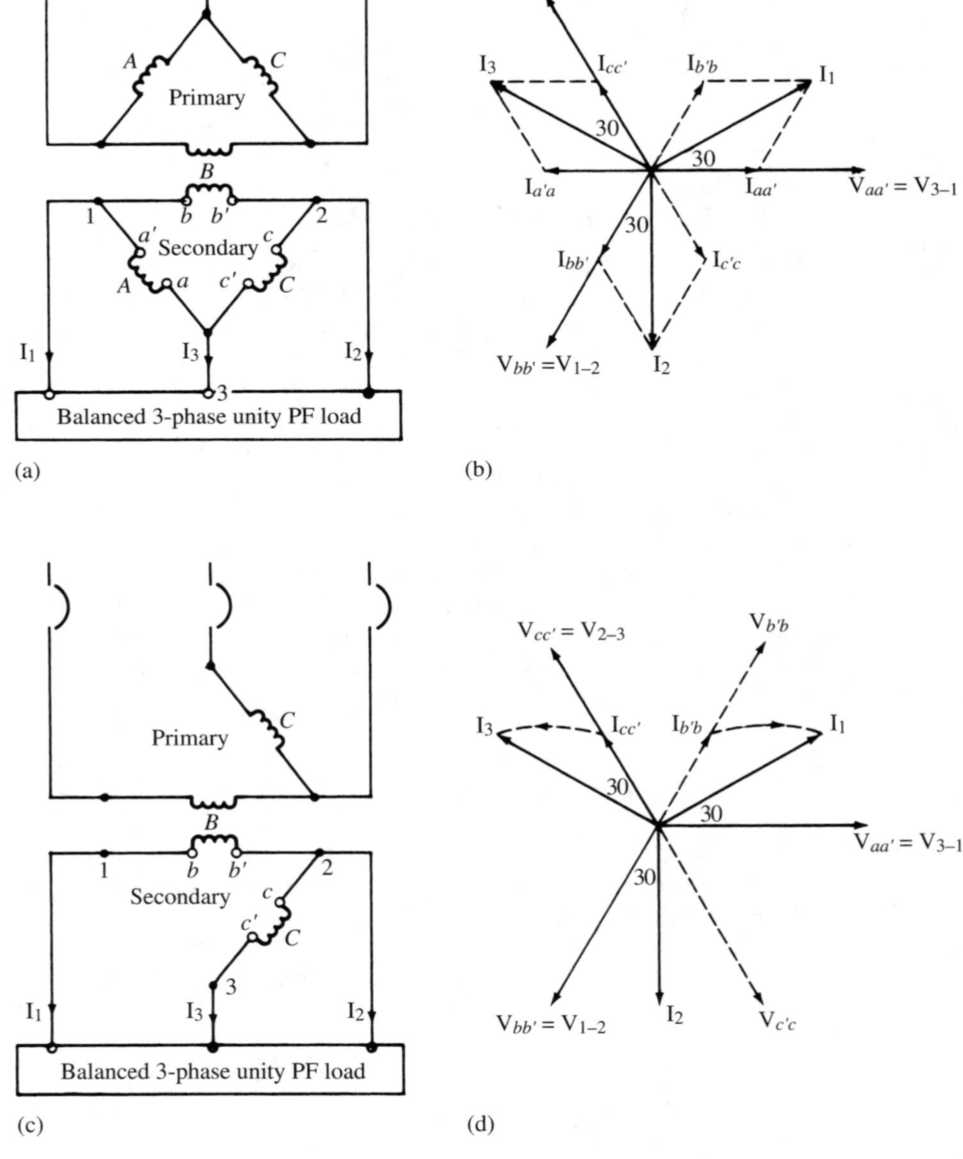

FIGURE 3.13
(a) Delta–delta bank; (b) phasor diagram for (a); (c) V–V bank; (d) phasor diagram for (c).

in Figure 3.13(c), however, the coil currents in the two remaining transformers must increase to equal the line currents. That is,

$$\mathbf{I}_{b'b} \Rightarrow \mathbf{I}_1 \qquad \mathbf{I}_{cc'} \Rightarrow \mathbf{I}_3$$

Figure 3.13(d) shows coil current $\mathbf{I}_{cc}$ increasing in magnitude and shifting its phase 30° to coincide with that of line current $\mathbf{I}_3$, and coil current $\mathbf{I}_{b'b}$ increasing in magnitude and shifting its phase 30° to coincide with that of line current $\mathbf{I}_1$. Thus, if a delta–delta bank is operating at rated load and one transformer is removed, the current in the two remaining transformer coils will increase to 1.73 times its normal rating. To prevent overheating and possible roasting of the windings when operating open-delta, the bank current and hence the bank apparent power must be rerated to reflect the lower kVA capacity. Thus,

$$I_{\text{V-V,rated}} = \frac{I_{\Delta\text{-}\Delta\text{,rated}}}{\sqrt{3}} = 0.577 \times I_{\Delta\text{-}\Delta\text{,rated}} \qquad \textbf{(3–12)}$$

Connecting the transformer bank open-delta did not change the three line voltages. Hence, the bank rating when connected open-delta is

$$S_{\text{V-V,rated}} = \frac{S_{\Delta\text{-}\Delta\text{,rated}}}{\sqrt{3}} = 0.577 \times S_{\Delta\text{-}\Delta\text{,rated}} \qquad \textbf{(3–13)}$$

EXAMPLE 3.8 Three 25-kVA, 480—120-V, single-phase, 60-Hz transformers are connected delta–delta. The total load on the bank is 50-kVA. A fault in one transformer required its removal, and the bank is operating in open-delta. Determine the maximum allowable power that the open-delta bank can handle without overheating.

Solution

The capacity of the delta–delta bank is

$$25 \times 3 = 75 \text{ kVA}$$

The capacity of the bank when operating open-delta is

$$75 \times 0.577 = \underline{43.3 \text{ kVA}}$$

EXAMPLE 3.9 It is desirable to use two transformers in open-delta to supply a balanced three-phase load that draws 50-kW at 120-V and 0.9 F_P lagging. The input voltage to the transformer bank is 450-V and 60-Hz. Determine the minimum power rating required for each transformer.

Solution

$$P = \sqrt{3}\, E_{\text{line}} I_{\text{line}} F_P \quad \Rightarrow \quad 50{,}000 = \sqrt{3} \times 120 \times I_{\text{line}} \times 0.9$$

$$I_{\text{line}} = 267.2918 \text{ A}$$

When operating open-delta, the transformer phase current equals the line current. Thus, the minimum apparent power rating of each transformer is

$$\frac{120 \times 267.2918}{1000} = \underline{32.1 \text{ kVA}}$$

3.11 THREE-PHASE TRANSFORMERS

Three-phase transformers have all three phases wound on a single magnetic core as shown in Figure 3.14(a) for shell-type construction, and in Figure 3.14(b) for core-type construction. The core-type transformer is simpler in construction, and limits third-harmonic fluxes and hence third-harmonic voltages to a relatively small value.[6]

[6] See Reference **[6]** for an analysis of third harmonics in three-phase transformers.

FIGURE 3.14
Basic construction of three-phase transformers: (a) shell type; (b) core type.

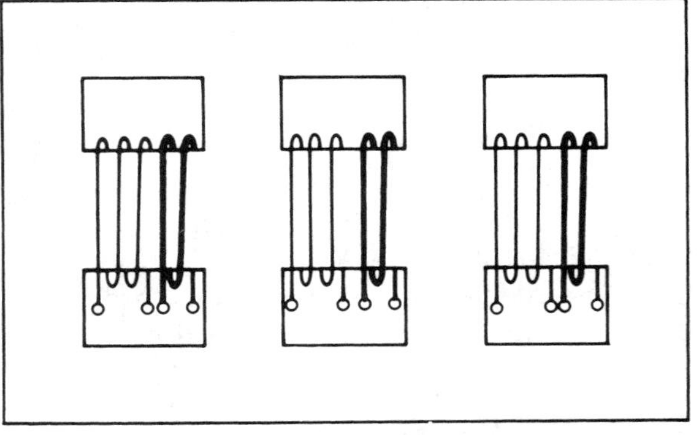

(a)

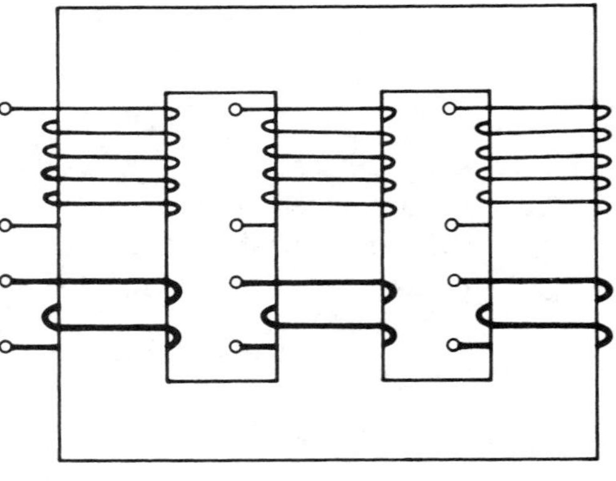

(b)

Three-phase transformers use much less material than three single-phase transformers for the same three-phase power and voltage ratings. Hence, they weigh less and cost much less to produce. Furthermore, since all three phases are in one tank, the wye or delta connections can be made internally, reducing the number of external high-voltage connections from six to three.

The principal disadvantage of a three-phase transformer, compared with its three-transformer counterpart, is that failure of one phase puts the entire transformer out of service. A decision on whether to use a three-phase transformer or three single-phase transformers, however, depends on many factors, including initial cost, cost of operation, cost of spares, cost of repairs, cost of downtime, space requirement, and need for continued operation if one phase is disabled.

3.12 BEWARE THE 30° PHASE SHIFT WHEN PARALLELING THREE-PHASE TRANSFORMER BANKS

There is an angular displacement, called *phase shift,* between the corresponding primary and secondary *line voltages* in the Y–Δ bank and in the Δ–Y bank, as shown in Figure 3.15(b), with the low voltage lagging the high voltage by 30°. There is no angular displacement between corresponding primary and secondary line voltages in a Y–Y bank, Δ–Δ bank, or a V–V bank. *Because of the phase shift inherent in* Y–Δ *and* Δ–Y *banks, they must not be paralleled with* Y–Y, Δ–Δ, *or* V–V *banks;* to do so would cause large circulating currents and severe overheating of the windings, even in the no-load condition **[5]**. Only banks with the same phase shift should be operated in parallel. Note that the bank ratio (ratio of line voltages) for Y–Y, Δ–Δ, or V–V banks is equal to the respective turns ratios. This may be deduced from Figures 3.12(a), (b), and (e).

EXAMPLE 3.10 Assume that a 60-Hz, 50-kVA, delta–delta transformer bank is accidentally paralleled with a 60-Hz, 50-kVA, delta–wye bank. Both have a bank ratio of 2400—240-V. The equivalent impedances referred to the respective secondaries are $\mathbf{Z}_\Delta = 0.1106\angle 60.2^\circ\ \Omega$ and $\mathbf{Z}_Y = 0.0369\angle 60.2^\circ\ \Omega$. Assuming the load switch is open, as shown in Figure 3.15(a), determine (a) the magnitude of the current circulating in each transformer secondary, (b) express the current in (a) in percent of rated transformer current for each transformer.

Solution

(a) The example is solved using loop (mesh) analysis. The respective secondary phase voltages are:[7]

$$\mathbf{E}_{ab} = 240\angle 0^\circ \text{ V} \qquad \mathbf{E}_{aN} = 138.6\angle -30^\circ \text{ V}$$
$$\mathbf{E}_{bc} = 240\angle -120^\circ \text{ V} \qquad \mathbf{E}_{bN} = 138.6\angle -150^\circ \text{ V}$$
$$\mathbf{E}_{ca} = 240\angle 120^\circ \text{ V} \qquad \mathbf{E}_{cN} = 138.6\angle 90^\circ \text{ V}$$

---

[7]See Sections A.7 and A.9 in Appendix A for voltage relationships in wye and delta connections.

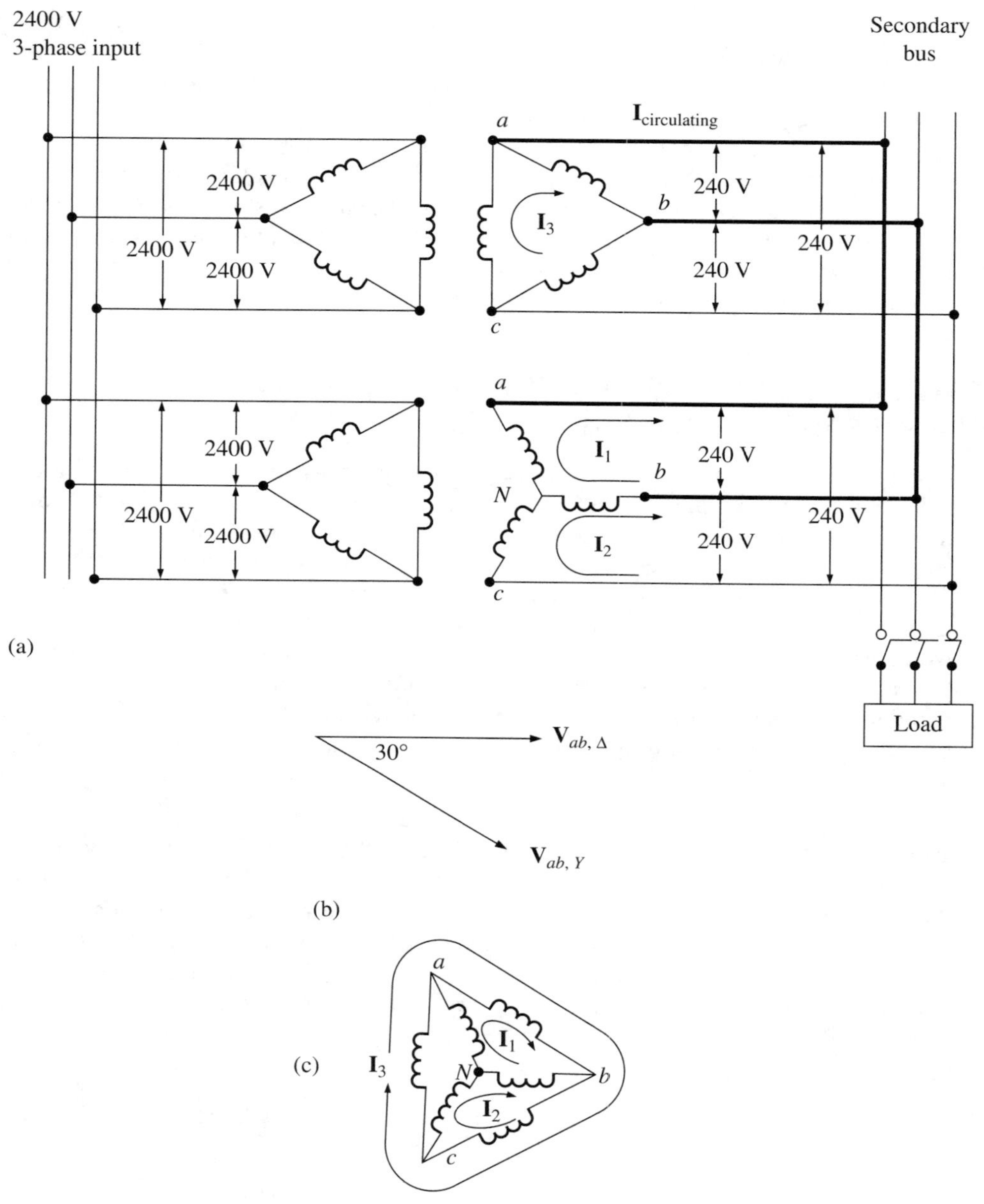

FIGURE 3.15
Paralleled three-phase transformer banks, whose corresponding output voltages are 30° out of phase: (a) circuit diagram; (b) phasor diagram showing one set of corresponding output voltages; (c) simplified circuit.

The loops selected in Figure 3.15(a) are as follows: loop 1 (*bNab*), loop 2 (*cNbc*), and loop 3 (*abc*). Since the load switch is open, the circuit diagram may be simplified as shown in Figure 3.15(c). Applying loop analysis to the three loops and taking into consideration the 30° phase shift shown in Figure 3.15(b),

$$\text{Loop 1} \quad (\mathbf{E}_{bN} + \mathbf{E}_{Na})\angle{-30°} + \mathbf{E}_{ab} = [2\mathbf{Z}_Y + \mathbf{Z}_\Delta]\mathbf{I}_1 - \mathbf{Z}_Y\mathbf{I}_2 + \mathbf{Z}_\Delta\mathbf{I}_3$$

$$\text{Loop 2} \quad (\mathbf{E}_{cN} + \mathbf{E}_{Nb})\angle{-30°} + \mathbf{E}_{bc} = -\mathbf{Z}_Y\mathbf{I}_1 + [2\mathbf{Z}_Y + \mathbf{Z}_\Delta]\mathbf{I}_2 + \mathbf{Z}_\Delta\mathbf{I}_3$$

$$\text{Loop 3} \quad \mathbf{E}_{ab} + \mathbf{E}_{bc} + \mathbf{E}_{ca} = \mathbf{Z}_\Delta\mathbf{I}_1 + \mathbf{Z}_\Delta\mathbf{I}_2 + 3\mathbf{Z}_\Delta\mathbf{I}_3$$

Substituting voltages and impedances, simplifying, and solving,

$$-240\angle{0° - 30°} + 240\angle{0°} = 0.1844\angle{60.2°}\,\mathbf{I}_1 - 0.0369\angle{60.2°}\,\mathbf{I}_2 + 0.1106\angle{60.2°}\,\mathbf{I}_3$$

$$-240\angle{-120° - 30°} + 240\angle{-120°} = -0.0369\angle{60.2°}\,\mathbf{I}_1 + 0.1844\angle{60.2°}\,\mathbf{I}_2 + 0.1106\angle{60.2°}\,\mathbf{I}_3$$

$$0 = 0.1106\angle{60.2°}\,\mathbf{I}_1 + 0.1106\angle{60.2°}\,\mathbf{I}_2 + 0.3318\angle{60.2°}\,\mathbf{I}_3$$

$$I_Y = I_1 = \underline{973\text{ A}} \qquad I_\Delta = I_3 = \underline{562\text{ A}}$$

(b)

$$I_{\text{phase}\Delta} = \frac{50000/3}{240} = 69.44\text{ A} \qquad I_{\text{phase Y}} = \frac{50000/3}{240/\sqrt{3}} = 120.28\text{ A}$$

$$I_\Delta = \frac{562 - 69.44}{69.44} = 7.09 \quad \text{or} \quad \underline{709\% I_{\Delta rated}}$$

$$I_Y = \frac{973 - 120.28}{120.28} = 7.09 \quad \text{or} \quad \underline{709\% I_{Y rated}}$$

As indicated in this example, paralleling three-phase banks whose corresponding line voltages are 30° out of phase will cause a very high circulating current between banks. The effect is similar to that caused by a short circuit.

3.13 HARMONIC SUPPRESSION IN THREE-PHASE CONNECTIONS

As previously discussed in Section 3.9, the magnetizing current that produces a sinusoidal flux and hence a sinusoidal output voltage is itself nonsinusoidal, containing many odd harmonic components. Suppressing any one of the harmonic components will result in a nonsinusoidal flux and hence a nonsinusoidal secondary voltage.

Figure 3.16(a) shows a wye-connected generator supplying a wye–wye transformer bank, with the neutral of the transformer bank connected to the neutral of the generator. The fundamental and third-harmonic components of the magnetizing currents for phases *A, B,* and *C* are shown in Figure 3.16(b); the three phases are separated vertically for easier viewing and are plotted on the *fundamental time axis.* The corresponding phasor diagrams are shown in Figure 3.16(c). Note that *the waves representing the respective fundamentals are 120° apart, but the corresponding third harmonics are in phase with each other.*

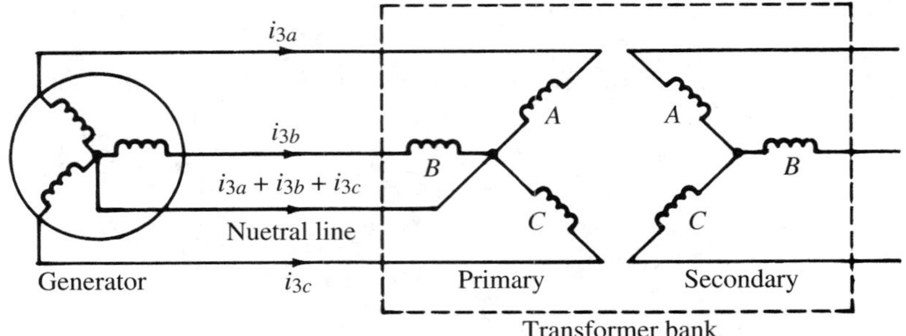

(a)

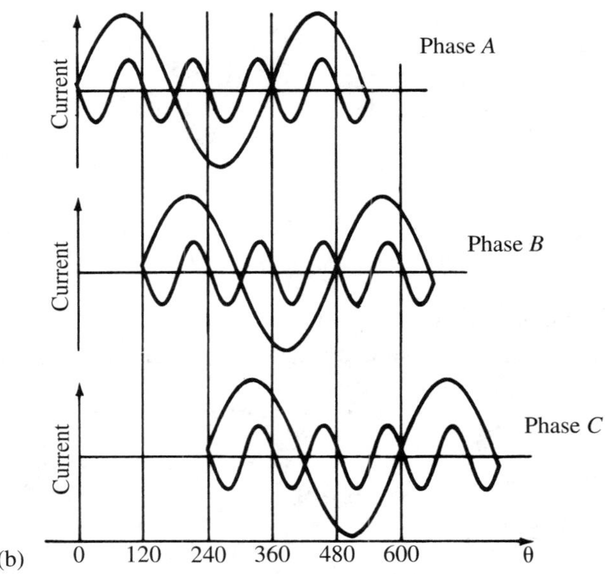

(b)

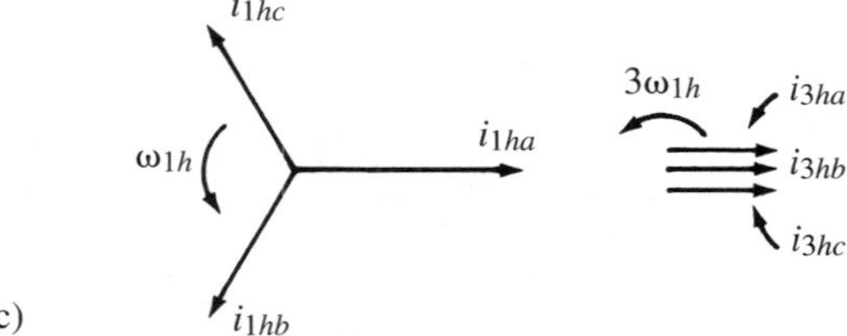

(c)

FIGURE 3.16
(a) Wye–wye bank with neutral connection to primary; (b) waves of fundamental and third harmonic; (c) phasors of fundamental and third harmonic.

Third-harmonic currents and their multiples, called *triplen harmonics,* have zero phase sequence and thus are not three-phase quantities. The three third harmonics (one from each transformer) are in phase with each other, the three ninth harmonics (one from each transformer) are in phase with each other, etc. Nontriplen harmonics, however, such as the second, fourth, fifth, and seventh, are three-phase quantities and must be treated as such.

Since the respective triplen harmonic currents of each phase of a three-phase wye-connected transformer bank are all in phase, all going in or all going out, they require a neutral line to the wye-connected source, as shown in Figure 3.16(a): only the third-harmonic currents are indicated. *If the neutral is not connected, the third-harmonic current will be suppressed, the flux will not be sinusoidal, and the resultant secondary voltage will not be sinusoidal.* The secondary output will have an appreciable third-harmonic voltage that may result in a resonance rise in voltage and overcurrent due to partial series resonance between the capacitive reactance of the lines and the leakage reactance of the transformer at the third-harmonic frequency. For this reason, a wye–wye bank without a line connecting the neutral of the wye primary to the neutral of a wye source, is not desirable for distribution systems.

The Delta Path for Triplen Harmonic Currents

Figure 3.17(a) shows a delta-wye transformer bank connected to a wye-connected source. Since there can be no neutral connection to a delta, the triplen harmonic currents will be suppressed. This will cause the flux in each transformer core to be nonsinusoidal, giving rise to induced emfs that are nonsinusoidal. The most significant component of this nonsinusoidal-induced emf is a third-harmonic voltage that causes a third-harmonic current to circulate in the delta. This is shown in Figure 3.17(a). The circulating third-harmonic current in the delta provides the missing component of flux that enables a sinusoidal output voltage. Other triplen harmonics, such as the ninth and fifteenth, are also circulating in the delta. However, their magnitudes are quite small compared to the magnitude of the third harmonic.

A similar phenomenon occurs with a wye–delta bank that has no neutral line to the generator. This is shown in Figure 3.17(b). In this case, however, the path for triplen harmonic currents is the delta-connected secondary.

The delta–wye bank, shown in Figure 3.17(a), is the preferred choice for industrial power systems; it provides a neutral for single-phase loads, a three-phase output, and a closed path that contains and isolates the triplen harmonic currents. The neutral is also used as a ground connection for safety.

Wye–Wye–Delta

In certain applications, where a wye–wye connection with no neutral tie to the source or load is desired, a third coil, called a *tertiary coil,* is wound on each of the three transformer cores. The three tertiary coils are connected in delta, as shown in Figure 3.17(c). The tertiary coils provide a path for triplen harmonic currents, thus enabling a sinusoidal flux and sinusoidal output voltage from the secondary.

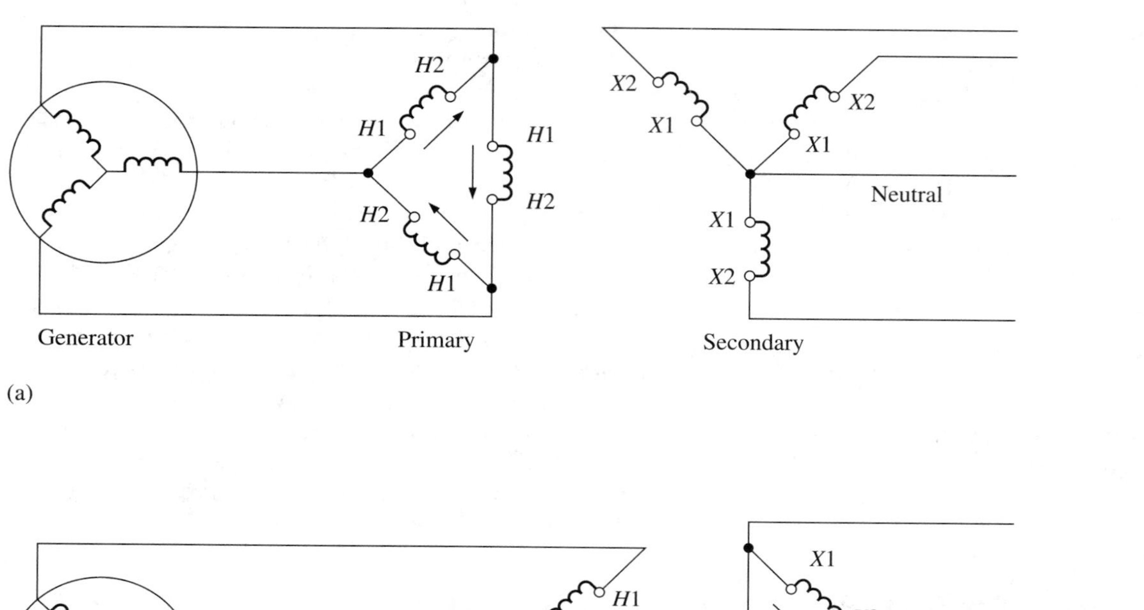

(a)

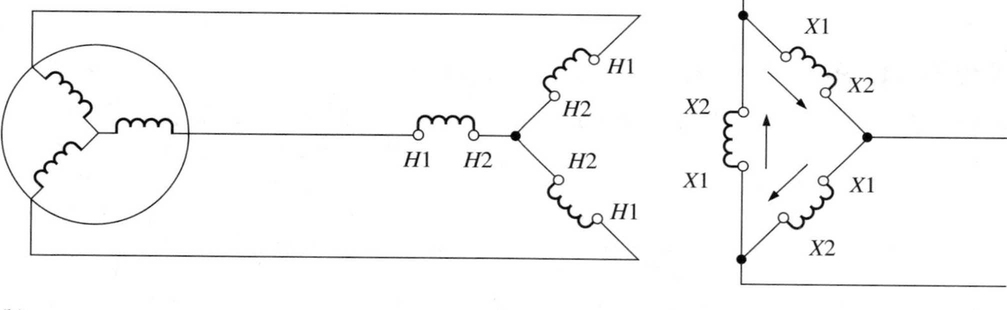

(b)

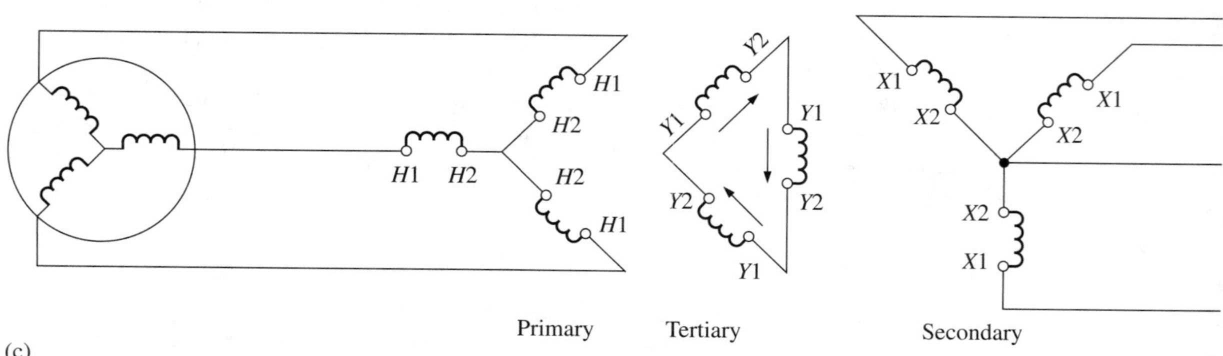

(c)

FIGURE 3.17
Third-harmonic currents: (a) delta–wye connection; (b) wye–delta connection; (c) wye–wye–delta.

3.14 INSTRUMENT TRANSFORMERS

Instrument transformers are used to transform high currents and high voltages to low values for instrumentation and control. Instrument potential-transformers are used in voltage measurements, and instrument current-transformers are used in current measurements. Both types also serve to insulate the low-voltage instruments from the high-voltage system.

The instrument potential-transformer or PT is a very accurate two-winding transformer, whose primary is connected across the voltage it is to measure and whose secondary is connected to a voltmeter or high-impedance relay. The operation of an instrument potential-transformer is essentially the same as that for all other two-winding transformers previously discussed.

An instrument current-transformer, also called a CT, is used to step down a relatively high current to some lower value for the operation of instruments and relays. It is also used in high-voltage circuits to isolate current-measuring instruments and relays from the high-voltage line. The primary of the CT is connected in series with the load circuit, and the secondary is connected to the instruments and/or relays, called the *burden.*

A representative *window-type* CT is shown in Figure 3.18(a). The primary consists of a single power-line conductor looped through the window of a ferromagnetic toroid. The secondary has many turns of insulated wire wound around the toroid.

The alternating flux produced by current in the power-line conductor [primary in Figure 3.18(a)] induces current in the closed secondary circuit that is approximately proportional to the primary current. The actual CT ratio, as stamped on the nameplate, is generally expressed as a current ratio with respect to a 5-A secondary. Two examples of CT ratios are 100 : 5 and 10000 : 5, indicating that a 5-A ammeter or 5-A relay is the required burden.

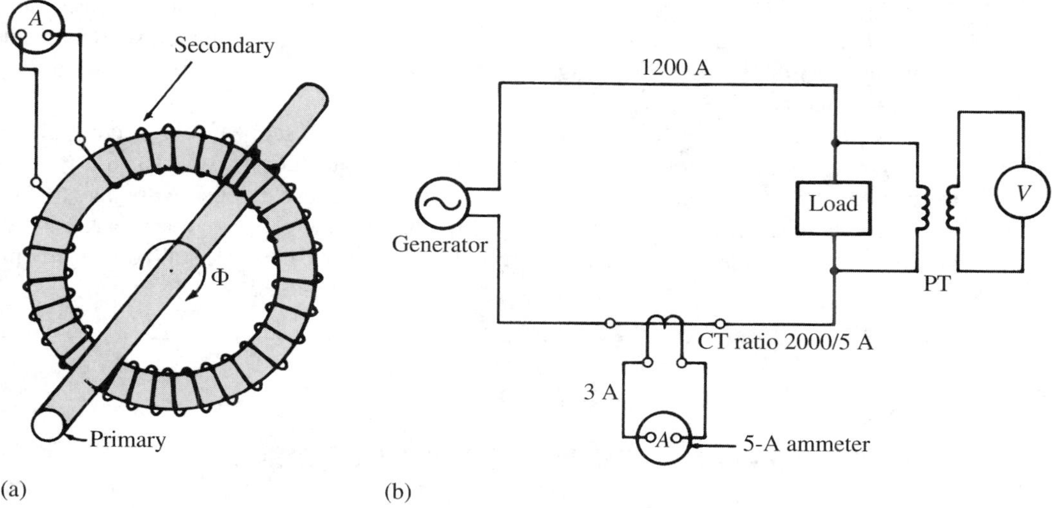

FIGURE 3.18
Current transformer: (a) window type; (b) circuit connections.

Figure 3.18(b) shows the circuit connections for metering a 1200-A load with a 2000/5-A current transformer. The current to the ammeter is

$$I_{\text{meter}} = 1200 \times \frac{5}{2000} = 3 \text{ A}$$

The secondary winding of a current transformer must be connected to a burden, or it must be short circuited at its terminals. Opening the secondary circuit while current is in the primary may cause dangerously high voltages at the secondary terminals and may also permanently magnetize the transformer iron, introducing errors in the transformer ratio. *Hence, before disconnecting instruments and relays from a current transformer, the secondary must be shorted.*

Instrument Transformer Accuracy

The core losses and the nonlinearity of the exciting current, caused by the nonlinear ferromagnetic core, prevents the true ratio of primary to secondary current from being exactly equal to the inverse of the turns ratio and also causes a small phase shift. To compensate for the ratio error, manufacturers of CTs make the turns ratio slightly different from the nameplate ratio. Although it is generally not practical to construct a CT with built-in compensation for phase-angle error, it must be taken into consideration when applying current transformers to solid-state circuitry and power system relaying.

Polarity of Instrument Transformers

The relative instantaneous polarities of instrument transformer terminals are indicated with $\pm$ markings or H_1 and X_1 markings and are generally supplemented with a bright paint mark; letter H_1 denotes the high-side polarity terminal, and X_1 the corresponding low-side polarity terminal.

Wattmeters, varmeters, power-factor meters, synchroscopes, relays, etc., all require specific polarity connections. Wrong polarity connections will cause errors in the previously mentioned instrument readings and may cause extensive damage to AC generators and distribution systems that utilize protective relaying. For this reason, a polarity test should always be made on rebuilt, repaired, or newly installed instrument transformers to make sure that the polarity markings are indeed correct.[8]

[8] See Reference **[4]** for polarity testing of instrument transformers.

SUMMARY OF EQUATIONS FOR PROBLEM SOLVING

Autotransformers

$$S_{\text{at}} = (a + 1) \cdot S_{2w} \qquad \textbf{(3–3)}$$

Parallel Operation of Transformers

$$\mathbf{I}_{\text{circulating}} = \frac{\mathbf{E}_A - \mathbf{E}_B}{\mathbf{Z}_A + \mathbf{Z}_B} \qquad \textbf{(3–4)}$$

$$\mathbf{I}_k = \mathbf{I}_{\text{bank}} \cdot \frac{\mathbf{Y}_k}{\mathbf{Y}_P} \qquad \textbf{(3–5)}$$

$$\mathbf{Y}_P = \mathbf{Y}_A + \mathbf{Y}_B + \cdots + \mathbf{Y}_k + \cdots + \mathbf{Y}_n \qquad \textbf{(3–6)}$$

Three-phase Connections

Wye: $V_{\text{line}} = \sqrt{3} \cdot V_{\text{phase}} \qquad I_{\text{line}} = I_{\text{phase}}$

Delta: $I_{\text{line}} = \sqrt{3} \cdot I_{\text{phase}} \qquad V_{\text{line}} = V_{\text{phase}}$

$$I_{\text{V-V,rated}} = \frac{I_{\Delta\text{-}\Delta\text{,rated}}}{\sqrt{3}} = 0.577 \times I_{\Delta\text{-}\Delta\text{,rated}} \qquad \textbf{(3–12)}$$

$$S_{\text{V-V,rated}} = \frac{S_{\Delta\text{-}\Delta\text{,rated}}}{\sqrt{3}} = 0.577 \times S_{\Delta\text{-}\Delta\text{,rated}} \qquad \textbf{(3–13)}$$

SPECIFIC REFERENCES KEYED TO TEXT

1. Bewley, L. V. *Alternating Current Machinery.* Macmillan, New York, 1949.
2. Huber, W. J. Selection and coordination criteria for current limiting fuses. *IEEE Trans. Industry Applications,* Vol. IA-13, No. 6, Nov./Dec. 1977.
3. Hubert, C. I. *Electric Circuits AC/DC: An Integrated Approach.* McGraw-Hill, New York, 1982.
4. Hubert, C. I. *Preventive Maintenance of Electrical Equipment.* Prentice Hall, Upper Saddle River, NJ, 2002.
5. Lawrence, R. R. *Principles of Alternating Current Machinery.* McGraw-Hill, New York, 1940.
6. MIT E. E. Staff. *Magnetic Circuits and Transformers.* Wiley, New York, 1943.
7. Standards publication: Dry type transformers for general applications. NEMA Publication No. ST 20-1972. National Electrical Manufacturers Association Washington, DC.
8. Say, M. G. *Alternating Current Machines.* Wiley, New York, 1983.
9. Zalar, D. A. A guide to the application of surge arresters for transformer protection. *IEEE Trans. Industry Applications,* Vol. IA-15, No. 6, Nov./Dec. 1979.

GENERAL REFERENCES

Heathcote, Martin J. *JSP Transformer Book: A Practical Technology of the Power Transformer,* 12th ed. Oxford, Boston, 1998.

Blume, L. F. *Transformer Engineering.* Wiley, New York, 1938.

Test Code for Liquid Immersed Distribution, Power, and Regulating Transformers, and Guide for Short-Circuit Testing of Distribution and Power Transformers. ANSI C57.12.90-1980, American National Standards Institute, New York, 1980.

Westinghouse Staff. *Electrical Transmission and Distribution Reference Book.* Westinghouse Electric Corp., 1964.

REVIEW QUESTIONS

1. Differentiate between subtractive polarity and additive polarity as it pertains to transformers. Why is subtractive polarity preferred?
2. What is meant by the basic impulse level (BIL) of a transformer?
3. How does an autotransformer differ physically and electrically from a two-winding transformer?
4. Sketch the circuit for a step-down autotransformer connected to a load and show the relative magnitude of currents and voltages in the windings.
5. What are the advantages and disadvantages of an autotransformer with respect to a two-winding transformer?
6. If a 50-kVA transformer has a turns ratio of 20 : 1, what would be its power rating if it is reconnected as an autotransformer?
7. What is meant by utilization voltage?
8. What is a buck-boost transformer, and what are its applications?
9. Using a circuit diagram, explain the effect of different turns ratios on the performance of paralleled transformers.
10. Under what conditions must the nameplate-power-rating of a transformer be rerated to a lower value? Explain.
11. What is transformer in-rush current? What does it depend on? What is its approximate range of values? Approximately how long does it last? Why is this information important?
12. What are transformer harmonics, and how are they generated?
13. Explain what would happen if three-phase transformer banks that have a 30° phase difference between corresponding secondary line voltages are paralleled.
14. State and explain the adverse effect on a three-phase distribution system if the third-harmonic component of exciting current is suppressed.
15. What are the advantages and disadvantages of a three-phase transformer over an equivalent three-phase bank composed of three single-phase transformers?
16. Sketch a circuit showing a transformer feeding a load. Include current and potential transformers along with associated instruments in both primary and secondary circuits.

PROBLEMS

3–1/4 A 2300—450-V, 60-Hz autotransformer is used for step-down operation and supplies a load whose impedance is $2\angle 10^\circ\ \Omega$. Neglecting the internal impedance of the transformer, determine (a) the load current; (b) high-side current and transformed current; (c) sketch the circuit and indicate the magnitude and relative directions of the currents in (b).

3–2/4 An autotransformer with a 25 percent tap is connected to a 600-V generator for step-up operation. The high side is connected to a 100-kVA 0.80 lagging power-factor load. Sketch the circuit and determine (a) secondary voltage; (b) load current; (c) primary current; (d) current in that section of the coil that is embraced by the 600-V generator.

3–3/4 An autotransformer with a total of 600 turns has its secondary winding embracing 200 turns. The 600 turns are connected to a 60-Hz, 2400-V driving voltage. A load connected to the secondary draws 4.8-kVA at 0.6 power factor lagging. Neglecting losses and leakage effects, determine (a) secondary voltage; (b) secondary current; (c) primary current; (d) power conducted; (e) power transformed; (f) the instantaneous maximum value of the core flux.

3–4/4 A 100-kVA, 60-Hz, 440—240-V autotransformer supplies a load consisting of a 240-V, 8-kW heater and a 240-V, 60-Hz, 10-hp motor. The motor operates at 90 percent load, 86 percent power-factor lagging, and is 88 percent efficient at that load. Calculate (a) the total apparent power supplied by the transformer; (b) the apparent power conducted; (c) the apparent power transformed.

3–5/5 A 120-V, 60-Hz air conditioner is to be operated in a remote area where the voltage drop in the long transmission line results in a utilization voltage of 102-V. Determine (a) the required step-up voltage ratio for satisfactory performance; (b) the voltage ratio of a standard buck-boost transformer that most closely meets the requirements of the load; (c) the voltage at the load with the buck-boost transformer installed. (d) Sketch the appropriate connection diagram, and show the NEMA standard terminal markings.

3–6/5 An electric boiler rated at 50-kW, 240-V, and 60-Hz is to be operated from a 60-Hz system whose utilization voltage is 269.5-V. Determine (a) the required step-down voltage ratio for satisfactory performance; (b) the voltage ratio of a standard buck-boost transformer that most closely approaches the requirement in (a); (c) the voltage at the load with the buck-boost transformer installed. (d) Sketch the appropriate connection diagram for the voltage transformation and show the NEMA standard terminal markings. (e) Determine the input and output currents.

3–7/5 A lighting installation consisting of 300 lamps, each rated at 300 W and 120-V, must be operated from a system whose utilization voltage is 132-V. (a) Since the overvoltage will shorten the life of the lamps, select a standard buck-boost transformer that will effectively match the utilization voltage to the load. (b) Sketch the appropriate connection diagram for the voltage transformation and show the NEMA standard terminal markings. (c) Calculate the input and output currents of the buck-boost transformer.

3–8/6 Two 50-kVA, 60-Hz transformers have the following voltage ratios and equivalent low-side impedances:

Transformer	Voltage Ratios	$R_{eq,Ls}$ (Ω)	$X_{eq,Ls}$ (Ω)
A	4800—482	0.0688	0.1449
B	4800—470	0.0629	0.1634

The transformers are connected in parallel and operated from a 4800-V, 60-Hz system. Calculate the circulating current.

3–9/6 A 75-kVA, 60-Hz, 4800—432-V transformer *A* is connected in parallel with a similar transformer *B*, whose exact ratio is unknown. The transformers are operating in the step-down mode and have a circulating current of $37.32\angle{-63.37°}$ A. The respective impedances as determined from a short-circuit test, and referred to the low side, are

$$\mathbf{Z}_{eq,A} = 0.0799\angle{62°}\Omega \qquad \mathbf{Z}_{eq,B} = 0.0676\angle{65°}\ \Omega$$

Determine the voltage ratio of transformer *B*.

3–10/7 Two 2400—240-V, 60-Hz, 100-kVA transformers *A* and *B* are operating in parallel and supplying 150-kW at 0.8 lagging power factor to a distribution system. The turns ratios are the same, and the equivalent impedances referred to the high side are $(0.869 + j2.38)\Omega$, and $(0.853 + j3.21)\Omega$, respectively. Determine the high-side current drawn by each transformer if the incoming voltage is 2470-V.

3–11/7 Two 4800—480-V, 167-kVA transformers are operated in parallel and supply a 480-V, 200-kVA, 0.72 lagging power-factor load. The percent impedance of each transformer is

$$\mathbf{Z}_A = (1.11 + j3.76)\% \qquad \mathbf{Z}_B = (1.46 + j4.81)\%$$

Sketch the equivalent circuit and determine (a) the total load current; (b) the current supplied by each transformer secondary.

3–12/7 Two 7200—240-V, 75-kVA transformers are to be operated in parallel. The per-unit impedances of the two transformers are

$$\mathbf{Z}_{A,PU} = 0.0121 + j0.0551 \qquad \mathbf{Z}_{B,PU} = 0.0201 + j0.0382$$

Determine the current supplied by each transformer secondary as a percentage of the total load current.

3–13/7 Three single-phase, 2400—120-V, 60-Hz, 200-kVA transformers are to be operated in parallel and supply a 500-kVA unity power-factor load. The percent resistance and percent reactance of the respective transformers are

Transformer	%*R*	%*X*
A	1.30	3.62
B	1.20	4.02
C	1.23	5.31

Sketch the connection diagram, the equivalent circuit, and determine the percentage of the total load taken by each transformer.

3–14/7 Three 7200—600-V, 500-kVA transformers are operating in parallel from a 7200-V source. The percent impedances, as indicated on the transformer nameplates, are

$$\mathbf{Z}_A = 5.34\% \qquad \mathbf{Z}_B = 6.08\% \qquad \mathbf{Z}_C = 4.24\%$$

What percent of the total current is supplied by transformer *B*?

3–15/7 The following 2400—240-V, 60-Hz transformers are to be operated as a parallel bank:

Transformer	kVA	Nameplate Impedance
A	50	3.53%
B	75	2.48%

Assume percent resistance is negligible. Can the bank be operated at its combined rating of 125-kVA without overheating? Show all work.

3–16/7 Can parallel operation of the following 2400—480-V transformers supply a 400-kVA, 0.8 lagging power-factor load without overloading any of the transformers? Show all work.

Transformer	kVA	Nameplate Impedance
A	100	3.68%
B	167	4.02%
C	250	4.25%

Assume percent resistance is negligible.

3–17/10 A delta–wye transformer bank is supplying a balanced three-phase, 500-kVA, 0.8 lagging power-factor load. The input voltage to the high side (delta) is 2400-V at 60-Hz. The turns ratio of each transformer is 6.9. Sketch the circuit and determine the line voltage, phase voltage, line current, and phase current for both the high side and low side.

3–18/10 A three-phase, 60-Hz, 75-hp, 890-rpm induction motor receives power from a 2300—550-V delta–delta transformer bank. The motor is operating at three-quarters load, and at that load has an efficiency of 89 percent and a power factor of 0.84 lagging. Sketch the circuit and, neglecting transformer losses, determine (a) the input power to the transformer bank; (b) the load on each of the remaining transformers if an open circuit occurs in one transformer; (c) the line current to the motor when operating closed delta; (d) repeat (c) for open delta.

3–19/10 Two 4160—450-V transformers are to be purchased to supply a 450-V, 90-kW, 0.75 lagging power-factor three-phase load. The transformers are to be connected in open delta. Specify the minimum power rating required for each transformer.

3–20/10 Three single-phase transformers are used to supply a total of 750-kVA at 450-V to a balanced three-phase load. The three-phase input to the bank is 2400-V. Determine (a) the bank ratio and the required transformer ratio if delta–wye connected; (b) the bank ratio and the required transformer ratio if delta–delta connected. (c) If each transformer is rated at 400-kVA, and the bank is delta–delta connected, will the bank be able to safely carry the load if one transformer is disconnected?

3–21/12 A bank of three single-phase 500-kVA transformers, connected delta–delta, was accidentally paralleled with a 400-kVA delta–wye bank. Although there was no load connected to the paralleled transformers, the resultant high circulating current between the transformer banks caused the tie breaker to trip. Both banks have a bank ratio of 7200—240-V. The impedance of transformers in the delta–delta bank is 2.2 percent, and is 3.1 percent in the delta–wye bank, both calculated on their respective bases. Determine the magnitude of the current that circulated between banks until the breaker tripped.

3–22/12 A 200-kVA, 60-Hz, 4600—460 Y/266-V, three-phase transformer (Δ–Y) was accidentally paralleled with a 200-kVA, 4600—460-V, 60-Hz, delta–delta, three-phase transformer. The per-unit impedance per phase is $0.0448\angle 72.33°$ for the delta–delta bank, and is $0.0420\angle 68.42°$ for the delta–wye bank, both calculated on their respective bases. Determine the magnitude of the line current circulating between banks.

4

Principles of Three-Phase Induction Machines

4.1 INTRODUCTION

Induction machines represent a class of rotating apparatus that includes induction motors, induction generators, induction frequency converters, induction phase converters, and electromagnetic slip couplings. Induction motors can be used effectively in all motor applications, except where very high torque or very fine adjustable speed control is required. Induction motors can range in size from fractional horsepower to more than 100,000 horsepower. They are more rugged, require less maintenance, and are less expensive than direct-current motors of equal power and speed ratings.

The induction motor was invented by Nikola Tesla (1856–1943) in 1888. It requires no electrical connections to the rotating member; the transfer of energy from the stationary member to the rotating member is by means of electromagnetic induction. A rotating magnetic field, produced by a stationary winding (called the stator), induces an alternating emf and current in the rotor. The resultant interaction of the induced rotor current with the rotating field of the stationary winding produces motor torque.

The torque-speed characteristic of an induction motor is directly related to the resistance and reactance of the rotor. Hence, different torque-speed characteristics may be obtained by designing rotor circuits with different ratios of rotor resistance to rotor reactance.

4.2 INDUCTION-MOTOR ACTION

An elementary three-phase two-pole induction motor is shown in Figure 4.1(a). The stator (stationary member) consists of three "blocks of iron" spaced 120° apart. The three coils wound around the iron blocks are connected in wye and energized from a three-phase system. The rotor consists of a laminated steel core containing conductors that are joined at the ends to form a cage similar to that used for exercising squirrels; hence, the name squirrel-cage rotor. When the stator windings are energized

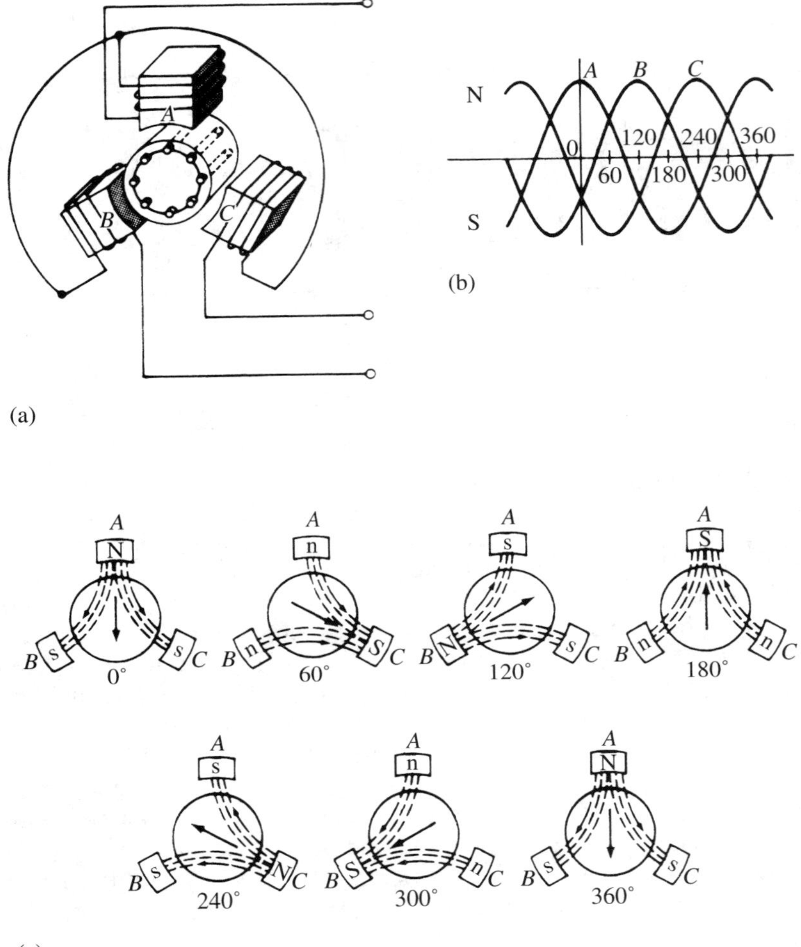

FIGURE 4.1
(a) Elementary three-phase induction motor; (b) three-phase flux waves; (c) instantaneous direction of resultant stator flux.

from a three-phase system, the currents in the three coils reach their maximum values at different instants. Since the three currents are displaced from each other by 120 electrical degrees, their respective flux contributions will also be displaced by 120°, as shown in Figure 4.1(b). Figure 4.1(c) is keyed to Figure 4.1(b), and shows the instantaneous direction of stator flux as it passes through the rotor at different instants of time. For example, at zero degrees, phase *A* is a maximum north pole, while phases *B* and *C* are weak south poles; at 60° phase *C* is a strong south pole, while phases *A* and *B* are weak north poles; at 120° phase *B* is a strong north pole, while phases *A* and *C* are weak south poles, and so forth. The large arrows indicate the instantaneous di-

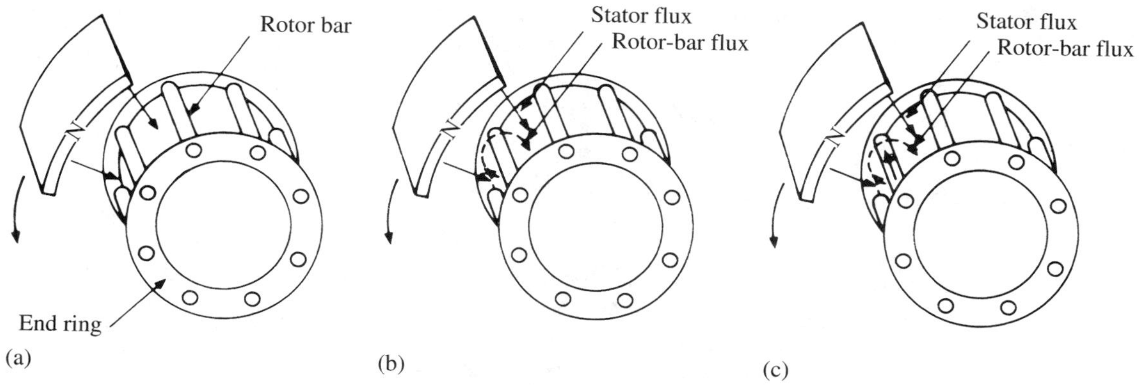

FIGURE 4.2
(a) Rotating field sweeping a rotor bar; (b) direction of flux generated around rotor bar; (c) direction of rotor-bar current.

rection of the resultant flux. The different angular positions assumed by the resultant flux vector show the plane of the flux to be revolving in a counterclockwise (CCW) direction. Although the flux generated by each coil is only an alternating flux, the combined flux contributions of the three staggered coils, carrying currents at appropriate sequential phase angles, produce a two-pole rotating flux. It is the rotating flux, not the alternating flux, that produces induction-motor action.

The rotating flux (also called *rotating field*) produced by the three-phase currents in the stationary coils, may be likened to the rotating field produced by a magnet sweeping around the rotor, as shown in Figure 4.2(a). The rotating magnetic field "cuts" the rotor bars (conductors) in its CCW sweep around the rotor. The speed of the rotating field is called the *synchronous speed.*

In accordance with Lenz's law, the voltage, current, and flux generated by the relative motion between a conductor and a magnetic field will be in a direction to oppose the relative motion. Hence, to satisfy Lenz's law, the conductors must develop a mechanical force or thrust in the same direction as the rotating flux (CCW). For this to happen, "flux bunching" must occur on the right side of the conductor; thus, the generated flux due to rotor-bar current must be clockwise (CW) as shown in Figure 4.2(b). The direction of rotor-bar current that produces this CW flux is determined by the right-hand rule and is shown in Figure 4.2(c).

4.3 REVERSAL OF ROTATION

The direction of rotation of an induction motor is dependent on the direction of rotation of the stator flux, which in turn is dependent on the phase sequence of the applied voltage. *Interchanging any two of the three line-leads to a three-phase induction motor will reverse the phase sequence, thus reversing the rotation of the motor.* As shown in Figure 4.1, the phase sequence *ABC* causes a CCW rotation of the magnetic field. Likewise, phase sequence *CBA* will cause a clockwise rotation. This can be illustrated by substituting letter *C* for letter *A* and letter *A* for letter *C* in Figure 4.1(b), and then resketching the corresponding flux diagrams in Figure 4.1(c). The resketching is left as an exercise for the student.

4.4 INDUCTION-MOTOR CONSTRUCTION

A cutaway view of a practical three-phase induction motor is shown in Figure 4.3. The stator core is an assembly of thin laminations stamped from silicon-alloy sheet steel; the use of silicon steel for the magnetic material minimizes hysteresis losses. The laminations are coated with oxide or varnish to minimize eddy-current losses.

Insulated coils are set in slots within the stator core. The overlapping coils are connected in series or parallel arrangements to form phase groups, and the phase groups are connected wye or delta. The connections, wye or delta, series or parallel, are dictated by voltage and current requirements.

FIGURE 4.3
Cutaway view of a three-phase induction motor (Courtesy Siemens Energy and Automation)

The rotors are of two basic types: squirrel cage and wound rotor. Small squirrel-cage rotors, such as that shown in Figure 4.4(a), use a slotted core of laminated steel into which molten aluminum is cast to form the conductors, end rings, and fan blades. Larger squirrel-cage rotors, as shown in Figure 4.4(b), use brass bars and brass end rings that are brazed together to form the squirrel cage. There is no insulation between the iron core and the conductors, and none is needed; the current induced in the rotor is contained within the circuit formed by the conductors and end rings, also called end connections. Skewing the rotor slots, as shown in Figure 4.4(a), helps avoid crawling (locking in at subsynchronous speeds) and reduces vibration.

A wound-rotor induction motor, shown in Figure 4.5, uses insulated coils that are set in slots and connected in a wye arrangement. The rotor circuit is completed through a set of slip rings, carbon brushes, and a wye-connected rheostat. The three-phase rheostat is composed of three rheostats connected in wye; a common lever is used to simultaneously adjust all three rheostat arms. Moving the rheostat to the "zero-resistance" position, extreme left in Figure 4.5, shorts the resistors and simulates a squirrel-cage motor. The rheostat is used to adjust starting torque and running speed.[1]

The transfer of energy from the stator to the rotor, whether squirrel cage or wound rotor, is by means of electromagnetic induction and occurs in a manner similar to that in a transformer. For that reason, the stator is often referred to as the primary and the rotor as the secondary. Since the energy to do work is transferred electromagnetically across the air gap between the stator and the rotor, the air gap is made quite small so as to offer minimum reluctance.

Each coil of an induction motor stator spans a portion of the stator circumference equal to or slightly less than the pole pitch; the pole pitch is equal to the stator circumference divided by the number of stator poles, and it may be expressed in terms of stator slots or stator arc. For example, each coil of a four-pole stator spans one-quarter or less of the circumference. If the coil span (also called coil pitch) is equal to the pole pitch it is called a full-pitch winding; if the span is less than full pitch, it is called a fractional-pitch winding. Figure 4.6 shows the coil span for full-pitch four-pole and eight-pole stator windings.[2] The three black arcs represent the end view of three stator coils, each representing one phase. The angles in Figure 4.6 are in mechanical degrees.

4.5 SYNCHRONOUS SPEED

The speed of the rotating flux, called *synchronous speed,* is directly proportional to the frequency of the supply voltage and inversely proportional to the number of pairs of poles; poles only occur in pairs. Expressed mathematically,

$$n_s = \frac{f_s}{P/2} = \frac{2 \times f_s}{P} \quad \text{r/s}$$

$$n_s = \frac{120 \times f_s}{P} \quad \text{r/min} \qquad \textbf{(4–1)}$$

[1]See Section 5.9, Chapter 5, for application of rheostats to wound-rotor motors.

[2]See Appendix B for more details on stator windings.

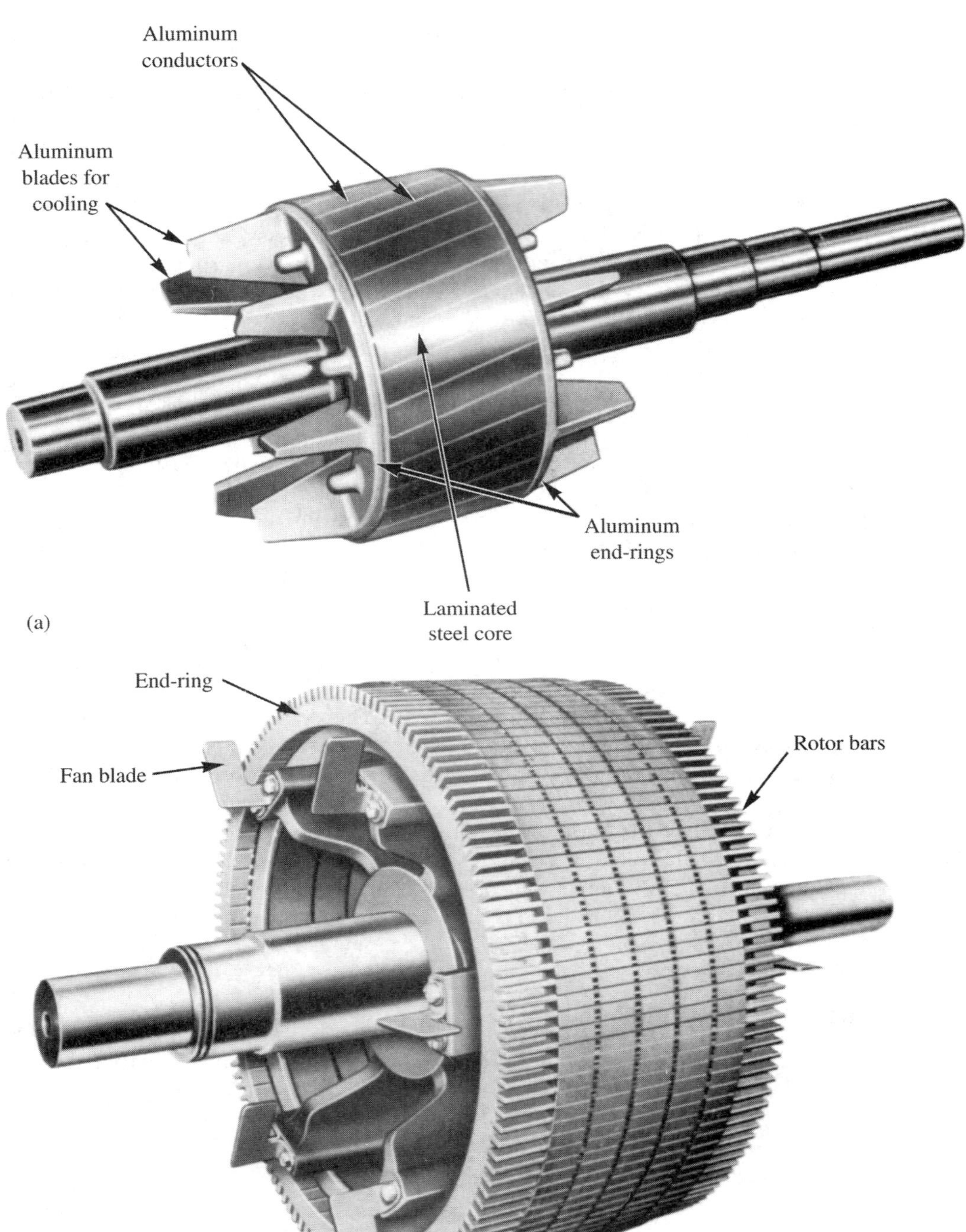

FIGURE 4.4
Squirrel-cage rotor: (a) cast-aluminum conductors; (b) brazed conductors and end rings. (Courtesy Siemens Energy and Automation)

FIGURE 4.5
Wound-rotor induction motor showing rheostat connections. (Courtesy Dresser-Rand, Electric Machinery)

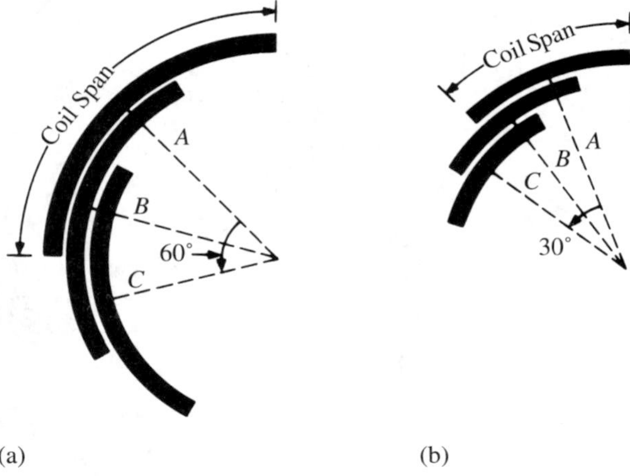

FIGURE 4.6
Coil span for (a) four-pole winding; (b) eight-pole winding.

where: f_s = frequency of the three-phase supply
n_s = synchronous speed
P = number of poles formed by the stator winding

The relationship between the speed of the rotating flux and the number of stator poles may be visualized by comparing the *mechanical degrees* of circular arc traveled by the flux in motors with different numbers of stator poles. This is shown in Figure 4.6, where the circular arc traveled by the rotating field of a four-pole motor is twice that of an eight-pole motor, assuming the same frequency and same time period; the centerline of flux rotates 60° (phase A to phase C) for a four-pole winding, and 30° (A to C) for an eight-pole winding.

Increasing the frequency of the supply voltage increases the frequency of the current in the stator coils, causing the flux to rotate at a proportionately higher speed.

The synchronous speed of an induction motor operating from a fixed-frequency system can be changed by changing the number of poles in the stator, by using a frequency converter to change the frequency, or both.

EXAMPLE 4.1 Determine the synchronous speed of a six-pole 460-V 60-Hz induction motor if the frequency is reduced to 85 percent of its rated value.

Solution

$$n_s = \frac{120 \times f_s}{P} = \frac{120(60 \times 0.85)}{6} = 1020 \text{ r/min}$$

4.6 MULTISPEED FIXED-FREQUENCY POLE-CHANGING MOTORS

Pole changing may be accomplished by using separate windings for each speed, or by reconnecting the windings of specially designed machines called *consequent-pole* motors.[3] When two separate windings are used, the machine is called a two-speed two-winding motor. Three separate windings, each arranged for a different number of poles, result in a three-speed three-winding motor. Pole arrangements of two, four, and six poles provide synchronous speeds of 3600, 1800, and 1200 r/min, respectively, from a 60-Hz system.

4.7 SLIP AND ITS EFFECT ON ROTOR FREQUENCY AND VOLTAGE

The difference between the speed of the rotating flux and the speed of the rotor is called *slip speed,* and the ratio of slip speed to synchronous speed is called *slip.* Expressed in equation form:

$$n = n_s - n_r \tag{4–2}$$

$$s = \frac{n_s - n_r}{n_s} \tag{4–3}$$

where:
n = slip speed (r/min)
n_s = synchronous speed (r/min)
n_r = rotor speed (r/min)
s = slip (pu)

The slip, as expressed in Eq. (4–3), is called *per-unit slip.*[4] The slip depends on the mechanical load connected to the rotor shaft (assuming a constant supply voltage and a constant supply frequency). Increasing the shaft load decreases the rotor speed, thus increasing the slip.

If the rotor is blocked to prevent turning, $n_r = 0$, and Eq. (4–3) reduces to

$$s = \frac{n_s - 0}{n_s} = 1$$

Releasing the brake allows the rotor to accelerate. The slip decreases with acceleration and approaches zero when all mechanical load is removed.

If operating with no shaft load, and the windage and friction are sufficiently small, the very low relative motion between the rotor and the rotating flux of the stator may cause the rotor to become magnetized along an axis of minimum reluctance. If this occurs, the rotor will lock in synchronism with the rotating flux of the stator; the slip will be zero, no induction motor torque will be developed, and the motor will act

[3]See Appendix B.

[4]If slip is given in percent, it must be divided by 100 to obtain the per-unit value before substituting into equations.

as a reluctance-synchronous motor.[5] The application of a small shaft load will cause it to pull out of synchronism, however, and induction-motor action will again occur. Solving Eq. (4–3) for n_r expresses the rotor speed in terms of slip:

$$n_r = n_s(1 - s) \tag{4–4}$$

Effect of Slip on Rotor Frequency

The frequency of the voltage induced in a rotor loop by a rotating magnetic field is given by[6]

$$f_r = \frac{P \times n}{120} \tag{4–5}$$

where: f_r = rotor frequency (Hz)
P = number of stator poles
n = slip speed (r/min)

Substituting Eq. (4–2) into Eq. (4–5)

$$f_r = \frac{P(n_s - n_r)}{120} \tag{4–6}$$

From Eq. (4–3)

$$n_s - n_r = sn_s$$

Substituting into Eq. (4–6)

$$f_r = \frac{sPn_s}{120} \tag{4–7}$$

If the rotor is blocked so that it cannot turn, $s = 1$, and Eq. (4–7) becomes

$$f_{BR} = \frac{Pn_s}{120} \tag{4–8}$$

where f_{BR} = frequency of voltage generated in the blocked rotor. Substituting Eq. (4–8) into Eq. (4–7) results in the general expression for rotor frequency in terms of slip and blocked-rotor frequency. Thus,

$$f_r = sf_{BR} \tag{4–9}$$

At blocked rotor, also called locked rotor, there is no relative motion between rotor and stator, the slip is 1.0, and the frequency of the voltage generated in the rotor is identical to the frequency of the applied stator voltage. That is,

$$f_{BR} = f_{stator}$$

---

[5]See Section 7.2, Chapter 7.

[6]See Section 1.15, Chapter 1.

Effect of Slip on Rotor Voltage

Referring to Figure 4.2, the voltage generated in a rotor loop (formed by two rotor bars and the end connections) as it is swept by the rotating stator flux is given by[7]

$$E_r = 4.44 N f_r \Phi_{\max}$$

Substituting Eq. (4–9) into Eq. (1–25),

$$E_r = 4.44 N s f_{\mathrm{BR}} \Phi_{\max} \qquad \textbf{(4–10)}$$

At blocked rotor, $s = 1$ and Eq. (4–10) becomes

$$E_{\mathrm{BR}} = 4.44 N f_{\mathrm{BR}} \Phi_{\max} \qquad \textbf{(4–11)}$$

Substituting Eq. (4–11) into Eq. (4–10),

$$E_r = s E_{\mathrm{BR}} \qquad \textbf{(4–12)}$$

Equation (4–12) is the general expression for the voltage induced in a rotor loop at any rotor speed, in terms of blocked-rotor voltage and slip.

[7]See Section 1.12, Chapter 1.

EXAMPLE 4.2 The frequency and induced voltage in the rotor of a certain six-pole wound-rotor induction motor, whose shaft is blocked, are 60 Hz and 100 V, respectively. Determine the corresponding values when the rotor is running at 1100 r/min.

Solution

$$n_s = \frac{120 f_s}{P} = \frac{120 \times 60}{6} = \underline{1200\text{ r/min}}$$

$$s = \frac{n_s - n_r}{n_s} = \frac{1200 - 1100}{1200} = \underline{0.0833}$$

$$f_r = s f_{\mathrm{BR}} = 0.0833 \times 60 = \underline{5.0\text{ Hz}}$$

$$E_r = s E_{\mathrm{BR}} = 0.0833 \times 100 = \underline{8.33\text{ V}}$$

4.8 EQUIVALENT CIRCUIT OF AN INDUCTION-MOTOR ROTOR

A feeling for the characteristic behavior of a three-phase induction motor may be obtained by manipulation and analysis of an equivalent-circuit model representing one phase of an induction-motor rotor, such as that shown in Figure 4.7(a). Note, however, that the power and torque developed by a three-phase motor is three times that developed by one of its phases. For purposes of simplicity in analysis, it will be assumed that the stator is an ideal stator, in that it produces a rotating magnetic field of constant amplitude and constant speed, and that it has no core losses, no copper losses, and no

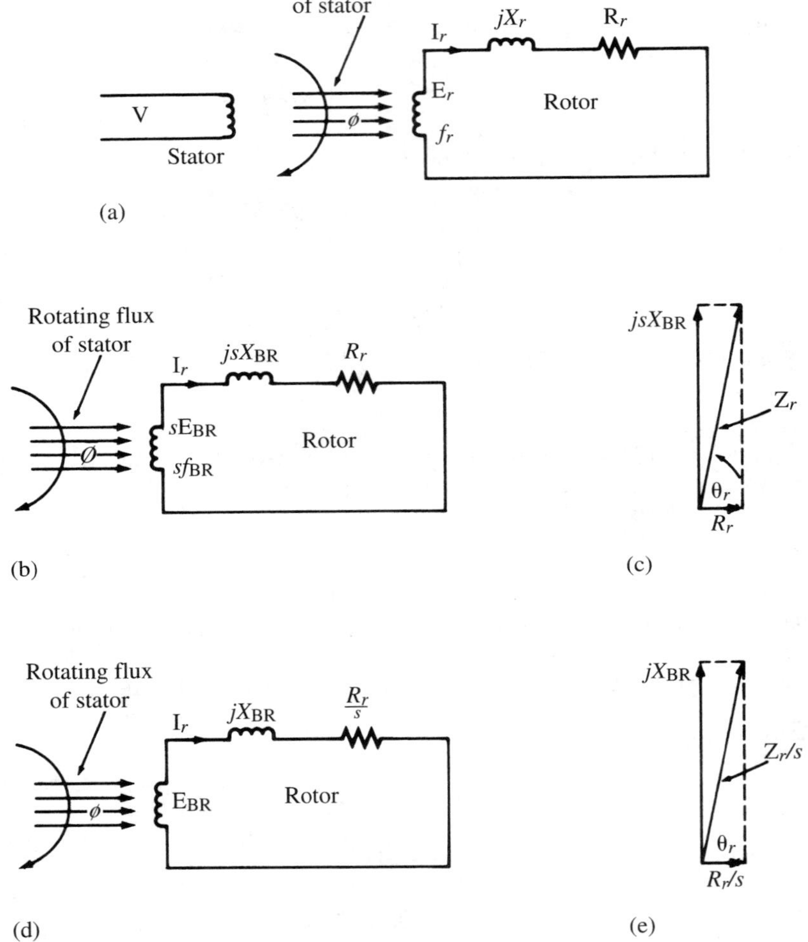

FIGURE 4.7
Equivalent circuits and corresponding impedance diagrams for an induction motor with an imaginary (ideal) stator and a real rotor.

voltage drops.[8] The rotor is represented by an electrically isolated closed circuit containing resistance and reactance acted on by an induced rotor voltage E_r. The rotor voltage is generated at frequency f_r by the rotating flux of the stator. The model, shown in Figure 4.7(a), represents one phase of a wound rotor, or one phase of an equivalent wound rotor if the rotor is squirrel cage.

---

[8]The effect of stator parameters (resistance and reactance) on induction-motor performance is discussed in Section 5.4, Chapter 5.

The rotor resistance is dependent on the length, cross-sectional area, resistivity, and skin effect of the rotor conductors, as well as external rheostat resistance if it is a wound rotor like that shown in Figure 4.5. The inductive reactance X_r of the rotor, called *leakage reactance,* is caused by leakage flux and is dependent on the shape of the rotor conductors, its depth in the iron core, the frequency of the rotor voltage, and the length of the air gap between the rotor iron and the stator iron.[9]

The leakage reactance of the rotor, expressed in terms of rotor frequency and rotor inductance, is

$$X_r = 2\pi f_r L_r \qquad \textbf{(4–13)}$$

Substituting Eq. (4–9) into Eq. (4–13) and simplifying,

$$X_r = 2\pi(sf_{\mathrm{BR}})L_r = s(2\pi f_{\mathrm{BR}} L_r)$$

$$X_r = sX_{\mathrm{BR}} \qquad \textbf{(4–14)}$$

Replacing X_r, E_r, and f_r in Figure 4.7(a) with their equivalent values in terms of slip results in Figure 4.7(b). The rotor impedance, as determined from the associated impedance diagram in Figure 4.7(c), is

$$\mathbf{Z}_r = R_r + jsX_{\mathrm{BR}} \qquad \textbf{(4–15)}$$

Applying Ohm's law to the rotor circuit in Figure 4.7(b),

$$\mathbf{I}_r = \frac{s\mathbf{E}_{\mathrm{BR}}}{\mathbf{Z}_r} = \frac{s\mathbf{E}_{\mathrm{BR}}}{R_r + jsX_{\mathrm{BR}}} \qquad \textbf{(4–16)}$$

Dividing both numerator and denominator by s,

$$\mathbf{I}_r = \frac{\mathbf{E}_{\mathrm{BR}}}{\mathbf{Z}_r/s} = \frac{\mathbf{E}_{\mathrm{BR}}}{R_r/s + jX_{\mathrm{BR}}} \qquad \textbf{(4–17)}$$

A modified equivalent series circuit and associated impedance diagram, corresponding to Eq. (4–17), are shown in Figures 4–7(d) and (e), respectively. The constant blocked-rotor voltage in Figure 4.7(d), combined with an equivalent rotor resistance that varies with the slip, provides a convenient tool for analysis of induction-motor behavior.

Expressing the rotor current in terms of magnitude and phase angle,[10]

$$\mathbf{I}_r = \frac{E_{\mathrm{BR}}\angle 0^\circ}{(Z_r/s)\angle \theta_r} = \frac{E_{\mathrm{BR}}}{Z_r/s}\angle -\theta_r$$

The magnitude of the rotor current is

$$I_r = \frac{E_{\mathrm{BR}}}{Z_r/s} \qquad \textbf{(4–18)}$$

---

[9]Leakage reactance, which also occurs in transformers, is explained in Section 2.9, Chapter 2.

[10]For convenience, the phase angle of $\mathbf{E}_{\mathrm{BR}}$ is assumed to be zero degrees.

Expressing Z_r/s and θ_r in terms of their components, as shown in Figure 4.7(e),

$$I_r = \frac{E_{BR}}{\sqrt{(R_r/s)^2 + X_{BR}^2}} \quad \textbf{(4–19)}$$

$$\theta_r = \tan^{-1}\left(\frac{X_{BR}}{R_r/s}\right) \quad \textbf{(4–20)}$$

EXAMPLE 4.3 The rotor of a certain 25-hp, six-pole, 60-Hz induction motor has equivalent resistance and equivalent reactance per phase of 0.10 Ω and 0.54 Ω, respectively. The blocked-rotor voltage/phase (E_{BR}) is 150 V. If the rotor is turning at 1164 r/min, determine (a) synchronous speed; (b) slip; (c) rotor impedance; (d) rotor current; (e) rotor current if changing the shaft load resulted in 1.24 percent slip; (f) speed for the conditions in (e).

Solution

(a) $n_s = \frac{120f}{P} = \frac{120 \times 60}{6} = \underline{1200 \text{ r/min}}$

(b) $s = \frac{n_s - n_r}{n_s} = \frac{1200 - 1164}{1200} = \underline{0.030}$

(c) $\mathbf{Z}_r = \frac{R_r}{s} + jX_{BR} = \frac{0.010}{0.03} + j0.54 = 3.3768\angle 9.20° \Rightarrow \underline{3.38\angle 9.20° \ \Omega}$

(d) $\mathbf{I}_r = \frac{\mathbf{E}_{BR}}{\mathbf{Z}_r} = \frac{150\angle 0°}{3.3768\angle 9.20°} = 44.421\angle -9.2° \Rightarrow \underline{44.4\angle -9.2° \text{ A}}$

(e) $\mathbf{Z}_r = \frac{R_r}{s} + jX_{BR} = \frac{0.10}{0.0124} + j0.54 = 8.08257\angle 3.83 \Rightarrow \underline{8.08\angle 3.83° \ \Omega}$

$\mathbf{I}_r = \frac{\mathbf{E}_{BR}}{\mathbf{Z}_r} = \frac{150\angle 0°}{8.08257\angle 3.83°} = 18.558\angle -3.83° \Rightarrow \underline{18.6\angle -3.83° \text{ A}}$

(f) $n_r = n_s(1-s) = 1200(1-0.0124) = \underline{1185 \text{ r/min}}$

4.9 LOCUS OF THE ROTOR CURRENT

The changes that take place in rotor-impedance angle θ_r, and rotor-current magnitude I_r, as an unloaded induction motor accelerates from standstill (blocked rotor) to synchronous speed, are shown in Figure 4.8(a). The curves are plots of Eqs. (4–20) and (4–19), respectively. *Note that the rotor current and the rotor impedance angle have their greatest values at blocked rotor, both decrease in value as the rotor accelerates, and both approach zero as the rotor approaches synchronous speed.* Note also that, *for low values of slip* ($s < 0.05$) *the rotor current is proportional to the slip.*

A phasor diagram representing the magnitude and phase angle of the rotor current for values of slip from $s = 1$ to $s = 0$ is shown in Figure 4.8(b). As indicated, the

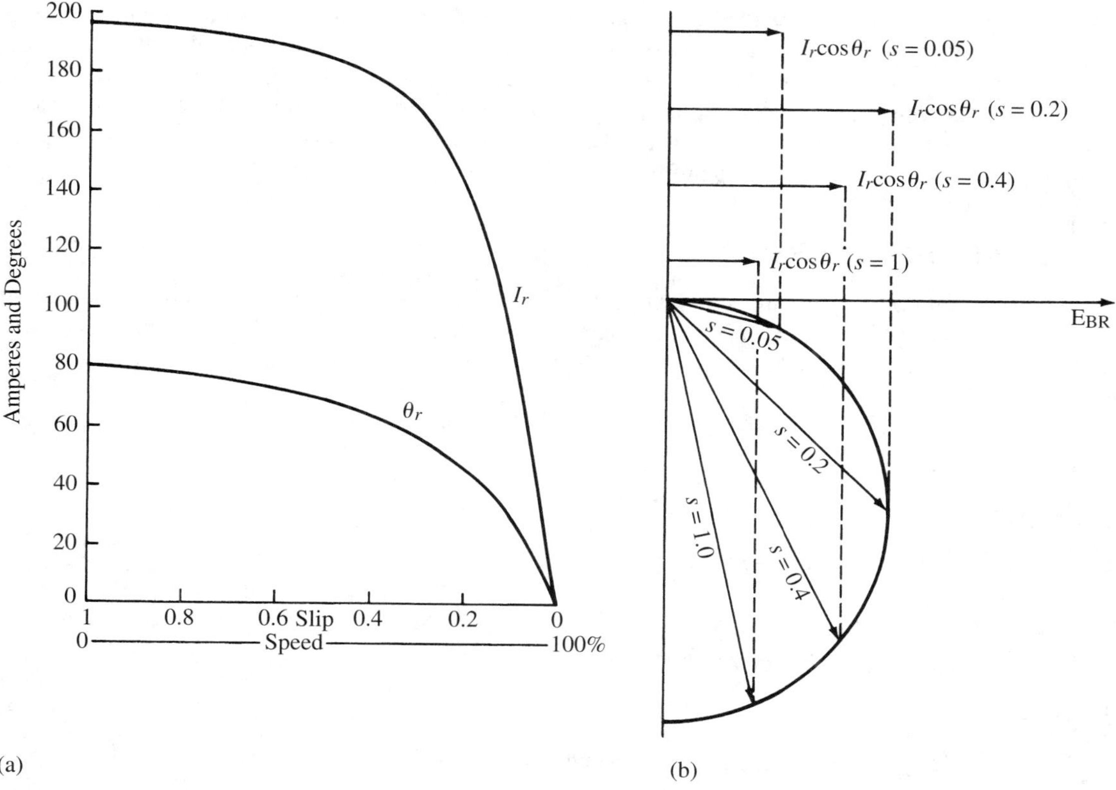

FIGURE 4.8
(a) Rotor current and rotor-impedance angle vs. speed for a representative induction motor; (b) locus of rotor-current phasor.

current phasor changes in both magnitude and phase angle as the machine accelerates from blocked rotor ($s = 1$) to synchronous speed ($s = 0$). *Note that the locus of the current phasor is a semicircle.* Proof of its semicircle character is obtained by expressing, Z_r/s in terms of $\sin\theta_r$, and then substituting into Eq. (4–18). Thus, from Figure 4.7(e),

$$\frac{Z_r}{s} = \frac{X_{BR}}{\sin\theta_r} \qquad \textbf{(4–21)}$$

Substituting Eq. (4–21) into Eq. (4–18) and simplifying,

$$I_r = \frac{E_{BR}}{X_{BR}}\sin\theta_r \qquad \textbf{(4–22)}$$

Equation (4–22) is the polar equation for a circle that is tangent to the horizontal axis at the origin and whose diameter is E_{BR}/X_{BR}.

Although the "circle diagram" in Figure 4.8(b) is for a machine with an ideal stator, it does provide a good picture of the relative changes that take place in rotor current and rotor phase angle during motor acceleration, motor loading, and motor unloading for *all* induction motors. Furthermore, since the energy supplied to the rotor is transferred magnetically across the air gap, changes in rotor current will cause corresponding changes in stator current.

4.10 AIR-GAP POWER

The power transferred electromagnetically across the air gap between the stator and the rotor is called *air-gap power* or *gap power.* From Figure 4.7(d), the gap power per phase in complex form is[11]

$$\mathbf{S}_{\text{gap}} = \mathbf{E}_{\text{BR}}\mathbf{I}_r^* \qquad \textbf{(4–23)}$$

where

$$\mathbf{E}_{\text{BR}} = E_{\text{BR}}\underline{/0^\circ} \qquad \mathbf{I}_r = I_r\underline{/-\theta_r}$$

Substituting $\mathbf{E}_{\text{BR}}$ and the conjugate of $\mathbf{I}_r$ into Eq. (4–23),

$$\mathbf{S}_{\text{gap}} = E_{\text{BR}}\underline{/0^\circ} \cdot (I_r\underline{/-\theta_r})^* = E_{\text{BR}}\underline{/0^\circ} \cdot (I_r\underline{/\theta_r}) = E_{\text{BR}}I_r\underline{/\theta_r}$$

Converting to rectangular form,

$$\mathbf{S}_{\text{gap}} = E_{\text{BR}}I_r \cos\theta_r + jE_{\text{BR}}I_r \sin\theta_r \qquad \textbf{(4–24)}$$

The active and reactive components of gap power in Eq. (4–24) are

$$P_{\text{gap}} = E_{\text{BR}}I_r \cos\theta_r \qquad \text{(Active power)} \qquad \textbf{(4–25)}$$

$$Q_{\text{gap}} = E_{\text{BR}}I_r \sin\theta_r \qquad \text{(Reactive power)} \qquad \textbf{(4–26)}$$

where: E_{BR} = blocked rotor voltage
I_r = magnitude of rotor current
θ_r = rotor impedance angle
$\cos\theta_r$ = power factor of rotor

Active component P_{gap} supplies the shaft power output as well as friction, windage, and heat losses in the rotor.

Reactive component Q_{gap} supplies the reactive power for the alternating magnetic field about the rotor current. Component Q_{gap} is not dissipated; it follows a sinusoidal pattern as it "see-saws" across the gap between the stator and the rotor.

Components P_{gap} and Q_{gap} may be represented in a power diagram as two sides of a right triangle whose diagonal is S_{gap} as shown in Figure 4.9. From the geometry of the power diagram,

$$S_{\text{gap}} = P_{\text{gap}} + jQ_{\text{gap}} \qquad \textbf{(4–27)}$$

[11]See Appendix A.5 for a review of complex power, also called phasor power.

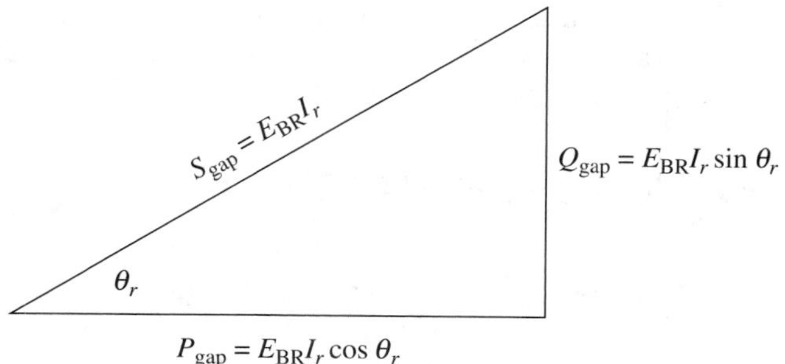

FIGURE 4.9
Power diagram for air-gap power.

The magnitude of S_{gap} is

$$S_{gap} = \sqrt{P_{gap}^2 + Q_{gap}^2} \tag{4–28}$$

EXAMPLE 4.4 For the motor operating at 1164 r/min in Example 4.3, determine the total three-phase apparent power crossing the air gap, its active and reactive components, and the rotor power factor.

Solution

$$\mathbf{S}_{gap} = 3E_{BR}I_r^* = 3 \times 150\angle 0° \times (44.421\angle -9.2°)^*$$

$$\mathbf{S}_{gap} = 3 \times 150\angle 0° \times 44.421\angle +9.2° = 19{,}989\angle 9.2° \text{ VA}$$

Converting to rectangular form,

$$\mathbf{S}_{gap} = (19{,}732 + j3197) \text{ VA}$$

$$P_{gap} = \underline{19{,}732 \text{ W}} \qquad Q_{gap} = \underline{3196 \text{ var}}$$

$$F_P = \cos 9.2° = \underline{0.99}$$

Interpreting the Circle Diagram

Blocked-rotor voltage E_{BR} is assumed constant because it is proportional to an assumed constant flux density. Thus, referring to Eq. (4–25), P_{gap} is proportional to $I_r \cos\theta_r$. This is shown in Figure 4.8(b) as the projection of the current phasor on the horizontal axis.

Assume the motor is partly loaded and operating at slip $s = 0.05$. Increasing the shaft load on the motor causes it to slow down and the slip to increase. The increased slip increases the magnitude and phase angle of the rotor current. This causes the current phasor in Figure 4.8(b) to elongate and rotate clockwise to a position that results in an $I_r \cos\theta_r$ sufficient to carry the load. Severely overloading the motor can cause the current phasor to rotate into a region where the load component of motor current $I_r \cos \theta_r$ no longer increases; in fact, it decreases. This is called a *breakdown condition,* causing rapid deceleration and very high, damaging currents.

4.11 MECHANICAL POWER AND DEVELOPED TORQUE

This section deals with the derivation of equations to be used for solving problems relating to slip, mechanical power developed, rotor power losses, shaft power out, and developed torque.

Most of the electrical power transferred across the air gap from stator to rotor is converted to mechanical power; the remainder is expended as I^2R heat-power losses in the rotor conductors. Expressed as an equation,

$$P_{\text{gap}} = P_{\text{mech}} + P_{\text{rcl}} \qquad \text{W} \tag{4–29}$$

where P_{rcl} = rotor conductor losses.

Examination of the *equivalent circuit* for one phase of the rotor in Figure 4.7(d), however, indicates that there is no provision for mechanical power. In fact, with respect to Figure 4.7(d), all of the air-gap power is dissipated as heat losses in the *equivalent resistance R_r/s;* the reactance X_{BR} draws no active power. Thus, the total air-gap power delivered to the rotor for all three phases expressed in terms of R_r/s is

$$P_{\text{gap}} = \frac{3I_r^2 R_r}{s} \qquad \text{W} \tag{4–30}$$

$$P_{\text{gap}} = \frac{P_{\text{rcl}}}{s} \qquad \text{W} \tag{4–31}$$

Equation (4–30) expresses the heat power expended in *equivalent rotor resistance* R_r/s. The *real rotor resistance,* as shown in Figure 4.7(a), however, is R_r. Thus, the actual heat power expended in the *real rotor conductors* for all three phases is

$$P_{\text{rcl}} = 3I_r^2 R_r \qquad \text{W} \tag{4–32}$$

Substituting Eqs. (4–30) and (4–32) into Eq. (4–29),

$$\frac{3I_r^2 R_r}{s} = P_{\text{mech}} + 3I_r^2 R_r \tag{4–33}$$

Solving Eq. (4–33) for P_{mech}, and rearranging terms,

$$P_{\text{mech}} = \frac{3I_r^2 R_r(1 - s)}{s} \qquad \text{W} \tag{4–34}$$

Substituting Eq. (4–30) into Eq. (4–34),

$$P_{\text{mech}} = P_{\text{gap}}(1 - s) \qquad \text{W} \tag{4–35}$$

Equation (4–35) represents the total mechanical power developed at slip s.

Steinmetz Equivalent Circuit

A modified equivalent circuit that is used extensively in induction-motor analysis is obtained by substituting Eq. (4–34) into Eq. (4–33), and dividing through by $3I_r^2$:

$$\frac{3I_r^2R_r}{s} = \frac{3I_r^2R_r(1 - s)}{s} + 3I_r^2R_r$$

$$\frac{R_r}{s} = \frac{R_r(1 - s)}{s} + R_r \qquad \textbf{(4–36)}$$

Equation (4–36) indicates that the equivalent resistance R_r/s of Figure 4.7(d) can be split into two series-connected components as shown in Figure 4.10,[12]

where: R_r = actual resistance per phase of the rotor windings (Ω)
$R_r(1 - s)/s$ = equivalent resistance per phase that expends energy at a rate equal to the mechanical power produced (Ω)

Developed Torque

From Eq. (4–4),

$$\frac{n_r}{n_s} = (1 - s)$$

Substituting into Eq. (4–34) and simplifying results in the following equation for mechanical power developed in the rotor of a three-phase motor in terms of rotor speed:

$$P_{\text{mech}} = \frac{3I_r^2R_rn_r}{sn_s} \quad \text{W} \qquad \textbf{(4–37)}$$

Motor nameplates and motor data, supplied by manufacturers and the National Electrical Manufacturers Association (NEMA), are expressed in hp, r/min, and lb-ft

[12] The equivalent-circuit model of the rotor shown in Figure 4.10 was developed by Charles Proteus Steinmetz.

FIGURE 4.10
Equivalent circuit of rotor, with R_r/s split into actual rotor resistance and an equivalent mechanical component.

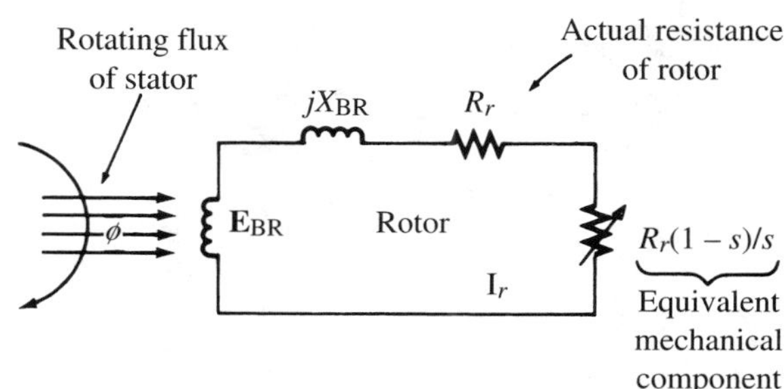

torque. Hence, these units will be used exclusively in all electric-motor problems.[13] Thus, converting Eq. (4–37) to horse-power,

$$P_{\text{mech}} = \frac{3I_r^2 R_r n_r}{746 s n_s} \quad \text{hp} \tag{4–38}$$

The basic equation that relates horsepower to developed torque and rotor speed is

$$P_{\text{mech}} = \frac{T_D n_r}{5252} \quad \text{hp} \tag{4–39}$$

where: T_D = developed torque (lb-ft)
n_r = shaft speed (r/min)
P_{mech} = mechanical power developed in rotor (hp)

Substituting Eq. (4–39) into Eq. (4–38) and solving for T_D,

$$\frac{T_D n_r}{5252} = \frac{3I_r^2 R_r n_r}{746 s n_s}$$

$$T_D = \frac{7.04 \times 3I_r^2 R_r}{s n_s} = \frac{21.12 I_r^2 R_r}{s n_s} \quad \text{lb-ft} \tag{4–40}$$

Substituting Eq. (4–30) into Eq. (4–40),

$$T_D = \frac{7.04 P_{\text{gap}}}{n_s} \quad \text{lb-ft} \tag{4–41}$$

Equations (4–40) and (4–41) represent the pound-foot torque developed in the rotor of a three-phase induction motor.

[13] Nameplates on machines used in the United States indicate r/min as RPM. Data on nameplates of foreign motors are expressed in SI units: rad/s, watts, and newton · meters (N · m).

EXAMPLE 4.5 A three-phase, 460-V, 25-hp, 60-Hz, four-pole induction motor operating at reduced load requires 14.58-kW input to the rotor. The rotor copper losses are 263 W, and the combined friction, windage, and stray power losses are 197 W. Determine (a) shaft speed; (b) mechanical power developed; (c) developed torque.

Solution

(a)

$$P_{\text{gap}} = \frac{P_{\text{rcl}}}{s} \quad \Rightarrow \quad 14{,}580 = \frac{263}{s}$$

$$s = 0.018$$

$$n_s = \frac{120 f_s}{P} = \frac{120 \times 60}{4} = 1800 \text{ r/min}$$

$$n_r = n_s(1 - s) = 1800(1 - 0.018) = \underline{1767.6 \text{ r/min}}$$

(b) $$P_{\text{mech}} = P_{\text{gap}} - P_{\text{rcl}} = 14{,}580 - 263 = \underline{14{,}317\ \text{W}}$$

Expressed in terms of horsepower,

$$P_{\text{mech}} = \frac{14{,}317}{746} = \underline{19.19\ \text{hp}}$$

(c) $$P_{\text{mech}} = \frac{T_D n_r}{5252} \quad \Rightarrow \quad T_D = \frac{5252 P_{\text{mech}}}{n_r}$$

$$T_D = \frac{5252 \times 19.19}{1767.6} = \underline{57.0\ \text{lb-ft}}$$

Or using Eq. (4–41),

$$T_D = \frac{7.04\, P_{\text{gap}}}{n_s} = \frac{7.04 \times 14{,}580}{1800} = \underline{57.0\ \text{lb-ft}}$$

4.12 TORQUE-SPEED CHARACTERISTIC

The developed torque, as expressed in Eq. (4–40), is a function of two variables: rotor current and slip. Substituting current equation (4–19) into torque equation (4–40) results in the expression for torque as a function of only one variable, slip. Making this substitution,

$$T_D = \frac{21.12 R_r}{s n_s} \cdot \left[\frac{E_{\text{BR}}}{\sqrt{(R_r/s)^2 + X_{\text{BR}}^2}}\right]^2$$

$$T_D = \frac{21.12 R_r E_{\text{BR}}^2}{s n_s [(R_r/s)^2 + X_{\text{BR}}^2]} \quad \text{lb-ft} \qquad \textbf{(4–42)}$$

A plot of Eq. (4–42), called the torque-speed characteristic of an induction motor, is shown in Figure 4.11. The inset is the locus of the rotor-current phasor extracted from Figure 4.8(b); it is used to show the correlation between rotor current and developed torque for an induction motor with an ideal stator. Although derived for a machine with an ideal stator, the torque-speed curve and the discussions associated with it are representative of the general behavior of real machines and, as such, provides the reader with a feel for what takes place in both normal and overload operation.

Locked-Rotor Torque

If the rotor is at rest, mechanical inertia prevents rotation at the instant voltage is applied to the stator; in effect, the motor behaves as though the rotor is locked. The torque developed at the locked-rotor stage is the turning moment produced by the interaction of the rotor current and the stator flux when the shaft is at rest and a three-phase voltage is applied to the stator. Locked-rotor torque, also called blocked-rotor torque, or static torque, may be calculated from Eq. (4–42) with $s = 1.0$.

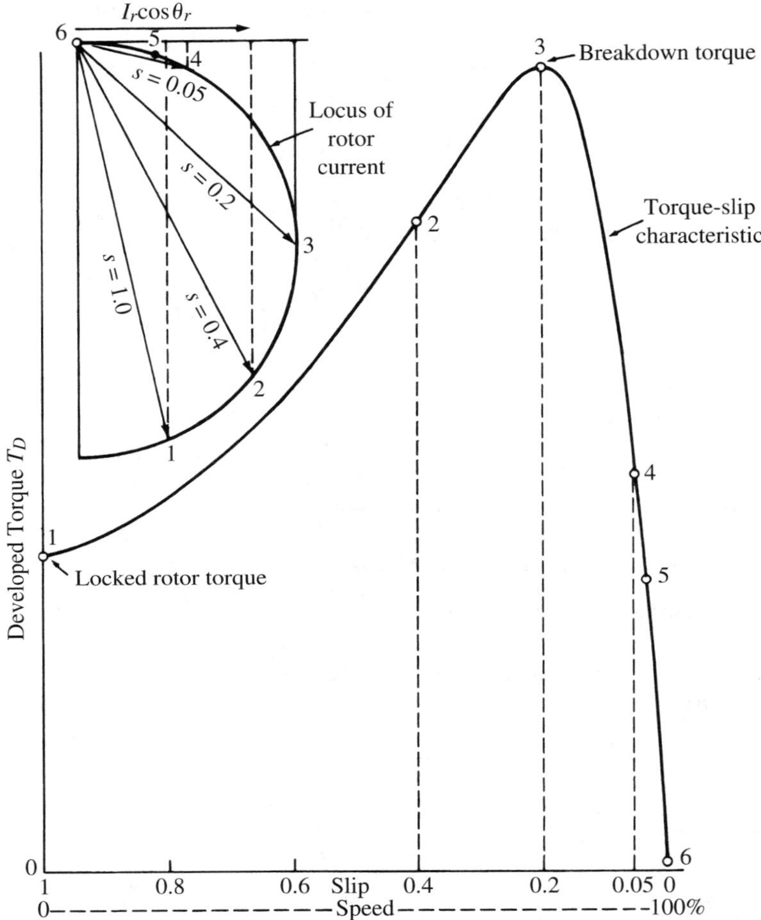

FIGURE 4.11
Representative torque-speed characteristic of an induction motor and corresponding locus of rotor current phasor.

The locked-rotor torque may vary somewhat with different standstill positions of the rotor with respect to the stator. Since this can affect starting, the locked-rotor torque data supplied by a manufacturer is the smallest of the locked rotor values obtained for different rotor positions. Poorly designed motors, or motors rewound for different speeds, or repaired by cutting out defective coils may have significantly lower values of locked-rotor torque at certain angular positions.

If the load torque on the shaft is equal to or exceeds the locked-rotor torque, the motor will not start. Should this occur, and protective devices do not clear the machine from the line, the motor will burn out! Locked-rotor conditions are indicated by point 1 in Figure 4.11.

Acceleration

Assuming the load torque on the shaft is less than the locked-rotor torque, the rotor will accelerate. As the machine accelerates from its standstill position, the slip decreases, causing the magnitude and phase angle of rotor current to decrease. This is shown in Figures 4.11 and 4.8, where $I_r \cos \theta_r$ increases from its low value at blocked rotor to some maximum value, called the breakdown value, and then decreases again with further acceleration. Note the correlation between developed torque (T_D) and $I_r \cos \theta_r$ in Figure 4.11.

Approaching Synchronous Speed

If the shaft is lightly loaded, the rotor speed approaches that of the rotating flux. The current becomes quite small, and even though it is almost in phase with the induced emf, the very low value of $I_r \cos \theta_r$ results in a very low T_D. If there is no load on the shaft, the rotor may sometimes lock in synchronism with the rotating flux. This becomes possible when the relative motion between the two is so small that the rotor iron becomes magnetized along some axis of minimum reluctance, and locks in synchronism with the rotating flux:[14] Under such conditions, the machine no longer develops induction-motor torque; the slip is zero and no current appears in the rotor. The reluctance torque is very small, however, and a light load on the shaft will pull it out of synchronism.

Behavior During Loading and Breakdown

Assume the induction motor portrayed in Figure 4.11 is operating at no load (point 6 on the curve and on the phasor diagram). For this condition, the load torque is essentially windage and friction. As shaft load is applied, the load torque becomes greater than the developed torque, and the motor slows down; the resultant increase in slip causes an increase in $I_r \cos \theta_r$, which in turn causes an increase in the developed torque. If rated load torque is applied to the shaft, the motor will decelerate until the increase in developed torque caused by the increase in slip equals the load torque on the shaft plus windage, friction, and stray load. The motor will then operate at the steady-state speed indicated by point 5.

Further increases in shaft load (overload) cause additional deceleration, accompanied by increases in $I_r \cos \theta_r$, and thus increases in developed torque. If, however, the load torque on the shaft is increased to a value greater than the maximum torque that the machine can develop (point 3), the machine will "break down"; increases in slip, due to increases in shaft load above the breakdown value, cause a rapid decrease in $I_r \cos \theta_r$, and hence a rapid decrease in developed torque. The machine will suffer a sharp drop in speed and may stop. The very high current associated with a high slip will burn out the motor windings unless protective devices remove the machine from

[14]This is called reluctance-motor action, and is discussed in Section 7.2, Chapter 7.

the line. *The breakdown torque is defined as the maximum torque that a motor can develop while being loaded (at rated voltage and rated frequency) without suffering an abrupt drop in speed.*

Although an induction motor can be operated momentarily at overloads up to the breakdown point, it cannot do so continuously without overheating and causing severe damage to both stator and rotor. To prevent damage if a sustained overload occurs, motor control circuits use overload relays and/or solid-state devices to trip the machine from the line.

No-Load Conditions

If there is no load on the shaft, the rotor will run at or near synchronous speed and the rotor current will be at or near zero. Under such conditions, the line current drawn by the stator will be only enough to produce the rotating magnetic field and supply the friction, windage, and iron losses. Thus, in a way, the no-load current drawn by the stator of an induction motor is similar to the exciting current of a transformer that supplies only the transformer flux and iron losses.

Neglecting the induction motor "exciting current," the stator (pri) current will be directly proportional to the rotor (sec) current.[15] Increasing the shaft load increases the rotor current, causing a proportional increase in stator current:

$$I_{\text{stator}} \propto I_{\text{rotor}} \tag{4–43}$$

4.13 PARASITIC TORQUES

The periodic variation of magnetic-circuit reluctance, caused by rotor and stator slots, results in a nonsinusoidal space distribution of the rotating flux. Analysis of this rotating flux pattern shows it to consist of a number of rotating fields of different speeds called *space harmonics.* The first harmonic, called the *fundamental,* runs at a speed corresponding to the number of poles in the actual winding. The fifth space-harmonic rotates backward at one-fifth the speed of the fundamental, the seventh space-harmonic rotates forward at one-seventh the speed of the fundamental, and so forth. There are no even space harmonics and no third harmonics or its multiples. Although the fundamental dominates, the component torques produced by the fifth and seventh harmonics, called *parasitic torques* or *harmonic torques,* can cause undesirable bumps and dips in the motor torque-speed characteristic during acceleration, and may even cause the rotor to lock in at some subsynchronous speed and "crawl." Figure 4.12 shows the effect of parasitic torques on the torque-speed characteristic. The presence of significant dips in the torque-speed characteristic of an induction motor may indicate a defective design, a damaged rotor, or improper repair of a damaged stator **[4], [5]**.

---

[15]The complete equivalent circuit of an induction motor, including stator and rotor windings, is similar to the equivalent circuit of a transformer. See Section 5.4, Chapter 5, for more details.

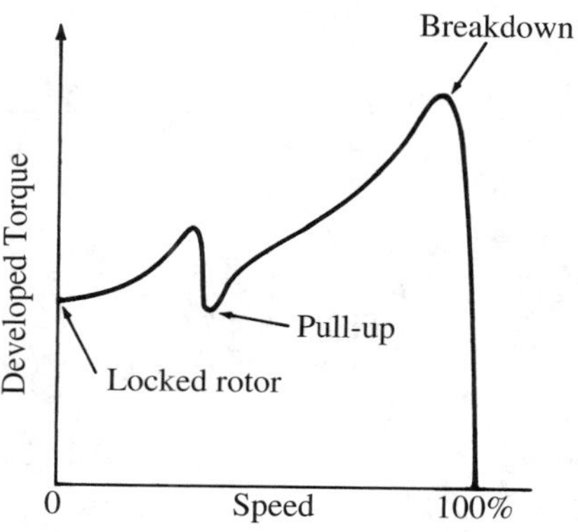

FIGURE 4.12
Effect of parasitic torques on the torque-speed characteristic of an induction motor.

4.14 PULL-UP TORQUE

The *pull-up torque* of an induction motor is the minimum torque developed by the motor during the period of acceleration from rest to the speed at which breakdown torque occurs. For the torque-speed characteristic in Figure 4.12, the pull-up torque is the value of torque at the bottom of the dip caused by a parasitic torque. If the pull-up torque is less than the load torque on the shaft, the motor will not accelerate past the pull-up point.

4.15 LOSSES, EFFICIENCY, AND POWER FACTOR

Calculations involving overall motor efficiency must take into account the losses that occur in both the stator and the rotor. The stator losses include all hysteresis losses and eddy-current losses in stator and rotor (called core losses), and I^2R losses in the stator winding (called stator conductor losses or stator copper losses). Thus, given the input power to the stator, and the stator losses, the net power crossing the air gap is

$$P_{gap} = P_{in} - P_{core} - P_{scl} \quad \text{W} \tag{4-44}$$

where: P_{in} = total 3-phase power input to stator
P_{core} = core loss
P_{scl} = stator conductor loss

Figure 4.13 shows the flow of power from stator input to shaft output, and accounts for the losses in both stator and rotor. The *power-flow diagram* is a useful adjunct to problem solving in that it often suggests a convenient method of solution. As indicated in the power-flow diagram, the total power loss for the motor is

$$P_{loss} = P_{scl} + P_{core} + P_{rcl} + P_{f,w} + P_{stray} \quad \text{W} \tag{4-45}$$

The friction losses (f) are due to bearing friction plus friction between carbon brushes and slip rings if a wound rotor motor, and the windage losses (w) are due to the shaft-mounted cooling fan plus other air disturbances caused by rotation; the friction and

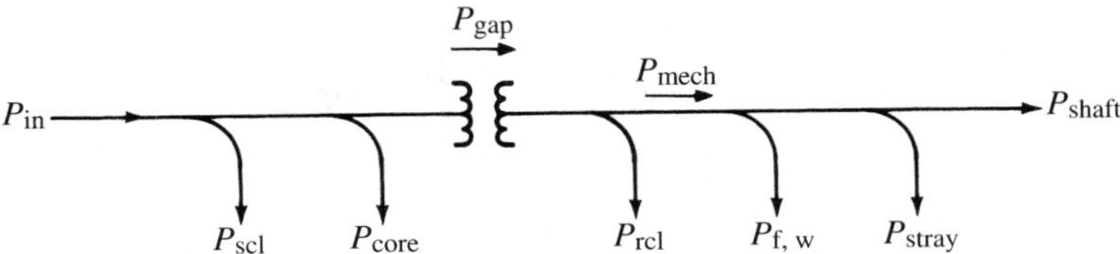

FIGURE 4.13
Power-flow diagram showing flow of power from stator input to shaft output.

windage losses are essentially constant for the normal load range of machines that conform to NEMA designs *A, B,* and *C*. Such machines change speed only slightly from no load to 115 percent rated load.[16]

The stray load loss is a collection of small losses not otherwise accounted for that vary with the load and, for calculation purposes, are assumed to be proportional to the square of the rotor current **[1]**. Expressed mathematically,

$$P_{\text{stray}} \propto I_r^2 \quad \textbf{(4–46)}$$

The stray losses include eddy-current losses in the stator conductors due to stator slot-leakage flux, losses in the end turns, end shields, and other parts in the end region due to stray flux, losses in the rotor due to harmonics produced by the stator load current, etc. **[2]**.

Typical magnitudes of induction-motor losses, expressed as a percentage of the total loss, and the factors affecting these losses, are given in Table 4.1 for three-phase four-pole NEMA design *B* motors between 1 and 125 hp **[3]**. NEMA design motors are discussed in Chapter 5.

[16]NEMA design classifications and their industrial applications are discussed in Chapter 5.

TABLE 4.1
Typical induction motor losses for four-pole motors

Losses	Percent of Total Losses	Factors Affecting These Losses
P_{scl}	35–40	Stator conductor size
P_{rcl}	15–25	Rotor conductor size
P_{core}	15–25	Type and quantity of magnetic material
P_{stray}	10–15	Primarily manufacturing and design methods
$P_{f,w}$	5–10	Selection and design of fans and bearings

Source: J. Keinz, and R. Houlton, NEMA Nominal Efficiency—What Is It and Why? *IEEE Trans. Industry and Applications,* Vol. IA-17, No. 5, Sept./Oct., 1981. © 1981 IEEE. Reprinted by permission.

Useful Shaft-Power Output and Shaft Torque

The useful shaft-power output is equal to the total mechanical power developed from all three phases minus friction, windage, and stray power losses.

$$P_{\text{shaft}} = P_{\text{mech}} - P_{f,w} - P_{\text{stray}} \qquad \text{W} \tag{4–47}$$

Shaft torque is the output torque of the motor. It is the torque transmitted to the load, and may be determined from

$$P_{\text{shaft}} = \frac{T_{\text{shaft}} \cdot n_r}{5252} \tag{4–48}$$

Efficiency

The efficiency of an induction motor is equal to the ratio of the useful power out to the total power in. Expressed as an equation

$$\eta = \frac{P_{\text{shaft}}}{P_{\text{in}}} \tag{4–49}$$

The efficiency in Eq. (4–49) is expressed in decimal form and generally called *per-unit efficiency.*

Power Factor

Power factor is the ratio of active power to apparent power. Thus, for the induction motor, the power factor is

$$F_P = \frac{P_{\text{in}}}{S_{\text{in}}} \tag{4–50}$$

$$S_{\text{in}} = \sqrt{3}\, V_{\text{line}} I_{\text{line}}$$

where: F_P = power factor (pu)
P_{in} = active power (W)
S_{in} = apparent power (VA)

Note the difference between power factor and efficiency.

EXAMPLE 4.6 A three-phase, 230-V, 60-Hz, 100-hp, six-pole induction motor operating at rated conditions has an efficiency of 91.0 percent and draws a line current of 248 A. The core loss, stator copper loss, and rotor conductor loss are 1697 W, 2803 W, and 1549 W, respectively. Determine (a) power input; (b) total losses; (c) air-gap power; (d) shaft speed; (e) power factor; (f) combined windage, friction, and stray load loss; (g) shaft torque.

Solution

The power-flow diagram is shown in Figure 4.14.

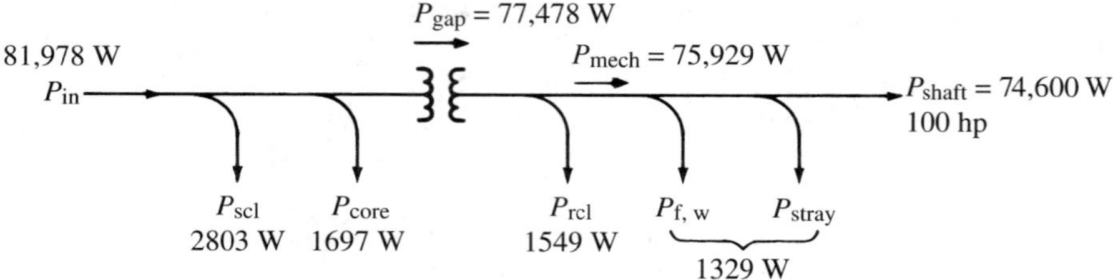

FIGURE 4.14
Power-flow diagram for Example 4.6.

(a)

$$\eta = \frac{P_{shaft}}{P_{in}} \quad \Rightarrow \quad 0.910 = \frac{100 \times 746}{P_{in}}$$

$$P_{in} = \underline{81{,}978 \text{ W}}$$

(b)

$$P_{loss} = P_{in} - P_{shaft} = 81{,}978 - 100 \times 746 = \underline{7378 \text{ W}}$$

(c) From Figure 4.14,

$$P_{gap} = P_{in} - P_{core} - P_{scl} = 81{,}978 - 1697 - 2803 = \underline{77{,}478 \text{ W}}$$

(d)

$$P_{gap} = \frac{P_{rcl}}{s} \quad \Rightarrow \quad 77{,}478 = \frac{1549}{s}$$

$$s = 0.0200$$

$$n_s = \frac{120f}{P} = \frac{120 \times 60}{6} = 1200 \text{ r/min}$$

$$n_r = n_s(1 - s) = 1200(1 - 0.0200) = \underline{1176 \text{ r/min}}$$

(e)

$$S = \sqrt{3}\, V_{line} I_{line} = \sqrt{3} \times 230 \times 248 = 98{,}796 \text{ VA}$$

$$F_P = \frac{P_{in}}{S_{in}} = \frac{81{,}978}{98{,}796} = \underline{0.83}$$

(f)

$$P_{loss} = P_{core} + P_{scl} + P_{rcl} + P_{w,f} + P_{stray}$$

$$7378 = 1697 + 2803 + 1549 + P_{w,f} + P_{stray}$$

$$P_{w,f} + P_{stray} = \underline{1329 \text{ W}}$$

(g) From Eq. (4–39)

$$T_{shaft} = \frac{5252 \times P_{shaft}}{n_r} = \frac{5252 \times 100}{1176} = \underline{446.6 \text{ lb-ft}}$$

SUMMARY OF EQUATIONS FOR PROBLEM SOLVING

$$n_s = \frac{f_s}{P/2} = \frac{2 \times f_s}{P} \quad \text{r/s} \qquad n_s = \frac{120 \times f_s}{P} \quad \text{r/min} \tag{4–1}$$

$$n = n_s - n_r \tag{4–2}$$

$$s = \frac{n_s - n_r}{n_s} \tag{4–3}$$

$$n_r = n_s(1 - s) \tag{4–4}$$

$$f_r = \frac{P \times n}{120} \tag{4–5}$$

$$f_r = \frac{P(n_s - n_r)}{120} \tag{4–6}$$

$$f_r = \frac{sPn_s}{120} \tag{4–7}$$

$$f_{BR} = \frac{Pn_s}{120} \tag{4–8}$$

$$f_r = sf_{BR} \tag{4–9}$$

$$E_r = sE_{BR} \tag{4–12}$$

$$\mathbf{I}_r = \frac{s\mathbf{E}_{BR}}{\mathbf{Z}_r} = \frac{s\mathbf{E}_{BR}}{R_r + jsX_{BR}} \tag{4–16}$$

$$I_r = \frac{E_{BR}}{\sqrt{(R_r/s)^2 + X_{BR}^2}} \tag{4–19}$$

$$\theta_r = \tan^{-1}\left(\frac{X_{BR}}{R_r/s}\right) \tag{4–20}$$

$$P_{gap} = E_{BR} I_r \cos \theta_r \tag{4–25}$$

$$P_{gap} = P_{mech} + P_{rcl} \quad \text{W} \tag{4–29}$$

$$P_{gap} = \frac{3I_r^2 R_r}{s} \qquad P_{gap} = \frac{P_{rcl}}{s} \quad \text{W} \tag{4–30, 4–31}$$

$$P_{rcl} = 3I_r^2 R_r \quad \text{W} \tag{4–32}$$

$$P_{mech} = P_{gap}(1 - s) \quad \text{W} \tag{4–35}$$

$$P_{\text{mech}} = \frac{3I_r^2 R_r n_r}{sn_s} \quad \text{W} \qquad P_{\text{mech}} = \frac{T_D n_r}{5252} \quad \text{hp} \qquad \textbf{(4–37, 4–39)}$$

$$T_D = \frac{21.12 I_r^2 R_r}{sn_s} \quad \text{lb-ft} \qquad T_D = \frac{7.04 P_{\text{gap}}}{n_s} \quad \text{lb-ft} \qquad \textbf{(4–40, 4–41)}$$

$$P_{\text{loss}} = P_{\text{scl}} + P_{\text{core}} + P_{\text{rcl}} + P_{f,w} + P_{\text{stray}} \quad \text{W} \qquad \textbf{(4–45)}$$

$$P_{\text{shaft}} = P_{\text{mech}} - P_{f,w} - P_{\text{stray}} \quad \text{W} \qquad \textbf{(4–47)}$$

$$\eta = \frac{P_{\text{shaft}}}{P_{\text{in}}} \qquad F_P = \frac{P_{\text{in}}}{S_{\text{in}}} \qquad S_{\text{in}} = \sqrt{3}\, V_{\text{line}} I_{\text{line}} \qquad \textbf{(4–49, 4–50)}$$

SPECIFIC REFERENCES KEYED TO TEXT

1. Institute of Electrical and Electronic Engineers, *Standard Test Procedure for Polyphase Induction Motors and Generators.* IEEE STD 112-1996, IEEE, New York, 1996.
2. Jimoh, A. A., R. D. Findlay, and M. Poloujadoff. Stray losses in induction machines, part I, definition, origin and measurement; part II, calculation and reduction. *IEEE Trans. Power Apparatus and Systems,* Vol. PAS-104, No. 6, June 1985, pp. 1500–1512.
3. Keinz, J., and R. Houlton, NEMA nominal efficiency—What is it and why? *IEEE Trans. Industry and Applications,* Vol. 1A-17, No. 5, Sept./Oct. 1981.
4. Liwschitz, J. M., M. Garik, and C. C. Whipple. *Alternating Current Machines.* Van Nostrand, New York, 1961.
5. Say, M. *Alternating Current Machines.* Halsted Press, New York, 1984.

REVIEW QUESTIONS

1. Explain how a rotating flux is produced in the stator of a three-phase induction motor.
2. With the aid of suitable sketches, explain how a rotating flux causes a squirrel-cage rotor to rotate.
3. What is phase sequence and how does it affect the operation of an induction motor?
4. Make two separate sketches showing the line connections to a three-phase induction motor for different directions of rotation.
5. State two reasons for skewing rotor slots.
6. (a) Differentiate between a squirrel-cage motor and a wound-rotor motor. (b) How is the speed of a wound-rotor motor adjusted?
7. Differentiate between synchronous speed, rotor speed, slip speed, and slip.
8. What two methods are used to change the synchronous speed of a three-phase induction motor?

9. Explain how slip affects rotor frequency and rotor voltage.
10. (a) Draw the circle diagram for the rotor of an induction motor. (b) Using the circle diagram as an aid to your analysis, explain the changes that take place in air-gap power as the rotor accelerates from standstill to near synchronous speed.
11. Differentiate between air-gap power, mechanical power developed, and shaft power out.
12. (a) Sketch the equivalent circuit for an induction-motor rotor and the related impedance diagram. (b) Determine from the impedance diagram the magnitude and phase angle of the rotor impedance in terms of its components.
13. (a) Draw the circle diagram for the rotor of an induction motor. (b) Using the circle diagram as an aid to your analysis, explain the changes that take place in air-gap power as the rotor accelerates from standstill to near synchronous speed.
14. (a) Sketch a representative torque-slip characteristic of a squirrel-cage induction motor and circle the points corresponding to locked rotor, breakdown torque, and rated torque. (b) Sketch the circle diagram for the rotor and draw the current phasors corresponding to the points circled in (a). (c) Using the sketches as an aid to your analysis, explain in detail the behavior of an induction motor as the machine is loaded from no load, to full load, to breakdown; assume that the machine had accelerated to rated speed before loading. Include in your analysis the reasons for changes in motor torque with increased shaft load.
15. What causes parasitic torques and what adverse effect can they have on induction-motor operation?
16. Differentiate between locked-rotor torque, pull-up torque, and breakdown torque.
17. Differentiate between efficiency and power factor.
18. List the types of losses in an induction motor and state the factors affecting these losses.
19. Sketch the power-flow diagram for an induction motor and show the relationship between power in, air-gap power, shaft power out.

PROBLEMS

4–1/7 A four-pole, 60-Hz, 10-hp, 460-V, three-phase induction motor operates at 1750 r/min when fully loaded and at its rated frequency and rated voltage. Determine (a) synchronous speed; (b) slip speed; (c) per-unit slip.

4–2/7 A 100-hp, 16-pole, 460-V, three-phase, 60-Hz induction motor has a slip of 2.4 percent when running at rated conditions. Determine (a) synchronous speed; (b) rotor speed; (c) rotor frequency.

4–3/7 A 60-Hz, four-pole, 450-V, three-phase induction motor operating at rated conditions has a speed of 1775 r/min. Determine (a) synchronous speed; (b) slip; (c) slip speed; (d) rotor frequency.

4–4/7 A 200-hp, 2300-V, three-phase, 60-Hz, wound-rotor induction motor has a blocked-rotor voltage of 104-V. The shaft speed and slip speed, when operating at rated load, are 1775 r/min and 25 r/min, respectively. Determine (a) number of poles; (b) slip; (c) rotor frequency; and (d) rotor voltage at slip speed.

4–5/7 A six-pole three-phase induction motor is operating at 480 r/min from a 25-Hz, 230-V supply. The voltage induced in the rotor when blocked is 90 V. Determine (a) slip speed; (b) rotor frequency and rotor voltage at 480 r/min.

4–6/7 A 100-hp, three-phase induction motor, operating at rated load, runs at 423 r/min when connected across a 450-V, 60-Hz supply. The slip at this load is 0.06. Determine (a) synchronous speed; (b) number of stator poles; (c) rotor frequency.

4–7/7 A four-pole/eight-pole, multispeed, 60-Hz, 10-hp, 240-V, three-phase induction motor operating with four poles runs at 1750 r/min when fully loaded and at its rated voltage and frequency. Determine (a) slip speed; (b) percent slip; (c) the synchronous speed if operating in the eight-pole mode and at 20 percent rated frequency.

4–8/11 A 20-hp, 230-V, 60-Hz, four-pole, three-phase induction motor operating at rated load has a rotor copper loss of 331 W, and a combined friction, windage, and stray power loss of 249 W. Determine (a) mechanical power developed; (b) air-gap power; (c) shaft speed; (d) shaft torque.

4–9/11 A 12-pole, 50-Hz, 20-hp, 220-V, squirrel-cage motor operating at rated conditions runs at 480 r/min, is 85 percent efficient, and has a power factor of 0.73 lagging. Determine (a) synchronous speed; (b) slip; (c) line current; (d) rated torque; (e) rotor frequency.

4–10/11 A three-phase, 230-V, 30-hp, 50-Hz, six-pole induction motor is operating with a shaft load that requires 21.3kW of input to the rotor. The rotor copper losses are 1.05 kW, and the combined friction, windage, and stray power losses for this load are 300 W. Determine (a) shaft speed; (b) mechanical power developed; (c) developed torque; (d) shaft torque; (e) percent of rated horsepower load that the machine is required to deliver.

4–11/15 A 30-hp, three-phase, 12-pole, 460-V, 60-Hz induction motor operating at reduced load draws a line current of 35 A, and has an efficiency and power factor of 90 and 79 percent, respectively. The stator conductor loss, rotor conductor loss, and core loss are 837 W, 485 W, and 375 W, respectively. Sketch the power-flow diagram, enter known values, and determine (a) input power; (b) shaft horsepower; (c) total losses; (d) rotor speed; (e) shaft torque; (f) combined windage, friction, and stray load loss.

4–12/15 A three-phase 5000-hp, 4000-V, 60-Hz, four-pole induction motor is operating at 4130 V, 60 Hz, and 67 percent rated load. The breakdown of losses for this load are as follows: stator conductors, 12.4 kW; rotor conductors, 9.92 kW; core, 12.44 kW; stray power, 10.2 kW; friction and windage, 18.2 kW. Sketch the power-flow diagram, enter known values, and determine (a) shaft speed; (b) shaft torque; (c) developed torque; (d) input power to the stator; (e) overall efficiency.

4–13/15 A 10-pole, 125-hp, 575-V, 60-Hz, three-phase induction motor operating at rated conditions draws a line current of 125 A and has an overall efficiency of 93 percent. The core loss, stator conductor loss, and rotor conductor loss are 1053 W, 2527 W, and 1755 W, respectively. Sketch the power-flow diagram, substitute values, and determine (a) shaft speed; (b) developed torque; (c) shaft torque; (d) power factor; (e) combined windage, friction, and stray power loss.

4–14/15 A 40-hp, 50-Hz, 2300-V, eight-pole induction motor is operating at 80 percent rated load and 6 percent reduced voltage. The efficiency and power factor for these conditions are 85 and 90 percent, respectively. The combined windage, friction, and stray power losses are 1011 W, the rotor conductor losses are 969 W, and the stator conductor losses are 1559 W. Sketch the power-flow diagram, enter values, and determine (a) mechanical power developed; (b) shaft speed; (c) shaft torque; (d) slip speed; (e) line current; (f) core loss.

4–15/15 A three-phase, 5-hp, 60-Hz, 115-V, four-pole induction motor operating at rated voltage, rated frequency, and 125 percent rated load has an efficiency of 85.4 percent. The stator conductor loss, rotor conductor loss, and core loss are 223.2 W, 153 W, and 114.8 W, respectively. Sketch the power-flow diagram, enter the given data, and determine (a) shaft speed; (b) shaft torque; (c) loss in torque due to the combined friction, windage, and stray power.

4–16/15 A three-phase, 50-hp, 230-V, 60-Hz, four-pole induction motor is operating at rated load, rated voltage, and rated frequency. Assume a system overload results in a 5 percent drop in frequency, and a 7 percent drop in voltage. To help reduce the system load, the shaft load is reduced to 70 percent rated horsepower, resulting in a line current of 100 A. Assume the losses for the new operating conditions are as follows: stator conductor loss, 1015 W; rotor conductor loss, 696 W; core loss, 522 W; and the combined windage, friction, and stray power loss is 667 W. Sketch the power-flow diagram, enter given data, and determine (a) percent efficiency; (b) speed; (c) shaft torque; (d) power factor.

4–17/15 A three-phase, 25-hp, 230-V, 60-Hz, two-pole induction motor drives a load that demands a constant torque regardless of speed (constant load torque). The machine is operating at rated voltage, rated frequency, and its rated speed of 3575 r/min. Determine the shaft horsepower, speed, and efficiency if the frequency drops to 54 Hz. The power factor and line current for the new conditions are 89 percent and 55 A, respectively, and the respective stator conductor loss, rotor conductor loss, and core loss, are 992.7 W, 496 W, and 546 W, respectively.

5

Classification, Performance, Applications, and Operation of Three-Phase Induction Machines

5.1 INTRODUCTION

Selecting the best induction motor for a specific application requires consideration of many factors and often presents a complex problem that requires sound judgment and considerable experience. To extract the optimum performance from a driven machine, the motor must be selected to match as closely as possible the operating characteristics of the load. To do this, a host of questions must be answered. What are the power, torque, and speed characteristics of the driven load? Must the speed be constant, adjustable, or inherently variable? Is the machine to be operated on continuous, short-time, or intermittent duty? What are the external conditions under which the motor will be required to operate? What about the ambient temperature in which the machine is to operate? Perhaps special insulation is required. What type of control is needed—manual, magnetic, or solid-state; full voltage or reduced voltage? What are the voltage and frequency constraints?

In an effort to assist the purchaser in selecting and obtaining the proper motor for the particular application, the National Electrical Manufacturers Association (NEMA) developed product standards for motors that include frame dimensions, voltage and frequency, power ratings, service factors, temperature rises, and performance characteristics. The benefits derived from these standards are greater availability of motors, a sounder basis for accurate comparison of machines, prompter repair service, and shorter delivery time.

NEMA data stamped on motor nameplates provide a wealth of information on motor operation, characteristics, and applications. Properly operated, within the bounds of its nameplate ratings, the machine will provide many years of efficient and reliable service. However, when operating at off-rated frequency, off-rated voltage, overloaded, in wrong ambient, etc., the performance of the machine will be different, with the amount of deviation from the expected normal operation depending on the percent variation in voltage, frequency, temperature, and so forth.

Sustained operation with unbalanced line voltages can cause a decrease in locked-rotor torque and breakdown torque, as well as severe overheating with a high probability of shortened life, unless the motor is derated as specified by NEMA for the particular unbalanced conditions.

High in-rush current associated with every full-voltage start (or attempted full-voltage start) causes severe thermal and mechanical stresses on rotor and stator components, as well as causing large voltage drops in the distribution system. Reduction of high in-rush current may be accomplished through various starting methods that use current-limiting impedances, autotransformers, reconnection of windings, solid-state starting, and the like.

The expected motor current, developed torque, and speed may be calculated for specific sets of conditions using the resistance and reactance of the motor windings. Simplifying approximations for "normal running" and blocked-rotor conditions make calculations relatively easy. Induction-motor parameters are available from the manufacturer or may be approximated through appropriate electrical tests.

A very interesting aspect of induction motors is their application as an induction generator. Driven by wind turbines, gas turbines, and the like, they range in size from a few kilowatts to more than 10 MW and are used in the sequential production of two forms of energy: process steam and electrical energy.

5.2 CLASSIFICATION AND PERFORMANCE CHARACTERISTICS OF NEMA-DESIGN SQUIRREL-CAGE INDUCTION MOTORS

The National Electrical Manufacturers Association standardized five basic design categories of induction motors to match the torque-speed requirements of the most common types of mechanical loads. Representative torque-speed characteristics of four of these basic designs are shown in Figure 5.1. Note that the design *C* motor has its maximum torque occur at blocked rotor ($s = 1$), the design *D* motor has its maximum torque occur at (or near) blocked rotor, and the design *A* and design *B* motors have their respective maximum torques occur at a slip of approximately 0.15. The design *E* (energy efficient)[1] motor, not shown, has a torque-speed characteristic somewhat similar to that of the design *B* motor shown in Figure 5.1.

---

[1]The Energy Policy Act of 1992 (EPACT92) requires that general-purpose, foot-mounted, T-frame, continuous-duty, single-speed, NEMA design A and B induction motors of two, four, and six poles, manufactured after October 4, 1997, have a NEMA nominal efficiency stamped on the motor nameplate. Motor ratings between 1 and 200 hp, 230/460 V, 60-Hz, are affected. Tables of nominal efficiencies for different horsepower and pole arrangements are provided in Reference **[9]**.

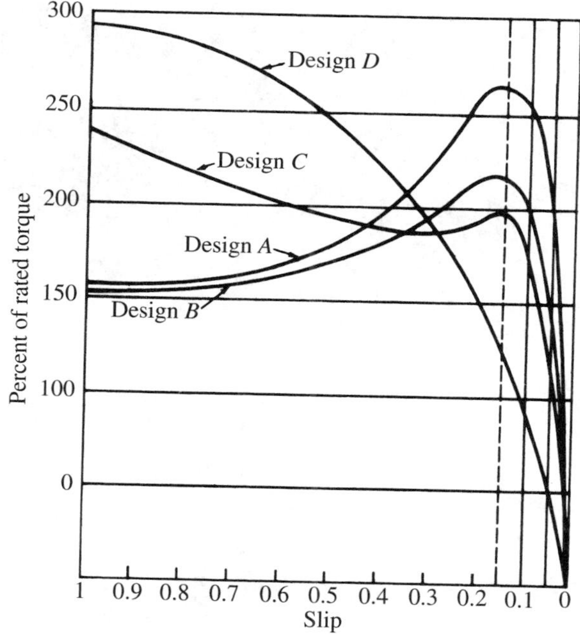

FIGURE 5.1
Torque-speed characteristics of basic NEMA-design squirrel-cage induction motors.

The characteristic curves in Figure 5.1 are "ideal" curves in that they do not include the effect of parasitic torques.[2] Parasitic torques cause dips in the torque-speed characteristic and are always present to some extent. The magnitude and location of these dips cannot be determined from the motor parameters. Hence, if this information is critical to a particular application, the manufacturer should be contacted for the actual test characteristics of the specific motor.

The different torque-speed characteristics shown in Figure 5.1 are obtained by selecting the proper combination of rotor and stator resistance and rotor and stator leakage reactance, with the rotor parameters playing the dominant role.

Representative cross sections of the three most commonly used NEMA-design squirrel-cage rotors are shown in Figure 5.2. The design *D* rotor has relatively high-resistance, low-reactance rotor bars close to the surface. Design *B* rotors and design *A* (not shown) have low-resistance rotor bars that extend deeper into the iron, resulting in low R_r and high X_{BR}. The design *C* rotor combines the features of both design *B* and design *D* rotors; it has high resistance with low reactance at the surface bars, and low resistance with high reactance at the deeper bars. Design *E* motors utilize thinner laminations of low-loss steel to minimize eddy-current losses; longer cores for low flux density to improve power factor and minimize rated current; larger cross-section conductors in the rotor and stator to reduce I^2R losses; special design low-loss cooling fans and bearings to reduce windage and friction loss; and special design winding configurations to minimize stray load losses.

[2]See parasitic torques in Section 4.12, Chapter 4.

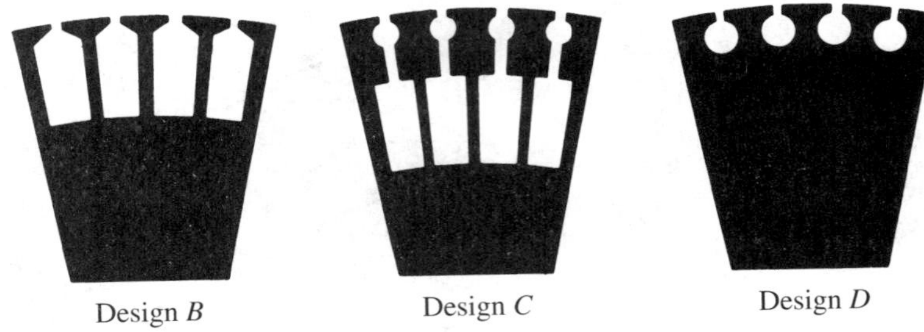

FIGURE 5.2
Representative cross sections of some NEMA-design rotors.

Induction Motor Applications

NEMA design induction motors are available for almost all applications:

The design *B* motor serves as the basis for comparison of motor performance with other designs. It has the broadest field of application and is used to drive centrifugal pumps, fans, blowers, and machine tools.

The design *A* motor has essentially the same characteristics as the design *B*, except for a somewhat higher breakdown torque. Since its starting current is higher, however, its field of application is limited.

The design *C* motor has a higher locked-rotor torque, but a lower breakdown torque than the design *B*. The higher starting torque makes it suitable for driving plunger pumps, vibrating screens, and compressors without unloading devices. The starting current and slip at rated torque are essentially the same as for the design *B*.

The design *D* motor has a very high locked-rotor torque and a high slip. Its principal field of application is in high-inertia loads such as flywheel-equipped punch presses, elevators, and hoists.

The design *E* motor is a high-efficiency motor that is used to drive centrifugal pumps, fans, blowers, and machine tools. However, except for isolated cases, the locked-rotor torque, breakdown torque, and pull-up torque of design *E* motors are somewhat lower than that of design *B* motors for the same power and synchronous speed ratings. Furthermore, the locked-rotor current (starting current) of design *E* motors is significantly higher than that of design *B* motors for the same power and synchronous speed ratings.

5.3 NEMA TABLES [9]

The *minimum* values of locked-rotor torque, breakdown torque, and pull-up torque, as specified for NEMA-design squirrel-cage medium-size induction motors with continuous ratings, are given in Table 5.1 through 5.7, respectively, for specific horsepower, frequency, and synchronous speed ratings. These minimum torque values are expressed

TABLE 5.1

Minimum locked-rotor torque, in percent of full-load torque, of single-speed, 60–50-Hz, polyphase, squirrel-cage, continuous-rated, medium motors with rated voltage and frequency applied for NEMA designs *A, B, C,* and *D.*

	Synchronous Speed (rpm)							
	60 Hz	3600	1800	1200	900	720	600	514
hp	50 Hz	3000	1500	1000	750	—	—	—
Designs A and B								
$\frac{1}{2}$		—	—	—	140	140	115	110
$\frac{3}{4}$		—	—	175	135	135	115	110
1		—	275	170	135	135	115	110
$1\frac{1}{2}$		175	250	165	130	130	115	110
2		170	235	160	130	125	115	110
3		160	215	155	130	125	115	110
5		150	185	150	130	125	115	110
$7\frac{1}{2}$		140	175	150	125	120	115	110
10		135	165	150	125	120	115	110
15		130	160	140	125	120	115	110
20		130	150	135	125	120	115	110
25		130	150	135	125	120	115	110
30		130	150	135	125	120	115	110
40		125	140	135	125	120	115	110
50		120	140	135	125	120	115	110
60		120	140	135	125	120	115	110
75		105	140	135	125	120	115	110
100		105	125	125	125	120	115	110
125		100	110	125	120	115	115	110
150		100	110	120	120	115	115	—
200		100	100	120	120	115	—	—
250		70	80	100	100	—	—	—
300		70	80	100	—	—	—	—
350		70	80	100	—	—	—	—
400		70	80	—	—	—	—	—
450		70	80	—	—	—	—	—
500		70	80	—	—	—	—	—
Design C								
1			285	255	225			
1.5			285	250	225			
2			285	250	225			
3			270	250	225			
5			255	250	225			
$7\frac{1}{2}$			250	225	200			
10			250	225	200			
15			225	210	200			
20–200, inclusive			200	200	200			

Design D: 150 hp and smaller with 4, 6, and 8 poles, 275 percent full-load torque.

Source: Reprinted by permission of the National Electrical Manufacturers Association from *NEMA Standards Publication MG 1-1998, Motors & Generators.* Copyright 1999 by NEMA, Washington, DC.

TABLE 5.2
Minimum locked-rotor torque, in percent of full-load torque, of single-speed, 60–50-Hz, polyphase, squirrel-cage, continuous-rated, medium motors with rated voltage and frequency applied for NEMA design *E*.

		Synchronous Speed (rpm)			
hp	60 Hz 50 Hz	3600 3000	1800 1500	1200 1000	900 750
$\frac{1}{2}$		190	200	170	150
$\frac{3}{4}$		190	200	170	150
1		180	190	170	150
$1\frac{1}{2}$		180	190	160	140
2		180	190	160	140
3		170	180	160	140
5		160	170	150	130
$7\frac{1}{2}$		150	160	150	130
10		150	160	150	130
15		140	150	140	120
20		140	150	140	120
25		130	140	140	120
30		130	140	140	120
40		120	130	130	120
50		120	130	130	120
60		110	120	120	110
75		110	120	120	110
100		100	110	110	100
125		100	110	110	100
150		90	100	100	90
200		90	100	100	90
250		80	90	90	90
300		80	90	90	
350		75	75	75	
400		75	75		
450		75	75		
500		75	75		

Source: Reprinted by permission of the National Electrical Manufacturers Association from *NEMA Standards Publication MG 1-1998, Motors & Generators*. Copyright 1999 by NEMA, Washington, DC.

as a *percent of rated torque,* and assume that rated voltage and rated frequency are applied to the stator.

To determine the minimum values of locked-rotor torque, breakdown torque, and pull-up torque for a specific machine, calculate rated torque from the nameplate data, and then multiply it by the respective percentages in the NEMA tables. Although the torque values obtained from the tables are minimum values, *for application considerations, it is best to assume that the minimum values are the actual values.*

TABLE 5.3
Minimum breakdown torque, in percent of full-load torque, of single-speed, 60–50-Hz, polyphase, squirrel-cage, continuous-rated, medium motors with rated voltage and frequency applied for NEMA designs *A*, *B*, and *C*.

	Synchronous Speed (rpm)							
hp	60 Hz 50 Hz	3600 3000	1800 1500	1200 1000	900 750	720 —	600 —	514 —
Designs A and B								
$\frac{1}{2}$		—	—	—	225	200	200	200
$\frac{3}{4}$		—	—	275	220	200	200	200
1		—	300	265	215	200	200	200
$1\frac{1}{2}$		250	280	250	210	200	200	200
2		240	270	240	210	200	200	200
3		230	250	230	205	200	200	200
5		215	225	215	205	200	200	200
$7\frac{1}{2}$		200	215	205	200	200	200	200
10–125, inclusive		200	200	200	200	200	200	200
150		200	200	200	200	200	200	—
200		200	200	200	200	200	—	—
250		175	175	175	175	—	—	—
300–350		175	175	175	—	—	—	—
400–500, inclusive		175	175	—	—	—	—	—
Design C								
3			200	225	200			
5			200	200	200			
$7\frac{1}{2}$–20			200	190	190			
25–200, inclusive			190	190	190			

Source: Reprinted by permission of the National Electrical Manufacturers Association from *NEMA Standards Publication MG 1-1998, Motors & Generators.* Copyright 1999 by NEMA, Washington, DC.

EXAMPLE 5.1 Determine the minimum values of locked-rotor torque, breakdown torque, and pull-up torque that can be expected from a three-phase, 10-hp, 460-V, 60-Hz, six-pole, NEMA design *C* motor whose rated speed is 1150 r/min.

Solution

$$n_s = \frac{120f}{P} = \frac{120(60)}{6} = 1200 \text{ r/min}$$

$$\text{hp} = \frac{Tn}{5252}$$

$$10 = \frac{T(1150)}{5252}$$

$$T_{\text{rated}} = \underline{45.67 \text{ lb-ft}}$$

TABLE 5.4
Minimum breakdown torque, in percent of full-load torque, of single-speed, 60–50-Hz, polyphase, squirrel-cage, continuous-rated, medium motors with rated voltage and frequency applied for NEMA design *E*.

	Synchronous Speed (rpm)				
	60 Hz	3600	1800	1200	900
hp	50 Hz	3000	1500	1000	750
$\frac{1}{2}$		200	200	170	160
$\frac{3}{4}$		200	200	170	160
1		200	200	180	170
$1\frac{1}{2}$		200	200	190	180
2		200	200	190	180
3		200	200	190	180
5		200	200	190	180
$7\frac{1}{2}$		200	200	190	180
10		200	200	180	170
15		200	200	180	170
20		200	200	180	170
25		190	190	180	170
30		190	190	180	170
40		190	190	180	170
50		190	190	180	170
60		180	180	170	170
75		180	180	170	170
100		180	180	170	160
125		180	180	170	160
150		170	170	170	160
200		170	170	170	160
250		170	170	160	160
300		170	170	160	—
350		160	160	160	—
400		160	160	—	—
450		160	160	—	—
500		160	160	—	—

Source: Reprinted by permission of the National Electrical Manufacturers Association from *NEMA Standards Publication MG 1-1998, Motors & Generators*. Copyright 1999 by NEMA, Washington, DC.

TABLE 5.5
Minimum pull-up torque, in percent of full-load torque of single-speed, 60–50-Hz, polyphase, squirrel-cage, continuous-rated, medium motors with rated voltage and frequency applied for NEMA designs *A* and *B*.

	Synchronous Speed (rpm)							
hp	60 Hz 50 Hz	3600 3000	1800 1500	1200 1000	900 750	720 —	600 —	514 —
$\frac{1}{2}$		—	—	—	100	100	100	100
$\frac{3}{4}$		—	—	120	100	100	100	100
1		—	190	120	100	100	100	100
$1\frac{1}{2}$		120	175	115	100	100	100	100
2		120	165	110	100	100	100	100
3		110	150	110	100	100	100	100
5		105	130	105	100	100	100	100
$7\frac{1}{2}$		100	120	105	100	100	100	100
10		100	115	105	100	100	100	100
15		100	110	100	100	100	100	100
20		100	105	100	100	100	100	100
25		100	105	100	100	100	100	100
30		100	105	100	100	100	100	100
40		100	100	100	100	100	100	100
50		100	100	100	100	100	100	100
60		100	100	100	100	100	100	100
75		95	100	100	100	100	100	100
100		95	100	100	100	100	100	100
125		90	100	100	100	100	100	100
150		90	100	100	100	100	100	—
200		90	90	100	100	100	—	—
250		65	75	90	90	—	—	—
300		65	75	90	—	—	—	—
350		65	75	90	—	—	—	—
400		65	75	—	—	—	—	—
450		65	75	—	—	—	—	—
500		65	75	—	—	—	—	—

Source: Reprinted by permission of the National Electrical Manufacturers Association from *NEMA Standards Publication MG 1-1998, Motors & Generators*. Copyright 1999 by NEMA, Washington, DC.

From Table 5.1, a 10-hp, design *C* motor with a synchronous speed of 1200 r/min should have a minimum locked-rotor torque equal to 225 percent full-load torque. Thus,

$$T_{\text{locked rotor}} = 2.25(45.67) = \underline{102.8 \text{ lb-ft}}$$

From Table 5.3, the minimum breakdown torque is 190%.

$$T_{\text{breakdown}} = 1.90(45.67) = \underline{86.8 \text{ lb-ft}}$$

From Table 5.6, the minimum pull-up torque is 165%.

$$T_{\text{pull-up}} = 1.65(45.67) = \underline{75.4 \text{ lb-ft}}$$

TABLE 5.6
Minimum pull-up torque, in percent of full-load torque of single speed, 60–50-Hz, polyphase, squirrel-cage, continuous-rated, medium motors with rated voltage and frequency applied for NEMA design *C*.

	Synchronous Speed (rpm)		
	60 Hz 1800	1200	900
hp	50 Hz 1500	1000	750
1	195	180	165
$1\frac{1}{2}$	195	175	160
2	195	175	160
3	180	175	160
5	180	175	160
$7\frac{1}{2}$	175	165	150
10	175	165	150
15	165	150	140
20	165	150	140
25	150	150	140
30	150	150	140
40	150	150	140
50	150	150	140
60	140	140	140
75	140	140	140
100	140	140	140
125	140	140	140
150	140	140	140
200	140	140	140

Source: Reprinted by permission of the National Electrical Manufacturers Association from *NEMA Standards Publication MG 1-1998, Motors & Generators.* Copyright 1999 by NEMA, Washington, DC.

The Upgrading Problem [12]

Before replacing a design *B* motor with a design *E* motor of the same horsepower and synchronous speed ratings, be sure to check the NEMA tables to see if the design *E* motor has sufficient torque to start and accelerate the load. The following is a comparison of the significant points on the motor speed-torque curves for design *B* and design *E* motors, both rated 60 hp with a synchronous speed of 1800 rpm:

Minimum Torque in Percent of Rated Torque from NEMA Tables

NEMA Design	Locked Rotor	Breakdown	Pull-Up
B	140	200	100
E	120	180	90

TABLE 5.7
Minimum pull-up torque, in percent of full-load torque of single-speed, 60–50-Hz, polyphase, squirrel-cage, continuous-rated, medium motors with rated voltage and frequency applied for NEMA design *E*.

		Synchronous Speed (rpm)			
	60 Hz	3600	1800	1200	900
hp	50 Hz	3000	1500	1000	750
$\frac{1}{2}$		130	140	120	110
$\frac{3}{4}$		130	140	120	110
1		120	130	120	110
$1\frac{1}{2}$		120	130	110	100
2		120	130	110	100
3		110	120	110	100
5		110	120	110	100
$7\frac{1}{2}$		100	110	110	100
10		100	110	110	100
15		100	110	100	90
20		100	110	100	90
25		90	100	100	90
30		90	100	100	90
40		90	100	100	90
50		90	100	100	90
60		80	90	90	80
75		80	90	90	80
100		70	80	80	70
125		70	80	80	70
150		70	80	80	70
200		70	80	80	70
250		60	70	70	70
300		60	70	70	—
350		60	60	60	—
400		60	60	—	—
450		60	60	—	—
500		60	60	—	—

Source: Reprinted by permission of the National Electrical Manufacturers Association from *NEMA Standards Publication MG 1-1998, Motors & Generators*. Copyright 1999 by NEMA, Washington, DC.

Note that, for the same horsepower and speed ratings (60 hp, 1800 rpm), the design *E* motor has lower minimum critical torques than the design *B* motor. This may cause problems. For the given load, the motor must be able to develop sufficient locked-rotor torque to start, sufficient pull-up torque to accelerate, and sufficient breakdown torque to handle any peak loads. It would also be wise to check with the manufacturer of the motor for their recommendations.

5.4 MOTOR PERFORMANCE AS A FUNCTION OF MACHINE PARAMETERS, SLIP, AND STATOR VOLTAGE

Motor performance as a function of machine parameters, slip, and applied stator voltage requires analysis of the complete equivalent-circuit model of an induction motor, including both rotor and stator circuits, as shown in Figure 5.3. The rotor circuit is identical to that previously shown in Figure 4.7(d) in Chapter 4. The stator circuit includes stator resistance R_s, stator leakage reactance X_s, resistance R_{fe}, which accounts for hysteresis and eddy-current losses in the iron, and magnetizing reactance X_M, which accounts for the magnetizing component of the exciting current.

The circuit model shown in Figure 5.3 is similar to that of a transformer (see Figure 2.9 in Chapter 2), where the resistance and leakage reactance are separated from the respective primary and secondary windings, leaving an ideal transformer between the two. Because of its similarity, the equivalent-circuit reductions previously developed for transformers may be adapted to induction motors. Thus, Figure 5.3 may be reduced to the simple series–parallel circuit shown in Figure 5.4, with all parameters referred to the stator. The relationship between the actual parameters shown in Figure 5.3 and the parameters referred to the stator (all per phase), as shown in Figure 5.4, are:

$$R_2 = a^2R_r = R_r \text{ referred to the stator}$$
$$X_2 = a^2X_{\text{BR}} = X_{\text{BR}} \text{ referred to the stator}$$
$$\mathbf{I}_2 = \mathbf{I}_r/a = \mathbf{I}_r \text{ referred to the stator}$$
$$\mathbf{E}_2 = \mathbf{E}_s = a\mathbf{E}_{\text{BR}} = \mathbf{E}_{\text{BR}} \text{ referred to the stator}$$
$$a = N_s/N_r = \text{ratio of stator turns per phase to rotor turns per phase}^3$$
$$R_{\text{fe}} = \text{equivalent resistance per phase that accounts for the core loss}$$
$$X_M = \text{equivalent reactance per phase that accounts for the magnetizing current}$$

[3] For squirrel-cage rotors, the ratio is stator turns per phase to equivalent wound-rotor turns per phase.

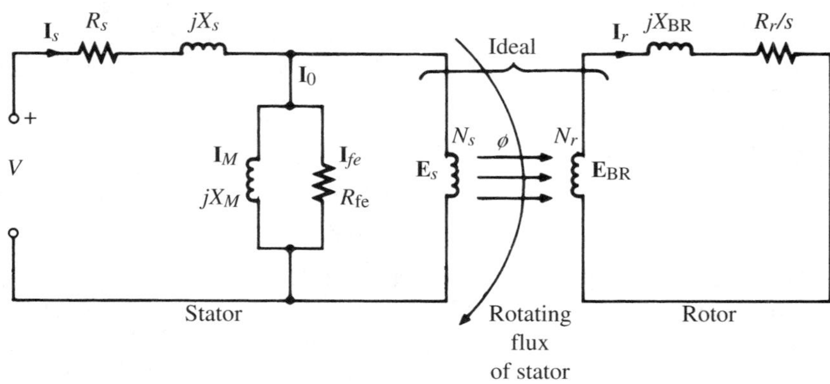

FIGURE 5.3
Equivalent-circuit model of an induction motor showing rotor and stator as separate circuits.

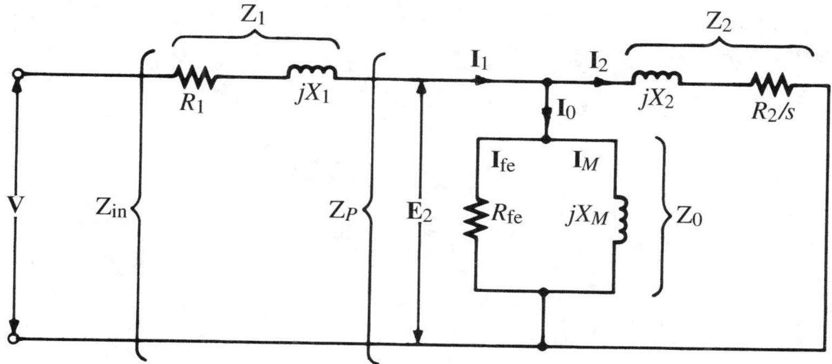

FIGURE 5.4
Equivalent series–parallel model of an induction motor with all parameters referred to the stator.

$\mathbf{I}_0$ = exciting current (no-load current) per phase
R_r = *actual* rotor resistance per phase
$\mathbf{I}_{fe}$ = core loss component of exciting current
$\mathbf{I}_M$ = magnetizing component of exciting current
X_{BR} = *actual* blocked-rotor reactance per phase
$\mathbf{I}_r$ = *actual* rotor current per phase
$\mathbf{V}$ = *actual* voltage per phase applied to the stator
$\mathbf{I}_1$ = *actual* stator current per phase

Power, Torque, Speed, Losses, and Efficiency

Stator–conductor losses, rotor–conductor losses, core losses, air-gap power, mechanical power developed, developed torque, shaft horsepower, shaft torque, speed, and efficiency may be readily determined by first solving the equivalent-circuit shown in Figure 5.4 for $\mathbf{I}_1$, $\mathbf{I}_2$, and $\mathbf{E}_2$, and then substituting into the appropriate equations previously developed in Chapter 4. Thus, from Figure 5.4,

$$\mathbf{Z}_2 = \frac{R_2}{s} + jX_2 \qquad \mathbf{Z}_0 = \frac{R_{fe} \cdot jX_M}{R_{fe} + jX_M}$$

$$\mathbf{Z}_P = \frac{\mathbf{Z}_2 \cdot \mathbf{Z}_0}{\mathbf{Z}_2 + \mathbf{Z}_0} \qquad \mathbf{Z}_{in} = \mathbf{Z}_1 + \mathbf{Z}_P$$

$$\mathbf{I}_1 = \frac{\mathbf{V}}{\mathbf{Z}_{in}} \qquad \mathbf{E}_2 = \mathbf{I}_1 \cdot \mathbf{Z}_P \qquad \mathbf{I}_2 = \frac{\mathbf{E}_2}{\mathbf{Z}_2}$$

Note: Parameter data supplied by manufacturers, or in technical papers, or as given in professional licensing examinations sometimes omit R_{fe}. To solve problems when R_{fe} is not given, simply equate $\mathbf{Z}_0 = jX_M$. The resultant error will be relatively small.

Applying the equations previously developed in Chapter 4 to the equivalent-circuit values used in Figure 5.4,

$$P_{\text{scl}} = 3 \cdot I_1^2 R_1 \qquad P_{\text{rcl}} = 3 \cdot I_2^2 R_2$$

$$P_{\text{gap}} = P_{\text{rcl}} \cdot \frac{1}{s} \qquad P_{\text{mech}} = P_{\text{rcl}} \cdot \frac{1-s}{s}$$

$$P_{\text{shaft}} = \frac{P_{\text{mech}} - \text{P}_{f,w} - P_{\text{stray}}}{746} \quad \text{hp}$$

From Eq. (4–40) of Chapter 4,

$$T_D = \frac{21.12 \cdot I_2^2 R_2}{s \cdot n_s} \quad \text{lb-ft} \tag{5–1}$$

The core loss expressed in terms of R_{fe} in Figure 5.4 is

$$P_{\text{core}} = \frac{3E_2^2}{R_{\text{fe}}} \tag{5–2}$$

EXAMPLE 5.2 A 60-Hz, 15-hp, 460-V, three-phase, six-pole, wye-connected induction motor is driving a centrifugal pump at 1185 r/min. The combined friction, windage, and stray power losses are 166 W, and the motor parameters (in ohms per phase) referred to the stator are:

$$R_1 = 0.200 \qquad R_2 = 0.250 \qquad X_M = 42.0$$
$$X_1 = 1.20 \qquad X_2 = 1.29 \qquad R_{\text{fe}} = 317$$

Determine (a) slip; (b) line current; (c) apparent power, active power, reactive power, and power factor of the motor; (d) equivalent rotor current; (e) stator copper loss; (f) rotor copper loss; (g) core loss; (h) air-gap power; (i) mechanical power developed; (j) developed torque; (k) shaft horsepower; (l) shaft torque; (m) efficiency. (n) Sketch the power-flow diagram.

Solution

(a)

$$n_s = \frac{120f}{P} = \frac{120 \times 60}{6} = 1200 \text{ r/min}$$

$$s = \frac{n_s - n_r}{n_s} = \frac{1200 - 1185}{1200} = \underline{0.0125}$$

(b) Referring to Figure 5.4,

$$\mathbf{Z}_2 = \frac{R_2}{s} + jX_2 = \frac{0.250}{0.0125} + j1.29 = 20 + j1.29 = \underline{20.0416\angle 3.6905°}\ \Omega$$

$$\mathbf{Z}_0 = \frac{R_{\text{fe}} \cdot jX_M}{R_{\text{fe}} + jX_M} = \frac{317(42.0\angle 90°)}{317 + j42.0} = 41.6361\angle 82.4527° = 5.4687 + j41.2754\ \Omega$$

$$\mathbf{Z}_P = \frac{\mathbf{Z}_2 \cdot \mathbf{Z}_0}{\mathbf{Z}_2 + \mathbf{Z}_0} = \frac{(20.0416\underline{/3.6905^\circ})(41.6361\underline{/82.4527^\circ})}{(20 + j1.29) + (5.4687 + j41.2754)}$$

$$\mathbf{Z}_P = 16.8226\underline{/27.0370^\circ} = 14.9841 + j7.6470\ \Omega$$

$$\mathbf{Z}_{\text{in}} = \mathbf{Z}_1 + \mathbf{Z}_P = (0.200 + j1.20) + (14.9841 + j7.6470) = 17.5735\underline{/30.2271^\circ}\ \Omega$$

$$\mathbf{I}_1 = \frac{\mathbf{V}}{\mathbf{Z}_{\text{in}}} = \frac{(460/\sqrt{3})\underline{/0^\circ}}{17.5735\underline{/30.2271^\circ}} = \underline{15.1126\underline{/-30.2271^\circ}\ \text{A}}$$

(c)

$$\mathbf{S} = 3\mathbf{V}\mathbf{I}_1^* = 3(460/\sqrt{3})\underline{/0^\circ} \times 15.1126\underline{/+30.2271^\circ}$$

$$\mathbf{S} = 12{,}040.857\underline{/30.2271^\circ} = 10{,}403.7 + j6061.7\ \text{VA}$$

Thus,

$$P_{\text{in}} = \underline{10{,}404\ \text{W}} \quad \Rightarrow \quad \underline{10.4\ \text{kW}}$$

$$Q_{\text{in}} = \underline{6062\ \text{var}} \quad \Rightarrow \quad \underline{6.06\ \text{kvar}}$$

$$S_{\text{in}} = \underline{12{,}041\ \text{VA}} \quad \Rightarrow \quad \underline{12.0\ \text{kVA}}$$

$$F_P = \cos(30.23^\circ) = 0.864 \quad \text{or} \quad \underline{86.4\%}$$

(d) From Figure 5.4,

$$\mathbf{E}_2 = \mathbf{I}_1\mathbf{Z}_P = (15.1126\underline{/-30.2271^\circ})(16.8226\underline{/27.037^\circ}) = 254.2332\underline{/-3.1901^\circ}\ \text{V}$$

$$\mathbf{I}_2 = \frac{\mathbf{E}_2}{\mathbf{Z}_2} = \frac{254.2332\underline{/-3.1901^\circ}}{20.0416\underline{/3.6905^\circ}} = \underline{12.6853\underline{/-6.8806^\circ}\ \text{A}}$$

(e)

$$P_{\text{scl}} = 3I_1^2R_1 = 3(15.1126)^2(0.20) = 137.03 \quad \Rightarrow \quad \underline{137\ \text{W}}$$

(f)

$$P_{\text{rcl}} = 3I_2^2R_2 = 3(12.6853)^2(0.25) = 120.69 \quad \Rightarrow \quad \underline{121\ \text{W}}$$

(g)

$$P_{\text{core}} = 3\left(\frac{E_2^2}{R_{\text{fe}}}\right) = 3\frac{(254.2332)^2}{317} = 611.68 \quad \Rightarrow \quad \underline{612\ \text{W}}$$

(h)

$$P_{\text{gap}} = \frac{P_{\text{rcl}}}{s} = \frac{120.6876}{0.0125} = 9655.20 \quad \Rightarrow \quad \underline{9655\ \text{W}}$$

(i)

$$P_{\text{mech}} = \frac{P_{\text{rcl}}(1 - s)}{s} = \frac{120.6876(1 - 0.0125)}{0.0125} = 9534.3 \quad \Rightarrow \quad \underline{9534\ \text{W}}$$

(j)

$$T_D = \frac{21.12 \cdot I_2^2R_2}{s \cdot n_s} = \frac{21.12(12.6853)^2(0.25)}{0.0125 \times 1200} = \underline{56.64\ \text{lb-ft}}$$

(k)

$$\text{Loss} = P_{\text{scl}} + P_{\text{rcl}} + P_{\text{core}} + P_{f,w} + P_{\text{stray}}$$

$$\text{Loss} = 137.03 + 120.69 + 611.68 + 166 = 1035\ \text{W}$$

$$P_{\text{shaft}} = \frac{P_{\text{in}} - \text{loss}}{746} = \frac{10{,}404 - 1035}{746} = \underline{12.56\ \text{hp}}$$

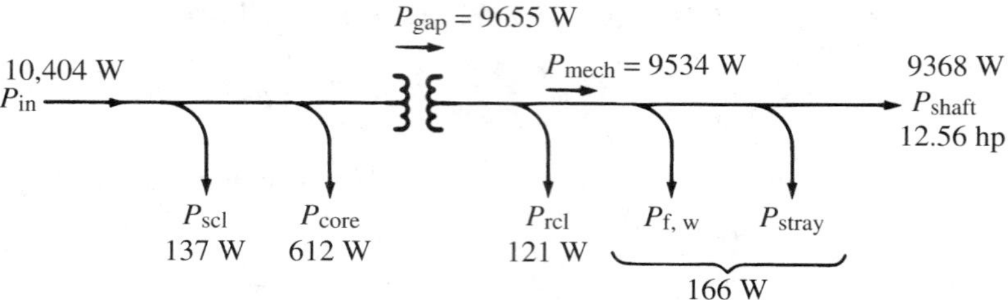

FIGURE 5.5
Power-flow diagram for Example 5.2.

(l)

$$\text{hp} = \frac{Tn}{5252}$$

$$12.56 = \frac{T(1185)}{5252}$$

$$T = \underline{55.7 \text{ lb-ft}}$$

(m)

$$\eta = \frac{P_{out}}{P_{in}} = \frac{12.56 \times 746}{10{,}404} = 0.900 \quad \text{or} \quad \underline{90.0\%}$$

(n) The power-flow diagram is shown in Figure 5.5.

5.5 SHAPING THE TORQUE-SPEED CHARACTERISTIC

The maximum torque that an induction motor can develop, for a given applied voltage and frequency, is dependent on the relative magnitudes of R_1, X_1, and X_2, and is independent of rotor resistance R_2. The slip at which this maximum torque occurs, however, is directly proportional to R_2. These two very significant relationships are not readily apparent from Eq. (5–1).

To show how T_D is related to the motor parameters requires solving for $\mathbf{I}_2$ in Figure 5.4, and then substituting the result into Eq. (5–1). Unfortunately, the resultant messy mathematical expression would completely obscure the basic relationships being sought. A simplified mathematical expression that is easier to interpret may be obtained by shifting the exciting current components shown in Figure 5.4 to the input terminals, as shown in Figure 5.6(a). Although the *approximate equivalent circuit* shown in Figure 5.6(a) is very useful for developing a reasonably accurate expression that clearly shows how rotor and stator parameters affect the value of $T_{D,\max}$, and the value of slip at which $T_{D,\max}$ occurs, it should not be used in place of Figure 5.4 for precise current, power, and efficiency calculations.

Solving the circuit in Figure 5.6(a) for $\mathbf{I}_2$,

$$\mathbf{I}_2 \cong \frac{\mathbf{V}}{R_1 + jX_1 + R_2/s + jX_2} \qquad \textbf{(5–3)}$$

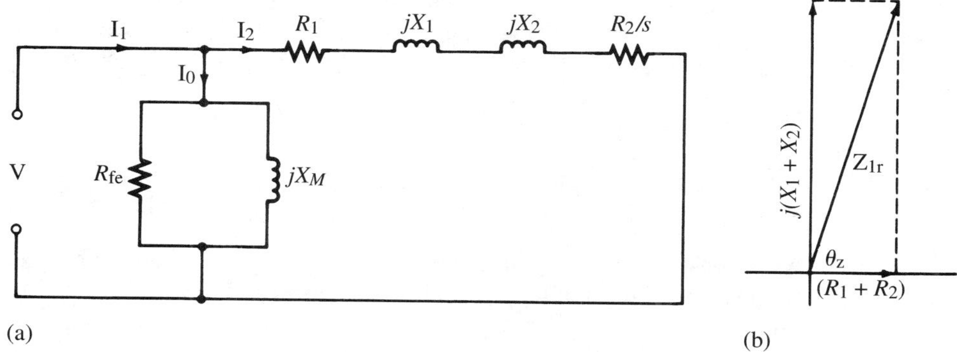

FIGURE 5.6
(a) Approximate equivalent circuit of an induction motor; (b) phasor diagram for locked-rotor conditions ($s = 1$).

$$|\mathbf{I}_2| \cong \frac{|\mathbf{V}|}{\sqrt{[(R_1 + R_2/s)^2 + (X_1 + X_2)^2]}} \tag{5–4}$$

Substituting Eq. (5–4) into Eq. (5–1),

$$T_D \cong \frac{21.12V^2R_2/s}{[(R_1 + R_2/s)^2 + (X_1 + X_2)^2]n_s} \tag{5–5}$$

As indicated in Eq. (5–5), for a given slip and given machine parameters, the developed torque is proportional to the square of the applied stator voltage. That is,

$$T_D \propto V^2 \tag{5–6}$$

This very useful relationship has significant applications in motor starting and motor breakdown problems.

Slip at Which Maximum Torque Occurs

Inspection of the torque-slip curves in Figure 5.1 shows that at breakdown the slope of the curve is zero. Hence, the slip at which maximum torque occurs may be determined by taking the derivative of Eq. (5–5) with respect to s, and then solving for the value of s that makes the slope (derivative) equal to zero. The resulting mathematical expression is

$$s_{T_{D,\max}} \cong \frac{R_2}{\sqrt{R_1^2 + (X_1 + X_2)^2}} \tag{5–7}$$

Note that the slip at which $T_{D,\max}$ occurs is directly proportional to rotor resistance R_2. Thus, in those applications that require a very high starting torque (e.g., elevators or hoists), the rotor circuit is designed with sufficient resistance so that $T_{D,\max}$ occurs at blocked rotor ($s = 1$). The value of R_2 required for this condition is obtained by substituting 1.0 for the slip in Eq. (5–7) and solving for R_2. Thus, for $T_{D,\max}$ to occur at blocked rotor,

$$R_{2,s=1} = \sqrt{R_1^2 + (X_1 + X_2)^2} \tag{5–8}$$

Maximum Torque

An expression for $T_{D,\max}$ in terms of induction-motor parameters may be obtained by solving Eq. (5–7) for R_2/s, substituting into Eq. (5–5), and simplifying. Thus, from Eq. (5–7),

$$\frac{R_2}{s} = \sqrt{R_1^2 + (X_1 + X_2)^2} \tag{5–9}$$

Substituting Eq. (5–9) into Eq. (5–5) and simplifying,

$$T_{D,\max} \cong \frac{21.12V^2\sqrt{R_1^2 + (X_1 + X_2)^2}}{n_s\left[(R_1 + \sqrt{R_1^2 + (X_1 + X_2)^2})^2 + (X_1 + X_2)^2\right]}$$

$$T_{D,\max} \cong \frac{21.12V^2}{2n_s\left[\sqrt{R_1^2 + (X_1 + X_2)^2} + R_1\right]} \tag{5–10}$$

As indicated in Eq. (5–10), the maximum torque (breakdown torque) that a given induction motor can develop is independent of rotor resistance. *Changing the value of rotor resistance will change the slip at which* $T_{D,max}$ *occurs, but will not change the value of* $T_{D,max}$.

EXAMPLE 5.3 A three-phase, 40-hp, 460-V, four-pole, 60-Hz squirrel-cage induction motor has a rated speed of 1751 r/min, and the following parameters expressed in ohms:

$$R_1 = 0.102 \qquad R_2 = 0.153 \qquad R_{fe} = 102.2$$
$$X_1 = 0.409 \qquad X_2 = 0.613 \qquad X_M = 7.665$$

Determine (a) the speed at which maximum torque is developed; (b) the maximum torque that the machine can develop; (c) rated shaft torque; (d) which NEMA design fits this motor.

Solution

(a)

$$s_{T_{D,\max}} = \frac{R_2}{\sqrt{R_1^2 + (X_1 + X_2)^2}} = \frac{0.153}{\sqrt{(0.102)^2 + (0.409 + 0.613)^2}}$$

$$s_{T_{D,\max}} = 0.1490$$

$$n_s = \frac{120f}{P} = \frac{120 \cdot 60}{4} = 1800 \text{ r/min}$$

$$n_r = n_s(1 - s) = 1800(1 - 0.1490) = \underline{1532 \text{ r/min}}$$

(b)

$$T_{D,\max} = \frac{21.12V^2}{2n_s\left[\sqrt{R_1^2 + (X_1 + X_2)^2} + R_1\right]}$$

$$T_{D,\max} = \frac{21.12(460/\sqrt{3})^2}{2(1800)[\sqrt{0.102^2 + (0.409 + 0.613)^2} + 0.102]}$$

$$T_{D,\max} = \underline{366.5 \text{ lb-ft}}$$

(c)

$$\text{hp} = \frac{Tn}{5252} \quad \Rightarrow \quad 40 = \frac{T \cdot 1751}{5252}$$

$$T_{D,\text{shaft}} = \underline{120.0 \text{ lb-ft}}$$

(d) Maximum torque is developed at a slip of 0.1490. This places the machine in the design *A* category, as seen from the curves in Figure 5.1.

EXAMPLE 5.4 A three-phase, 50-hp, 460-V, 60-Hz, four-pole, design *B* induction motor operating at rated load, rated voltage, and rated frequency has an operating speed of 1760 r/min. If to reduce a system overload, the utility drops the line voltage to 90 percent rated voltage, determine (a) the amount of torque that must be removed from the motor shaft in order to maintain 1760 r/min; (b) the expected minimum starting torque for the lower voltage; (c) the percent change in developed torque caused by the 10 percent drop in system voltage.

Solution

(a) At rated conditions,

$$\text{hp} = \frac{Tn}{5252} \quad \Rightarrow \quad 50 = T \times \frac{1760}{5252}$$

$$T_{\text{rated}} = 149.2 \text{ lb-ft}$$

Using proportionality (5–6), the developed torque at 1760 r/min and 90 percent rated voltage is

$$T_{D2} = T_{D1}\left(\frac{V_2}{V_1}\right)^2 = 149.2\left(\frac{460 \times 0.90}{460}\right)^2 = \underline{120.9 \text{ lb-ft}}$$

The required reduction in torque is

$$149.2 - 120.9 = \underline{28.3 \text{ lb-ft}}$$

(b) The expected minimum locked-rotor torque at rated voltage and rated frequency, as obtained from Table 5.1, is 140 percent rated torque.

$$T_{\text{lr}} = 1.40 \times 149.2 = 208.88 \text{ lb-ft}$$

The expected minimum starting torque at 90 percent rated voltage is

$$T_{\text{lr2}} = T_{\text{lr1}}\left(\frac{V_2}{V_1}\right)^2 = 208.88 \times \left(\frac{460 \times 0.90}{460}\right)^2 = \underline{169.2 \text{ lb-ft}}$$

(c) The percent change in torque caused by a 10 percent drop in system voltage is as follows. At 1760 r/min,

$$\frac{120.9 - 149.2}{149.2} = -0.19 \quad \text{or} \quad \underline{-19\%}$$

At locked rotor,

$$\frac{169.2 - 208.88}{208.88} = -0.19 \quad \text{or} \quad \underline{-19\%}$$

Note that a 10 percent drop in applied stator voltage results in a 19 percent drop in developed torque.

5.6 SOME USEFUL APPROXIMATIONS FOR NORMAL RUNNING AND OVERLOAD CONDITIONS OF SQUIRREL-CAGE MOTORS

Normal running conditions are defined as operating between no load and 15 percent overload with rated voltage and rated frequency. The 15 percent overload represents the permissible continuous overload for motors with a 1.15 service factor on their respective nameplates. When operating under these conditions, the slip is very small, usually <0.03, permitting very simplified mathematical approximations that are useful for solving many induction-motor problems.

Recall equations for rotor current and developed torque, Eqs. (5–4) and (5–5), respectively:

$$|\mathbf{I}_2| \cong \frac{|\mathbf{V}|}{\sqrt{[(R_1 + R_2/s)^2 + (X_1 + X_2)^2]}} \tag{5–4}$$

$$T_D \cong \frac{21.12V^2R_2/s}{[(R_1 + R_2/s)^2 + (X_1 + X_2)^2]n_s} \tag{5–5}$$

In Eqs. (5–4) and (5–5), for very low values of slip,

$$R_1 \ll \frac{R_2}{s} \gg (X_1 + X_2)$$

Thus, for values of $s \le 0.03$, the bracketed expression in the denominators of Eqs. (5–4) and (5–5) may be replaced by R_2/s without introducing any significant error.[4] That is,

$$\left[\left(R_1 + \frac{R_2}{s}\right)^2 + (X_1 + X_2)^2\right]_{s \le 0.03} \Rightarrow \left(\frac{R_2}{s}\right)^2$$

[4]A slip of $s \le 0.03$ is an arbitrary constraint that provides good approximations for most machines. Exact determinations, however, require the procedure outlined in Section 5.4, when calculating rotor current and developed torque.

Substituting into Eqs. (5–4) and (5–5) results in the following approximations:

$$\underset{s\leq 0.03}{I_2} \cong \frac{V}{R_2/s} = \frac{V\cdot s}{R_2} \tag{5–11}$$

$$\underset{s\leq 0.03}{T_D} \cong \frac{21.12V^2R_2/s}{(R_2/s)^2 n_s} = \frac{21.12V^2\cdot s}{R_2 n_s} \tag{5–12}$$

Examination of Eqs. (5–11) and (5–12) shows that for $s \leqslant 0.03$, both I_2 and T_D are directly proportional to the slip. Thus, expressed as a proportion, and assuming rated voltage and rated frequency,

$$\underset{s\leq 0.03}{I_2} \propto s \tag{5–13}$$

$$\underset{s\leq 0.03}{T_D} \propto s \tag{5–14}$$

Graphical justification for proportionalities (5–13) and (5–14) is illustrated in Figure 5.7. The curves show the behavior of rotor current and rotor torque *as the motor is loaded from its running no-load condition.*[5]

The approximations developed in this section may also be applied to problems involving overloads of up to 150 percent or more of rated torque, providing the initial and final conditions of rotor current and developed torque are known to lie on (or close to) the linear section of the respective curve. It is important, however, to note that when operating under high overload conditions, rapid and severe heating of the motor will occur; sustained overload operation at high overloads will cause severe damage to both rotor and stator.

[5]The curves in Figure 5.7 were plotted using the motor parameters in Example 5.2.

EXAMPLE 5.5

A 575-V, 100-hp, 60-Hz, 12-pole, wye-connected squirrel-cage motor operating at rated torque load and 591.1 r/min draws a line current of 89.2 A at rated voltage and rated frequency. The motor parameters expressed in ohms are

$$R_1 = 0.060 \qquad R_2 = 0.055 \qquad R_{\text{fe}} = 67.0$$
$$X_1 = 0.034 \qquad X_2 = 0.034 \qquad X_M = 11.22$$

If an increase in shaft load causes T_D to increase by 25 percent (an obvious overload), determine for the new conditions (a) shaft speed; (b) rotor current referred to the stator. Assume the overload operation is known to lie on the linear sections of the respective torque-slip and current-slip curves.

Solution

(a)

$$n_s = \frac{120f}{P} = 120 \times \frac{60}{12} = 600 \text{ r/min}$$

At rated load,

$$s = \frac{n_s - n_r}{n_s} = \frac{600 - 591.1}{600} = 0.01483$$

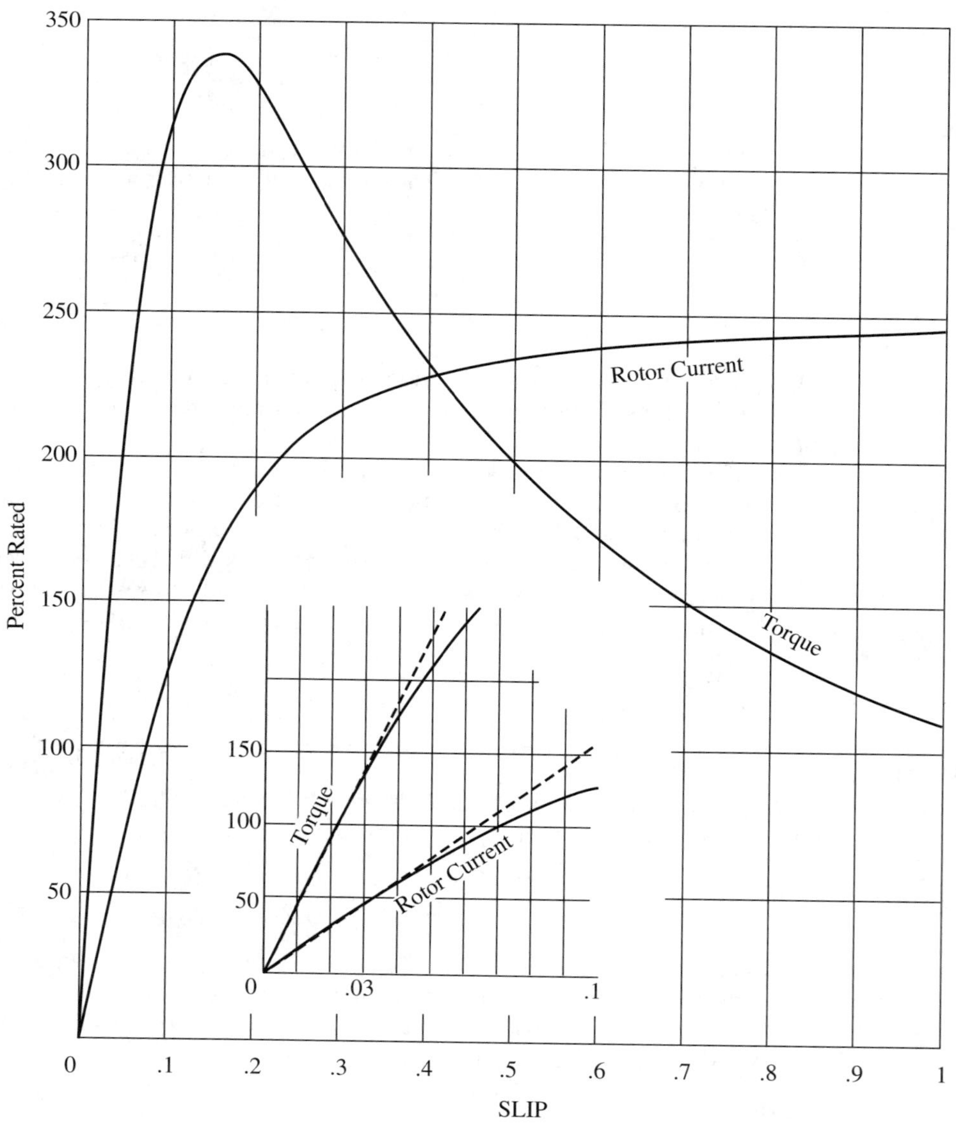

FIGURE 5.7
Rotor current and developed torque vs. slip.

From proportionality (5–14),

$$s_2 = s_1 \cdot \frac{T_{D2}}{T_{D1}} = 0.01483 \times \frac{1.25 \times T_{D1}}{T_{D1}} = 0.01854$$

$$n_r = n_s(1 - s) = 600(1 - 0.01854) = \underline{589 \text{ r/min}}$$

(b)

$$\underset{s \leq 0.03}{I_2} = \frac{V \cdot s}{R_2} = \frac{(575/\sqrt{3}) \cdot (0.01854)}{0.055} = \underline{112 \text{ A}}$$

5.7 NEMA CONSTRAINTS ON VOLTAGE AND FREQUENCY

NEMA-rated induction motors are expected to operate successfully at rated load, as long as variations in applied voltage and applied frequency do not exceed the following constraints **[9]**:

1. A voltage variation of up to ±10 percent rated voltage while operating at rated frequency.
2. A frequency variation of ±5 percent rated frequency, while operating at rated voltage.
3. A combined variation in voltage and frequency, with the sum of the *absolute* values of the respective variations not exceeding 10 percent providing the frequency does not exceed 5 percent of rated frequency.

Note: In accordance with constraint 3, a 9 percent *rise* in system voltage, accompanied by a 4 percent *drop* in system frequency would cause a combined variation of 9% + 4% = 13%. Adding algebraically 9% + (−4%) = 5% is *incorrect!* The calculation must be an *addition of the absolute values.* Off-standard frequency and off-standard voltage each have an adverse effect on motor performance, and *the adverse effects of a decrease in frequency do not offset the adverse effects of an increase in voltage,* and vice versa.

5.8 EFFECT OF OFF-RATED VOLTAGE AND OFF-RATED FREQUENCY ON INDUCTION-MOTOR PERFORMANCE

Induction-motor speed, current, and developed torque are a function of the source frequency and voltage. Thus, a significant deviation from rated motor frequency can have serious adverse effects on motor operation. Large interconnected utilities have a relatively stable frequency and stable voltage. However, isolated power plants, such as those found on ships, offshore drilling rigs, and in certain rural areas, may at times experience both off-rated voltage and off-rated frequency.

Effect on Running Torque

The effect of different frequencies and different voltages on the developed torque may be determined by expressing the synchronous speed in Eq. (5–12) in terms of frequency. Making the substitution,

$$\underset{s\leq 0.03}{T_D} \cong \frac{21.12V^2 \cdot s}{R_2(120f/P)} \tag{5–15}$$

Equation (5–15) indicates that for *all values of slip* $\leq$ *0.03 the developed torque is proportional to the slip, to the square of the applied voltage, and inversely proportional to the frequency.* That is,

$$\underset{s\leq 0.03}{T_D} \propto \frac{V^2 \cdot s}{f} \tag{5–16}$$

Note: Changes in friction, windage, and stray power losses are generally very small for ranges in slip ≤ 0.03. Hence, if these losses are not known, and $s \leq 0.03$, T_{shaft} may be substituted for T_D in proportionality (5–16) without introducing a significant error.

EXAMPLE 5.6 A 230-V, 20-hp, 60-Hz, six pole, three-phase induction motor driving a constant torque load at rated frequency, rated voltage, and rated horsepower has a speed of 1175 r/min, and an efficiency of 92.1 percent. Determine (a) the new operating speed if a system disturbance causes a 10 percent drop in voltage and a 6 percent drop in frequency; (b) the new shaft horsepower. Assume that windage, friction, and stray power losses are essentially constant.

Solution

(a)

$$V_2 = 0.90(230) = 207 \text{ V}$$

$$f_2 = 0.94(60) = 56.4 \text{ Hz}$$

$$n_{s1} = 120f_1/P = 120(60)/6 = 1200 \text{ r/min}$$

$$n_{s2} = 120f_2/P = 120(56.4)/6 = 1128 \text{ r/min}$$

$$s_1 = (n_{s1} - n_{r1})/n_{s1} = (1200 - 1175)/1200 = 0.02083$$

With a constant torque load and $s_1 \leq 0.03$,

$$\left[\frac{V^2 \cdot s}{f}\right]_1 = \left[\frac{V^2 \cdot s}{f}\right]_2$$

$$s_2 = s_1 \cdot \left(\frac{V_1}{V_2}\right)^2 \cdot \frac{f_2}{f_1} = 0.02083 \times \left(\frac{230}{0.90 \times 230}\right)^2 \times \frac{60 \times 0.94}{60}$$

$$s_2 = 0.02417$$

$$n_{r2} = n_{s2}(1 - s_2) = 1128(1 - 0.02417) = \underline{1101 \text{ r/min}}$$

(b) $$P = \frac{T \cdot n}{5252} \quad \Rightarrow \quad \frac{P_2}{P_1} = \frac{T_2 \cdot n_{r2}}{T_1 \cdot n_{r1}}$$

Thus, with a constant torque load, $T_2 = T_1$,

$$P_2 = P_1 \times \frac{T_2 \cdot n_{r2}}{T_1 \cdot n_{r1}} = 20 \times \frac{T_1 \cdot 1101}{T_1 \cdot 1175} = \underline{18.7 \text{ hp}}$$

Effect on Locked-Rotor Current

Calculations to determine locked-rotor current are based solely on the applied voltage and the locked-rotor input impedance. Thus, referring to the approximate equivalent circuit in Figure 5.6(a), and noting that $s = 1.0$, at blocked rotor,

$$\underset{s=1.0}{I_2} = \frac{V}{Z_{\text{lr}}} = \frac{V}{\sqrt{(R_1 + R_2)^2 + (X_1 + X_2)^2}} \qquad \textbf{(5–17)}$$

where: Z_{lr} = locked-rotor impedance (Ω)

Except for high-slip machines such as the design D motor, or other special high-slip designs, *most squirrel-cage induction motors are low-resistance, high-reactance machines, whose impedance angle at locked rotor is $\geq 75°$.* This is shown in Figure 5.6(b), where

$$\theta_z = \arctan\left(\frac{X_1 + X_2}{R_1 + R_2}\right) \qquad \textbf{(5–18)}$$

Note that, with angle $\theta_z \geq 75°$,

$$|\mathbf{Z}_{\text{lr}}| \cong X_1 + X_2$$

Hence, with $s = 1.0$, and $\theta_z \geq 75°$, Eq. (5–17) may be approximated as

$$\underset{s=1.0,\theta_z\geq 75°}{I_2} \cong \frac{V}{X_1 + X_2} \qquad \textbf{(5–19)}$$

Expressing the leakage reactance in terms of frequency, and factoring,

$$\underset{s=1.0,\theta_z\geq 75°}{I_2} \cong \frac{V}{f(2\pi L_1 + 2\pi L_2)} \qquad \textbf{(5–20)}$$

Equation (5–20) indicates that the locked-rotor current is directly proportional to the applied voltage and inversely proportional to the applied frequency. That is,

$$\underset{s=1.0,\theta_z\geq 75°}{I_2} \propto \frac{V}{f} \qquad \textbf{(5–21)}$$

Furthermore, from Figure 5.6(a),

$$\mathbf{I}_1 = \mathbf{I}_2 + \mathbf{I}_0$$

And, at locked rotor,

$$|\mathbf{I}_2| >> |\mathbf{I}_0|$$

Hence, at locked rotor, the stator current (I_1, defined as I_{lr}) is approximately equal to the rotor current (I_2),

$$I_{lr} = I_1 \cong I_2 \tag{5–22}$$

Thus rewriting Eq. (5–21),

$$\underset{s=1.0,\theta_z \geq 75^\circ}{I_{lr}} \propto \frac{V}{f} \tag{5–23}$$

EXAMPLE 5.7 A 20-hp, four-pole, three-phase, 230-V, 60-Hz, design *B,* wye-connected motor draws 151 A when started at rated voltage and rated frequency. Determine the expected locked-rotor line current if the motor is started from a 220-V, 50-Hz system.

Solution

Using proportionality (5–23),

$$\frac{I_{lr1}}{I_{lr2}} = \frac{V_1/f_1}{V_2/f_2} \qquad \Rightarrow \qquad I_{lr2} = I_{lr1} \cdot \frac{V_2/f_2}{V_1/f_1}$$

$$I_{lr2} = 151 \times \frac{220/50}{230/60} = \underline{173 \text{ A}}$$

Note: In those industries where there is a mix of 25-Hz and 60-Hz power systems, extreme care must be exercised to prevent the accidental connection of a 25-Hz motor to a 60-Hz system, and vice versa. For example, connecting a motor rated at 25 Hz, 425 r/min, to a 60-Hz system would cause a dangerous overspeed approximating $425(60/25) = 1020$ r/min.[6] Connecting a 60-Hz motor, whose locked-rotor current is 1085 A, to a 25-Hz system (with the same line voltage) will cause the locked-rotor current to be approximately $1085(60/25) = 2604$ A, resulting in a burned-out motor.

Effect on Locked-Rotor Torque

From the BLI rule as expressed in Eq. (1–12), Chapter 1, the force on a rotor conductor (and hence the torque) is proportional to the current in the conductor and the stator flux density in which it is immersed. Expressed mathematically,

$$T_D \propto B_{stator} I_{rotor}$$

Furthermore, the flux density is proportional to the stator current.

$$B_{stator} \propto I_{stator}$$

[6]See Table 5.11, Section 5.18, for allowable overspeeds of induction motors.

Thus,

$$T_D \propto I_{\text{stator}} I_{\text{rotor}}$$

Using the terminology in Figure 5.6(a), at locked rotor,

$$T_{\text{lr}} \propto I_1 I_2$$

However, from Eq. (5–22), at locked rotor,

$$I_{\text{lr}} = I_1 \cong I_2$$

Thus, at locked rotor,

$$T_{\text{lr}} \propto I_{\text{lr}}^2 \tag{5–23a}$$

Substituting proportionality (5–23) into proportionality (5–23a),

$$I_{\text{lr}} = \left[\frac{V}{f}\right]^2 \tag{5–23b}$$

Proportionality (5–23b) shows that the locked-rotor torque is directly proportional to the square of the applied voltage, and inversely proportional to the square of the frequency. Although the effects of magnetic saturation and conductor skin effect were not considered, proportionality (5–23b) is accurate enough for practical applications.

EXAMPLE 5.8 (a) From the NEMA tables, determine the expected minimum locked-rotor torque for a 75-hp, four-pole, 60-Hz, 240-V, 1750-rpm, design *E* motor. (b) Repeat (a) assuming system overloading made it mandatory to drop the voltage and frequency to 230 V and 58 Hz, respectively.

Solution

(a) From Table 5.2, the minimum locked-rotor torque is 120% rated.

$$\text{hp}_{\text{rated}} = \frac{T_{\text{rated}} \times n_{\text{rated}}}{5252} \quad \Rightarrow \quad T_{\text{rated}} = \frac{5252 \times \text{hp}_{\text{rated}}}{n_{\text{rated}}}$$

$$T_{\text{rated}} = \frac{5252 \times 75}{1750} = 225 \text{ lb-ft}$$

$$T_{\text{lr}} = 225 \times 1.20 = \underline{270 \text{ lb-ft}}$$

(b) Using proportionality (5–23b)

$$\frac{T_{\text{lr2}}}{T_{\text{lr1}}} = \frac{\left[\dfrac{V_2}{f_2}\right]^2}{\left[\dfrac{V_1}{f_1}\right]^2} \quad \Rightarrow \quad T_{\text{lr2}} = T_{\text{lr1}} \times \left[\frac{V_2}{f_2}\right]^2 \times \left[\frac{f_1}{V_1}\right]^2$$

$$T_{\text{lr2}} = 270 \times \left[\frac{230}{58}\right]^2 \times \left[\frac{60}{240}\right]^2 = \underline{265 \text{ ft-lb}}$$

Operating 60-Hz Motors on a 50-Hz System

Operating an induction motor at significantly below rated frequency, such as operating a 60-Hz motor at 50 Hz, causes a significant decrease in magnetizing reactance and, because of magnetic saturation effects, an out-of-proportion increase in magnetizing current. The net result is severe overheating of the motor windings. To prevent overheating, a reduction in applied frequency must be accompanied by a reduction in applied voltage. Simply stated, *the ratio of volts per hertz must be kept constant.* General-purpose, three-phase, 60-Hz, NEMA-design induction motors with two, four, six, or eight poles are capable of operating satisfactorily from 50-Hz systems, *provided the horsepower and voltage ratings at 50 Hz are 5/6 of the corresponding ratings at 60 Hz.* When operating in this manner, overheating will not occur, and the locked-rotor torque and breakdown torque at 50 Hz will be essentially the same as for 60-Hz operation **[9]**.

These relationships, expressed mathematically are:

$$V_{50} = \frac{5}{6} V_{60} \tag{5–23c}$$

$$\text{hp}_{50} = \frac{5}{6} \text{hp}_{60} \tag{5–23d}$$

$$\left[\frac{T \cdot n_r}{5252}\right]_{50} = \frac{5}{6}\left[\frac{T \cdot n_r}{5252}\right]_{60} \tag{5–23e}$$

Substituting proportionality (5–16) and simplifying,

$$\left.\left[\frac{V^2 \cdot s \cdot n_r}{f}\right]_{50} = \frac{5}{6}\left[\frac{V^2 \cdot s \cdot n_r}{f}\right]_{60}\right\} s \le 0.03 \tag{5–23f}$$

EXAMPLE 5.9 A 40-hp, 460-V, 60-Hz, four-pole, three-phase, design *B* induction motor operating at rated voltage, rated frequency, and rated horsepower runs at 1770 r/min and draws a line current of 52.0 A. If the machine is operated at 5/6 rated horsepower from a 385-V, 50-Hz supply, determine (a) shaft r/min; (b) slip.

Solution

(a)

$$n_{s,60} = \frac{120(60)}{4} = 1800 \text{ r/min}$$

$$n_{s,50} = \frac{120(50)}{4} = 1500 \text{ r/min}$$

$$s_{60} = \frac{n_s - n_r}{n_s} = \frac{1800 - 1770}{1800} = 0.01667$$

$$\text{hp}_{50} = \frac{5}{6} \text{hp}_{60}$$

$$\left[\frac{T \cdot n_r}{5252}\right]_{50} = \frac{5}{6}\left[\frac{T \cdot n_r}{5252}\right]_{60}$$

Substituting proportionality (5–16) and simplifying,

$$\left[\frac{V^2 \cdot s \cdot n_r}{f}\right]_{50} = \frac{5}{6}\left[\frac{V^2 \cdot s \cdot n_r}{f}\right]_{60} \Big\} \; s \leq 0.03$$

Substituting given and calculated values and solving for s_{50},

$$\left[\frac{(385)^2 \cdot s_{50} \cdot n_{r,50}}{50}\right] = \frac{5}{6}\left[\frac{(460)^2(0.01667)(1770)}{60}\right]$$

$$s_{50} = \frac{29.251}{n_{r,50}}$$

From Eq. (4–3),

$$s_{50} = \frac{1500 - n_{r,50}}{1500}$$

Equating and solving for $n_{r,50}$

$$\frac{29.251}{n_{r,50}} = \left[\frac{1500 - n_{r,50}}{1500}\right]$$

$$n_{r,50}^2 - 1500n_{r,50} + 43{,}876.5 = 0$$

Using the quadratic formula,

$$n_{r,50} = \frac{1500 \pm \sqrt{(-1500)^2 - 4(43{,}876.5)}}{2}$$

$$n_{r,50} = \underline{1470 \text{ r/min}} \qquad \underbrace{29.8 \text{ r/min}}_{\text{not valid}}$$

Proportionality (5–16) is valid for $s \leq 0.03$. The slip at 29.8 r/min, however, is $(1500 - 29.8)/1500 = 0.980$. Hence, the 29.8 r/min is not valid.

(b)

$$s_{50} = \frac{1500 - 1470}{1500} = \underline{0.020}$$

5.9 WOUND-ROTOR INDUCTION MOTOR

A wound-rotor induction motor, shown in Figure 5.8 and Figure 4.5, uses a wound rotor in place of a squirrel-cage rotor. The wound rotor has a regular three-phase winding similar to that of the stator and is wound with the same number of poles. The phases are usually wye connected and terminate at the slip rings. A wye-connected rheostat with a common lever is used to adjust the resistance of the rotor circuit. The rheostat provides speed control, torque adjustment at locked rotor, and current limiting during starting and acceleration.

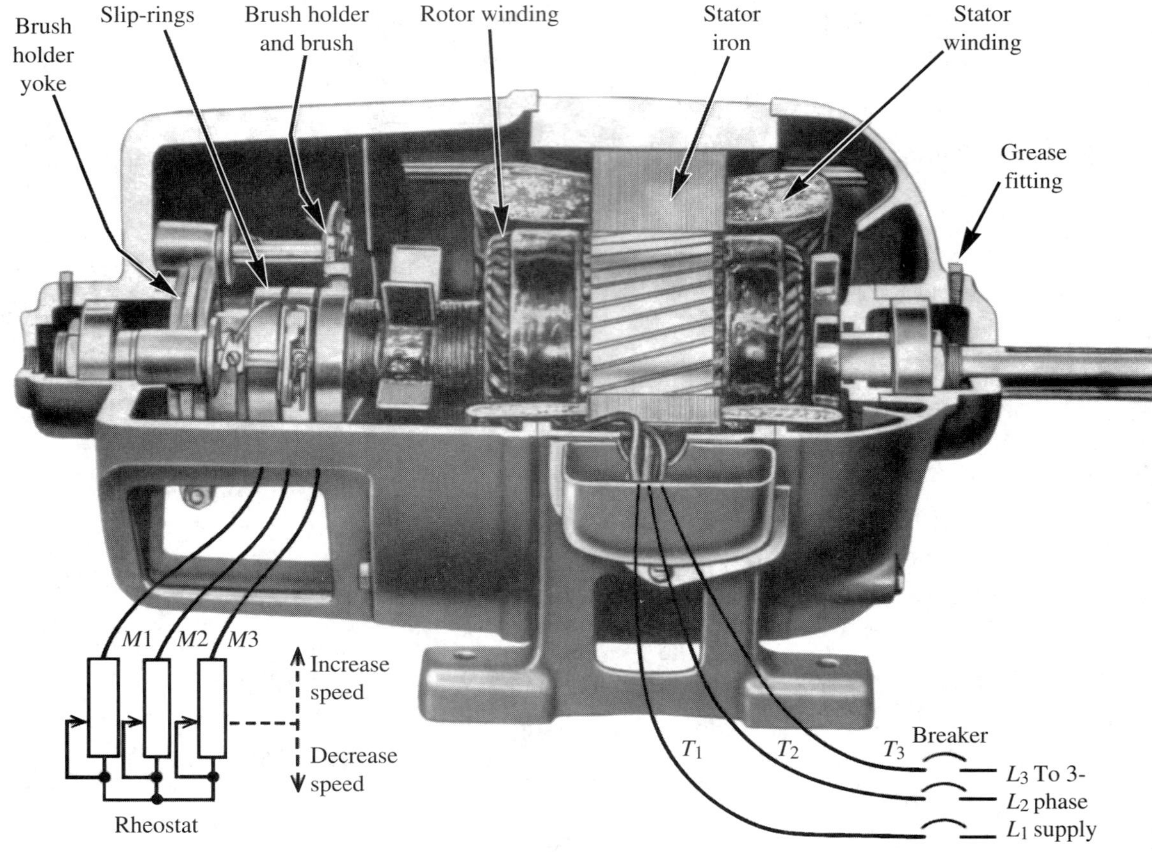

FIGURE 5.8
Cutaway view of a wound-rotor motor. (Courtesy Magnetek Louis Allis Company)

Figure 5.9 illustrates a typical family of torque-slip curves that may be obtained through rotor-rheostat adjustment. Curve 5 is obtained with maximum rheostat resistance in the rotor circuit, and curve 1 is obtained with the rheostat shorted ($R_{\text{rheo}} = 0\ \Omega$). Changing the rheostat setting changes the slip at which $T_{D,\max}$ occurs, but does not change the value of $T_{D,\max}$. Note that increasing the resistance of the rheostat causes $T_{D,\max}$ to shift to the left. A rheostat setting that results in curve 3 causes $T_{D,\max}$ to occur at locked rotor ($s = 1$). Higher values of rotor circuit resistance, such as that represented by curve 4 and curve 5, cause $T_{D,\max}$ to occur at values of slip greater than 1.0.

A slip greater than 1.0 can only occur during *plugging* operations. Plugging is the electrical reversal of a motor before it comes to rest and is accomplished by interchanging any two of the three line leads going to the stator. The characteristics when plugging are shown with broken lines in Figure 5.9. When in the plugging mode, n_r is negative with respect to synchronous speed. Hence, the slip when plugging is greater than 1.0.

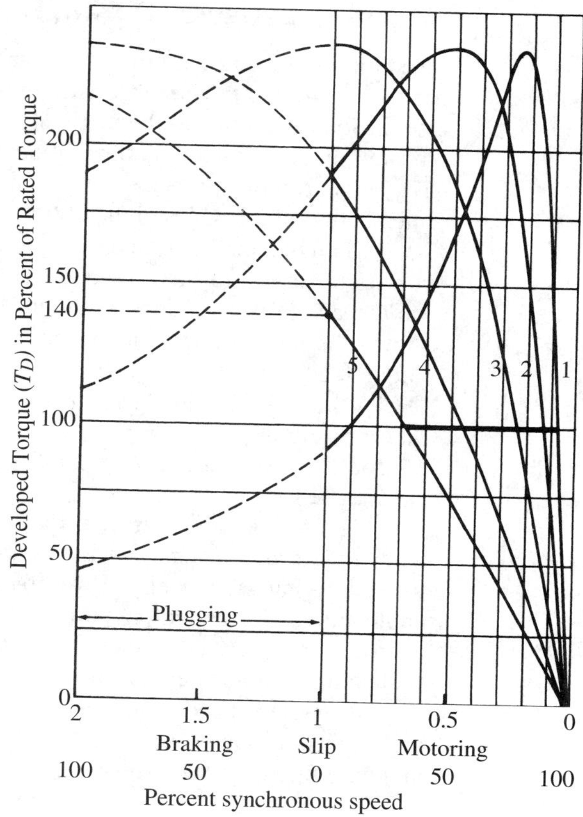

FIGURE 5.9
Family of torque-slip (speed) curves for a representative wound-rotor motor.

Although plugging is sometimes used to provide a fast stop or a fast reversal of squirrel-cage or wound-rotor motors, damage due to overheating may occur unless the motor and/or control circuit are specifically designed to prevent excessive current during such operations. The high transient torques produced by plugging may also damage the driven equipment.

The slip at which $T_{D,\max}$ occurs may be determined from the machine parameters by modifying Eq. (5–7) to include the rheostat resistance. The resultant equation is:

$$s_{T_{D,\max}} = \frac{R_2 + R'_{\text{rheo}}}{\sqrt{R_1^2 + (X_1 + X_2)^2}} \tag{5–24}$$

where R'_{rheo} = rotor-rheostat resistance referred to the stator. That is,

$$R'_{\text{rheo}} = a^2 R_{\text{rheo}} \tag{5–24a}$$

where: a = stator-to-rotor turns-ratio

As indicated in Eq. (5–24), the slip at which $T_{D,\max}$ occurs is directly proportional to the per phase value of the rotor circuit resistance. Expressed as a proportion,

$$s_{T_{D,\max}} \propto (R_2 + R'_{\text{rheo}}) \tag{5–25}$$

Machine Behavior During Rheostat Adjustment

Assume the machine represented in Figure 5.9 is at rest ($s = 1$), the rheostat is adjusted to provide the characteristic indicated by curve 5, a constant 100-percent-rated torque load is applied to the motor shaft, and full voltage is applied to the stator. As indicated on curve 5, the motor will develop 140-percent-rated torque, and since $T_D > T_{load}$, the rotor will accelerate from standstill to a speed corresponding to the intersection of curve 5 and the 100-percent-rated torque line.

Reducing the rheostat resistance causes the torque-slip characteristic to shift from curve 5 to curve 4 to curve 3, etc., resulting in higher speeds. The heavy line connecting the respective intersections of the torque-slip curves with the 100 percent torque line shows the speed range for the particular motor, rheostat settings, and 100 percent torque load.

With no shaft load, and negligible windage and friction, the speed for every rheostat setting will be approximately synchronous speed, this is indicated in Figure 5.9 for $T_D \cong 0$. *Speed changes through rheostat control can be obtained only if the machine is loaded.*

The torque-slip characteristics of wound-rotor motors make them adaptable to loads requiring constant-torque variable-speed drives, high starting torques, and relatively low starting currents. Blowers, hoists, compressors, and stokers are some of its applications. Prolonged operations at below 50 percent synchronous speed should be avoided, however, as the reduced ventilation at slow speed may overheat the machine. If such operation is necessary, auxiliary cooling is required.

The rated speed as indicated on the nameplate of a wound-rotor induction motor is its speed at rated load, rated voltage, rated frequency, rated operating temperature, and with the rotor rheostat shorted.

EXAMPLE 5.10

A family of torque-slip curves for a wye-connected, 400-hp, 2300-V, 14-pole, 60-Hz, wound-rotor induction motor is shown in Figure 5.10. Curves *A* and *D* indicate the extremes of rheostat adjustment. Determine (a) the range of rotor speeds available by rheostat adjustment, assuming 100 percent rated torque load on the shaft; (b) the rheostat resistance required to obtain 260 percent rated torque when starting. The ratio of stator turns per phase to rotor turns per phase is 3.8, and the motor parameters, in ohms/phase, are

$$R_1 = 0.403 \qquad R_2 = 0.317$$
$$X_1 = 1.32 \qquad X_2 = 1.32 \qquad X_M = 35.46$$

Solution

(a)

$$n_s = \frac{120f}{P} = 120 \times \frac{60}{14} = 514.29 \text{ r/min}$$

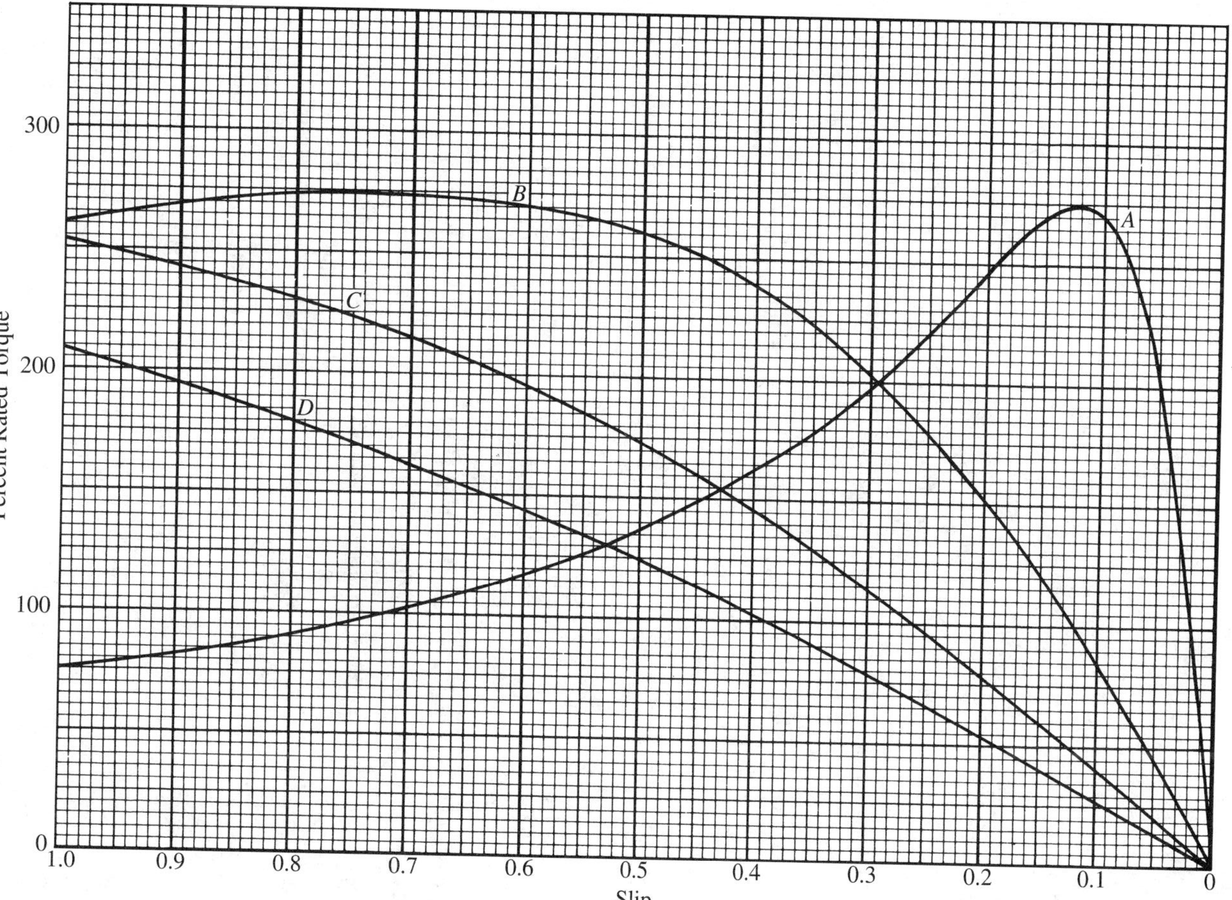

FIGURE 5.10
Torque-speed curves for Example 5.10.

From the intersection of the curves with the 100 percent torque line, low speed occurs at $s \cong 0.395$, and high speed occurs at $s \cong 0.02$. Thus, for low speed (curve D),

$$n_r = n_s(1 - s)$$
$$n_r = 514.29(1 - 0.395)$$
$$n_r \approx \underline{311 \text{ r/min}}$$

For high speed (curve A),

$$n_r = n_s(1 - s)$$
$$n_r = 514.29(1 - 0.02)$$
$$n_r \approx \underline{504 \text{ r/min}}$$

(b) Curve B has a locked-rotor torque of 260 percent. A rheostat setting that results in curve B will cause $T_{D,\max}$ to occur at $s \approx 0.74$ and is as follows:

$$s_{T_{D,max}} = \frac{R_2 + R'_{\text{rheo}}}{\sqrt{R_1^2 + (X_1 + X_2)^2}} \quad \Rightarrow \quad R'_{\text{rheo}} = s_{T_{D,\max}} \cdot \sqrt{R_1^2 + (X_1 + X_2)^2} - R_2$$

$$R'_{\text{rheo}} = 0.74 \times \sqrt{(0.403)^2 + (1.32 + 1.32)^2} - 0.317 = 1.66\ \Omega$$

From Eq. (5–24a),

$$R_{\text{rheo}} = \frac{R'_{\text{rheo}}}{a^2} = \frac{1.66}{(3.8)^2} = \underline{0.115\ \Omega/\text{phase}}$$

5.10 NORMAL RUNNING AND OVERLOAD CONDITIONS FOR WOUND-ROTOR INDUCTION MOTORS

Examination of the torque-slip curves for the wound-rotor motors in Figures 5.9 and 5.10 shows them to be essentially linear from zero torque to approximately 110 percent torque for all values of rheostat resistance. Hence, the simplified equations developed for solving squirrel-cage motor problems, modified to include rheostat resistance, can be used for solving wound-rotor motor problems. Modifying Eqs. (5–11) and (5–12),

$$\underset{s \le 0.03}{I_2} \cong \frac{V \cdot s}{R_2 + R'_{\text{rheo}}} \tag{5–26}$$

$$\underset{s \le 0.03}{T_D} \cong \frac{21.12V^2 \cdot s}{(R_2 + R'_{\text{rheo}})n_s} \tag{5–27}$$

Examination of Eqs. (5–26) and (5–27) shows that for $s \le 0.03$, both I_2 and T_D are directly proportional to the slip and inversely proportional to $(R_2 + R'_{\text{rheo}})$. Because of the added resistance in the rotor circuit, however, the constraint for $s \le 0.03$ for squirrel-cage motors does not necessarily apply for all rheostat settings of a wound-rotor motor. For example, curve 5 in Figure 5.9 is essentially linear from $s = 0$ to $s = 0.50$. Thus, expressed as a proportion, and assuming rated voltage, rated frequency, and operation in the linear region,

$$\underset{\text{linear}}{I_2} \propto \frac{s}{R_2 + R'_{\text{rheo}}} \tag{5–28}$$

$$\underset{\text{linear}}{T_D} \propto \frac{s}{R_2 + R'_{\text{rheo}}} \tag{5–29}$$

EXAMPLE 5.11 A three-phase, wye-connected, 400-hp, four-pole, 380-V, 50-Hz, wound-rotor induction motor operating at rated conditions with the rheostat shorted has a slip of 0.0159. The machine parameters expressed in ohms are

$$R_1 = 0.00536 \qquad R_2 = 0.00613 \qquad R_{\text{fe}} = 7.66$$
$$X_1 = 0.0383 \qquad X_2 = 0.0383 \qquad X_M = 0.5743$$

Determine (a) the rotor frequency; (b) the slip at which $T_{D,\max}$ occurs; (c) the rotor speed at one-half rated torque load; (d) the rheostat resistance per phase required to operate the machine at 1000 r/min and one-half rated torque load (assume the motor is operating on the linear section of the curve and the stator to rotor turns ratio is 2.0); (e) rated torque.

Solution

(a)

$$f_r = sf_{\text{BR}} = 0.0159 \times 50 = \underline{0.795 \text{ Hz}}$$

(b)

$$s_{T_{D,\max}} = \frac{R_2 + R'_{\text{rheo}}}{\sqrt{R_1^2 + (X_1 + X_2)^2}} = \frac{0.00613 + 0}{\sqrt{0.00536^2 + (0.0383 + 0.0383)^2}}$$

$$s_{T_{D,\max}} = \underline{0.0798}$$

(c) The loss in torque due to friction, windage, and stray load is a small percentage of developed torque. Hence, little error is introduced if Eq. (5–27) (for developed torque) is used to determine the load torque. Thus, using proportionality (5–29) to set up a ratio,

$$\frac{T_{D1}}{T_{D2}} = \frac{[s/(R_2 + R'_{\text{rheo}})]_1}{[s/(R_2 + R'_{\text{rheo}})]_2} \quad \Rightarrow \quad s_2 = s_1 \times \frac{T_{D2}}{T_{D1}} \times \frac{[(R_2 + R'_{\text{rheo}})]_2}{[(R_2 + R'_{\text{rheo}})]_1}$$

$$s = 0.0159 \times \frac{0.5T_{\text{rated}}}{T_{\text{rated}}} \times \frac{0.00613}{0.00613} = 0.00795$$

$$n_s = \frac{120f}{p} = \frac{120(50)}{4} = 1500 \text{ r/min}$$

$$n_r = n_s(1 - s) = 1500(1 - 0.00795) = \underline{1488 \text{ r/min}}$$

(d)

$$s = \frac{n_s - n_r}{n_s} = \frac{1500 - 1000}{1500} = 0.3333$$

$$\frac{T_{D1}}{T_{D2}} = \frac{[s/(R_2 + R'_{\text{rheo}})]_1}{[s/(R_2 + R'_{\text{rheo}})]_2} \quad \Rightarrow \quad R'_{\text{rheo},2} = \frac{s_2}{s_1} \times \frac{T_{D1}}{T_{D2}} \times [R_2 + R'_{\text{rheo}}]_1 - R_2$$

$$R'_{\text{rheo},2} = \frac{0.3333}{0.0159} \times \frac{T_{\text{rated}}}{0.5T_{\text{rated}}} \times (0.00613 + 0.0) - 0.00613 = \underline{0.2509\ \Omega}$$

$$R_{\text{rheo}} = \frac{R'_{\text{rheo}}}{a^2} = \frac{0.2509}{2^2} = \underline{0.0627\ \Omega/\text{phase}}$$

(e)

$$n_r = n_s(1 - s) = 1500(1 - 0.0159) = 1476.2 \text{ r/min}$$

$$\text{hp} = \frac{Tn}{5252} \quad \Rightarrow \quad 400 = \frac{T(1476.2)}{5252}$$

$$T = \underline{1423 \text{ lb-ft}}$$

EXAMPLE 5.12 A 50-hp, 10-pole, 60-Hz, 575-V, wye-connected, three-phase, wound-rotor induction motor, operating with the rotor rheostat in the circuit develops its maximum torque at 45 percent slip. Determine the percentage increase or decrease in rotor circuit resistance required to cause $T_{D,\text{max}}$ to occur at 80 percent slip.

Solution

From proportionality (5–25),

$$\frac{s_{T_{D,\max 1}}}{s_{T_{D,\max 2}}} = \frac{[R_2 + R'_{\text{rheo}}]_1}{[R_2 + R'_{\text{rheo}}]_2} \quad \Rightarrow \quad \left[\frac{0.45}{0.80}\right] = \frac{[R_2 + R'_{\text{rheo}}]_1}{[R_2 + R'_{\text{rheo}}]_2}$$

$$[R_2 + R'_{\text{rheo}}]_2 = 1.78[R_2 + R'_{\text{rheo}}]_1 = \underline{78\% \text{ increase}}$$

Thus, a 78 percent increase in equivalent rotor circuit resistance is required to cause $T_{D,\text{max}}$ to occur at 80 percent slip.

5.11 MOTOR NAMEPLATE DATA

Nameplate data offer very pertinent information on the limits, operating range, and general characteristics of electrical apparatus. Interpretation of these data and adherence to their specifications are vital to the successful operating and servicing of such equipment. Correspondence with the manufacturer should always be accompanied by the complete nameplate data of the apparatus. Figure 5.11 illustrates a typical nameplate for an induction motor.

The nameplate lists the *rated operating conditions* of the motor as guaranteed by the manufacturer. For example, for the three-phase motor represented by the nameplate in Figure 5.11, if exactly 460 V, 60 Hz, and three phases are supplied to the stator, and the motor is located in a 40°C ambient region, with the shaft loaded to exactly

FIGURE 5.11
Induction-motor nameplate. (Courtesy Reliance Electric Company)

150 hp, the motor will run at approximately 1785 r/min, draw a line current of approximately 163 A, and have a guaranteed efficiency of 95.8 percent.

Motors rarely, if ever, operate at the exact rated conditions specified by the manufacturer, however. The utilization voltage (voltage at the apparatus) rarely corresponds exactly to the motor nameplate voltage, the motor rarely operates at exactly rated horsepower, and the ambient temperature is rarely 40°C. Although the system frequency most often matches the rated frequency of the motor, there are instances, especially in isolated generator systems (offshore drilling rigs or ships), where the frequency is subject to change.

The nameplate acts as a guide to motor applications, and satisfactory performance is assured if the applied voltage is approximately rated voltage, the frequency is approximately rated frequency, the shaft load does not exceed the service factor rating, and the temperature of the ambient is within the limits indicated on the nameplate. The horsepower rating for each speed of a multispeed motor is based on the type of industrial application: constant horsepower, constant torque, or variable torque.[7]

Nominal Efficiency[8]

The *nominal efficiency* indicated on the nameplate is the average efficiency of a large number of motors of the same design.

---

[7]See Appendix C.

[8]See Reference **[9]** for tables of nominal and minimum efficiencies of NEMA-design induction motors.

Guaranteed Efficiency

The *guaranteed efficiency* is the minimum efficiency to be expected when operating at rated nameplate values.

Design Letter

The *design letter* indicates the NEMA-design characteristics of the machine and serves to direct the reader to the minimum values of locked-rotor torque, breakdown torque, and pull-up torque that may be expected from the machine (see Tables 5.1 through 5.7 in Section 5.3).

Service Factor

The *service factor* (S.F.) of a motor is a multiplier that, when multiplied by the rated power, indicates the permissible loading, provided that the voltage and frequency are maintained at the value specified on the nameplate. Note, however, that if induction motors are operated at a service factor greater than 1.0, the efficiency, power factor, and speed will be different from those at rated load.

Insulation Class

The letter designating *insulation class* specifies the maximum allowable temperature rise above the temperature of the cooling medium for motor windings, based on a maximum ambient temperature of 40°C. *All winding temperatures are to be determined by winding resistance measurement.* A list of maximum allowable winding-temperature rise above the ambient temperature for different classes of insulation systems used in medium single-phase and polyphase induction motors is given in Table 5.8. The maximum winding temperatures are for continuous-duty motors, or motors with short-time ratings of 5, 15, 30, and 60 min. *Note:* Table 5.8 is not valid for ambient temperatures above 40°C. See References **[9]** and **[5]** for more detailed information regarding ambient temperatures above 40°C, high altitude, etc. The motor whose nameplate is shown in Figure 5.11, has Class F insulation, a service factor of 1.15, and is rated for continuous duty. Thus, based on a 40°C ambient, the maximum allowable temperature rise is 115°C, as obtained from Table 5.8.

Frame Number

The *frame number,* for example, 445T in Figure 5.11, determines the critical mounting dimensions of the motor, including shaft length, shaft diameter, shaft height, location of mounting holes, etc. Mounting dimensions for specific frame sizes are given in Reference **[9]**.

TABLE 5.8
Maximum allowable temperature rise for medium single-phase and polyphase induction motors in °C, based on a maximum ambient temperature of 40°C[a]

Class of insulation system (see MG 1-1.65)	A	B	F[b]	H[b,c]
Time rating (shall be continuous or any short-time rating given in MG 1-10.36)				
Temperature rise (based on a maximum ambient temperature of 40°C), °C				
1. Windings, by resistance method				
(a) Motors with 1.0 service factor other than those given in items 1(c) and 1(d)	60	80	105	125
(b) All motors with 1.15 or higher service factor	70	90	115	—
(c) Totally enclosed nonventilated motors with 1.0 service factor	65	85	110	135
(d) Motors with encapsulated windings and with 1.0 service factor, all enclosures	65	85	110	—
2. The temperatures attained by cores, squirrel-cage windings, commutators, collector rings, and miscellaneous parts (such as brushholders, brushes, pole tips, uninsulated shading coils) shall not injure the insulation or the machine in any respect.				

[a]Reproduced by permission of the National Electrical Manufacturers Association from *NEMA Standards Publication MG 1-1998, Motors & Generators*. Copyright 1999 by NEMA, Washington, DC.
[b]Where a Class F or H insulation system is used, special consideration should be given to bearing temperatures, lubrication, etc. (Footnote approved as Authorized Engineering Information.)
[c]This column applies to polyphase induction motors only.

Enclosure

The *enclosure* (TEFC in Figure 5.11) indicates that the motor is totally enclosed and is fan cooled by an external fan mounted on the motor shaft.

Code Letter

The *code letter* provides a means for determining the expected locked-rotor *in-rush current* to the stator when starting the motor with rated voltage and rated frequency applied directly to the stator terminals. The code letter directs the reader to a table of locked-rotor kVA per horsepower, from which the in-rush current may be calculated (see Table 5.9).

5.12 LOCKED-ROTOR IN-RUSH CURRENT

At locked rotor each phase of an induction motor acts as a simple *R–L* series circuit. Closing the switch to such a circuit causes a combined transient and steady-state current **[3]**.

TABLE 5.9
NEMA code letters for locked-rotor kVA per horsepower[a]

Code Letter	kVA/hp[b]	Code Letter	kVA/hp[b]
A	0.0–3.15	K	8.0–9.0
B	3.15–3.55	L	9.0–10.0
C	3.55–4.0	M	10.0–11.2
D	4.0–4.5	N	11.2–12.5
E	4.5–5.0	P	12.5–14.0
F	5.0–5.6	R	14.0–16.0
G	5.6–6.3	S	16.0–18.0
H	6.3–7.1	T	18.0–20.0
J	7.1–8.0	U	20.0–22.4
		V	22.4 and up

[a]Reprinted by permission of the National Electrical Manufacturers Association from *NEMA Standards Publication MG 1-1998, Motors & Generators.* Copyright 1999 by NEMA, Washington, DC.
[b]Locked kVA per horsepower range includes the lower figure up to, but not including, the higher figure. For example, 3.14 is designated by letter A, and 3.15 by letter B.

Thus, the locked-rotor in-rush current to an induction motor consists of a steady-state component ($i_{\text{lr,ss}}$), called the *normal in-rush current,* and a transient component ($i_{\text{lr,tr}}$) that decays to insignificance in a very short time. Unless otherwise specified, the term *locked-rotor current* always refers to the steady-state component.

The steady-state component for a specific motor may be obtained from the manufacturer or approximated from data usually stamped on the motor nameplate. The steady-state component may also be calculated from the machine parameters and the rated voltage as outlined in Section 5.4. Using machine parameters and solving on a per-phase basis,

$$\mathbf{I}_{\text{lr,ss}} = \left[\frac{\mathbf{V}_{\text{phase}}}{\mathbf{Z}_{\text{in}}}\right]_{s=1.0} \tag{5–30}$$

Expressed in the time domain as a sinusoidal current,

$$i_{\text{lr,ss}} = \sqrt{2}\left|\frac{\mathbf{V}_{\text{phase}}}{\mathbf{Z}_{\text{in}}}\right|_{s=1.0} \sin(2\pi f t - \theta_z) \tag{5–31}$$

The transient component of locked-rotor in-rush current per phase may be approximated by the following exponential function:

$$i_{\text{lr,tr}} = A\varepsilon^{-(R/L)t} \tag{5–32}$$

where: $R = Z_{\text{in}} \cos \theta_z$
$L = (Z_{\text{in}} \sin \theta_z)/(2\pi f)$

A = coefficient whose value depends on the magnitude and phase angle of the applied voltage at the instant the switch is closed:

$$Z_{\text{in}}\angle\theta_z = Z_{\text{in}}\cos\theta_z + jZ_{\text{in}}\sin\theta_z = R + jX_L$$

Combining the steady-state component and the transient component,

$$i_{\text{lr}} = i_{\text{lr,ss}} + i_{\text{lr,tr}} \tag{5–33}$$

$$i_{\text{lr}} = \sqrt{2}\left|\frac{\mathbf{V}_{\text{phase}}}{\mathbf{Z}_{\text{in}}}\right|_{s=1.0}\sin(2\pi ft - \theta_z) + A\varepsilon^{-(R/L)t} \tag{5–34}$$

Since the three-phase voltages are 120° apart, their voltage zeros occur at different instants of time. Hence, the transient behavior indicated by Eq. (5–34) is per phase.

If the motor is at rest and the switch is closed at the instant that the applied voltage has its maximum value, the transient component of locked-rotor current will be zero. If the switch is closed at the instant that the voltage wave is passing through zero, the transient component will be a maximum, as shown in Figure 5.12 for one phase of a representative motor.[9]

In the preceding discussion of locked-rotor in-rush current, and in Figure 5.12, it was assumed that the rotor was blocked and could not rotate. For normal motor operation, with relatively low inertia loads, however, the decrease in slip during acceleration would cause a rapid reduction in locked-rotor current, decaying to rated current at rated load in 2 s to 5 s or less.

[9]See Reference **[3]** for transients in driven systems.

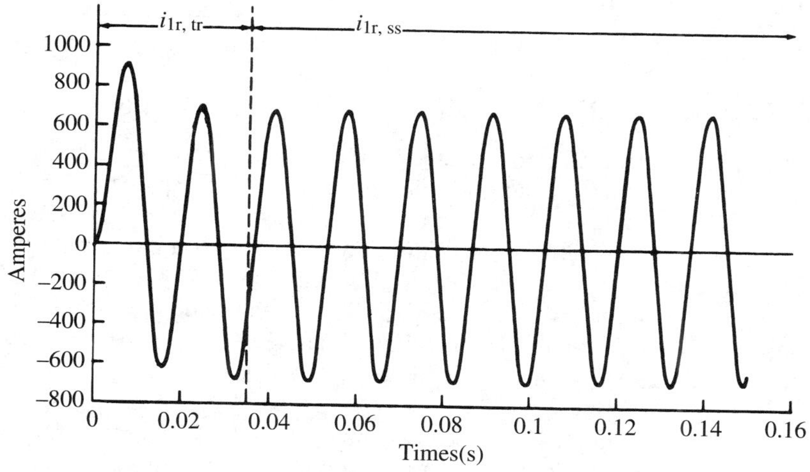

FIGURE 5.12
Locked-rotor in-rush current.

EXAMPLE 5.13 For the motor whose nameplate is shown in Figure 5.11, determine the expected in-rush current.

Solution

The motor is a NEMA design *B* machine rated at 150 hp at 460 V, 60 Hz. It has a rated current of 163 A, a nominal efficiency of 96.2 percent, and a Code G designation for locked-rotor kVA/hp. Referring to Table 5.9, the expected range of starting kVA/hp for a Code G machine is

$$5.6 \leq \text{kVA/hp} < 6.3$$

The expected range of locked-rotor current can be determined from the following apparent-power equation:

$$\text{kVA/hp} \times \text{hp} \times 1000 = \sqrt{3} \times V_{\text{line}} \times I_{\text{line}}$$

Thus, for the lower limit:

$$5.6 \times 150 \times 1000 = \sqrt{3} \times 460 I_{\text{line}}$$

$$I_{\text{lr,ss}} = 1054 \text{ A}$$

For the upper limit:

$$6.3 \times 150 \times 1000 = \sqrt{3} \times 460 I_{\text{line}}$$

$$I_{\text{lr,ss}} = 1186 \text{ A}$$

Thus, the expected range of in-rush current at locked rotor, with rated voltage and rated frequency applied to the stator, is

$$\underline{1054 \text{ A}} \leq I_{\text{lr,ss}} < \underline{1186 \text{ A}}$$

This is a lot of current for a motor whose full-load rating is 163 A.

5.13 EFFECT OF NUMBER OF STARTS ON MOTOR LIFE

The high in-rush current associated with every start, or attempted start, causes severe thermal and mechanical stresses on rotor and stator components. The effects of these stresses are cumulative and adversely affect the service life of the machine. Large motors (above 50 hp) and motors with high inertia loads are particularly susceptible to damage caused by frequent starting. For this reason, manufacturers of large motors often provide information on the permissible number of starts, and the required waiting period between starts. The number of starts and cooling periods between starts are based on the inertia (WK^2) of the motor and load, the type of load, and the temperature of the machine **[9]**.

Operating and maintenance personnel should avoid unnecessary starts; repeated attempts at starting a motor after tripping due to fault or overload can cause extensive damage to the machine. The reason for tripping or failure to start must be determined and corrected before attempting a restart **[4]**.

5.14 RECLOSING OUT-OF-PHASE SCENARIO

When the stator of an induction motor is disconnected from the line, the closed circuit formed by the rotor bars and end rings prevents the quick collapse of flux. The flux in the revolving rotor induces a three-phase voltage in the stator windings that appears at the stator terminals. This *residual voltage* decays at a rate determined by the open-circuit time constant of the motor. In an elapsed time equal to one time constant, the residual voltage will decay to 36.8 percent of line voltage. Open-circuit time constants of 0.3 s are not uncommon.

If after a power interruption the motor is reconnected to the power line, with the residual voltage out of phase with the line voltage, the in-rush current will be higher than if the motor were started from a stopped position. The worst possible condition would be reclosure immediately after a power interruption, with the residual voltage almost equal to the line voltage and nearly 180° out of phase; the in-rush current would be almost double the locked-rotor value, and severe damage to the motor could occur. If the machine is large, a power blackout may occur.

Reclosing with high residual voltage and out of phase is likely to occur when switching rapidly from a failed power supply to a standby or emergency supply. Rapid and safe reclosing to alternate power supplies may be accomplished with an in-phase monitor, which measures the phase angle between the source voltage and the motor residual voltage, and initiates reclosure when the phase angle approaches zero.

Permanently connecting a voltmeter across the motor terminals and manually reclosing when the residual voltage drops below 20 percent or less of rated voltage is another method that will enable a relatively quick restart while avoiding an abnormally high in-rush current. If reclosed at 20 percent voltage and 180° out of phase, the net voltage applied to the stator windings will be only 120 percent rated voltage. *Note:* In those applications where capacitors for power-factor improvement are connected directly across the motor terminals, the residual voltage will take a much longer time to decay; the capacitors cause the motor to act as a self-excited induction generator (see Section 5.18).

5.15 EFFECT OF UNBALANCED LINE VOLTAGES ON INDUCTION MOTOR PERFORMANCE

If the three line voltages supplied to a three-phase induction motor are not equal, they not only cause unequal phase currents in the rotor and stator windings but the percentage current unbalance may be 6 to 10 times larger than the percentage voltage unbalance. The resultant increase in I^2R losses will overheat the insulation, shortening its life. Unbalanced voltages also cause a decrease in locked-rotor and breakdown torque. Thus, in those applications where there is only a small margin between the locked-rotor torque and the load torque, severe voltage unbalance may prevent the motor from starting. The full-load speed of running motors is reduced slightly by voltage unbalance.

Percent voltage unbalance is defined by NEMA as the maximum line voltage deviation from the average value of the three line voltages, times 100. Expressed as an equation,

$$\%\text{UBV} = \frac{V_{\text{max dev}}}{V_{\text{avg}}} \cdot 100 \tag{5–35}$$

where: $\%\text{UBV}$ = percent unbalanced voltage
$V_{\text{avg}} = (V_1 + V_2 + V_3)/3$
$V_{\text{max dev.}}$ = maximum volt deviation between a line voltage and V_{avg}

Note: Voltage measurements, taken for the purpose of determining voltage unbalance, should be made as close as possible to the motor terminals, and the readings should be taken with a digital voltmeter for greater accuracy.

Empirical results obtained from laboratory tests indicate that the percentage increase in motor temperature, due to voltage unbalance, is approximately equal to two times the square of the percentage voltage unbalance **[2]**. Expressed as an equation,

$$\%\Delta T \cong 2(\%\text{UBV})^2 \tag{5–36}$$

where $\%\Delta T$ = percent increase in motor temperature. Thus, assuming operation at rated load, the expected temperature rise caused by voltage unbalance is

$$T_{\text{UBV}} \cong T_{\text{rated}} \cdot \left(1 + \frac{\%\Delta T}{100}\right) \tag{5–37}$$

where: T_{rated} = expected temperature rise from Table 5.8 (C°)
T_{UBV} = expected temperature rise due to voltage unbalance (C°)

The effect of higher operating temperatures on the life of electrical insulation may be approximated by the *ten-degree rule.* This rule, developed by A. M. Montsinger in a classic study, demonstrated that insulation life is approximately halved with each 10°C increase in temperature, and conversely is doubled with each 10°C decrease in temperature **[1]**. Although not precise, this ten-degree rule has been substantiated over time, and does provide a rough approximation of the effect of motor temperature on insulation life. Expressing the ten-degree rule in equation form.

$$\text{RL} \cong \frac{1}{2^{(\delta T/10)}} \tag{5–38}$$

where: RL = relative life of insulation
$\delta T = T_{\text{UBV}} - T_{\text{rated}}$

Thus, a motor with an expected life of 20 years operating with the same rated load, but with a phase unbalance that causes $\delta T = 15°\text{C}$, would from Eq. (5–38), have a relative life of RL = 0.35. The expected life of the insulation, assuming continued operation at the higher temperature, would be $0.35 \times 20 = 7$ years.

In those applications where voltage unbalance cannot be corrected, the motor should be *derated,* that is, operated at a lower horsepower. The derating curve shown

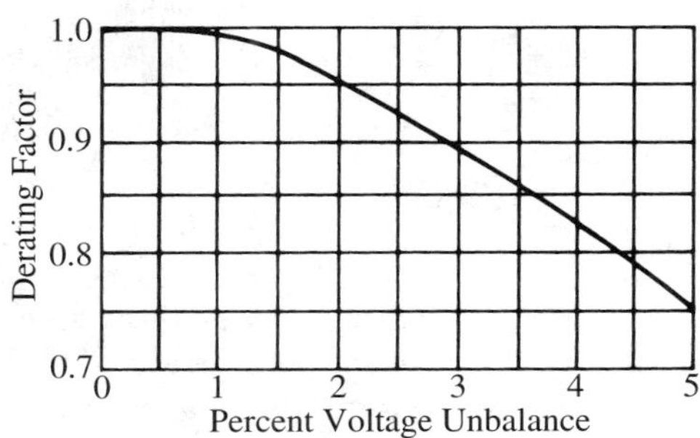

FIGURE 5.13
Derating curve for induction motors. (Courtesy National Electrical Manufacturers Association. From *NEMA Standards Publication MG1-1998.* Copyright 1999 by NEMA)

in Figure 5.13 should be used to determine the required derating **[9]**. A 1 percent unbalance will not cause significant problems. Operating a motor with a voltage unbalance greater than 5 percent is not recommended, however.

EXAMPLE 5.14

A 30-hp, design *B*, 460-V, 60-Hz, four-pole, totally enclosed nonventilated induction motor with Class F insulation, and a service factor of 1.0, is to be operated at rated power from an unbalanced three-phase system. The expected life of the motor under normal operating conditions is 20 years. If the three line-to-line voltages are 460 V, 455 V, and 440 V, determine (a) the percent voltage unbalance; (b) the expected approximate temperature rise if operating at rated load in a 40°C ambient; (c) the expected insulation life; (d) the required derating of the motor to prevent shortening insulation life.

Solution

(a)

$$V_{\text{avg}} = \frac{460 + 455 + 440}{3} = 451.667 \text{ V}$$

The voltage deviations from the average are:

$$|460 - 451.667| = 8.333 \text{ V}$$
$$|455 - 451.667| = 3.333 \text{ V}$$
$$|440 - 451.667| = 11.333 \text{ V}$$

$$\%\text{UBV} = \frac{V_{\text{max dev}}}{V_{\text{avg}}} \cdot 100 = \frac{11.667}{451.667} \cdot 100 = 2.5831 \quad \Rightarrow \quad \underline{2.58}$$

(b)

$$\%\Delta T \cong 2(\%\text{UBV})^2 = 2(2.5831)^2 = 13.344 \quad \Rightarrow \quad \underline{13.34}$$

The rated temperature rise (from Table 5.8) for a totally enclosed nonventilated motor with Class F insulation and 1.0 service factor is 110°C. Thus,

$$T_{\text{UBV}} \cong T_{\text{rated}} \cdot \left(1 + \frac{\%\Delta T}{100}\right) = 110\left(1 + \frac{13.344}{100}\right) = 124.6784 \quad \Rightarrow \quad \underline{125°\text{C}}$$

(c) $\delta T = T_{\text{UBV}} - T_{\text{rated}} = 124.6784 - 110 = 14.6784°\text{C}$

$$\text{RL} \cong \frac{1}{2^{(\delta T/10)}} = \frac{1}{2^{(14.6784/10)}} = 0.361$$

Thus, if operating with a 2.58 percent voltage unbalance, the expected life is $0.361 \times 20 \approx$ 7.2 years.

(d) From the derating curve, the derating factor for a 2.58 percent voltage unbalance is ≈ 0.92. Hence, to prevent excessive heating and a shortened life, the shaft horsepower output should be limited to $30 \times 0.92 =$ 27.6 hp.

5.16 PER-UNIT VALUES OF INDUCTION-MOTOR PARAMETERS

Manufacturers of electrical machinery often publish induction-motor parameters as per-unit impedance values instead of actual impedance values. Per-unit values are useful in machine design in that a wide range of machine sizes have approximately the same values. This makes it easy to detect gross errors in design calculations and also provides a convenient means for comparing the relative performance of machines without regard to machine size. For example, machines with higher per-unit rotor resistance will have $T_{D,\max}$ occur at a greater slip.

Per-unit values of induction-motor parameters are defined as a ratio of the actual value of the respective parameters divided by a common base impedance. The base impedance for induction motors is calculated from *output power* rather than input power to avoid having to make assumptions of power factor and efficiency **[6]**.

Defining base values for a three-phase induction motor:

$$P_{\text{base}} = \text{base power/phase} = \text{rated shaft power output} \div 3\ (\text{W})$$
$$V_{\text{base}} = \text{base voltage} = \text{rated voltage/phase}\ (\text{V})$$
$$I_{\text{base}} = \text{base current} = \text{rated current/phase}\ (\text{A})$$

$$I_{\text{base}} = \frac{P_{\text{base}}}{V_{\text{base}}} \tag{5–39}$$

$$Z_{\text{base}} = \frac{V_{\text{base}}}{I_{\text{base}}} \tag{5–40}$$

Expressing the motor parameters in terms of per-unit values:

$$\left.\begin{aligned} r_1 &= \frac{R_1}{Z_{\text{base}}} & r_2 &= \frac{R_2}{Z_{\text{base}}} & r_{\text{fe}} &= \frac{R_{\text{fe}}}{Z_{\text{base}}} \\ x_1 &= \frac{X_1}{Z_{\text{base}}} & x_2 &= \frac{X_2}{Z_{\text{base}}} & x_M &= \frac{X_M}{Z_{\text{base}}} \end{aligned}\right\} \tag{5–41}$$

Note: Z_{base} may also be expressed in terms of P_{base} as derived below:

$$Z_{\text{base}} = \frac{V_{\text{base}}}{I_{\text{base}}} = \frac{V_{\text{base}}}{P_{\text{base}}/V_{\text{base}}} = \frac{V_{\text{base}}^2}{P_{\text{base}}} \tag{5–41a}$$

EXAMPLE 5.15 Given the following per-unit values for a 50-hp, 460-V, six-pole, 60-Hz, wye-connected induction motor:

$$r_1 = 0.021 \qquad r_2 = 0.020 \qquad r_{fe} = 20.0$$
$$x_1 = 0.100 \qquad x_2 = 0.0178 \qquad x_M = 3.68$$

Determine the machine parameters in ohms.

Solution

$$V_{base} = \frac{460}{\sqrt{3}} = 265.58 \text{ V}$$

$$P_{base} = 50 \times 746 \div 3 = 12{,}433.33 \text{ W}$$

$$Z_{base} = \frac{V_{base}^2}{P_{base}} = \frac{(265.58)^2}{12{,}433.33} = 5.67\ \Omega$$

From equation set (5–41),

$$R_1 = r_1 Z_{base} = 0.021(5.67) = \underline{0.119\ \Omega}$$
$$X_1 = x_1 Z_{base} = 0.100(5.67) = \underline{0.567\ \Omega}$$
$$R_2 = r_2 Z_{base} = 0.020(5.67) = \underline{0.113\ \Omega}$$
$$X_2 = x_2 Z_{base} = 0.0178(5.67) = \underline{0.101\ \Omega}$$
$$R_{fe} = r_{fe} Z_{base} = 20.0(5.67) = \underline{113.40\ \Omega}$$
$$X_M = x_M Z_{base} = 3.68(5.67) = \underline{20.87\ \Omega}$$

5.17 DETERMINATION OF INDUCTION-MOTOR PARAMETERS

In those cases where induction motor parameters are not readily available from the manufacturer, they can be approximated from a DC test, a no-load test, and a blocked-rotor test **[7]**.

DC Test

The purpose of the DC test is to determine R_1. This is accomplished by connecting any two stator leads to a variable-voltage DC source as shown in Figure 5.14(a). The DC source is adjusted to provide approximately rated stator current, and the resistance between the two stator leads is determined from voltmeter and ammeter readings. Thus, from Figure 5.14(a),

$$R_{DC} = \frac{V_{DC}}{I_{DC}}$$

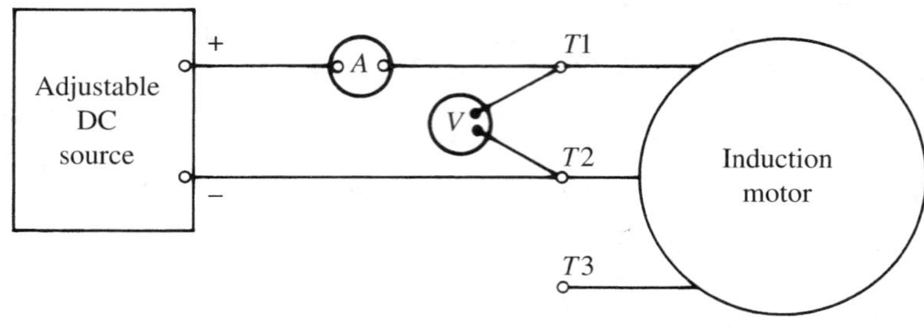

(a)

(b)

(c)

FIGURE 5.14
Circuits for DC test to determine parameter R_1.

If the stator is wye connected,[10] as shown in Figure 5.14(b),

$$R_{DC} = 2R_{1,\text{wye}}$$

$$R_{1,\text{wye}} = \frac{R_{DC}}{2} \tag{5–42}$$

If the stator is delta connected, as shown in Figure 5.14(c),

$$R_{DC} = \frac{R_{1\Delta} \cdot 2R_{1\Delta}}{R_{1\Delta} + 2R_{1\Delta}} = \frac{2}{3}R_{1\Delta}$$

$$R_{1\Delta} = 1.5R_{DC} \tag{5–43}$$

Blocked-Rotor Test

The blocked-rotor test is used to determine X_1 and X_2. When combined with data from the DC test, it also determines R_2.

The test is performed by blocking the rotor so that it cannot turn, and measuring the line voltage, line current, and three-phase power input to the stator. Connections for the test are shown in Figure 5.15. An adjustable voltage AC supply (not shown) is

---

[10]If there is no indication as to whether the connections are wye or delta, assume an equivalent wye connection and proceed as outlined.

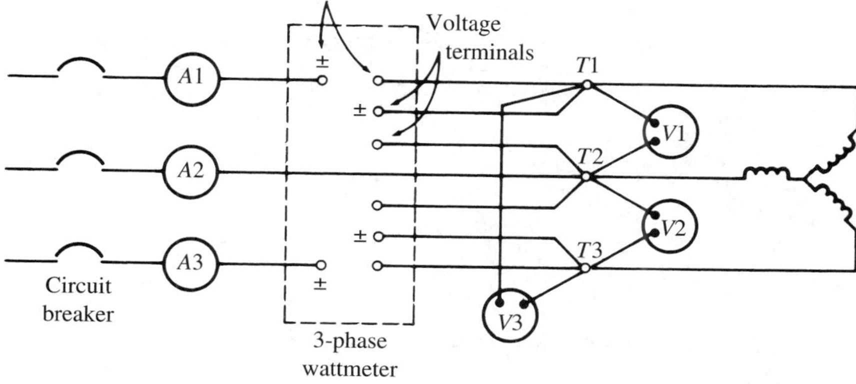

FIGURE 5.15
Basic circuit for blocked-rotor test and no-load test.

used to adjust the blocked-rotor current to approximately rated current. If instrument transformers and single-phase wattmeters are used, the effect of transformer ratios and the direction of the wattmeter readings (whether positive or negative) must be considered **[3]**.

Since the exciting current (I_0) at blocked rotor is considerably less than the rotor current (I_2), the exciting current may be neglected, enabling a simplification of the equivalent circuit. This is shown in Figure 5.16, where X_M and R_{fe} are drawn with dotted lines and omitted when making blocked-rotor calculations.

To minimize errors that would otherwise be introduced by magnetic saturation and resistance skin effect, if tested at rated voltage and rated frequency, *the IEEE test code recommends that the blocked-rotor test be made using 25 percent rated frequency with the test voltage adjusted to obtain approximately rated current* **[7]**. Thus, a 60-Hz motor would use a 15-Hz test voltage. The total reactance calculated from the 15-Hz test is then corrected to 60 Hz by multiplying by 60/15. The total resistance calculated

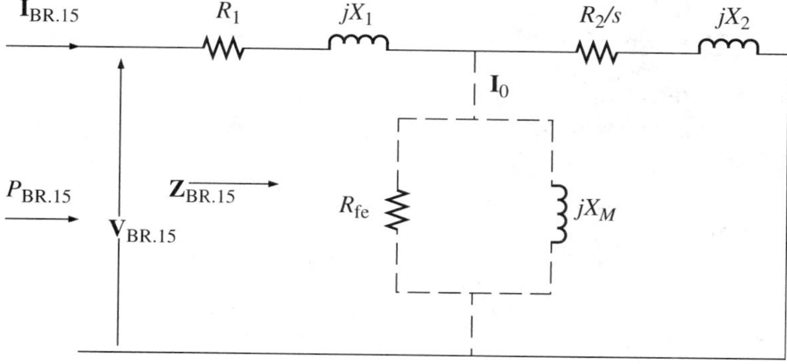

FIGURE 5.16
Equivalent circuit per phase for blocked-rotor test.

from the 15-Hz test is essentially correct, however, and must not be adjusted. Thus, referring to Figure 5.16, where all values are per phase,

$$R_1 + R_2 = R_{BR,15} \tag{5–44}$$

$$Z_{BR,15} = \frac{V_{BR,15}}{I_{BR,15}} \tag{5–45}$$

$$R_{BR,15} = \frac{P_{BR,15}}{I^2_{BR,15}} \tag{5–46}$$

Resistance R_2 is obtained from $R_{BR,15}$ by substituting R_1 from the DC test into Eq. (5–44). Thus,

$$R_2 = R_{BR,15} - R_1 \tag{5–47}$$

From Figure 5.16,

$$\begin{aligned} Z_{BR,15} &= \sqrt{R^2_{BR,15} + X^2_{BR,15}} \\ X_{BR,15} &= \sqrt{Z^2_{BR,15} - R^2_{BR,15}} \end{aligned} \tag{5–48}$$

Converting $X_{BR,15}$ to 60 Hz,

$$X_{BR,60} = \frac{60}{15} X_{BR,15} \tag{5–49}$$

where:

$$X_{BR,60} = X_1 + X_2 \tag{5–50}$$

If the NEMA-design letter of the induction motor is known, the division of blocked-rotor reactance between X_1 and X_2, as shown in Eq. (5–50), may be determined from Table 5.10 **[7]**. If the NEMA design letter is not known, an equal division between X_1 and X_2 is generally assumed.

No-Load Test

The no-load test is used to determine the magnetizing reactance X_M and the combined core, friction, and windage losses. These losses are essentially constant for all load conditions.

TABLE 5.10
Division of blocked-rotor reactance for NEMA-design motors

	A, D	*B*	*C*	Wound Rotor
X_1	$0.5X_{BR}$	$0.4X_{BR}$	$0.3X_{BR}$	$0.5X_{BR}$
X_2	$0.5X_{BR}$	$0.6X_{BR}$	$0.7X_{BR}$	$0.5X_{BR}$

The connections for the no-load test are identical to those shown in Figure 5.15 for the blocked-rotor test. The only difference is in the operation of the test: For the no-load test, the rotor is unblocked and allowed to run unloaded at rated voltage and rated frequency.

At no load, the operating speed is very close to synchronous speed; the slip is ≈ 0, causing the current in the R_2/s branch to be very small. For this reason, the R_2/s branch is drawn with dotted lines, as shown in Figure 5.17, and omitted from the no-load current calculations. Furthermore, since $I_M >> I_{fe}$, $I_0 \approx I_M$; thus, the R_{fe} branch is also drawn with dotted lines, as in the figure, and omitted from the no-load current calculations.

Referring to the approximate equivalent circuit shown in Figure 5.17 for the no-load test, the apparent power input per phase is

$$S_{NL} = V_{NL} I_{NL}$$

The reactive power per phase is determined from

$$S_{NL} = \sqrt{P_{NL}^2 + Q_{NL}^2} \tag{5–51}$$

Solving for Q_{NL},

$$Q_{NL} = \sqrt{S_{NL}^2 - P_{NL}^2} \tag{5–52}$$

Expressing the reactive power in terms of current and reactance, and solving for the equivalent reactance at no load,

$$Q_{NL} = I_{NL}^2 X_{NL} \tag{5–53}$$

Thus,

$$X_{NL} = \frac{Q_{NL}}{I_{NL}^2} \tag{5–54}$$

where, as indicated in Figure 5.17,

$$X_{NL} = X_1 + X_M \tag{5–55}$$

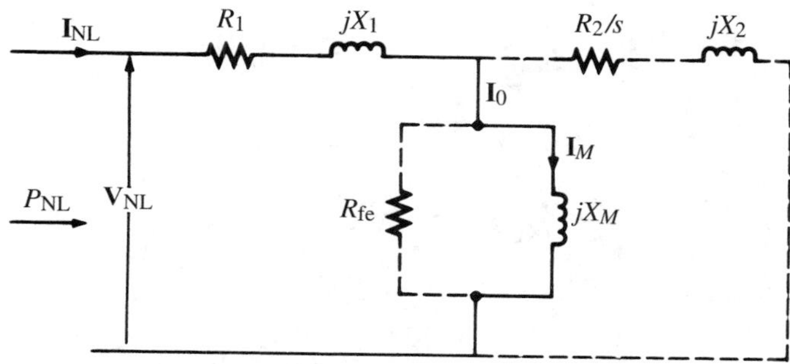

FIGURE 5.17
Equivalent circuit per phase for no-load test.

Substituting X_1 as determined from the blocked-rotor test, into Eq. (5–55), permits the determination of X_M.

The input power per phase at no load includes the core loss, stator copper loss, windage loss, and friction loss (all per phase). That is,

$$P_{NL} = I_{NL}^2 R_1 + P_{core} + P_{w,f} \tag{5-56}$$

Separation of friction and windage losses from the no-load loss may be accomplished by plotting the no-load power vs. voltage squared for low values of voltage and then extrapolating to zero voltage [7].

EXAMPLE 5.16 The following data were obtained from no-load, blocked-rotor, and DC tests of a three-phase, wye-connected, 40-hp, 60-Hz, 460-V, design *B* induction motor whose rated current is 57.8 A. The blocked-rotor test was made at 15 Hz.

Blocked Rotor	No-Load	DC
$V_{line} = 36.2$ V	$V_{line} = 460.0$ V	$V_{DC} = 12.0$ V
$I_{line} = 58.0$ A	$I_{line} = 32.7$ A	$I_{DC} = 59.0$ A
$P_{3\ phase} = 2573.4$ W	$P_{3\ phase} = 4664.4$ W	

(a) Determine R_1, X_1, R_2, X_2, X_M, and the combined core, friction, and windage loss.

(b) Express the no-load current as a percent of rated current.

Solution

(a) Converting the AC test data to corresponding phase values for a wye-connected motor,

$$P_{BR,15} = \frac{2573.4}{3} = 857.80 \text{ W}$$

$$V_{BR,15} = \frac{36.2}{\sqrt{3}} = 20.90 \text{ V}$$

$$I_{BR,15} = 58.0 \text{ A}$$

$$P_{NL} = \frac{4664.4}{3} = 1554.80 \text{ W}$$

$$V_{NL} = \frac{460}{\sqrt{3}} = 265.581 \text{ V}$$

$$I_{NL} = 32.7 \text{ A}$$

Determination of R_1:

$$R_{DC} = \frac{V_{DC}}{I_{DC}} = \frac{12.0}{59.0} = 0.2034\ \Omega$$

$$R_{1,wye} = \frac{R_{DC}}{2} = \frac{0.2034}{2} = \underline{0.102\ \Omega/\text{phase}}$$

Determination of R_2:

$$Z_{BR,15} = \frac{V_{BR,15}}{I_{BR,15}} = \frac{20.90}{58.0} = 0.3603\ \Omega$$

$$R_{BR,15} = \frac{P_{BR,15}}{I^2_{BR,15}} = \frac{857.8}{(58)^2} = 0.2550\ \Omega/\text{phase}$$

$$R_2 = R_{BR,15} - R_{1,wye} = 0.2550 - 0.102 = \underline{0.153\ \Omega/\text{phase}}$$

Determination of X_1 and X_2:

$$X_{BR,15} = \sqrt{Z^2_{BR,15} - R^2_{BR,15}} = \sqrt{(0.3603)^2 - (0.255)^2} = 0.2545\ \Omega$$

$$X_{BR,60} = \frac{60}{15} X_{BR,15} = \frac{60}{15}(0.2545) = 1.0182\ \Omega$$

From Table 5.10, for a design B machine,

$$X_1 = 0.4X_{BR,60} = 0.4(1.0182) = \underline{0.4073\ \Omega/\text{phase}}$$
$$X_2 = 0.6X_{BR,60} = 0.6(1.0182) = \underline{0.6109\ \Omega/\text{phase}}$$

Determination of X_M:

$$S_{NL} = V_{NL}I_{NL} = 265.581(32.7) = 8684.50\ \text{VA}$$

$$Q_{NL} = \sqrt{S^2_{NL} - P^2_{NL}} = \sqrt{(8684.50)^2 - (1554.8)^2} = 8544.19\ \text{var}$$

$$X_{NL} = \frac{Q_{NL}}{I^2_{NL}} = \frac{8544.19}{(32.7)^2} = 7.99\ \Omega$$

$$X_{NL} = X_1 + X_M \quad \Rightarrow \quad 7.99 = 0.4073 + X_M$$

$$X_M = \underline{7.58\ \Omega/\text{phase}}$$

Determination of combined friction, windage, and core loss:

$$P_{NL} = I^2_{NL}R_{1,wye} + P_{core} + P_{f,w}$$
$$1554.8 = (32.7)^2(0.102) + P_{core} + P_{f,w}$$
$$P_{core} + P_{f,w} = \underline{1446\ \text{W/phase}}$$

(b)

$$\%\text{I}_{NL} = \frac{\text{I}_{NL}}{\text{I}_{rated}} \times 100 = \frac{32.7}{57.8} = \underline{56.6\%}$$

Note: The no-load current (exciting current) of a three-phase induction motor is large, generally 40% or higher in terms of rated current.

5.18 INDUCTION GENERATORS

Induction generators have the same basic construction as squirrel-cage induction motors. In fact, all induction motors can be operated very effectively as induction generators by driving them at a speed greater than synchronous speed. In induction generator applications, however, where the machine is not started as a motor, and hence does not require a high starting torque, induction generators are generally designed with lower resistance values to provide a lower slip and a higher efficiency at rated load.

Induction generators are suitable for operation by wind turbines, hydraulic turbines, steam turbines, and gas engines powered by natural gas or biogas. They can range in size from a few kilowatts to 10 MW or higher, and are used extensively in cogeneration operations. *Cogeneration* is the sequential production of two forms of energy, usually steam for process operations and electricity for plant use and for sale to utilities.

Motor-to-Generator Transition

Figure 5.18 shows an induction motor connected to a three-phase system, with its shaft mechanically coupled to a steam turbine. Assume the turbine valve is closed so that no steam enters the turbine, and the motor is started at full voltage by closing the circuit breaker. The induction motor accelerates and drives the turbine at a speed somewhat less than the synchronous speed of the rotating stator flux. Gradually opening the turbine valve causes a gradual buildup of turbine torque, adding to that developed by the induction motor, resulting in an increase in rotor speed. When the speed of the turbine-motor set reaches the synchronous speed of the stator, the slip becomes zero, R_2/s becomes infinite, rotor current I_2 becomes zero, and no motor torque is developed. At zero slip, the induction machine is neither a motor nor a generator; it is "floating" on the bus. The only current in the stator is the exciting current that supplies the rotating magnetic field and the iron losses.

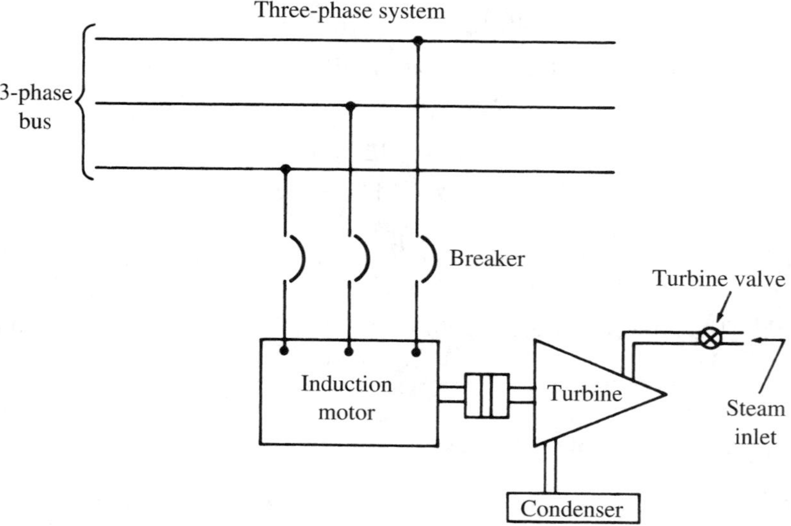

FIGURE 5.18
Induction motor coupled to a turbine for induction-generator operation.

Prime movers, such as turbines, diesels, and waterwheels, can only supply active power. Hence, when connected to an energized system, the magnetizing current I_M is supplied by the power system whether the machine operates as a motor or a generator.

The speed of the rotating flux is independent of the rotor speed; it is a function of the number of stator poles and the frequency of the power system to which the stator is connected. *Increasing the rotor speed above synchronous by increasing the energy input to the turbine will cause the slip to become negative,* reversing the direction of the air-gap power ($P_{gap} = P_{rcl}/s$). Thus, instead of active power being transferred across the air gap from the stator to the rotor, as in motor action, a negative slip causes the power to be transferred in the reverse direction, from the rotor to the stator, and on into the distribution system.

Figure 5.19 shows plots of air-gap power, developed torque, and line current vs. rotor speed for induction-machine operation, as a motor below synchronous speed and as a generator above synchronous speed. The curves were plotted from calculated data using the four-pole, 40-hp, 460-V, 60-Hz motor in Example 5.3. Note the smooth transition of power from motor action to generator action as the shaft speed is raised above synchronous speed.

When operating as an induction generator, the interaction of the magnetic flux of the stator and the magnetic flux of the rotor produce a countertorque in opposition to the driving torque of the prime mover. As the speed of the prime mover is increased, the increase in electrical power supplied to the distribution system by the induction generator causes an increase in countertorque. The countertorque increases with increasing speed until it attains a maximum value called the *pushover torque.* The pushover torque when operating as an induction generator corresponds to the breakdown torque when operating as an induction motor.

As shown in Figure 5.19, increasing the prime-mover speed beyond the pushover point causes the power output to decrease. This decreases the load (countertorque) on the prime mover, causing a rapid increase in speed. The resultant overspeed can damage both the induction machine and the prime mover. Automatic closure of the turbine valve is necessary to prevent damaging overspeed when overloading past the pushover point. A similar effect occurs when an induction generator is loaded and the breaker connecting the generator to the power system is tripped. The sudden loss of load causes the turbine to overspeed.

To assure some degree of protection against damage due to accidental overspeed, NEMA specifications require that squirrel-cage and wound-rotor motors (except crane motors) be so constructed that, in an emergency, they will be able to withstand without mechanical injury the respective overspeeds listed in Table 5.11 **[9]**.

Induction-Generator Starting and Operation with Other Three-Phase Sources

Normal procedure for starting induction generators to be paralleled with the bus is to use the prime mover to accelerate the machine to synchronous speed or slightly above, and then close the circuit breaker. No special care is required, since it generates no measurable voltage until the breaker is closed; closing the breaker causes the three-phase bus to supply the magnetizing current necessary to establish the rotating field.

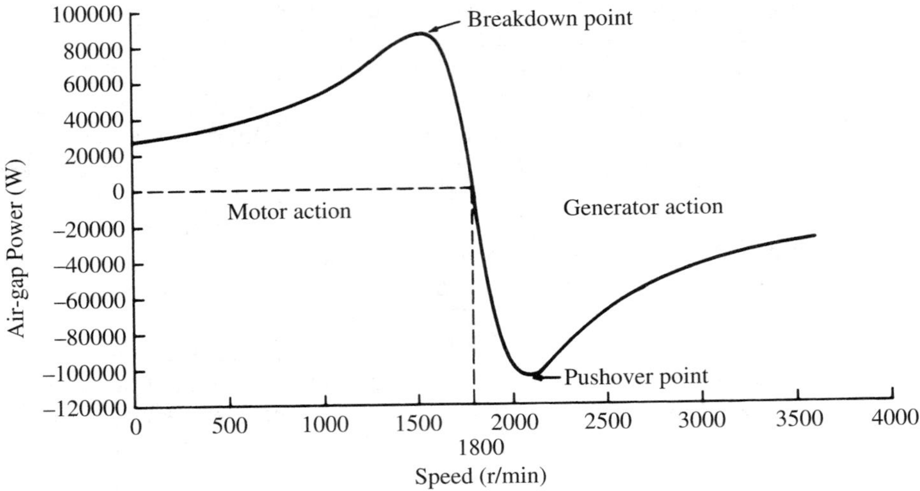

(a)

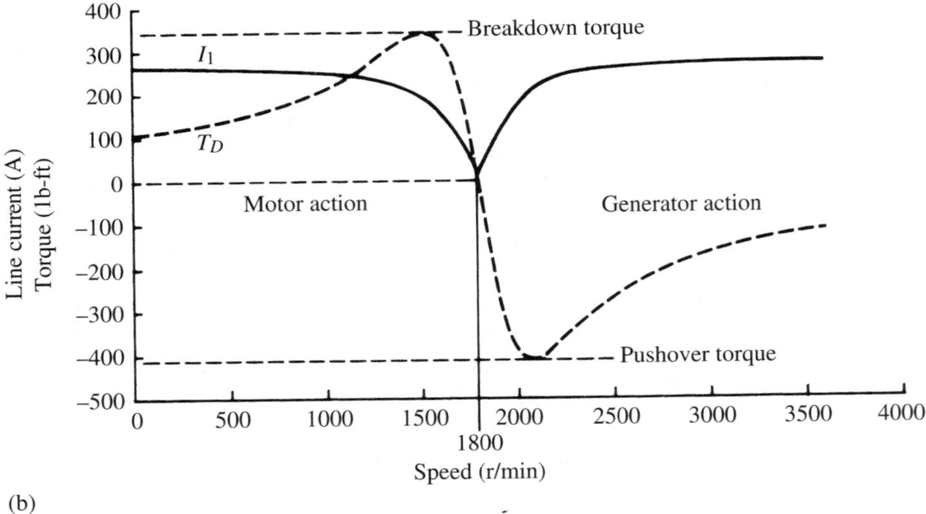

(b)

FIGURE 5.19
Plots of air-gap power, developed torque, and line current vs. rotor speed for induction-machine operation.

The voltage and frequency of an induction generator that is connected to an energized power system cannot be adjusted; its frequency and voltage are the respective frequency and voltage of the electrical system to which it is connected. The only control over the induction generator is load control through speed adjustment of the prime mover.

TABLE 5.11
Allowable emergency overspeed of squirrel-cage and wound-rotor motors

hp	Synchronous Speed (r/min)	Percent above Synchronous Speed
≤200	1201 and over	25
	1200 and below	50
250–500,	1801 and over	20
inclusive	1800 and below	25

EXAMPLE 5.17 A three-phase, six-pole, 460-V, 60-Hz induction motor rated at 15-hp, 1182 r/min is driven by a turbine at 1215 r/min, as shown in Figure 5.18. The equivalent circuit diagram is shown in Figure 5.20 and the motor parameters (in ohms) are:

$$R_1 = 0.200 \qquad R_2 = 0.250 \qquad R_{fe} = 317$$
$$X_1 = 1.20 \qquad X_2 = 1.29 \qquad X_M = 42.0$$

Determine the active power that the motor, driven as an induction generator, delivers to the system. *Note:* This is the same machine driven as a motor in Example 5.2.

Solution

$$s = \frac{n_s - n_r}{n_s} = \frac{1200 - 1215}{1200} = -0.0125$$

$$Z_2 = \frac{R_2}{s} + jX_2 = -0.20 + j1.29 = 20.0416\underline{/176.30°}\ \Omega$$

Note: The apparent equivalent resistance is negative ($R_2/s = -20$) when operating as an induction generator.

$$\mathbf{Z}_0 = \frac{R_{fe} \cdot jX_M}{R_{fe} + jX_M} = \frac{317(42.0\underline{/90°})}{317 + j42.0} = 41.6361\underline{/82.4527°} = 5.4687 + j41.2754 \quad \Omega$$

$$Z_P = \frac{Z_2 \cdot Z_0}{Z_2 + Z_0} = \frac{(20.0416\underline{/176.30°})(41.6361\underline{/82.4527°})}{(-0.20 + j1.29) + 5.4687 + j41.2754)}$$

$$Z_P = 18.5527\underline{/149.91°} = -16.0530 + j9.301 \quad \Omega$$

$$Z_{in} = Z_1 + Z_P = (0.2 + j1.2) + (-16.0530 + j9.301) = 19.0153\underline{/146.4802°} \quad \Omega$$

$$I_1 = \frac{V}{Z_{in}} = \frac{460/\sqrt{3}\underline{/0°}}{19.0153\underline{/146.4802°}} = 13.9667\underline{/-146.4802°} \quad \text{A}$$

$$S = 3VI_1^* = 3(460/\sqrt{3}\underline{/0°})(13.9667\underline{/-146.4802°})^* = 11{,}127.87\underline{/146.48°} \quad \text{VA}$$

$$S = -9277 + j6145 \quad \text{VA}$$

$$P = \underline{-9277\ \text{W}}$$

The negative sign indicates power flow *from* the induction machine *to* the electrical distribution system.

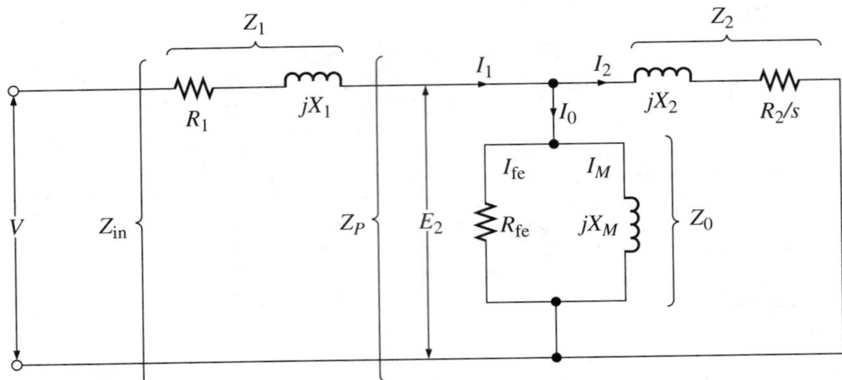

FIGURE 5.20
Equivalent circuit for Example 5.17.

Isolated-Generator Operation

The buildup of voltage in an induction generator that is isolated from other power sources requires the use of capacitors in parallel with the stator windings, as shown in Figure 5.21. The capacitors provide the magnetizing current that is necessary for the

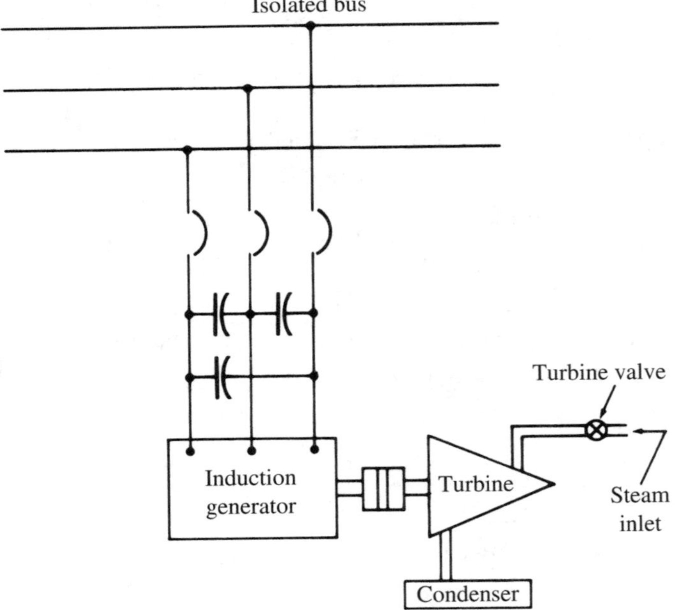

FIGURE 5.21
Capacitors in parallel with stator windings for single induction-generator operation.

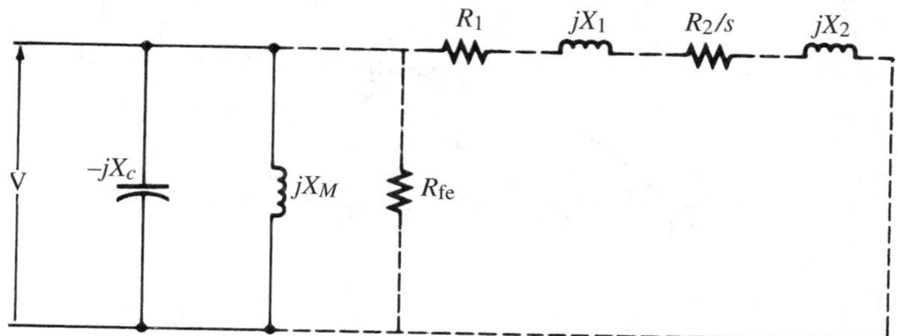

FIGURE 5.22
Approximate equivalent circuit for a self-excited induction generator.

buildup of a rotating magnetic field. Induction generators that do not depend on the power system to establish a rotating magnetic field are called *self-excited* generators.

The theory of voltage buildup is explained with the aid of the approximate equivalent-circuit diagram in Figure 5.22, which shows a capacitor connected across the output terminals. Resistance R_{fe}, shown with dotted lines, is relatively high compared to X_M and can be ignored, since it has very little effect on voltage buildup. Furthermore, with no load on the machine, the slip is zero, causing R_2/s to be infinite, and no current appears in the branch formed by R_1, jX_1, R_2/s, and jX_2, so this branch is also drawn with dotted lines and ignored.

When the prime mover is started, the residual magnetism[11] in the revolving rotor generates a low voltage in magnetizing reactance X_M. This low voltage appears across the capacitor, causing a small current in the circuit formed by X_M and the capacitor. The small current in X_M causes voltage $I_M X_M$ that appears at the terminals as output voltage V; this voltage is higher than the residual voltage, causing a higher capacitor current, which also appears in X_M, causing a further voltage buildup, and so forth.

The process of voltage buildup is illustrated in Figure 5.23(a), which shows the magnetization curve and capacitance line for a representative induction generator. The magnetization curve is a plot of voltage vs. current for magnetizing reactance X_M, and is obtained by running the induction machine as a motor with no load, as shown in Figure 5.23(b), and plotting applied voltage vs. current as the voltage is raised in steps from zero voltage to rated voltage. The nonlinearity of the magnetization curve is caused by magnetic saturation effects in the stator and rotor iron. The capacitance line is an Ohm's law plot of voltage vs. current for the capacitor when measured alone, as shown in Figure 5.23(c). For the capacitor line shown in Figure 5.23(a), $V = IX_C$, where X_C is the slope of the line.

---

[11]An induction generator that has lost its residual magnetism will not build up voltage if operating as a single generator. Switching on a capacitor bank with some residual voltage, however, or running the machine briefly as a motor, will be sufficient to build up flux.

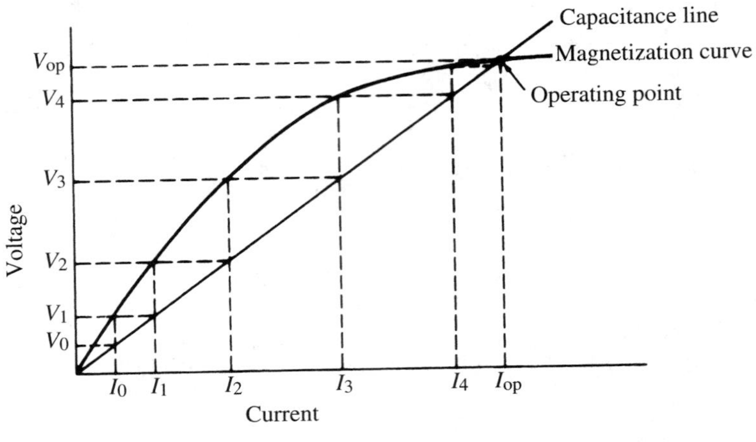

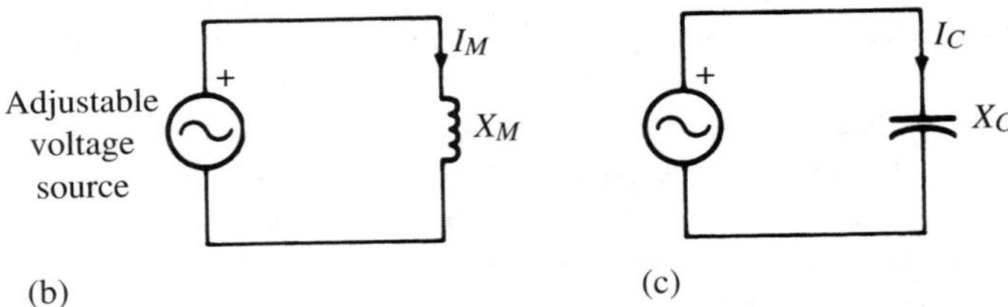

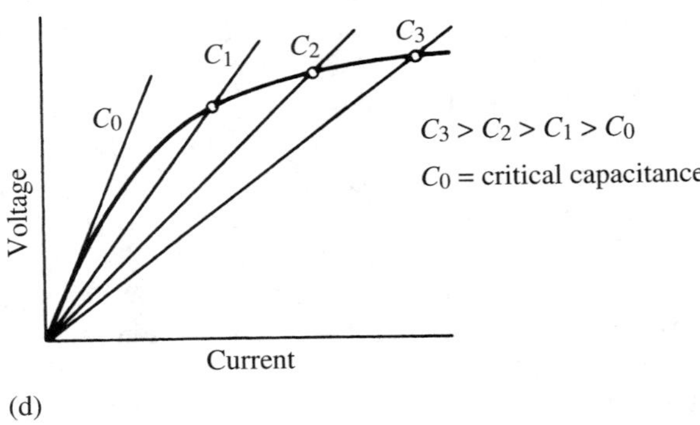

FIGURE 5.23
(a) Process of voltage buildup in a self-excited induction generator; (b) circuit for obtaining magnetization curve; (c) circuit for obtaining capacitance line; (d) operating points for several values of capacitance.

Voltage Buildup

Referring to Figure 5.23(a), the voltage V_0 due to residual magnetism causes current I_0 in the capacitor. Current I_0 also appears in reactance X_M, which as indicated on the magnetizing curve, causes the voltage across it to rise to V_1. Voltage V_1 across the capacitor causes higher current I_1, which in turn causes the reactance voltage to rise to V_2, and so forth. The point of intersection of the magnetization curve and the capacitance line is the no-load operating point of the induction generator. Although described as a step process, the actual buildup is a smooth and fairly rapid rise to the operating point.

Adjustment of the no-load voltage is accomplished by changing the slope of the capacitance line. Since X_C is the slope of the capacitance line, and $X_C = 1/(2\pi fC)$, increasing the capacitance decreases the slope and therefore increases the voltage. The operating points for several values of capacitance are shown in Figure 5.23(d). The value of capacitance that causes the capacitance line to be tangent to the magnetization curve is called the critical capacitance. Values of capacitance less than the critical capacitance C_0 will prevent the buildup of voltage.

The frequency of an induction generator when operating self-excited may be determined from the capacitive reactance at the operating point. Thus, from Figure 5.23(a) and Ohm's law,

$$X_C = \frac{V_{\text{op}}}{I_{\text{op}}} \quad \Rightarrow \quad \frac{1}{2\pi fC} = \frac{V_{\text{op}}}{I_{\text{op}}}$$

$$f = \frac{I_{\text{op}}}{V_{\text{op}}} \cdot \frac{1}{2\pi C} \tag{5–57}$$

The principal disadvantage of self-excited induction generators is the rapid falloff in voltage when load is applied, requiring significantly higher capacitance with higher and more lagging power-factor loads. Despite this drawback, however, self-excited induction generators are very effective for battery-charging systems and other applications where radical changes in terminal voltage are not disturbing to equipment operation.

5.19 DYNAMIC BRAKING OF INDUCTION MOTORS

Dynamic braking is the slowing down of a machine by converting the kinetic energy stored in the rotating mass to heat energy in the rotor and/or stator windings. To do this, the motor is switched from the line to a braking circuit that causes the motor to behave as a generator with a connected load; the load is the resistance of the rotor and/or stator windings. Since the only source of energy is the rotating parts of the induction motor and the driven equipment, the machine will slow down. Dynamic braking of induction motors may be accomplished by DC injection and/or capacitive braking. Note, however, that this type of braking does not provide holding torque at the end of the braking period. Hence, where required, as in hoists or other applications where rolling is not permitted after a stop, a mechanical brake must be used to hold the shaft.

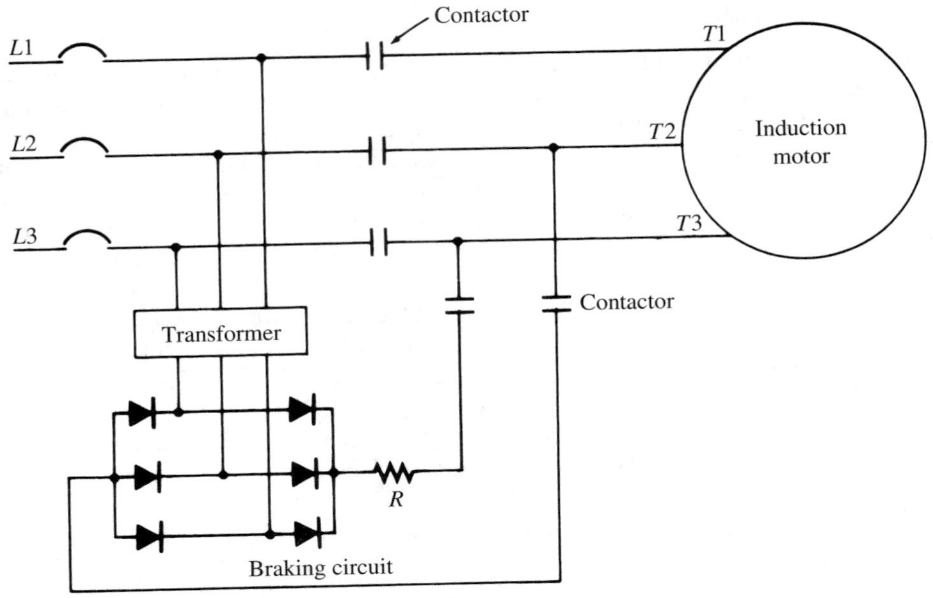

FIGURE 5.24
Dynamic braking using DC injection.

DC Injection

In DC injection the motor is disconnected from the line, and a DC source supplied by a rectifier is then connected to any two terminals of the stator through a current-limiting resistor, as shown in Figure 5.24. The direct current in the stator winding sets up a stationary magnetic field that generates a voltage in the windings of the spinning rotor. The resultant current in the closed loops formed by the squirrel cage (or wound rotor) dissipates the rotational energy as I^2R losses, rapidly slowing the rotor.

The rate of deceleration by DC injection may be adjusted by adjusting resistor R in Figure 5.24, by using a variable ratio transformer, or by using a thyristor (SCR) control circuit in place of the resistor **[10]**.

Capacitor Braking

In capacitor braking, the motor is disconnected from the line, and a capacitor bank is then connected to the stator terminals, as shown in Figure 5.25. When braking, the motor behaves as a self-excited induction generator as described in Section 5.18. During capacitive braking, the rotational energy is dissipated as I^2R losses in both stator and rotor windings. The braking effect can be increased by adding a resistor load, as shown with broken lines in Figure 5.25.

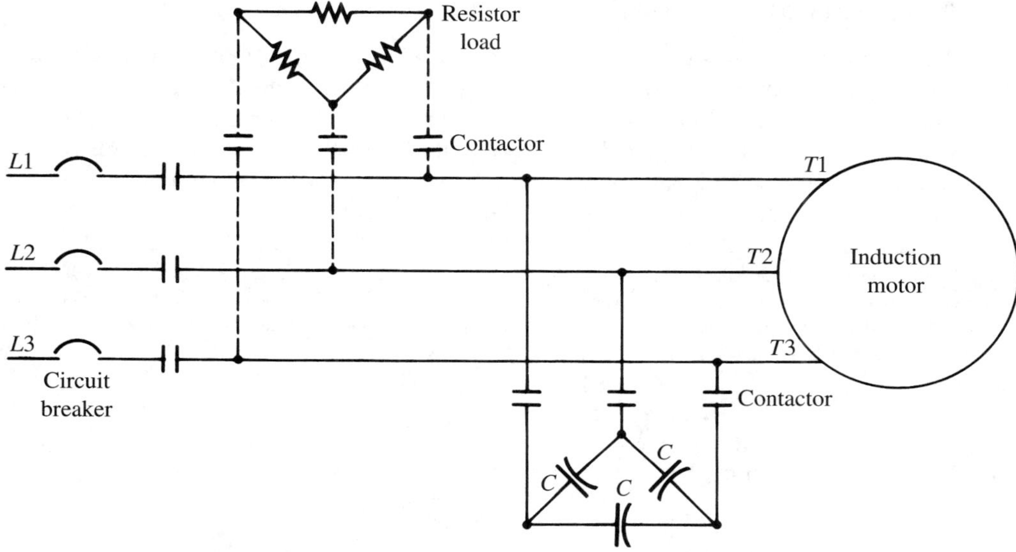

FIGURE 5.25
Dynamic braking using capacitors.

5.20 INDUCTION-MOTOR STARTING

Induction motors of almost any horsepower may be started by connecting them across full voltage, as shown in Figure 5.26, and most are started that way. In many cases, however, the high *in-rush current* associated with full-voltage starting can cause large voltage dips in the distribution system; lights may dim or flicker, unprotected control systems may drop out due to low voltage, and unprotected computers may go off line or lose data. Furthermore, the impact torque that occurs when starting at full voltage can, if high enough, damage gears and other components of the driven equipment.

The methods commonly used for reducing in-rush current are reduced voltage starting using autotransformers, current limiting through wye-delta connections of stator windings, part-winding connections, series impedance, and solid-state control.

FIGURE 5.26
Full-voltage starting.

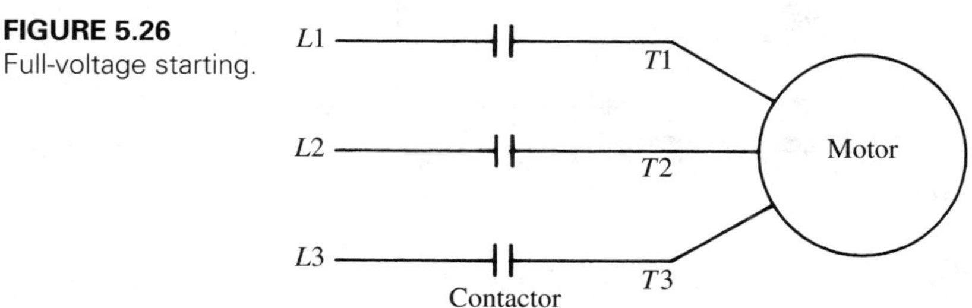

Autotransformer Starting

Autotransformer starting, shown in Figure 5.27(a), uses a wye-connected autotransformer (called a compensator) with taps that provide 50, 65, and 80 percent of full voltage.[12] The motor is started at reduced voltage by closing the *S* contacts. When near rated speed, the *S* contacts are opened and the *R* contacts are closed, connecting full voltage across the stator. The big advantage of autotransformer starting is that it draws less line current for the same amount of motor-starting current than does any other starter.

---

[12]Delta-connected autotransformers are not used for reduced-voltage starting because ratios higher than 2 to 1 cannot be obtained, and the required power rating for ratios less than 2 to 1 is greater than that for wye-connected and open-delta autotransformers with the same power output. Furthermore, for ratios other than 2 to 1, there is a phase shift between primary and secondary **[11]**.

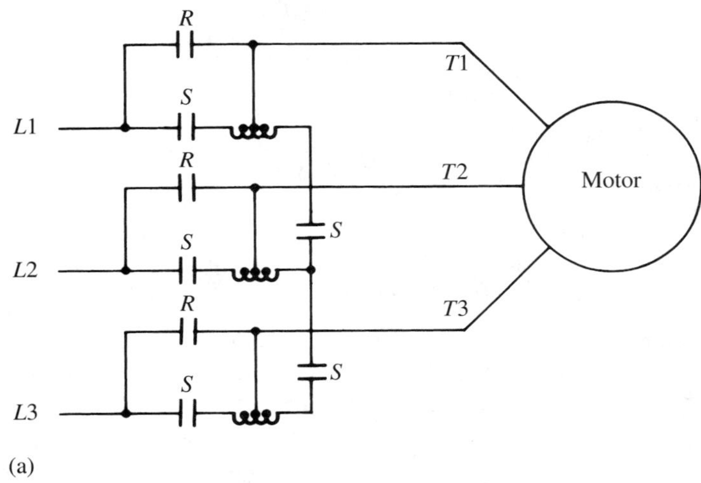

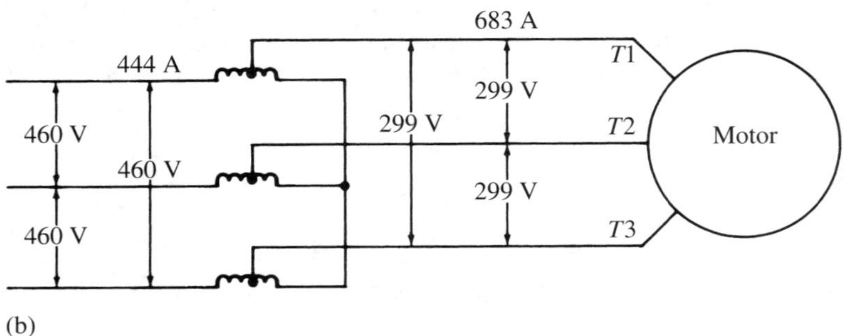

FIGURE 5.27
(a) Circuit for autotransformer starting; (b) current and voltage distributions for Example 5.18.

EXAMPLE 5.18 A three-phase, 125-hp, 460-V, 156-A, 60-Hz, six-pole 1141 r/min, design *B* motor with code letter H is to be started at reduced voltage using an autotransformer with a 65 percent tap. Determine (a) locked-rotor torque and expected average in-rush current to the stator if the motor is started at rated voltage; (b) repeat part (a) assuming the motor is started at reduced voltage using an autotransformer with a 65 percent tap; (c) the in-rush line current when starting at reduced voltage.

Solution

(a)

$$P = \frac{Tn}{5252} \quad \Rightarrow \quad 125 = \frac{T(1141)}{5252}$$

$$T_{\text{rated}} = 575.37 \text{ lb-ft}$$

From Table 5.1 the minimum locked-rotor torque for a 125-hp design *B*, six-pole motor is 125 percent rated torque.

$$T_{\text{lr,460}} = 575.37 \times 1.25 = \underline{719.2 \text{ lb-ft}}$$

From Table 5.9 the average locked-rotor kVA/hp that may be expected for a code H motor is

$$\frac{6.3 + 7.1}{2} = 6.70 \text{ kVA/hp}$$

The expected average in-rush current to the stator is

$$\text{kVA/hp} \times 1000 \times \text{hp} = \sqrt{3}\, V_{\text{line}} I_{\text{line}}$$

$$6.70 \times 1000 \times 125 = \sqrt{3} \times 460 \times I_{\text{lr,460}}$$

$$I_{\text{lr,460}} = \underline{1051 \text{ A}}$$

(b) With the 65 percent tap, the voltage impressed across the stator at locked rotor is

$$V_2 = 0.65 \times 460 = 299 \text{ V}$$

Since the motor input impedance is constant at locked rotor, the average in-rush current to the stator will be proportional to the voltage applied to the stator. Thus,

$$I = 1051 \times 0.65 = 683 \text{ A}$$

The locked-rotor torque is proportional to the square of the voltage. Thus,

$$T_2 = T_1 \cdot \left(\frac{V_2}{V_1}\right)^2 = 719.2 \times \left(\frac{0.65 \times 460}{460}\right)^2 = \underline{303.9 \text{ lb-ft}}$$

The current and torque curves for starting at rated voltage and 65 percent rated voltage are shown in Figures 5.28(a) and (b), respectively. The transition from low voltage to

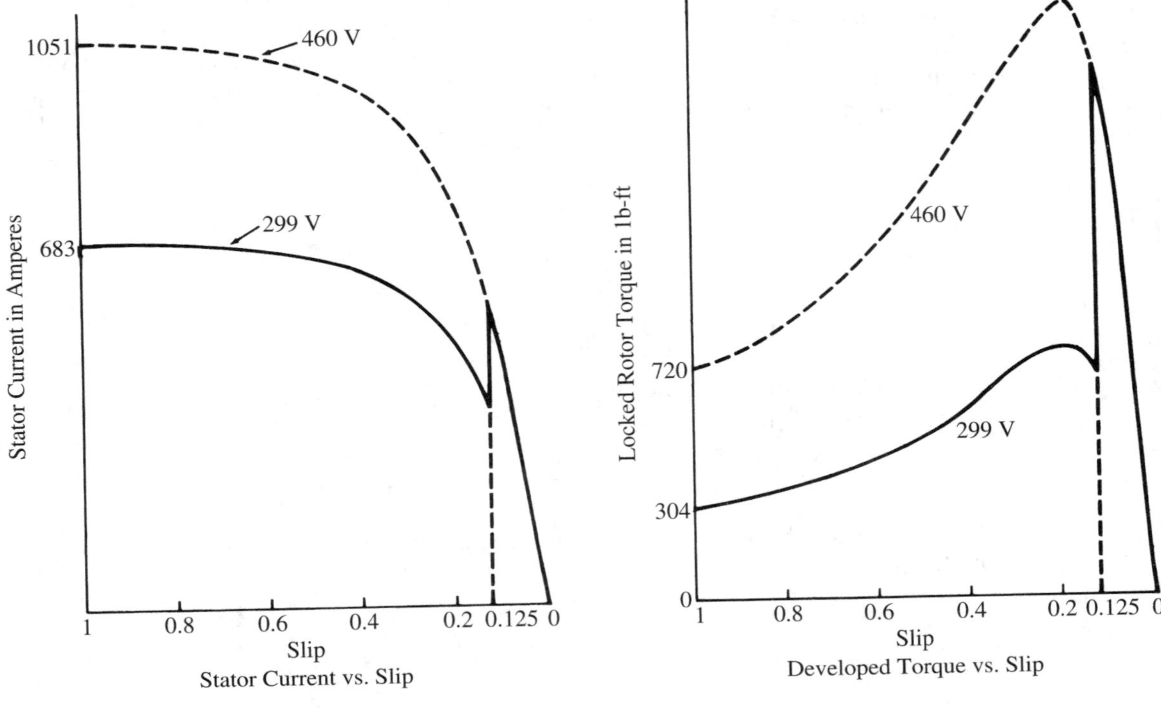

FIGURE 5.28
Current and torque curves for Example 5.18.

rated voltage for this starter is assumed to occur at $s = 0.125$. This corresponds to a shaft speed of

$$n_r = n_s(1 - s) = 1200(1 - 0.125) = 1050 \text{ r/min}$$

The solid lines indicate the overall behavior of the machine when using autotransformer starting.

(c) The bank ratio for a wye-connected autotransformer equals the turns ratio. Thus,

$$a = \frac{V_{\text{HS,line}}}{V_{\text{LS,line}}} = \frac{V_{\text{HS,line}}}{0.65 \cdot V_{\text{HS,line}}} = \frac{1}{0.65}$$

$$I_{\text{HS}} = \frac{I_{\text{LS}}}{a} = 683 \times 0.65 = \underline{444 \text{ A}}$$

Wye-Delta Starting

The circuit for wye-delta starting, also called star-delta starting, is shown in Figure 5.29. The three phases of the stator are connected in wye during start-up, and then

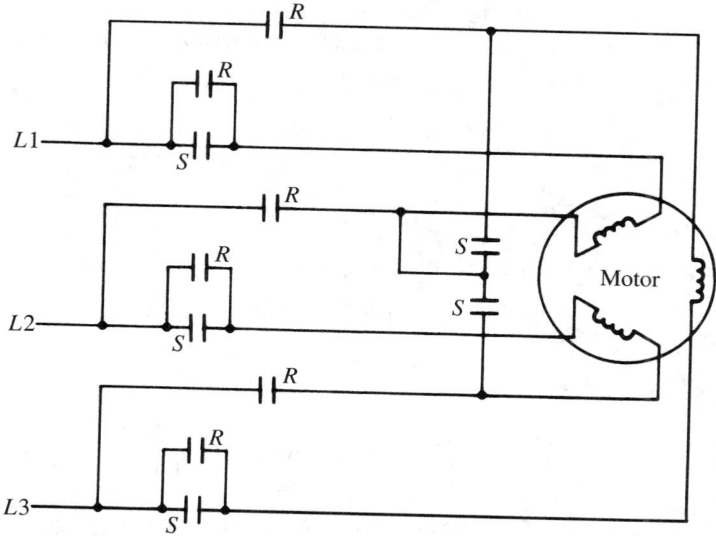

FIGURE 5.29
Wye-delta starter.

reconnected in delta when the starting current is sufficiently decreased. All motors connected for wye-delta starting are actually delta-connected machines that are only wye connected for starting purposes. The S contacts provide the wye connection, and the R contacts provide the delta connection. When connected in wye, the voltage across each phase of the stator winding is $V_{\text{line}}/\sqrt{3}$.

EXAMPLE 5.19 A 60-hp, 460-V, 60-Hz, 77-A, three-phase, 1750 r/min, design B motor has a locked-rotor impedance of $0.547\underline{/69.1^\circ}$ Ω/phase. Assuming the machine is connected for wye-delta starting, determine (a) the locked-rotor current per phase and the expected minimum locked-rotor torque when starting; (b) the locked-rotor current per phase, assuming the motor is started delta connected; (c) the code letter.

Solution

(a) The voltage per phase when wye connected is

$$\frac{460}{\sqrt{3}} = 265.6 \text{ V}$$

The corresponding locked-rotor current per phase is

$$I_{\text{lr}} = \frac{V}{Z} = \frac{460/\sqrt{3}}{0.547} = \underline{485.5 \text{ A}}$$

The minimum locked-rotor torque at rated voltage and rated frequency for a 1750 r/min, 60-hp, design B machine, as determined from Table 5.1, is 140 percent full-load torque. Full-load torque at rated voltage is

$$\text{hp} = \frac{Tn}{5252} \qquad \Rightarrow \qquad 60 = \frac{T(1750)}{5252}$$

$$T_{\text{rated}} = 180 \text{ lb-ft}$$

The minimum expected locked-rotor torque at rated voltage (delta connected) is

$$T_{\text{lr}(460)} = 1.4 \times 180 = 252 \text{ lb-ft}$$

Thus, if wye connected,

$$\frac{T_2}{T_1} = \left(\frac{V_2}{V_1}\right)^2 \qquad \Rightarrow \qquad T_2 = 252 \times \left(\frac{460/\sqrt{3}}{460}\right)^2 = \underline{84 \text{ lb-ft}}$$

(b) If started delta connected (full-voltage starting), the locked-rotor current per phase would be

$$I_{\text{lr},\Delta} = \frac{V}{Z} = \frac{460}{0.547} = 840.95 \text{ A/phase}$$

The corresponding line current is

$$840.95 \times \sqrt{3} = \underline{1457 \text{ A}}$$

(c) The code letter is determined at rated voltage. Thus,

$$S_{\text{lr}} = \sqrt{3} \times 460 \times \frac{1457}{1000} = 1161 \text{ kVA}$$

$$\text{kVA/hp} = 1161/60 = 19.4$$

This corresponds to Code T in Table 5.9.
Note: The current and torque curves for wye-delta starting are similar to those in Figure 5.28 for autotransformer starting.

Series-Impedance Starting

The series-impedance starter, shown in Figure 5.30(a), uses a resistor or an inductor in series with each phase of the stator windings to limit the current during start-up. The running contacts (R) are open when starting, to limit the in-rush current, and are closed to short out the impedance when the motor is near rated speed. The ohmic value of the resistor or reactor is generally selected to provide approximately 70 percent

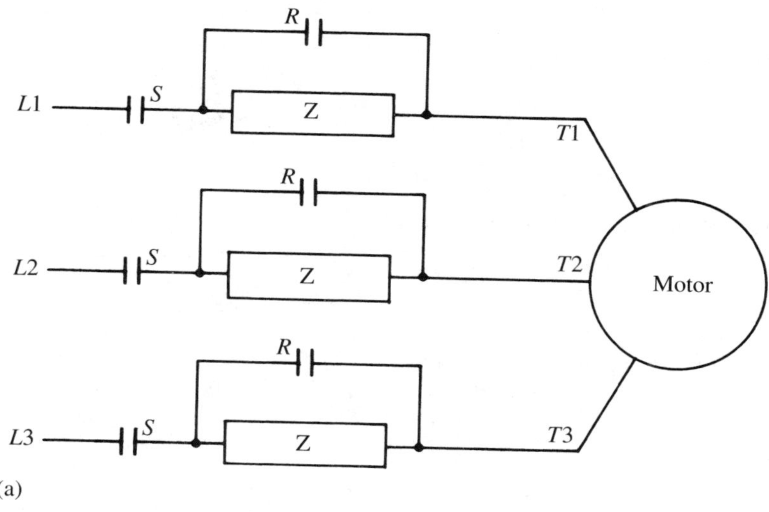

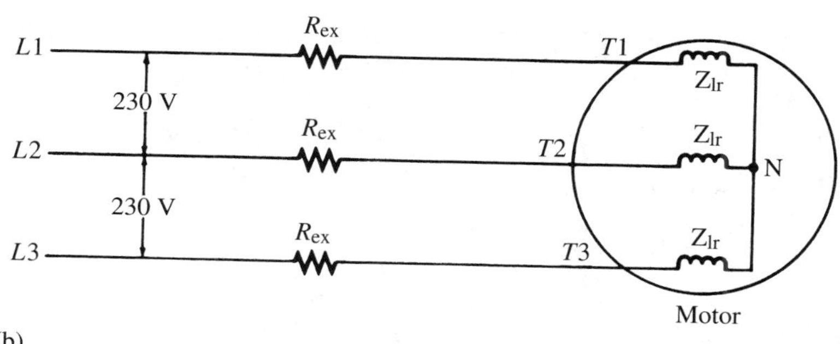

FIGURE 5.30
(a) Series-impedance starter; (b) circuit for Example 5.20.

rated voltage at the motor terminals when starting. The series-impedance starter provides smooth acceleration and is the simplest method for starting induction motors.

EXAMPLE 5.20 A 30-hp, wye-connected, three-phase, 230-V, 60-Hz, 78-A 1748 r/min, design *B* motor has a locked-rotor impedance per phase of $0.273\angle 69^\circ\ \Omega$ at rated temperature and frequency. The motor is to be started using series resistors in each line. Determine (a) the resistance of the resistors required to limit the locked-rotor current to three times rated current; (b) the *stator voltage* per phase at locked rotor; (c) the expected minimum locked-rotor torque when starting as a percent of rated torque.

Solution

(a) Referring to Figure 5.30(b), where R_{ex} is the external starting resistance, the impedance of one branch, from line to neutral, is

$$\mathbf{Z} = R_{ex} + \mathbf{Z}_{lr} = R_{ex} + 0.273\underline{/69^\circ}\ \Omega$$
$$\mathbf{Z} = R_{ex} + 0.0978 + j0.2549\ \Omega$$

The voltage from line to neutral is $230/\sqrt{3} = 132.79$ V, and the locked-rotor current is to be limited to

$$I_{lr} = 3 \times 78 = 234\ \text{A}$$

The three-phase stator and series-connected resistors present a balanced three-phase circuit. Hence, the magnitude of the current in the respective branches are equal. Applying Ohm's law to one branch,

$$I_{lr} = \frac{V_{branch}}{Z_{branch}} \quad \Rightarrow \quad 234 = \frac{132.79}{\sqrt{(R_{ex} + 0.0978)^2 + (0.2549)^2}}$$

$$R_{ex} = \underline{0.4093\ \Omega}$$

(b) The stator voltage per phase when starting is the voltage between $T1$ and N in Figure 5.30(b); it is the impedance drop across one phase of the motor winding:

$$V_{T1-N} = IZ_{lr} = 234 \times 0.273 = \underline{63.88\ \text{V}}$$

(c) The minimum locked-rotor torque at rated voltage and frequency from Table 5.1 is 150 percent rated torque.

$$T_{lr} = \underline{1.5\ T_{rated}}$$

Note: Calculations must always be made on the basis of the *actual voltage applied to the stator.* Thus,

$$\frac{1.5T_{rated}}{T_{63.88}} = \left[\frac{132.79}{63.88}\right]^2$$

$$T_{lr,63.88} = 0.347T_{rated} \quad \text{or} \quad 34.7\%\ T_{rated}$$

EXAMPLE 5.21

A certain 208-V, 7.5-hp, 60-Hz, four-pole, wye-connected, design B induction motor with Code letter H, has a rated current of 24 A at 1722 r/min. The motor is to be started using the series-impedance method with inductors in each line. Calculate the inductance and voltage rating of each series-connected inductor required to limit the starting current to approximately $2 \times I_{rated}$.

Solution

The circuit is similar to that shown in Figure 5.30(b), with R_{ex} replaced by jX_{ex}. A rough approximation of the motor locked-rotor impedance may be calculated from motor horsepower, code letter, and voltage rating. From Table 5.9, the average locked-rotor kVA/hp for Code letter H is

$$\frac{6.3 + 7.1}{2} = 6.7 \text{ kVA/hp}$$

The corresponding approximate locked-rotor current is

$$\text{kVA/hp} \times 1000 \times \text{hp} = \sqrt{3}\, V_{\text{line}} I_{\text{line}}$$
$$6.7 \times 1000 \times 7.5 = \sqrt{3} \times 208\, I_{\text{lr}}$$
$$I_{\text{lr}} = 139.5 \text{ A}$$

For a wye-connected stator, $I_{\text{phase}} = I_{\text{line}}$. Applying Ohm's law to one phase,

$$Z_{\text{lr}} \cong \frac{V_{\text{phase}}}{I_{\text{lr}}} = \frac{208/\sqrt{3}}{139.5} = \underline{0.861\ \Omega}$$

Assuming a phase angle of 70° for the locked-rotor impedance of a design B machine,[13]

$$\mathbf{Z}_{\text{lr}} \approx 0.861\angle 70^\circ \approx 0.294 + j0.809\ \Omega$$

The external inductor required to limit the current to $2 \times I_{\text{rated}}$ is determined from Ohm's law. Assuming the inductor has negligible resistance,

$$\mathbf{I} = \frac{\mathbf{V}}{\mathbf{Z}} = \frac{\mathbf{V}}{R_{\text{lr}} + jX_{\text{lr}} + jX_{\text{ex}}} \quad \Rightarrow \quad |\mathbf{I}| = \frac{|\mathbf{V}|}{\sqrt{R_{\text{lr}}^2 + (X_{\text{lr}} + X_{\text{ex}})^2}}$$

[13] Design A and design B machines have relatively low resistance and high reactance at locked rotor, resulting in an impedance phase angle of approximately 75°. The higher resistance to reactance ratio for design D and other high-slip machines at locked rotor results in a locked-rotor impedance angle of approximately 50°.

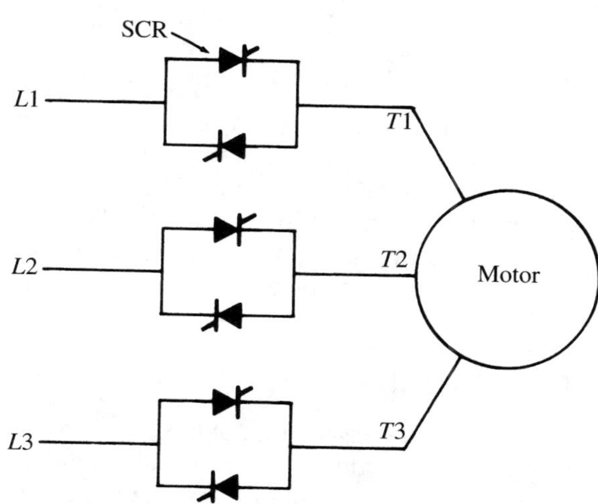

FIGURE 5.31
Solid-state starter.

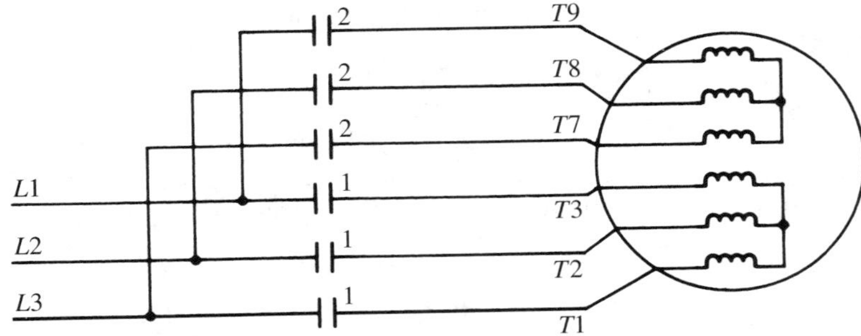

FIGURE 5.32
Part-winding starter.

$$2 \times 24 = \frac{208/\sqrt{3}}{\sqrt{(0.294)^2 + (0.809 + X_{ex})^2}}$$

$$X_{ex} = 1.675\ \Omega$$

$$X_{ex} = 2\pi f L \quad \Rightarrow \quad 1.675 = 2\pi 60 L \quad \Rightarrow \quad L = \underline{4.44\text{ mH}}$$

$$\text{Voltage rating} = IX_L = 2 \times 24 \times 1.675 = \underline{80.4\text{ V}}$$

Solid-State Starting

A solid-state starter, shown in Figure 5.31, uses back-to-back thyristors (SCRs) to limit the current. The control circuitry (not shown) allows a gradual buildup of current. The smooth buildup permits a soft start with no impact loading and no significant voltage dips. Solid-state starters can be designed to incorporate many special features, such as speed control, power factor control, protection against overload, and single phasing.

Part-Winding Method

The part-winding method uses a stator with two identical three-phase windings, each capable of supplying one-half of the rated power. The power circuit for starting a part-winding motor is shown in Figure 5.32. Contacts 1 are closed first, energizing one winding. After a brief time delay, contacts 2 are closed, energizing both windings. The part-winding starter is one of the least expensive starters, but is limited to dual-voltage motors that are operated on the low-voltage connections.

5.21 MOTOR BRANCH CIRCUITS

When planning the installation of electric motors, consideration must be given to the ratings of disconnecting devices, branch-circuit overcurrent and ground-fault protection devices, motor running-overload protection devices, motor, motor controller, rheostat if wound rotor, and all connecting conductors. The correct selection of protective devices, cable sizes, and disconnecting devices for a particular motor application requires familiarization with Article 430 of the National Electric Code (NEC) **[8].** Adherence to the NEC requirements is absolutely essential; failure to do so may result in damage to equipment and injury to personnel.

SUMMARY OF EQUATIONS FOR PROBLEM SOLVING

Note: If wound-rotor, replace R_2 with $(R_2 + R'_{\text{rheo}})$.

Exact Solutions

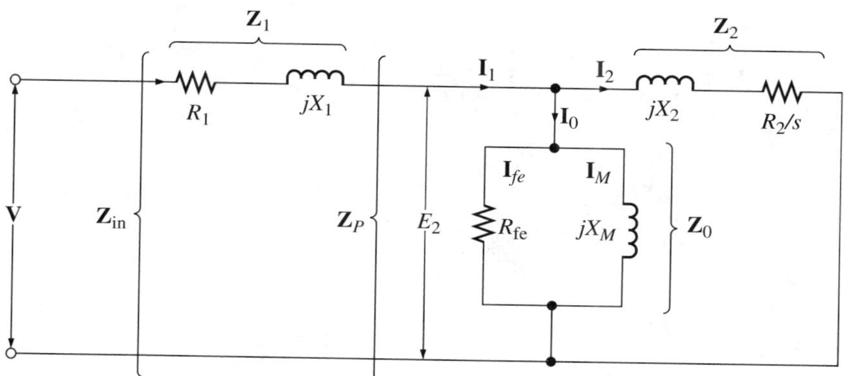

$$s = (n_s - n_r)/n_s \qquad n_r = n_s(1 - s)$$

$$\mathbf{Z}_2 = \frac{R_2}{s} + jX_2 \qquad \mathbf{Z}_0 = \frac{R_{\text{fe}} \cdot jX_M}{R_{\text{fe}} + jX_M}$$

$$\mathbf{Z}_P = \frac{\mathbf{Z}_2 \cdot \mathbf{Z}_0}{\mathbf{Z}_2 + \mathbf{Z}_0} \qquad \mathbf{Z}_{\text{in}} = \mathbf{Z}_1 + \mathbf{Z}_P$$

$$\mathbf{I}_1 = \frac{\mathbf{V}}{\mathbf{Z}_{\text{in}}} \qquad \mathbf{E}_2 = \mathbf{I}_1 \cdot \mathbf{Z}_P \qquad \mathbf{I}_2 = \frac{\mathbf{E}_2}{\mathbf{Z}_2}$$

$$P_{\text{scl}} = 3I_1^2R_1 \qquad P_{\text{rcl}} = 3I_2^2R_2 \qquad P_{\text{core}} = 3E_2^2/R_{\text{fe}}$$

$$P_{\text{gap}} = \frac{P_{\text{rcl}}}{s} \qquad P_{\text{mech}} = \frac{P_{\text{rcl}}(1 - s)}{s}$$

$$\text{Loss} = P_{\text{scl}} + P_{\text{rcl}} + P_{\text{core}} + P_{f,w} + P_{\text{stray}}$$

$$T_D = \frac{21.12 \cdot I_2^2R_2}{s \cdot n_s} \qquad \text{hp} = \frac{Tn}{5252} \tag{5-1}$$

Close Approximations Using Approximate Equivalent Circuit

Note: If wound rotor, replace R_2 with $(R_2 + R'_{\text{rheo}})$.

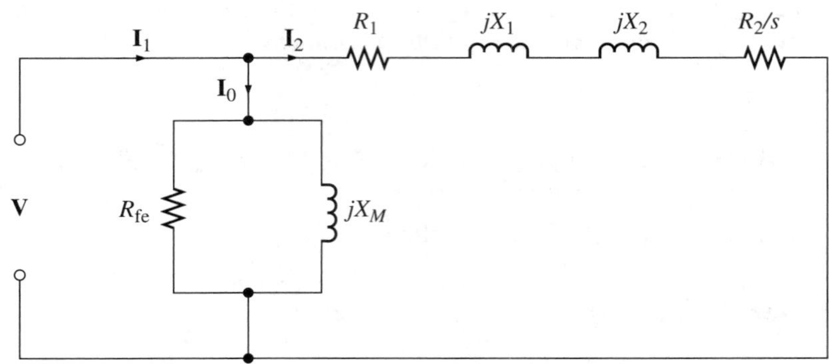

$$I_2 \cong \frac{V}{\sqrt{[(R_1 + R_2/s)^2 + (X_1 + X_2)^2]}} \tag{5–4}$$

$$T_D \cong \frac{21.12V^2 R_2/s}{[(R_1 + R_2/s)^2 + (X_1 + X_2)^2]n_s} \tag{5–5}$$

$$s_{T_{D,\max}} \cong \frac{R_2}{\sqrt{R_1^2 + (X_1 + X_2)^2}} \tag{5–7}$$

$$T_{D,\max} \cong \frac{21.12V^2}{2n_s[\sqrt{R_1^2 + (X_1 + X_2)^2} + R_1]} \tag{5–10}$$

Approximations for Normal Running and Overload Conditions

Squirrel Cage with $s \leq 0.03$

$$\underset{s \leq 0.03}{I_2} \cong \frac{V \cdot s}{R_2} \qquad \underset{s \leq 0.03}{T_D} \cong \frac{21.12V^2 \cdot s}{R_2 n_s} \tag{5–11, 5–12}$$

$$\underset{s \leq 0.03}{I_2} \propto s \qquad \underset{s \leq 0.03}{T_D} \propto s \tag{5–13, 5–14}$$

Wound-Rotor Motor

$$S_{T_{D,\max}} = \frac{R_2 + R'_{\text{rheo}}}{\sqrt{R_1^2 + (X_1 + X_2)^2}} \tag{5–24}$$

$$R'_{\text{rheo}} = a^2 \cdot R_{\text{rheo}} \tag{5–24a}$$

$$\underset{\text{linear}}{I_2} \propto \frac{s}{R_2 + R'_{\text{rheo}}} \qquad \underset{\text{linear}}{T_D} \propto \frac{s}{R_2 + R'_{\text{rheo}}} \tag{5–28, 5–29}$$

Approximations for Off-Rated Voltage and Frequency

$$\underset{s \leq 0.03}{T_D} \propto \frac{V^2 \cdot s}{f} \qquad \textbf{(5–16)}$$

$$\underset{s=1.0, \theta_z \geq 75°}{I_{lr}} \cong I_2 \propto \frac{V}{f} \qquad \theta_z = \arctan\left(\frac{X_1 + X_2}{R_1 + R_2}\right) \qquad \textbf{(5–23, 5–18)}$$

60-Hz Motors on 50-Hz Systems

$$V_{50} = \frac{5}{6} V_{60} \qquad \textbf{(5–23c)}$$

$$hp_{50} = \frac{5}{6} hp_{60} \qquad \textbf{(5–23d)}$$

$$\left[\frac{T \cdot n_r}{5252}\right]_{50} = \frac{5}{6}\left[\frac{T \cdot n_r}{5252}\right]_{60} \qquad \textbf{(5–23e)}$$

$$\left.\left[\frac{V^2 \cdot s \cdot n_r}{f}\right]_{50} = \frac{5}{6}\left[\frac{V^2 \cdot s \cdot n_r}{f}\right]_{60}\right\} s \leq 0.03 \qquad \textbf{(5–23f)}$$

In-Rush Current

$$\mathbf{I}_{lr,ss} = \left[\frac{\mathbf{V}_{phase}}{\mathbf{Z}_{in}}\right]_{s=1.0} \qquad \textbf{(5–30)}$$

$$i_{lr,ss} = \sqrt{2}\left|\frac{\mathbf{V}_{phase}}{\mathbf{Z}_{in}}\right|_{s=1.0} \sin(2\pi f t - \theta_z) \qquad \textbf{(5–31)}$$

Effect of Unbalanced Line Voltages

$$\%UBV = \frac{V_{max\ dev}}{V_{avg}} \cdot 100 \qquad \%\Delta T \cong 2(\%UBV)^2 \qquad \textbf{(5–35, 5–36)}$$

$$T_{UBV} \cong T_{rated} \cdot \left(1 + \frac{\%\Delta T}{100}\right) \qquad \delta T = T_{UBV} - T_{rated} \qquad \textbf{(5–37)}$$

$$RL \cong \frac{1}{2^{(\delta T/10)}} \qquad \textbf{(5–38)}$$

Per-Unit Conversions

P_{base} = rated three-phase shaft-power output in watts $\div$ 3
V_{base} = rated voltage/phase
I_{base} = base current = rated current/phase

$$I_{base} = \frac{P_{base}}{V_{base}} \qquad Z_{base} = \frac{V_{base}}{I_{base}} = \frac{V_{base}}{P_{base}/V_{base}} = \frac{V_{base}^2}{P_{base}} \tag{5–39, 5–41a}$$

$$\left.\begin{array}{lll} r_1 = \dfrac{R_1}{Z_{base}} & r_2 = \dfrac{R_2}{Z_{base}} & r_{fe} = \dfrac{R_{fe}}{Z_{base}} \\ x_1 = \dfrac{X_1}{Z_{base}} & x_2 = \dfrac{X_2}{Z_{base}} & x_M = \dfrac{X_M}{Z_{base}} \end{array}\right\} \tag{5–41}$$

Determination of Induction Motor Parameters

DC Test

$$R_{1,wye} = \frac{R_{DC}}{2} \qquad R_{1,\Delta} = 1.5R_{DC} \tag{5–42, 5–43}$$

Blocked-Rotor Test

$$Z_{BR,15} = \frac{V_{BR,15}}{I_{BR,15}} \qquad R_{BR,15} = \frac{P_{BR,15}}{I_{BR,15}^2} \tag{5–45, 5–46}$$

$$R_2 = R_{BR,15} - R_1 \qquad X_{BR,15} = \sqrt{Z_{BR,15}^2 - R_{BR,15}^2} \tag{5–47, 5–48}$$

$$X_{BR,60} = \tfrac{60}{15}X_{BR,15} \qquad X_{BR,60} = X_1 + X_2 \tag{5–49, 5–50}$$

No-Load Test

$$S_{NL} = V_{NL}I_{NL} \qquad Q_{NL} = \sqrt{S_{NL}^2 - P_{NL}^2} \qquad X_{NL} = \frac{Q_{NL}}{I_{NL}^2} \tag{5–52, 5–54}$$

$$X_{NL} = X_1 + X_M \qquad P_{NL} = I_{NL}^2 R_1 + P_{core} + P_{w,f} \tag{5–55, 5–56}$$

SPECIFIC REFERENCES KEYED TO TEXT

1. Brancato, E. L. Insulation aging. A critical and historical review. *IEEE Trans. Electrical Insulation,* Vol. EI-13, No. 4, Aug. 1978.
2. Brighton, R. J., Jr., and P. N. Ranade. Why overload relays do not always protect motors. *IEEE Trans. Industry Applications,* Vol. IA-18, No. 6, Nov./Dec. 1982.
3. Hubert, C. I. *Electric Circuits AC/DC: An Integrated Approach.* McGraw-Hill, New York, 1982.
4. Hubert, C. I. *Preventive Maintenance of Electrical Equipment.* Prentice Hall, Upper Saddle River, NJ, 2002.

5. Institute of Electrical and Electronics Engineers. *General Principles for Temperature Limits in the Rating of Electrical Equipment.* IEEE STD 1-2000, IEEE, New York, 2000.
6. Institute of Electrical and Electronics Engineers. *Standard Definitions of Basic Per-Unit Quantities for Alternating-Current Machines.* IEEE STD 86-1975, IEEE, New York, 1975.
7. Institute of Electrical and Electronics Engineers. *Standard Test Procedure for Polyphase Induction Motors and Generators.* IEEE STD 112-1996, IEEE, New York, 1996.
8. National Fire Protection Association. Motors circuits, and controllers. Article 430, National Electrical Code NFPA No. 70, NFPA, Quincy, MA, 1999.
9. National Electrical Manufacturers Association. *Motors and Generators.* Publication No. MG 1-1998, NEMA, Rosslyn, VA, 1999.
10. Shemanske, R. Electronic motor braking. *IEEE Trans. Industry Applications,* Vol. IA-19. No. 5 Sept./Oct. 1983.
11. Blume, L. F. *Transformer Engineering,* Wiley, New York, 1938.
12. DeDad, J. Design E motor: You may have problems. *Electrical Construction and Maintenance,* Sept. 1999, pages 36, 38.

GENERAL REFERENCES

Heathcote, M. *J. & P Transformer Book, A Practical Technology of the Transformer,* 12th ed. Newnes Butterworth-Heineman, Boston, 1998.

Lawrence, R. R. *Principles of Alternating Current Machinery.* McGraw-Hill, New York, 1940.

Matsch, L. W., and J. D. Morgan. *Electromagnetic and Electromechanical Machines.* Harper & Row, New York, 1986.

Smeaton, R. W. *Motor Application and Maintenance Handbook,* 2nd ed. McGraw-Hill, New York, 1987.

Wildi, T. *Electrical Power Technology.* Wiley, New York, 1981.

REVIEW QUESTIONS

1. Sketch (on one set of coordinate axes) the torque-speed characteristics of design *A, B, C,* and *D* motors, and state an application for each design.
2. Sketch representative cross sections of design *A, B, C,* and *D* rotors, and explain why the respective construction gives the desired characteristics.
3. Sketch the complete equivalent circuit of a squirrel-cage motor and label all components.
4. How does the developed torque vary with the applied stator voltage?
5. What effect does increasing rotor-circuit resistance have on the breakdown torque and on the slip at which breakdown occurs?

6. What is the approximate slip range for NEMA-design squirrel-cage motors operating between no-load and rated service factor load? Assume rated voltage and rated frequency.
7. What is the effect of off-rated voltage and off-rated frequency on running torque for $s \leq 0.03$?
8. Under what constraints may a 60-Hz motor be operated from a 50-Hz system?
9. What are the maximum permissible frequency and voltage variations that a NEMA-design motor can tolerate and still operate successfully at rated load?
10. (a) Sketch a family of torque-speed curves for a wound-rotor induction motor and indicate the relative values of rotor resistance for each curve. (b) If operating at no load, what effect does adjusting the rheostat have on the motor speed? (c) State an application for a wound-rotor motor.
11. Define utilization voltage, nominal efficiency, code letter, design letter, service factor, and insulation class, and state how this information can be used.
12. (a) What is in-rush current and what causes it? (b) Under what conditions is the transient in-rush a minimum and (c) a maximum?
13. State how to approximate in-rush current from nameplate data.
14. What specific adverse effect does in-rush current have on the service life of an induction motor?
15. (a) What is meant by "reclosing out of phase" and how does it affect an induction motor? (b) What can be done to avoid reclosing out of phase?
16. (a) How does NEMA define voltage unbalance? (b) What is the maximum permissible voltage unbalance?
17. (a) What adverse effects do unbalanced line voltages have on induction-motor performance? (b) If the voltage unbalance cannot be corrected, what should be done to protect the machine?
18. What is the ten-degree rule as it pertains to electrical insulation?
19. What are some of the advantages of expressing induction-motor parameters in terms of per-unit values?
20. What is an induction generator? How does it operate? What are its principal fields of application?
21. Sketch air-gap power vs. speed for an induction machine, showing its behavior as it goes from zero speed to twice synchronous speed. Indicate the breakdown point and the pushover point.
22. What determines the output voltage and frequency of an induction generator in parallel with other three-phase sources?
23. What is the normal procedure for starting and then paralleling an induction generator with a live three-phase bus?
24. Using suitable diagrams, explain how an isolated induction generator builds up voltage as a self-excited machine.
25. Explain how dynamic braking is accomplished, using (a) DC injection; (b) capacitors.
26. If induction motors of almost any power rating can be started by connecting them across full voltage, why use reduced voltage starters?

PROBLEMS

5–1/3 A three-phase, 230-V, 25-hp, 60-Hz, two-pole, NEMA design *A* induction motor runs at 3564 r/min when operating at rated conditions. Determine, in lb-ft, the minimum expected values of (a) locked-rotor torque; (b) breakdown torque; (c) pull-up torque.

5–2/3 A 50-Hz, design *B*, three-phase, 380-V, four-pole induction motor, with rated shaft output of 45 kW, has a speed of 1490 r/min. Determine, in lb-ft, the minimum expected values of (a) locked-rotor torque; (b) breakdown torque; (c) pull-up torque.

5–3/3 Determine, in lb-ft, the minimum values of locked-rotor torque, breakdown torque, and pull-up torque that can be expected from a 5-hp, four-pole, 440-V, 60-Hz, three-phase, design *C* induction motor whose rated speed is 1776 r/min.

5–4/3 (a) Prepare a table listing the NEMA minimum expected values of locked-rotor torque, breakdown torque, and pull-up torque for six-pole, 100-hp, 240-V, 60-Hz, three-phase induction motors, designs *A, B, C, D,* and *E.* (b) Referring to the table in part (a), under what conditions would it be safe to substitute a design *E* motor for design *B* motor.

5–5/4 A 25-hp, 60-Hz, 575-V, six-pole motor is operating at a slip of 0.030. The stray power loss and the combined windage and friction loss at this load are 230.5 W and 115.3 W, respectively. The motor is wye connected, and the motor parameters in ohms/phase are

$$R_1 = 0.3723 \qquad R_2 = 0.390 \qquad X_M = 26.59$$
$$X_1 = 1.434 \qquad X_2 = 2.151 \qquad R_{fe} = 354.6$$

Determine (a) motor input impedance per phase; (b) line current; (c) active, reactive, and apparent power and power factor; (d) equivalent rotor current; (e) stator copper loss; (f) rotor copper loss; (g) core loss; (h) air-gap power; (i) mechanical power developed; (j) developed torque; (k) shaft horsepower; (l) shaft torque; (m) efficiency. (n) Sketch the power-flow diagram and enter all data.

5–6/4 The shaft load on a 40-hp, 60-Hz, 460-V, four-pole induction motor is such as to cause the machine to operate at 1447 r/min. The stray load loss and combined windage and friction loss, when operating at this load, are 450 W and 220 W, respectively. The motor parameters in ohms/phase are

$$R_1 = 0.1418 \qquad R_2 = 1.10 \qquad X_M = 21.27$$
$$X_1 = 0.7273 \qquad X_2 = 0.7284 \qquad R_{fe} = 212.73$$

The motor is NEMA design *D* and wye connected. Determine (a) motor input impedance per phase; (b) line current; (c) active, reactive, and apparent power and power factor; (d) equivalent rotor current; (e) stator copper loss; (f) rotor copper loss; (g) core loss; (h) air-gap power; (i) mechanical power developed; (j) developed torque; (k) shaft horsepower; (l) shaft torque; (m) efficiency. (n) Sketch the power-flow diagram and enter all data. (o) If the speed at rated load is 1190 r/min, determine the expected minimum locked-rotor torque.

5–7/4 A three-phase, eight-pole induction motor, rated at 847 r/min, 30 hp, 60 Hz, 460 V, operating at reduced load, has a shaft speed of 880 r/min. The combined stray power loss, windage loss, and friction loss is 350 W. The motor parameters in ohms/phase are

$$R_1 = 0.1891 \qquad R_2 = 0.191 \qquad X_M = 14.18$$
$$X_1 = 1.338 \qquad X_2 = 0.5735 \qquad R_{fe} = 189.1$$

The motor is NEMA design *C* and wye connected. Determine (a) motor input impedance per phase; (b) line current; (c) active, reactive, and apparent power and power factor; (d) equivalent rotor current; (e) stator copper loss; (f) rotor copper loss; (g) core loss; (h) air-gap power; (i) mechanical power developed; (j) developed torque; (k) shaft horsepower; (l) shaft torque; (m) efficiency. (n) Sketch the power-flow diagram and enter all data. (o) Determine the expected minimum locked-rotor torque, breakdown torque, and pull-up torque.

5–8/5 For the motor in Problem 5–5/4, determine (a) the shaft speed at maximum torque; (b) the value of the maximum torque.

5–9/5 For the motor in Problem 5–7/4, determine (a) the shaft speed at maximum torque; and (b) the value of the maximum torque.

5–10/5 A three-phase, 60 Hz, 75-hp, 460-V, six-pole, design *C* induction motor, operating at rated conditions, has a slip of 0.041. Determine (a) the expected minimum locked-rotor torque; (b) the expected minimum locked-rotor torque if the stator is connected to a 400-V, 60-Hz supply; (c) the percent change in applied voltage; (d) the resultant percent change in locked-rotor torque.

5–11/5 A three-phase, wye-connected, design *A*, eight-pole motor rated at 50 Hz, 240 V, 20 hp has a slip of 1.80 percent when operating at rated conditions. The motor is to drive a constant-torque load of 145 lb-ft from a 208-V, 50-Hz supply. The breakaway torque required to get the motor started is 155 lb-ft. Will the motor start? Show all work!

5–12/6 A 75-hp, two-pole, wye-connected motor, operating from a 60-Hz, 2300-V line, is delivering 75.4 hp at 3500 r/min and 18.9 A. Determine (a) the new shaft speed if the torque load is reduced by 25 percent; (b) the rotor current; (c) the air-gap power. The motor parameters expressed in ohms per phase are

$$R_1 = 1.08 \qquad R_2 = 2.14 \qquad R_{\text{fe}} = 1892$$
$$X_1 = 8.14 \qquad X_2 = 3.24 \qquad X_M = 147.5$$

5–13/6 A 250-hp, four-pole, wye-connected motor, operating from a 60-Hz, 460-V line, is delivering 255 hp at 1777 r/min. If the shaft torque load is reduced by

25 percent, determine the approximate values of (a) slip; (b) shaft speed; (c) rotor current; (d) shaft horsepower. The motor parameters expressed in ohms/phase are:

$$R_1 = 0.0626 \qquad R_2 = 0.0118 \qquad R_{fe} = 32.25$$
$$X_1 = 0.027 \qquad X_2 = 0.040 \qquad X_M = 2.465$$

5–14/6 A 10-hp, 440-V, 60-Hz, two-pole, induction motor operating at rated load develops 15.5 lb-ft torque at 3492 r/min. The motor parameters expressed in ohms/phase are:

$$R_1 = 0.740 \qquad R_2 = 0.647 \qquad R_{fe} = \text{not known}$$
$$X_1 = 1.33 \qquad X_2 = 2.01 \qquad X_M = 77.6$$

If the load torque is reduced by 30 percent, determine the approximate values of (a) slip; (b) shaft speed; (c) rotor current; (d) shaft horsepower.

5–15/6 A 15-hp, 440-V, 60-Hz, six-pole motor operates at 1173 r/min when carrying rated load. The motor parameters expressed in ohms/phase are:

$$R_1 = 0.301 \qquad R_2 = 0.327 \qquad R_{fe} = 496$$
$$X_1 = 0.833 \qquad X_2 = 1.25 \qquad X_M = 30.3$$

If a 15 percent torque overload occurs, determine the approximate values of (a) speed; (b) rotor current; (c) shaft power.

5–16/8 A 150-hp, three-phase, wye-connected, 60-Hz, 4000-V, six-pole induction motor, operating at rated conditions from an isolated power system, has a shaft speed of 1175 r/min. A very heavy power demand on the electrical system caused the voltage and frequency to decrease by 15 percent and 5 percent, respectively. To compensate for this off-normal condition, the torque load on the motor was reduced to 82 percent of rated torque. Determine the operating speed for the new conditions.

5–17/8 A three-phase, wye-connected, 50-hp, 60-Hz, 460-V, four-pole induction motor, operating at rated conditions, has an efficiency, power factor, and slip of 89.6 percent, 79.5 percent, and 3.0 percent, respectively. Operating the motor from a 430-V, 55-Hz supply results in a shaft speed of 1750 r/min. Determine the resultant shaft horsepower for the new operating conditions. Assume the windage, friction, and stray load losses are the same.

5–18/8 A four-pole, 60-hp, 440-V, 60-Hz, 1760 r/min, three-phase induction motor driving a loaded conveyer develops a locked-rotor torque equal to 161 percent rated torque when rated voltage and rated frequency are applied. The *load torque* at start-up is 114 percent rated motor torque. To compensate for inertia and static friction, the developed torque at locked rotor must be at least 15 percent greater than the load torque. If a very heavy power demand on the electrical system causes the voltage to decrease by 15 percent and the frequency to decrease by 3 percent, will the motor start? Show all work.

5–19/8 A three-phase, 125-hp, 60-Hz, eight-pole, 575-V, design *B* induction motor is to be operated from a 50-Hz system. Determine (a) allowable voltage at 50 Hz; (b) allowable shaft load in horsepower; (c) new synchronous speed; (d) shaft speed assuming a slip of 2.1 percent; (e) shaft torque at 2.1 percent slip.

5–20/9 A three-phase, wye-connected, 25-hp, 60-Hz, 575-V, six-pole, wound-rotor motor, operating at nameplate conditions, runs at 1164 r/min with the rheostat shorted. The stator/rotor turns ratio is 2.15, and the motor parameters in ohms/phase are

$$R_1 = 0.3723 \qquad R_2 = 0.390 \qquad R_{fe} = 26.59$$
$$X_1 = 1.434 \qquad X_2 = 2.151 \qquad X_m = 354.6$$

Determine (a) slip at which $T_{D,\max}$ occurs; (b) $T_{D,\max}$; (c) the rheostat resistance/phase required to operate the machine at rated torque load and 1074 r/min.

5–21/9 A 40-hp, 60-Hz, 460-V, four-pole, wye-connected wound-rotor motor, with slip rings shorted, has its breakdown torque occur at 25 percent slip. The rotor impedance in ohms/phase referred to the stator is $0.158 + j0.623$, and the turns ratio is 1.28. Determine the rheostat resistance required to cause the breakdown torque to occur at 60 percent slip.

5–22/12 A 75-hp, 460-V, six-pole NEMA design *A* machine with Code letter H is operating at rated load, is 89 percent efficient, has a power factor of 84 percent, and has a slip of 2.36 percent. Determine (a) rated current; (b) the expected range of locked-rotor current with rated voltage and rated frequency applied.

5–23/12 A 30-hp, 60-Hz, 230-V, two-pole, design *E* motor, with Code letter C, operating at rated conditions has an efficiency of 91.8 percent and a power factor of 86.2 percent. Determine (a) rated current; (b) the expected range of locked-rotor current.

5–24/12 A 150-hp, 60-Hz, 460-V, four-pole, design *B* motor, with Code letter R, operating at rated conditions has an efficiency of 94.5 percent and a power factor of 86.2 percent. Determine (a) rated current; (b) the expected range of locked-rotor current.

5–25/15 A 60-hp, design *C*, 230-V, 60-Hz, six-pole motor with a 1.15 service factor and Class B insulation is operating at rated horsepower from an unbalanced three-phase system. The three line-to-line voltages are 232 V, 238 V, and 224 V. The machine is new and has an expected life of 20 years. Determine (a) the percent voltage unbalance; (b) the expected approximate temperature rise if operating at rated load in a 40°C ambient situation, with the percent unbalance in (a); (c) the expected insulation life; (d) the required rerating, if any, to prevent shortening the life of the insulation.

5–26/15 A three-phase, 30-hp, 460-V, 60-Hz, 1770 r/min, design *B* induction motor with Class F insulation and service factor 1.0 is operating at rated shaft load in a 40°C ambient situation, and has an expected 20-year life. A preventive maintenance check shows line-to-line voltages to be 449.2 V, 431.3 V, and

462.4 V. Determine (a) percent voltage unbalance; (b) expected temperature rise; (c) expected insulation life; (d) rerating of motor, if any to prevent shortening insulation life.

5–27/16 A three-phase, wye-connected 10-hp, 60-Hz, 230-V, four-pole induction motor has the following per-unit parameters:

$$R_1 = 0.0358 \qquad R_2 = 0.0264 \qquad R_{fe} = \text{not known}$$
$$X_1 = 0.0964 \qquad X_2 = 0.1450 \qquad X_M = 3.02$$

Determine the machine parameters in ohms/phase.

5–28/16 A three-phase, 460-V, wye-connected, 200-hp, 60-Hz, eight-pole, design *B*, squirrel-cage induction motor, has the following per-unit parameters:

$$R_1 = 0.011 \qquad R_2 = 0.011 \qquad R_{fe} = \text{not known}$$
$$X_1 = 0.123 \qquad X_2 = 0.210 \qquad X_M = 2.994$$

Determine the corresponding ohmic values.

5–29/16 A three-phase, wye-connected, 100-hp, 60-Hz, 440-V, 10-pole induction motor has the following parameters in ohms:

$$R_1 = 0.0864 \qquad R_2 = 0.078 \qquad R_{fe} = 110$$
$$X_1 = 0.146 \qquad X_2 = 0.218 \qquad X_M = 3.185$$

Determine the corresponding per-unit values.

5–30/17 A three-phase, design *B*, wye-connected, 25-hp, 575-V, 60-Hz induction motor, operating at rated conditions, draws a line current of 27 A. Data from a 15-Hz blocked-rotor test, a 60-Hz no-load test, and a DC test are:

Blocked Rotor	No-Load	DC
$V_{line} = 54.7$ V	$V_{line} = 575$ V	$V_{DC} = 20$ V
$I_{line} = 27.0$ A	$I_{line} = 11.8$ A	$I_{DC} = 27$ A
$P_{3\text{-phase}} = 1653$ W	$P_{3\text{-phase}} = 1264.5$ W	

Determine R_1, R_2, X_1, X_2, X_M, and the combined core, friction, and windage loss.

5–31/17 The following data were obtained from a no-load test, a 15-Hz blocked-rotor test, and a DC test on a three-phase, wye-connected, four-pole, 30-hp, 460-V, 60-Hz, 40-A, design *C* induction motor:

Blocked Rotor	No-Load	DC
$V_{line} = 42.39$ V	$V_{line} = 458.6$ V	$V_{DC} = 15.4$ V
$I_{line} = 40$ A	$I_{line} = 17.0$ A	$I_{DC} = 40.2$ A
$P_{3\text{-phase}} = 1828.8$ W	$P_{3\text{-phase}} = 1381.4$ W	

Determine R_1, R_2, X_1, X_2, X_M, and the combined core, friction, and windage loss.

5–32/17 A three-phase, design *A*, wye-connected, 15-hp, 460-V, 60-Hz induction motor draws a line current of 14 A when operating at rated conditions. A

60-Hz no-load test, a 15-Hz blocked-rotor test, and a DC test provide the following data:

Blocked Rotor	No-Load	DC
$V_{line} = 18.5$ V	$V_{line} = 459.8$ V	$V_{DC} = 5.6$ V
$I_{line} = 13.9$ A	$I_{line} = 6.2$ A	$I_{DC} = 14.0$ A
$P_{3\text{-phase}} = 264.6$ W	$P_{3\text{-phase}} = 799.5$ W	

Determine R_1, R_2, X_1, X_2, X_M, and the combined core, friction, and windage loss.

5–33/18 A 75-hp, 2300-V, 60-Hz, two-pole, wye-connected motor, driven at 3650 r/min by a steam turbine, is connected to a 2300-V, 60-Hz distribution system. The motor parameters in ohms are:

$$R_1 = 1.08 \qquad R_2 = 2.14 \qquad R_{fe} = 1892$$
$$X_1 = 8.14 \qquad X_2 = 3.24 \qquad X_M = 187.5$$

Determine the active power that the machine delivers to the system.

5–34/18 A 60-Hz, 15-hp, 460-V, six-pole, wye-connected, three-phase induction motor is connected to a 460-V distribution system and driven at 1210 r/min by a diesel engine. The motor parameters in ohms are:

$$R_1 = 0.200 \qquad R_2 = 0.250 \qquad X_M = 42.0$$
$$X_1 = 1.20 \qquad X_2 = 1.29 \qquad R_{fe} = 317$$

Determine the active power that the machine delivers to the system.

5–35/18 A three-phase, wye-connected, 400-hp, four-pole, 380-V, 50-Hz induction motor is driven by a wind turbine at 1515 r/min. The motor parameters in ohms are:

$$R_1 = 0.00536 \qquad R_2 = 0.00613 \qquad R_{fe} = 7.66$$
$$X_1 = 0.0383 \qquad X_2 = 0.0383 \qquad X_M = 0.5743$$

Determine the active power that the machine delivers to the system.

5–36/20 A 200-hp, 1150 r/min, 440-V, 60-Hz pump motor uses a wye-connected autotransformer for reduced voltage starting. The transformer has a 65 percent tap. The starting torque at rated voltage is 150 percent rated torque. Sketch a one-line diagram and determine the blocked-rotor torque when (a) starting at rated voltage; (b) starting at reduced voltage.

5–37/20 A three-phase 12-pole, 220-V, 60-Hz, 50-hp squirrel-cage motor, operating at rated load, has a speed of 595 r/min, an efficiency of 89 percent, and a power factor of 81 percent lagging. The locked-rotor current and locked-rotor torque at rated voltage and rated frequency are 725 A and 120 percent rated torque, respectively. Determine (a) rated line current; (b) rated torque; (c) minimum voltage required to obtain a blocked-rotor torque equal to 70 percent rated torque; (d) autotransformer bank ratio required to provide the minimum voltage in (c); (e) stator current at blocked rotor with autotransformer in circuit; (f) corresponding input current to the transformer when starting the motor at the voltage in (c).

5–38/20 A 50-hp, six-pole, three-phase, 450-V, 60-Hz, 1120 r/min induction motor, operating at rated conditions, has an efficiency of 91 percent and a power factor of 89 percent. When starting at rated voltage, the motor develops 170 percent rated torque and draws five times rated current. Determine (a) slip at rated conditions; (b) rated torque; (c) blocked-rotor torque; (d) rated line current. (e) Design a V–V transformer bank of the correct turns ratio that will limit the stator line current to 200 percent rated current. (f) Determine the primary line current for the conditions in (e); (g) motor line current when starting at the reduced voltage.

5–39/20 A three-phase, four-pole, 460-V, 60-Hz, 200-hp, design *B* induction motor has a locked-rotor torque equal to 125 percent rated torque when rated voltage is applied. The efficiency power factor and slip at rated conditions are 92, 82, and 2.0 percent, respectively. The locked-rotor current at rated voltage is 1450 A. The motor is to be used to drive a centrifugal pump whose specifications require the motor to have a minimum starting torque of 357 ft-lb. Determine (a) rated line current; (b) rated torque; (c) minimum stator voltage required to start the load; (d) required transformer bank ratio; (e) and then sketch the circuit showing the motor and wye-connected autotransformer.

5–40/20 A three-phase, design *C*, 25-hp, 60-Hz, 27-A, 575-V, six-pole motor has a rated speed of 1164 r/min. The locked-rotor impedance is $3.49 \angle 78.18^\circ$ Ω/phase, and the motor was designed for wye-delta starting. Determine (a) locked-rotor line current and phase current when wye starting; (b) locked-rotor line current and phase current when delta starting; (c) the expected minimum locked-rotor torque if delta connected; (d) the expected minimum locked-rotor torque if wye connected.

5–41/20 A four-pole, 30-hp, 60-Hz, 460-V, design *B*, wye-connected induction motor, operating at rated conditions, draws 40 A and has a slip of 2.89 percent. The locked rotor impedance is $1.93\angle 79.02^\circ$ Ω/phase. (a) Design a series-resistor starter that will limit the starting current to 200 percent rated current; (b) determine the stator voltage per phase at locked rotor; (c) determine the expected minimum locked-rotor torque when using the starting resistance.

5–42/20 A 75-hp, 60-Hz, 2300-V, 20-A, two-pole, wye-connected, design *B* motor with Code K delivers rated horsepower at 3500 r/min. Assume the phase angle of the locked-rotor impedance is 75°. Determine (a) the required reactance of a series-reactor starter that will limit the starting current to 350 percent of rated current; (b) the expected minimum locked-rotor torque as a percent of rated torque when starting with the series reactor.

6

Single-Phase Induction Motors

6.1 INTRODUCTION

Single-phase induction motors are used extensively in industrial, commercial, and domestic applications. They are used in clocks, refrigerators, freezers, fans, blowers, pumps, washing machines, machine tools, and range in size from a fraction of a horsepower to 15 hp.

Large single-phase induction motors are split-phase machines that have two separate windings physically displaced by 90 electrical degrees, and phase-splitting circuits that cause the current and associated flux of one winding to lag or lead the current and associated flux of the other winding. The net effect is the production of a rotating magnetic field that sweeps a squirrel-cage rotor, developing induction-motor action. Smaller single-phase induction motors use a much simpler device called a shading coil to provide the phase-splitting effect.

Phase-splitting circuits are also used to operate three-phase induction motors from a single-phase source. This enables larger motors to be operated in isolated areas where three-phase sources are not available. Such operation, however, must not be confused with a fault condition, called *single phasing,* where a three-phase motor operating from a three-phase source has an open circuit occur in one of the three lines. In this case, there is no phase splitter, and the resultant excessive current and vibration may damage the motor if protective devices do not function.

6.2 QUADRATURE FIELD THEORY AND INDUCTION-MOTOR ACTION

Single-phase induction motors are unique in that they cannot develop a rotating field, and hence cannot develop motor torque unless auxiliary methods are used to start the rotor turning. This is illustrated in Figure 6.1(a), where the stationary squirrel-cage rotor acts as the short-circuited secondary of a transformer. For the instant shown, the

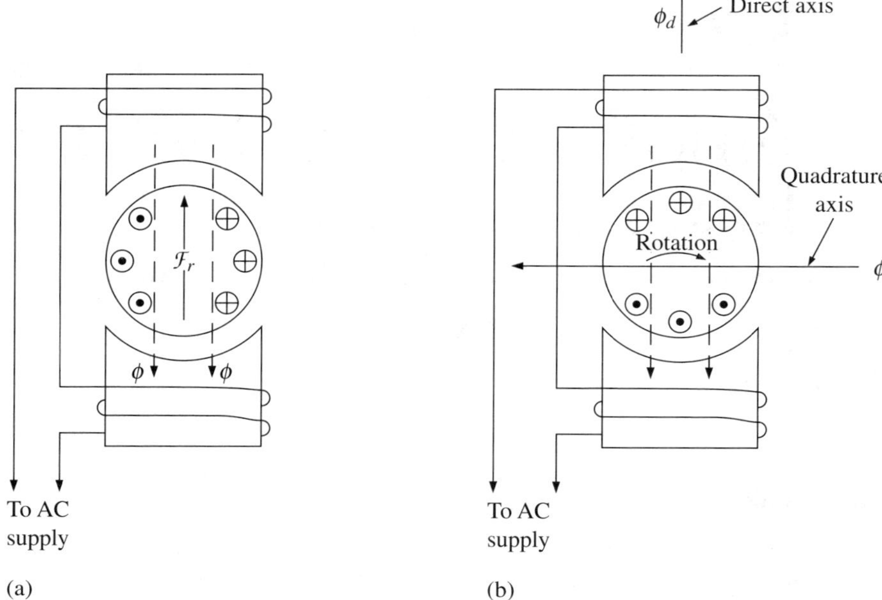

FIGURE 6.1
Instantaneous direction of current in rotor bars caused by (a) transformer action; (b) speed voltage.

stator flux, also called main-pole flux, is increasing in the downward direction, causing the rotor mmf $\mathscr{F}_r$ to be in the upward direction. Since the magnetic axis of the rotor is in line with the magnetic axis of the stator, no rotating field will be developed, and hence no motor action will be produced.

If the rotor is rotated mechanically (by hand or otherwise), as shown in Figure 6.1(b), however, a rotating field will be created, induction-motor action will occur, and the machine will accelerate to near synchronous speed. This behavior is explained by the quadrature field theory, or the double-revolving field theory **[5]**.

Quadrature Field Theory

Spinning the rotor mechanically, as shown in Figure 6.1(b), causes the rotor conductors to cut the flux of the main poles. For the instant shown the current in the rotor bars, caused by the induced speed-voltage, produces a magnetic field that is 90 electrical degrees from the centerline of the main-pole flux. The main pole flux is called the *direct-axis flux,* and the flux caused by the speed-voltage is called the *quadrature-axis flux.* The instantaneous directions of speed-voltage, associated current, and quadrature flux are determined from Lenz's law. As the rotor turns, the conductors shift their relative positions, but other conductors take their place, thus maintaining the quadrature flux. Note that *the flux produced by the stator winding alternates sinusoidally along the direct axis, and the induced flux in the rotor alternates sinusoidally along the quadrature axis.*

Figure 6.2(a) shows the phase relationship between the main-pole flux, speed-voltage, associated rotor current, and quadrature flux. Note that the high ratio of rotor reactance to rotor resistance causes the rotor current, and hence the quadrature flux, to lag the induced voltage by almost 90°.

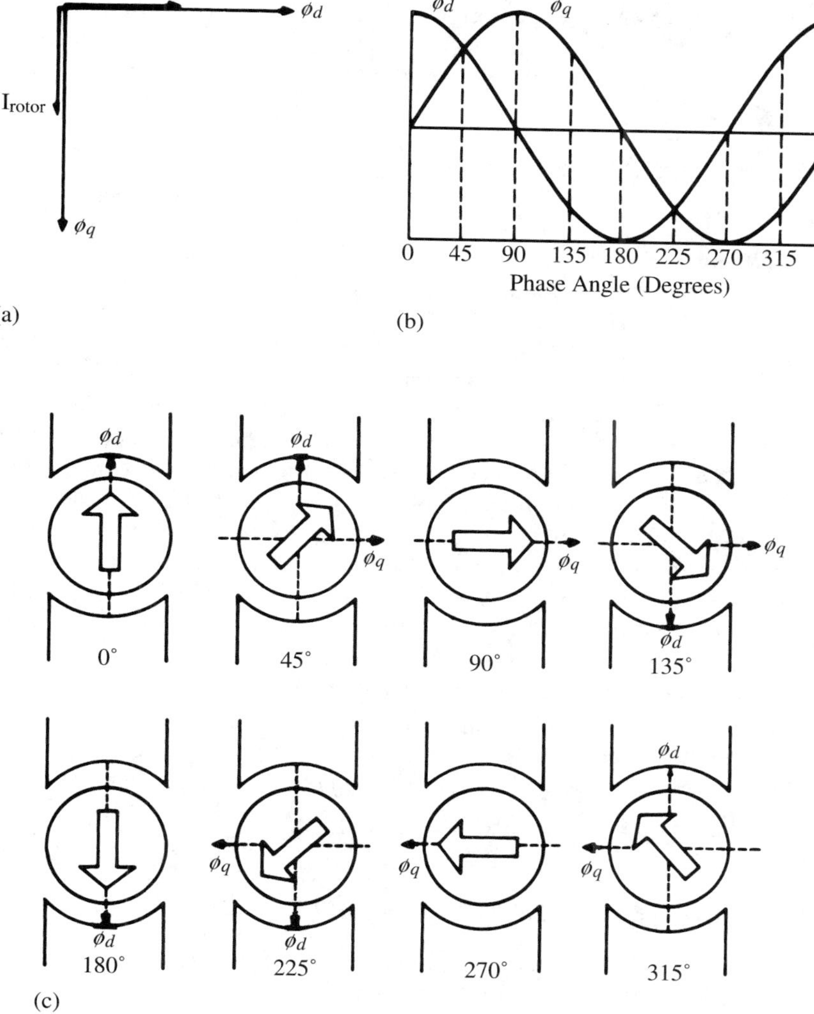

FIGURE 6.2
(a) Phase relationship between main-pole flux, speed-voltage, associated rotor current, and quadrature flux; (b) flux waves corresponding to phasor diagram in (a); (c) instantaneous directions of ϕ_d, ϕ_q, and resultant flux for corresponding phase angles in (b).

A speed-voltage is always in phase with the flux that caused it, and flux due to current is always in phase with the current that caused it. Thus, as shown in Figure 6.2(a), the quadrature flux lags the direct-axis flux by almost 90 electrical degrees. The flux waves corresponding to those in the phasor diagram are shown in Figure 6.2(b).

Figure 6.2(c) shows the instantaneous directions of the direct-axis flux, quadrature-axis flux, and resultant flux for the corresponding phase angles in Figure 6.2(b). The large arrows indicate the instantaneous direction of the resultant flux. As indicated in Figure 6.2(c), the resultant flux vector revolves in a clockwise (CW) direction. Manually spinning the rotor produces a quadrature field, which in combination with the direct-axis field, causes a rotating flux in the direction of the original spin. The rotating flux accelerates the rotor to near synchronous speed.

6.3 INDUCTION-MOTOR ACTION THROUGH PHASE SPLITTING

Initiating a rotating magnetic field from a single-phase source, without resorting to mechanical means, requires the use of two stator windings and a phase-splitting circuit. The physical layout of the windings for an elementary two-pole split-phase motor is shown in Figure 6.3(a), and the corresponding equivalent-circuit diagram is shown in Figure 6.3(b). The main winding supplies the direct-axis flux (ϕ_d), and an auxiliary winding displaced 90 electrical degrees from the main winding supplies the quadrature-axis flux (ϕ_q). The auxiliary winding is also called the *starting winding.*

The phase splitter is connected in series with the auxiliary winding and causes the current in the auxiliary winding to be out of phase with the current in the main winding. Since the magnetic field due to a current is in phase with the current that produces it, the quadrature field and main field will be out of phase, resulting in a rotating flux and induction-motor action.

Phase splitting may be accomplished through the use of capacitance or resistance. If accomplished through the use of capacitance, the motor is called a capacitor-start split-phase motor; if accomplished through the use of resistance, it is called a resistance-start split-phase motor. Regardless of the means used to start the rotor turning (be it phase splitting or mechanical action), however, *once it starts turning, self-excitation will maintain the quadrature field,* and the auxiliary winding with its phase splitter may be disconnected.

6.4 LOCKED-ROTOR TORQUE

The locked-rotor torque of a split-phase motor is proportional to the product of the magnitudes of the locked-rotor current in each winding times the sine of the angle of phase displacement between the two currents **[6]**. Expressed mathematically,

$$T_{\text{lr}} = k_{\text{sp}} \cdot I_{\text{mw}} \cdot I_{\text{aw}} \sin \alpha \tag{6–1}$$

$$\alpha = |\theta_{i,\text{mw}} - \theta_{i,\text{aw}}| \tag{6–2}$$

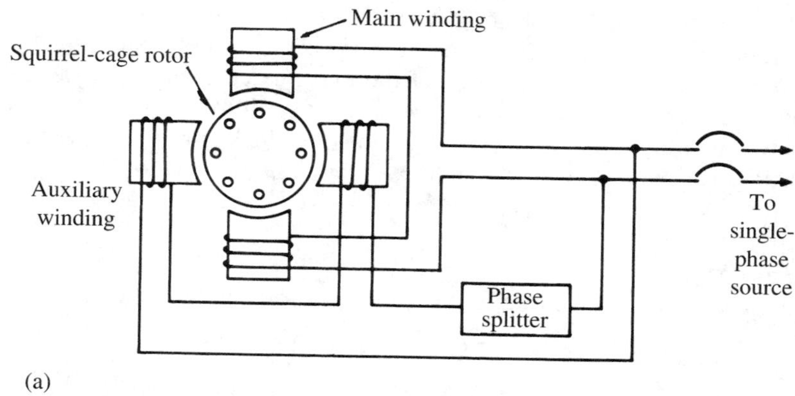

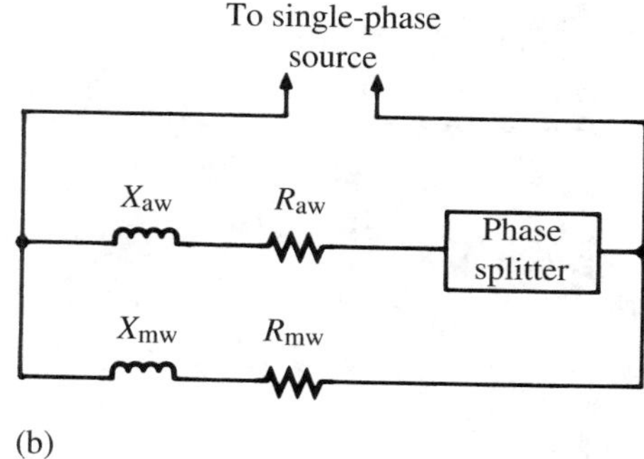

FIGURE 6.3
(a) Elementary two-pole single-phase motor with phase splitter; (b) equivalent-circuit diagram.

where:

k_{sp} = machine constant, split-phase motor
I_{aw} = current in auxiliary winding (A)
I_{mw} = current in main winding (A)
$\theta_{i,aw}$ = phase angle of current in auxiliary winding
$\theta_{i,mw}$ = phase angle of current in main winding
α = phase displacement between $\mathbf{I}_{aw}$ and $\mathbf{I}_{mw}$

EXAMPLE 6.1 The main and auxiliary windings of a hypothetical 120-V, 60-Hz, split-phase motor have the following locked-rotor parameters:

$$R_{mw} = 2.00\ \Omega \qquad X_{mw} = 3.50\ \Omega$$
$$R_{aw} = 9.15\ \Omega \qquad X_{aw} = 8.40\ \Omega$$

The motor is connected to a 120-V, 60-Hz system. Determine (a) locked-rotor current in each winding; (b) phase-displacement angle between the two currents; (c) locked-rotor torque in terms of the machine constant; (d) external resistance required in series with the auxiliary winding in order to obtain a 30° phase displacement between the currents in the two windings; (e) locked-rotor torque for the conditions in (d); (f) percent increase in locked-rotor torque due to the addition of external resistance.

Solution

(a) The circuit for the original condition is shown in Figure 6.4(a).

$$\mathbf{Z}_{mw} = 2.00 + j3.50 = 4.0311\underline{/60.2551^\circ}\ \Omega$$
$$\mathbf{Z}_{aw} = 9.15 + j8.40 = 12.4211\underline{/42.5530^\circ}\ \Omega$$

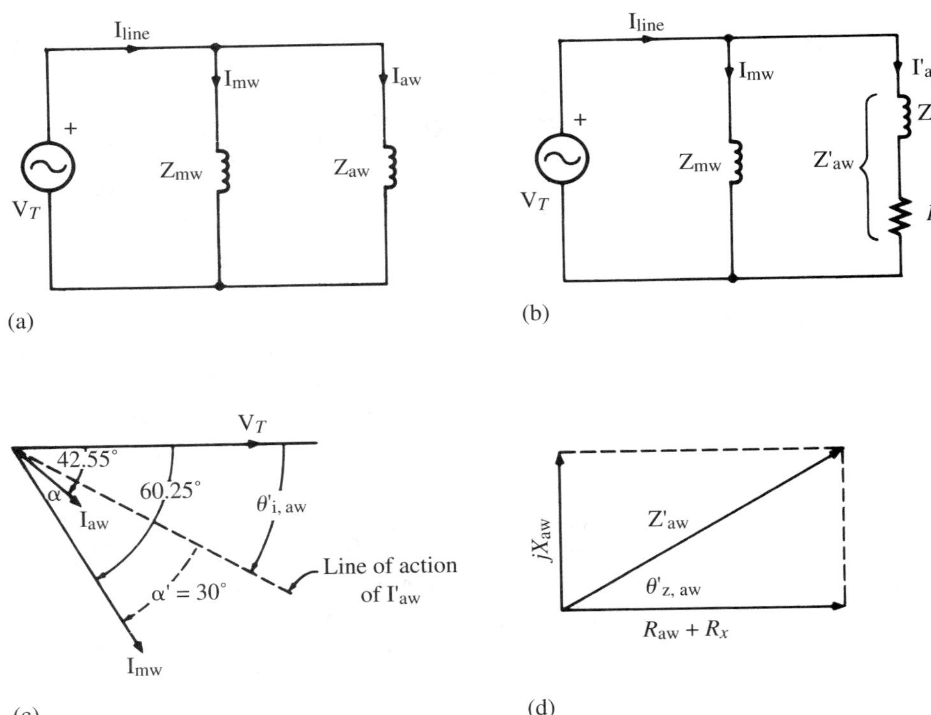

FIGURE 6.4
Diagrams for Example 6.1: (a) original circuit; (b) modified circuit; (c) phasor diagram for determining required phase angle of auxiliary current for new conditions; (d) impedance diagram for new auxiliary-circuit branch.

$$\mathbf{I}_{\text{mw}} = \frac{120\angle 0^\circ}{4.0311\angle 60.2551^\circ} = \underline{29.7688\angle -60.2551^\circ} \quad \Rightarrow \quad \underline{29.8\angle -60.3^\circ \text{ A}}$$

$$\mathbf{I}_{\text{aw}} = \frac{120\angle 0^\circ}{12.4211\angle 42.5530^\circ} = \underline{9.6610\angle -42.5530^\circ} \quad \Rightarrow \quad \underline{9.66\angle -42.6^\circ \text{ A}}$$

(b) $\alpha = |\theta_{i,\text{mw}} - \theta_{i,\text{aw}}| = |-60.2551 - (-42.5530)| = 17.7021^\circ \quad \Rightarrow \quad \underline{17.7^\circ}$

(c)

$$T_{\text{lr}} = k_{\text{sp}} \cdot I_{\text{mw}} \cdot I_{\text{aw}} \sin \alpha$$

$$T_{\text{lr}} = k_{\text{sp}} \times 29.7688 \times 9.6610 \times \sin 17.7021 = \underline{87.45 k_{\text{sp}}}$$

(d) The circuit for the new condition, with a resistor in series with the auxiliary winding, is shown in Figure 6.4(b). A phasor diagram showing the respective currents for the old condition and the desired location of the new auxiliary winding current $\mathbf{I}'_{\text{aw}}$ is shown in Figure 6.4(c). From Figure 6.4(c), the required phase angle for $\mathbf{I}'_{\text{aw}}$ is

$$\theta'_{i,\text{aw}} = -60.2551^\circ + 30^\circ = -30.2551^\circ$$

Applying Ohm's law to the auxiliary branch in Figure 6.4(b),

$$\mathbf{I}'_{\text{aw}} = \frac{\mathbf{V}_T}{\mathbf{Z}'_{\text{aw}}} \quad \Rightarrow \quad \underline{I'_{\text{aw}}\angle -30.2551^\circ} = \frac{V_T\angle 0^\circ}{Z'_{\text{aw}}\angle \theta'_{z,\text{aw}}}$$

$$\theta'_{z,\text{aw}} = 30.2551^\circ$$

From the impedance diagram for the new auxiliary-circuit branch shown in Figure 6.4(d),

$$\tan \theta'_{\text{aw}} = \frac{X_{\text{aw}}}{R_{\text{aw}} + R_x} \quad \Rightarrow \quad R_x = \frac{X_{\text{aw}}}{\tan(\theta'_{\text{aw}})} - R_{\text{aw}}$$

$$R_x = \frac{8.40}{\tan(30.2551^\circ)} - 9.15 = 5.2508 \quad \Rightarrow \quad \underline{5.25\ \Omega}$$

(e)

$$\mathbf{I}'_{\text{aw}} = \frac{\mathbf{V}_T}{\mathbf{Z}'_{\text{aw}}} = \frac{120\angle 0^\circ}{9.15 + 5.2508 + j8.40} = \underline{7.1979\angle -30.2551^\circ} \text{ A}$$

$$T_{\text{lr}} = k_{\text{sp}} \cdot I_{\text{mw}} \cdot I_{\text{aw}} \sin \alpha$$

$$T_{\text{lr}} = k_{\text{sp}} \times 29.7688 \times 7.1979 \times \sin 30^\circ = \underline{107.1 \cdot k_{\text{sp}}}$$

(f)

$$\frac{107.1 - 87.45}{87.45} \times 100 = \underline{22.5\% \text{ increase}}$$

Note: The added resistance in the auxiliary winding circuit decreased the auxiliary winding current, but increased the locked-rotor torque.

Graphical Analysis of Example 6.1

Because only the auxiliary winding has series-connected elements to provide phase splitting, the current in the main winding may be assumed constant, permitting Eq. (6–1) to be written as

$$T_{\text{lr}} \propto I_{\text{aw}} \sin \alpha \tag{6–3}$$

Graphs of I_{aw}, α, and ($I_{\text{aw}} \cdot \sin \alpha$) vs. the total resistance of the auxiliary winding circuit in Example 6.1, as R_x is increased from 0 Ω to 20 Ω, are shown in Figure 6.5. Note that the current in the auxiliary winding decreases with increasing resistance; angle α increases with increasing resistance; but the locked-rotor torque, represented by ($I_{\text{aw}} \sin \alpha$), reaches a peak value with an auxiliary circuit resistance of approximately 14.2 Ω, and then falls off with increasing resistance.

For every split-phase motor there is an optimum value of auxiliary-circuit resistance that will maximize the locked-rotor torque. The phase displacement (α) for this optimum value of resistance is generally between 25° and 30°.

6.5 PRACTICAL RESISTANCE-START SPLIT-PHASE MOTORS

The circuit diagram for a general-purpose resistance-start split-phase motor is shown in Figure 6.6(a). The auxiliary winding is wound with smaller diameter wire than the main winding, causing the auxiliary winding to have a higher ratio of resistance to reactance than the main winding. The switch in the auxiliary circuit is a magnetic relay, a solid-state switch, or a centrifugally operated switch. The centrifugally operated switch shown in the circuit diagram of Figure 6.6(a) is closed when the motor is at rest, and opens when the rotor is at approximately 75 to 80 percent synchronous speed. A solid-state switch, called a triac, is shown with broken lines in Figure 6.6(a); the switch closes when starting and is set to open at approximately 75 percent synchronous speed. A magnetic relay (not shown) is closed by the high motor-starting current, and springs open when acceleration of the motor reduces the current to approximately 80 percent of the locked-rotor current. A representative phasor diagram for the motor (when starting) is shown in Figure 6.6(b). A typical torque-speed characteristic for a resistance-start split-phase motor is shown in Figure 6.6(c).

The resistance-start split-phase motor is adaptable to loads such as centrifugal pumps, oil burners, blowers, and other loads of similar characteristics that require moderate torques and constant speed. This motor offers no means for speed control from a fixed-frequency source other than that obtainable by reconnecting for different pole arrangements.

A cutaway view of a general-purpose split-phase induction motor is shown in Figure 6.7. Note the centrifugal mechanism mounted on the squirrel-cage rotor and the associated switch mounted on the end bell.

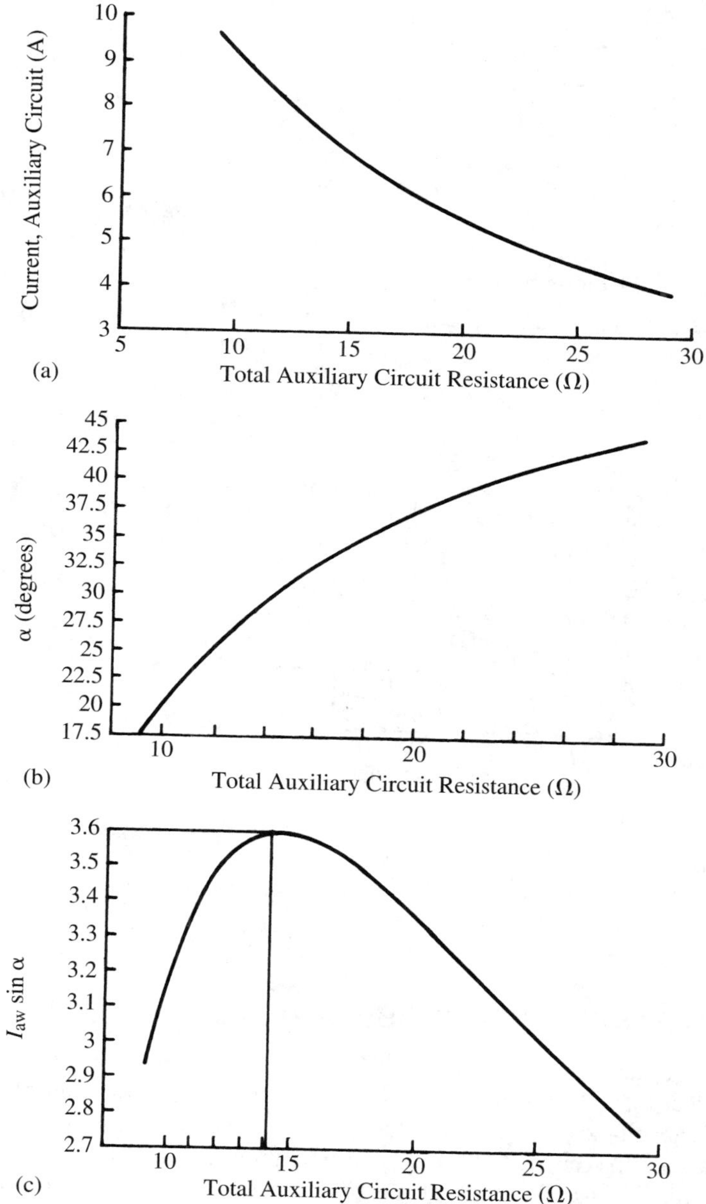

FIGURE 6.5
Graphs of auxiliary winding current, phase-displacement angle α, and locked-rotor torque represented by $I_{aw} \sin \alpha$, for the split-phase motor in Example 6.1.

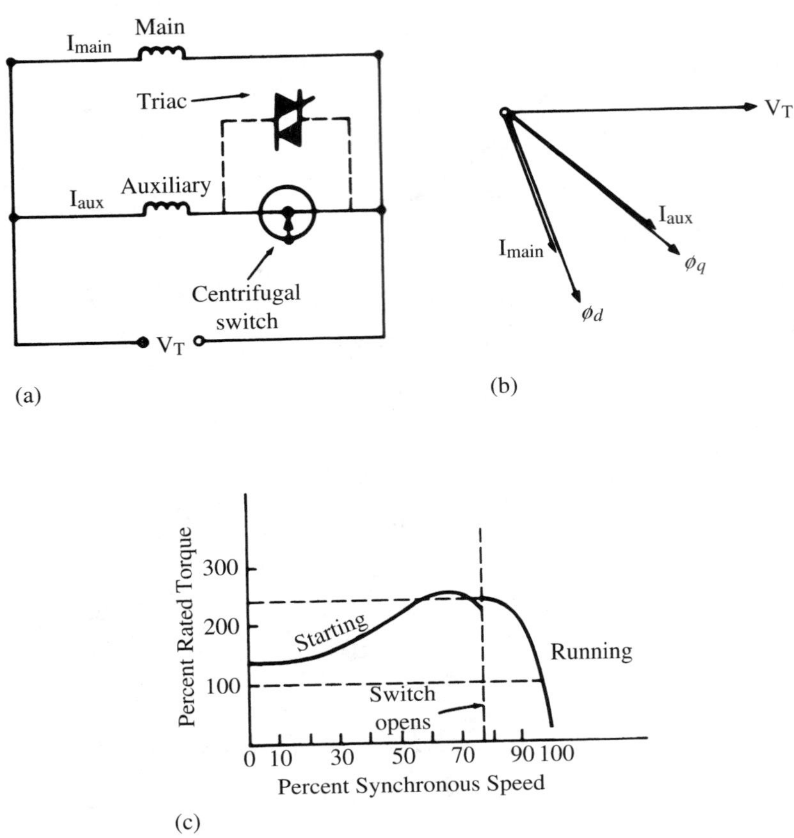

FIGURE 6.6
Resistance-start split-phase motor: (a) circuit diagram; (b) phasor diagram; (c) torque-speed characteristic.

6.6 CAPACITOR-START SPLIT-PHASE MOTORS

A capacitor-start split-phase motor develops a much larger $I_{aw} \sin \alpha$, and hence a much larger locked-rotor torque, than does the resistance-start split-phase motor. The value of capacitance that produces the greatest locked-rotor torque in a capacitor-start split-phase motor causes a phase-displacement angle α of between 75° and 88°, compared to the 25° to 30° phase-displacement angle of the resistance-start split-phase motor. The circuit diagram and phase relationships for the capacitor-start split-phase motor are shown in Figures 6.8(a) and (b), respectively.

A typical torque-speed characteristic for a capacitor-start split-phase induction motor is shown in Figure 6.8(c). The starting curve shows the motor characteristics with both auxiliary and main windings energized. The running curve shows the characteristic behavior after the auxiliary winding is disconnected. A comparison of this

FIGURE 6.7
Cutaway view of a split-phase motor. (Courtesy Magnetek-Century Electric Company)

characteristic with that of the resistance-start split-phase motor in Figure 6.6(c) shows that the running characteristics of both machines are essentially the same. The significant difference between the two machines is the starting torque; about 130 percent rated for the resistance-start split-phase motor, and 300 percent rated for the capacitance-start split-phase motor. The high starting torque and good speed regulation of the capacitor-start motor make it well suited for applications to stokers, compressors, reciprocating pumps, and other loads of similar characteristics. This motor offers no means for speed control from a fixed-frequency source other than that obtainable by reconnecting for a different number of poles.

Note that neither the resistance-start split-phase motor nor the capacitance-start split-phase motor can attain synchronous speed. The rotating flux is dependent on

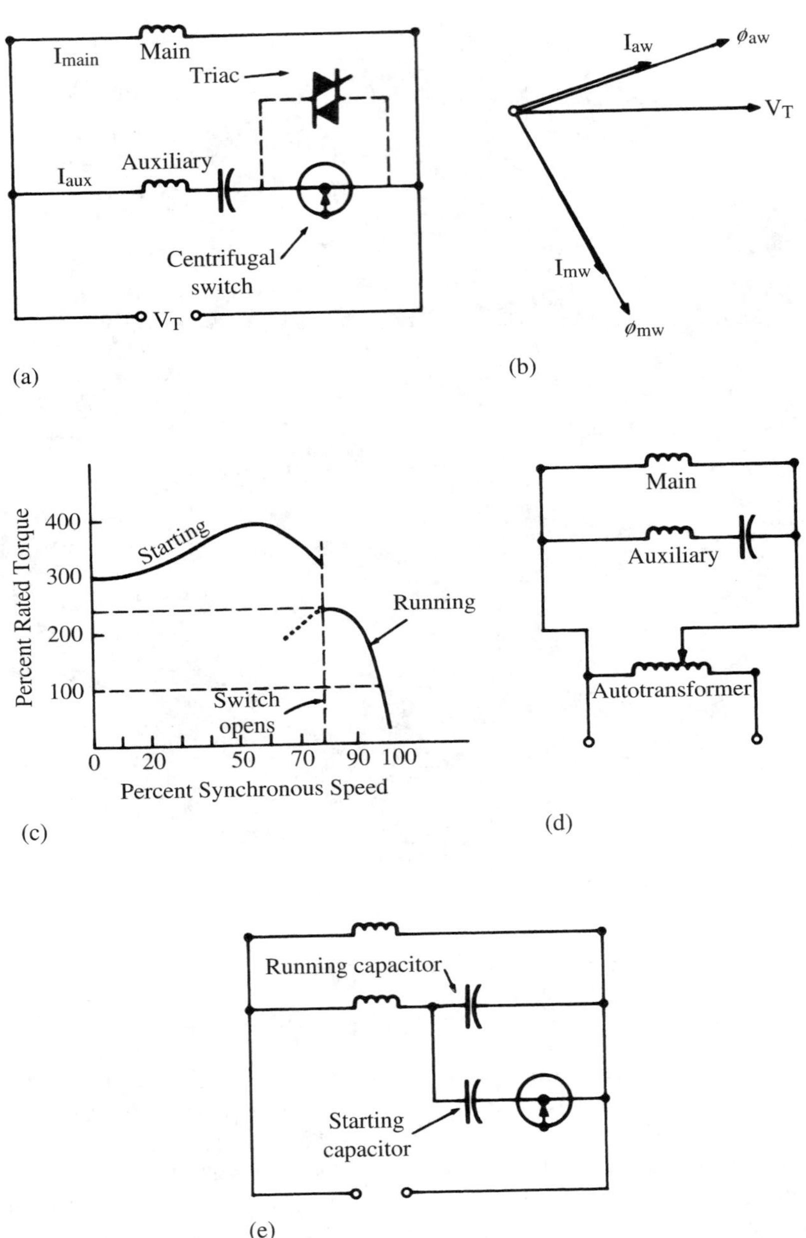

FIGURE 6.8
Capacitor motors: (a) circuit for capacitor-start motor; (b) phasor diagram corresponding to (a); (c) torque-speed characteristic for motor in (a); (d) permanent-split capacitor motor; (e) two-value capacitor motor.

induced current in the rotor to produce the quadrature field. Thus, as the rotor approaches synchronous speed, the speed-voltage induced in the rotor, the associated current in the rotor, and the quadrature flux approach zero. Hence, the accelerating torque will become zero at slightly below synchronous speed.

However, permanent-split capacitor motors and two-value capacitor motors (described below) are in effect two-phase motors, and at no load could attain synchronous speed.

Permanent-Split Capacitor Motor

A permanent-split capacitor motor utilizes a permanently connected auxiliary circuit containing a capacitor. There is no switch in the auxiliary circuit, and its operation is smoother and quieter than a capacitor-start or resistance-start motor of the same power rating. The value of capacitance for this type of motor is smaller than the one used in the capacitor-start motor and is a compromise between the best starting and best running performances. The primary field of application for a permanent-split capacitor motor is for shaft-mounted fans used in unit heaters and for ventilating fans. Its speed may be varied by a tapped or slide-wire autotransformer in the main line, as shown in Figure 6.8(d); by using an external resistor or reactor in series with the main winding or in series with both windings; or by adjusting the number of turns in the main winding through the use of taps and a selector switch or by solid-state control **[8]**.

Two-Value Capacitor Motor

A two-value capacitor motor, having the circuit shown in Figure 6.8(e), provides a greater amount of capacitance for starting than for running. This enables a greater locked-rotor torque than is obtainable with the permanent-split capacitor motor, and the reduced capacitance when running results in improved power factor, improved efficiency, and a higher breakdown torque **[8]**.

EXAMPLE 6.2 Using the given data for the split-phase motor windings in Example 6.1, determine (a) the capacitance required in series with the auxiliary winding in order to obtain a 90° phase displacement between the current in the main winding and the current in the auxiliary winding at locked rotor; (b) locked-rotor torque in terms of the machine constant.

Solution

(a) The winding impedances in Example 6.1 are

$$\mathbf{Z}_{\text{mw}} = 2.00 + j3.50 = 4.0311\underline{/60.2551^\circ}\ \Omega$$

$$\mathbf{Z}_{\text{aw}} = 9.15 + j8.40 = 12.4211\underline{/42.5530^\circ}\ \Omega$$

The circuit for the original conditions is shown in Figure 6.9(a).

$$\mathbf{I}_{mw} = \frac{120\angle 0^\circ}{4.0311\angle 60.2551^\circ} = \underline{29.7688\angle -60.2551^\circ \text{ A}}$$

$$\mathbf{I}_{aw} = \frac{120\angle 0^\circ}{12.4211\angle 42.5530^\circ} = \underline{9.6610\angle -42.5530^\circ \text{ A}}$$

The circuit diagram for the new condition (with a capacitor in series with the auxiliary winding) is shown in Figure 6.9(b), and a phasor diagram showing the respective

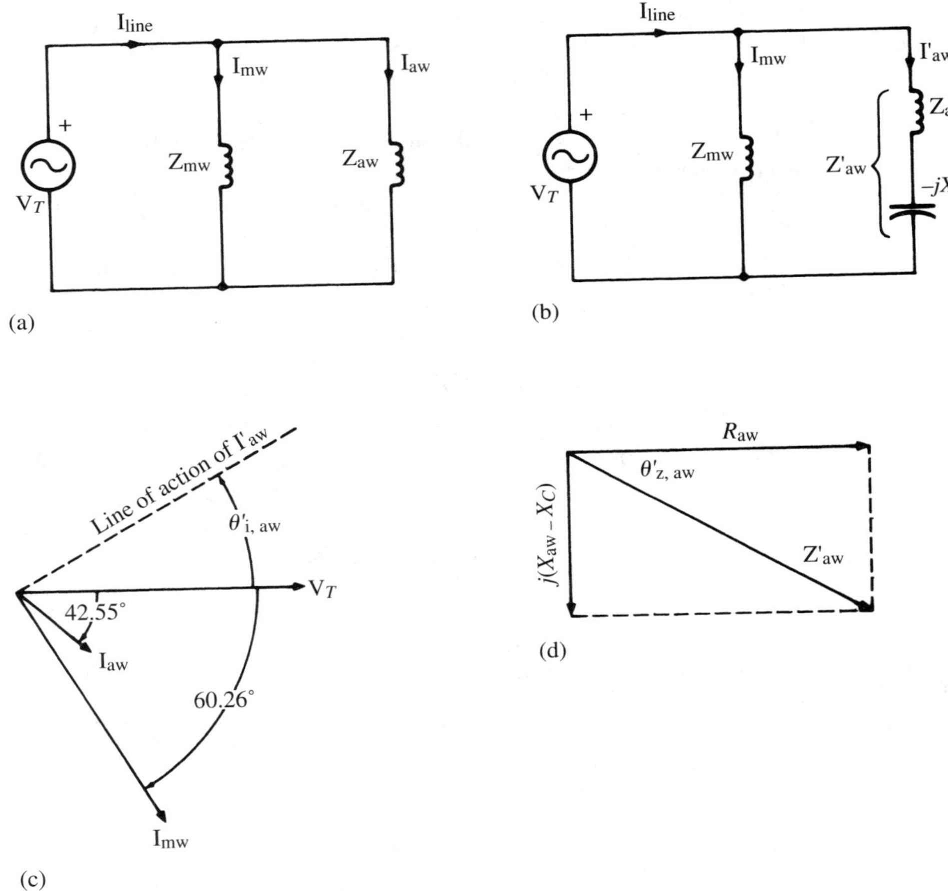

FIGURE 6.9
Diagrams for Example 6.2: (a) original circuit; (b) modified circuit; (c) phasor diagram for determining required phase angle of auxiliary current for new conditions; (d) impedance diagram for new auxiliary-circuit branch.

currents for the original condition and the desired location of the new auxiliary-winding current is shown in Figure 6.9(c). The required phase angle for $\mathbf{I}'_{\text{aw}}$ is

$$\theta'_{i,\text{aw}} = 90° - 60.26° = 29.74°$$

Applying Ohm's law to the auxiliary branch in Figure 6.9(b),

$$\mathbf{Z}'_{\text{aw}} = Z'_{\text{aw}}\underline{/\theta'_{z,\text{aw}}} = \frac{V_T\underline{/0°}}{I'_{\text{aw}}\underline{/29.74°}}$$

Thus,

$$\theta'_{z,\text{aw}} = -29.74°$$

From the impedance diagram, shown in Figure 6.9(d) for the new auxiliary circuit branch,

$$\tan(\theta'_{z,\text{aw}}) = \frac{X_{\text{aw}} - X_C}{R_{\text{aw}}} \quad \Rightarrow \quad X_C = X_{\text{aw}} - R_{\text{aw}} \cdot \tan(\theta'_{z,\text{aw}})$$

$$X_C = 8.40 - 9.15 \times \tan(-29.74°) = 13.628\ \Omega$$

$$C = \frac{1}{2\pi f X_C} = \frac{1}{2\pi 60 \times 13.628} = \underline{194.6\ \mu\text{F}}$$

(b)

$$\mathbf{I}'_{\text{aw}} = \frac{120\underline{/0°}}{9.15 + j8.40 - j13.628} = 11.387\underline{/29.74°}$$

$$T_{\text{lr}} = k_{\text{sp}} \cdot I_{\text{mw}} \cdot I_{\text{aw}} \sin\alpha$$

$$T_{\text{lr}} = k_{\text{sp}} \times 29.7688 \times 11.387 \times \sin 90° = \underline{338.9k_{\text{sp}}}$$

Note: The percent increase in locked-rotor torque obtained by capacitor start in Example 6.2, with respect to the locked-rotor torque obtained by resistor start in Example 6.1, is

$$\frac{338.9 - 107.1}{107.1} \times 100 = 216\%$$

Graphical Analysis of Example 6.2

Graphs of I_{aw}, α, and ($I_{\text{aw}} \sin\alpha$), plotted against X_C for Example 6.2 are shown in Figure 6.10. Note that the current in the auxiliary winding increases and then decreases with increasing capacitive reactance (resonance phenomena); angle α increases with increasing capacitive reactance; and the locked-rotor torque, represented by $I_{\text{aw}} \sin\alpha$, increases to some peak value, and then falls off with increasing capacitive reactance. Note that, for the given winding parameters, the optimum value of capacitive reactance that resulted in a phase-displacement angle of approximately 75° produced the greatest locked-rotor torque.

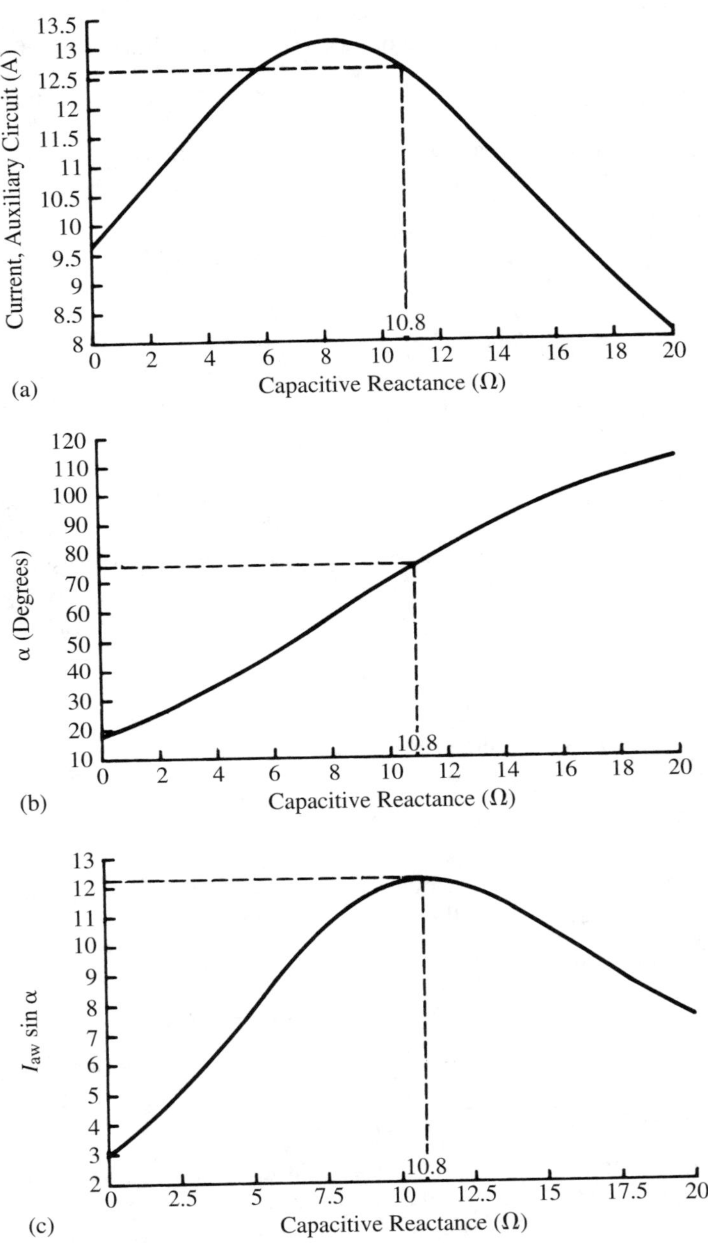

FIGURE 6.10
Graphs of auxiliary-winding current, phase-displacement angle α, and locked-rotor torque represented by $I_{aw} \sin \alpha$, for the capacitor-start motor in Example 6.2.

A comparison of the $I_{aw} \sin \alpha$ curve for the capacitor motor in Figure 6.10(c) with that of the split-phase motor in Figure 6.5(c) shows that with the same windings, phase shifting with capacitance can produce significantly greater locked-rotor torques than can phase shifting with resistance.

6.7 REVERSING SINGLE-PHASE INDUCTION MOTORS

Reversing a single-phase induction motor is accomplished by stopping the machine, interchanging the leads to the auxiliary circuit, and then restarting. This reverses the quadrature-axis flux, causing flux rotation to be in the opposite direction.

6.8 SHADED-POLE MOTORS

The shaded-pole motor, illustrated in Figure 6.11(a), utilizes a short-circuited coil or copper ring, called a *shading coil,* to provide starting torque. The shading coil is wound around a part of the pole face and acts as the short-circuited secondary of a transformer. In accordance with Lenz's law the current in the shading coil sets up a magnetomotive force (mmf) in a direction to oppose the mmf of the field that produced it. This causes a time delay in the buildup of flux in the shading portion of the pole face. The flux reaches its maximum in the unshaded portion of the pole face before it reaches its maximum in the shaded portion, thus causing a shifting flux in the direction from the unshaded part to the shaded part. The shifting flux "sweeping" the squirrel-cage rotor develops induction-motor action. Figure 6.11(b) shows a representative torque-speed characteristic for the shaded-pole motor. A shaded-pole motor cannot be reversed unless it has two sets of shading coils with two switches, one set for each direction of rotation. The principal field of application for shaded-pole motors is in clocks, record players, small fans, and other loads with similar characteristics.

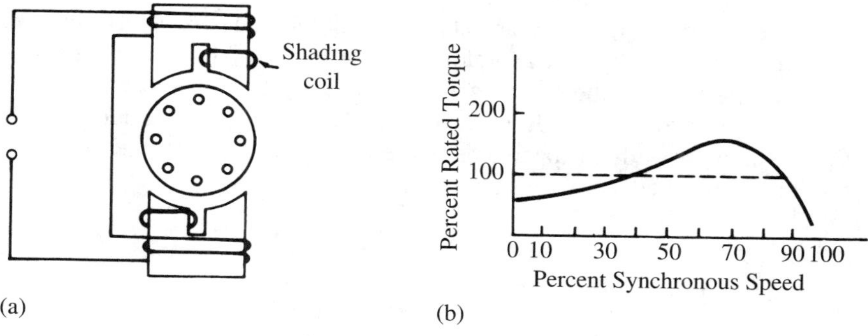

FIGURE 6.11
Shaded-pole motor: (a) construction; (b) torque-speed characteristic.

TABLE 6.1
Range of standard power ratings, single-phase motors

Motor	Power Range
Capacitor start	1 mhp to 10 hp
Resistance start	1 mhp to 10 hp
Two-value capacitor	1 mhp to 10 hp
Permanent-split capacitor	1 mhp to 1.5 hp
Shaded-pole	1 mhp to 1.5 hp

6.9 NEMA-STANDARD RATINGS FOR SINGLE-PHASE INDUCTION MOTORS

The standard frequency and voltage ratings for single-phase induction motors are

60 Hz—115 V and 230 V

50 Hz—110 V and 220 V

The ranges of NEMA-standard power ratings for single-phase induction motors are listed in Table 6.1.[1]

6.10 OPERATION OF THREE-PHASE MOTORS FROM SINGLE-PHASE LINES

In many developing countries, and in certain remote regions of the United States, and on isolated farms, the only source of power is a single-phase system. The power rating of single-phase motors, however, is generally limited to 10 hp. Hence, to fill the need for higher power requirements, three-phase motors are purchased and operated from a single-phase supply using capacitors for phase conversion.

Figure 6.12 shows the capacitor connections and switching arrangement required to obtain satisfactory starting and running characteristics when operating a three-phase motor from a single-phase source **[4]**. Contactor *S* is closed when starting, and opened when the rotor reaches approximately 80 percent synchronous speed.

An analytical study of 220-V, 60-Hz, three-phase induction motors, operating single phase with a capacitance phase converter, showed that the best starting performance (blocked rotor) was obtained with

$$C_1 + C_2 \approx 230\ \mu\text{F/hp} \tag{6–4}$$

The best running performance was obtained with

$$C_1 \approx 26.5\ \mu\text{F/hp} \tag{6–5}$$

---

[1]See Reference **[7]** for complete tables of power ratings, speed ratings, locked-rotor torque, and breakdown torque.

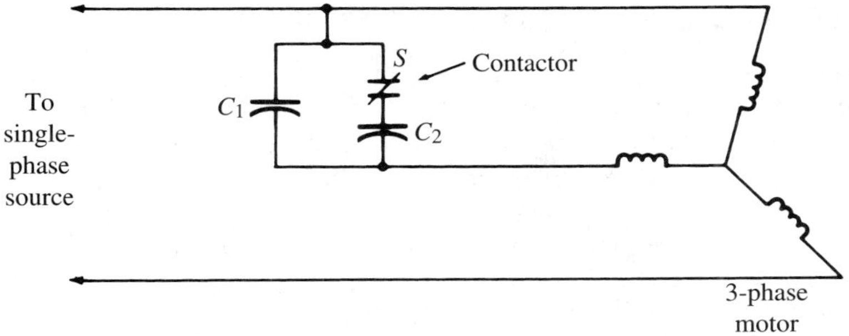

FIGURE 6.12
Capacitor connections and switching arrangement for operating a three-phase motor from a single-phase line.

Because of reduced breakdown torque and higher operating temperature, however, a three-phase motor operating single phase with a capacitance phase converter should be derated to approximately two-thirds of the three-phase rating. Thus,

$$P_{\text{rated }1\phi} = \frac{2}{3} P_{\text{rated }3\phi} \tag{6–6}$$

Three-phase wye-connected motors that have all six leads available, called an *open-wye motor,* provide greater flexibility for designing three-phase to single-phase conversion systems. Large open-wye motors such as a 100-hp, three-phase machine driving a rock crusher have been successfully operated from a single-phase source **[2]**.

EXAMPLE 6.3 A large exhaust fan for a farm building on a remote farm requires a 35-hp, 1175-rpm motor. The motor is to be operated from a 220-V, 60-Hz, single-phase system. Determine (a) the NEMA-standard horsepower rating of a three-phase, design *B* motor that will provide satisfactory service when used with a capacitance phase converter; (b) the required running capacitance; (c) the additional capacitance required for starting.

Solution

(a) From Eq. (6–6),

$$P_{\text{rated }3\phi} = 35 \times \frac{3}{2} = 52.5 \text{ hp}$$

From Table 5.1 in Chapter 5, the motor that most closely fits the requirement, without overloading, is a four-pole, 220-V, <u>60-hp</u> machine.

(b)

$$C_1 = 26.5\ \mu\text{F} \times 60 = \underline{1590\ \mu\text{F}}$$

(c)

$$C_1 + C_2 = 230\ \mu\text{F} \times 60 = 13{,}800\ \mu\text{F}$$

$$C_2 = (13{,}800 - 1590) = \underline{12{,}210\ \mu\text{F}}$$

6.11 SINGLE PHASING (A FAULT CONDITION)

Single phasing is a *fault condition* in which a three-phase motor is operating with one line open, as shown in Figure 6.13(b) and Figure 6.14(b). Although the three-phase motor will not start with one line open, if the motor is running when single phasing occurs, it will continue to run as long as the shaft load is less than 80 percent rated load and the remaining single-phase voltage is normal; rotation of the rotor produces a quadrature field that maintains rotation. *Single phasing is not the same as operating with a phase converter (as discussed in Section 6.10), and excessive vibration caused by single phasing may damage the motor and its connected load.*

Assuming the shaft load remains the same when single phasing as when operating three phase,

$$\sqrt{3} \cdot V_{\text{line}} \cdot I_{\text{line } 3\phi} \cdot F_{P3\phi} = V_{\text{line}} \cdot I_{\text{line } 1\phi} \cdot F_{P1\phi}$$

$$I_{\text{line } 1\phi} = \sqrt{3} \cdot I_{\text{line } 3\phi} \times \frac{F_{P3\phi}}{F_{P1\phi}} \tag{6–7}$$

where:

$I_{\text{line } 1\phi}$ = line current when single phasing
$I_{\text{line } 3\phi}$ = line current when operating three phase
$F_{P1\phi}$ = power factor when single phasing
$F_{P3\phi}$ = power factor when operating three phase

FIGURE 6.13
Wye-connected motor: (a) normal; (b) single phasing.

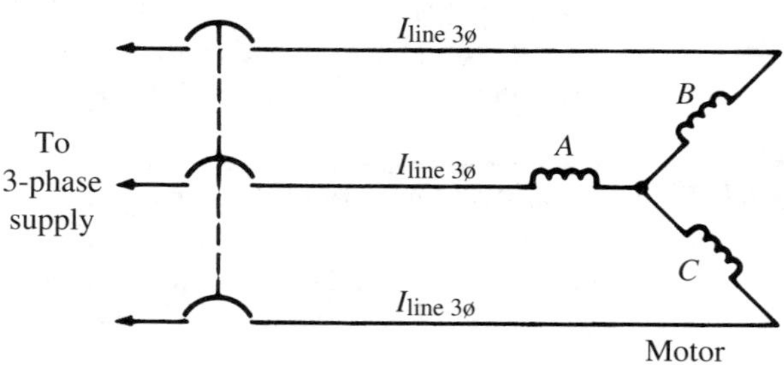

(a)

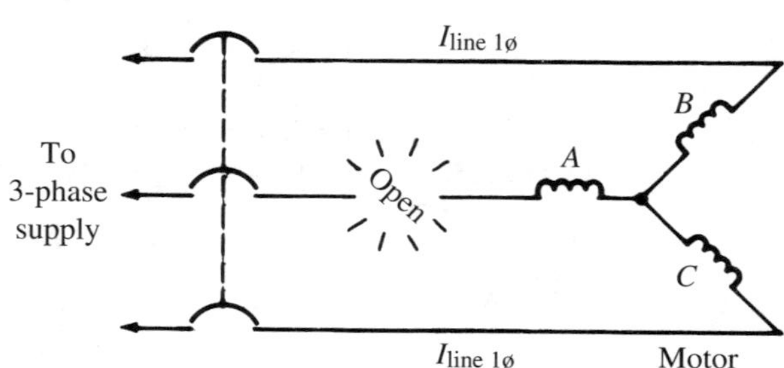

(b)

FIGURE 6.14
Delta-connected motor: (a) normal; (b) single phasing.

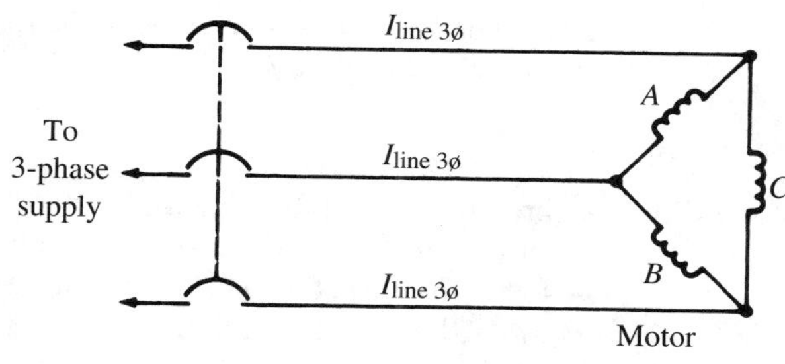

(a)

(b)

Thus, for the wye-connected stator in Figure 6.13 and for the delta-connected stator in Figure 6.14, when single phasing occurs, the current in the two remaining *lines* will be increased by 73 percent.[2] The distribution of current in the *stator windings* will depend on whether the windings are connected in wye or delta.

For the wye connection, shown in Figure 6.13(b), the phase current and line current are the same. Thus,

$$I_{YA} = 0 \qquad \textbf{(6–8)}$$
$$I_{YB} = I_{YC} = I_{\text{line } 1\phi}$$

For the delta connection, shown in Figure 6.14(b), the phase currents will divide in an inverse proportion to the paralleled impedances. Thus, the current distribution for the delta connection will be

$$I_{\Delta A} = I_{\Delta B} = \frac{1}{3} I_{\text{line } 1\phi} \qquad \textbf{(6–9)}$$
$$I_{\Delta C} = \frac{2}{3} I_{\text{line } 1\phi}$$

If single phasing occurs while operating at or near rated load, the increase in phase current will cause rapid heating of the windings, and therefore protective devices,

[2]A small reduction in power factor occurs when single phasing, resulting in a somewhat higher line current.

must be provided to trip the machines from the supply lines, or severe damage to stator and rotor winding may occur **[1],[3]**.

EXAMPLE 6.4 A three-phase, 2300-V, 350-hp, 78-A, 60-Hz, four-pole, wye-connected, design *B* motor operating at one-half rated load has an efficiency and power factor of 93.6 and 84.4 percent, respectively. Determine (a) motor line current and motor phase current; (b) motor line current and phase current if one line opens; assume the efficiency and power factor are essentially the same as before the open occurred; (c) the line and phase currents if the power factor when single phasing is 82.0 percent.

Solution

(a) $P_{\text{in}} = \sqrt{3}\, V_{\text{line}} I_{\text{line}} F_P \quad \Rightarrow \quad \dfrac{(350/2)746}{0.936} = \sqrt{3}(2300) \cdot I_{\text{line}}(0.844)$

$$I_{\text{line}} = \underline{41.5 \text{ A}}$$

$$I_{\text{phase}} = I_{\text{line}} = \underline{41.5 \text{ A}}$$

(b) When single phasing,

$$I_{\text{line }1\phi} = \sqrt{3} \times I_{\text{line }3\phi} \times \frac{F_{P3\phi}}{F_{P1\phi}} = \sqrt{3} \times 41.5 \times \frac{0.844}{0.844} = \underline{71.9 \text{ A}}$$

$$I_{\text{phase}} = I_{\text{line}} = \underline{71.9 \text{ A}}$$

(c)

$$I_{\text{line}} = 71.9 \times \frac{0.844}{0.820} = \underline{74.0 \text{ A}}$$

$$I_{\text{phase}} = I_{\text{line}} = \underline{74.0 \text{ A}}$$

Single phasing is a fault condition that should not be allowed to continue.

SUMMARY OF EQUATIONS FOR PROBLEM SOLVING

Locked-Rotor Conditions (Single-Phase Motors)

$$T_{\text{lr}} \propto I_{\text{mw}} \cdot I_{\text{aw}} \sin \alpha \qquad \alpha = |\theta_{i,\text{mw}} - \theta_{i,\text{aw}}| \qquad \textbf{(6–1, 6–2)}$$

Operating Three-Phase 220-V Motors From a Single-Phase Line

$$C_1 + C_2 \approx 230\ \mu\text{F/hp} \qquad \textbf{(6–4)}$$

$$C_1 \approx 26.5\ \mu\text{F/hp} \qquad \textbf{(6–5)}$$

$$P_{\text{rated }1\phi} = \frac{2}{3} P_{\text{rated }3\phi} \qquad \textbf{(6–6)}$$

Single Phasing (Fault Condition)

$$I_{\text{line }1\phi} = \sqrt{3} \cdot I_{\text{line }3\phi} \times \frac{F_{P3\phi}}{F_{P1\phi}} \qquad \textbf{(6–7)}$$

$$I_{YA} = 0 \qquad I_{YB} = I_{YC} = I_{\text{line }1\phi} \qquad \textbf{(6–8)}$$

$$I_{\Delta A} = I_{\Delta B} = \frac{1}{3} I_{\text{line }1\phi} \qquad I_{\Delta C} = \frac{2}{3} I_{\text{line }1\phi} \qquad \textbf{(6–9)}$$

SPECIFIC REFERENCES KEYED TO TEXT

1. Brighton, R. S., Jr., and P. N. Ranade. Why overload relays do not always protect motors. *IEEE Trans. Industry Applications,* Vol. IA-18, No. 6, Nov./Dec. 1982.
2. Elliott, K. C., Jr. Open-wye-type phase conversion systems. *IEEE Trans. Industry and General Applications,* Vol. IGA-6, No. 2, Mar./Apr. 1970.
3. Griffith, M. S. A penetrating gaze at one open phase: Analyzing the polyphase induction motor dilemma. *IEEE Trans. Industry Applications,* Vol. IA-13, No. 6, Nov./Dec. 1977.
4. Habermann, R., Jr. Single-phase operation of a three-phase induction motor with a single-phase converter. *Trans. AIEE-PAS,* 1954, p. 833.
5. Matsch, L. W., and J. D. Morgan. *Electromagnetic and Electromechanical Machines.* Harper & Row, New York, 1986.
6. McPherson, G. *An Introduction to Electrical Machines and Transformers.* Wiley, New York, 1981.
7. National Electrical Manufacturers Association. *Motors and Generators.* Publication No. MG 1-1998, NEMA, Rosslyn, VA, 1999.
8. Vienott, C. G., and J. E. Martin. *Fractional and Sub-fractional Horsepower Electric Motors.* McGraw-Hill, New York, 1986.

REVIEW QUESTIONS

1. Explain how a resistance-start split-phase induction motor develops a rotating magnetic field.
2. Explain how a capacitance-start split-phase induction motor develops a rotating magnetic field.
3. Explain how a shaded pole motor develops a rotating magnetic field.
4. A grindstone operated by a resistance-start split-phase motor will start only when spun by hand; it will accelerate to near rated speed and then go into a recycling state: slow-down, speed-up, slow-down, speed-up, etc. Determine the cause and explain the behavior.
5. A capacitor-start motor does not run when the breaker is closed; it hums loudly and starts to smoke. If the breaker is closed while the shaft is spun by hand, however, the machine comes up to speed and operates properly. What are the three probable faults?

PROBLEMS

6–1/4 An experimental 120-V, 1/4-hp, 60-Hz, split-phase motor has the following locked-rotor parameters referred to the respective main and auxiliary windings:

$$R_{\text{mw}} = 3.94\ \Omega \qquad X_{\text{mw}} = 4.20\ \Omega$$
$$R_{\text{aw}} = 8.42\ \Omega \qquad X_{\text{aw}} = 6.28\ \Omega$$

Determine (a) the amount of external resistance required in series with the auxiliary winding in order to obtain a 30° phase displacement between the current in the main winding and the current in the auxiliary winding; (b) the locked-rotor current drawn by each winding; (c) the line current at locked rotor.

6–2/4* (a) For the machine in Problem 6–1/4, tabulate and plot $I_{\text{aw}} \sin \alpha$ vs. total auxiliary circuit resistance for 30 equal increments of R_x ranging from 0.0 Ω to 15.0 Ω. (b) Determine from the curve the value of R_x that will provide maximum starting torque.

6–3/6 Using the given data in Problem 6–1/4, determine (a) the amount of capacitance required in series with the auxiliary winding in order to obtain a phase displacement of 80.6° between the current in the main winding and the current in the auxiliary winding; (b) line current at locked rotor with capacitance in the auxiliary-winding circuit.

6–4/6* For the machine in Problem 6–3/6, tabulate and plot $I_{\text{aw}} \sin \alpha$ vs. capacitive reactance for 30 equal increments of X_C ranging from 0.0 Ω to 20 Ω. (b) Determine from the curve the value of capacitive reactance that will provide maximum starting torque. (c) Determine the value of capacitance for (b). (d) Determine the line current at locked rotor.

6–5/10 A three-phase, 50-hp, 60-Hz, 220-V, delta-connected motor is to be operated from a 220-V single-phase system using a capacitance phase converter. Determine (a) the total capacitance required for starting; (b) the running capacitance; (c) the rerated motor horsepower.

6–6/10 A centrifugal pump that requires a 20-hp input at 1185 r/min must be driven from a 220-V, 60-Hz, single-phase system. (a) Select a three-phase motor (to be used with a capacitance phase converter) that most closely suits the desired characteristics of the load. (b) Determine the required running capacitance. (c) Determine the additional capacitance required for good starting characteristics.

6–7/11 A 460-V, 25-hp, 60-Hz, six-pole, delta-connected, design *B* motor with Code letter G operating at 75 percent rated output, has an efficiency, power factor, and slip of 94.4, 83.0, and 1.67 percent, respectively. (a) Determine motor line currents and motor phase currents. (b) Neglecting changes in power factor and slight changes in efficiency, determine the motor line currents and motor phase currents if a single-phase fault occurs.

---

*Solution by computer is recommended.

6–8/11 A three-phase, 4000-V, 150-hp, 890-r/min, 60-Hz, design *B*, wye-connected motor has a full-load current of 21 A and draws a locked-rotor current of 120 A. The motor is operating at 60 percent rated shaft output, and has an efficiency and power factor of 93.0 and 73.8 percent, respectively, at that load. A fault in the power supply causes an open in one of the three line leads feeding the motor. (a) Assuming the efficiency remains essentially the same, but the power factor drops to 70.1 percent, determine the motor line current and the current in each of the three motor phases. (b) If the motor is stopped, by opening its breaker, will it be able to restart with one line out? Explain.

7

Specialty Machines

7.1 INTRODUCTION

This chapter provides a brief introduction to machines with special applications. Machines, such as reluctance motors and hysteresis motors, are used for timing devices, tape recorders, turntables, and others with constant-speed requirements. They are used extensively in process industries, such as the man-made fiber industry, where many components of the process line must operate in synchronism.

Stepper motors are used in conjunction with pulse-driving circuits for precise positioning of mechanical systems. They are essential components of disk drives, printers, plotters, robots, and other applications that require accurate step-by-step positioning.

Linear induction motors are used to apply mechanical forces and to cause movement in a straight or curved line. They are used in conveyer systems, door openers, aircraft launchers, electromagnetic guns, liquid-metal pumps for nuclear reactors, high-speed rail transportation, and the like.

Universal motors have applications in low-power apparatus, such as vacuum cleaners, small power tools, and kitchen appliances.

7.2 RELUCTANCE MOTORS

The reluctance motor, also called a reluctance-synchronous motor, is an induction motor with a modified squirrel-cage rotor such as those shown in Figure 7.1. The *notches, flats,* or *barrier slots* provide equally spaced areas of high reluctance. The sections of rotor periphery between the high-reluctance areas are called salient poles; the number of salient poles must match the number of stator poles. The stator winding may be three phase, as discussed in Chapter 5, or single phase, as discussed in Chapter 6.

When the stator is energized, the rotor accelerates as a squirrel-cage induction motor. As the rotor approaches synchronous speed, the slip is very small and the rotating flux of the stator moves slowly past the salient poles of the rotor. At some critical speed, the low-reluctance paths provided by the salient poles cause them to *snap into synchronism* with the rotating flux of the stator. When this occurs, the slip is zero, induction-motor action ceases, and the rotor is pulled around by simple magnetic attraction called *reluctance torque.*

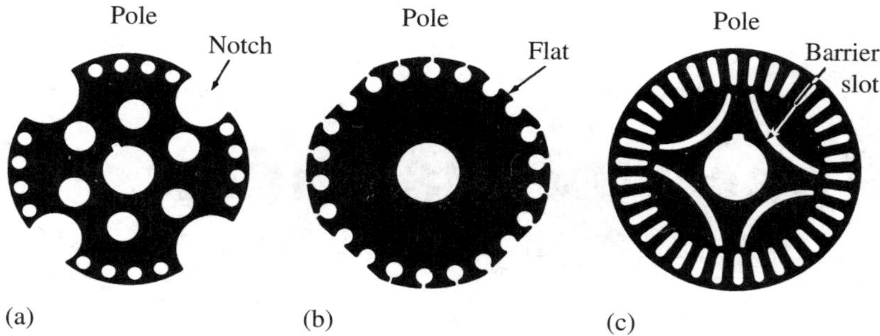

FIGURE 7.1
Types of rotor laminations used in reluctance motors. (Courtesy of Bodine Electric Company, Chicago, Illinois 60618)

Figure 7.2 shows simulated stroboscopic views of the reluctance rotor for different load conditions, as it operates in synchronism with the rotating stator flux. Assuming no shaft load and negligible rotational losses, the centerline of the rotor poles remains coincident with the centerline of the rotating poles produced by the stator, as shown in Figure 7.2(a).

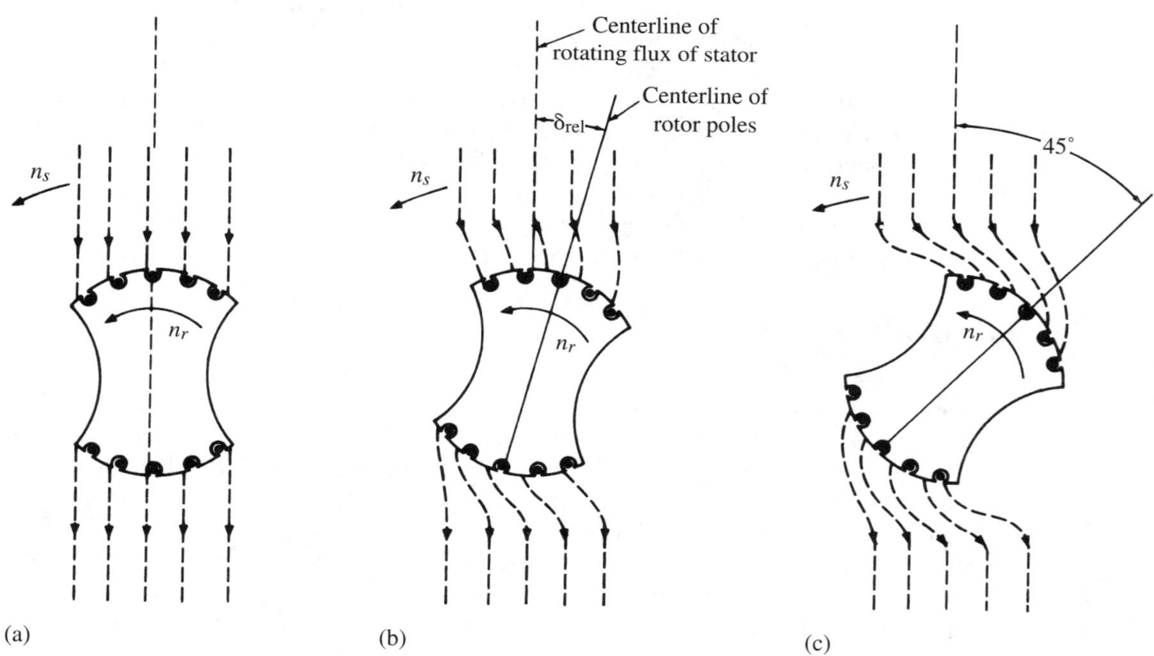

FIGURE 7.2
Simulated stroboscopic view of a reluctance motor, showing position of rotor with respect to the rotating field for (a) no load; (b) partial load; (c) maximum load.

With a step increase in shaft load, the rotor momentarily slows down, causing the salient poles of the rotor to lag the rotating poles of the stator. The angle of lag, called the *torque angle,* is shown in Figure 7.2(b). As soon as the transient slowdown for angular adjustment is completed, the rotor returns to synchronous speed at the new torque angle; the increase in reluctance torque caused by the increase in torque angle just balances the torque load on the shaft plus windage and friction. Reluctance torque increases with increasing torque angle, attaining its maximum value at $\delta_{rel} = 45°$, as shown in Figure 7.2(c).

If the shaft load is increased to a value that causes $\delta_{rel} > 45°$, the increased length of the magnetic flux path between the centerline of the rotating stator poles and the centerline of the corresponding rotor poles causes a decrease in flux and hence a decrease in magnetic attraction. In effect, the flux lines are "over-stretched," the rotor falls out of synchronism, and the machine runs as an induction motor at a slip speed determined by the amount of overload.

The average value of reluctance torque, expressed in terms of applied voltage, applied frequency, and torque angle is[1]

$$\underset{s=0}{T_{rel}} = K\left(\frac{V}{f}\right)^2 \times \sin(2\delta_{rel}) \qquad \textbf{(7–1)}$$

where:

T_{rel} = average value of reluctance torque
V = applied voltage (V)
f = line frequency (Hz)
δ_{rel} = torque angle (electric degrees)

Constant K takes into account the constants of the motor, such as reluctance, number of turns of wire, and units used.

As indicated in Eq. (7–1), and as shown in Figure 7.2, each increment increase in shaft load causes additional increases in the torque angle. Maximum reluctance torque, called the pull-out torque, occurs at $\delta_{rel} = 45°$.

The constant-speed characteristic of reluctance motors, their simple construction, ruggedness, and particularly their low cost make them desirable for the more economical timing devices, turntables, and tape recorders. They are also used in automatic processes such as are found in the food processing and packaging industries. Synthetic textile industries often require that as many as 90 to 100 machines, each driving different sections of the production line, be operated in synchronism over a wide range of speeds. Three-phase reluctance motors rated at 100 hp and more are used in larger power applications where the complexity and cost of "standard" synchronous motors are not warranted.[2]

---

[1]For a more detailed analysis and derivations see Reference **[3]**.

[2]"Standard" synchronous motors rotors have electromagnets that require a direct current source. See Chapter 8 for further details.

EXAMPLE 7.1 A certain 10-hp, four-pole, 240-V, 60-Hz reluctance motor, operating under rated load conditions, has a torque angle of 30°. Determine (a) torque load on the shaft; (b) torque angle if the voltage drops to 224 V. (c) Will the rotor pull out of synchronism?

Solution

(a)

$$n_s = \frac{120 \cdot f}{P} = \frac{120 \times 60}{4} = \underline{1800 \text{ r/min}}$$

$$P_{\text{shaft}} = \frac{T \cdot n}{5252} \quad \Rightarrow \quad 10 = \frac{T \times 1800}{5252}$$

$$T_{\text{rel}} = \underline{29.18 \text{ lb-ft}}$$

(b) With the same torque load and constant frequency,

$$[V^2 \sin(2\delta_{\text{rel}})]_1 = [V^2 \sin(2\delta_{\text{rel}})]_2$$

$$240^2 \sin(2 \times 30) = 224^2 \sin(2\delta_{\text{rel}})$$

$$\delta_{\text{rel}} = \underline{41.90°}$$

(c) Because δ_{rel} is less than 45°, the rotor will not pull out of synchronism.

7.3 HYSTERESIS MOTORS

The stator of an hysteresis motor (also called a *hysteresis-synchronous motor*) is the same as that for an induction motor. The rotor, however, consists of a smooth cylinder made of very hard permanent-magnet alloy material and a nonmagnetic support, as shown in Figure 7.3.

The principle of hysteresis-motor action is explained using the elementary hysteresis motor shown in Figure 7.4. The magnets represent the stator flux, which serves to induce opposite magnetic polarity in the hardened-alloy rotor. With the magnets stationary, as shown in Figure 7.4(a), the magnetic axis of the rotor poles is coincident with the magnetic axis of the stator.

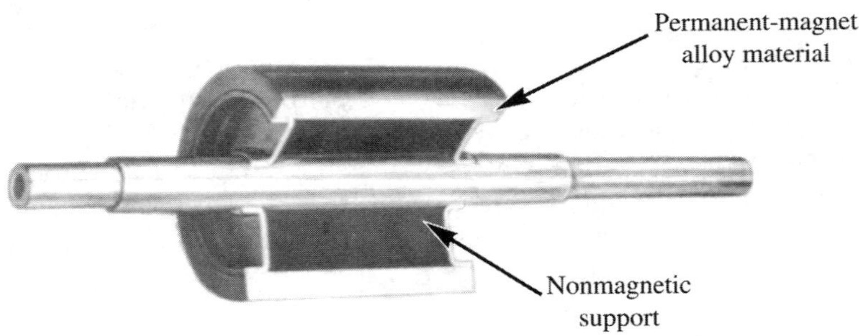

FIGURE 7.3
Rotor for hysteresis motor. (Courtesy of Bodine Electric Company, Chicago, Illinois 60618)

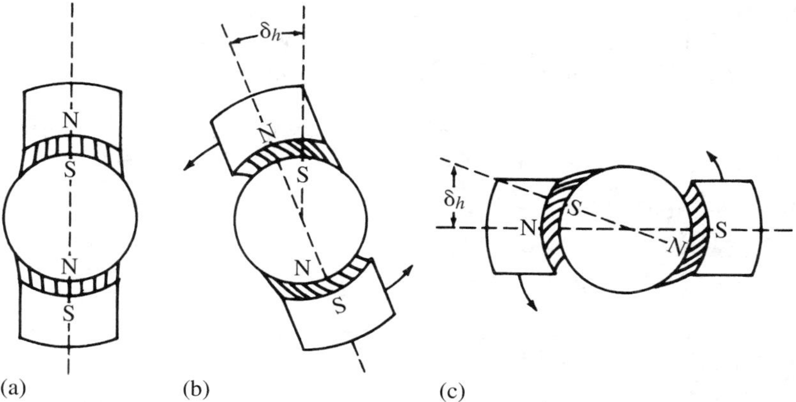

FIGURE 7.4
Hysteresis motor behavior: (a) magnets and rotor stationary; (b) and (c) rotor blocked and magnets rotating.

Spinning the "stator magnets," *with the rotor blocked,* as shown in Figures 7.4(b) and (c), provides a rotating magnetic field that exerts a torque on the induced magnetic poles of the rotor. As the stator poles rotate, the induced magnetic poles in the rotor are constantly reforming in new positions, following the rotating flux. Because of hysteresis, however, the *rotor poles always lag the stator poles by angle* δ_h. The constant lag-angle results in a constant force of attraction, and hence a constant accelerating torque. Releasing the rotor, and assuming no overload, the constant torque will accelerate the rotor to synchronous speed.

The energy transferred to the rotor by the rotating flux of the stator is in the form of hysteresis energy, and is expended as heat-power losses at locked rotor. From Section 1.7 of Chapter 1, the hysteresis power loss, expressed in terms of frequency and flux density, is

$$P_h = k_h f_r B_{max}^n \qquad \textbf{(1–11)}$$

where: f_r = frequency of flux reversal in rotor (Hz)
B_{max} = maximum value of flux density in air gap (T)
P_h = heat-power loss due to hysteresis (W)
k_h = constant

With the rotor free to turn, however, hysteresis power is converted to mechanical power. Expressing the mechanical power developed in the rotor during acceleration as a function of hysteresis loss and slip,

$$P_{mech} = P_h\left(\frac{1 - s}{s}\right) \qquad \textbf{(7–2)}$$

Equation (7–2) is similar to that for the mechanical power developed in an induction motor and is an extension of Eq. (4–34). Substituting Eq. (1–11) into Eq. (7–2), and expressing P_{mech} in terms of torque and rotor speed,

$$\frac{T_h n_r}{5252} = k_h f_r B_{max}^n \left(\frac{1 - s}{s}\right) \quad \textbf{(7–3)}$$

From Section 4.7 in Chapter 4,

$$n_r = n_s(1 - s) \quad \textbf{(4–4)}$$

$$f_r = sf_s \quad \textbf{(4–9)}$$

where: f_s = stator frequency (H_z)

Substituting Eqs. (4–4) and (4–9) into Eq. (7–3), and simplifying,

$$\frac{T_h n_s(1 - s)}{5252} = k_h \cdot s \cdot f_s B_{max}^n \left(\frac{1 - s}{s}\right)$$

$$T_h = \frac{5252 k_h f_s B_{max}^n}{n_s} \quad \textbf{(7–4)}$$

Expressing synchronous speed in terms of frequency, and substituting into Eq. (7–4),

$$n_s = \frac{120 \cdot f_s}{P}$$

$$T_h = \frac{5252 k_h f_s B_{max}^n}{(120 \cdot f_s)/P}$$

$$T_h = \frac{5252 k_h B_{max}^n}{120/P} \quad \textbf{(7–5)}$$

Equation (7–5) shows the hysteresis torque to be constant, independent of frequency and speed. In fact, *hysteresis torque depends solely on the area of the rotor's hysteresis loop.*

The hysteresis motor develops a constant hysteresis torque at all speeds from standstill up to but not including synchronous speed. At synchronous speed the rotor becomes magnetized along some random axis and pulled around by simple magnetic attraction. *The hysteresis power* P_h *is zero at synchronous speed, hysteresis torque is zero at synchronous speed, and the machine runs as a permanent-magnet synchronous motor, developing magnet torque.*

Figure 7.5 shows simulated stroboscopic views of the hysteresis rotor for different load conditions while operating at synchronous speed. Assuming no shaft load and negligible rotational losses, the centerline of induced rotor-magnets remains coincident with the centerline of the rotating poles produced by the stator, as shown in Figure 7.5(a).

With a step increase in shaft load, the rotor momentarily slows down, causing the induced rotor-magnets to lag the rotating poles of the stator by angle δ_{mag} as shown in Figure 7.5(b). As soon as the transient slowdown for angular adjustment is

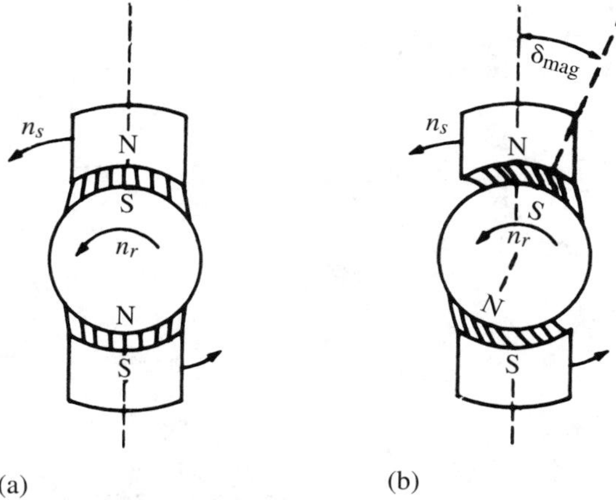

FIGURE 7.5
Simulated stroboscopic views of an hysteresis motor operating at synchronous speed: (a) no load; (b) partial load.

completed, the rotor returns to synchronous speed at the new torque angle; the increase in magnet torque caused by the increase in torque angle just balances the load torque on the shaft. *Note:* Hysteresis torque is also developed during the transient period when shaft load is increasing.

Magnet torque occurs only at synchronous speed and is proportional to the sine of the torque angle.[3] That is,

$$\underbrace{T_{\text{mag}}}_{s=0} \propto \sin(\delta_{\text{mag}}) \tag{7–6}$$

The magnet torque increases with increasing torque angle, attaining its maximum value at $\delta_{\text{mag}} = 90°$.

If the shaft load is increased to a value that causes $\delta_{\text{mag}} > 90°$, the rotor pulls out of synchronism, the magnet torque drops to zero, and although the machine develops hysteresis torque, it is not sufficient to carry the load torque that caused loss of synchronism.

A representative torque-speed curve for a hysteresis motor is shown in Figure 7.6. The dark vertical line indicates the normal range of operation while running at synchronous speed. Expressed mathematically,

$$\underbrace{T_{\text{mag}}}_{s=0} \leq T_h \tag{7–7}$$

$$\underbrace{T_{\text{mag, max}}}_{s=0} = T_h \tag{7–8}$$

Note: Maximum magnet torque, which precedes "pull-out," is equal to the hysteresis torque.

---

[3]Proportionality (7–6) may be derived from Eq. (8–14) for magnet power, developed for cylindrical-rotor synchronous machines in Chapter 8.

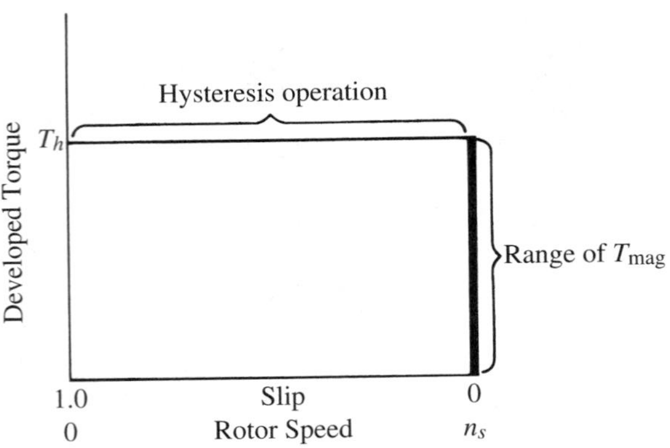

FIGURE 7.6
Representative torque-speed characteristic of an hysteresis motor.

Some Unique Features of the Hysteresis Motor

1. The constant hysteresis torque from locked rotor to synchronous speed permits the hysteresis motor to synchronize any load that it can accelerate; no other motor can perform in this manner.
2. The smooth rotor provides quiet operation. It does not suffer from magnetic pulsations caused by slots and/or salient poles that are present in the rotors of other motors.
3. The relatively high resistance and high reactance of the hysteresis rotor limit the starting current to approximately 150 percent rated current. This contrasts significantly with the reluctance rotor, whose low reactance and low resistance results in a locked-rotor current of approximately 600 percent rated current.

Although more expensive than a reluctance motor of equivalent power, the very quiet operation of the hysteresis motor makes it particularly desirable for driving phonographs and tape recorders. Furthermore, its ability to synchronize any load that it can accelerate and its smooth synchronization without any abrupt transition make it very adaptable to drives such as those used in synthetic fiber production, where snap-synchronization may cause product damage.

Hysteresis-Reluctance Motors

An hysteresis-reluctance motor combines the torque characteristic of an hysteresis motor with that of a reluctance motor. The characteristic is obtained by cutting reluctance notches in the hysteresis ring to produce a salient pole effect. The hysteresis-reluctance motor has applications in robotics, machine tools, and textile industries **[5]**.

7.4 STEPPER MOTORS

Stepper motors, also called *stepping motors* or *step motors,* are highly accurate pulse-driven motors that change their angular position in steps, in response to input pulses from digitally controlled systems. Stepper motors are used for precise positioning of mechanical systems and may be used without feedback. Examples of their applications include head positioning in computer disk drives, positioning of carriage, ribbon, print head, and paper feed in typewriters and printers, robots, and the like.

The step angle per input pulse depends on the construction of the stepper motor and the control system used. Steppers with a 45° step angle provide a *resolution* of 360/45 = 8 steps per revolution; steppers with a 1.8° step angle provide 360/1.8 = 200 steps per revolution; and so on. The total angle traveled by the rotor is equal to the step angle times the number of steps. Expressed in equation form,

$$\text{Resolution} = \text{steps/rev} = \frac{360°}{\beta} \tag{7–9}$$

$$\theta = \beta \times \text{steps} \tag{7–10}$$

where: β = step angle (deg/pulse)
θ = total angle traveled by rotor (deg)

The speed of a stepper motor is a function of the step angle and the *stepping frequency* (called the *pulse rate*). Thus,

$$n = \frac{\beta \times f_p}{360} \tag{7–11}$$

where: n = shaft speed (r/s)
f_p = stepping frequency (pulses/s)

EXAMPLE 7.2 A stepper motor has a 2.0° step angle. Determine (a) resolution; (b) number of steps required for the rotor to make 20.6 revolutions; (c) shaft speed if the stepping frequency is 1800 pulses/s.

Solution

(a)

$$\text{Steps/rev} = \frac{360}{2.0} = \underline{180}$$

(b)

$$\theta = \beta \times \text{Steps} \quad \Rightarrow \quad 20.6 \times 360 = 2.0 \times \text{steps}$$

$$\text{Steps} = \underline{3708}$$

(c)

$$n = \frac{\beta \times f_p}{360} = \frac{2.0 \times 1800}{360} = \underline{10 \text{ r/s}}$$

7.5 VARIABLE-RELUCTANCE STEPPER MOTORS

The variable reluctance stepper motor, shown in Figure 7.7, will be used to illustrate the simple construction and general principles of operation. The toothed stator and toothed rotor are constructed from soft steel that retains very little residual magnetism. Coils wound around the stator teeth provide the magnetic attraction that establishes

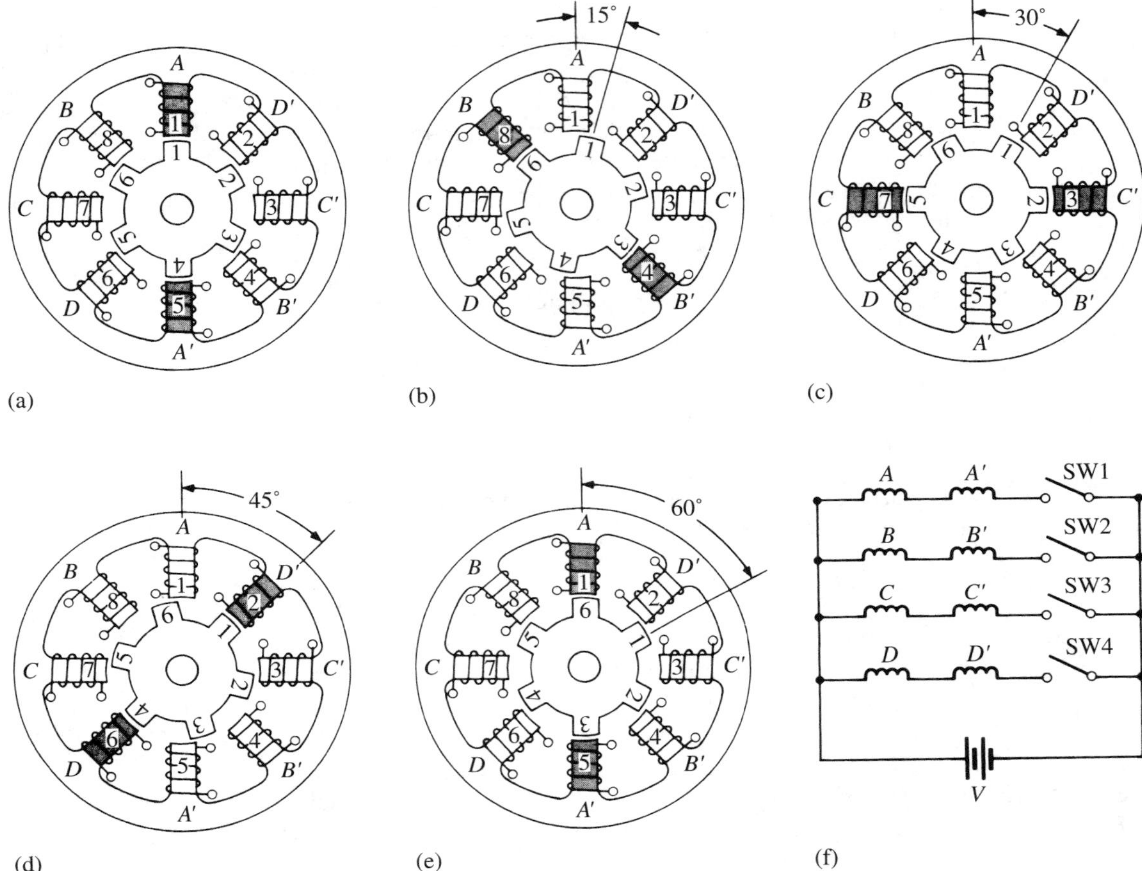

FIGURE 7.7
Variable-reluctance stepper motor showing different step positions corresponding to the switching sequence in (f). (Courtesy Superior Electric Company)

the rotor position. The motor shown in Figure 7.7 is called a *variable-reluctance stepper* because the reluctance of the magnetic circuit formed by the rotor and stator teeth varies with the angular position of the rotor. Energizing one or more stator coils causes the rotor to step forward (or backward) to a position that forms a path of least reluctance with the magnetized stator teeth.

A simple circuit arrangement for sequencing current to the stator coils is shown in Figure 7.7(f). The eight stator coils are connected in two-coil groups to form four separate circuits called *phases.* Each phase has its own independent switch. Although shown as mechanical switches in Figure 7.7(f), in actual practice, switching of phases is accomplished with solid-state control.

Figure 7.7(a) shows the position of the rotor with switch *SW*1 closed, energizing phase *A;* the rotor is in a position of minimum reluctance with rotor teeth 1 and 4 aligning with stator teeth 1 and 5, respectively. Closing switch *SW*2 and opening *SW*1 energizes phase *B,* causing rotor teeth 3 and 6 to align with stator teeth 4 and 8, respectively, as shown in Figure 7.7(b), for an angular step of 15°. Closing switch *SW*3 and opening switch *SW*2 energizes phase *C,* causing rotor teeth 2 and 5 to align with stator teeth 3 and 7, as shown in Figure 7.7(c). As each switch is closed, and the preceding one opened, the rotor moves an additional step angle of 15°. The stepping sequence, shown in Figures 7.7(a) to (e), follows the sequence of switches, repeating 1 through 4, over and over, until the desired number of revolutions or fraction of a revolution is achieved.

The direction of rotation for the 1-to-4 switching sequence, shown in Figure 7.7, resulted in clockwise (CW) stepping of the rotor. Reversing the sequence of pulses by closing the switches in the order of 4–3–2–1, will cause counterclockwise (CCW) stepping.

The relationship between step angle and the number of teeth in the rotor and the number of teeth in the stator is

$$\beta = \frac{|N_s - N_r|}{N_s \cdot N_r} \times 360 \qquad \textbf{(7–12)}$$

where: β = step angle in space degrees
N_s = number of teeth in stator core
N_r = number of teeth in rotor core

Thus, the step angle for the stepper motor in Figure 7.7 is

$$\beta = \frac{8 - 6}{8 \times 6} \times 360 = 15°$$

Half-Step Operation

Half-step operation is accomplished by using the modified pulsing sequence: *A, A & B, B, B, & C, C,* and so forth. This is illustrated in Figure 7.8 for three successive pulses. Energizing only phase *A* causes the rotor position shown in Figure 7.8(a); energizing phase *A* and phase *B* moves the rotor to the position shown in Figure 7.8(b); and energizing only phase *B* moves the rotor to the position shown in Figure 7.8(c). With each pulse the rotor moves 7.5° in a CW direction.

Microstepping

Microstepping utilizes two phases with unequal currents. For example, in Figure 7.8(b), instead of exciting phase *A* and phase *B* equally, the current in phase *A* is held constant while the current in phase *B* is increased in very small increments (pulses) until maximum current is reached. The current in phase *A* is then reduced to zero using the same very small increments. The resultant very small step angles, called

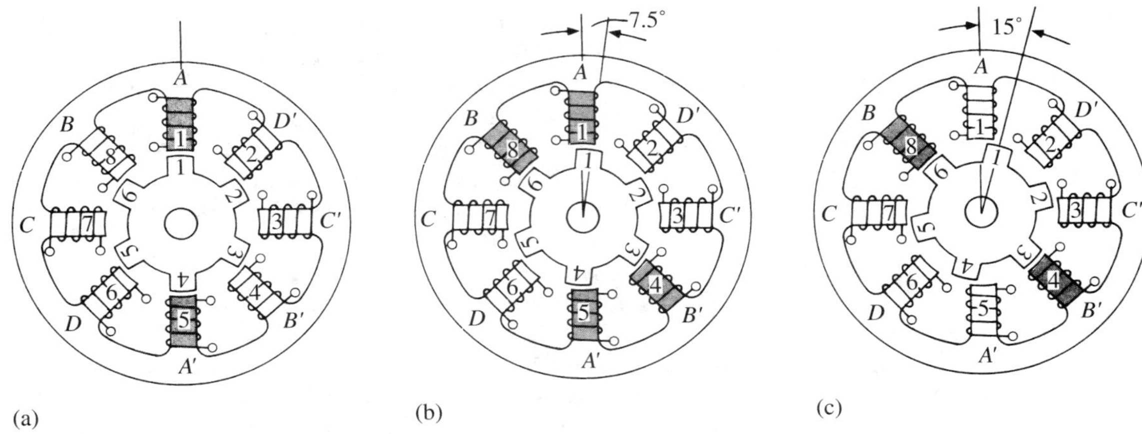

FIGURE 7.8
Variable-reluctance stepper motor sequenced for half-step operation, showing three successive pulses.

microsteps, provide smooth low-speed operation and high resolution. For example, a stepper motor with a resolution of 200 steps/rev (1.80° step angle) can with microstepping have a resolution of 20,000 steps/rev (0.0180° step angle).

Holding Torque

The holding torque, called *static torque,* is the maximum load-torque that can be applied to an *energized* stepper motor without having it slip poles.[4]

Assume that phase *A* is energized and the rotor is at rest, as indicated in Figure 7.9(a). If from this rest position the rotor is mechanically forced one 15° step clockwise, as shown in Figure 7.9(b), the resultant force of attraction on rotor teeth 1 and 4, by the respective magnetized stator teeth 1 and 5, develops a restoring torque. Releasing the rotor will cause it to flip back to the original position shown in Figure 7.9(a). Forcing the rotor 30° clockwise (two steps), while phase *A* is still energized, will result in the relative tooth positions shown in Figure 7.9(c). This is an unstable equilibrium position because rotor teeth 1 and 6 have equal forces of attraction; the net torque is zero. When released, the rotor can go in either direction: back two steps to the original position shown in Figure 7.9(a), or ahead an additional two steps to the position shown in Figure 7.9(d).

A plot of restoring torque vs. forced step displacement, called a *static-torque curve,* is shown in Figure 7.10. Note that maximum restoring torque is produced with one-step displacement. A two-step displacement has zero restoring force, but the rotor is unstable when released, and can move backwards to its original position, or move ahead for a total of four steps from its original position.

---

[4] The variable-reluctance stepper motor provides holding torque only while the windings are energized. Residual magnetism has no significant effect on the developed torque.

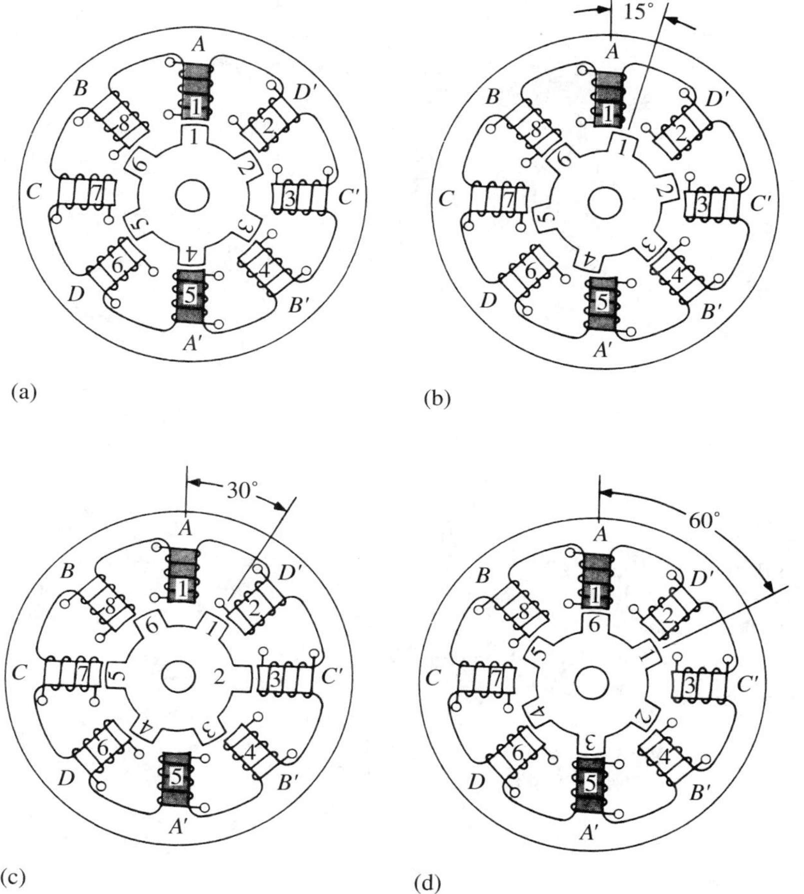

FIGURE 7.9
Rotor positions that develop restoring torque in a variable-reluctance stepper motor with phase *A* energized: (a) rotor at rest; (b) rotor forced 15° CW; (c) rotor forced 30° CW (unstable); (d) rotor flips to 60° position.

Step Accuracy

Step accuracy, expressed in percent, indicates the total error introduced by a stepper motor in a single step movement. The error is noncumulative in that it does not increase as additional steps are taken. Typical step accuracies for commercial stepper motors range from 1 to 10 percent of step size.

7.6 PERMANENT-MAGNET STEPPER MOTORS

A simplified diagram of a permanent-magnet stepper motor, shown in Figure 7.11, is used to illustrate the physical features common to this type of motor. The rotor, shown in Figure 7.11(b), has two toothed sections, separated by a permanent magnet. The two

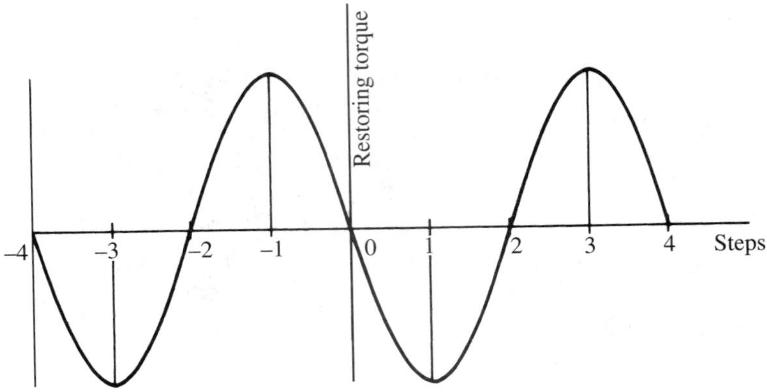

FIGURE 7.10
Typical static-torque curve for a stepper motor.

sections are offset from each other by one-half of a tooth pitch. The magnet provides opposite polarity to each section, developing five north poles in one section and five south poles in the other section. Figures 7.11(a) and (c) show the two end views of the combined stator and rotor. Note that all five north poles are on one end, and all south poles are on the other end. The stator coils shown in Figures 7.11(a) and (c) span both rotor sections. An axial view of the assembled stepper motor is shown in Figure 7.11(d).

Each rotor section contributes to the development of torque. In effect, the sections are in parallel. The net effect is that of a five-tooth rotor with a four-tooth stator. The step angle for the stepper in Figure 7.11 is

$$\beta = \frac{|N_s - N_r|}{N_s \times N_r} \times 360 = \frac{|4 - 5|}{4 \times 5} \times 360 = 18°$$

The principle of operation of a permanent-magnet stepper motor is developed using the circuit diagram, switching table, and the corresponding rotor positions in Figure 7.12. For simplicity, only the south section of the rotor is shown. The rotor positions are keyed to the switching sequence for CW rotation; phase *A* is energized by *SW*1, and phase *B* is energized by *SW*2. Half-stepping and microstepping may also be accomplished by using methods similar to those described for variable-reluctance steppers.

7.7 STEPPER-MOTOR DRIVE CIRCUITS

Figure 7.13 shows the general structure of a drive circuit for a variable-reluctance stepper motor such as the one shown in Figure 7.7. The input pulses that control the motor are filtered in block *F*, and passed on to bidirectional counter *C*. The output bits of the counter are fed to the input of decoder *D*. Electronic switches *S*1, *S*2, *S*3, and *S*4 energize and de-energize the windings in accordance with the input pulse commands.

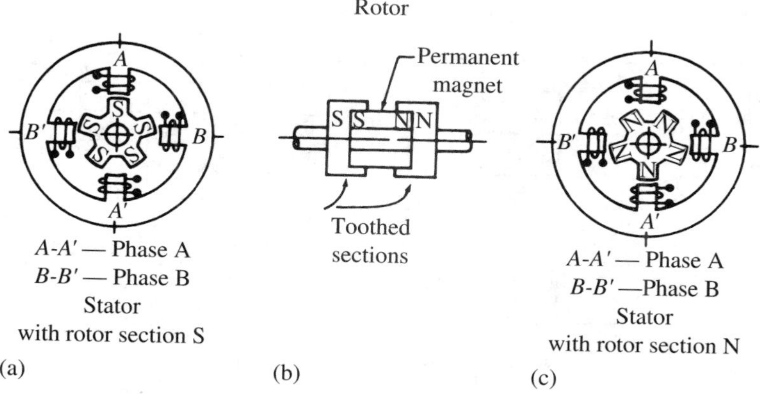

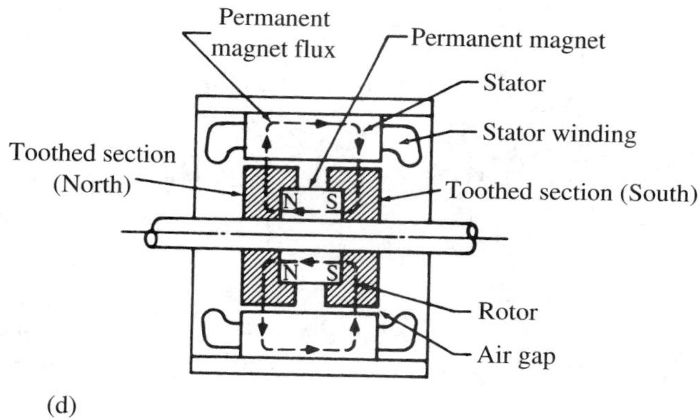

FIGURE 7.11
Permanent-magnet stepper motor: (a) stator and south section of rotor; (b) rotor; (c) stator and north section of rotor; (d) axial view of assembled motor. (Courtesy Superior Electric Company)

A typical solid-state switch is shown in Figure 7.14. Amplifier A converts the logic levels from the decoder to adequate voltage and current levels to turn transistor Q_A on and off. Resistor R_s is used to discharge the energy in the coil when Q_A is turned off.

The drive-circuit structure for a permanent-magnet stepper is essentially the same as far as the filter, counter, and decoder are concerned. The interconnection of phase windings, switches, and power supply, however, is significantly different because of the flux reversal requirement. Figure 7.15 shows one example of a switching arrangement that may be used for controlling a permanent-magnet stepper motor such as the one shown in Figure 7.12. The diodes provide a path for the decaying phase-currents whenever a conducting transistor is turned off. For example, referring to Figure 7.15(b), Q_1 is on and current i_1 is supplying phase A. When Q_1 is turned off, the

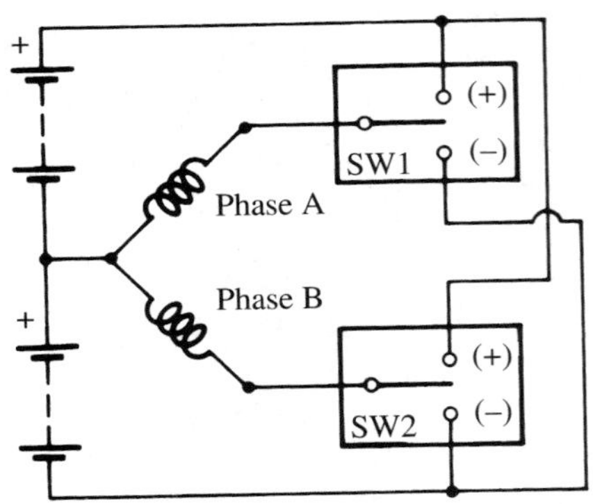

Step	CW Rotation		CCW Rotation	
	SW1	SW2	SW1	SW2
1	off	(–)	off	(–)
2	(+)	off	(–)	off
3	off	(+)	off	(+)
4	(–)	off	(+)	off
1	off	(–)	off	(–)

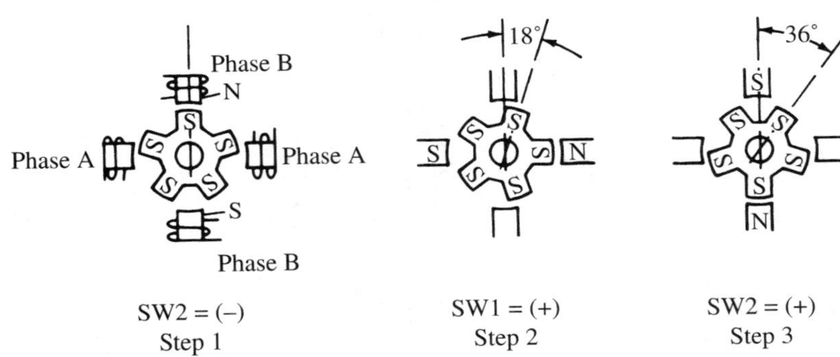

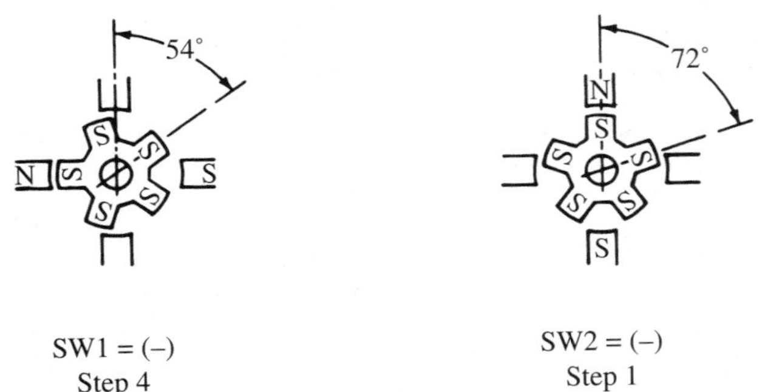

FIGURE 7.12
Circuit diagram of a permanent-magnet stepper motor with rotor positions keyed to switching sequence for CW rotation. (Courtesy Superior Electric Company)

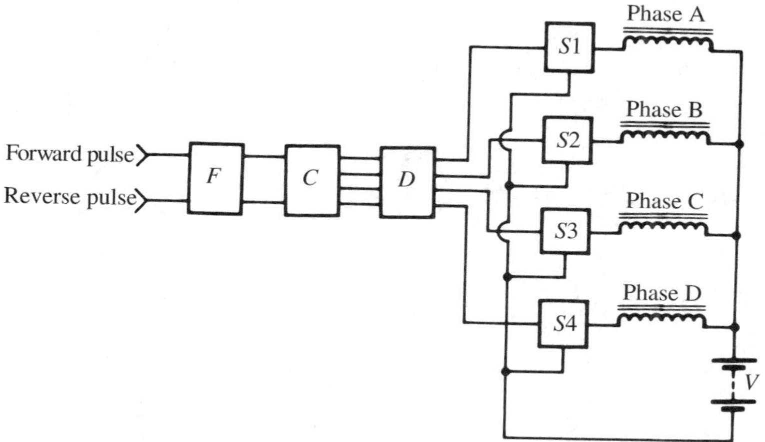

FIGURE 7.13
Drive circuit for a variable-reluctance stepper motor. (Courtesy Superior Electric Company)

inductance of phase A would prevent the immediate decay of current in it. As shown in Figure 7.15(c), however, the presence of diode D_2 provides a discharge path through an opposing V_2, quickly squelching the current. The main disadvantage of this switching arrangement is the requirement for two power supplies [7].

7.8 LINEAR INDUCTION MOTOR

The principle of linear-induction-motor (LIM) operation is the same as that for the three-phase induction motor discussed in Chapter 4 and the single-phase induction motor discussed in Chapter 6. Instead of using a rotating field to sweep a squirrel-cage rotor, however, the LIM uses a linearly moving field to sweep a conducting rail or the conductors of a ladder-type cage. This is illustrated in Figure 7.16(a), where a magnetic field, moving at velocity U_s, is shown sweeping the rungs of an aluminum ladder.

FIGURE 7.14
Solid-state switching circuit.

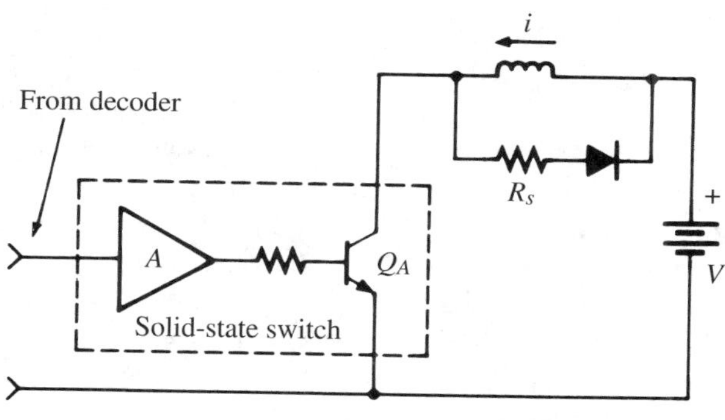

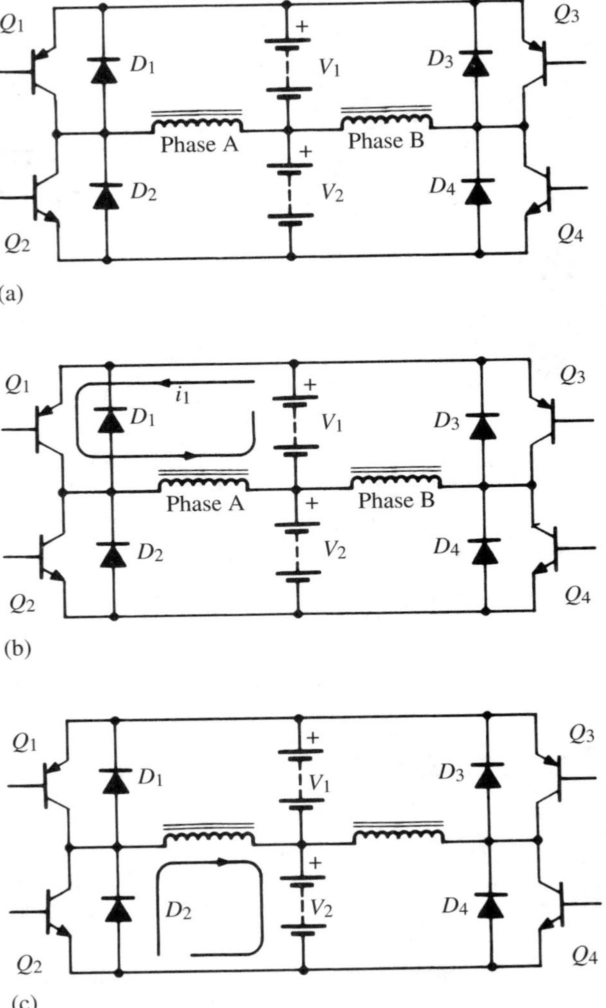

FIGURE 7.15
Drive circuit for a permanent-magnet stepper motor: (a) unenergized circuit; (b) Q_1 is in on state; (c) Q_1 off, diode D_1 provides discharge path. (Courtesy Superior Electric Company)

In accordance with Lenz's law, the voltage, current, and flux generated by the relative motion between a conductor and a magnetic field will be in a direction to oppose the relative motion. Hence, to satisfy Lenz's law the rungs must develop a mechanical force in the same direction as the sweeping magnetic field. For this to happen, "flux bunching" must occur on the left side of the conducting rung, as shown in Figure 7.16(b). The direction of induced current in the rung, as determined by the right-hand rule, is shown in Figure 7.16(c). In keeping with induction-motor nomenclature, the circuit that produces the sweeping flux is called the *primary,* and the conducting sheet, rail, or ladder is called the *secondary.* Thus, if the primary is fixed, the secondary (ladder) will move to the right. If the primary is free to move and the secondary is blocked, the primary will move to the left.

FIGURE 7.16
Diagrams of an elementary linear induction motor: (a) magnetic field sweeping rungs of ladder; (b) direction of flux around rung as determined by Lenz's law; (c) direction of current in rung as determined by right-hand rule.

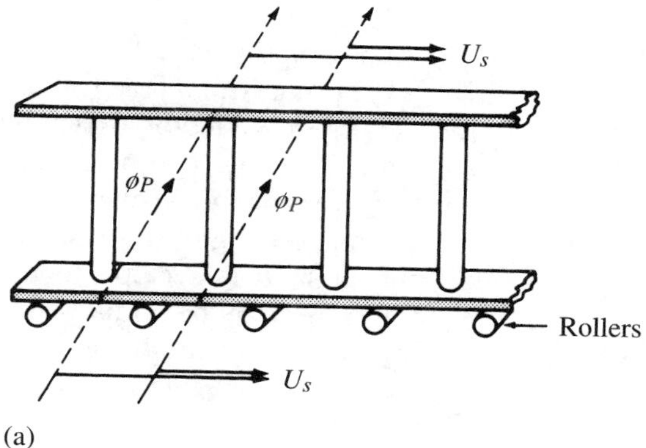

(a)

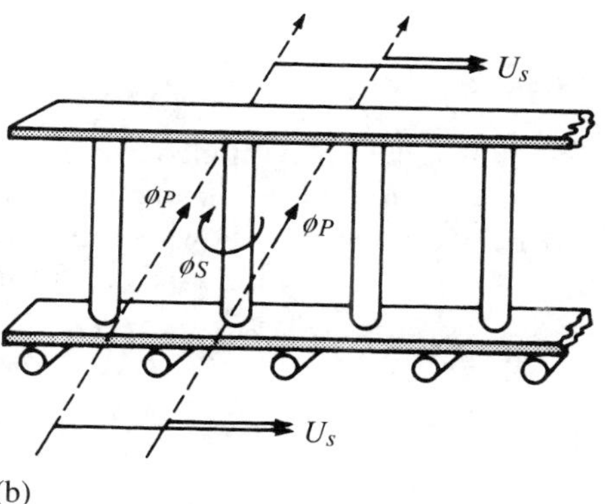

(b)

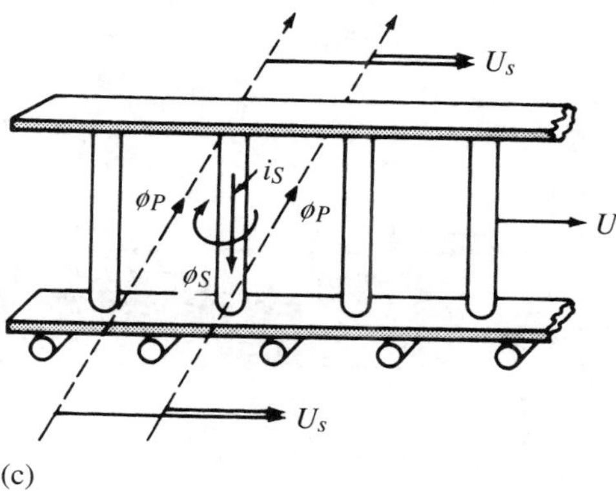

(c)

Reversal of a linear induction motor is accomplished by reversing the phase sequence of the primary voltage.

The primary of a practical three-phase LIM is wound in a manner similar to that for a squirrel-cage motor, except that the windings are laid in a straight line, as shown in Figure 7.17 for a two-pole machine with one coil per phase per pole; connections between poles are not shown. The secondary is a conducting sheet or rail of copper or aluminum. Although the conducting rail has no squirrel-cage bars, the induced eddy currents in the rail develop a force in a direction to oppose the relative motion **[2],[6]**.

The speed of the primary flux, called *synchronous speed,* is a function of the frequency of the applied voltage and the span of the primary coils (pole pitch).[5] In one cycle of applied voltage, the magnetic field travels a linear distance equal to two pole pitches **[4]**. Thus,

$$U_s = 2\tau f \tag{7–13}$$

where: U_s = synchronous speed (m/s)
τ = pole pitch (m)
f = supply frequency (Hz)

Unlike the rotary induction motor, *the synchronous speed of a linear induction motor is not dependent on the number of poles,* and any number of poles may be used, odd or even. More poles, however, increase the total force developed. Expressing the speed of the secondary in terms of synchronous speed and slip,

$$U = U_s(1 - s) \tag{7–14}$$

and

$$s = \frac{U_s - U}{U_s} \tag{7–15}$$

---

[5]See Appendix B for more details on machine windings, including a developed view of the stator of a squirrel-cage motor that simulates a three-phase linear motor.

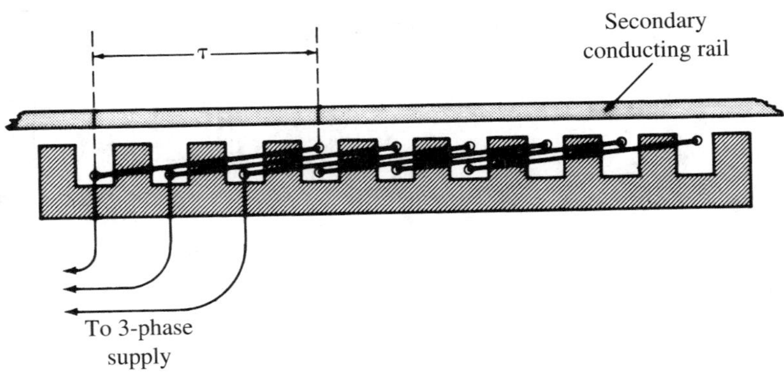

FIGURE 7.17
Linear induction motor showing conducting rail and pole pitch.

where: s = slip
U = speed of secondary (m/s)

The relatively large air gap in a LIM causes its slip at rated load to be much higher than that for a corresponding squirrel-cage induction motor.

Large three-phase LIMs have applications in traveling cranes, aircraft and missile launching, traction motors for high-speed railroads, conveyers, liquid-metal pumps, and other high-power applications where linear motion is required. Smaller linear induction motors that use a split-phase-type primary have applications in door closers and other low-power positioning requirements.

EXAMPLE 7.3 A three-phase moving-rail LIM has three poles and a pole pitch of 0.24 m. The LIM is operated from a 50-Hz system. Determine (a) the synchronous speed; (b) the rail speed assuming a slip of 16.7 percent.

Solution

(a) $$U_s = 2\tau f = 2 \times 0.24 \times 50 = \underline{24 \text{ m/s}}$$

(b) $$U = U_s(1 - s) = 24(1 - 0.167) = \underline{20.0 \text{ m/s}}$$

7.9 UNIVERSAL MOTOR

A universal motor is a relatively small *series motor* specially designed to operate at approximately the same speed and output power on either DC or single-phase AC at a frequency of 60 Hz or less, and an rms voltage that is approximately equal to the DC voltage. Such motors have NEMA-standard ratings of from 0.010 hp to 1.0 hp at a rated speed of 5000 r/min or above.

An elementary universal motor, shown in Figures 7.18(a) and (b), has its rotating part, called the *armature,* connected in series with the *series-field* winding. The rotating commutator and stationary brushes constitute a rotary switch that reverses the direction of the current in the armature coil as the coil rotates. The equivalent-circuit diagram is shown in Figure 7.18(c).

The direction of developed torque, and hence the direction of armature rotation, is independent of the polarity of the AC source. This is shown in Figures 7.18(a) and (b) for the respective alternations of the AC source. The direction of the mechanical force exerted on each conductor is determined by the flux bunching rule.

The torque developed by the universal motor is proportional to the flux density of the series field and the current in the armature conductors. That is,

$$T_D \propto B_p I_a \tag{7–16}$$

where: T_D = developed torque
B_p = flux density due to current in series-field winding
I_a = armature current

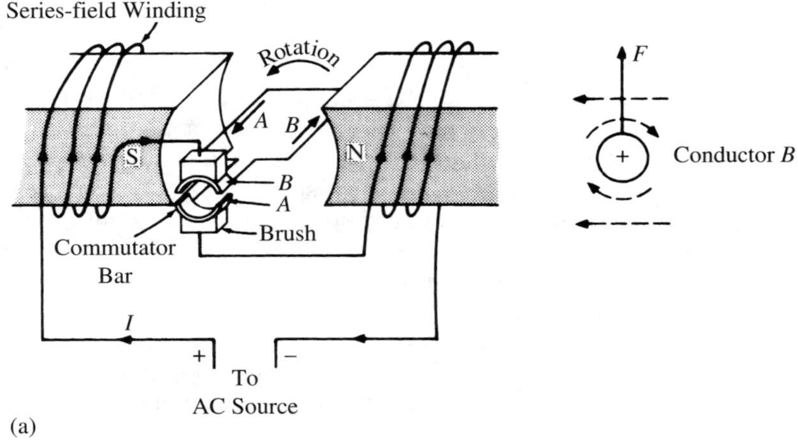

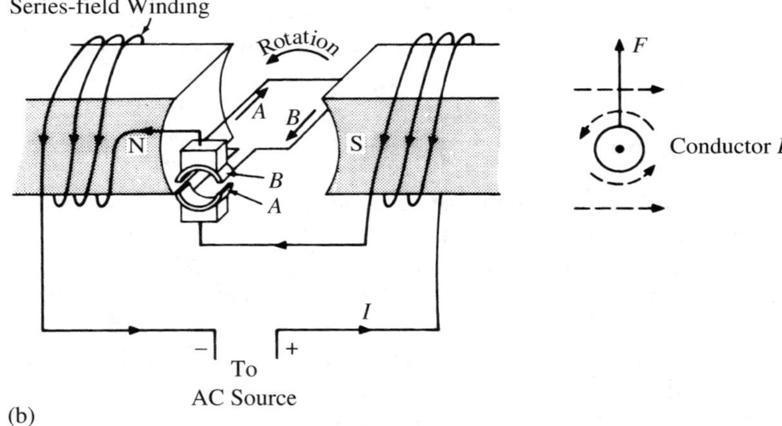

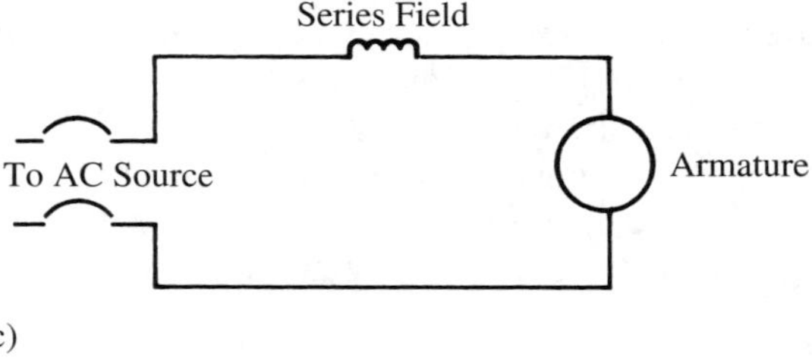

FIGURE 7.18
Elementary universal motor: sketches (a) and (b) show the same direction of rotation regardless of polarity of voltage source; (c) equivalent circuit.

As indicated in Figure 7.18(c), however, the current in the series field is the armature current. Hence, neglecting magnetic saturation effects,

$$B_p \propto I_a$$

Substituting into Eq. (7–16),

$$T_D \propto I_a^2 \tag{7–17}$$

As indicated in Eq. (7–17), the torque developed by a universal motor is approximately proportional to the square of the armature current.[6]

Universal motors can develop higher torques, can accelerate to higher speeds, and have a higher power-to-weight ratio than induction motors of the same power ratings.

Reversing the direction of rotation of a universal motor is accomplished by reversing the direction of current in the series field or in the armature, but not in both. Speed adjustment is accomplished by using an autotransformer or solid-state control to reduce the voltage applied to the motor; reducing the applied voltage reduces the armature current, which reduces the developed torque, and hence reduces the speed.

The torque and speed characteristics of the universal motor are essentially the same whether operating on AC or DC. Furthermore, because of its relatively small dimensions, no-load speeds in excess of 12,000 r/min are achieved without damage. Universal motors have applications in vacuum cleaners, portable power tools, and kitchen appliances.

Series motors, operating from a 25-Hz single-phase system are used for traction purposes on some electrified railroads **[1]**.

[6]More detailed theory and characteristics of commutator-type motors are given in Chapters 10 and 11.

SUMMARY OF EQUATIONS FOR PROBLEM SOLVING

Reluctance Motor

$$\underbrace{T_{\text{rel}}}_{s=0} = K\left(\frac{V}{f}\right)^2 \times \sin(2\delta_{\text{rel}}) \tag{7–1}$$

Hysteresis Motor

$$T_h = \frac{5252 k_h B_{\max}^n}{120/P} \tag{7–5}$$

$$\underbrace{T_{\text{mag}}}_{s=0} \propto \sin(\delta_{\text{mag}}) \tag{7–6}$$

$$\underbrace{T_{\text{mag}}}_{s=0} \leq T_h \tag{7–7}$$

$$\underbrace{T_{\text{mag, max}}}_{s=0} = T_h \tag{7–8}$$

Stepper Motors

$$\text{steps/rev} = \frac{360°}{\beta} \qquad \theta = \beta \times \text{steps} \tag{7–9, 7–10}$$

$$n = \frac{\beta \times f_p}{360} \qquad \beta = \frac{|N_s - N_r|}{N_s \cdot N_r} \times 360 \tag{7–11, 7–12}$$

Linear Induction Motors

$$U_s = 2\tau f \qquad U = U_s(1 - s) \tag{7–13, 7–14}$$

$$s = \frac{U_s - U}{U_s} \tag{7–15}$$

Series Motor

$$T_D \propto I_a^2 \tag{7–17}$$

SPECIFIC REFERENCES KEYED TO TEXT

1. Jones, A. J. Amtrack's Richmond static frequency converter project. *IEEE Vehicular Technology Society News,* May 2000, pages 4–10.
2. Laithwaite, E. R. Linear electric machines—A personal view. *Proc. IEEE,* Vol. 63, No. 2, Feb. 1975.
3. Mablekos, V. E. *Electric Machine Theory for Power Engineers.* Harper & Row, New York, 1980.
4. Nasar, S. A., and I. Boldea. *Linear Motion Machines.* Wiley, New York, 1976.
5. Rahman, M. A., and A. M. Osheiba. Steady-state performance analysis of polyphase hysteresis-reluctance motors. *IEEE Trans. Industry Applications,* Vol. IA-21, No. 4, May/June 1985.
6. Say, M. G. *Alternating Current Machines.* Wiley, New York, 1978.
7. Superior Electric Company. Step motor systems. Tech. Paper A-1-A.

REVIEW QUESTIONS

1. Explain the principle of reluctance-motor operation: how it starts, accelerates, and synchronizes.
2. At what stages of reluctance-motor operation (locked rotor, acceleration, steady-state synchronous speed, loading and unloading at synchronous speed) does the machine develop (a) induction-motor torque; (b) reluctance-motor torque?
3. Explain the principle of hysteresis-motor operation: how it starts, accelerates, and synchronizes.
4. At what stages of hysteresis-motor operation (locked rotor, acceleration, steady-state synchronous speed, loading and unloading at synchronous speed) does the machine develop hysteresis torque?
5. (a) Explain the principle of stepper-motor operation. (b) How does it differ in operation from a squirrel-cage induction motor?
6. Explain how half-stepping and microstepping are accomplished.
7. Will the overall accuracy of a stepper motor be greater at 100 steps than at 10 steps? Explain.
8. Explain the principle of linear-induction-motor operation, and state how a LIM may be reversed.
9. What determines the synchronous speed of a LIM? How does this compare with the factors that determine the synchronous speed of a squirrel-cage induction motor?
10. Explain why the torque developed by a universal motor varies as the square of the armature current.
11. Using suitable diagrams, show how the direction of rotation of a universal motor is reversed.
12. How may the speed of a universal motor be adjusted?

PROBLEMS

7–1/2 A reluctance motor operating at rated voltage and frequency has a shaft load that results in a torque angle of 25°. How much additional torque load (expressed in percent) can be added to the shaft without causing loss of synchronism?

7–2/2 A reluctance motor operating at 240 V, 60 Hz, has a torque angle of 15.6°. The motor is supplied by an isolated generator. Heavy demands on the generator cause a 10 percent drop in voltage and a 5 percent drop in generator speed. Assuming a constant torque load on the motor shaft, determine the new torque angle.

7–3/3 Rated torque load for a certain hysteresis motor occurs at a torque angle of 36°. Determine (a) percent rated torque the motor can carry before pulling out of synchronism; (b) hysteresis torque developed if the motor pulls out of synchronism.

7–4/3 Determine the magnet torque developed by a hysteresis motor, in percent of locked-rotor torque, for $\delta_{mag} = 60°$.

7–5/4 A stepper motor with a step angle of 1.8° is driven at a rate of 6000 pulses/s. Determine (a) resolution; (b) rotor speed; (c) number of pulses required to rotate 46.8°.

7–6/4 Determine the pulse rate required to obtain 3600 r/min from a stepper motor with a resolution of 500 steps/rev.

7–7/4 A stepper motor that makes 500 steps/rev is operated in half-step mode. Determine (a) resolution; (b) number of steps required to turn the rotor 75.6°.

7–8/4 Determine the required resolution for a stepper motor that is to operate at a frequency of 4000 pulse/s, and travel 137.52° in 0.0191 s.

7–9/8 A three-phase, four-pole LIM with a pole pitch of 32 cm is operated from a 240-V, 60-Hz system. When loaded, the motor runs at 30.72 m/s. Determine the slip.

7–10/8 (a) Determine the required pole pitch for a two-pole, 60-Hz, 450-V LIM that will provide a synchronous speed of 40.6 m/s. (b) Assuming a 32 percent slip, determine the speed of the secondary.

8

Synchronous Motors

8.1 INTRODUCTION

Synchronous motors are used principally in large power applications because of their high operating efficiency, reliability, controllable power factor, and relatively low sensitivity to voltage dips. They are constant-speed machines with applications in mills, refineries, power plants, and the like, to drive pumps, compressors, fans, pulverizers, and other large loads, and to assist in power-factor correction. One very interesting application is the 44-MW, 10-kV, 60-Hz, 50-pole, 144 r/min synchronous motors used to drive the *Queen Elizabeth II* passenger ship. A solid-state volts/hertz drive circuit provides speed control through frequency adjustment.

Synchronous machines designed specifically for power-factor control have no external shafts and are called *synchronous condensers.* They "float" on the bus, supplying reactive power to the system. The direction of the reactive power, and hence the power factor of the system, is adjusted by changing the field excitation of the machine.

Unless otherwise specified, when discussing the behavior of individual motors and/or generators, it will be assumed that the machine is connected to a power source of unlimited capacity and zero impedance, called an *infinite bus.* The terminal voltage and frequency of the infinite bus remain constant and are unaffected by any power drawn from or supplied to the infinite bus. Large power systems in the United States and other highly industrialized countries may be considered to approximate those of the infinite bus.

8.2 CONSTRUCTION

The stator of a three-phase synchronous motor, called the *armature,* is identical to that of a three-phase induction motor, and when energized from a three-phase supply develops a rotating field in the same manner as described for induction motors.

Coils wound (or mounted) on the rotor are energized from a DC source to form electromagnets that lock in synchronism with poles of opposite polarity produced by the rotating flux of the stator. The number of rotor poles is made equal to the number of stator poles.

The rotor of a two-pole 60-Hz motor, such as the cylindrical rotor shown in Figure 8.1, is constructed from a solid-steel forging so as to withstand the large centrifugal stresses inherent in high-speed operation. *Cylindrical rotors* (also called *round rotors*) cannot accelerate high-inertia loads, however, and are thus limited in application to pumps, fans, blowers, and other loads with similar low-starting-torque requirements.

Synchronous motors built for use with high-inertia low-speed loads have many poles projecting radially outward from a steel *spider,* as shown in Figure 8.2. These *salient poles* are bolted or keyed to the spider, and the spider is keyed to the shaft.

The squirrel-cage winding, called a *pole-face* winding, *amortisseur* winding, or *damper* winding, shown in Figure 8.2, is used to accelerate the rotor to near synchronous speed.[1] In those applications where high accelerating torque is required throughout the accelerating period, the rotor is designed with a double squirrel-cage winding, similar to that in a NEMA design *C* induction motor. The damper winding also helps suppress oscillations (called *hunting*) caused by torque pulsations when driving loads such as reciprocating compressors. Currents generated in the damper windings during such rotor oscillations provide a countertorque that serves to dampen the oscillations.

The *excitation windings,* also called *magnet windings, field coils,* or *field windings,* form the rotor poles. The individual field coils are connected in series or series–parallel combinations, in a manner that provides alternate north and south poles. The rotor circuit terminates at slip rings, also called collector rings, and graphite

[1]Some salient-pole designs obtain starting and accelerating torque from eddy currents and hysteresis losses in massive solid-steel pole plates that form the faces of the rotor poles. The eddy currents provide induction-motor torque, and the hysteresis effect provides hysteresis torque.

FIGURE 8.1
Two-pole cylindrical rotor for high-speed synchronous-machine applications. (Courtesy TECO Westinghouse)

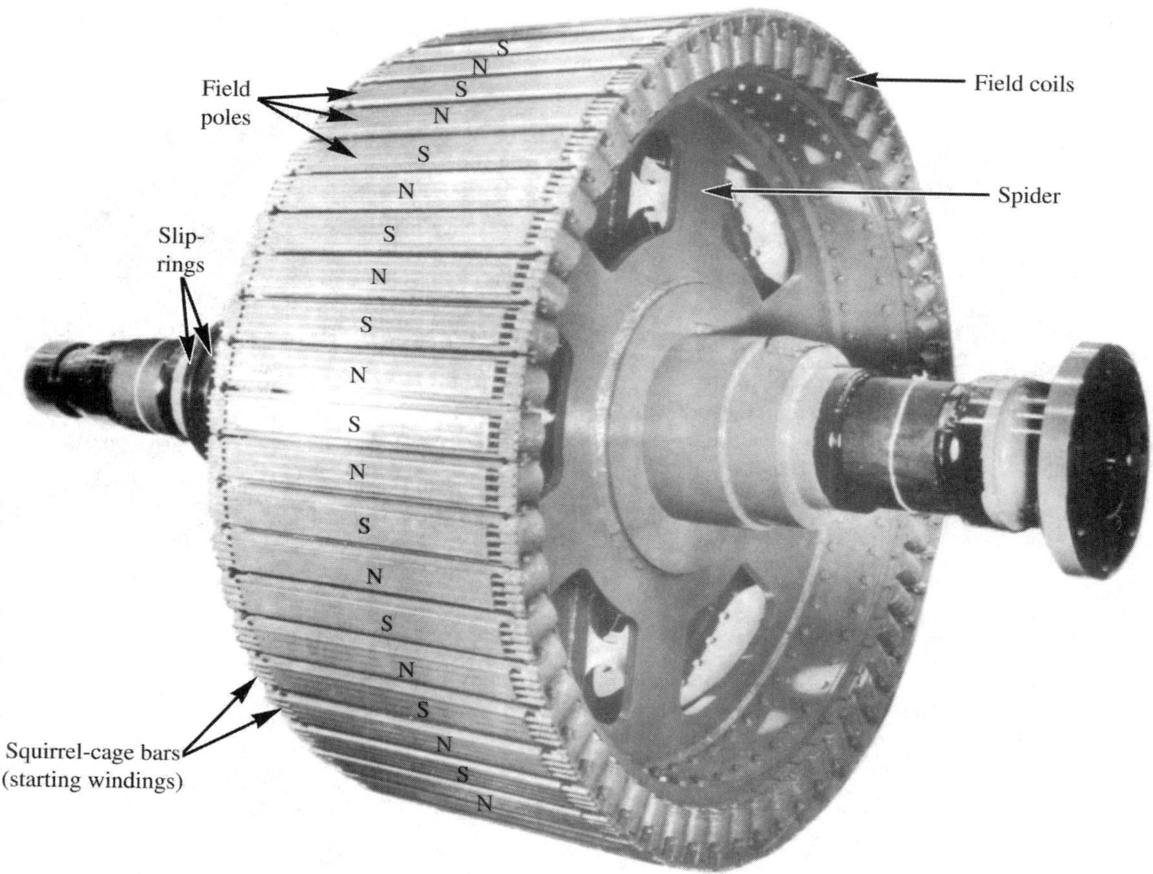

FIGURE 8.2
Salient-pole rotor for slow-speed synchronous-machine applications. (Courtesy TECO Westinghouse)

brushes pressed against the rings provide the connection between the field windings and the DC excitation source.

A salient-pole rotor with a shaft-mounted DC exciter is shown in Figure 8.3. Direct current from the exciter armature is supplied to the field windings by means of carbon brushes (not shown) riding on the commutator that connects to carbon brushes riding on the slip rings.

A salient-pole rotor equipped with a brushless excitation system is shown in Figure 8.4. Brushless excitation is provided by a small three-phase exciter armature, a three-phase rectifier, and control circuitry, all mounted on the same shaft.

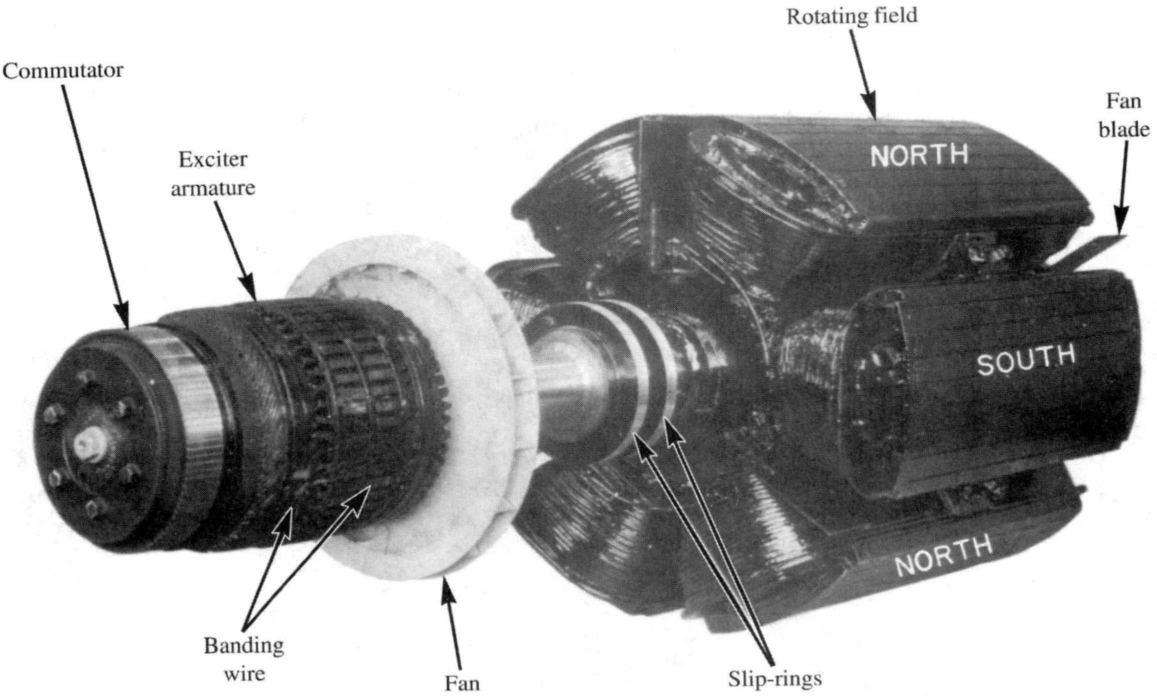

FIGURE 8.3
Salient-pole rotor with shaft-mounted DC exciter. (Courtesy GE Industrial Systems)

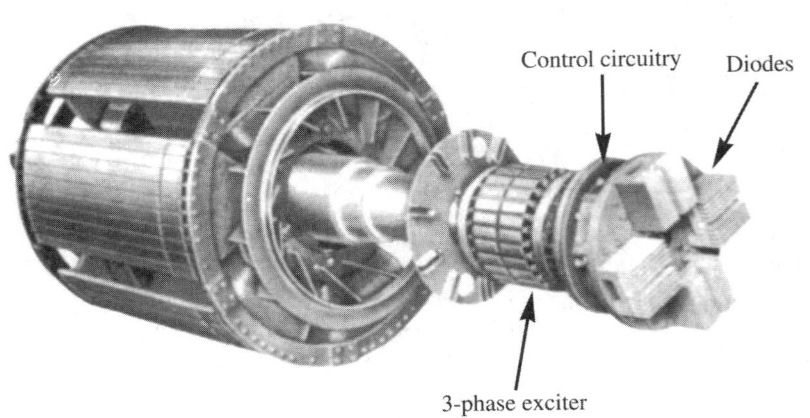

FIGURE 8.4
Salient-pole rotor with a brushless excitation system. (Courtesy Dresser Rand, Electric Machinery)

8.3 SYNCHRONOUS MOTOR STARTING

When starting, the rotor is allowed to accelerate to its maximum speed as an induction motor using its built-in damper windings.[2] At this speed the slip is very small and the rotating flux of the stator moves very slowly relative to the revolving rotor. Direct current is then applied to the magnet windings, forming alternate north and south poles that "lock" in rotational synchronism with the corresponding opposite poles of the rotating flux, the slip is zero, all induction-motor action ceases, and the machine operates at synchronous speed. The synchronous speed, as developed in Section 4.5 in Chapter 4, is expressed in terms of the stator frequency and the number of poles:

$$n_s = \frac{120 \cdot f}{P} \qquad \text{r/min} \qquad \textbf{(4–1)}$$

If the rotor magnets are energized before the machine reaches its maximum speed as an induction motor, the rotor may not synchronize and severe vibration will occur; every time a pole of the rotating stator flux passes a rotor pole, alternate attraction and repulsion occurs. Such out-of-step operation, called *pole slipping,* also causes cyclic current surges and torque pulses at slip frequency in the armature windings.

The adverse effects of frequent starting, or operating with unbalanced three-phase line voltages, are the same as for induction motors.[3]

A simplified circuit diagram for rotor and stator connections is shown in Figure 8.5. A varistor, or resistor and switch, connected across the field leads during locked rotor and acceleration, prevents a high induced emf in the field windings, and the induced current in the circuit formed by the field windings and external resistor provides additional induction-motor torque.

The circuit for a brushless excitation system is shown in Figure 8.6. A frequency-sensitive solid-state control circuit monitors the frequency of the emf induced in the motor field winding by the rotating flux of the stator. The frequency of the emf is the same as that in the squirrel-cage winding and is a function of the frequency of the applied stator voltage and the slip. That is,

$$f_r = sf_s \qquad \textbf{(8–1)}$$

At locked rotor $s = 1.0$, causing the rotor frequency to equal the applied stator frequency. As the rotor accelerates, the slip becomes smaller, and the frequency decreases.

When the frequency-sensitive circuit observes the "high frequency" generated in the rotor coils at locked rotor, and during high-slip acceleration, it closes SCR-2 and opens SCR-1; opening SCR-1 blocks the exciter current, and closing SCR-2 connects the discharge resistor across the field windings. At near-synchronous speed, the rotor

---

[2]Although some synchronous motors are accelerated by eddy-current torque and hysteresis torque, most are accelerated by induction-motor torque through damper windings.

[3]In Chapter 5, see Section 5.13 for frequent starting, and Section 5.15 for unbalanced voltages.

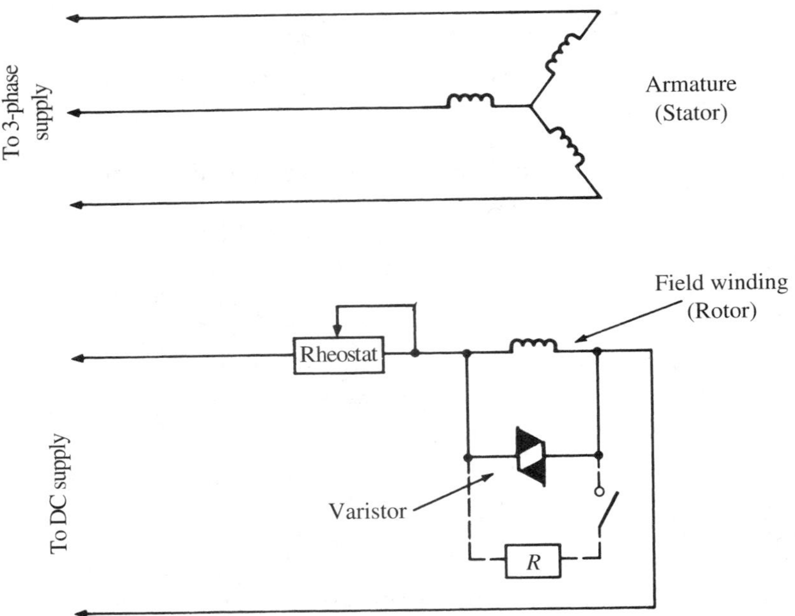

FIGURE 8.5
Simplified circuit diagrams for rotor and stator connections of a synchronous motor.

frequency is very low (approaching zero), and the control circuit opens SCR-2 and closes SCR-1; opening SCR-2 disconnects the discharge resistor, and closing SCR-1 admits current to the field windings. The solid-state switching is programmed to close SCR-1 at an instant that will ensure that the rotor poles will be facing stator poles of opposite polarity, thus preventing pole slipping.

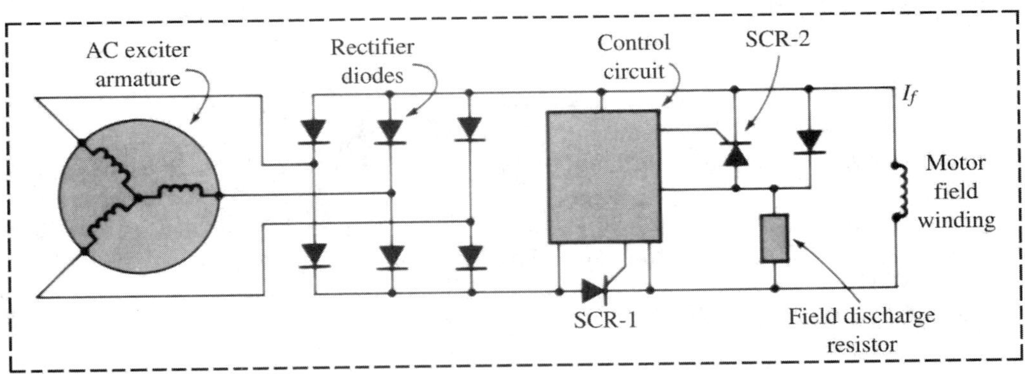

FIGURE 8.6
Circuit diagram of a brushless excitation system for a synchronous machine. (Courtesy Dresser Rand, Electric Machinery)

Synchronous motors without built-in starting components are not energized until they have been accelerated to approximately synchronous speed by an auxiliary motor or turbine. No load is connected to a synchronous motor during the acceleration period. When near synchronous speed, the magnet windings are energized, and the field current adjusted to obtain an induced stator voltage approximately equal to the supply voltage. The circuit breaker connecting the stator to the three-phase supply is closed when the induced stator voltage is in phase with the supply voltage. This procedure, called paralleling, is discussed in Section 9.6 of Chapter 9.

Reversing a Synchronous Motor

The direction of rotation of a synchronous motor is determined by its starting direction, as initiated by induction-motor action. Thus, to reverse the direction of a three-phase synchronous motor, it is necessary to first stop the motor and then reverse the phase sequence of the three-phase connections at the stator. Reversing the current to the field windings will not affect the direction of rotation.

8.4 SHAFT LOAD, POWER ANGLE, AND DEVELOPED TORQUE

Although the rotor of a synchronous motor rotates in synchronism with the rotating flux of the stator, increases in shaft load cause the rotor magnets to change their angular position with respect to the rotating flux. This displacement angle can be seen by viewing the rotor with a strobe light synchronized with the stator frequency.

Figure 8.7 shows a "strobe view" of a salient-pole motor operating at synchronous speed with no load on the shaft. The direction of rotation is counterclockwise (CCW). As the machine is loaded, the rotor changes its relative position with respect to the rotating flux of the stator, lagging behind it by angle δ. Angle δ, expressed in electrical degrees, is called the *power angle, load angle,* or *torque angle.* The broken line on the south pole in Figure 8.7 shows the position of the rotating magnets relative to the rotating flux of the stator for a representative load.

A synchronous motor operates at the same average speed for all values of load from no load to peak load. When the load on a synchronous motor is increased, the motor slows down just enough to allow the rotor to change its angular position in relation to the rotating flux of the stator, and then goes back to synchronous speed. Similarly, when the load is removed, it accelerates just enough to cause the rotor to decrease its angle of lag in relation to the rotating flux, and then goes back to synchronous speed. When the peak load that the machine can handle is exceeded, the rotor pulls out of synchronism.

The torque developed by all synchronous motors has two components: a reluctance-torque component and a magnet-torque component. The *reluctance-torque component* is due to the normal characteristic of magnetic materials in a magnetic field to align themselves so that the reluctance of the magnetic circuit is a minimum[4]

---

[4]See Section 7.2, Chapter 7.

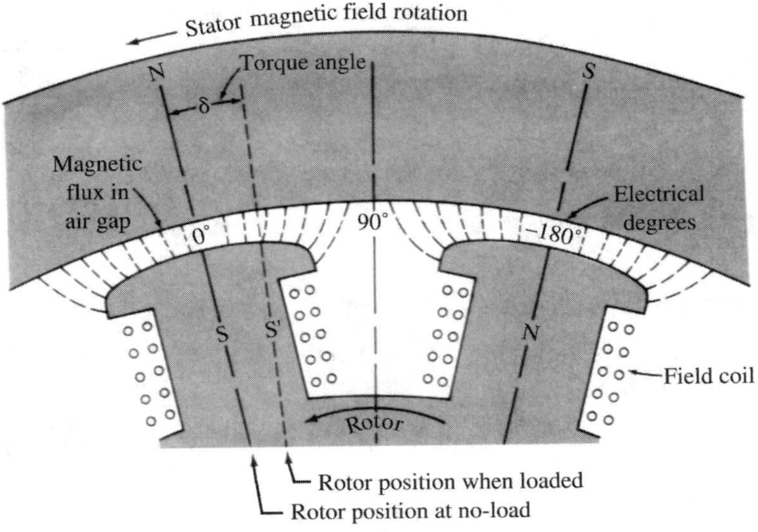

FIGURE 8.7
Strobe view of a salient-pole motor operating at synchronous speed with no load on the shaft. (Courtesy Dresser Rand, Electric Machinery)

The *magnet-torque component* is due to the magnetic attraction between the field poles (magnets) on the rotor and the corresponding opposite poles of the rotating stator flux.

The analysis of synchronous-motor behavior in this chatper is based on magnet-torque alone. This is justified for cylindrical-rotor motors because the smooth rotor surface provides an insignificant reluctance torque. It is also justified for salient-pole motors operating from 50 percent rated load to above 100 percent rated load, with power factors ranging from unity to leading; the reluctance torque for such loads is significantly smaller than the magnet torque. The performance of lightly loaded salient-pole motors is discussed in Section 8.13.

8.5 COUNTER-EMF AND ARMATURE-REACTION VOLTAGE

The air-gap flux in a synchronous motor includes a rotating field flux (Φ_f) due to DC current (I_f) in the rotating magnets, and a rotating armature flux, called *armature-reaction flux* (Φ_{ar}), caused by the three-phase armature currents in the stator windings. The magnitude and phase angle of the armature-reaction flux is a function of the magnitude and phase angle of the armature current. Both the magnet flux and the armature-reaction flux rotate in the same direction, and at synchronous speed with respect to the armature windings.

The rotating fluxes Φ_f and Φ_{ar} generate speed voltages in the stator conductors, as visualized in Figure 8.8 for one armature phase of a three-phase synchronous

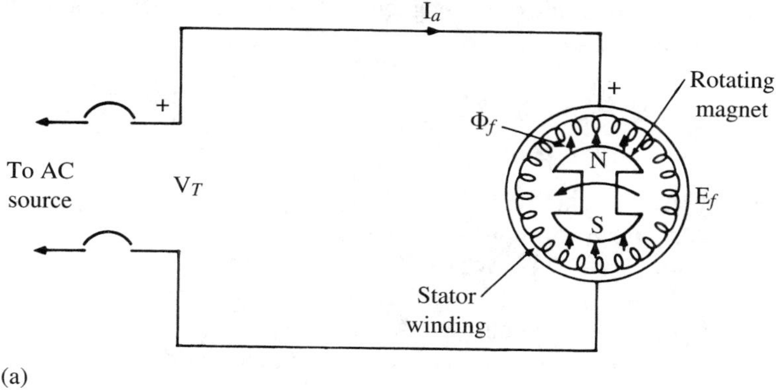

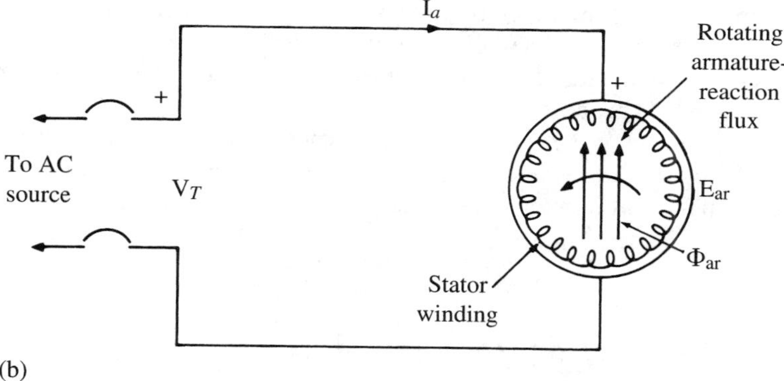

FIGURE 8.8
Separate circuits showing emfs generated by component magnetic fields for one phase of a synchronous motor: (a) due to rotating magnets; (b) due to rotating armature-reaction flux.

motor. Although shown as separate diagrams for ease of visualization, both Φ_f and Φ_{ar} rotate within the same three-phase stator winding.

Counter-emf

The rotor magnets sweeping the stator conductors, as shown in Figure 8.8(a), generate a speed-voltage, called a counter-emf (cemf) or *excitation voltage,* that acts in opposition to the applied voltage. The speed-voltage is proportional to the field flux and the speed of rotation. Expressed in equation form:

$$E_f = n_s \Phi_f k_f \tag{8–2}$$

Expressing the flux in terms of magnetomotive force (mmf) and the reluctance of the magnetic circuit,

$$E_f = n_s \cdot \frac{N_f I_f}{\mathcal{R}} \cdot k_f \tag{8–3}$$

where: E_f = excitation voltage/phase (V)
n_s = synchronous speed (r/min)
Φ_f = pole flux (Wb)
N_f = number of turns of conductor in field coil
I_f = DC field current (A)
$\mathcal{R}$ = reluctance of magnetic circuit (A-t/Wb)
k_f = constant

The speed of a synchronous motor is equal to the speed of the rotating flux, and is therefore constant for a given frequency. Thus, as indicated in Eq. (8–3), the excitation voltage is a function of the field current alone. Note, however, that because of magnetic saturation effects, the reluctance of the magnetic circuit is not constant. Hence, Φ_f and E_f are not proportional to I_f.

The cemf plays a significant role in synchronous-motor operation, and as will be explained later, may be less than, equal to, or greater than the applied stator voltage. Its adjustment, through changes in field current, is also used to correct system power factor.

Armature-Reaction Voltage

The rotating armature-reaction flux sweeping the stator conductors, as visualized in Figure 8.8(b), generates a speed voltage called the *armature-reaction voltage.* The armature-reaction speed voltage, expressed in terms of armature-reaction flux is

$$E_{\text{ar}} = n_s \Phi_{\text{ar}} k_a \tag{8–4}$$

where: E_{ar} = armature-reaction voltage (V)
Φ_{ar} = armature-reaction flux (Wb)
n_s = synchronous speed (r/min)
k_a = constant

Neglecting the effects of magnetic saturation, the armature-reaction flux is proportional to the armature current. Hence, the armature-reaction voltage may be conveniently expressed in terms of the armature current and a fictitious *armature-reaction reactance.* Thus,

$$\mathbf{E}_{\text{ar}} = \mathbf{I}_a \cdot jX_{\text{ar}} \tag{8–5}$$

where: X_{ar} = armature-reaction reactance (Ω/phase).

8.6 EQUIVALENT-CIRCUIT MODEL AND PHASOR DIAGRAM OF A SYNCHRONOUS-MOTOR ARMATURE

The equivalent-circuit model for one armature phase of a *cylindrical rotor* synchronous motor is shown in Figure 8.9(a). All values are per phase. Applying Kirchhoff's voltage law to Figure 8.9(a),

$$\mathbf{V}_T = \mathbf{I}_a R_a + \mathbf{I}_a jX_\ell + \mathbf{I}_a jX_{\text{ar}} + \mathbf{E}_f \quad \textbf{(8–6)}$$

Combining reactances,

$$X_s = X_\ell + X_{\text{ar}} \quad \textbf{(8–7)}$$

Substituting Eq. (8–7) into Eq. (8–6)

$$\mathbf{V}_T = \mathbf{E}_f + \mathbf{I}_a(R_a + jX_s) \quad \textbf{(8–8)}$$

$$\mathbf{V}_T = \mathbf{E}_f + \mathbf{I}_a \mathbf{Z}_s \quad \textbf{(8–9)}$$

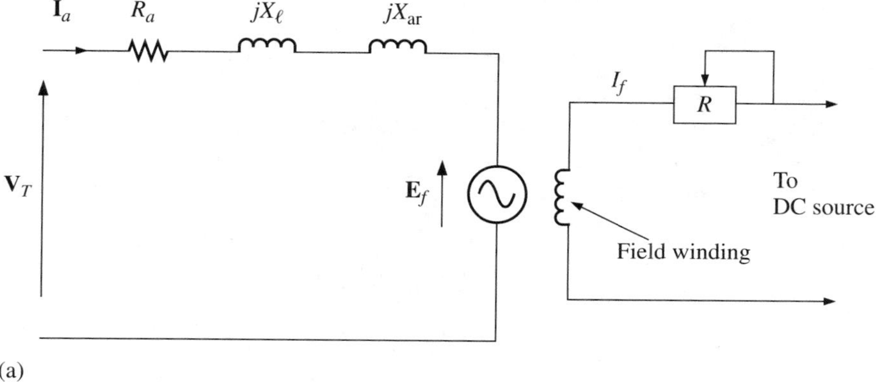

(a)

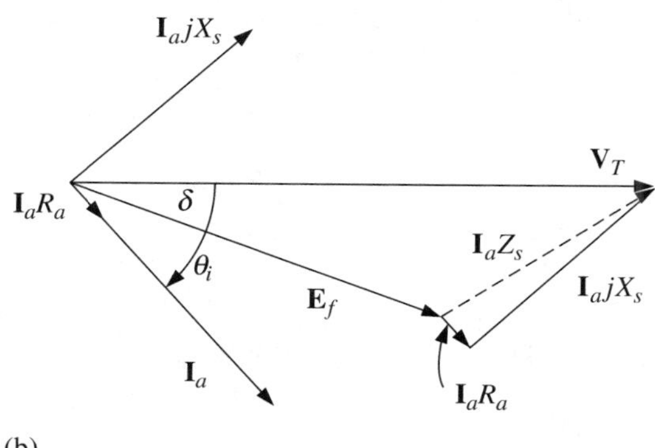

(b)

FIGURE 8.9
(a) Equivalent-circuit model for one phase of a synchronous-motor armature; (b) phasor diagram corresponding to the equivalent-circuit model in (a).

where: R_a = armature resistance (Ω/phase)
X_ℓ = armature leakage reactance (Ω/phase)
X_s = synchronous reactance (Ω/phase)
Z_s = synchronous impedance (Ω/phase)
V_T = applied voltage/phase (V)

A phasor diagram showing the component phasors and the tip-to-tail determination of $\mathbf{V}_T$ is shown in Figure 8.9(b). *The phase angle δ of the excitation voltage in Figure 8.9(b) is equal to the torque angle in Figure 8.7.* The torque angle is also called the *load angle* or *power angle.*

8.7 SYNCHRONOUS-MOTOR POWER EQUATION (MAGNET POWER)

Except for very small machines, the armature resistance of a synchronous motor is relatively insignificant compared to its synchronous reactance, enabling Eq. (8–9) to be approximated as

$$\mathbf{V}_T = \mathbf{E}_f + \mathbf{I}_a jX_s \quad \textbf{(8–10)}$$

The equivalent-circuit and phasor diagram corresponding to Eq. (8–10) are shown in Figure 8.10 and are normally used for the analysis of synchronous-motor behavior, as the motor responds to changes in load and/or changes in field excitation.

From the *geometry* of the phasor diagram,

$$I_a X_s \cos\theta_i = -E_f \sin\delta \quad \textbf{(8–11)}$$

Multiplying through by V_T and rearranging terms,

$$V_T I_a \cos\theta_i = \frac{-V_T E_f}{X_s}\sin\delta \quad \textbf{(8–12)}$$

Since the left side of Eq. (8–12) is an expression for active *power input,* the magnet power/phase developed by the synchronous motor can be expressed as

$$P_{\text{in},1\phi} = V_T I_a \cos\theta_i \quad \textbf{(8–13)}$$

or

$$P_{\text{in},1\phi} = \frac{-V_T E_f}{X_s}\sin\delta \quad \textbf{(8–14)}$$

Thus, for a three-phase synchronous motor,

$$P_{\text{in}} = 3 \times V_T I_a \cos\theta_i \quad \textbf{(8–15)}$$

or

$$P_{\text{in}} = 3 \times \frac{-V_T E_f}{X_s}\sin\delta \quad \textbf{(8–16)}$$

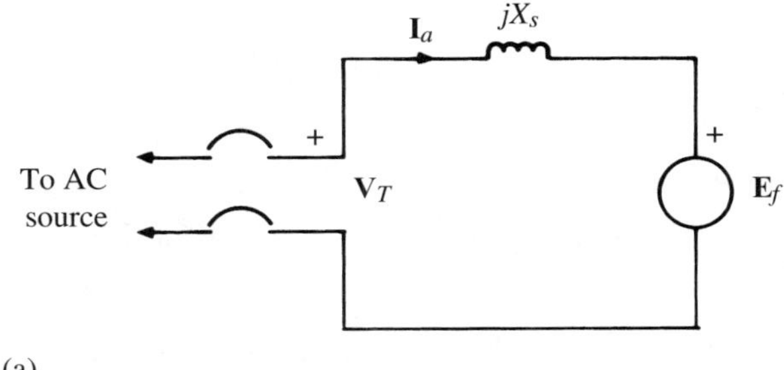

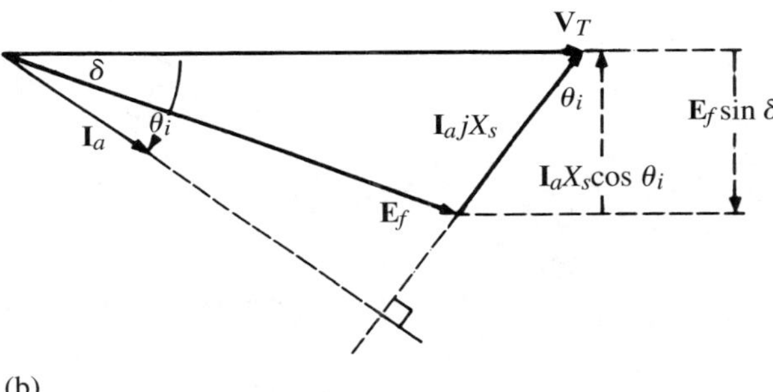

FIGURE 8.10
(a) Equivalent-circuit; (b) phasor diagram for a synchronous motor, assuming armature resistance is negligible.

Equation (8–14), called the *synchronous-machine power equation,* expresses the magnet power/phase developed by a cylindrical-rotor motor, in terms of its excitation voltage and power angle. Assuming a constant source voltage and constant frequency, Eqs. (8–13) and (8–14) may be expressed as proportionalities that are very useful in synchronous-machine analysis:

$$P \propto I_a \cos\theta_i \quad \textbf{(8–17)}$$

$$P \propto E_f \sin\delta \quad \textbf{(8–18)}$$

EXAMPLE 8.1 A 100-hp, three-phase, wye-connected, 60-Hz, 460-V, four-pole, cylindrical-rotor synchronous motor is operating at rated conditions and 80 percent power-factor leading. The efficiency, excluding field and stator losses, is 96 percent, and the synchronous reactance is 2.72 Ω/phase. Determine (a) developed torque; (b) armature current; (c) excitation voltage; (d) power angle; (e) maximum torque (also called pull-out torque).

Solution

(a)

$$n_s = \frac{120f}{P} = \frac{120 \times 60}{4} = 1800 \text{ r/min}$$

$$P_{\text{mech}} = \frac{Tn}{5252} = \quad \Rightarrow \quad T_{\text{dev}} = \frac{5252 \cdot P_{\text{mech}}}{n}$$

$$T_{\text{dev}} = \frac{5252 \times 100/0.96}{1800} = \underline{304 \text{ lb-ft}}$$

(b)

$$S = \frac{P_{\text{shaft}} \times 746}{\eta \times F_P} = \frac{100 \times 746}{0.96 \times 0.80} = 97{,}135 \text{ VA}$$

The power-factor angle is negative for a leading power factor.[5]

$$\theta = -\cos^{-1}0.80 = -36.87°$$

$$V_{1\phi} = \frac{460}{\sqrt{3}} = 265.581 \text{ V}$$

$$\mathbf{S}_{1\phi} = \mathbf{V}_T\mathbf{I}_a^* \quad \Rightarrow \quad \frac{97{,}135}{3}\angle{-36.87°} = 265.581\angle{0°} \times \mathbf{I}_a^*$$

$$\mathbf{I}_a^* = 121.92\angle{-36.87°} \quad \Rightarrow \quad \underline{\mathbf{I}_a = 121.92\angle{36.87°} \text{ A}}$$

(c)

$$\mathbf{E}_f = \mathbf{V}_T - \mathbf{I}_a jX_s = 265.581\angle{0°} - (121.92\angle{36.87°})(2.72\angle{90°})$$

$$\mathbf{E}_f = 265.58 - 331.62\angle{126.87°}$$

$$\mathbf{E}_f = 534.96\angle{-29.73°} \quad \Rightarrow \quad \underline{535\angle{-29.7°} \text{ V}}$$

(d)

$$\delta = \underline{-29.7°}$$

(e) Pull-out torque occurs at $\delta = -90°$.

$$P_{\text{in}} = 3 \cdot \frac{-V_T E_f}{X_s} \cdot \sin\delta = 3 \cdot \frac{-265.581 \times 534.96}{2.72} \cdot \sin(-90°)$$

$$P_{\text{in}} = 156{,}700 \text{ W}$$

$$P = \frac{Tn}{5252} \quad \Rightarrow \quad T_{\text{pull-out}} = \frac{5252 \cdot P}{n}$$

$$T_{\text{pull-out}} = \frac{5252 \times 156{,}700}{746 \times 1800} = \underline{613 \text{ lb-ft}}$$

8.8 EFFECT OF CHANGES IN SHAFT LOAD ON ARMATURE CURRENT, POWER ANGLE, AND POWER FACTOR

The effect that changes in shaft load have on armature current, power angle, and power factor are illustrated in Figure 8.11; the applied stator voltage, frequency, and

[5]See Sections A.5 and A.12 in Appendix A for the relationship between leading and lagging power factor and the sign of the power-factor angle.

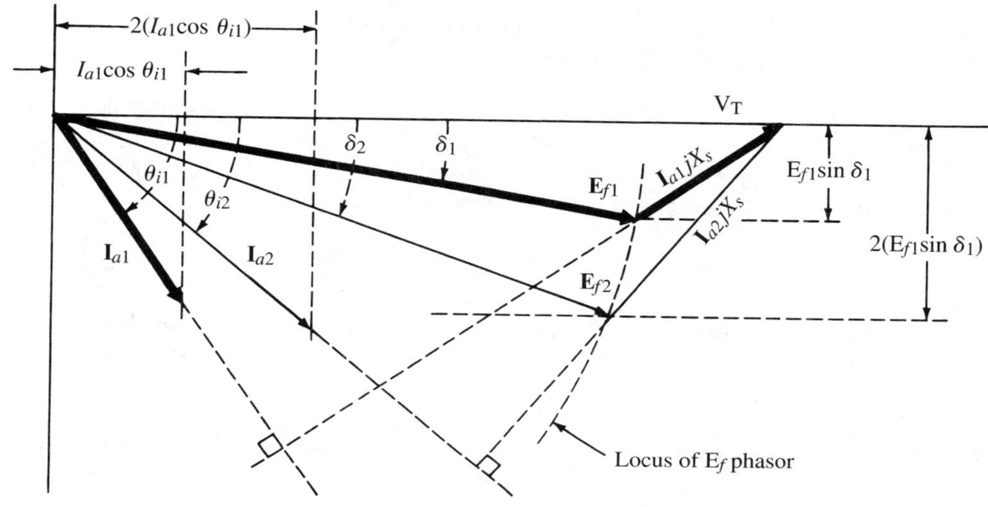

FIGURE 8.11
Phasor diagram showing effect of changes in shaft load on armature current, power angle, and power factor of a synchronous motor.

field excitation are assumed constant. The heavy lines indicate the initial load conditions, and the light lines indicate the new steady-state conditions that correspond to doubling the shaft load. *Note:* In accordance with Eqs. (8–17) and (8–18), doubling the shaft load doubles both $I_a \cos\theta_i$ and $E_f \sin\delta$. When adjusting phasor diagrams to show new steady-state conditions, the line of action of the new $\mathbf{I}_a jX_s$ phasor must be perpendicular to the new $\mathbf{I}_a$ phasor. Furthermore, as shown in Figure 8.11, if the excitation is not changed, increasing the shaft load causes the locus of the $\mathbf{E}_f$ phasor to describe a circular arc, increasing its phase angle with increasing shaft load. Note also that an increase in shaft load is also accompanied by a decrease in θ_i resulting in an increase in power factor.

As additional load is placed on the machine, the rotor continues to increase its lag relative to the rotating flux, thereby increasing both the angle of lag of the cemf phasor and the magnitude of the stator current. During all this loading, however, except for the transient conditions whereby the rotor assumes a new position in relation to the rotating flux, the average speed of the machine does not change. Finally, with increased loading, a point is reached at which a further increase in δ fails to cause a corresponding increase in motor torque, and the rotor pulls out of synchronism. The point of maximum torque occurs at a power angle of approximately 90° for a cylindrical-rotor machine, as is evidenced in Eq. (8–16). The critical value of torque that causes a synchronous motor to pull out of synchronism is called the *pull-out torque*.

8.9 EFFECT OF CHANGES IN FIELD EXCITATION ON SYNCHRONOUS-MOTOR PERFORMANCE

Intuitively we can expect that increasing the strength of the magnets will increase the magnetic attraction, and thereby cause the rotor magnets to have a closer alignment with the corresponding opposite poles of the rotating stator flux; the result is a smaller power angle. Proof of this behavior can be seen in Eq. (8–16). Assuming a constant shaft load, the steady-state value of $E_f \sin \delta$ must be constant. A step increase in E_f will cause a transient increase in $E_f \sin \delta$, and the rotor will accelerate. As the rotor changes its angular position, δ decreases until $E_f \sin \delta$ has the same steady-state value as before, at which time the rotor is again operating at synchronous speed. The change in angular position of the rotor magnets relative to the rotating flux of the stator occurs in a fraction of a second.

The effect of changes in field excitation on armature current, power angle, and power factor of a synchronous motor operating with a *constant shaft load,* from a constant voltage, constant frequency supply, is illustrated in Figure 8.12. From Eq. (8–18), for a constant shaft load,

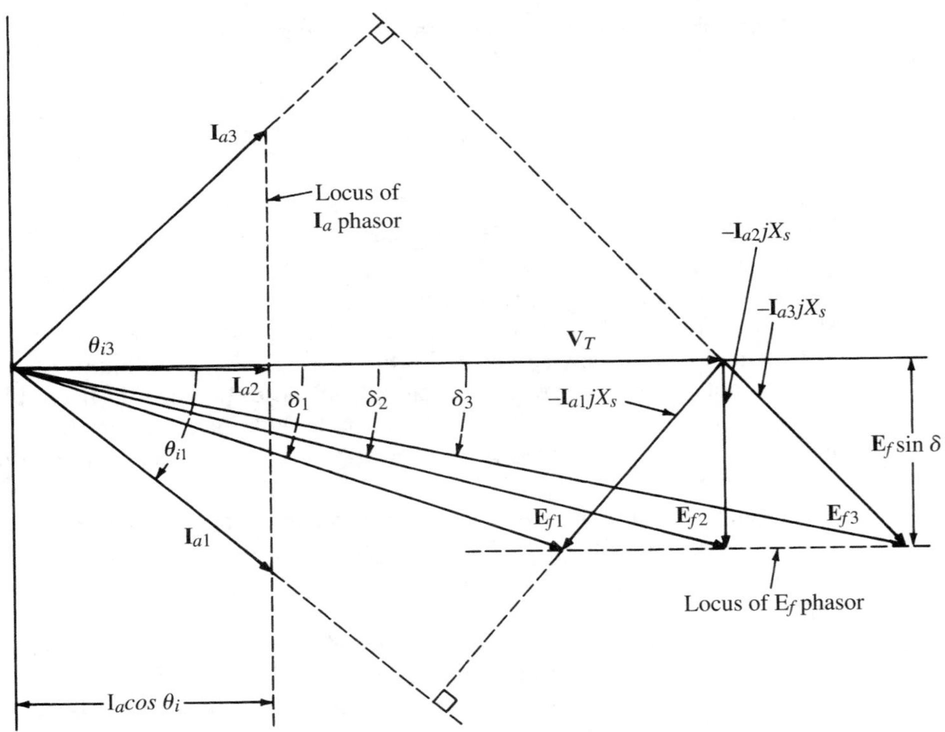

FIGURE 8.12
Phasor diagram showing the effect of changes in field excitation on armature current, power angle, and power factor of a synchronous motor.

$$E_{f1}\sin\delta_1 = E_{f2}\sin\delta_2 = E_{f3}\sin\delta_3 = E_f\sin\delta \qquad \textbf{(8–19)}$$

This is shown in Figure 8.12, where the locus of the tip of the $\mathbf{E}_f$ phasor is a straight line parallel to the $\mathbf{V}_T$ phasor. Similarly, from Eq. (8–17), for a constant shaft load,

$$I_{a1}\cos\theta_{i1} = I_{a2}\cos\theta_{i2} = I_{a3}\cos\theta_{i3} = I_a\cos\theta_i \qquad \textbf{(8–20)}$$

This is shown in Figure 8.12, where the locus of the tip of the $\mathbf{I}_a$ phasor is a line perpendicular to the $\mathbf{V}_T$ phasor.

Note that increasing the excitation from $\mathbf{E}_{f1}$ to $\mathbf{E}_{f3}$ in Figure 8.12 caused the angle of the current phasor (and hence the power factor) to go from lagging to leading. The value of field excitation that results in unity power factor is called *normal excitation.* Excitation greater than normal is called overexcitation, and excitation less than normal is called underexcitation. Furthermore, as indicated in Figure 8.12, when operating in the overexcited mode, $|\mathbf{E}_f| > |\mathbf{V}_T|$.

8.10 V CURVES

Curves of armature current vs. field current or armature current vs. excitation voltage are called V curves, and are shown in Figure 8.13 for representative synchronous-motor loads. The curves are related to the phasor diagram in Figure 8.12, and illustrate the effect of different values of field excitation on armature current and power factor for specific shaft loads. Note that increases in shaft load require increases in field excitation in order to maintain unity power factor.

The locus of the left endpoints of the V curves in Figure 8.13 represents the *stability limit* ($\delta = -90°$). Any reduction in excitation below the stability limit for a particular load will cause the rotor to pull out of synchronism.

The constant-load V curves shown in Figure 8.13 may be plotted from laboratory data (I_a vs. I_f) as I_f is varied, or graphically by plotting $|\mathbf{I}_a|$ vs. $|\mathbf{E}_f|$ from a family of phasor diagrams such as that shown in Figure 8.12, or from the following mathematical expression for the V curves[6]:

$$I_a = \frac{1}{X_s}\cdot\left[E_f^2 + V_T^2 - 2\sqrt{E_f^2V_T^2 - X_s^2P_{\text{in},1\phi}^2}\right]^{0.5} \qquad \textbf{(8–21)}$$

Equation (8–21) is based on the geometry of the phasor diagram and assumes R_a is negligible. *Note:* If the developed torque is less than the shaft load plus windage and friction, instability will occur, and the expression under the square root sign will be negative.

The family of V curves shown in Figure 8.13 represents computer plots of Eq. (8–21), assuming a three-phase, 40-hp, 220-V, 60-Hz synchronous motor with a synchronous reactance of 1.27 Ω/phase.

---

[6]See Reference **[1]** for the derivation of Eq. (8–21) and the general nature of V curves.

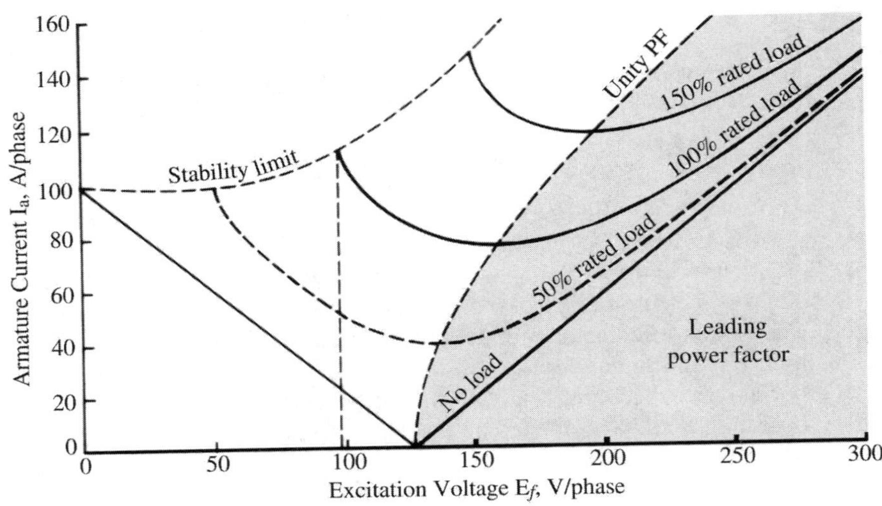

FIGURE 8.13
Family of representative V curves for a synchronous motor.

EXAMPLE 8.2 Referring to the V curve for 100 percent load in Figure 8.13, determine (a) the minimum value of excitation that will maintain synchronism. (b) Repeat (a) using Eq. (8–16). (c) Repeat (a) using Eq. (8–21). (d) Determine the power angle if the field excitation voltage is increased to 175 percent of the stability limit determined in (c).

Solution

(a) Estimation of minimum excitation voltage from the graph is

$$E_f \approx \underline{98\text{ V}}$$

(b) From Eq. (8–16), and neglecting losses,

$$P_{\text{in}} = 3 \times \frac{-V_T E_f}{X_s} \sin\delta \quad \Rightarrow \quad E_f = \frac{-P_{\text{in}} \cdot X_s}{3 \cdot V_T \sin\delta}$$

$$E_f = \frac{-40 \times 746 \times 1.27}{3 \times 220/\sqrt{3} \times \sin(-90°)} \approx \underline{99\text{ V}}$$

(c) From Eq. (8–21), the value of E_f that causes the expression under the square-root sign to equal zero is the minimum excitation that will maintain synchronism. Thus for,

$$E_f^2 V_T^2 - X_s^2 P_{\text{in},1\phi}^2 = 0$$

$$E_f = \frac{X_s P_{\text{in},1\phi}}{V_T} = \frac{1.27 \times 40 \times 746/3}{220/\sqrt{3}} \approx \underline{99\text{ V}}$$

(d) From Eq. (8–18),

$$1.75E_{f1} \sin \delta_2 = E_{f1} \sin(-90°)$$
$$\delta_2 = \underline{-35°}$$

8.11 SYNCHRONOUS-MOTOR LOSSES AND EFFICIENCY

A power-flow diagram that illustrates the flow of power through a synchronous motor, from stator and rotor input to shaft output, is shown in Figure 8.14. As indicated in the power-flow diagram, the total power loss for the motor is

$$P_{\text{loss}} = P_{\text{scl}} + P_{\text{core}} + P_{\text{fcl}} + P_{f,w} + P_{\text{stray}} \quad \text{W} \tag{8–22}$$

where:
P_{scl} = stator-conductor loss
P_{fcl} = field-conductor loss
P_{core} = core loss
$P_{f,w}$ = friction and windage loss
P_{stray} = stray power loss

Except for the transient conditions that occur when the field current is increased or decreased (magnetic energy stored or released), the total energy input to the field coils is constant and all of it is converted to I^2R losses in the copper conductors.

Efficiency

The overall efficiency of a synchronous motor is given by

$$\eta = \frac{P_{\text{shaft}}}{P_{\text{in}} + P_{\text{field}}} = \frac{P_{\text{shaft}}}{P_{\text{shaft}} + P_{\text{loss}}} \tag{8–23}$$

The nameplates of synchronous motors and manufacturers' specification sheets generally provide only the overall efficiency for rated load conditions. Hence, only the total losses may be determined. The separation of losses into the components listed in Eq. (8–22) requires a very involved test procedure **[2]**. If the armature copper losses and field copper losses can be calculated, however, they should be subtracted from the total input power to obtain a closer approximation of the mechanical power developed.

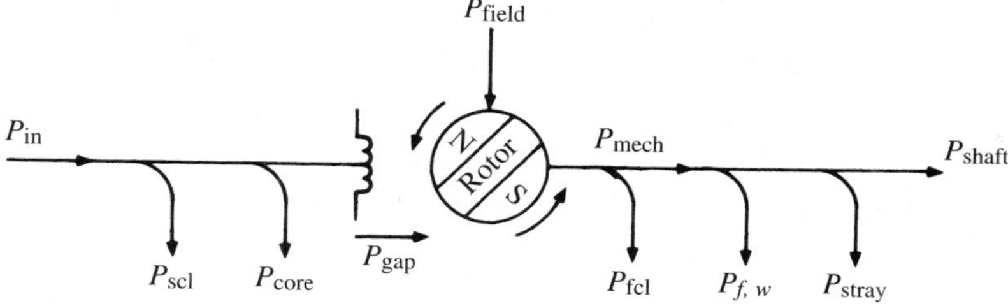

FIGURE 8.14
Power-flow diagram for a synchronous motor.

8.12 USING SYNCHRONOUS MOTORS TO IMPROVE THE SYSTEM POWER FACTOR

In industrial applications where both synchronous motors and induction motors are used, the synchronous motor is usually operated at a leading power factor to compensate for the lagging power factor of the induction motors. In those cases where a synchronous motor is operated without load, merely for the purpose of improving the power factor of a system, the machine is called a *synchronous condenser*. Machines specifically designed for this application are built without external shafts.

EXAMPLE 8.3 A three-phase, 60-Hz, 460-V system supplies the following loads:

1. A six-pole, 60-Hz, 400-hp, three-phase, wye-connected induction motor, operating at three-quarters rated load with an efficiency of 95.8 percent and a power factor of 89.1 percent.
2. A 50-kW, delta-connected, three-phase resistance heater.
3. A 300-hp, 60-Hz, four-pole, wye-connected, cylindrical-rotor synchronous motor, operating at one-half rated load, with a torque angle of $-16.4°$.

Neglecting copper losses, the synchronous motor is operating at 96 percent efficiency, and its synchronous reactance is 0.667 Ω/phase. Determine (a) system active power; (b) power factor of the synchronous motor; (c) system power factor; (d) percent change in synchronous motor field current required to adjust the system power factor to unity (neglect saturation effects); (e) power angle of the synchronous motor for the conditions in (d).

Solution

The circuit diagram is shown in Figure 8.15(a) and the power diagram in Figure 8.15(b).

(a)

$$P_{\text{ind mot}} = \frac{400 \times 0.75 \times 746}{0.958} = 233{,}611.7 \text{ W}$$

$$P_{\text{heater}} = 50{,}000 \text{ W}$$

$$P_{\text{syn mot}} = \frac{300 \times 0.5 \times 746}{0.96} = 116{,}562.5 \text{ W}$$

$$P_{\text{system}} = P_{\text{ind mot}} + P_{\text{heater}} + P_{\text{syn mot}} = 400{,}174.19 \quad \Rightarrow \quad \underline{400.2 \text{ kW}}$$

(b)

$$V_T/\text{phase} = 460/\sqrt{3} = 265.581 \text{ V}$$

$$P_{\text{in}} = 3 \times \frac{-V_T E_f}{X_s} \sin\delta \quad \Rightarrow \quad E_f = \frac{-P_{\text{in}} \cdot X_s}{3 \cdot V_T \sin\delta}$$

$$E_f = \frac{-116{,}562.5 \times 0.667}{3 \times 265.581 \times \sin(-16.4°)} = 345.614 \text{ V}$$

$$\mathbf{E}_f = 345.614\underline{/-16.4°} \text{ V}$$

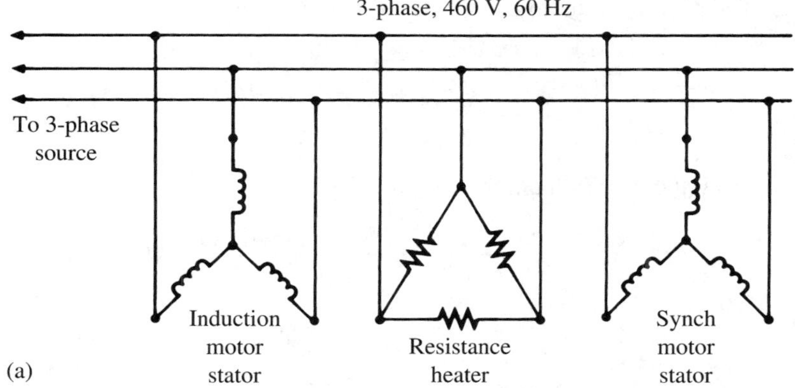

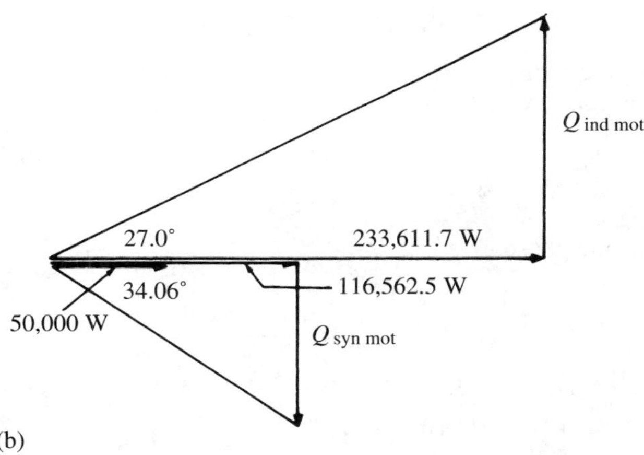

FIGURE 8.15
(a) Circuit diagram for Example 8.3; (b) power diagram.

$$\mathbf{E}_f = \mathbf{V}_T - \mathbf{I}_a jX_s$$

$$345.614\underline{/-16.4^\circ} = 265.581\underline{/0^\circ} - \mathbf{I}_a \times 0.667\underline{/90^\circ}$$

$$117.79\underline{/-55.94^\circ} = -\mathbf{I}_a \times 0.667\underline{/90^\circ}$$

$$\mathbf{I}_a = 176.6\underline{/34.06^\circ}\ \text{A}$$

$$\theta = (\theta_v - \theta_i) = (0 - 34.06) = -34.06^\circ$$

$$F_P = \cos(-34.06^\circ) = \underline{0.828\ \text{leading}}^7$$

---

[7]See Sections A.5 and A.12 in Appendix A for the relationship between the leading and lagging power factor and the sign of the power-factor angle.

(c) The power-factor angles for the respective loads are

$$\theta_{\text{ind mot}} = \cos^{-1}0.891 = 27.0°$$
$$\theta_{\text{heater}} = \cos^{-1}1.0 = 0.0°$$
$$\theta_{\text{syn mot}} = -34.06°$$

From the power diagram in Figure 8.15(b),

$$\tan 27.0° = \frac{Q_{\text{ind mot}}}{233{,}611.7} \quad \Rightarrow \quad Q_{\text{ind mot}} = 119{,}035.3 \text{ var}$$

$$\tan(-34.06°) = \frac{Q_{\text{syn mot}}}{116{,}562.5} \quad \Rightarrow \quad Q_{\text{syn mot}} = -78{,}800.1 \text{ var}$$

$$Q_{\text{sys}} = 119{,}035.3 - 78{,}800.1 = 40{,}235.2 \text{ var}$$
$$\mathbf{S}_{\text{sys}} = P + jQ = 400{,}174.19 + j40{,}235.2 = \underline{402{,}191\angle 5.74° \text{ VA}}$$
$$F_{P,\text{sys}} = \cos(5.74°) = \underline{0.995 \text{ lagging}}$$

(d) To obtain unity power factor, the synchronous motor must supply an additional $-40{,}235.2$ var. Thus,

$$\mathbf{S}_{\text{syn mot}} = 116{,}562.5 - j(78{,}800.1 + 40{,}235.2)$$
$$\mathbf{S}_{\text{syn mot}} = \underline{166{,}602\angle -45.6° \text{ VA}}$$

For one phase,

$$\mathbf{S}_{\text{syn mot}} = \frac{166{,}602}{3}\angle -45.6° = \underline{55{,}534.0\angle -45.6° \text{ VA}}$$

$$\mathbf{S}_{\text{syn mot}} = \mathbf{V}_T\mathbf{I}_a^* \quad \Rightarrow \quad 55{,}534.0\angle -45.6° = (265.58\angle 0°)\mathbf{I}_a^*$$
$$\mathbf{I}_a^* = 209.10\angle -45.60°$$
$$\mathbf{I}_a = 209.10\angle 45.60°$$
$$\mathbf{E}_f = \mathbf{V}_T - \mathbf{I}_a jX_s = 265.581\angle 0° - (209.10\angle 45.60°)(0.667\angle 90°)$$
$$\mathbf{E}_f = 265.581 + 99.65 - j97.58 = \underline{378.04\angle -14.96° \text{ V}}$$

Neglecting magnetic saturation, $E_f \propto \Phi_f \propto I_f$

$$\Delta E_f = \frac{378.04 - 345.614}{345.614} \times 100 = \underline{9.38\%}$$

(e) $$\delta = \underline{-14.96°}$$

8.13 SALIENT-POLE MOTOR

The rotor poles and associated interpolar spaces in the magnetic circuit of a salient-pole motor (such as that previously shown in Figure 8.7) present a cyclic variation of reluctance to the armature-reaction flux sweeping the stator conductors: low reluctance along each pole axis and high reluctance along each interpolar space. The net effect on the armature winding is the same as would be produced by two component waves of armature-reaction flux, 90° apart, sweeping the stator conductors. The component of armature-reaction flux acting along the axis of each field pole is called the *direct-axis* component, and the component of armature-reaction flux acting in the interpolar space between alternate north and south poles is called the *quadrature-axis* component. These component fluxes generate voltage drops in the armature that may be expressed in terms of armature current and fictitious armature-reaction reactances. The associated power equation, expressed in terms of these reactances, includes a magnet component and a reluctance component[8]:

$$P_{\text{salient},1\phi} = \underbrace{\frac{-V_T E_f}{X_d} \sin \delta}_{\text{magnet power}} - \underbrace{V_T^2 \left[\frac{X_d - X_q}{2X_d X_q}\right] \sin 2\delta}_{\text{reluctance power}} \qquad \textbf{(8–24)}$$

where: X_d = direct-axis synchronous reactance (Ω/phase)
X_q = quadrature-axis synchronous reactance (Ω/phase)

Normal Operation

Within its normal operating range (half-rated load to rated load), and normal excitation, the excitation voltage (E_f) in Eq. (8–24) is high enough to cause $P_{\text{mag}} >> P_{\text{rel}}$. Thus, for normal operation,

$$P_{\text{salient},1\phi} \approx \frac{-V_T E_f}{X_d} \sin \delta \qquad \textbf{(8–25)}$$

Note: If the interpolar space between north and south poles is reduced to zero, the rotor becomes a cylindrical rotor, $X_d = X_q = X_s$, the reluctance component of power = 0, and Eq. (8–24) reduces to Eq. (8–25).

Because the reluctance component of power is independent of the excitation voltage, low values of excitation will significantly reduce the magnet component without affecting the reluctance component. Thus, if the motor is unloaded, or very lightly loaded, and the field circuit is opened, the excitation voltage will drop to a very small residual value. This will cause the magnet power in Eq. (8–24) to drop to an insignificant value, and the motor will operate as a *reluctance-synchronous motor*. The reluctance-torque contribution of a salient-pole motor makes it more stable at low excitation and/or sudden overloads than the cylindrical-rotor machine.

Plots of reluctance torque and magnet torque vs. δ are shown superimposed on a strobe view of the rotor in Figure 8.16. Curve *A* represents reluctance torque, curve *B* represents magnet torque, and curve *C* is the algebraic summation of curve *A* and curve *B*. The broken sections indicate regions of unstable operation for the component torques and for the resultant torque.

---

[8]See References **[3]** and **[4]** for in-depth derivations.

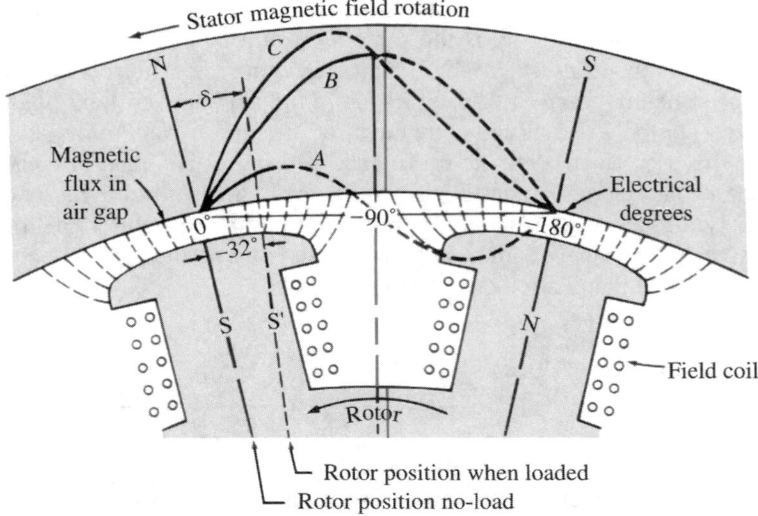

FIGURE 8.16
Plots of reluctance torque, magnet torque, and total torque vs. power angle shown superimposed on a strobe view of the rotor. (Courtesy Dresser Rand, Electric Machinery)

As indicated in Eq. (8–24), and shown in Figure 8.16, the maximum value of reluctance torque occurs when the shaft load is sufficient to cause the rotor to lag its no-load position by 45° ($\delta = -45°$). The magnet torque has its maximum value at $\delta = -90°$. The net torque for a salient-pole motor, due to combined reluctance and magnet contributions, has its maximum value, called the pull-out torque, at $\delta \approx -70°$. Rated torque for the representative motor shown in Figure 8.16 occurs at $\delta \approx -32°$.

EXAMPLE 8.4 A three-phase, 200-hp, 60-Hz, 2300-V, 900 r/min, salient-pole synchronous motor has direct-axis and quadrature-axis reactances equal to 36.66 Ω/phase and 23.33 Ω/phase, respectively. Neglecting losses, determine (a) the developed torque if the field current is adjusted so that the excitation voltage is equal to two times the applied stator voltage, and the power angle is −18°; (b) the developed torque in percent of rated torque, if the load is increased until maximum reluctance torque occurs.

Solution

(a) $$P_{\text{mag},1\phi} = \frac{-V_T E_f}{X_d} \sin\delta = \frac{-(2300/\sqrt{3})(2 \times 2300/\sqrt{3})}{36.66} \sin(-18°)$$

$$P_{\text{mag},1\phi} = 29{,}727.2 \text{ W}$$

$$P_{rel,1\phi} = -V_T^2\left[\frac{X_d - X_q}{2X_dX_q}\right]\sin 2\delta = -(2300/\sqrt{3})^2\left[\frac{36.66-23.33}{2\times 36.66\times 23.33}\right]\sin(-36°)$$

$$P_{rel,1\phi} = 8076.9 \text{ W}$$

$$P_{salient,3\phi} = 3(P_{mag,1\phi} + P_{rel,1\phi}) = 3(29{,}727.2 + 8076.9) = 113{,}412.3 \text{ W}$$

$$P_{salient,3\phi} = \frac{113{,}412.3}{746} = 152.0 \text{ hp}$$

$$P = \frac{Tn}{5252} \quad \Rightarrow \quad T = \frac{5252 \cdot P}{n}$$

$$T = \frac{5252 \times 152.0}{900} = \underline{887 \text{ lb-ft}}$$

(b) The reluctance torque has its maximum value of $\delta = -45°$.

$$P_{mag,1\phi} = \frac{-V_T \cdot E_f \sin\delta}{X_d}$$

$$P_{mag,1\phi} = \frac{-(2300/\sqrt{3})(2\times 2300/\sqrt{3})}{36.66}\sin(-45°) = 68{,}023.1 \text{ W}$$

$$P_{rel,1\phi} = -V_T^2 \cdot \left[\frac{X_d - X_q}{2X_dX_q}\right]\sin 2\delta$$

$$P_{rel,1\phi} = -(2300/\sqrt{3})^2\left[\frac{36.66 - 23.33}{2\times 36.66\times 23.33}\right]\sin(-90°) = 13{,}741.3 \text{ W}$$

$$P_{salient,3\phi} = 3(68{,}023.1 + 13{,}741.3) = 245{,}293.2 \text{ W}$$

$$P_{salient,3\phi} = \frac{245{,}293.2}{746} = 328.8 \text{ hp}$$

Because the speed of a synchronous motor is constant,

$$\text{Percent rated torque} = \frac{328.8}{200}\times 100 = \underline{164\%}$$

This is an excessive overload, and if the machine is operated at this load for more than several minutes, rapid overheating of armature and field windings will occur.

8.14 PULL-IN TORQUE AND MOMENT OF INERTIA

The pull-in torque is the torque required to pull the rotor into synchronism with the rotating field of the stator. The amortisseur winding accelerates the rotor to its maximum speed as an induction motor, at which time application of current to the rotor magnets develops additional torque to pull the rotor into synchronism. If the load inertia is too high, however, the motor may not develop enough torque to synchronize or may take excessive time. In either event, the resultant pole slipping will cause high armature

TABLE 8.1
Synchronous-motor torques in percent of rated torque[a] (minimum values)

Speed (rpm)	hp	Power Factor	Torques: Locked Rotor	Pull-in (Based on Normal Wk^2 of Load)[b,c]	Pull-out[c]
500 to 1800	200 and below	1.0	100	100	150
	150 and below	0.8	100	100	175
	250 to 1000	1.0	60	60	150
	200 to 1000	0.8	60	60	175
	1250 and larger	1.0	40	60	150
		0.8	40	60	175
450 and below	All ratings	1.0	40	30	150
		0.8	40	30	200

[a] Reproduced by permission of the National Electrical Manufacturers Association from *NEMA Standards Publication MG 1-1998, Motors and Generators,* Copyright 1999 by NEMA, Washington, DC.
[b] Values of normal Wk^2 of load are given in *Motors and Generators* 1-21.12.
[c] With rated excitation current applied.

current pulses and violent torque pulses that may damage the winding. To ensure quick synchronization when the field current is applied, the moment of inertia (Wk^2) of the connected load must not exceed the normal values recommended for the specific horsepower and speed rating of the motor.

Table 8.1 lists the minimum values of locked-rotor torque, pull-in torque, and pull-out torque for NEMA-standard *salient-pole* synchronous motors for unity and 0.8 *leading* power factors.[9] The motors must be capable of delivering the minimum pull-out torque for at least one minute. The normal values of Wk^2, on which the pull-in torque values in Table 8.1 are based, are given in extensive NEMA tables **[5]**.

8.15 SPEED CONTROL OF SYNCHRONOUS MOTORS

The speed of a synchronous motor is directly proportional to the stator frequency and inversely proportional to the number of poles:

$$n_r = n_s = \frac{120f}{P}$$

Because of the nature of its construction, however, pole changing of the rotor is not practical. Hence, the only way to change the speed of a synchronous motor is to change the frequency of the applied voltage. Furthermore, as discussed in Section 5.8

[9] Corresponding values of torque for cylindrical-rotor synchronous motors are subject to individual negotiation between manufacturer and user.

in Chapter 5, to avoid overheating of the stator winding, the ratio of volts/hertz must be kept constant. This may be accomplished by means of adjustable frequency generators with volts/hertz regulators, or by using solid-state devices exclusively. With solid-state control, the fixed frequency AC supply is converted to adjustable voltage DC, and then inverted to adjustable frequency AC, or by using a cycloconverter (see Section 13.16).

8.16 DYNAMIC BRAKING

Synchronous motors may be decelerated rapidly by converting the energy stored in the moving parts to electrical energy and dissipating it as heat in resistors. To do this, the motor armature is disconnected from the three-phase supply lines and connected across a three-phase resistor bank. The field is maintained at full strength by the exciter. The motor acts as a generator and feeds current to the resistor bank, dissipating energy at a rate equal to $3 \times I^2R$. Resistance R is the resistance of each armature phase and series-connected external resistor.

SUMMARY OF EQUATIONS FOR PROBLEM SOLVING

$$n_r = n_s = \frac{120 \cdot f}{P} \tag{4–1}$$

$$E_f = n_s \Phi_f k_f \tag{8–2}$$

$$\mathbf{V}_T = \mathbf{I}_a\mathbf{R}_a + \mathbf{I}_a jX_\ell + \mathbf{I}_a jX_{\text{ar}} + \mathbf{E}_f \tag{8–6}$$

$$\mathbf{V}_T = \mathbf{E}_f + \mathbf{I}_a jX_s \tag{8–10}$$

$$P_{\text{in},1\phi} = V_T I_a \cos\theta_i \tag{8–13}$$

$$P_{\text{in},1\phi} = \frac{-V_T E_f}{X_s} \sin\delta \tag{8–14}$$

$$I_a = \frac{1}{X_s} \cdot \left[E_f^2 + V_T^2 - 2\sqrt{E_f^2 V_T^2 - X_s^2 P_{\text{in},1\phi}^2}\right]^{0.5} \tag{8–21}$$

$$P_{\text{salient},1\phi} = \underbrace{\frac{-V_T E_f}{X_d} \sin\delta}_{\text{magnet power}} - \underbrace{V_T^2\left[\frac{X_d - X_q}{2X_d X_q}\right] \sin 2\delta}_{\text{reluctance power}} \tag{8–24}$$

$$P = \frac{Tn}{5252}$$

SPECIFIC REFERENCES KEYED TO TEXT

1. Bewley, L. V. *Alternating Current Machinery.* Macmillan, New York, 1949.
2. Institute of Electrical and Electronics Engineers. *Test Procedures for Synchronous Machines.* IEEE STD 115-1983, IEEE, New York, 1995.
3. Mablekos, V. E. *Electric Machine Theory for Power Engineers.* Harper & Row, New York, 1980.
4. McPherson, G. *An Introduction to Electrical Machines and Transformers.* Wiley, New York, 1981.
5. National Electrical Manufacturers Association. *Motors and Generators.* Publication No. MG 1-1998, NEMA, Rosslyn, VA, 1999.

GENERAL REFERENCE

Smeaton, R. W. *Motor Application and Maintenance Handbook,* 2nd ed. McGraw-Hill, New York, 1987

REVIEW QUESTIONS

1. Discuss two designs of salient-pole rotors that provide starting and accelerating torque.
2. How does an amortisseur winding reduce hunting caused by pulsating loads?
3. State how a synchronous motor is started, stopped, and reversed.
4. How can the speed of a synchronous motor be adjusted?
5. What are the two components of synchronous-motor torque? What are they due to?
6. Explain why it is possible for the cemf of a synchronous motor to be greater than the applied voltage.
7. What is meant by the *torque angle* of a synchronous motor? What factors affect the magnitude of this angle?
8. What effect does increasing the excitation of a fully loaded synchronous motor, from its stability limit to its maximum excitation, have on its armature current? Explain.
9. What effect does increasing the excitation of a fully loaded synchronous motor, from its stability limit to its maximum excitation, have on its pull-out torque? Explain.
10. What effect does increasing the excitation of a fully loaded synchronous motor, from its stability limit to its maximum excitation, have on its power factor? Explain.
11. Will increasing the shaft load cause the speed of a synchronous motor to operate at a new lower steady-state speed? Explain.
12. Using a phasor diagram, explain how a synchronous motor adjusts its input power to accommodate an increase in shaft load.

13. Differentiate between pull-in torque, pull-out torque, and locked-rotor torque.
14. What effect does load inertia have on pull-in torque, pull-out torque, and locked-rotor torque?
15. Explain the dynamic braking process for synchronous motors.

PROBLEMS

8–1/3 Determine the speed of a 40-pole synchronous motor operating from a three-phase, 50-Hz, 4600-V system.

8–2/3 Determine the frequency required to operate a 16-pole, 480-V synchronous motor at 225 r/min.

8–3/3 A three-phase 50-hp, 2300-V, 60-Hz synchronous motor is operating at 90 r/min. Determine the number of poles in the rotor.

8–4/7 A two-pole, three-phase, 1000-hp, 2300-V, wye-connected synchronous motor operating at rated power, rated voltage, and rated frequency has a power factor of 0.80 leading, an efficiency of 96.5 percent, and a power angle of $-23°$. The synchronous reactance is 4.65 Ω/phase. Determine (a) line current; (b) cemf/phase.

8–5/7 A 4000-hp, 13,200 V, 60-Hz, two-pole, three-phase, wye-connected cylindrical-rotor synchronous motor, operating at rated load and 0.84 power-factor leading, has an efficiency (excluding field losses and armature resistance losses) of 96.5 percent. The synchronous reactance per phase is 49.33 Ω. Sketch the phasor diagram and determine (a) rated torque; (b) armature current; (c) excitation voltage; (d) power angle; (e) pull-out torque.

8–6/7 A 2500-hp, 6600-V, 60-Hz, 3600 r/min, three-phase, wye-connected synchronous motor is operating at a power angle of $-28.5°$ and a cemf of 4500 V/phase. The synchronous reactance is 15.8 Ω/phase. Determine (a) the developed torque; (b) the shaft-power out, assuming a mechanical efficiency of 94.3 percent; (c) power factor; (d) pull-out torque.

8–7/7 A 200-hp, 460-V, four-pole, 60-Hz, three-phase, wye-connected synchronous motor, operating at rated load, has an efficiency of 94 percent and a power factor of 80 percent leading. Assuming the synchronous reactance is 1.16 Ω/phase, determine (a) the excitation voltage and power angle; (b) the pull-out torque; (c) the pull-out torque if the excitation voltage is cut in half.

8–8/8 The phasor diagram for a synchronous motor operating at 0.5 rated load is shown in Figure 8.17. Using the same applied voltage phasor, construct a new phasor diagram that shows the new conditions, assuming the shaft load is increased to its rated value. Show all construction lines.

8–9/9 Given the phasor diagram for the synchronous motor shown in Figure 8.18, neglecting magnetic saturation, and showing all construction lines, construct a new phasor diagram for each of the following specified conditions: (a) field current unchanged, shaft load doubled; (b) shaft load unchanged, field current doubled; (c) shaft load doubled, field current doubled.

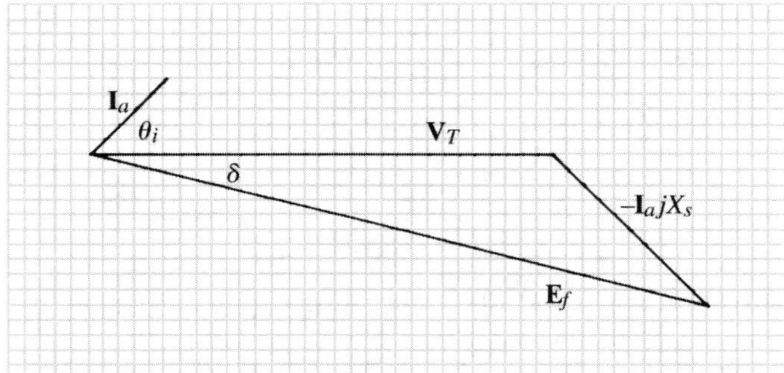

FIGURE 8.17
Phasor diagram for Problem 8–8/8.

8–10/9 Given the phasor diagram for the synchronous motor shown in Figure 8.18, neglecting magnetic saturation, and showing all construction lines, construct a new phasor diagram for each of the following specified conditions: (a) field current doubled, shaft load unchanged; (b) shaft load doubled, field current cut in half; (c) shaft load cut in half, field current halved.

8–11/10* (a) A 15,000-hp, 13,200-V, 60-Hz, 1800 r/min, wye-connected, cylindrical-rotor synchronous motor has a synchronous reactance of 14.95 Ω/phase. Plot a family of V curves corresponding to rated load, 75 percent rated load, and 50 percent rated load. (b) Sketch the stability limit and unity power-factor locus. (c) Approximate from the curves the minimum excitation voltage that will just maintain synchronism for each load condition.

8–12/10 A 250-hp, 575-V, three-phase, 60-Hz, two-pole, wye-connected synchronous motor operating at one-half rated load has the V curve shown in

---

*Solution by computer is recommended.

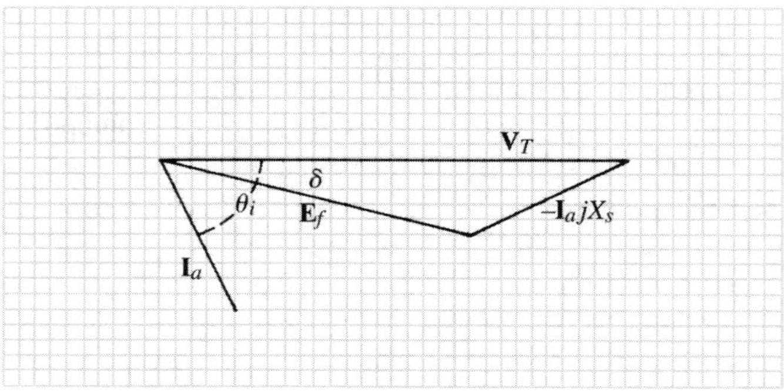

FIGURE 8.18
Phasor diagram for Problems 8–9/9 and 8–10/9.

Figure 8.19. (a) Approximate (from the curve) the phase values of excitation voltage and armature current for operation at unity power factor. (b) Approximate the stability limit. (c) Using the nameplate voltage and data obtained in (a), make a scale drawing of the corresponding phasor diagram. (d) Using the phasor diagram, determine the power angle and the synchronous reactance drop. (e) Determine the synchronous reactance.

8–13/10 The V curve shown in Figure 8.20 is for a 460-V, 350-hp, 60-Hz, 1800 r/min, cylindrical-rotor synchronous motor. The machine is operating at less than rated load, with an excitation voltage of 500 V/phase. The synchronous reactance is 0.65 Ω/phase. Sketch the corresponding phasor diagram to scale, and determine: (a) power angle; (b) active and reactive components of input power; (c) power factor; (d) developed torque.

8–14/12 A three-phase, 60-Hz, 460-V system supplies an induction motor that draws 50 kVA at 0.84 power factor, a 30-kW incandescent lighting load, and a 125-hp, six-pole, wye-connected synchronous motor, with a synchronous impedance of 1.45 Ω/phase, that draws 80 kW. Determine (a) the synchronous-motor power factor required to operate the system at unity power factor; (b) the excitation voltage required for the conditions in (a); (c) the synchronous-motor power angle for the conditions in (a).

8–15/12 Repeat Problem 8–14/12, assuming the system is to be operated at 0.80 leading power factor.

8–16/12 A 460-V, three-phase, 60-Hz system supplies power to a 10-kVA, delta-connected resistor load; a 50-kVA, 0.86 power factor induction motor, and

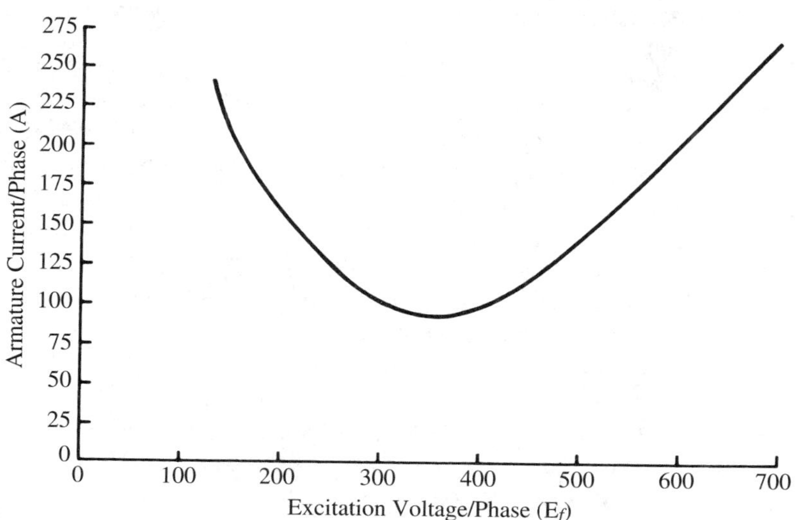

FIGURE 8.19
V curve for Problem 8–12/10.

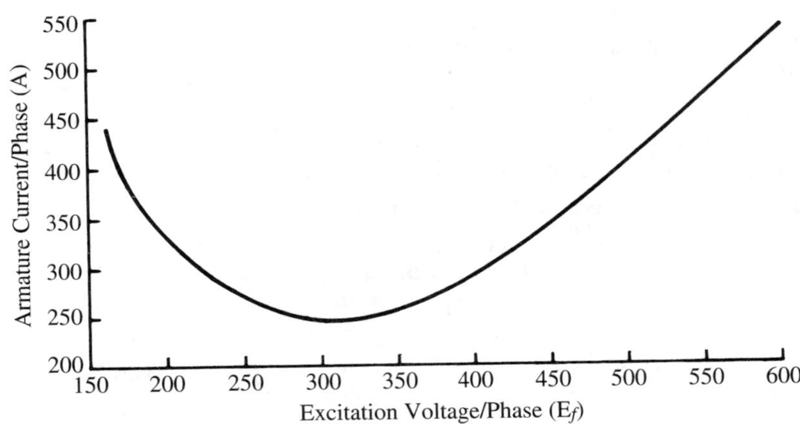

FIGURE 8.20.
V curve for Problem 8–13/10.

a wye-connected synchronous motor that is operating at 0.62 power factor leading, and has a synchronous impedance of 6.05 Ω/phase. If the system is operating at unity power factor, determine (a) active and reactive components of synchronous motor power; (b) synchronous motor current; (c) excitation line voltage.

8–17/12 A three-phase, 60-Hz, 575-V system, operating at 1.5 MW and 0.92 power factor lagging, includes a 500-hp, 575-V, 1200 r/min synchronous motor operating at rated load, 0.84 power-factor leading, and an efficiency (discounting field losses) of 96.2 percent. The synchronous reactance is 0.567 Ω/phase. Determine (a) system power factor if the synchronous motor is disconnected; (b) system power factor if the synchronous motor is unloaded but remains connected to the bus and its excitation remains unchanged. Assume the no-load losses are negligible.

8–18/13 A 3500-hp, 60-Hz, 4000-V, 450 r/min, wye-connected synchronous motor has a direct-axis synchronous reactance of 5.76 Ω/phase, and a quadrature-axis synchronous reactance of 4.80 Ω/phase. The motor is operating at rated conditions with a power angle and efficiency of −34.6° and 97.1 percent, respectively. Determine (a) excitation voltage; (b) magnet component of torque; (c) reluctance component of torque; (d) percent of total power contributed by the reluctance component.

8–19/14 A 20-pole, 700-hp, 2300-V, 60-Hz, 0.8 power-factor wye-connected synchronous motor has a direct-axis reactance of 8.91 Ω/phase and a quadrature-axis reactance of 6.48 Ω/phase. Using Table 8.1, determine (a) minimum expected locked-rotor torque; (b) minimum expected pull-in torque, assuming normal Wk^2; (c) minimum expected pull-out torque, assuming rated field current is applied.

9

Synchronous Generators (Alternators)

9.1 INTRODUCTION

Synchronous generators, also called *alternators* or *AC generators,* are the principal sources of electrical power throughout the world, and range in size from a fraction of a kVA to 1500 MVA.

The bulk of electric power is generated with high-speed steam turbines driving cylindrical rotor machines, and low-speed hydraulic turbines driving salient-pole machines. Steam for steam turbines is obtained from fossil fuels, nuclear fuel, and in some cases from geothermal energy. Water for hydraulic turbines is obtained from dammed rivers or pumped storage (water pumped into high-elevation reservoirs for later use).

Diesel-driven generators and gas-turbine-driven generators are used for large and small isolated loads, for utility standby power to handle peak demand, and for applications in remote pumping stations, ships, drilling rigs, and the like.

The construction of synchronous generators is essentially the same as for synchronous motors described in Chapter 8. It is interesting to note that synchronous generators driven by hydraulic turbines and connected to large power systems are sometimes operated as synchronous motors during periods of least usage; through a switching arrangement, the synchronous generator is driven as a synchronous motor in the reverse direction, pumping water back from a lower reservoir to a higher reservoir. Such storing of water for later use is called *pumped storage.*

This chapter starts with a simple transition from synchronous-motor action (developed in Chapter 8) to synchronous-generator action. This is followed by the general procedure for safe and efficient paralleling (and shutdown) of synchronous generators with other machines or with an "infinite bus." The theory of load transfer between machines, power-factor control, and the avoidance of potential operational problems are stressed. Voltage regulation, determination of parameters, and generator efficiency complete the chapter.

9.2 MOTOR-TO-GENERATOR TRANSITION

Figure 9.1(a) shows an infinite bus connected to a synchronous motor whose shaft is mechanically coupled to a water pump and a steam turbine. The design of the turbine is such that when steam hits the turbine blades, the torque exerted by the turbine will be in the same direction as the synchronous-motor torque. The equivalent circuit for one phase is shown in Figure 9.1(b).

With the turbine valve closed, so that no steam enters the turbine, the synchronous motor will drive the pump and the turbine at synchronous speed.[1] The phasor diagram for synchronous-motor action is shown in Figure 9.2(a) for a cylindrical-rotor machine. The corresponding phasor equation is

$$\mathbf{V}_T = \mathbf{E}_f + \mathbf{I}_a jX_s \tag{9–1}$$

where: $\mathbf{V}_T$ = terminal voltage/phase (V)
$\mathbf{E}_f$ = excitation voltage/phase (V)
$\mathbf{I}_a$ = armature current/phase (A)
X_s = synchronous reactance/phase (Ω)

[1] A vacuum is maintained on the turbine to prevent windage from overheating the turbine blades.

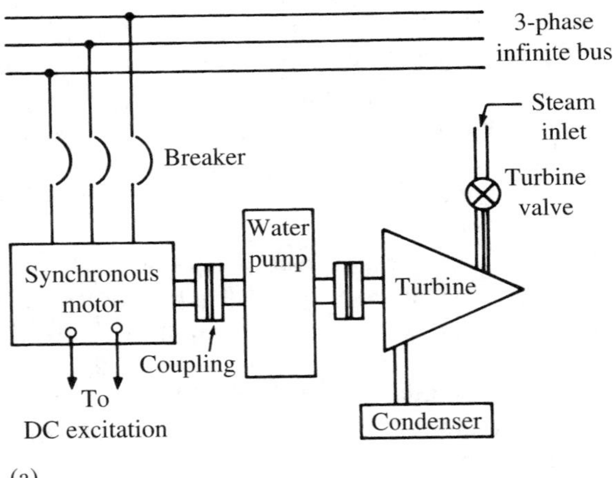

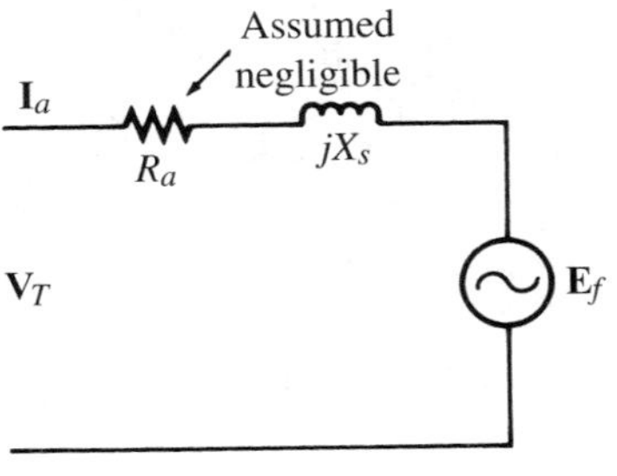

FIGURE 9.1
(a) Synchronous motor mechanically coupled to a water pump and a steam turbine; (b) equivalent circuit for one phase of the synchronous-motor armature.

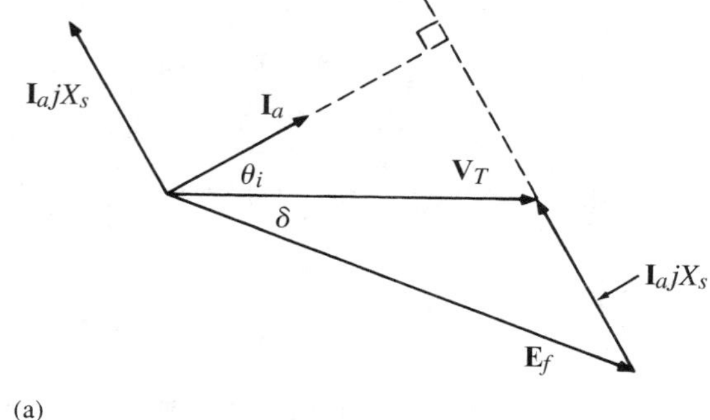

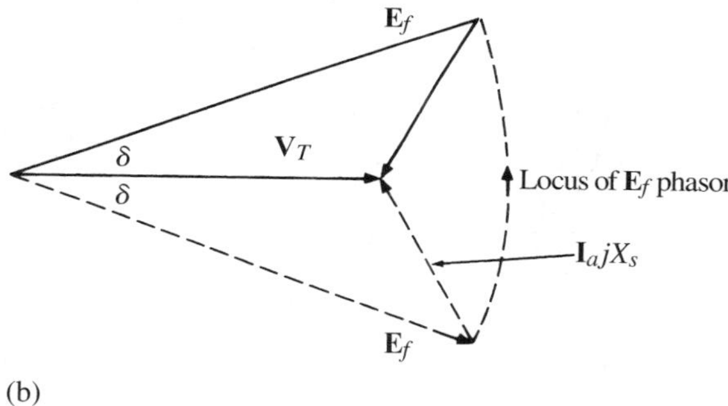

FIGURE 9.2
(a) Phasor diagram for one phase of a synchronous machine acting as a motor; (b) phasor diagram showing transition from motor action to generator action.

Figure 9.2(b) illustrates how admitting steam to the turbine affects the power angle of a synchronous machine. Admitting a small amount of steam to the turbine by partially opening the turbine valve causes the turbine to take part of the pump load, resulting in a decrease in the synchronous-motor power angle.

If the flow of steam is increased to a value that causes the turbine to carry the entire pump load, the synchronous motor will operate at no load and its power angle will be zero. Further increases in steam will cause the power angle to increase in the positive direction. Since the field current was not changed, the locus of the excitation voltage describes a circular arc. Except for transient changes in speed as the power angle in Figure 9.2(b) changes, the machine continues to run at synchronous speed.

The synchronous machine absorbs power and thus provides motor action when δ is negative, operates at no load when δ is zero, and (as will be shown) provides generator action, delivering power to the three-phase bus when δ is positive.

Strobe views of a two-pole rotor, and the corresponding phasor diagrams for motoring, for no load, and for generating are shown in Figure 9.3. When the synchronous machine is operating as a motor (δ negative), as shown in Figure 9.3(a), it is driven by the bus voltage, and the excitation voltage is a cemf (voltage drop). *However, when operating as a generator (δ positive),* as shown in Figure 9.3(c), *the excitation voltage becomes the source voltage, and the active component of current* ($I_a \cos \theta_i$) *is reversed.*

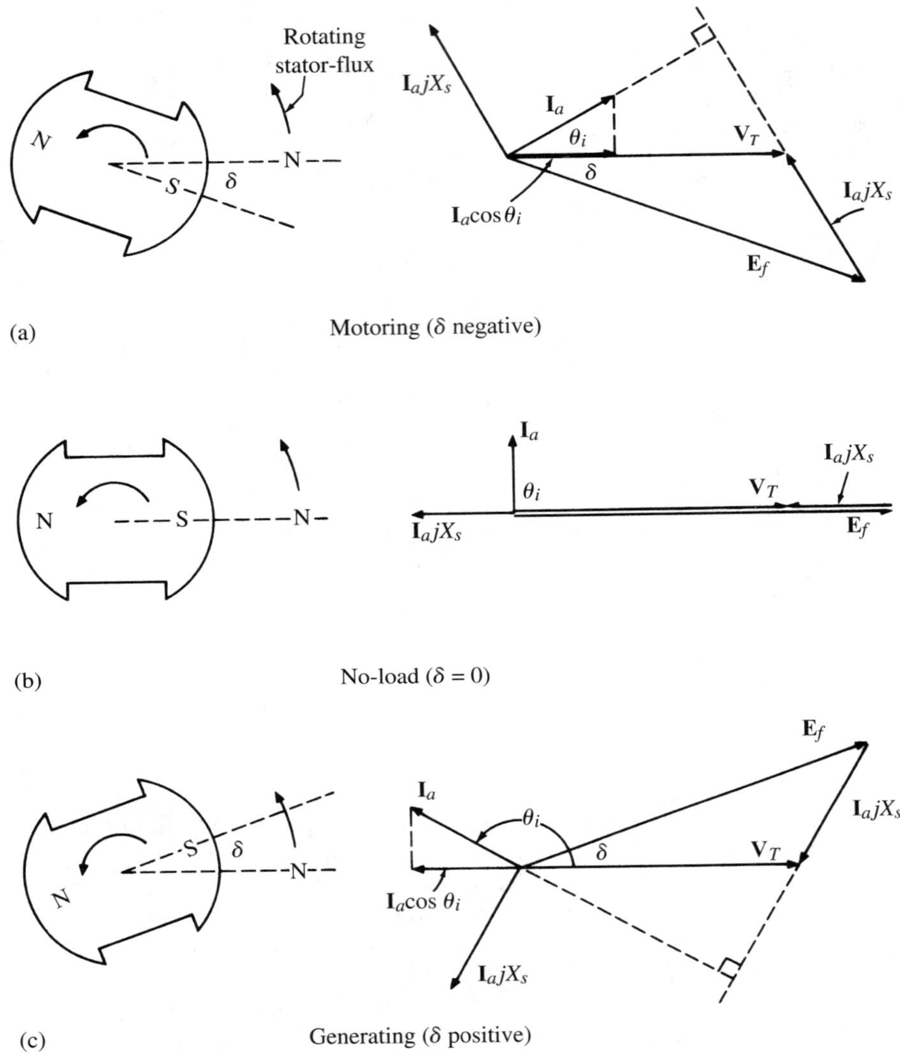

FIGURE 9.3
Strobe views of a two-pole rotor and the corresponding phasor diagrams for (a) motoring, (b) no load, and (c) generating.

Although the phasor diagram shown in Figure 9.3(c) is correct for a synchronous generator, it is not used in that form when analyzing generator behavior because one would have to contend with a negative power factor.[2] The reason for this awkward situation is that the changes in the phasor diagrams, as the machine went from motor ac-

---

[2]The power factor would have a negative value because $\theta_v = 0$, $\theta_i > 90°$: $\theta = (\theta_v - \theta_i)$ and $F_P = \cos\theta$.

tion in Figure 9.3(a) to generator action in Figure 9.3(c), were based on an assumed direction of *electric power in* for both motor and generator action.

Phasor Diagram of a Synchronous Generator

When thinking in terms of generator operation and *power out,* as shown in Figure 9.4(a), the excitation voltage $\mathbf{E}_f$ becomes the source voltage and the bus voltage $\mathbf{V}_T$ becomes a voltage drop. Thus, in terms of Kirchhoff's voltage law,

$$\mathbf{E}_f = \mathbf{V}_T + \mathbf{I}_a jX_s$$

Rearranging terms,

$$\mathbf{V}_T = \mathbf{E}_f - \mathbf{I}_a jX_s \qquad \textbf{(9–1a)}$$

where: $\mathbf{V}_T$ = terminal voltage/phase (V)
$\mathbf{E}_f$ = excitation voltage/phase (V)
$\mathbf{I}_a$ = armature current/phase (A)
X_s = synchronous reactance/phase (Ω)

As previously developed in Section 8.5 of Chapter 8, the excitation voltage is related to the pole flux and field current in the following manner:

$$E_f = n_s \Phi_f k_f \qquad \textbf{(9–2)}$$

FIGURE 9.4
(a) Equivalent circuit; (b) preferred phasor diagram for generator action.

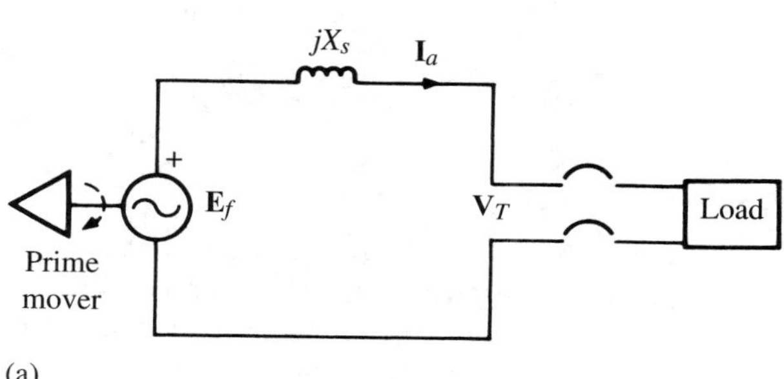

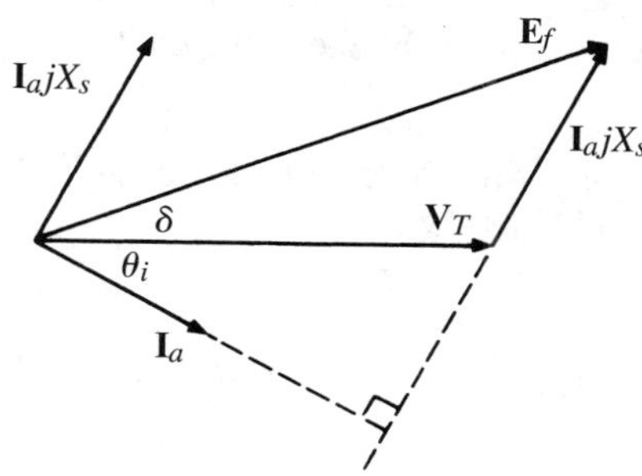

$$E_f = n_s \cdot \frac{N_f I_f}{\mathcal{R}} \cdot k_f \tag{9–3}$$

where:
n_s = synchronous speed (r/min)
Φ_f = flux/magnet pole (Wb)
k_f = constant
N_f = turns of wire/pole
I_f = DC field current (A)
$\mathcal{R}$ = reluctance of magnetic circuit (A-t/Wb)

Note that the difference between Eq. (8–10) for motor action and Eq. (9–1) for generator action is that the armature current is reversed. The phasor diagram corresponding to Eq. (9–1) for generator action is shown in Figure 9.4(b).

9.3 SYNCHRONOUS-GENERATOR POWER EQUATION

The equations developed in Section 8.7 for electric power input to a synchronous machine acting as a motor apply equally well to the synchronous machine acting as a generator. The only difference is the *direction* of electric power: *in* for a motor and *out* for a generator. Thus,

$$P_{\text{in},1\phi} = \frac{-V_T E_f}{X_s} \sin\delta \qquad \text{(motor action)} \tag{8–14}$$

$$P_{\text{out},1\phi} = \frac{+V_T E_f}{X_s} \sin\delta \qquad \text{(generator action)} \tag{9–4}$$

EXAMPLE 9.1 A certain three-phase, 460-V, two-pole, 60-Hz, wye-connected synchronous alternator, with a synchronous reactance of 1.26 Ω/phase, is connected to an infinite bus. The power angle, when supplying 112 kW to the bus, is 25°. Neglecting losses, determine (a) turbine torque supplied to the alternator; (b) excitation voltage; (c) active and reactive components of apparent power; (d) power factor. (e) Neglecting saturation effects, determine the excitation voltage if the field current is reduced to 85 percent of its value in (a). (f) Determine the turbine speed.

Solution

(a) Neglecting losses, the power input from the turbine to the synchronous machine is 112,000 W.

$$\text{hp} = \frac{Tn}{5252} \qquad \Rightarrow \qquad T = \frac{\text{hp} \times 5252}{n}$$

$$T = \frac{112{,}000 \times 5252}{746 \times 3600} = \underline{219.0 \text{ lb-ft}}$$

(b)

$$V_T = \frac{460}{\sqrt{3}} = 265.581 \text{ V/phase}$$

$$P_{out} = 3 \times \frac{V_T E_f}{X_s} \sin\delta \quad \Rightarrow \quad E_f = \frac{P_{out} \cdot X_s}{3 \cdot V_T \cdot \sin\delta}$$

$$E_f = \frac{112{,}000 \times 1.26}{3 \times 265.581 \times \sin(25°)} = \underline{419.1 \text{ V/phase}}$$

(c) Using Eq. (9–1),

$$\mathbf{V}_T = \mathbf{E}_f - \mathbf{I}_a jX_s \quad \Rightarrow \quad 265.581\angle 0° = 419.1\angle 25° - \mathbf{I}_a \times 1.26\angle 90°$$

$$114.253 + j177.119 = \mathbf{I}_a \times 1.26\angle 90°$$

$$\mathbf{I}_a = 167.279\angle -32.824° \text{ A}$$

$$\mathbf{S} = 3 \times \mathbf{V}_T \mathbf{I}_a^* = 3 \times 265.581\angle 0° \times 167.279\angle 32.824°$$

$$\mathbf{S} = 133{,}278.3\angle 32.824° = 111{,}990 + j72{,}245 \text{ VA}$$

$$P = 112 \text{ kW}$$

$$Q = 72.2 \text{ kvar lagging}$$

(d)

$$F_P = \cos(32.824°) = \underline{0.84 \text{ lagging}}$$

(e) From Eq. (9–3), and neglecting saturation effects, $E_f \propto I_f$. Thus,

$$E_f = 0.85 \times 419.1 = \underline{356.2 \text{ V/phase}}$$

(f)

$$n_s = \frac{120f}{P} = \frac{120 \times 60}{2} = \underline{3600 \text{ r/min}}$$

Salient-Pole Generator

The equation for *electric power in,* previously discussed for salient-pole motors in Section 8.13, also applies to salient-pole generators. Thus, in terms of *electric power out,*

$$P_{out,1\phi} = \frac{V_T E_f}{X_d} \sin\delta + V_T^2 \left[\frac{X_d - X_q}{2X_d X_q}\right] \sin 2\delta \tag{9–5}$$

A plot of Eq. (9–5) for electric *power out* is shown in Figure 9.5 for a representative salient-pole synchronous machine connected to an infinite bus. Reading from left to right, the plot shows a smooth transition from motor action to generator action as load torque on the motor shaft is removed and turbine torque is applied. Although the curve is plotted from −180° to +180°, loading beyond the maximum values will cause the machine to pull out of synchronism, whether it is operating as a motor or as a generator. Maximum power is called *pull-out power.*

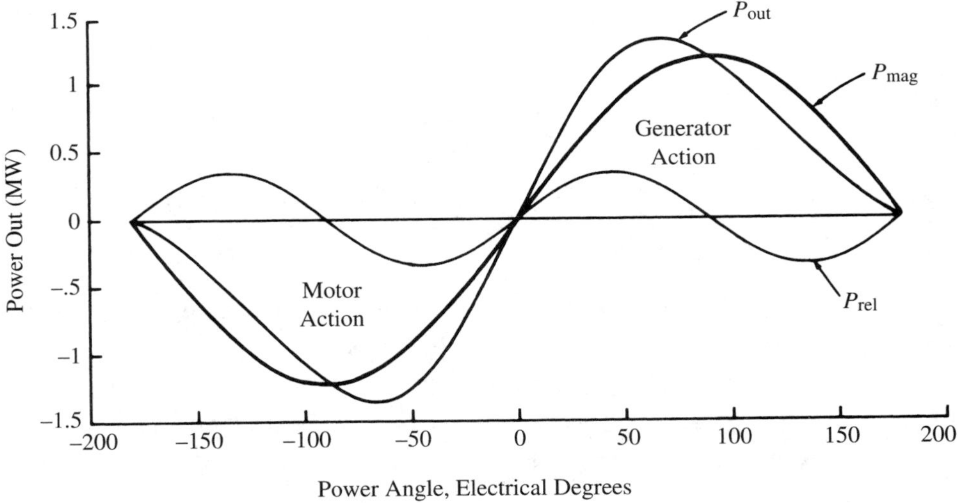

FIGURE 9.5
Plot of *electric power out* for a representative synchronous machine, showing a smooth transition from motor action to generator action.

9.4 GENERATOR LOADING AND COUNTERTORQUE

When load is connected to a generator, the interaction of the armature current with the pole flux develops a torque in opposition to the driving torque of the prime mover.[3] This opposing torque, called *countertorque,* causes the machine to slow down. As the machine decelerates, its speed-governor responds to admit more energy to the prime mover. When the energy input to the prime mover is equal to the load on the generator plus losses, the system is in equilibrium and the machine operates at the new lower speed. If the prime mover has a constant-speed governor (isochronous), however, the machine will initially drop in speed, and as sufficient energy is admitted to the prime mover, the speed will return to its initial no-load value.

9.5 LOAD, POWER FACTOR, AND THE PRIME MOVER

The voltage at the terminals of a generator is affected considerably by the power factor of the load. Hence, the rated voltage of an alternator, as indicated on the nameplate, is always given for rated kVA at a specific power factor and field current.

The multitude of different loads in distribution systems—resistive, inductive, and capacitive—generally results in an average system power factor of approximately 80 percent lagging. Hence, the terminal voltage for rated kVA at 80 percent power

[3]See Section 1.11, Chapter 1.

factor is generally indicated on the nameplate of most general-service alternators. If an alternator is designed to supply only a unity power-factor load, however, the generator voltage indicated on the nameplate will be the value for rated kVA at unity power factor.

The energy supplied to an alternator by its prime mover must be adequate to carry the load that the generator is called on to deliver. Although the generator can deliver the same kVA at different power factors, the prime mover supplies only active power. Hence, if the kVA demand from the generator is always at 80 percent power factor or less, the maximum active power that the prime mover will be required to supply will be equal to the rated kVA of the machine multiplied by a power factor of 0.80 or less. Under such circumstances, it is not necessary to install a prime mover large enough to supply rated kVA at unity power factor.

9.6 PARALLELING SYNCHRONOUS GENERATORS

Two or more generators operating in parallel provide economy of operation and flexibility for scheduling routine maintenance. As increases in oncoming load approaches the rated load of the machine or machines already on the bus, additional machines may be paralleled to share the load. Similarly, as the bus load decreases, one or more machines are taken off the line to allow the remaining units to operate at higher efficiencies. A simplified one-line diagram[4] for a two-generator system designed for parallel operation is shown in Figure 9.6. Generator *A* is connected to the bus and is supplying

[4]Each line from generator to breaker to load represents the three line leads of a three-phase generator. Three slash marks indicate three conductors/line (IEEE STD 315-1975).

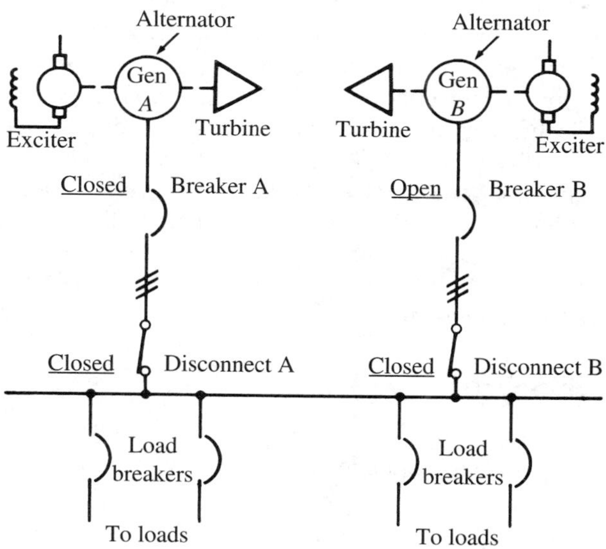

FIGURE 9.6
Simplified one-line diagram for a two-generator system.

the entire load. Generator *B* is the incoming machine; its breaker is open. The disconnect switch is closed prior to starting the prime mover.

Machines that are to be paralleled must have the same phase sequence.[5] Any attempt to parallel an alternator with another already on the bus, but with opposite phase sequence, may not only black out the plant, but may also cause serious damage to the machines and associated equipment. Most modern plants have automatic synchronizers that correctly synchronize a generator with the bus. Knowledge of manual synchronization is necessary for nonautomatic plants, however, and for plants with defects in the automatic synchronizing equipment.

Procedure for Paralleling

The voltage of the incoming machine should be adjusted to obtain a value approximately equal to the bus voltage. Although a few volts higher is preferred, a few volts lower will not adversely affect the operation. The frequency of the incoming machine should be adjusted to be a fraction of a hertz higher than the bus frequency. Then assuming that the bus-disconnect switch is closed, the circuit breaker should be closed at the instant the voltage wave of the incoming machine is in phase with the bus voltage. The no-load speed (and hence the no-load frequency) of a synchronous generator is changed by remote control from the generator panel. A control switch on the generator panel actuates a servomotor that raises or lowers the no-load speed setting of the governor.

The frequency of the incoming machine should be a fraction of hertz higher than the bus frequency to ensure its taking load at the instant it is paralleled. If its frequency is slightly lower than the bus frequency, it will be motorized at the instant the circuit breaker is closed. Alternators in parallel are locked in rotational synchronism. Hence, an incoming machine with a frequency lower than the bus frequency will be driven as a motor by the other alternators on the bus.

Although it is possible to adjust the frequency of the incoming machine to the same frequency as the bus, and to be in phase with the bus, the incoming machine will "float" on the line when paralleled. It will not be driven as a motor, nor will it take load. Hence, there is no advantage to this procedure. Since an alternator is paralleled for the specific purpose of taking some of the bus load, it is advantageous to adjust it to a fraction of a hertz higher than the bus frequency prior to paralleling. This will cause the voltage wave of the incoming machine to slide slowly past the voltage wave of the bus, as illustrated in Figure 9.7(a); the voltage wave representing the incoming machine is shown slightly higher than the bus voltage. The incoming machine should be paralleled with the bus at an instant when its voltage wave coincides with that of the bus.

Synchroscope

A synchroscope, shown in Figure 9.7(b), indicates the instantaneous angle of phase displacement (error angle) between two voltage waves. If the voltage waves have different

---

[5]See Appendix A.10.

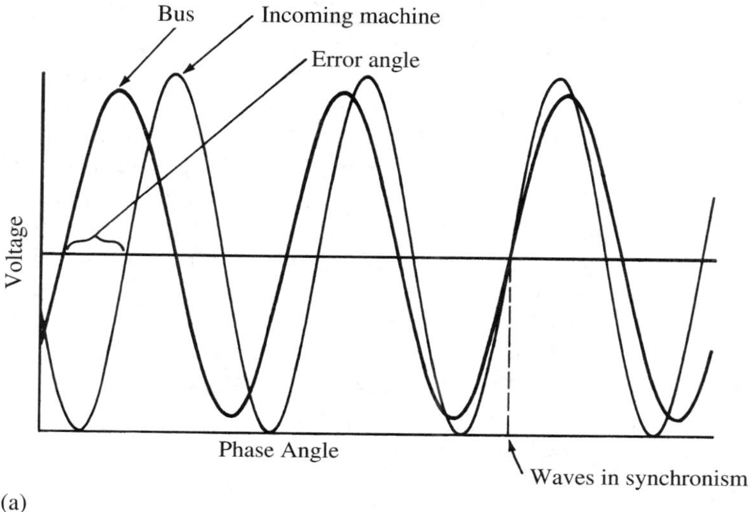

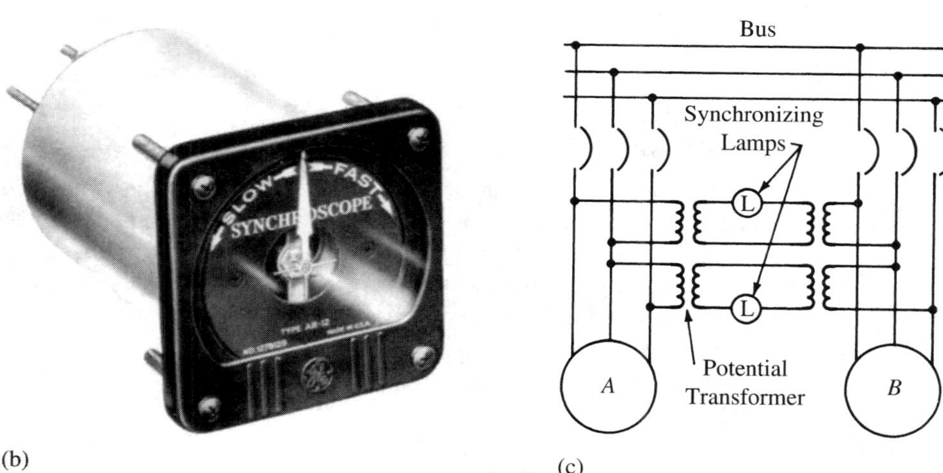

FIGURE 9.7
(a) Voltage waves of bus and incoming generator, with incoming machine at a slightly higher voltage and a slightly higher frequency; (b) synchroscope (Courtesy GE Industrial Systems); (c) circuit showing synchronizing lamp connections.

frequencies, the angle between the two waves, called the *error angle,* will always be changing, causing the pointer to revolve. When properly connected, a synchroscope indicates the electrical speed of the incoming machine with respect to the bus. If the synchroscope revolves in the direction marked SLOW, the incoming machine has a lower frequency than that of the bus. If the synchroscope revolves in the direction marked

FAST, the incoming machine has a higher frequency than that of the bus. If the frequency of the incoming machine is equal to the bus frequency, the pointer will not revolve, and the position of the pointer will indicate the error angle. If this happens, the prime mover of the incoming machine should have its speed increased by adjusting its governor control. *The incoming machine should be paralleled at the instant the synchroscope pointer enters the zero-degree position while revolving slowly in the FAST direction.* Once paralleled, the synchroscope pointer will no longer revolve, but will stay in the zero-degree position.

Because of the slight delay caused by human reaction time and breaker closing time, however, it is good practice to start the breaker closing operation one or two degrees *before* the synchroscope pointer reaches the zero-degree position.

Closing the breaker when the error angle is greater than 10° will cause a high transient synchronizing current between the incoming generator and the bus, which may cause automatic tripping of the generator breaker. If the bus is already heavily loaded, and the breaker is closed when the error angle approaches 180°, the entire system may be blacked out. Furthermore, the high transient current associated with an extreme out-of-phase attempt at synchronization causes excessive oscillating torques on both the stator and rotor windings that may result in immediate damage, or at best, a decrease in its service life.

Synchronizing Lamps

Synchronizing lamps are used to indicate the phase sequence of the incoming machine with respect to the bus and as a check on the synchroscope.[6] The lamp connections are shown in Figure 9.7(c). If both generator and bus have the same phase sequence, both lamps will go bright and dark in unison. If, however, the incoming machine is of opposite phase sequence with respect to the bus, the lamps will alternate repeatedly—one bright and one dark.

Assuming the phase sequence is correct, both lamps will be dark when the synchroscope is at approximately zero degrees. *Note:* If the lamps and the synchroscope are in conflict, and the lamps go bright and dark in unison, the lamps are correct.

Synchronizing with lamps has the disadvantage of a dark period that may extend over 15° or more of error angle. Hence, when this method is used, the circuit breaker should be closed midway through the dark period. Another disadvantage is that it does not indicate whether the incoming machine is fast or slow.

A functional diagram for a two-generator system supplying a common bus load is shown in Figure 9.8; it illustrates the basic instrumentation required for effective operation. A plant with only one generator, synchronized with an infinite bus, would omit panel 3, but would provide protective relaying on a substitute panel.

---

[6]Phase sequence may also be checked with a small handheld phase-sequence indicator or with a small three-phase motor **[5]**.

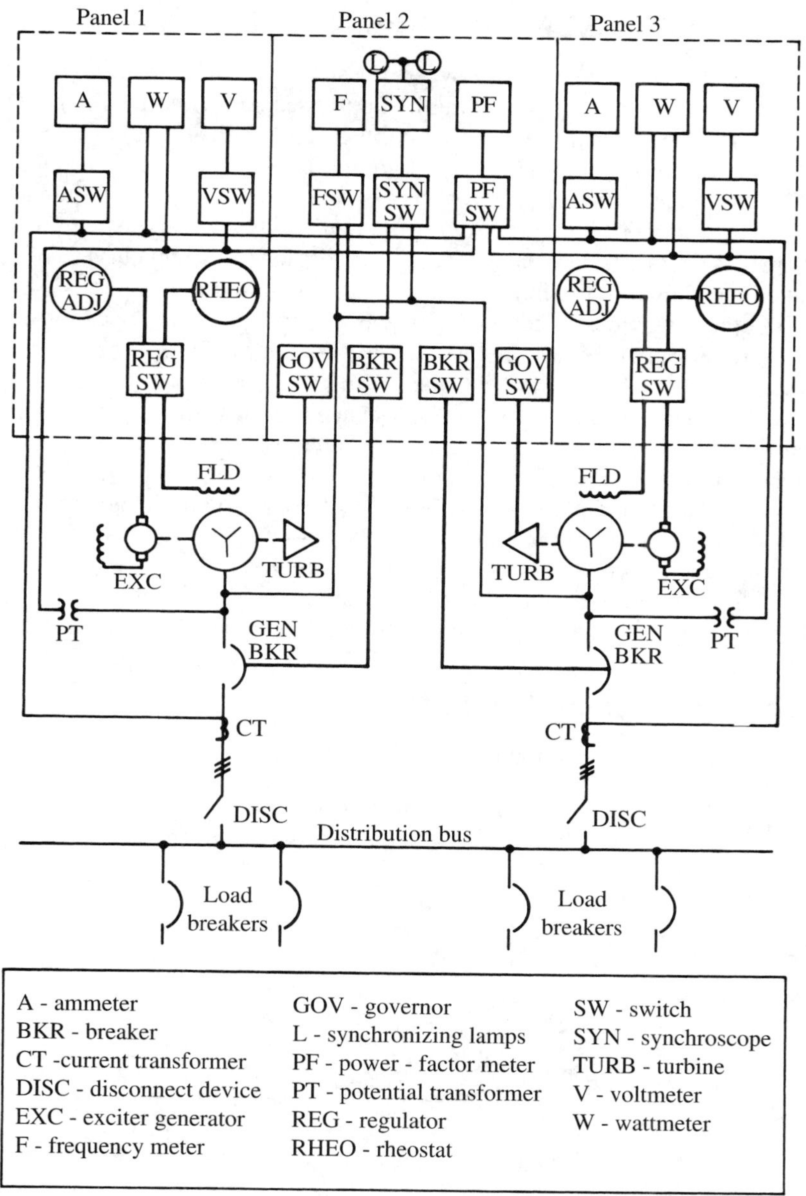

FIGURE 9.8
Functional diagram for a two-generator system supplying a common bus load.

9.7 PRIME-MOVER GOVERNOR CHARACTERISTICS

The transfer of active power between alternators in parallel is accomplished by adjustment of the no-load speed setting of the respective prime-mover governors, and the transfer of reactive power is accomplished by adjustment of the respective field rheostats or voltage regulators (see Figure 9.8).

A typical prime-mover governor characteristic, shown in Figure 9.9, is a plot of prime-mover speed (or generator frequency) vs. active power. Although usually drawn as a straight line, the actual characteristic has a slight curve. The drooping characteristic shown in the figure provides inherent stability of operation when paralleled with other machines. Machines with zero droop, called *isochronous* machines, are inherently unstable when operated in parallel; they are subject to unexpected load swings, unless electrically controlled with solid-state regulators.

The no-load speed setting (and hence the no-load frequency setting) of a synchronous generator can be changed by remote control from the generator panel by using a remote-control switch, shown as GOV SW in Figure 9.8. The switch actuates a servomotor that repositions the no-load speed setting of the governor, raising or lowering the characteristic without changing its slope. Curves for different no-load speed settings are shown with broken lines in Figure 9.9.

The governor parameters that determine the division of active power between alternators in parallel are governor speed regulation and governor droop.

Governor Speed Regulation

Governor speed regulation is defined as

$$\text{GSR} = \frac{n_{\text{nl}} - n_{\text{rated}}}{n_{\text{rated}}} = \frac{f_{\text{nl}} - f_{\text{rated}}}{f_{\text{rated}}} \tag{9–6}$$

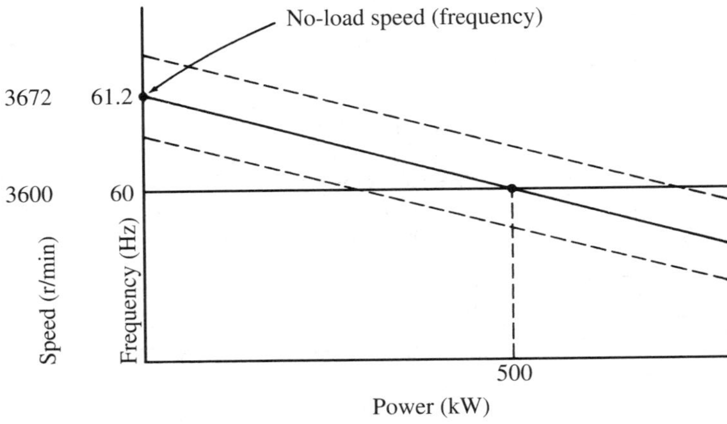

FIGURE 9.9
Typical prime-mover governor characteristics.

where: n_{rated} = rated speed (r/min)
n_{nl} = no-load speed (r/min)
f_{rated} = rated frequency (Hz)
f_{nl} = no-load frequency (Hz)

Governor Droop

Governor droop (GD) or droop rate is defined as the ratio of the change in frequency to the corresponding change in active power:

$$\text{GD} = \frac{\Delta f}{\Delta P} = \frac{f_{\text{nl}} - f_{\text{rated}}}{P_{\text{rated}}} \qquad \textbf{(9–7)}$$

The governor droop may be expressed in hertz/watt, hertz/kilowatt, or hertz/megawatt, as appropriate. The droop of a properly functioning governor is constant for a given droop setting. It is independent of the no-load speed setting and is unaffected by parallel operation.[7]

[7]Wear or other defects in the governing mechanism can affect the droop.

EXAMPLE 9.2 Referring to Figure 9.9, the governor characteristic drawn with a solid line represents the rated operating conditions for a 500-kW, 460-V, 60 Hz, three-phase, two-pole synchronous generator. Determine (a) speed regulation; (b) governor droop.

Solution

(a)

$$\text{GSR} = \frac{f_{\text{nl}} - f_{\text{rated}}}{f_{\text{rated}}} = \frac{61.2 - 60}{60} = \underline{0.02}$$

(b)

$$\text{GD} = \frac{\Delta_f}{\Delta_P} = \frac{61.2 - 60}{500} = 0.00240 \text{ Hz/kW} \quad \text{or} \quad \underline{2.4 \text{ Hz/MW}}$$

9.8 DIVISION OF ACTIVE POWER BETWEEN ALTERNATORS IN PARALLEL

The division of oncoming bus load between alternators in parallel is determined inherently by the droops of the respective governors, and cannot be readily changed by the operator. The shifting of bus load between machines to provide a desirable distribution is easily accomplished, however, by changing the no-load speed settings of the respective governors. *To shift some load from one generator to another requires an energy shift between prime movers.* The generator that is to take a greater share of the existing bus load must have more energy admitted to its prime mover; this is accomplished by raising the no-load speed setting of its governor. Similarly, the turbine generator that is to lose some load must have less energy admitted to its prime mover; this is accomplished by lowering the no-load speed setting of its governor.

Machines With Identical Governor Characteristics

Figure 9.10 illustrates graphically the transfer of load between two turbine generators with identical droop characteristics. Machine *B* carries a load of 150 kW at 60 Hz. Machine *A* has just been paralleled and is carrying no load. The initial conditions are indicated by subscript 1 and the solid lines. To transfer some load from machine *B* to machine *A* without permanently changing the frequency of the system requires adjustment of both governors.

Assume for the sake of argument that one-third of the load on machine *B* is to be transferred to machine *A*; that is, machine *B* is to carry 100 kW and machine *A* is to carry 50 kW. The governor control switch of machine *A* is turned to RAISE and held there until one-half the amount of load that is to be transferred is shifted, namely, 25 kW. This adjustment raises the no-load speed setting, and hence raises the entire characteristic of machine *A* without changing its droop. Since 25 kW was removed from machine *B*, it also increases in speed. Its characteristic, however, has not been shifted. The system is now operating at some higher frequency f_2, and the characteristic of machine *B* intersects the new and higher frequency line at 125 kW, as denoted by subscript 2. To complete the transfer of load, the governor control switch of machine *B* is moved to LOWER and held there until the remaining 25 kW is transferred. The characteristic of machine *B* is thereby lowered, and the frequency of the system comes back to its original 60-Hz value. This is indicated by subscript 3, where machine *A* now takes 50 kW and machine *B* takes 100 kW. Throughout the transition period the slope of the curves was not changed. The entire characteristic of each machine was moved either up or down.

During the transition period, when more energy was admitted to the prime mover of machine *A*, the energy balance of the system was disturbed; more energy was

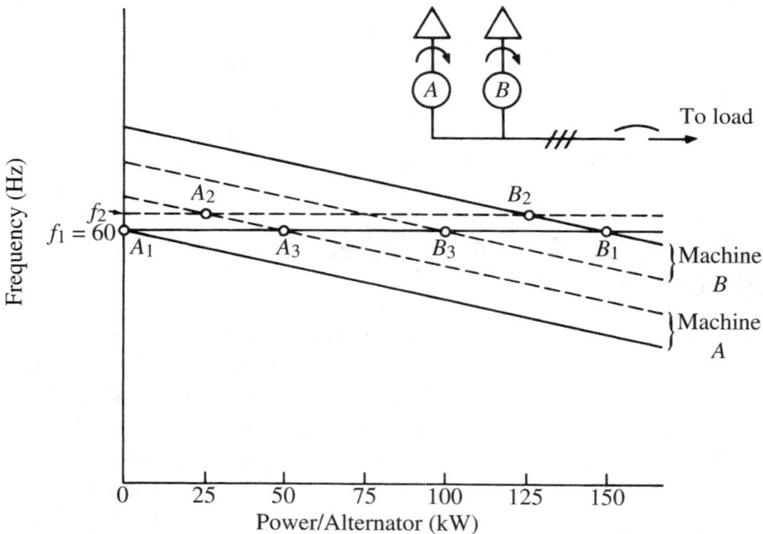

FIGURE 9.10
Graphical illustration of load transfer between machines with identical governor characteristics.

fed into the turbine generators than was going out. The total input was greater than the output, resulting in an increase in system speed and frequency. Prime-mover *A* increased its speed because of the direct increase in energy input brought about by raising the no-load speed setting of its governor. Prime-mover *B* increased its speed because of a reduction in its load. Except for the brief interval when the power angles are changing, both machines maintain identical speeds. Machines in parallel cannot run at different electrical speeds; if one increases in speed, so must the other. Like two engines driving a bull gear through identical pinions and flexible couplings, both machines are locked in rotational synchronism; but the engine with the greater fuel input will do more work than the other.

In actual practice, the division of active power between machines is accomplished with very little transient change in frequency, by making very small governor adjustments (alternately on each machine) until the desired transfer is accomplished.

If the governor control switches are located on a common synchronizing panel, both controls may be operated simultaneously, holding one on RAISE and the other on LOWER until the desired transfer is accomplished. This may be a little tricky, however.

Machines with identical governor droops (droop rate) will divide all increases in bus load and all decreases in bus load equally between them, regardless of the power ratings of the individual machines and the number of machines in parallel. For example, assume that the system indicated by subscript 3 in Figure 9.10 has an additional 50 kW added to the bus load. The system will operate at a new lower frequency, and the additional load will be divided equally between both machines. This is shown in Figure 9.11, where at 60 Hz, generator *A* is supplying 50 kW and generator *B* is supplying 100 kW. The additional load caused the frequency of the system to drop to a value that resulted in generator *A* delivering 75 kW and generator *B* delivering 125 kW. The new distribution is indicated by subscript 4 and the intersection of the new frequency line with the governor characteristics.

Machines With Dissimilar Governor Characteristics

Synchronous-generator sets with dissimilar governor characteristics do not divide increases or decreases in bus load equally. *The machine with the least droop assumes a greater portion of the change in bus load.* This is shown in Figure 9.12, where three paralleled generators, each with a different governor droop, are taking equal shares of the bus load at frequency f_1. The broken lines show the effect that different governor droops have on the distribution of oncoming bus load; the new lower frequency f_2 caused by the oncoming bus load causes the machine with the least droop to take a greater share of the oncoming load. Thus, if generators with different power ratings are to be operated in parallel, the respective governor droops should be adjusted so that load distribution between generators will be in proportion to their respective power ratings.

Machines With Isochronous Governors

The extreme case of dissimilar governor droops is illustrated in Figure 9.13. Generator *A* has a drooping governor characteristic and generator *B* is isochronous. The loads on machines *A* and *B* are 40 kW and 120 kW, respectively, and are indicated in Figure 9.13 by subscript 1. To transfer 60 kW from generator *B* to generator *A*, the governor

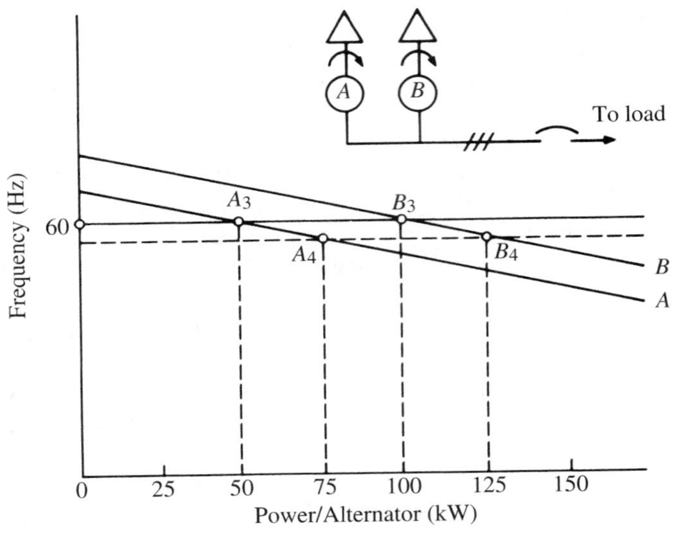

FIGURE 9.11
Effect of oncoming load on frequency and load division, for paralleled machines that have identical governor droops.

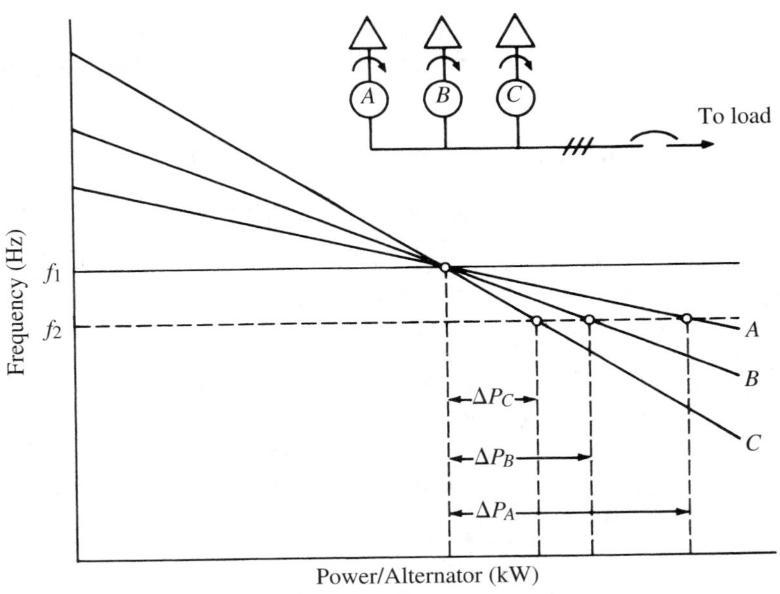

FIGURE 9.12
Effect of oncoming load on frequency and load division, for paralleled machines that have different governor droops.

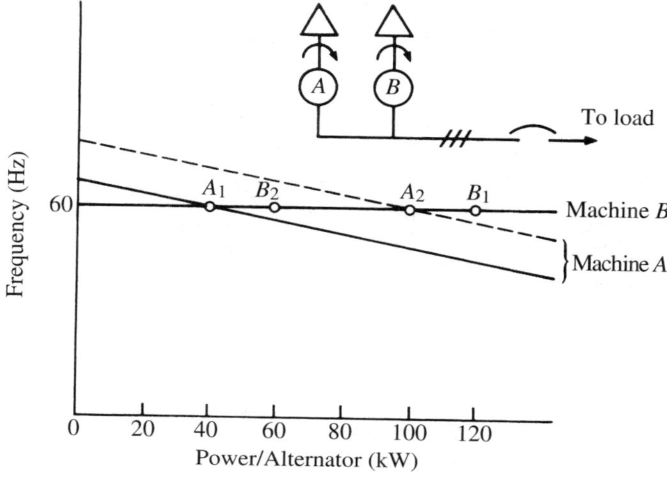

FIGURE 9.13
Effect of governor adjustment in a constant frequency system.

control switch of generator A should be turned to RAISE and held there until the entire transfer is completed. The governor control switch of generator B should not be used, since adjustment of its governor would affect the frequency of the system. The transfer of load must be made by adjusting only the governor control of the machine with droop. The final conditions are indicated by subscript 2.

For a given no-load speed setting of its governor, *an alternator with a drooping characteristic can carry only one value of load at a given frequency.* An isochronous machine, however, can carry any value of load from no load to rated load at the given frequency. Thus, if an isochronous machine is in parallel with a droop machine, any increase or decrease in bus load will be absorbed by the isochronous machine.

Inherent Instability of Two or More Isochronous Machines in Parallel

When two or more alternators with isochronous governors are in parallel, their operation is very unstable unless controlled by special-purpose electronic governors. The slightest change in frequency may cause a considerable interchange of load between machines. This is illustrated in Figure 9.14. Since no two governors are identical in the true sense of the word, it may be assumed that generator A has zero droop and generator B has a very minute droop.

The initial conditions are illustrated by subscript 1, where both generators are shown as having equal shares of the bus load. If a slight decrease in system speed occurs, due perhaps to very slight irregularities in the isochronous governing mechanism, a considerable portion of the bus load will be thrown on generator B, and generator A will then be lightly loaded or driven as a motor; the new condition is shown by subscript 2. Likewise, a slight increase in speed would shift the load in the other direction, as indicated by subscript 3.

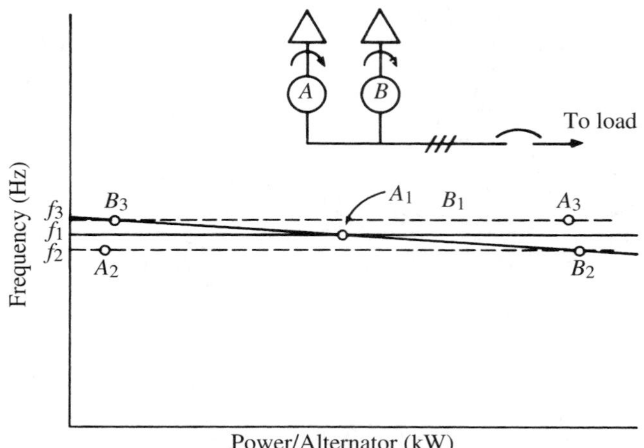

FIGURE 9.14
Graphic illustration of the inherent instability of isochronous machines in parallel.

The parallel operation of isochronous machines requires fast-acting electronic governors. Mechanical governors do not respond fast enough to avoid wild load swings **[1]**.

9.9 MOTORING OF ALTERNATORS

When the energy input to the prime mover of an alternator is insufficient to drive it at its synchronous speed, it will not only lose its load but will be driven by the other alternators on the bus at the synchronous speed of the system. Such "motoring" is automatically detected by a reverse-power relay, also called a *power-directional relay,* which trips the circuit breaker of the affected machine. Magnetic damping action on a rotating disk provides a time delay (adjustable) to prevent a transient load swing from tripping the generator breaker. Sustained motoring will trip the breaker, however, after expiration of the time delay. The only other indication of motoring is the backward deflection of the wattmeter pointer. If observed early enough, the operator may correct the condition by admitting more energy to the prime mover of the motorized machine. Sustained motoring is undesirable, because it results in a useless expenditure of energy. Furthermore, in the case of steam turbines, windage losses during motorization can overheat the turbine blades.

9.10 GENERAL PROCEDURE FOR SAFE SHUTDOWN OF AC GENERATORS IN PARALLEL WITH OTHER MACHINES

For safe shutdown of AC generators in parallel with other machines, these steps should be followed:

1. Using governor and field controls, gradually reduce the active and reactive power of the generator to essentially zero (see Section 9.12).

2. When the wattmeter indicates zero, or reverses (motoring), trip the generator breaker.
3. Switch the automatic voltage regulator to manual.
4. Reduce the voltage to its minimum value and then trip the exciter breaker.
5. Trip the turbine.

Note: In some large steam-turbine-generator applications (1000 MVA and above) the manufacturer may recommend a different sequence for shutdown to provide an increased margin against potential overspeed **[4]**.

9.11 CHARACTERISTIC TRIANGLE AS A TOOL FOR SOLVING LOAD DISTRIBUTION PROBLEMS BETWEEN ALTERNATORS IN PARALLEL

The solution of problems involving load distribution between generators may be accomplished in a straightforward manner by defining a *characteristic triangle* for each machine, and then using similar triangles to determine a solution.

Referring to Figure 9.15, the characteristic triangle is formed by the governor characteristic and the rated frequency line, with the intersection occurring at rated frequency and rated active power. The characteristic triangle is fixed for a given governor droop, and it does not change with changes in load or changes in the no-load speed setting of the governor.

An increase in load causes a decrease in frequency, as shown in Figure 9.15, and the intersection of the new frequency line with the governor characteristic establishes

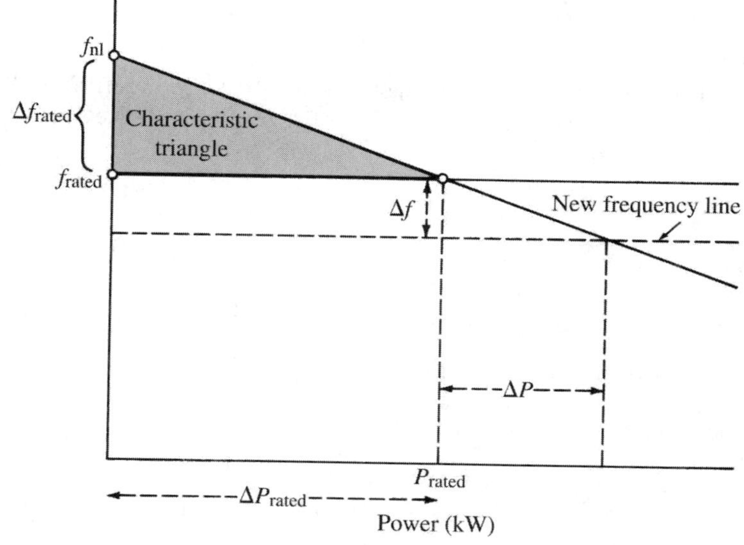

FIGURE 9.15
Characteristic triangle formed by the governor characteristic and the rated frequency line.

a new triangle similar to the characteristic triangle. Thus, from the geometry of similar triangles, and Eq. (9–7)

$$GD = \frac{f_{nl} - f_{rated}}{P_{rated}} = \frac{\Delta f}{\Delta P} \quad \textbf{(9–8)}$$

From Eq. (9–6),

$$f_{nl} - f_{rated} = GSR \times f_{rated}$$

Substituting into Eq. (9–8),

$$GD = \frac{GSR \times f_{rated}}{P_{rated}} = \frac{\Delta f}{\Delta P} \quad \textbf{(9–9)}$$

where: GD = governor droop
GSR = governor speed regulation
Δf = change in frequency due to change in load
ΔP = change in load

Since the governor droop is unaffected by increases or decreases in load, Eq. (9–9) holds true for all operating frequencies and for all no-load speed settings of the governor.

EXAMPLE 9.3 Two generators, *A* and *B*, are in parallel and sharing a total bus load of 300 kW at 60 Hz. Both generators are rated at 460 V, 500 kW, 60 Hz, and have 2.0 percent speed regulation. Generator *A* is carrying 100 kW, and generator *B* is carrying 200 kW. Assuming generator *A* is tripped off the bus, determine the frequency of (a) generator *A;* (b) generator *B;* (c) bus.

Solution

The characteristic curves for both machines are shown in Figure 9.16. With generator *A* tripped off the bus, its frequency rises to f_{a2}; generator *B* takes all the load, causing its frequency to drop to f_{b2}.

(a) For machine *A*,

$$\frac{GSR \times f_{rated}}{P_{rated}} = \frac{\Delta f_a}{\Delta P_a} \quad \Rightarrow \quad \frac{0.020 \times 60}{500} = \frac{\Delta f_a}{100}$$

$$\Delta f_a = 0.24 \text{ Hz}$$

Thus, $f_a = 60 + 0.24 = \underline{60.24 \text{ Hz.}}$

(b) Since both machines are identical,

$$\Delta f_b = 0.24 \text{ Hz}$$
$$f_b = 60 - 0.24 = \underline{59.76 \text{ Hz}}$$

(c) The bus frequency is the frequency of generator *B*.

$$f_{bus} = \underline{59.76 \text{ Hz}}$$

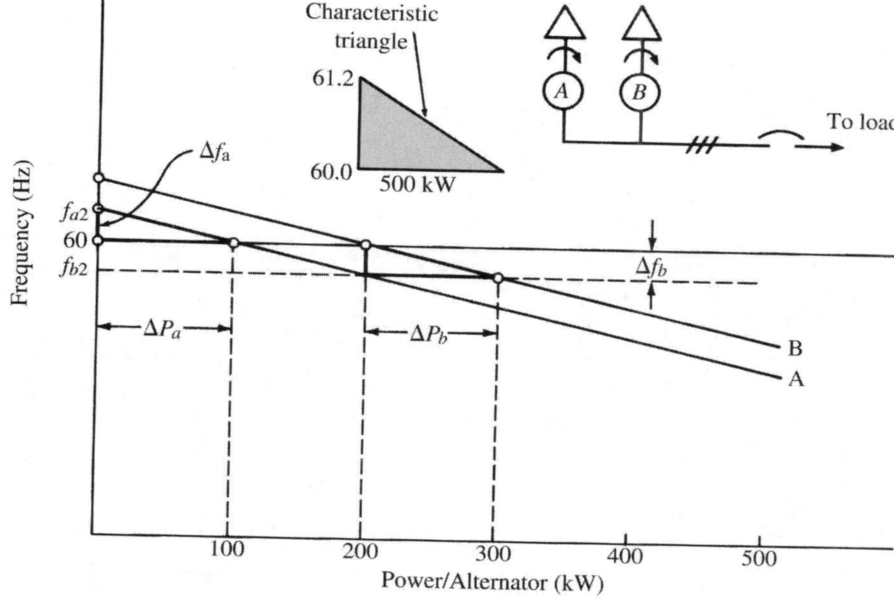

FIGURE 9.16
Governor characteristics for Example 9.3.

EXAMPLE 9.4 A 500-kW, 60-Hz, 2300-V, six-pole alternator *A* is paralleled with a 300-kW, 60-Hz, 2300-V, four-pole machine *B*. Both machines have a speed regulation of 2.43 percent. The machines are carrying equal shares of a 400-kW bus load at a frequency of 60.5 Hz. If the bus load increases to a total of 500 kW, determine (a) operating frequency; (b) load carried by each machine.

Solution

The characteristic curves are shown in Figure 9.17.

(a)

$$\frac{\text{GSR} \times f_{\text{rated}}}{P_{\text{rated}}} = \frac{\Delta f}{\Delta P}$$

Machine *A*	**Machine *B***
$\dfrac{0.0243 \times 60}{500} = \dfrac{\Delta f}{\Delta P_a}$	$\dfrac{0.0243 \times 60}{300} = \dfrac{\Delta f}{\Delta P_b}$
$\Delta P_a = 342.936\Delta f$	$\Delta P_b = 205.761\ \Delta f$

$$\Delta P_a + \Delta P_b = [342.936 + 205.761]\Delta f$$

$$500 - 400 = 548.697\Delta f$$

$$\Delta f = 0.182\text{ Hz}$$

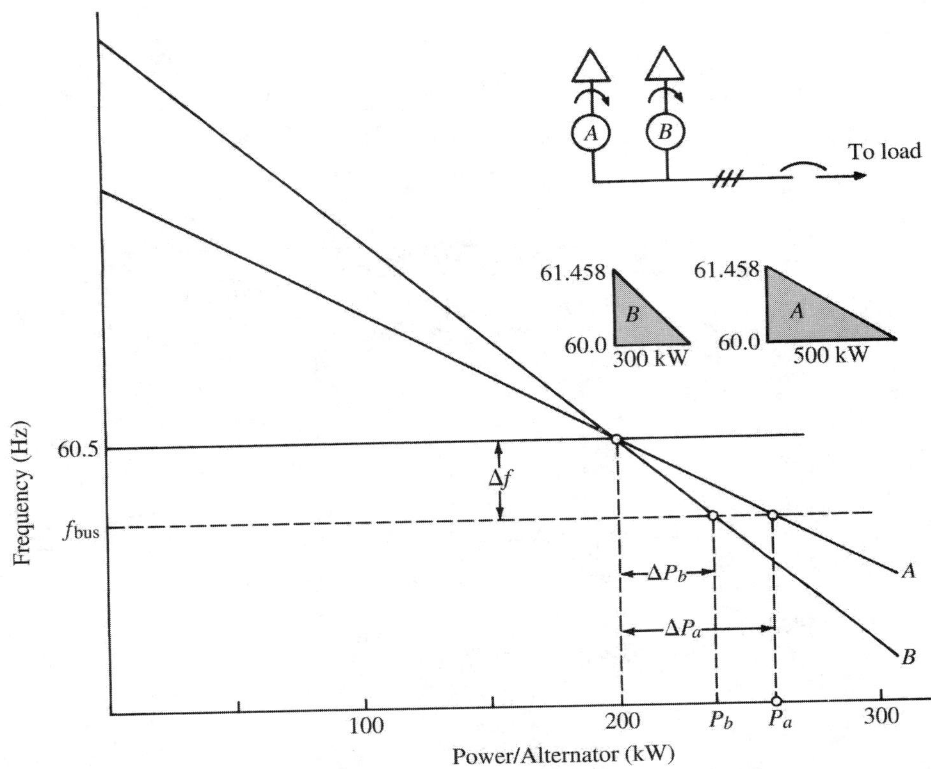

FIGURE 9.17
Governor characteristics for Example 9.4.

Referring to Figure 9.17,

$$f_{\text{bus}} = 60.5 - 0.182 = \underline{60.318 \text{ Hz}}$$

(b)

$$\Delta P_a = 342.936\Delta f = 342.936 \times 0.182 = 62.414 \text{ kW}$$

$$\Delta P_b = 100 - 62.414 = 37.586 \text{ kW}$$

$$P_a = 200 + 62.41 = \underline{262.41 \text{ kW}}$$

$$P_b = 200 + 37.59 = \underline{237.59 \text{ kW}}$$

EXAMPLE 9.5 A 1000-kW, 60-Hz synchronous alternator A is in parallel with a 600-kW, 60-Hz alternator B. Both machines have identical governor droops of 0.0008 Hz/kW. Machine A takes two-thirds of a 900-kW bus load at 60.2 Hz. If an additional 720-kW load is connected to the bus, determine (a) bus frequency; (b) load on each machine.

Solution

(a) Since both machines have identical governor droops, they will take equal shares of the additional bus load. Thus,

$$\Delta P_a = \Delta P_b = \frac{720}{2} = 360 \text{ kW}$$

$$\text{GD} = \frac{\Delta f}{\Delta P_a} \quad \Rightarrow \quad 0.0008 = \frac{\Delta f}{360}$$

$$\Delta f = 0.288 \text{ Hz}$$

$$f_{\text{bus}} = 60.2 - 0.288 = \underline{59.91 \text{ Hz}}$$

(b)

$$P_a = \frac{2}{3} \times 900 + 360 = \underline{960 \text{ kW}}$$

$$P_b = \frac{1}{3} \times 900 + 360 = \underline{660 \text{ kW}}$$

Note: Machine *B* is operating at a 10 percent overload.

EXAMPLE 9.6 Three diesel-driven 60-Hz alternators *A, C,* and *B* are in parallel and taking equal shares of a 210-kW, 60-Hz bus load. The ratings of the alternators are 500 kW, 200 kW, and 300 kW, respectively. The circuit is shown in Figure 9.18(a), and the governor characteristics are shown in Figure 9.18(b). If a three-phase, 440-kW resistance load and a three-phase, 60-Hz induction motor that draws 200 kVA at 0.80 power factor are added to the bus, determine (a) the system kilowatts; (b) the system frequency; (c) the kilowatt loads carried by each machine.

Solution

(a) The increase in bus load is

$$\Delta P_{\text{bus}} = 440 + 0.8 \times 200 = 600 \text{ kW}$$

$$P_{\text{syst}} = 210 + 600 = \underline{810 \text{ kW}}$$

(b) The respective governor droops obtained from the initial loadings shown in Figure 9.18(b) are

$$\text{GD}_a = \frac{\Delta f_a}{\Delta P_a} = \frac{60.2 - 60}{70} = 0.002857 \text{ Hz/kW}$$

$$\text{GD}_b = \frac{\Delta f_b}{\Delta P_b} = \frac{60.4 - 60}{70} = 0.005714 \text{ Hz/kW}$$

$$\text{GD}_c = \frac{\Delta f_c}{\Delta P_c} = \frac{60.6 - 60}{70} = 0.008571 \text{ Hz/kW}$$

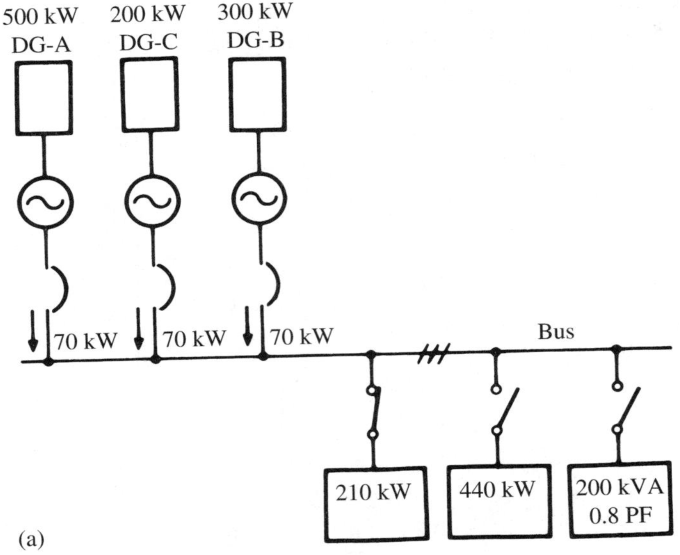

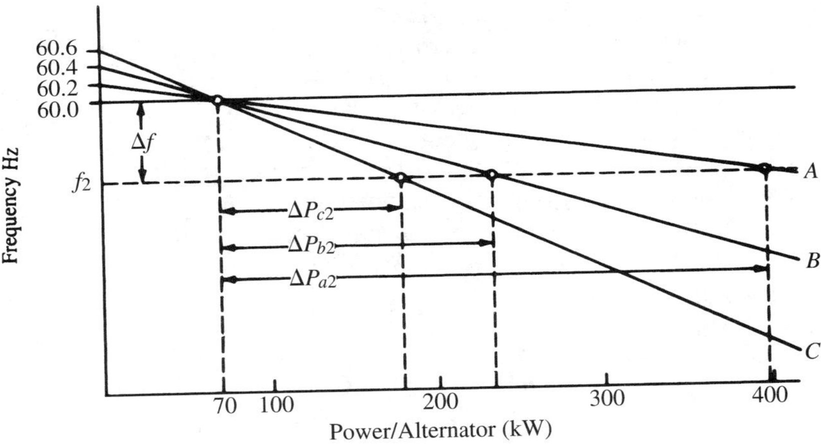

FIGURE 9.18
Governor characteristics for Example 9.6.

The increased load causes the bus frequency to drop to f_2. Referring to Figure 9.18(b), using similar triangles, and noting that Δf is the same for all machines,

$$\frac{\Delta f}{\Delta P_{a2}} = 0.002857 \qquad \frac{\Delta f}{\Delta P_{b2}} = 0.005714 \qquad \frac{\Delta f}{\Delta P_{c2}} = 0.008571$$

$$\Delta P_{a2} = 350\Delta f \qquad \Delta P_{b2} = 175\Delta f \qquad \Delta P_{c2} = 116.6667\,\Delta f$$

$$\Delta P_{a2} + \Delta P_{b2} + \Delta P_{c2} = 600$$

$$(350 + 175 + 116.6667)\Delta f = 600 \quad \Rightarrow \quad \Delta f = 0.9351 \text{ Hz}$$

$$f_2 = 60 - \Delta f = 60 - 0.9351 = \underline{59.06 \text{ Hz}}$$

(c)

$$P_{a2} = 70 + \Delta P_{a2} = 70 + 350 \times 0.9351 = \underline{397.3 \text{ kW}}$$

$$P_{b2} = 70 + \Delta P_{b2} = 70 + 175 \times 0.9351 = \underline{233.6 \text{ kW}}$$

$$P_{c2} = 70 + \Delta P_{c2} = 70 + 116.6667 \times 0.9351 = \underline{179.1 \text{ kW}}$$

9.12 DIVISION OF REACTIVE POWER BETWEEN ALTERNATORS IN PARALLEL

Adjustment of the no-load speed setting of the respective governors to balance the active power between paralleled alternators does not balance the reactive power. Balancing reactive power between machines requires adjustment of the respective field currents.

The effect of active power transfer on reactive power loading will be discussed for the two identical paralleled generators shown in Figure 9.19; generator 1 is carrying the entire bus load, and generator 2 is floating on the bus. The phasor diagrams corresponding to the initial load conditions are shown with *solid lines* in Figures 9.20(a) and (b). Current $\mathbf{I}_{a1}$ supplied by generator 1 is the bus current. The drawings in Figure 9.20 are constructed to scale.

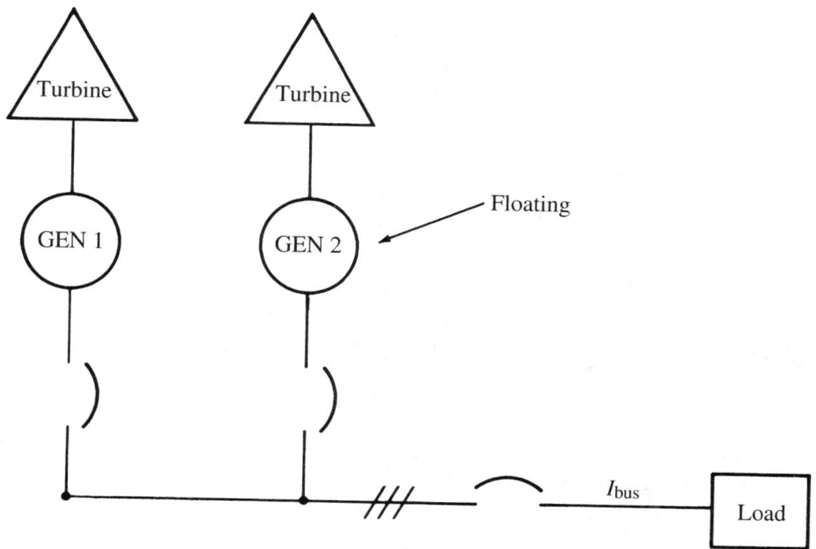

FIGURE 9.19
Two generators in parallel, one floating on the bus.

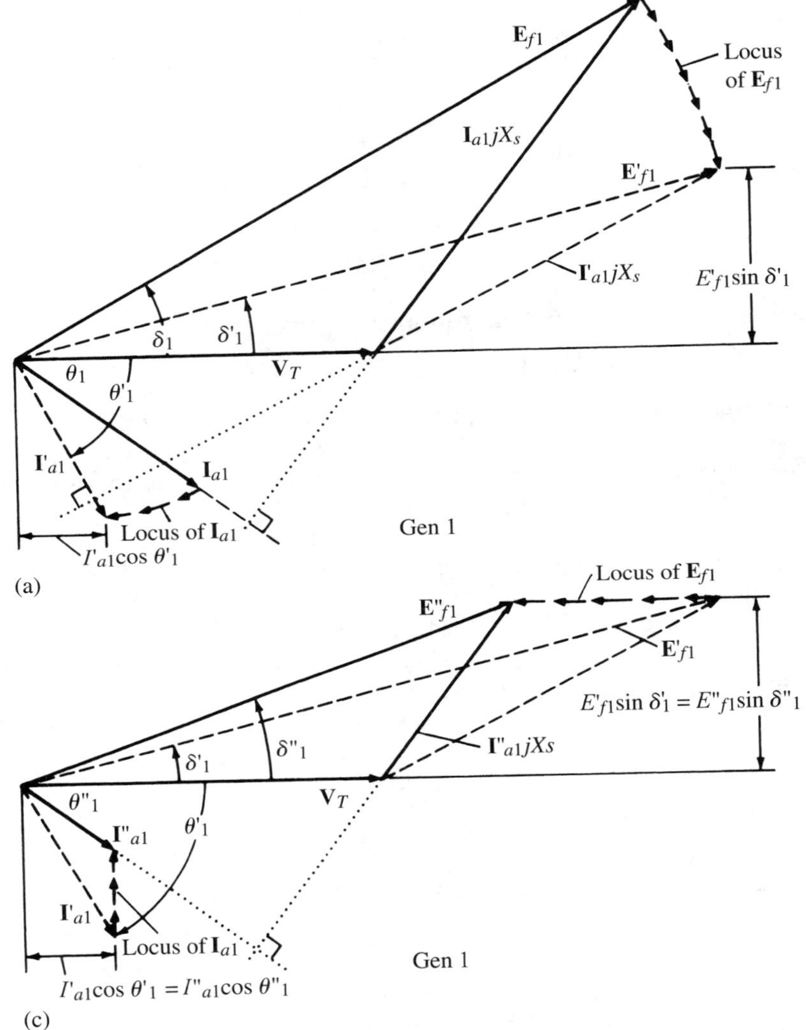

FIGURE 9.20
Phasor diagrams: (a) and (b) show the effect of governor adjustments on the transfer of active power; (c) and (d) the effect of field-current adjustments on the transfer of reactive power.

Balancing the active power between generators by simultaneously adjusting the no-load speed settings of the respective governors causes the power angle of generator 2 to increase, and that of generator 1 to decrease. The new conditions are shown by broken lines and single-primed variables in Figure 9.20(a) and (b). For the new conditions,

$$\frac{V_T E'_{f1}}{X_{s1}} \sin \delta'_1 = \frac{V_T E'_{f2}}{X_{s2}} \sin \delta'_2$$

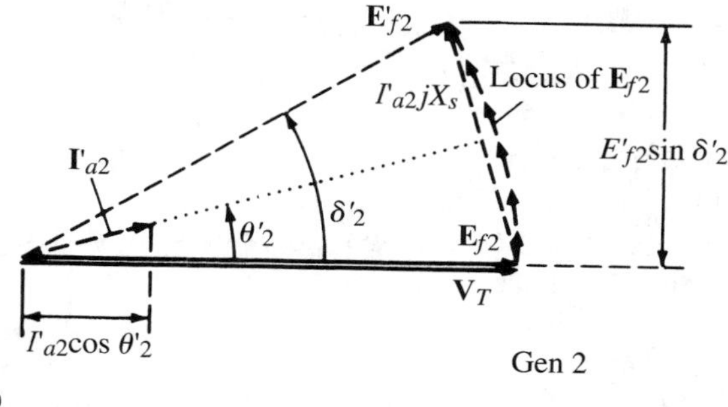

(b)

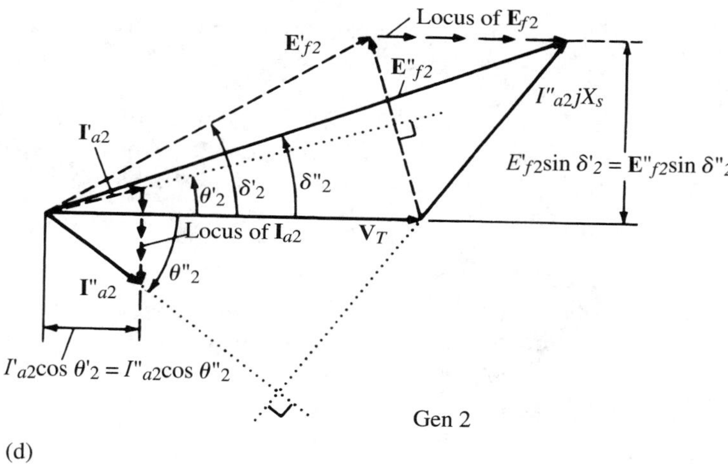

(d)

FIGURE 9.20 ***(Continued)***

Since the machines are assumed to be identical, $X_{s1} = X_{s2}$, and since they are in parallel V_T is the same for both machines. Hence,

$$E'_{f1} \sin \delta'_1 = E'_{f2} \sin \delta'_2$$

The equivalence is shown in Figures 9.20(a) and (b). Furthermore, with both machines supplying equal active power to the load,

$$V_T I'_{a1} \cos \theta'_1 = V_T I'_{a2} \cos \theta'_2$$
$$I'_{a1} \cos \theta'_1 = I'_{a2} \cos \theta'_2$$

This equivalence is also shown on the drawings in Figures 9.20(a) and (b). *Note:*

1. The armature current $\mathbf{I}'_{a1}$ in generator 1 is lagging, but the armature current $\mathbf{I}'_{a2}$ in generator 2 is leading, indicating a difference in power factor, and hence a difference in reactive power distribution between machines.

2. Although not shown in Figures 9.20(b), (c), and (d), the magnitude and phase angle of the bus current do not change, and remain equal to $\mathbf{I}_{a1}$ [shown in Figure 9.20(a)].

To balance the reactive power between machines (for the example illustrated in Figure 9.20), the field excitation of each machine must be adjusted so that their respective armature currents are equal in magnitude and their respective phase angles equal that of the bus current. To accomplish this, the field rheostat (or voltage regulator control) of the machine that accepted active power load (generator 2) should be turned in the "raise voltage" direction, and that of the machine that gave up some active power load (generator 1) should be turned in the "lower voltage" direction, until the power-factor meter (or var meter) indicates the same for both machines.

The phasor diagrams corresponding to the final steady-state condition for balanced active power and balanced reactive power between machines is shown with solid lines and double-primed variables in Figures 9.20(c) and (d).

The locus of the current and emf phasors in Figure 9.20 illustrates the transition process that takes place as active and reactive power is balanced between identical machines.

The alternators represented in Figure 9.20 and the associated discussion are for identical machines that are paralleled and adjusted to take equal shares of the total active and reactive power. If the machines are not identical, the respective synchronous reactances will be different and must be accounted for.

EXAMPLE 9.7

A three-phase distribution bus supplies a 500-kW unity power-factor load, and several induction motors that draw a total of 200 kVA at 85.2 percent power factor, as shown in Figure 9.21(a). Two alternators (A and B) supplying the bus take equal shares of the active power load. Determine (a) the active and reactive components of the bus load. (b) If the power factor of generator A is 0.94 lagging, determine the reactive power supplied by each machine.

Solution

(a)

$$P_{\text{bus}} = 500 + 200 \times 0.852 = \underline{670.40 \text{ kW}}$$

$$\theta_{\text{motors}} = \cos^{-1}(0.852) = 31.57°$$

$$Q_{\text{bus}} = Q_{\text{motors}} = 200 \times \sin(31.57) = \underline{104.7 \text{ kvar}}$$

(b)

$$P_a = \frac{670.40}{2} = 335.2 \text{ kW} \qquad \theta_a = \cos^{-1}(0.94) = 19.95°$$

From the power diagram shown in Figure 9.21(b),

$$\tan(19.95°) = \frac{Q_a}{335.2} \quad \Rightarrow \quad Q_a = \underline{121.7 \text{ kvar}}$$

$$Q_b = Q_{\text{bus}} - Q_a = 104.7 - 121.7 = \underline{-17.0 \text{ kvar}}$$

The negative sign indicates a leading power factor.

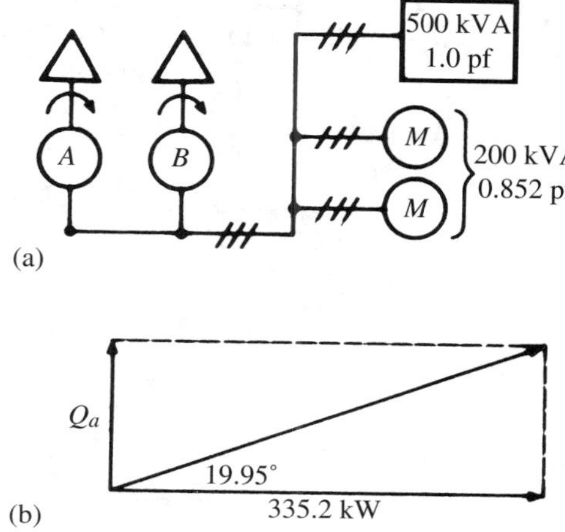

FIGURE 9.21
(a) One-line diagram and (b) power diagram for Example 9.7.

9.13 ACCIDENTAL LOSS OF FIELD EXCITATION

Accidental loss of field excitation supplied to a synchronous machine in parallel with other machines will cause the affected unit to pull out of synchronism. The resultant reduction in its load will cause its prime mover to accelerate and drive the alternator as an induction generator.[8] The effect will be rapid overheating of the rotor forging by eddy currents, with possible damage to rotor wedges that hold the rotor end turns in place. If protective equipment, such as a loss-of-field relay, does not respond to trip the generator breaker, and the field excitation cannot be immediately restored, the generator breaker must be tripped by the operator.

Furthermore, the large amount of reactive power drawn by the affected machine as it operates asynchronously (as an induction generator) will cause overheating of the stator windings. On complete loss of excitation, the rotor may overheat in as little as 10 seconds **[2]**.

9.14 PER-UNIT VALUES OF SYNCHRONOUS MACHINE PARAMETERS

Manufacturers of electrical machinery generally publish synchronous machine parameters as per-unit impedance instead of actual impedance values. Per-unit values are useful in machine design in that a wide range of machine sizes have approximately the same values. This makes it easy to detect gross errors in design calculations, and provides a convenient means of comparing the relative performance of machines without regard to machine size.

---

[8]See Section 5.18 on induction generators.

Per-unit parameters are defined as a ratio of the actual value of the respective parameters divided by a common base impedance. The base values for a synchronous machine are defined as follows:

$$S_{\text{base}} = \text{rated apparent power/phase (VA)}$$
$$V_{\text{base}} = \text{rated voltage/phase (V)}$$

$$I_{\text{base}} = \frac{S_{\text{base}}}{V_{\text{base}}} \quad \text{A} \tag{9–10}$$

$$Z_{\text{base}} = \frac{V_{\text{base}}}{I_{\text{base}}} = \frac{V_{\text{base}}^2}{S_{\text{base}}} \quad \Omega \tag{9–11}$$

Using Z_{base} as defined in Eq. (9–11), per-unit parameters are

$$R_{\text{pu}} = \frac{R_{\text{a}}}{Z_{\text{base}}} \qquad X_{\text{pu}} = \frac{X_{\text{s}}}{Z_{\text{base}}} \qquad \mathbf{Z}_{\text{pu}} = R_{\text{pu}} + jX_{\text{pu}} \tag{9–12}$$

EXAMPLE 9.8 Determine the per-unit impedance of a three-phase, wye-connected, 100-kVA, 480-V, 60-Hz synchronous generator whose synchronous impedance is (0.0800 + j2.300) Ω.

Solution

$$S_{\text{base}} = \frac{100{,}000}{3}\ \text{VA} \qquad V_{\text{base}} = \frac{480}{\sqrt{3}}\ \text{V}$$

$$Z_{\text{base}} = \frac{V_{\text{base}}^2}{S_{\text{base}}} = \frac{(480/\sqrt{3})^2}{100{,}000/3} = 2.304\ \Omega$$

$$R_{\text{pu}} = \frac{R_a}{Z_{\text{base}}} = \frac{0.0800}{2.304} = \underline{0.0347} \qquad X_{\text{pu}} = \frac{X_s}{Z_{\text{base}}} = \frac{2.300}{2.304} = \underline{0.9983}$$

$$\mathbf{Z}_{\text{pu}} = R_{\text{pu}} + jX_{\text{pu}} = 0.0347 + j0.9983 = \underline{\underline{0.9989\angle 88.01^\circ}}$$

9.15 VOLTAGE REGULATION

The voltage regulation of an alternator is the percentage change in terminal voltage from no load to rated load, with respect to rated voltage, with the field current held constant. The voltages are all phase voltages or all line voltages.

$$\text{VR} = \frac{V_{\text{nl}} - V_{\text{rated}}}{V_{\text{rated}}} \times 100 \tag{9–13}$$

where: VR = voltage regulation (%)
V_{rated} = voltage indicated on nameplate of machine (V)
V_{nl} = no-load voltage (open-circuit voltage) (V)

Note 1. *V_{nl} is the voltage measured after tripping the generator breaker while the generator is operating at rated load and rated voltage.*

Note 2. V_{nl} *is not the excitation voltage.* Substituting E_f for V_{nl} in Eq. (9–13) results in gross errors and should not be used as an approximate value for V_{nl} because E_f does not account for the effect of magnetic saturation.

Note 3. The voltage regulation will be positive for unity power-factor and lagging power-factor loads (usual operating conditions), but for leading power-factor loads the regulation may be negative.

The voltage regulation of an alternator is a figure of merit used for comparison with other machines. Machines with high values of voltage regulation have relatively large values of synchronous reactance and suffer large drops in output voltage with lagging power-factor loads. Thus, the voltage regulation of a machine is an indicator of the change in field current required to maintain system voltage, when going from no load to rated load at some specific power factor. Furthermore, when operating in parallel, alternators with wide differences in voltage regulation will not divide the reactive power load in proportion to their apparent power ratings.

The no-load voltage for a specific power factor may be determined by operating the machine at rated load conditions, then removing the load and observing the no-load voltage. Since this is generally not a practical method, especially for very large machines, other methods have been developed to determine the no-load voltage without having to load and unload the machine. The IEEE standard for determining the no-load voltage, as given in Reference **[6]**, is an exact but complicated and time-consuming process.

The Approximate Method

A graphical method that provides a reasonable approximation of the no-load voltage may be obtained by reference to the actual magnetization curve, and a linearized magnetization curve drawn from the origin through the rated voltage point on the magnetization curve, as shown in Figure 9.22. See References **[3],[8]** for a more detailed analysis.

Defining points on the field-current axis of the magnetization curve,

$$I_{f2} = I_{f1} = \Delta I_f$$

where:

I_{f1} = field current required to produce rated armature voltage at no load

I_{f2} = field current required to obtain a value of V_{nl} that will result in rated voltage when rated load is applied

ΔI_f = increment increase in field current required to compensate for the synchronous reactance drop and magnetic saturation

ΔV = voltage drop due to magnetic saturation

To determine the no-load voltage using the approximate method:

1. Enter the value of E_f calculated from Eq. (9–1) on the linearized magnetization curve (point *a*) in Figure 9.22.
2. Project vertically from the linearized curve to the actual magnetization curve (point *b*). Then project horizontally to the voltage axis and read V_{nl}.

EXAMPLE 9.9 A hypothetical three-phase, two-pole, 60-Hz, 1000-kVA, 4.8-kV, wye-connected synchronous generator has a synchronous reactance of 13.80 Ω/phase. The machine is operating at rated kVA, rated voltage, and a power factor of 90.0 percent lagging.

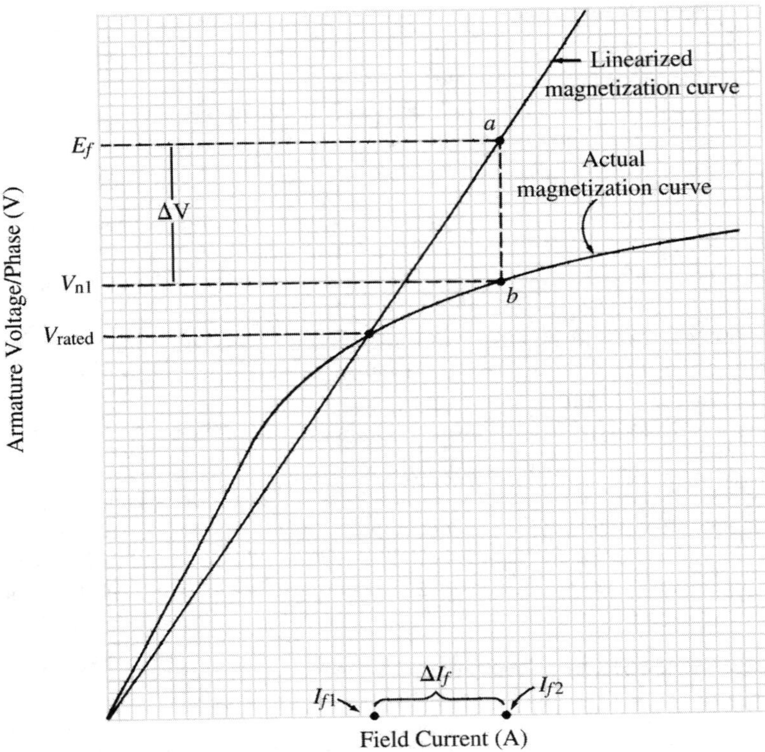

FIGURE 9.22
Magnetization curve and its linear approximation for a representative synchronous generator.

Determine (a) excitation voltage; (b) power angle; (c) no-load voltage, assuming the field current is not changed; (d) voltage regulation; (e) no-load voltage if the field current is reduced to 80 percent of its value at rated load.

Solution

$$V_T = \frac{4800}{\sqrt{3}} = 2771.28 \text{ V} \qquad \theta = \cos^{-1}(0.900) = 25.84^\circ$$

$$\mathbf{S} = \mathbf{V}_T \mathbf{I}_a^* \quad \Rightarrow \quad \frac{1000 \times 10^3}{3} \underline{/25.84^\circ} = (2771.28\underline{/0^\circ}) \cdot \mathbf{I}_a^*$$

$$\mathbf{I}_a^* = 120.28\underline{/25.84^\circ} \quad \Rightarrow \quad \mathbf{I}_a = 120.28\underline{/-25.84^\circ}$$

(a) The circuit diagram for one phase is shown in Figure 9.23(a).

$$\mathbf{E}_f = \mathbf{V}_T + \mathbf{I}_a jX_s = 2771.28\underline{/0^\circ} + (120.28\underline{/-25.84^\circ})(13.80\underline{/90^\circ})$$

$$\mathbf{E}_f = 3494.8 + j1493.9 = 3800.7\underline{/23.1^\circ} \text{ V}$$

$$E_f = \underline{3801 \text{ V}}$$

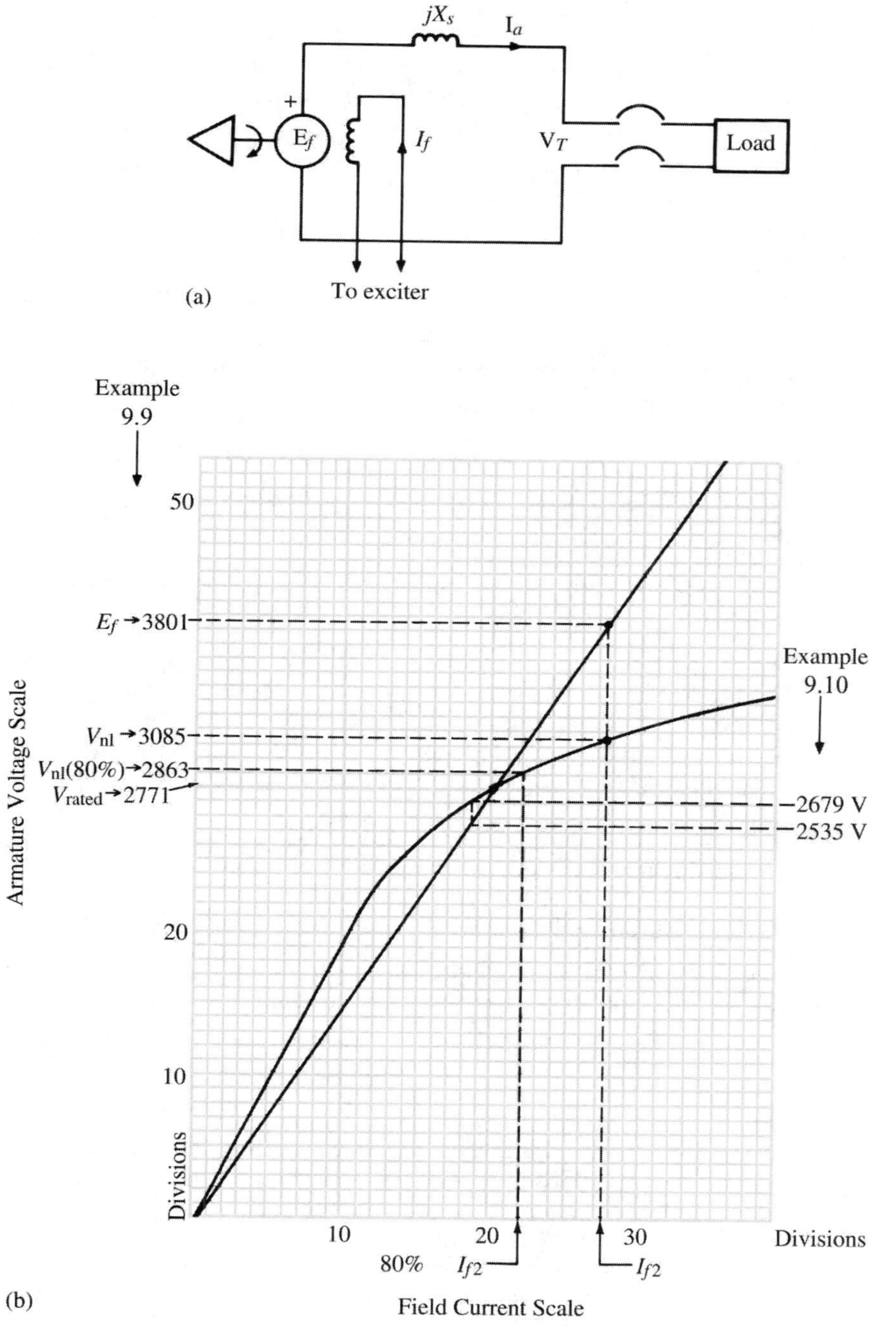

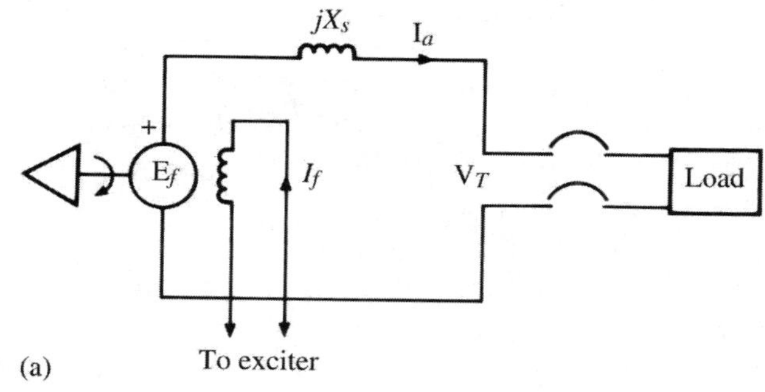

(a)

Example 9.9

50

$E_f \rightarrow 3801$

Example 9.10

$V_{nl} \rightarrow 3085$

$V_{nl}(80\%) \rightarrow 2863$

$V_{rated} \rightarrow 2771$

–2679 V

–2535 V

Armature Voltage Scale

20

10

Divisions

10 20 30 Divisions

80% I_{f2} I_{f2}

(b)

Field Current Scale

FIGURE 9.23
Equivalent circuit and magnetization curve for Examples 9.9 and 9.10.

(b) $$\delta = \underline{23.1°}$$

(c) The no-load voltage is obtained graphically from Figure 9.23(b). The linearized curve is drawn to intersect the magnetization curve at a rated phase voltage of 2771 V. This scales the voltage axis at

$$2271 \text{ V} \div 30 \text{ scale divisions} = 92.37 \text{ V/div}$$

Thus, voltage E_f is located at

$$\frac{3801 \text{ V}}{92.37 \text{ V/div}} = 41.2 \text{ divisions on the linearized curve}$$

Projecting vertically from the 3801-V point on the linearized curve to the magnetization curve determines V_{nl}.

$$V_{nl} = 33.4 \text{ div} \times 92.37 \text{ V/div} \approx \underline{3085 \text{ V}}$$

(d) $$\text{VR} = \frac{V_{nl} - V_{rated}}{V_{rated}} \times 100 = \frac{3085 - 2771}{2771} \times 100 \approx \underline{11\%}$$

Note: If E_f were erroneously substituted for V_{nl}, the calculated voltage regulation would be

$$\frac{3801 - 2771}{2771} \times 100 = 37\%$$

Such erroneous substitution ignores the limits that magnetic saturation imposes on the magnitude of the induced voltage.

(e) Rated field current (I_{f2}) is indicated by 27.5 div on the field-current scale. Thus, referring to Figure 9.23(b),

$$80\%\ I_{f2} \text{ corresponds to } 0.80 \times 27.5 = 22 \text{ div on the field-current scale}$$

Projecting up from the field-current axis to the magnetization curve determines the no-load voltage at 80 percent rated field current.

$$V_{nl} = 31 \text{ div} \times 92.37 \text{ V/div} \approx \underline{2863 \text{ V}}$$

EXAMPLE 9.10 Repeat Example 9.9 assuming a 90.0 percent leading power factor.

Solution

(a) $$\mathbf{V}_T = 2771.28\angle 0° \text{ V} \qquad \mathbf{I}_a = 120.28\angle 25.84°$$

$$\mathbf{E}_f = \mathbf{V}_T + \mathbf{I}_a jX_s = 2771.28 + (120.28\angle 25.84°)(13.80\angle 90°)$$

$$\mathbf{E}_f = 2047.81 + j1493.89 = 2534.81\angle 36.11°$$

$$E_f = \underline{2535 \text{ V}}$$

(b) $$\delta = \underline{36.1^\circ}$$

(c) Referring to Figure 9.23(b), voltage E_f is located at

$$\frac{2535}{92.37} \approx 27.4 \text{ div on the linearized curve (shown with dotted lines)}$$

Projecting vertically from the 2535 point on the linearized curve to the magnetization curve determines the no-load voltage. Thus,

$$V_{\text{nl}} = 29 \times 92.37 \approx \underline{2679 \text{ V}}$$

(d) $$\text{VR} = \frac{V_{\text{nl}} - V_{\text{rated}}}{V_{\text{rated}}} \times 100 = \frac{2679 - 2771}{2771} \times 100 \approx \underline{-3.32\%}$$

Note: The leading power-factor condition resulted in a negative voltage regulation.

9.16 DETERMINATION OF SYNCHRONOUS MACHINE PARAMETERS

The synchronous reactance depends on the degree of magnetic saturation, and will have different values for different loads and different field currents. The value of synchronous reactance supplied by the manufacturer for voltage-regulation calculations, and for analysis involving rated conditions, *is based on a field current that provides rated open-circuit voltage and an associated short-circuit current* **[6],[9]**.

In those cases where synchronous machine parameters are not readily available from the manufacturer, they can be *approximated* from a DC test, an open-circuit test, and a short-circuit test.

DC Test

The purpose of the DC test is to determine R_a. This is accomplished by connecting any two stator leads to a low-voltage DC source, as shown in Figure 9.24(a). The rheostat is adjusted to provide readable values of current and voltage, and the resistance between the two stator leads is determined from Ohm's law. Thus, from Figures 9.24(a) and (b) for a wye connection,

$$R_{\text{DC}} = \frac{V_{\text{DC}}}{I_{\text{DC}}} \qquad \textbf{(9–14)}$$

$$R_{\text{Y}} = \frac{R_{\text{DC}}}{2} \qquad \textbf{(9–15)}$$

If the stator is delta connected as shown in Figure 9.24(c),

$$R_{\text{DC}} = \frac{R_\Delta \cdot 2R_\Delta}{R_\Delta + 2R_\Delta} \quad \Rightarrow \quad R_\Delta = 1.5R_{\text{DC}} \qquad \textbf{(9–16)}$$

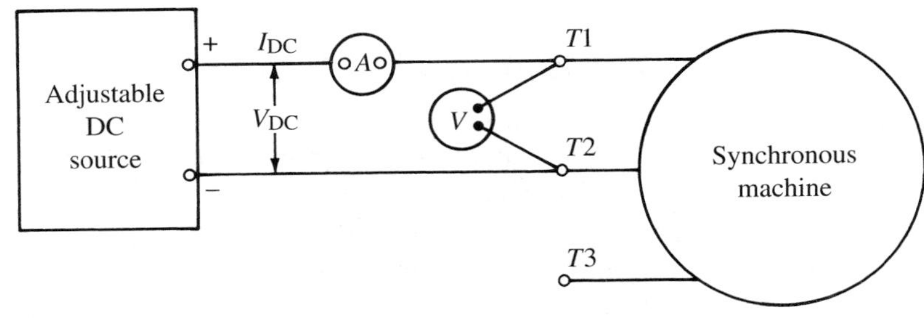

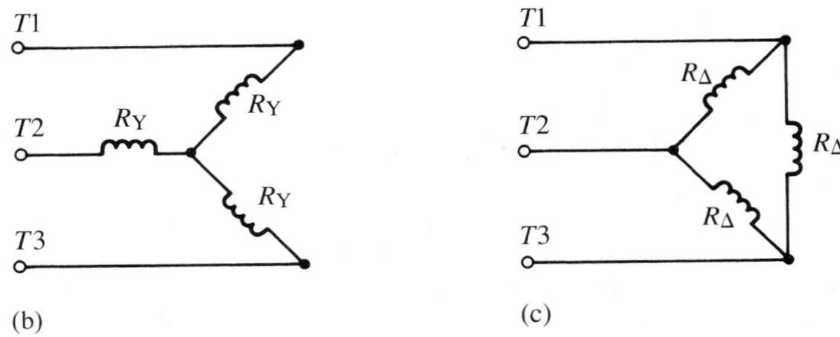

FIGURE 9.24
(a) DC test circuit for determining the equivalent resistance of a synchronous machine; (b) wye-connected stator; (c) delta-connected stator.

Note: To account for skin effect, the values of resistance obtained by Eqs. (9–15) and (9–16) should be multiplied by a skin-effect factor; although the value varies with different machines, a factor of 1.2 is a fair approximation. Thus,

$$R_a \approx 1.2 \times R_Y \quad \text{or} \quad R_a \approx 1.2 \times R_\Delta \tag{9–16a}$$

as appropriate.

Open-Circuit Test

In the open-circuit test, the machine is driven at rated speed by its prime mover, and the field current is adjusted to obtain rated voltage at no load. The circuit is shown in Figure 9.25(a). The data set obtained from the open-circuit test is

$$I_{f,OC} \qquad V_{OC,line}$$

where: $I_{f,OC}$ = field current that produces $V_{OC,line}$ (A)
$V_{OC,line}$ = open-circuit line voltage (V)

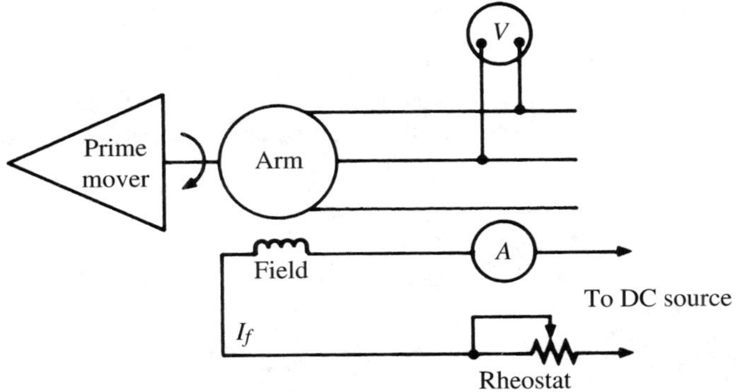

(a)

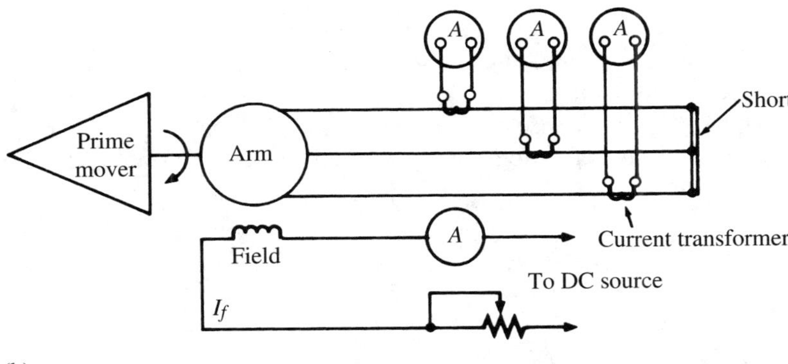

(b)

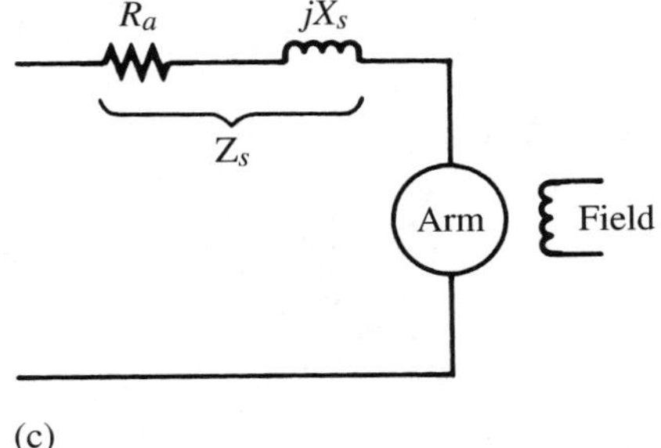

(c)

FIGURE 9.25
(a) Connections for open-circuit test; (b) connections for short-circuit test; (c) equivalent circuit for one phase.

Short-Circuit Test

With the three stator leads short circuited, as shown in Figure 9.25(b), the machine is driven at rated speed by its prime mover, and the field current slowly increased from its lowest value to the value it had in the open-circuit test. The data set obtained from the short-circuit test is

$$I_{f,\text{OC}} \qquad I_{\text{SC,line}}$$

where $I_{\text{SC,line}}$ is the average of the three line currents (A).

Determining Synchronous Reactance

The ratio of *phase voltage* obtained in the open-circuit test, to the *phase current* obtained in the short-circuit test, determines the synchronous impedance/phase.

$$Z_s = \frac{V_{\text{OC,phase}}}{I_{\text{SC,phase}}} \quad \Omega \tag{9–17}$$

where Z_s is the synchronous impedance/phase (Ω).

Substituting the appropriate R_a from equation set (9–16a) and Z_s from Eq. (9–17) into the series-impedance equation determines the synchronous reactance. Thus, referring to the equivalent circuit in Figure 9.25(c), for one phase of the synchronous machine stator,

$$\mathbf{Z}_s = R_a + jX_s \qquad \Rightarrow \qquad X_s = \sqrt{Z_s^2 - R_a^2} \tag{9–18}$$

Note: In those cases where $R_a \ll Z_s$, the armature resistance is neglected and $X_s \approx Z_s$.

Short-Circuit Ratio (SCR)

The short-circuit ratio of a synchronous generator is the ratio of field current required to obtain rated voltage at no load, to the field current required to obtain rated current when the armature is short circuited:

$$\text{SCR} = \frac{I_f \text{ for rated volts at no load}}{I_f \text{ for rated } I_a \text{ when shorted}} \tag{9–19}$$

The short-circuit ratio may also be determined from the reciprocal of the *per-unit* synchronous reactance:

$$\text{SCR} = \frac{1}{X_{\text{pu}}} \tag{9–20}$$

The short-circuit ratio is a measure of a machine's sensitivity to changes in loading. A machine with a high SCR is larger physically, weighs more, and costs more, but has a smaller voltage regulation than machines with lower SCRs. Machines with lower SCRs have larger synchronous reactance, and therefore require a voltage-regulating system that provides rapid and large changes in field current for small changes in load.

EXAMPLE 9.11 The short-circuit, open-circuit, and DC test data for a three-phase, wye-connected, 50-kVA, 240-V, 60-Hz synchronous alternator are

$$V_{\text{OC,line}} = 240.0\text{ V} \qquad I_{\text{SC,line}} = 115.65\text{ A}$$
$$V_{\text{DC}} = 10.35\text{ V} \qquad I_{\text{DC}} = 52.80\text{ A}$$

Determine (a) equivalent armature resistance; (b) synchronous reactance; (c) short-circuit ratio.

Solution

(a)

$$R_{\text{DC}} = \frac{V_{\text{DC}}}{I_{\text{DC}}} = \frac{10.35}{52.8} = 0.1960\ \Omega \qquad R_{\text{Y}} = \frac{R_{\text{DC}}}{2} = \frac{0.1960}{2} = 0.098\ \Omega$$

$$R_a \approx 1.2 \times R_{\text{Y}} = 1.2 \times 0.098 = \underline{0.1176\ \Omega}$$

(b)

$$Z_s = \frac{V_{\text{OC,phase}}}{I_{\text{SC,phase}}} = \frac{240/\sqrt{3}}{115.65} = 1.1981\ \Omega$$

$$X_s = \sqrt{Z_s^2 - R_a^2} = \sqrt{(1.1981)^2 - (0.1176)^2} = \underline{1.1923\ \Omega}$$

(c)

$$S_{\text{base}} = \frac{50{,}000}{3} \qquad V_{\text{base}} = \frac{240}{\sqrt{3}}$$

$$Z_{\text{base}} = \frac{V_{\text{base}}^2}{S_{\text{base}}} = \frac{(240/\sqrt{3})^2}{50{,}000/3} = 1.1520\ \Omega$$

$$X_{\text{pu}} = \frac{X_s}{Z_{\text{base}}} = \frac{1.1923}{1.1520} = 1.035$$

$$\text{SCR} = \frac{1}{X_{\text{pu}}} = \frac{1}{1.035} = \underline{0.966}$$

9.17 LOSSES, EFFICIENCY, AND COOLING OF AC GENERATORS

Losses in synchronous generators are the same as those presented for synchronous motors in Section 8.11. A power-flow diagram that illustrates the flow of power through a synchronous generator is shown in Figure 9.26. As indicated in the power-flow diagram, the total loss for the synchronous generator is

$$P_{\text{loss}} = P_{\text{stray}} + P_{f,w} + P_{\text{core}} + P_{\text{fcl}} + P_{\text{scl}} \tag{9–21}$$

The overall efficiency of a synchronous generator is given by

$$\eta = \frac{P_{\text{out}}}{P_{\text{in}} + P_{\text{field}}} = \frac{P_{\text{out}}}{P_{\text{out}} + P_{\text{losses}}} \tag{9–22}$$

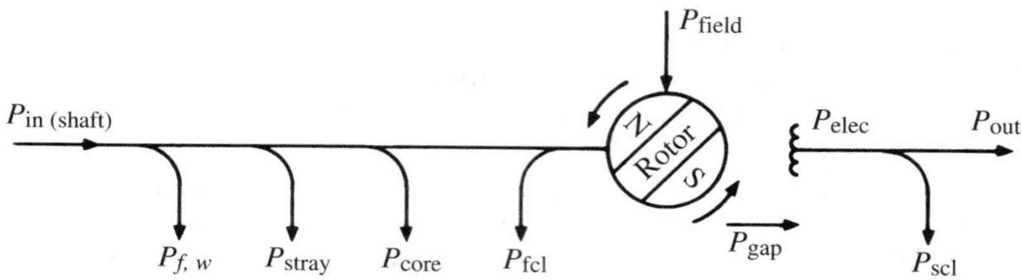

FIGURE 9.26
Power-flow diagram for a synchronous generator.

The field-power requirements for a synchronous generator vary from 1 kW for an alternator with a 25-kW output to 50 kW for a machine with a 7-MW output **[11]**. Since all power input to the field windings is a heat-power loss, many different cooling methods have evolved.

Cooling of Alternator Windings

The heat generated by the I^2R losses in the copper, and the hysteresis and eddy-current losses in the iron, are removed by ventilation. The simplest method, frequently used on small machines, uses an open housing and a fan mounted on the shaft to circulate ambient air through the windings. When the ambient temperature is relatively high, however, or the air is contaminated with moisture and dust, a totally enclosed air recirculating and cooling system is installed, as shown in Figure 9.27. The cooler, also called a heat exchanger, consists of a bank of finned tubes over which the hot air passes. Cold water circulating through the tubes absorbs the heat. The heat liberated from the windings is transferred to the air, which then passes through the cooler. A fan circulates the same air through the machine, absorbing heat from the windings and transferring it to the cooling water through the heat exchanger.

Hydrogen is often used as the cooling agent for large electrical machines. It has a thermal conductivity approximately seven times that of air, and because of its low density requires only one-tenth the amount of energy to blow it through the machine. The very considerable reduction in windage loss, coupled with the smaller physical dimensions of the hydrogen-cooled machine, more than offsets the extra cost of explosion-proof housings, complex shaft seals, and hydrogen-control equipment. The windings of some large alternators are cooled by circulating liquids through hollow tubular conductors that carry both the coolant and the current. The direct contact of coolant with the conductors results in excellent heat removal.

The new thrust in synchronous generator design is for highly efficient superconducting generators that use field coils wound with superconducting material. Prototypes of superconducting AC generators have been constructed and tested by several manufacturers **[7]**,**[10]**.

FIGURE 9.27
Installing a heat exchanger on top of the stator of a synchronous generator. (Courtesy GE Industrial Systems)

SUMMARY OF EQUATIONS FOR PROBLEM SOLVING

$$f = \frac{P \cdot n_s}{120} \quad \text{Hz} \tag{4–1}$$

$$E_f = n_s \Phi_f k_f \tag{9–2}$$

$$\mathbf{V}_T = \mathbf{E}_f - \mathbf{I}_a jX_s \tag{9–1a}$$

$$P_{\text{in},1\phi} = \frac{-V_T E_f}{X_s} \sin \delta \qquad \text{(motor action)} \tag{8–14}$$

$$P_{\text{out},1\phi} = \frac{+V_T E_f}{X_s} \sin\delta \qquad \text{(generator action)} \tag{9–4}$$

$$P_{\text{out},1\phi} = \frac{V_T E_f}{X_d} \sin\delta + V_T^2 \left[\frac{X_d - X_q}{2X_d X_q}\right] \sin 2\delta \qquad \text{W} \tag{9–5}$$

Governor Characteristics

$$\text{GSR} = \frac{n_{\text{nl}} - n_{\text{rated}}}{n_{\text{rated}}} = \frac{f_{\text{nl}} - f_{\text{rated}}}{f_{\text{rated}}} \tag{9–6}$$

$$\text{GD} = \frac{f_{\text{nl}} - f_{\text{rated}}}{P_{\text{rated}}} = \frac{\Delta f}{\Delta P} \tag{9–8}$$

$$\text{GD} = \frac{\text{GSR} \times f_{\text{rated}}}{P_{\text{rated}}} = \frac{\Delta f}{\Delta P} \tag{9–9}$$

Per-Unit Values of Synchronous Machine Parameters

S_{base} = rated apparent power/phase (VA)

V_{base} = rated voltage/phase (V)

$$I_{\text{base}} = \frac{S_{\text{base}}}{V_{\text{base}}} \qquad \text{A} \tag{9–10}$$

$$Z_{\text{base}} = \frac{V_{\text{base}}}{I_{\text{base}}} = \frac{V_{\text{base}}^2}{S_{\text{base}}} \qquad \Omega \tag{9–11}$$

$$R_{\text{pu}} = \frac{R_a}{Z_{\text{base}}} \qquad X_{\text{pu}} = \frac{X_s}{Z_{\text{base}}} \qquad \mathbf{Z}_{\text{pu}} = R_{\text{pu}} + jX_{\text{pu}} \tag{9–12}$$

Voltage Regulation

$$\text{VR} = \frac{V_{\text{n1}} - V_{\text{rated}}}{V_{\text{rated}}} \times 100 \qquad \text{percent} \tag{9–13}$$

Determination of Parameters

$$R_{\text{DC}} = \frac{V_{\text{DC}}}{I_{\text{DC}}} \qquad \text{R}_{\text{Y}} = \frac{\text{R}_{\text{DC}}}{2} \qquad R_{\Delta} = 1.5R_{\text{DC}} \qquad \Omega \tag{9–14, 9–15, 9–16}$$

$$Z_s = \frac{V_{\text{OC,phase}}}{I_{\text{SC,phase}}} \qquad X_s = \sqrt{Z_s^2 - R_a^2} \qquad \Omega \tag{9–17, 9–18}$$

$$\text{SCR} = \frac{I_f \text{ for rated volts at no load}}{I_f \text{ for rated } I_a \text{ when shorted}} = \frac{1}{X_{\text{pu}}} \tag{9–20}$$

Losses and Efficiency

$$P_{\text{loss}} = P_{\text{stray}} + P_{f,w} + P_{\text{core}} + P_{\text{fcl}} + P_{\text{scl}} \quad \text{W} \tag{9–21}$$

$$\eta = \frac{P_{\text{out}}}{P_{\text{in}} + P_{\text{field}}} = \frac{P_{\text{out}}}{P_{\text{out}} + P_{\text{losses}}} \tag{9–22}$$

SPECIFIC REFERENCES KEYED TO TEXT

1. Daley, J. M. Design considerations for operating on-site generators in parallel with utility service. *IEEE Trans. Industry Applications,* Vol. IA-21, No. 1, Jan./Feb. 1985, pp. 69–80.
2. Darron, H. G., J. L. Koepfinger, J. R. Mather, and P. A. Rusche. The influence of generator loss of excitation on bulk power system reliability. *IEEE Trans. Power Apparatus and Systems,* Vol. PAS-94, No. 5, Sept./Oct. 1975, pp. 1473–1483.
3. Del Toro, V. *Electric Machines and Power Systems,* Sec. 5–4. Prentice Hall, Upper Saddle River, NJ, 1985.
4. General Electric Co. *Generator Instructions,* GEK-75511B, 1985.
5. Hubert, C. I. *Preventive Maintenance of Electrical Equipment.* Prentice Hall, Upper Saddle River, NJ, 2002.
6. Institute of Electrical and Electronic Engineers. *Test Procedures for Synchronous Machines,* IEEE Standard 115-1995, IEEE, New York, 1995.
7. Kumagai, M., T. Tanaka, K. Ito, Y. Watanabe, K. Sato, and Y. Gocho. Development of superconducting AC generator. *IEEE Trans. Energy Conversion,* Vol. EC-1, No. 4, Dec. 1986, pp. 122–129.
8. Mablekos, Van E. *Electric Machine Theory for Power Engineers.* Harper & Row, New York, 1980.
9. McPherson, G. *An Introduction to Electrical Machines and Transformers.* Wiley, New York, 1981.
10. Nasar, S. A. *Handbook of Electrical Machines,* McGraw-Hill, New York, 1987.
11. National Electrical Manufacturers Association. *Motors and Generators,* Publication No. MG-1-1998. NEMA, Rosslyn, VA, 1999.

REVIEW QUESTIONS

1. Using a two-pole synchronous machine and a corresponding phasor diagram, explain the motor-to-generator transition.
2. What would cause a synchronous generator to fail to build up voltage?
3. Assume that a turbine generator is operating alone at rated load when an accidental "dead-short" occurs at its three-phase output terminals. Will the machine come to a quick stop, slow down, or overspeed? Explain.
4. Explain why an alternator develops a countertorque when a load is connected to its armature.

5. Why is the rated voltage of an alternator, as indicated on the nameplate, always given for rated kVA at a specific power factor and frequency?
6. A 450-V, three-phase, 600-kW, 60-Hz generator *A* is to be paralleled with another identical machine *B*. Machine *B* is on the bus and carrying a load of 300 kVA at 60 Hz, 450 V, and 0.80 power-factor lagging. State the procedure you would follow in order to parallel machine *A* with machine *B*. Include in your statements the apparatus operated, meters observed, and the specific values of voltage and frequency for the incoming machine. Assume the prime mover has been started.
7. Explain why alternators in parallel cannot operate at different frequencies.
8. If the pointer of a synchroscope stops revolving and remains at the 30° position, what does it signify? What must be done to enable correct paralleling?
9. Explain the inherent dangers of attempting to parallel alternators when the phase angle between the two is relatively large. What is the maximum allowable error angle?
10. What methods can be used to determine the phase sequence of an incoming alternator? How is opposite phase sequence indicated?
11. How is the transfer of active power accomplished between alternators in parallel?
12. A 500-kVA, three-phase, 60-Hz, 450-V alternator is supplying a bus load of 400 kW at 0.90 power factor. State in outline form the procedure to be followed in order to have each machine take an equal share of the active power load. Mention the apparatus adjusted and the instruments observed as well as the values of voltage, frequency, and power of each machine during the transition and when the change is completed.
13. Two identical alternators with slightly drooping governor characteristics share an equal portion of the bus load at 60 Hz. (a) Sketch representative governor droop characteristics of each machine on one set of coordinate axis and draw the 60-Hz line. (b) With dotted lines appropriately drawn and lettered to indicate the transition steps, show and explain how a portion of the load of one machine may be transferred to the other machine. The bus must operate at 60 Hz when the transfer is completed.
14. Given two alternators with drooping governor characteristics, explain why admitting more energy to the prime mover of one machine results in an increase in the frequency of both machines.
15. Assume two alternators *A* and *B* are in parallel and carrying identical loads. Machine *A* has 2 percent speed regulation, and machine *B* has 3 percent speed regulation. Sketch the characteristic curves and indicate (a) the relative load division when the bus load increases; (b) the relative load division when the bus load decreases.
16. Explain the behavior of two alternators in parallel, one with governor droop and the other adjusted for isochronous operation, when the load on the bus increases and when it decreases.
17. Explain why two alternators with isochronous governors are inherently unstable when operating in parallel.
18. What is motoring? Is it harmful? How can it be detected and corrected?

19. Using phasor diagrams, explain why the transfer of active power between alternators in parallel causes a change in the division of reactive power between them.
20. How can the transfer of reactive power be accomplished between alternators in parallel? What instruments are used to observe this transfer?
21. Explain why adjustment of the field excitation of alternators in parallel affects the power factor of the machines.
22. Explain with the aid of a phasor diagram how it is possible for one alternator to operate at a leading power factor and another in parallel with it to operate at a lagging power factor when the bus power factor is unity.
23. Explain with the aid of a phasor diagram how it is possible for the individual power factors of two paralleled alternators and the power factor of the bus to be all lagging and all different.
24. (a) State in outline form the correct procedure for paralleling a 2000-V, 50-Hz, 1000-kVA alternator with another on the bus. Assume both machines have identical characteristics. Include the approximate values of the voltage and frequency of the incoming machine just prior to synchronizing. (b) Assuming the bus load is 800 kW at 0.8 power-factor lagging, what adjustments must be made to each machine in order that both machines will divide equally the active power of the bus? (c) What adjustments must be made to each machine in order that both machines will divide equally the reactive power of the bus?
25. What are the losses in an AC generator? How is the heat dissipated?

PROBLEMS

9–1/2 Determine the speed required to obtain a frequency of 50 Hz from a four-pole alternator.

9–2/2 A certain three-phase alternator operating at no load has an induced emf of 2460 V at 60 Hz. Determine the voltage and frequency if the pole flux and rotor speed are each increased by 10 percent.

9–3/2 An alternator operating at no load has a generated emf of 346.4 V/phase and a frequency of 60 Hz. If the pole flux is decreased by 15 percent and the speed is increased by 6.8 percent, determine (a) induced voltage; (b) frequency.

9–4/3 A 2400-V, 60-Hz, three-phase, six-pole, wye-connected synchronous generator is connected to an infinite bus and is supplying 350 kW at a power angle of 28.2°. The stator has a synchronous reactance of 12.2 Ω/phase. Neglecting losses, determine (a) input torque to the alternator; (b) excitation voltage per phase; (c) armature current; (d) active and reactive components of power; (e) power factor.

9–5/3 A four-pole, 600-V synchronous generator is connected to an infinite bus that supplies a 60-Hz, 600-V, 2000-kVA, 80.4 percent power-factor load. The generator is driven by a steam turbine that delivers 1955 lb-ft of torque to the generator shaft, causing a power angle of 36.4°. The synchronous reactance is 1.06 Ω/phase. Neglecting losses, determine (a) mechanical power input to the generator rotor; (b) excitation voltage per phase; (c) armature current; (d) active and reactive components of apparent power delivered to the bus; (e) generator power factor.

9–6/3 A six-pole, 75-kVA, 340-V, 60-Hz, wye-connected diesel generator is supplying a bus load of 54.5 kVA at 78.9 percent power-factor lagging, 220 V, and 60 Hz. The armature has a synchronous impedance of $0.18 + j0.92$ Ω/phase. Determine (a) armature current; (b) excitation voltage per phase; (c) power angle; (d) shaft torque supplied by the diesel engine (neglect losses).

9–7/11 The ratings of two turbine generators *A* and *B* are 300-kW, 3.5 percent speed regulation and 600-kW, 2.5 percent speed regulation, respectively. The machines are operating in parallel at 60 Hz, 480 V. Machine *A* is supplying 260 kW, and machine *B* is supplying 590 kW. (a) Sketch a one-line diagram for the system. (b) On the same coordinate axes sketch the approximate governor-droop characteristics of both machines and label the curves. (c) Opening of a distribution breaker results in a remaining bus load of 150 kW. Determine the new frequency and the active power load carried by each machine.

9–8/11 Two 600-kW, 60-Hz, diesel-driven synchronous generators *A* and *B* have governor speed regulations of 2.0 and 5.0 percent, respectively. Both machines are in parallel and supplying equal shares of a 1000-kW bus load at 57 Hz. (a) Sketch the approximate governor characteristics for both machines on one set of coordinate axes, and indicate the operating frequency. Label both curves. (b) On the same diagram, approximate a new operating condition that assumes the load on the bus decreases to a total of 400 kW. (c) Determine the new frequency and the new load distribution for the conditions in (b).

9–9/11 A 700-kW, 60-Hz generator *A*, with 2.0 percent speed regulation, is operating in parallel with generator *B* of equal kilowatt rating, but whose speed regulation is 6.0 percent. The total bus load of 1000 kVA at 60 Hz, 2400 V, and 80.6 percent power-factor lagging is divided equally between the two machines. If a 200-kVA, 60.0 percent power-factor load is disconnected from the bus, determine (a) the new operating frequency; (b) the active power load on each machine.

9–10/11 Three 600-kW, 60-Hz, 480-V synchronous machines are in parallel, each supplying the following loads: generator *A*, 200 kW; generator *B*, 100 kW; and generator *C*, 300 kW. The speed regulations for machines *A*, *B*, and *C* are 2.0, 2.0, and 3.0 percent, respectively. The system frequency is 60 Hz. If the system load increases to 2000 kVA at 70.0 percent power-factor lagging, determine the new operating frequency and active power load carried by each machine.

9–11/11 Three 25-Hz turbine generators *A*, *B*, and *C* are connected in parallel and operating at 25 Hz, 2400 V. Generator *A* is rated at 600 kW and has a speed regulation of 2.0 percent. Generator *B* is rated at 500 kW and has a speed regulation of 1.5 percent. Generator *C* has a rating of 1000 kW and has a speed regulation of 4.0 percent. The loads on machines *A, B,* and *C* are 200 kW, 300 kW, and 400 kW, respectively. If an additional 800 kW is connected to the bus, determine (a) the new frequency; (b) the load on each generator.

9–12/12 Two synchronous machines *A* and *B* are in parallel and taking equal shares of the bus load at 480 V, 60 Hz. The power factor of the bus is 76.6 percent, and the power angle of each machine is 20.0°. (a) Using a straight edge and protractor draw (on graph paper) separate phasor diagrams for each machine. (b) Assuming the governor controls of both machines are adjusted so that machine *A* takes 25 percent of the bus load, construct the new conditions on the phasor diagrams. Show all construction lines and determine the new power factor of each machine.

9–13/12 A three-phase turbine generator in parallel with an infinite bus has a power angle of 27°, a power factor of 94.2 percent, and a line voltage of 600 V. (a) Using a straight edge, protractor, and graph paper, draw the phasor diagram to scale. (b) Assume the turbine is tripped, the generator excitation is not changed, and a mechanical load equal to 60 percent of the load it delivered when acting as a generator is applied to the shaft. Show the changes on the phasor diagram and determine the new power factor and the new power angle.

9–14/12 Two identical 480-V, 60-Hz diesel generators DG 1 and DG 2 are in parallel and are taking equal shares of the total bus load of 3000 kW at 80.2 percent power-factor lagging. (a) Sketch separate phasor diagrams showing the original conditions for each machine. (b) The governor controls of both machines are manipulated so that DG 1 takes two-thirds of the bus load. Assume the bus voltage, frequency, and field excitation do not change. Using ruler and compass, construct to scale, on the diagrams in (a), the new conditions for the respective machines. Show all construction lines, and using a protractor, determine the new power angle and new power factor of each machine.

9–15/12 A turbine generator TG 1 is paralleled with an infinite bus and is carrying 100 kW at 60.2 percent power-factor lagging. (a) Sketch the phasor diagram for the given conditions. (b) Assume the governor of TG 1 is adjusted so as to double the energy input to its prime mover. Using dotted lines, construct the new conditions on the phasor diagram in (a) and determine the new power angle. (c) Assume the conditions in (b) are further modified by a 25 percent increase in the pole flux of TG 1. Using broken lines and the same phasor diagram construct the new conditions and determine the corresponding power angle.

9–16/12 Two 450-V, three-phase, 60-Hz, 600-kW synchronous machines *A* and *B* are operating in parallel and share equally the active and reactive power of the bus. The bus load is 1000 kVA, operating at a power factor of 80.4 percent lagging. If the field excitation of alternator *A* is adjusted to cause its power factor to increase to 85.0 percent lagging, at what power factor will the other machine operate?

9–17/12 Two alternators *A* and *B* are in parallel supplying a 560-kVA, 480-V, 60-Hz bus load at a power factor of 82.8 percent lagging. Machine *A* is carrying 75 percent of the active power load at a power factor of 92.4 percent lagging. Determine the active and reactive power components supplied by each machine.

9–18/12 Two alternators *A* and *B* are in parallel and supplying a 600-V, 60-Hz bus load consisting of a 270-kVA unity power-factor load, a group of induction motors totaling 420 kVA at 89.4 percent power factor, and a synchronous motor that draws 300 kVA at 92.3 percent power-factor leading. If machine *A* is carrying 60 percent of the total active power load at a power factor of 70.4 percent lagging, determine the active and reactive components supplied by machine *B*.

9–19/12 Three identical 1000-kW diesel generating sets operating in parallel take equal shares of a 1500-kW bus load at 450 V, 60 Hz, and 80.0 percent power-factor lagging. The governor speed regulation of the three machines are adjusted to 1.0, 2.0, and 3.0 percent for prime movers *A, B,* and *C,* respectively. (a) Sketch representative governor droops of the three machines on one set of coordinate axes, and label the curves. (b) Calculate the new operating frequency and active power load on each generator if the total bus load is increased to 1850 kW. (c) If the system power factor for the new conditions is 95.0 percent lagging, machine *A* has a power factor of 90.0 percent lagging, and machine *B* has a power factor of 60.0 percent lagging, what is the power factor of machine *C*?

9–20/14 Determine the per-unit resistance and per-unit synchronous reactance of a 37.5-kVA, 480-V, 60-Hz, wye-connected alternator whose synchronous impedance is $1.47 + j7.68\ \Omega$/phase.

9–21/14 A 5000-kVA, 13,800-V, 60-Hz, wye-connected alternator has a synchronous reactance of 55.2 Ω/phase. Its resistance is negligible. Determine the per-unit impedance.

9–22/15 A 250-kVA, 480-V, three-phase, four-pole, 60-Hz synchronous generator with a synchronous reactance of 0.99 Ω/phase is operating at rated conditions and 83.2 percent power-factor lagging, and has the magnetization curve shown in Figure 9.28. Determine (a) excitation voltage; (b) power angle; (c) open-circuit phase voltage; (d) voltage regulation; (e) no-load voltage if the field current is reduced to 60 percent of its value at rated load.

9–23/15 Repeat Problem 9–22/15 assuming that the power factor is 83.2 percent leading.

9–24/15 A 1000-kVA, 4800-V, 60-Hz, 3600-r/min, wye-connected alternator with a synchronous reactance of 14.2 Ω/phase is supplying 600 kVA at 4800 V to a bus load operating at a power factor of 95.2 percent lagging. Assume the

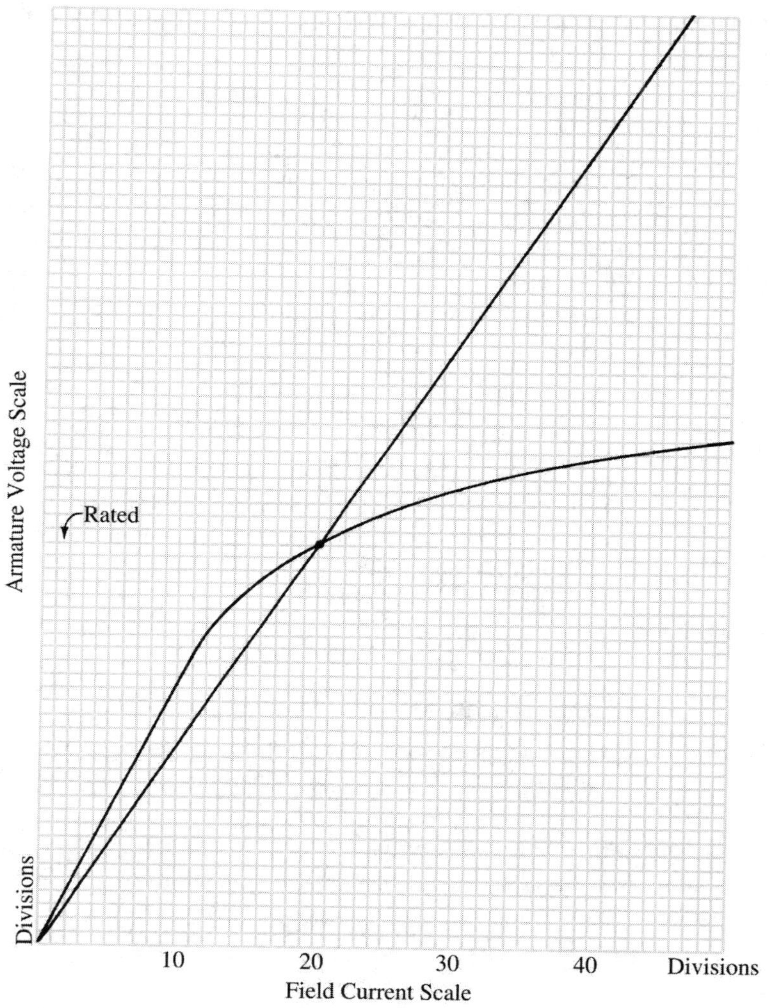

FIGURE 9.28
Magnetization curve for Problems 9–22, 9–23, 9–24, 9–25, 9–26, and 9–27.

curves in Figure 9.28 apply. Determine the open-circuit *line voltage* of the generator if a sudden short circuit tripped the generator breaker.

9–25/15 A 150-kVA, 240-V, delta-connected, 60-Hz synchronous generator has a synchronous impedance of $0.094 + j0.320\ \Omega$/phase, and is operating at rated conditions and 75.2 percent power-factor lagging. Assume the curves in Figure 9.28 apply. Determine the voltage regulation.

9–26/15 Repeat Problem 9–25/15 assuming operation is at rated conditions and unity power factor.

9–27/15 A 250-kVA, 450-V, 60-Hz, three-phase, wye-connected alternator has a synchronous impedance of $0.05 + j0.24\ \Omega$/phase. The machine is operating in parallel with other machines and is supplying its rated kVA at rated voltage, rated frequency, and 85.0 percent power-factor lagging. Assume the curves in Figure 9.28 apply. Sketch the equivalent circuit and corresponding phasor diagram for one phase, and determine the generator *line voltage* if the circuit breaker is tripped.

9–28/16 The short-circuit, open-circuit, and DC test data for a wye-connected, 25-kVA, 240-V, 60-Hz alternator are

$$V_{OC} = 240.0\text{ V} \qquad I_{SC} = 60.2\text{ A}$$
$$V_{DC} = 120.6\text{ V} \qquad I_{DC} = 50.4\text{ A}$$

Determine (a) equivalent armature resistance; (b) synchronous reactance; (c) synchronous impedance; (d) short-circuit ratio.

9–29/16 Data from tests performed to determine the parameters of a 200-kVA, 480-V, 60-Hz, three-phase, wye-connected alternator are

$$V_{OC} = 480.0\text{ V} \qquad I_{SC} = 209.9\text{ A}$$
$$V_{DC} = 91.9\text{ V} \qquad I_{DC} = 72.8\text{ A}$$

Determine (a) synchronous impedance; (b) short-circuit ratio.

9–30/16 A 350-kVA, 600-V, three-phase, four-pole, wye-connected alternator has a short-circuit ratio of 0.87, and its equivalent resistance is $0.644\ \Omega$/phase. Determine the synchronous impedance.

9–31/16 The short-circuit, open-circuit, and DC test data for a three-phase, 125-kVA, 480-V, delta-connected, synchronous generator are

$$V_{OC} = 480\text{ V} \qquad I_{SC} = 519.6\text{ A}$$
$$V_{DC} = 24.0\text{ V} \qquad I_{DC} = 85.6\text{ A}$$

Determine (a) equivalent resistance; (b) synchronous reactance; (c) synchronous impedance.

10

Principles of Direct-Current Machines

10.1 INTRODUCTION

The DC motor is the most versatile of all rotating electrical machines. Its speed may be easily adjusted in very fine increments ranging from standstill to rated speed and above, and if not properly controlled, may reach speeds high enough to cause destruction by centrifugal force. Direct-current motors can develop rated torque at all speeds from standstill (locked-rotor) to rated speed, and the torque that it can develop at standstill is many times greater than the torque developed by an AC motor of equal power and speed ratings.

Direct-current motors are used in a wide variety of industrial drives, such as robots, machine tools, petrochemical, pulp, paper and steel mills, oil drilling rigs, and mining. They are also used extensively in automotive systems and railroads.

Direct-current generators, however, once the mainstay of electric power for large and small industrial plants, are being increasingly replaced by solid-state devices that convert available AC to direct current for DC drive systems and other DC applications.

10.2 FLUX DISTRIBUTION AND GENERATED VOLTAGE IN AN ELEMENTARY DC MACHINE

The flux distribution in the air gap of an elementary two-pole DC machine is shown in Figure 10.1. Except for the interpolar region, the pole face of a DC machine is shaped to provide a uniform flux distribution per space degree in the air-gap region between the pole face and the armature. Although there is some fringing of flux in the interpolar region, there is no flux entering or leaving the armature in the magnetic-neutral plane (plane of zero flux).

Figure 10.2(a) shows several positions of an armature coil as it revolves counterclockwise (CCW) within the magnetic field, and Figure 10.2(b) shows the corresponding variation of flux through the window of the armature coil plotted against electrical degrees.

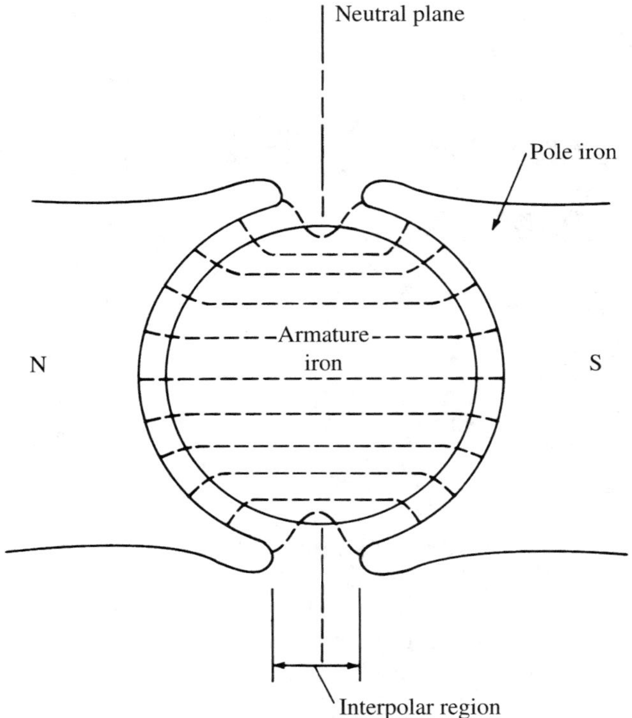

FIGURE 10.1
Flux distribution in the air gap of an elementary DC machine.

At zero degrees all of the flux that crosses the air gap passes through the armature coil. As the coil rotates from the zero-degree position, the amount of flux passing through the window decreases linearly with angular position until at 90° the flux through the window is zero. When past the 90° position, the flux passes through the opposite face of the coil. The flux increases linearly to its maximum value in the opposite direction at 180°, then decreases, reaching zero at 270°, and so forth. Except for small changes in flux in the interpolar region, the graph of flux through the coil window as a function of rotation angle is triangular in shape.

In accordance with Faraday's law, the magnitude of the voltage generated in the armature coil is equal to the number of turns of wire in the coil times the rate of change of flux through the coil window. Furthermore, in accordance with Lenz's law, the voltage will be generated in a direction to oppose the rotation that caused it. The combined *Faraday–Lenz* relationship is expressed as

$$e = -N_a \frac{d\phi}{dt} \qquad \textbf{(10–1)}$$

The minus sign in Eq. (10–1) is the application of Lenz's law to the Faraday equation.

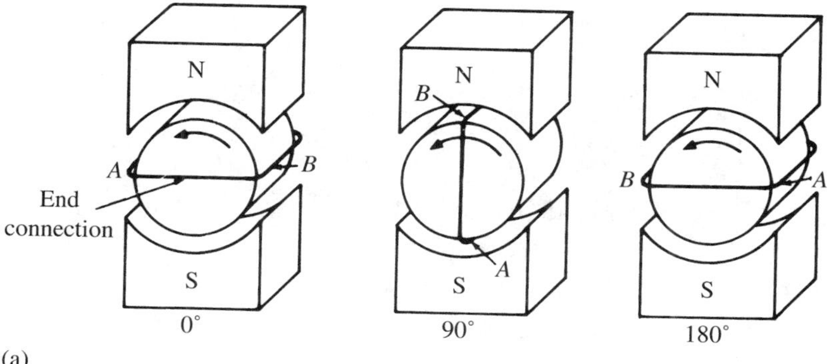

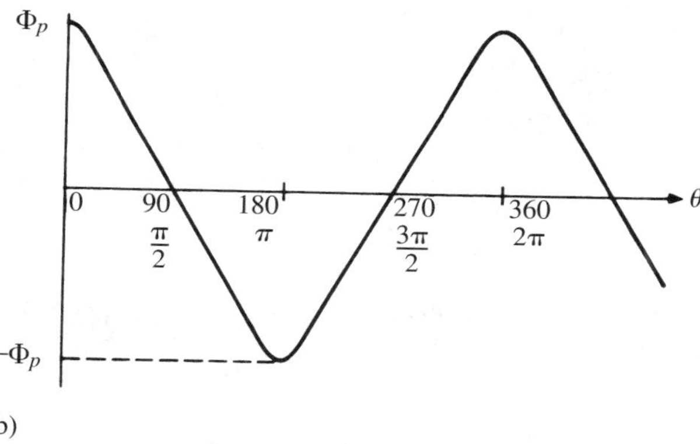

FIGURE 10.2
(a) Representative positions of an armature coil; (b) plot of flux through the window of an armature coil vs. angular position of coil.

Applying the Faraday–Lenz relationship to the triangular-shaped flux wave in Figure 10.2(b) results in the rectangular-shaped voltage wave shown in Figure 10.3. The angular velocity of the coil as it rotates at constant speed is

$$\omega = \frac{d\theta}{dt} \qquad \Rightarrow \qquad dt = \frac{d\theta}{\omega} \tag{10–2}$$

Substituting Eq. (10–2) into Eq. (10–1),

$$e = -N_a \omega \frac{d\phi}{d\theta} \tag{10–3}$$

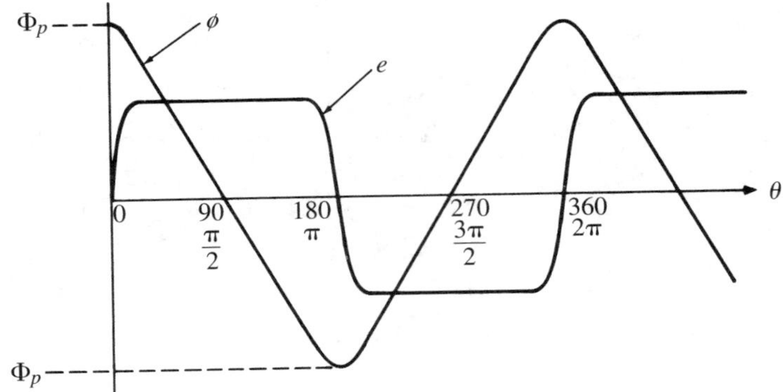

FIGURE 10.3
Rectangular voltage wave generated in an armature coil of a DC machine by a triangular-shaped flux wave through its window.

Referring to Figure 10.3, the slope of the voltage wave for the first one-half cycle of the flux wave is essentially constant and approximately equal to

$$\frac{d\phi}{d\theta} = -\frac{\Delta\phi}{\Delta\theta} = -\frac{2\Phi_p}{\pi} \tag{10–4}$$

Substituting Eq. (10–4) into Eq. (10–3) results in the average value of one-half cycle of the rectangular voltage wave:

$$E_a = \frac{2\omega N_a \Phi_p}{\pi} \tag{10–5}$$

Rectifying the rectangular-shaped voltage wave by solid-state or mechanical methods results in the wave shown in Figure 10.4, and Eq. (10–5) applies to the entire wave. Equation (10–5) can be expressed in terms of r/min by making the following substitutions:

$$\omega = 2\pi f \tag{10–6}$$

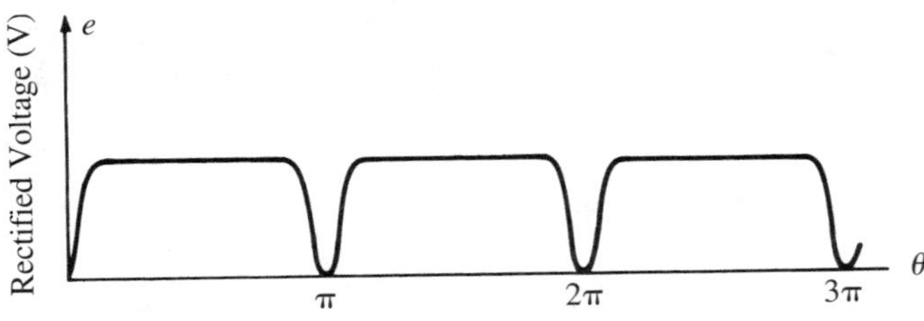

FIGURE 10.4
Rectified voltage wave.

$$f = \frac{Pn}{120} \tag{10–7}$$

Thus,

$$E_a = \frac{nPN_a\Phi_p}{30} \tag{10–8}$$

where: E_a = voltage induced in armature (avg) (V)
f = frequency (Hz)
n = rotational speed (r/min)
P = number of field poles
Φ_p = pole flux (Wb)
N_a = number of turns of conductor in armature

Each turn of wire in an armature coil has two conductors (coil sides). No voltage is generated in the end connections; the end connections (also called end turns) do not "cut" flux (see Figure 10.2). Thus, in terms of armature conductors,

$$N_a = \frac{z_a}{2} \tag{10–9}$$

where z_a = the total number of armature conductors.

A practical machine has many coils of wire that are distributed around the armature in series–parallel arrangements, with a minimum of two parallel paths. An expression for the average value of induced emf, in terms of total armature conductors, is obtained by substituting Eq. (10–9) into Eq. (10–8). Making the substitution,

$$E_a = \frac{nPz_a\Phi_p}{60a} \tag{10–10}$$

where: a = number of parallel paths
z_a = total number of armature conductors

The number of parallel paths and the number of series-connected conductors required for a given kilowatt rating are determined by the system voltage; low-voltage, high-current machines have more parallel paths, each with fewer series-connected conductors than high-voltage, low-current machines.

Substituting k_G for the constants in Eq. (10–10),

$$k_G = \frac{z_aP}{60a} \tag{10–11}$$

$$E_a = n\Phi_pk_G \tag{10–12}$$

EXAMPLE 10.1 A six-pole, 50-kW, DC machine operating at 1180 r/min has a generated emf of 136.8 V. If the speed is reduced to 75 percent of its original value, and the pole flux is doubled, determine (a) induced emf; (b) frequency of the rectangular voltage wave in the armature winding.

Solution

(a) From Eq. (10–12),

$$\frac{E_1}{E_2} = \frac{[n\Phi_p]_1}{[n\Phi_p]_2} \qquad \Rightarrow \qquad E_2 = E_1 \times \frac{[n\Phi_p]_2}{[n\Phi_p]_1}$$

$$E_2 = 136.8 \times \frac{0.75n \times 2\Phi_p}{n \times \Phi_p} = \underline{205.2\text{ V}}$$

(b)

$$f = \frac{Pn}{120} = \frac{6 \times 1180 \times 0.75}{120} = \underline{44.25\text{ Hz}}$$

10.3 COMMUTATION

The rectangular-shaped voltage wave generated within a DC armature coil is changed to a unidirectional voltage in the load circuit by means of a mechanical rectifier, called a *commutator,* mounted on the armature shaft. This is illustrated in Figure 10.5 for an elementary two-pole DC machine with one armature coil and a two-bar commutator. Connections to the external terminals are made via small stationary blocks of graphite, called *brushes,* that are pressed against the commutator by springs.

As shown in Figure 10.5, the generated voltage within the armature coil changes direction every 180 electrical degrees of rotation, but the voltage in the external circuit remains in the same direction. The rotating commutator and stationary brushes constitute a rotary switch that provides a switching action, called *commutation,* that switches the internal alternating voltage and current of an AC generator to direct voltage and direct current in the external circuit.

When the coil is rotating through the neutral plane, as shown in Figures 10.5(a) and (c), it is shorted by the brushes. Since the coil sides are not cutting flux, however, no armature voltage is generated, and no short-circuit current occurs.

A practical machine has many coils distributed around the armature, and the coils pass through the neutral plane one at a time. As one coil moves into the neutral plane another moves out, producing an essentially constant voltage.

10.4 CONSTRUCTION

A cutaway view of a DC machine is shown in Figure 10.6. The shunt field coils and series field coils[1] are wound around the same pole iron and provide the specific machine characteristics. Some DC machines omit the shunt field coils and use permanent magnets for the pole iron; in such cases, the series field would be wound around the magnets.

[1]Series fields are used in compound and series machines to shape motor and generator characteristics. See Chapters 11 and 12, respectively, for further details.

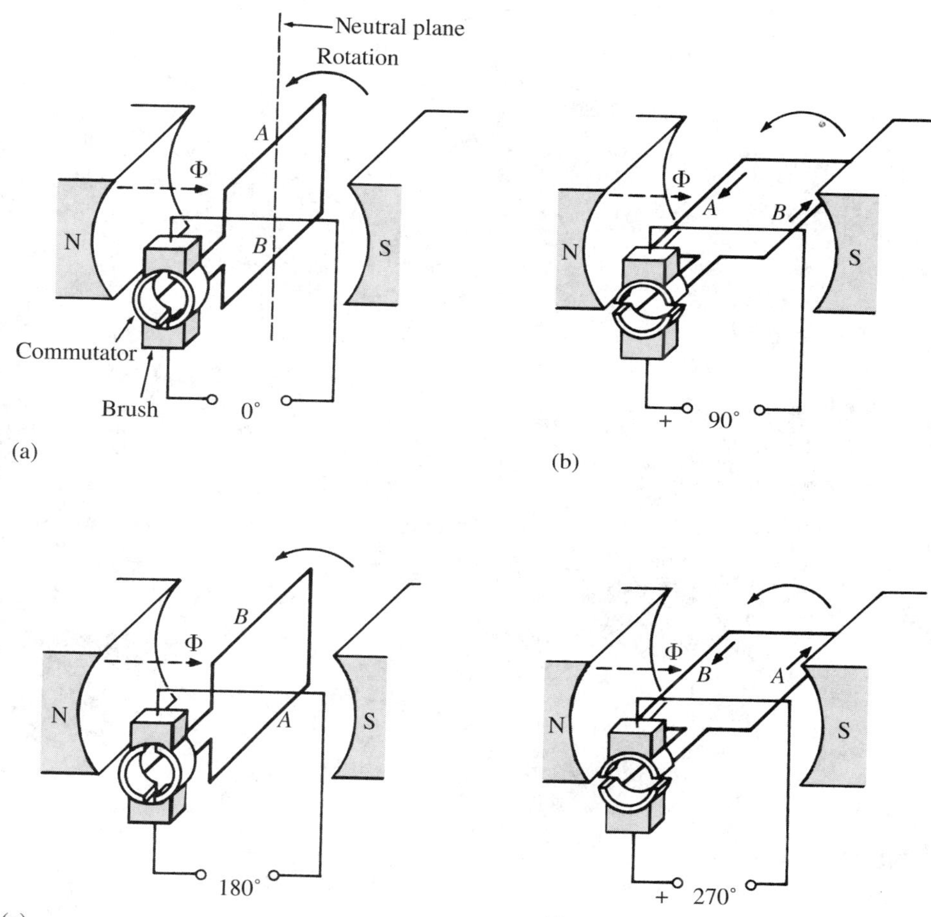

FIGURE 10.5
Sketches showing the commutation process illustrated with a one-coil armature.

Small poles, called *interpoles* or *commutating poles,* are located between the main field poles. The interpoles are used to minimize sparking between the brush and the commutator, which would otherwise occur when the machine is loaded. The interpole iron and main-pole iron are bolted to a yoke or frame of cast steel. Graphite or metal–graphite brushes provide the connection between the rotating commutator and the external load; they are designed to slide freely in metallic brush-boxes, and are connected to the box by a short flexible copper conductor called a *pigtail* or *shunt.*

The commutator is composed of alternate sections of copper bars and mica separators, clamped together with mica-insulated vee-rings. The number of commutator bars is determined by the number of coils in the armature, the number of poles, and the type of winding.

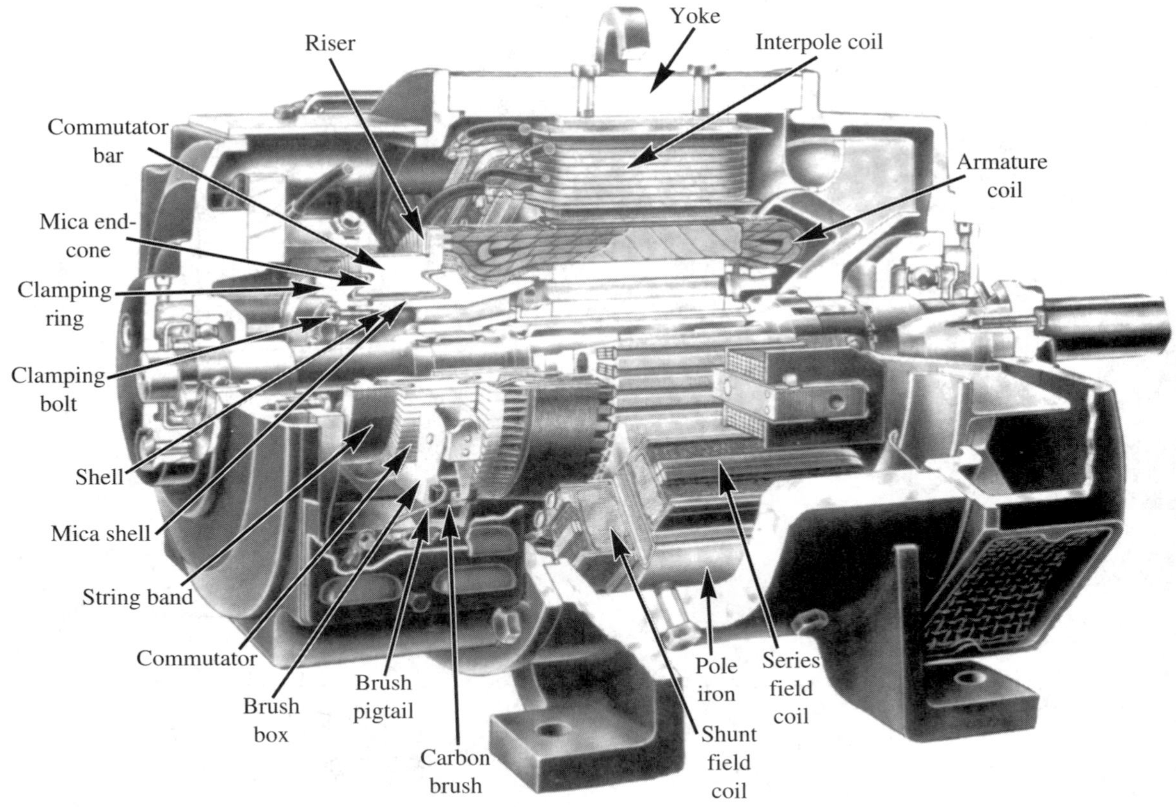

FIGURE 10.6
Cutaway view of a DC machine. (Courtesy Reliance Electric Company)

10.5 LAYOUT OF A SIMPLE ARMATURE WINDING

Figure 10.7 illustrates the layout of a simple armature winding (called a *lap winding*) for an eight-slot, eight-coil armature designed for operation with a two-pole field. Figure 10.7(a) shows the layout of the armature winding and its connections to the commutator, Figure 10.7(b) shows the physical layout of the top and bottom conductors in the armature slots, and Figure 10.7(c) is an elementary diagram that shows how the brushes form two parallel paths in the armature. The letters T and B refer to the top and bottom conductors in a single slot. For example, as seen in Figure 10.7(a), one side of coil 1 is in the top of slot 1, and the other side is in the bottom of slot 5.[2]

[2]For more information on armature windings, see the General References section at the end of this chapter.

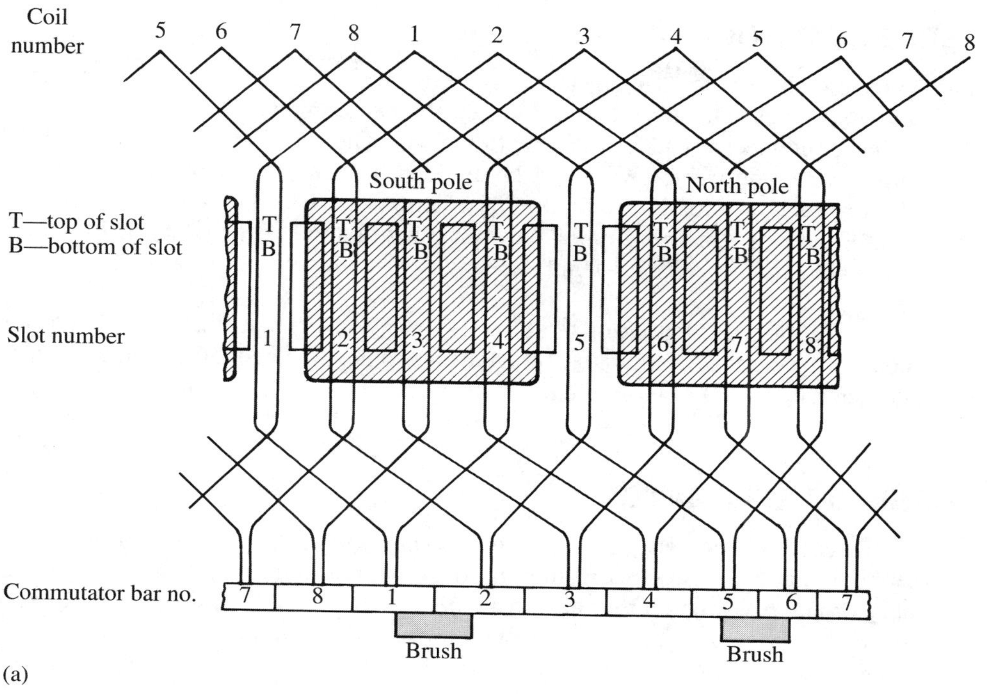

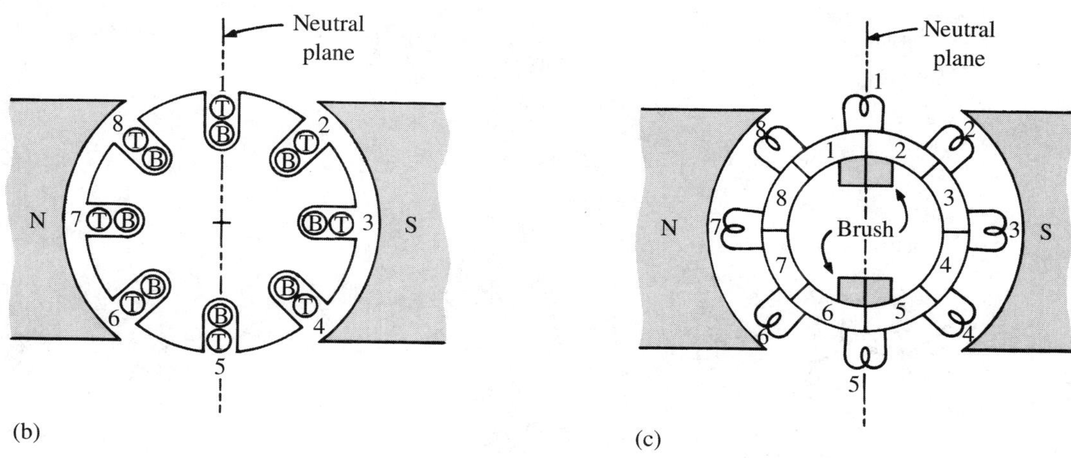

FIGURE 10.7
Layout of a simple armature winding: (a) coil distribution and commutator connections; (b) end view of an armature showing top and bottom conductors; (c) simplified connection diagram showing neutral plane and brush position.

10.6 BRUSH POSITION

As shown in Figure 10.7(c), each brush will short circuit every coil in turn as the armature rotates. This occurs at the instant the two commutator bars to which the coil ends are connected make contact with a brush. The shorting of armature coils by the brushes is undesirable but cannot be avoided. Hence, the brushes are positioned by the manufacturer to short circuit coils when they are in the magnetic neutral plane. This is called the *neutral setting.* Any other position will result in short circuiting coils while they are generating an emf, causing overheating of the coils and sparking at the brushes.

The physical layout of the brushes in any given machine depends on the coil connections to the commutator. When the brush position of a DC generator or motor is in doubt, electrical tests should be made to locate the correct neutral setting **[1],[2]**.

10.7 BASIC DC GENERATOR

The basic DC generator, called a *shunt generator,*[3] has its field winding connected either in parallel with the armature, or to a separate source of excitation such as a battery or another generator (called an *exciter*). The circuit diagram for a separately excited shunt generator that uses a battery to supply the field current is shown in Figure 10.8. The resistance of the armature winding and the resistance of the field winding are shown as R_a and R_f, respectively.

---

[3] The name *shunt generator* or *shunt motor* originally referred to a DC generator or DC motor whose field winding was always connected in parallel with the armature.

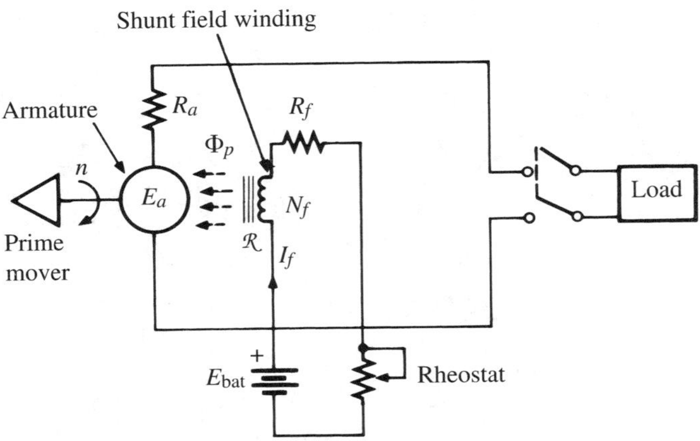

FIGURE 10.8
Equivalent-circuit diagram for a separately excited shunt generator.

Neglecting any residual magnetism, which is designed to be negligible for a separately excited machine, the pole flux can be determined from

$$\Phi_p = \frac{N_f I_f}{\mathcal{R}} \qquad \textbf{(10–13)}$$

where: I_f = field current (A)
N_f = number of turns of wire in a field coil
$\mathcal{R}$ = reluctance of magnetic circuit (A-t/Wb)
Φ_p = pole flux (Wb)

Substituting Eq. (10–13) into Eq. (10–12),

$$E_a = \frac{nN_f I_f k_G}{\mathcal{R}} \qquad \textbf{(10–14)}$$

Note that the reluctance $\mathcal{R}$ of the ferromagnetic material is not constant. Hence, Eq. (10–14) is nonlinear.

The current in the field windings as determined from Ohm's law is

$$I_f = \frac{E_{\text{bat}}}{R_f + R_{\text{rheo}}} \qquad \textbf{(10–15)}$$

where: I_f = field current (A)
E_{bat} = battery voltage (V)
R_f = resistance of field winding, including all poles (Ω)
R_{rheo} = resistance of rheostat, at setting (Ω)

Reducing the rheostat resistance increases the field current, which in turn increases the pole flux, causing an increase in the armature voltage.

The relationship between field current and induced armature voltage is shown graphically in Figure 10.9 for a representative DC machine. The curve, called the *magnetization curve* or *open-circuit characteristic,* is a plot of E_a vs. I_f as expressed in Eq. (10–14), assuming constant speed and no load connected to the armature terminals.

EXAMPLE 10.2 Assume that the resistance of the armature winding and the resistance of the field winding of the generator shown in Figure 10.8 are 0.014 Ω and 10.4 Ω, respectively, and that the magnetization curve is the one shown in Figure 10.9. Determine the rheostat setting required to obtain an induced emf of 290 V. The battery voltage is 240 V.

Solution

The required field current, as obtained from the magnetization curve in Figure 10.9, is ≈8.9 A. Applying Ohm's law to the shunt field circuit,

$$I_f = \frac{E_{\text{bat}}}{R_f + R_{\text{rheo}}} \quad \Rightarrow \quad R_{\text{rheo}} = \frac{E_{\text{bat}}}{I_f} - R_f$$

$$R_{\text{rheo}} = \frac{240}{8.9} - 10.4 = \underline{16.57}$$

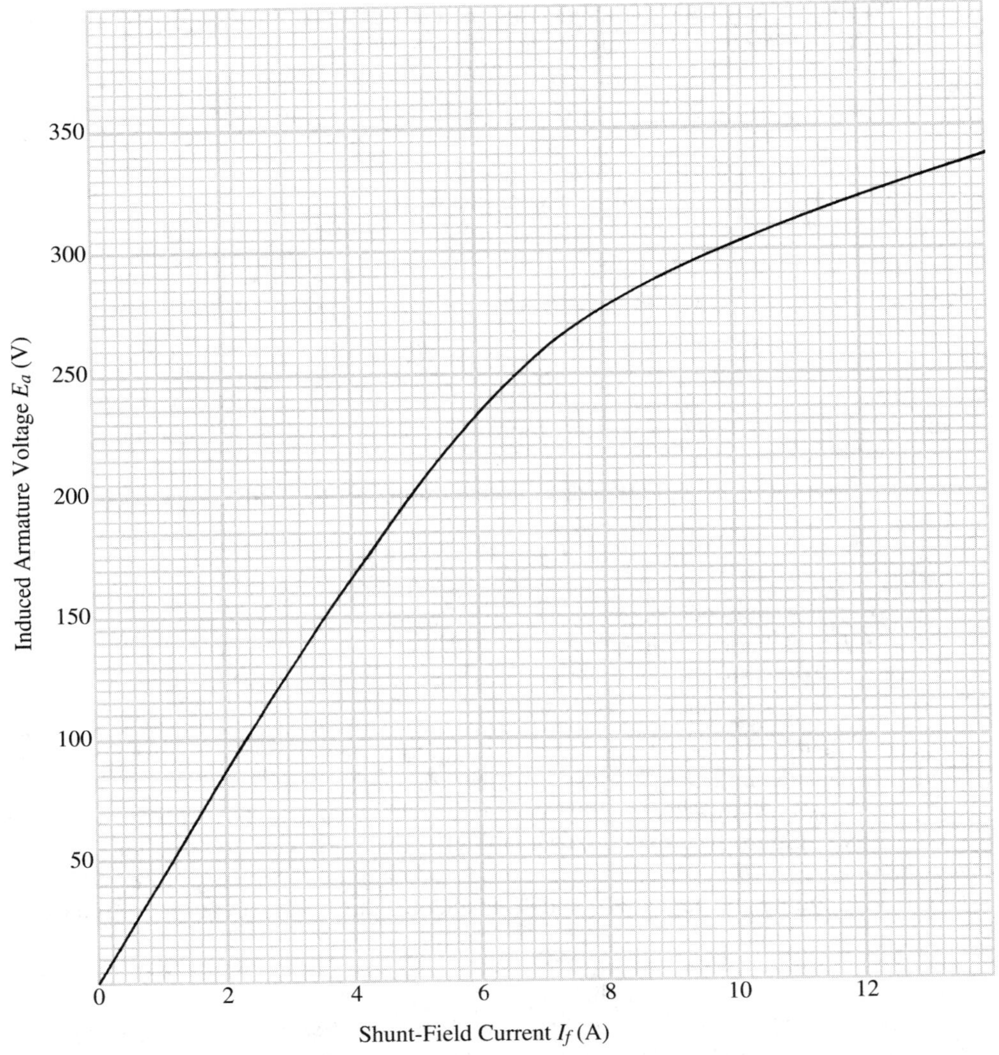

FIGURE 10.9
Magnetization curve for a representative separately excited shunt generator.

10.8 VOLTAGE REGULATION

The voltage regulation of a DC generator is the percent change in terminal voltage from no load to rated load, with respect to rated voltage:

$$\text{VR} = \frac{V_{\text{nl}} - V_{\text{rated}}}{V_{\text{rated}}} \times 100 \qquad \textbf{(10–16)}$$

where: VR = voltage regulation (percent)
V_{nl} = no-load (open-circuit) voltage (V)
V_{rated} = voltage indicated on nameplate of machine (V)

The no-load voltage may be obtained by operating the machine at its nameplate conditions, then removing the load and observing the no-load voltage. Although this is an exact method, it is not practical for very large machines.

An analytical method that provides a good approximation of the no-load voltage may be determined from the machine parameters and the associated magnetization curve (see Chapter 12).

EXAMPLE 10.3 A 100-kW, 1800 r/min generator operating at rated load has a terminal voltage of 240 V. If the voltage regulation is 2.3 percent, determine the no-load voltage.

Solution

$$\text{VR} = \frac{V_{nl} - V_{rated}}{V_{rated}} \quad \Rightarrow \quad V_{nl} = V_{rated} \times (1 + \text{VR})$$

$$V_{nl} = 240 \times (1 + 0.023) = \underline{245.5 \text{ V}}$$

10.9 GENERATOR-TO-MOTOR TRANSITION AND VICE VERSA

Generator-to-motor transition and vice versa are illustrated in Figure 10.10 using an elementary DC machine. For simplicity, only one armature coil is shown, and the magnetic field is supplied by permanent magnets. The armature is connected to a battery through the DC bus. The prime mover, which may be a steam turbine, diesel engine, etc., is mechanically coupled to the armature through a clutch. The clutch shown in Figure 10.10 is used for illustrative purposes. In actual practice the generator is directly coupled to the prime mover.

Generator Action

In Figure 10.10(a), the clutch is closed and the prime mover is shown driving the armature at a speed that causes its emf to be greater than the battery voltage. Since $E_a > E_{bat}$, the machine behaves as a generator, feeding current to the bus, charging the battery. The direction of current in the armature conductors, for the given magnetic polarity and direction of armature rotation, is determined by Lenz's law and the flux bunching rule. The equivalent circuit for generator action is shown in Figure 10.10(b). Resistor R represents the resistance of the armature windings, battery, and connecting cables.

Motor Action

Disengaging the prime mover by opening the clutch causes the armature speed and hence the generated emf to decrease. When the speed drops to a value that causes

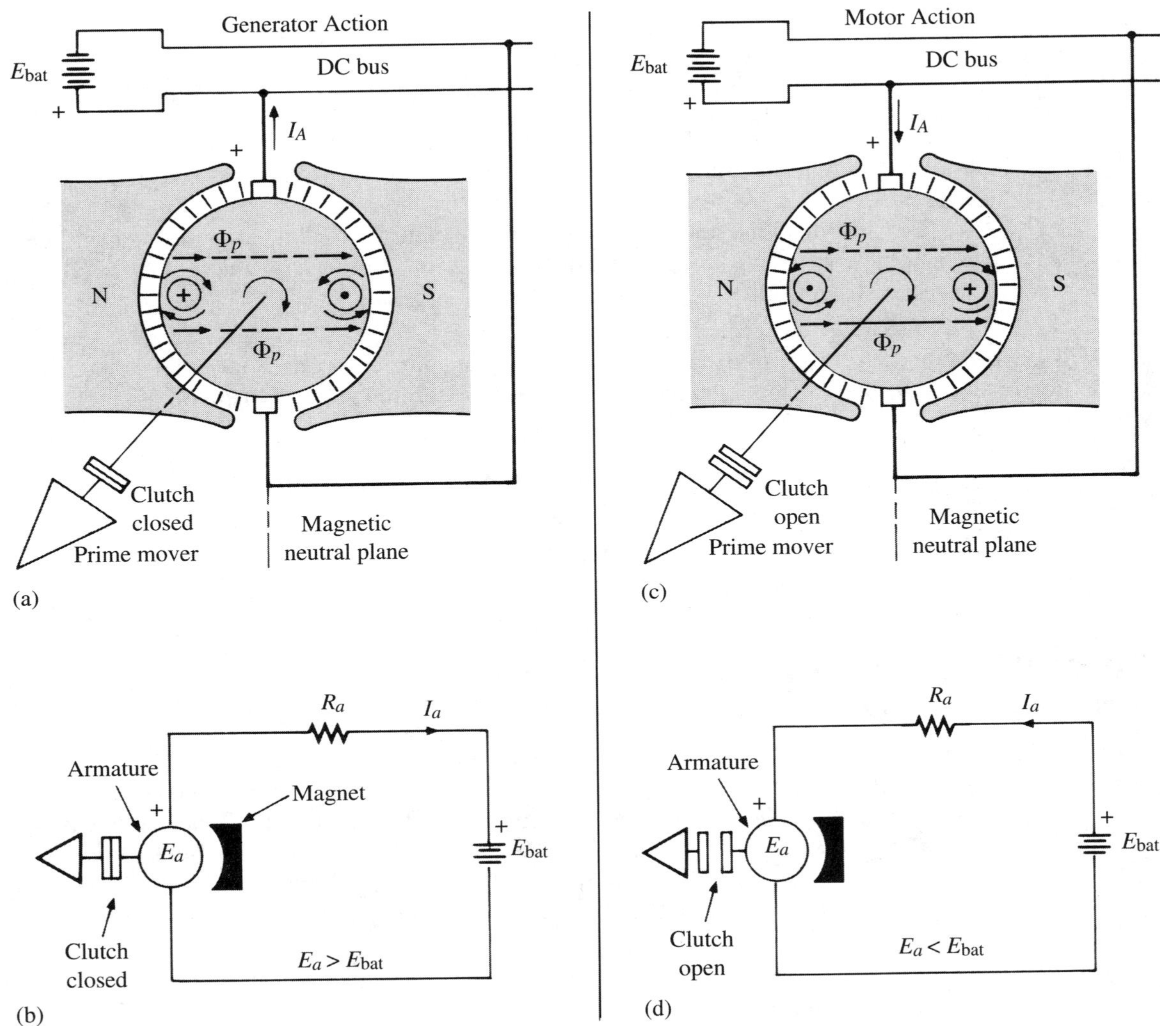

FIGURE 10.10
Generator-to-motor transition: (a) and (b) generator action; (c) and (d) motor action.

$E_a < E_{bat}$, the battery will act as the source, reversing the current in the armature and driving the machine as a motor. This is shown in Figure 10.10(c). Note that the rotation of the armature when motoring is in the same direction as it was when driven by the prime mover. The direction of rotation may be verified by Lenz's law and the flux bunching rule. Furthermore, as indicated in Figures 10.10(b) and (d), *the direction of the induced emf* (E_a) *is the same whether the machine is operating as a motor or as a generator;* this must be so, because in both cases the magnet polarity and the direction of rotation are the same.

All rotating machines (generators and motors) generate voltage *and* develop torque. When acting as a generator, as shown in Figures 10.10(a) and (b), the machine generates voltage, and if supplying current, develops a countertorque in opposition to the driving torque. When acting as a motor, as shown in Figures 10.10(c) and (d), the machine develops a driving torque, and if the shaft is free to turn, generates a counter-voltage (called a counter-emf or cemf) in opposition to the driving voltage. Furthermore, the machine will rotate in the same direction whether motoring or acting as a generator; the only outward indication of a change in the operating condition is the reversal of armature current, as would be indicated on an ammeter.

10.10 REVERSING THE DIRECTION OF ROTATION OF A DC MOTOR

The direction of rotation of a direct-current motor may be reversed by reversing either the current in the armature or the polarity of the field, but not both. Figures 10.11(a) and (b) illustrate how reversing the armature current reverses the direction of rotation. Figures 10.11(c) and (d) illustrate how reversing the polarity of the field reverses the direction of rotation. The direction of rotation in each case may be determined by the flux bunching rule, as shown on the inset in each figure.

10.11 DEVELOPED TORQUE

The relationship between developed torque, conductor current, and flux density, derived in Section 1.10 of Chapter 1 for an elementary motor, applies to all motors. Thus, by extension,

$$T_D = B_p I_a k_M \qquad \textbf{(10–17)}$$

where: B_p = flux density in air gap produced by shunt field poles (T)
I_a = armature current (A)
k_M = constant

Constant k_M depends on the design of the motor and includes the number of turns, effective length of armature conductors, number of poles, type of internal circuitry, and units used.

As indicated in Eq. (10–17) the torque developed by a DC motor is proportional to the flux density in the air gap and the current in the armature.[4]

10.12 BASIC DC MOTOR

The equivalent-circuit diagram for the basic DC motor, called a shunt motor, is shown in Figure 10.12. The shunt field winding is connected across the incoming lines, and it provides the flux that interacts with the armature current to produce the torque that results in motor action.

[4]The required flux density in the air gap of a DC machine depends on its power and speed ratings, and varies from a low of approximately 0.5 T to slightly over 1.0 T for large machines **[3]**.

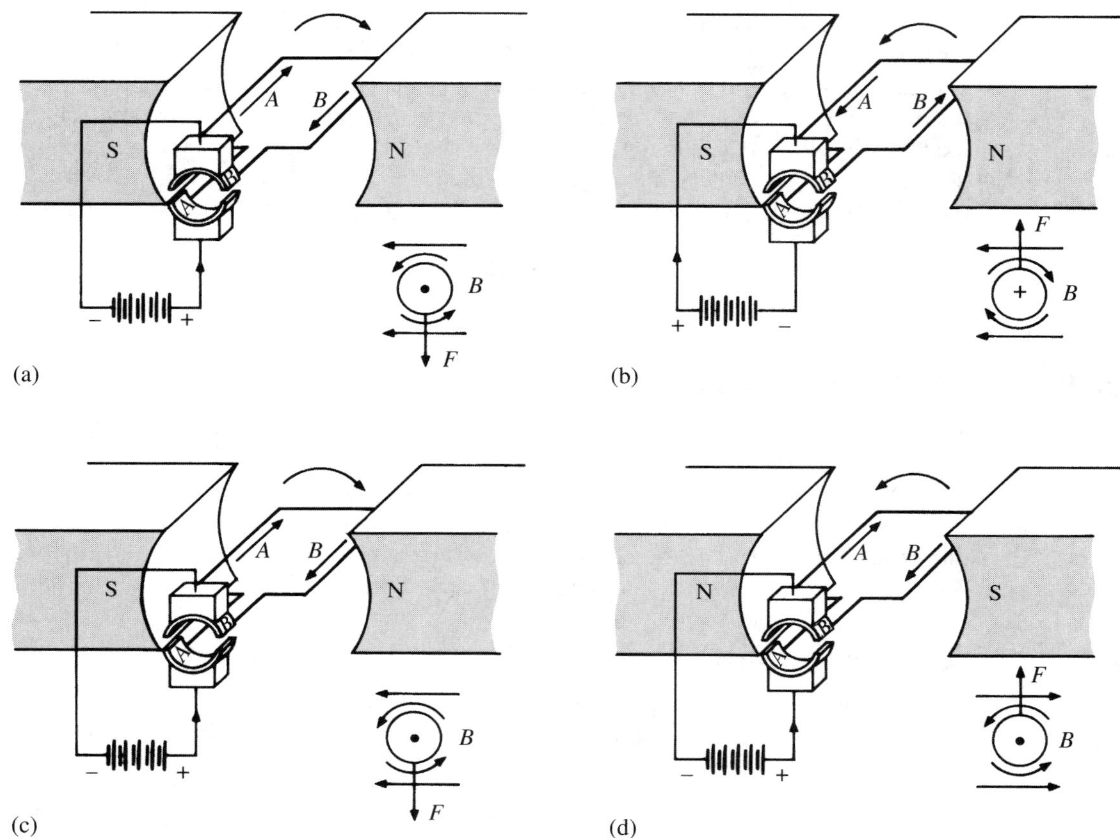

FIGURE 10.11
Reversing a direct-current motor by (a) and (b) reversing the armature current; (c) and (d) reversing the polarity of the field.

The shunt field does no useful work. It merely provides the necessary medium for the armature conductors to push against when developing rotary motion. It is similar to the roadbed against which the wheels of an automobile push; the roadbed does no work, but without it the car could not move. All of the energy supplied to the field is expended as I^2R losses in the field windings. This may be easily shown by connecting a wattmeter to measure the power drawn by the shunt field. Observation of the wattmeter will indicate no change in the power input to the field as the machine is loaded. A wattmeter placed in the armature circuit, however, will indicate proportional increases in power with increased shaft load. Assuming the developed torque is sufficient to cause rotation, a cemf will be generated in the armature that is proportional to the armature speed and the air-gap flux. That is,

$$E_a = n\Phi_p k_G \quad \textbf{(10–18)}$$

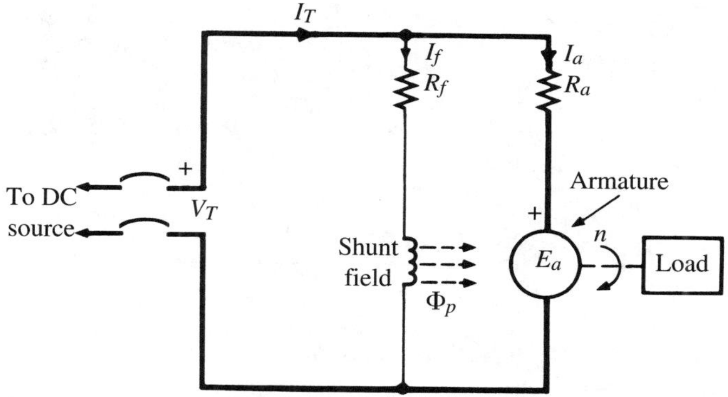

FIGURE 10.12
Equivalent-circuit diagram for a basic DC motor.

where Φ_p = flux in air gap produced by shunt field poles (Wb). Solving Eq. (10–18) for armature speed,

$$n = \left.\frac{E_a}{\Phi_p k_G}\right|_{\Phi_p \neq 0} \qquad \textbf{(10–19)}$$

Within the constraints indicated in Eq. (10–19), the speed of a DC motor is inversely proportional to the pole flux.

Note: If $\Phi_p = 0$, then the motor torque as indicated in Eq. (10–17) will be zero, and the armature will not turn.

Applying Kirchhoff's voltage law to the armature circuit in Figure 10.12 and solving for the current,

$$V_T = I_a R_a + E_a$$

$$I_a = \frac{V_T - E_a}{R_a}$$

EXAMPLE 10.4 A 240-V, 20-hp, 850 r/min shunt motor draws 72 A when operating at rated conditions. The respective resistances of the armature and shunt field are 0.242 Ω and 95.2 Ω. Determine the percent reduction in field flux required to obtain a speed of 1650 r/min, while drawing an armature current of 50.4 A.

Solution

$$I_{f1} = \frac{240}{95.2} = 2.52 \text{ A}$$

$$I_{a1} = I_T - I_{f1} = 72 - 2.52 = 69.48 \text{ A}$$

$$E_{a1} = V_T - I_{a1}R_a = 240 - 69.48 \times 0.242 = 223.19 \text{ V}$$

$$E_{a2} = V_T - I_{a2}R_a = 240 - 50.4 \times 0.242 = 227.80 \text{ V}$$

$$n_2 = n_1 \cdot \frac{[E_a/\Phi_p]_2}{[E_a/\Phi_p]_1} \qquad \Rightarrow \qquad \Phi_{p2} = \frac{n_1}{n_2} \times \frac{E_{a2}}{E_{a1}} \times \Phi_{p1}$$

$$\Phi_{p2} = \frac{850}{1650} \times \frac{227.80}{223.19} \times \Phi_{p1} = 0.5258\Phi_{p1}$$

$$\frac{0.5258\Phi_{p1} - \Phi_{p1}}{\Phi_{p1}} \times 100 = \underline{-47.4\%}$$

10.13 DYNAMIC BEHAVIOR WHEN LOADING AND UNLOADING A DC MOTOR

Figure 10.13 illustrates the dynamic behavior of the speed, cemf, armature current, and developed torque when a constant-torque load is applied to the shaft of a shunt motor; the circled numbers indicate the sequence of events. With the motor running, and no load on the shaft, the developed torque just balances the friction, windage, and small stray load loss. Adding shaft load (at time t_1) causes $T_{\text{load}} > T_D$ and the armature slows down. The decrease in speed decreases the cemf, causing an increase in armature current and hence an increase in the developed torque. The machine continues to slow down, increasing its armature current and thus developing greater torque, until T_D becomes equal to the torque load on the shaft. When this occurs (time t_2), deceleration ceases and the machine operates at a lower steady-state speed and a higher armature current.

Similarly, if the shaft load is reduced, $T_D > T_{\text{load}}$, and the armature accelerates. As the speed increases, the increase in cemf causes the armature current to decrease, decreasing the developed torque. The machine continues to accelerate, decreasing its armature current, developing less and less torque, until $T_D = T_{\text{load}}$. When this occurs, there is no further acceleration, and the machine operates at a higher steady-state speed and a lower armature current.

When adding or removing load, the resultant unbalanced torque causes deceleration or acceleration, respectively. This change in speed, with its attendant lowering or raising of the cemf, automatically adjusts the current in a direction to balance the opposing torques.

10.14 SPEED REGULATION

The speed regulation of a direct-current motor is the percent change in speed from no load to rated load with respect to rated speed when operating at rated voltage and rated temperature from a constant-voltage source:

$$SR = \frac{n_{\text{nl}} - n_{\text{rated}}}{n_{\text{rated}}} \times 100 \text{ percent} \qquad \textbf{(10–20)}$$

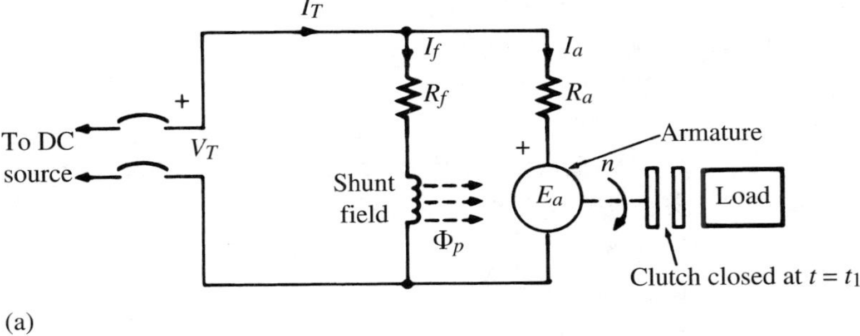

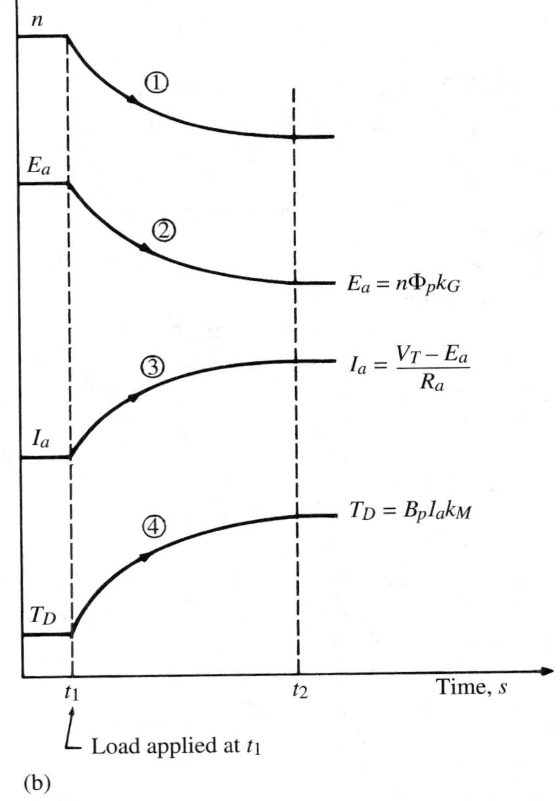

FIGURE 10.13
Dynamic behavior of a shunt motor when a constant-torque load is applied to the shaft: (a) equivalent circuit; (b) behavior of motor variables.

where:

$$n_{nl} = \text{no-load speed (r/min)}$$
$$n_{rated} = \text{rated speed (r/min)}$$
$$SR = \text{speed regulation}$$

EXAMPLE 10.5 A 120-V, 1750 r/min, 5-hp motor, operating at rated conditions, has a speed regulation of 4.0 percent. Determine the no-load speed.

Solution

From Eq. (10–20),

$$S_{nl} = n_{rated}\left(\frac{SR}{100} + 1\right) = 1750(0.04 + 1) = \underline{1820\text{ r/min}}$$

10.15 EFFECT OF ARMATURE INDUCTANCE ON COMMUTATION WHEN A DC MACHINE IS SUPPLYING A LOAD

Figure 10.14 illustrates the rectifying process that takes place when no load is connected to the armature. The arrows indicate only the voltage direction; there is no current in the armature coils unless a load is connected to the armature. As the armature rotates, the induced voltage in coil 2 changes from a clockwise direction when under the south pole (part a), to zero voltage in the commutating zone[5] (part b), to a counterclockwise direction when under the north pole (part c). The *commutating zone,* shown in Figure 10.14(a) is that portion of the commutator circumference where the armature coil is shorted by the brush.

When a load is connected to a DC machine, the commutator attempts to reverse the coil current as well as the coil voltage while the coil is passing through the commutation zone. The presence of coil inductance delays the change in coil current, however, preventing it from dropping to zero and reversing during the very brief commutation period.[6] *The commutation period is the duration of time that the armature coil is shorted by the brush.* The effect of armature inductance on the commutation process is illustrated in Figure 10.15.

Figure 10.15(a) shows coil 2 and coil 3 feeding current to the load via the commutator and brush. Figure 10.15(b) shows coil 2 moving through the commutating zone; the field poles no longer induce an emf in the coil, the coil is shorted by the brush, and the current starts to decrease. The decreasing current, however, causes an emf of self-induction which, in accordance with Lenz's law, delays the change. Hence, coil 2 rotates into the position shown in Figure 10.15(c) without having had sufficient time for its current to drop to zero. As commutator bar 3 slides off the brush,

---

[5]Any voltage produced by flux fringing into the commutation zone at no load is insignificant.

[6]If each pair of armature slots contains more than one coil, the inductive effect is increased by mutual inductance between the coils.

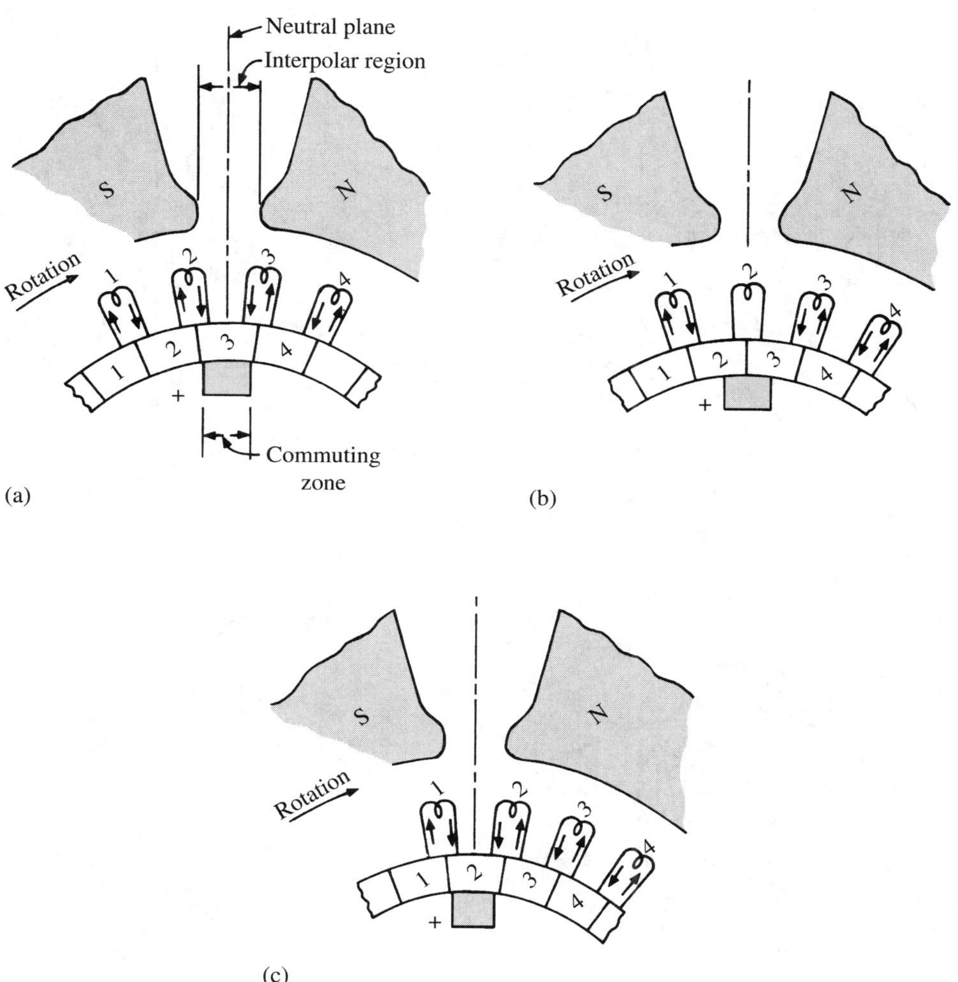

FIGURE 10.14
Commutation process under no-load conditions. The sequence of sketches shows the reversal of voltage in coil 2 as it enters and leaves the commutating zone.

the current in coil 2 is suddenly forced to zero, generating a high emf of self-induction that causes an arc to jump between commutator bar 3 and the brush. If allowed to continue, such arcing will severely damage both the commutator and the brushes, making it impossible to operate the machine unless corrective measures are taken.

Successful operation of a DC machine requires that during the very brief commutation period the coil current reverses from full value in one direction to full value in the opposite direction. Ideal commutation occurs when the current changes direction midway through the commutation period, as shown in Figure 10.15(d). Thus for ideal commutation, the coil current will be in phase with the coil voltage.

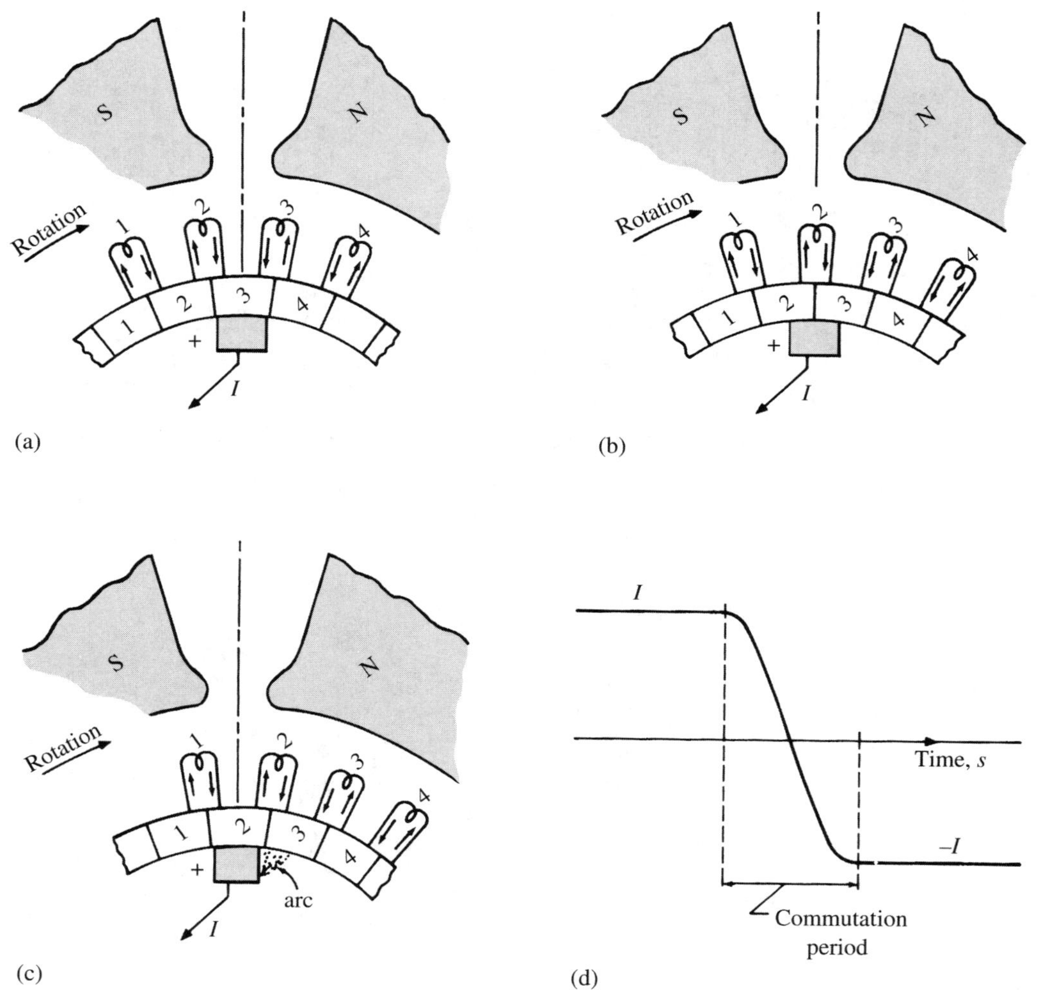

FIGURE 10.15
Commutation process under load conditions: (a), (b), and (c) show delay in current reversal and resultant arc caused by coil inductance; (d) graph of ideal commutation.

10.16 INTERPOLES

To eliminate sparking caused by the emf of self-induction, narrow poles called *interpoles* or *commutating poles* are installed in the neutral plane of DC machines. The interpoles, shown in the cutaway view of a DC machine in Figure 10.6, affect only the coil undergoing commutation. The interpole windings are designed to generate a neutralizing voltage of sufficient magnitude to force the reversal of current in each armature coil as it moves through the interpolar region. Ideally, the current in the coil undergoing commutation should start decreasing when it enters the commutating zone, become zero when in the neutral plane, and then rise in the opposite direction, reaching full value at the moment it leaves the commutating zone.

The interpole windings are connected in series with the armature, as shown in Figure 10.16 for an elementary machine, so that their strength will be proportional to the armature current. The interpole winding forms part of the armature circuit and should not be disconnected or reversed. Incorrect connections will result in worse sparking than if no interpoles were used. The polarity of the interpoles with respect to the polarity of the field poles is shown in Figure 10.16 for the specific direction of rotation, for both motors and generators.

Since an armature coil spans approximately 180 electrical degrees, both sides of the coil will enter the magnetic neutral plane at the same time. Hence, instead of using two interpoles to induce a neutralizing voltage in each coil as it enters the neutral plane, one interpole with more turns may be used.

Interpole Saturation

When operating at rated load or below, the interpole flux remains in proportion to the armature current, and as a result, the neutralizing emf will always be equal and opposite to the emf of self-induction. When operating well above rated load, however, the interpole iron saturates, and its flux is no longer proportional to the armature current. The interpole can no longer generate an opposing emf in proportion to the emf of self-induction, and a relatively large current will appear in the coils short circuited by the brushes, causing sparking.

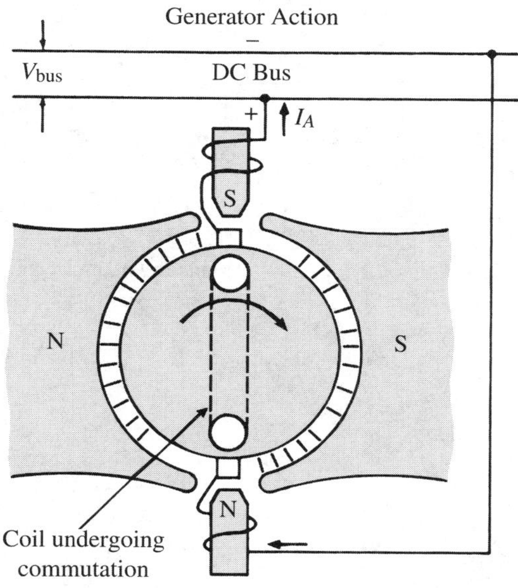

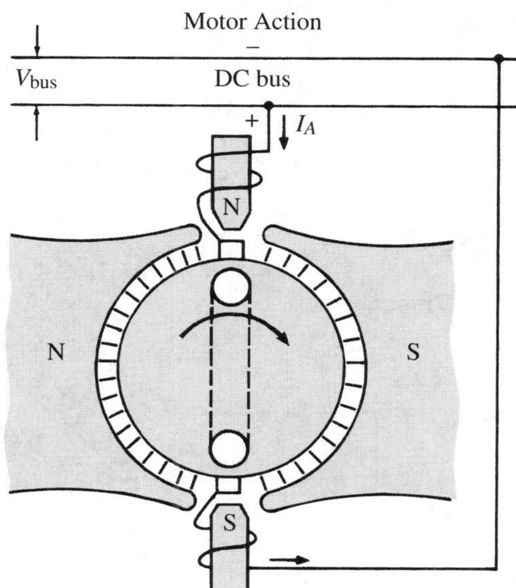

FIGURE 10.16
Location of interpoles, their connections, and their polarity for generators and motors.

10.17 ARMATURE REACTION

When a generator or motor is loaded, the current in the armature coils develops a magnetomotive force of its own that interacts with the magnetomotive force of the field poles, disturbing the uniform flux distribution in the air gap. This behavior is called armature reaction.

The effect of armature reaction on the flux distribution is shown in Figure 10.17(a) for generator action and in Figure 10.17(b) for motor action. The accompanying vector diagrams are useful only for showing the direction of flux shift due to the respective armature mmfs. Although only one armature coil is shown in Figure 10.17, the many coils of a practical machine are distributed uniformly around the armature as previously shown in Figure 10.7. Thus, when an armature coil rotates from the position shown in Figure 10.17, another coil takes its place, and the armature mmf remains fixed in space.

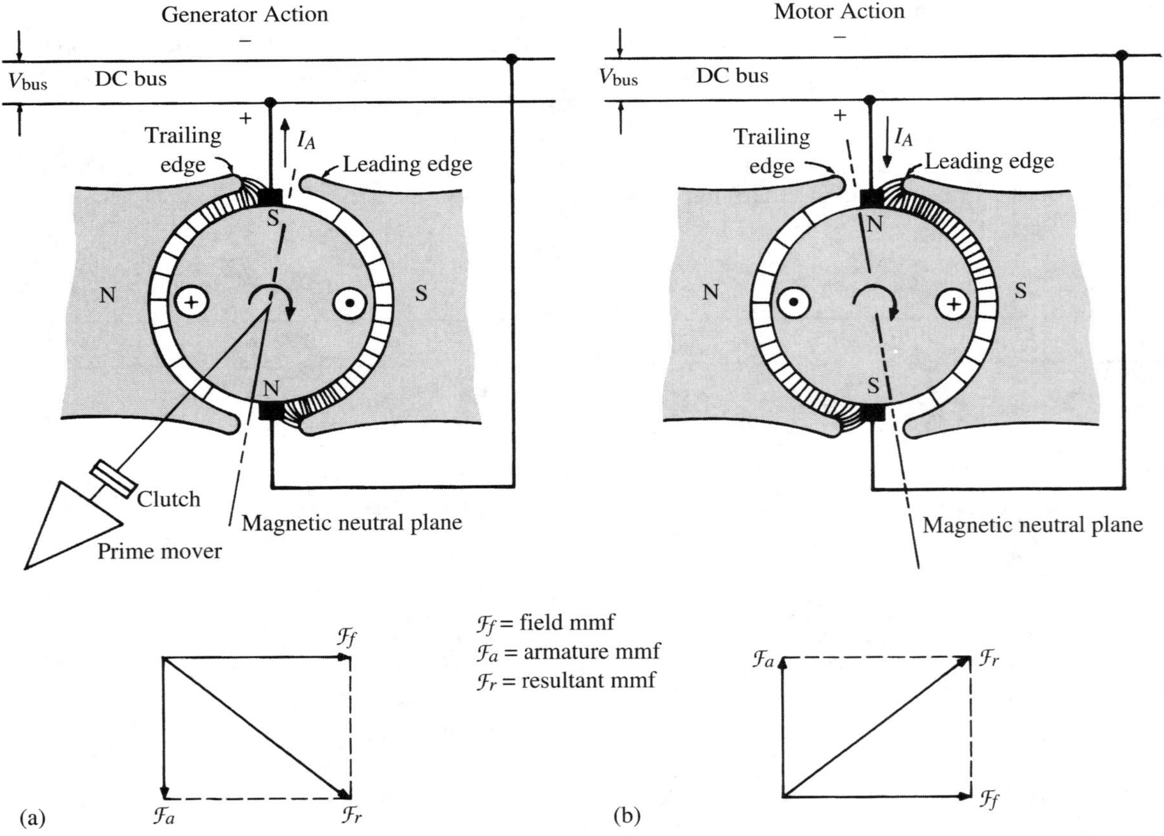

FIGURE 10.17
Effect of armature reaction in DC generators and motors. Note leakage of flux into interpolar region.

Armature Reaction in a DC Generator

As shown in Figure 10.17(a), the armature reaction mmf in a generator causes the neutral plane to shift its axis in the direction of rotation. The net result is an increase in flux at the *trailing edges* of the poles, a decrease in flux at the leading edges,[7] and some flux fringing into the interpolar region. Because of the nonlinear nature of the pole iron, however, the increase in flux at the trailing edges is less than the decrease in flux at the leading edges, resulting in a net reduction in total pole flux. *Although the resultant mmf vector* $\mathcal{F}_r$ *shown in Figure 10.17 indicates an increase in magnetization, the shift of the flux axis due to the armature mmf, combined with magnetic saturation, results in a demagnetizing effect.*

Armature reaction has two adverse effects on the performance of a DC generator:

1. Flux in the interpolar region causes a voltage to be generated in the coil undergoing commutation, causing arcing at the brushes.
2. The net reduction in total pole flux results in an undesirable reduction in generated voltage.

Armature Reaction in a Motor

As shown in Figure 10.17(b), the armature reaction mmf in a motor causes the neutral plane to shift its axis in a direction opposite to the direction of rotation. The net result is an increase in flux at the *leading edges* of the poles, a decrease in flux at the trailing edges, and some fringing of flux into the interpolar region. This has two adverse effects on the performance of a DC motor:

1. Flux in the interpolar region causes a voltage to be generated in the coil undergoing commutation, causing arcing at the brushes.
2. The net reduction in total pole flux results in an undesirable increase in motor speed.

Armature Reaction and Interpoles

The degree to which armature reaction affects the performance of a DC machine, whether as a motor or as a generator, is dependent on the magnitude of the armature current. Increased loading increases the armature mmf, thus increasing the adverse effects.

If the interpoles are made sufficiently strong by the addition of more turns, they will produce additional opposing mmf to nullify the effect of the armature mmf in the interpolar region and yet remain equally effective in opposing the emf of self-induction in the coils undergoing commutation. Although they do eliminate or reduce sparking, interpoles do not prevent the shift of flux across the pole face brought about by armature reaction; they merely prevent flux from fringing into the interpolar region. The total field flux is still reduced by saturation at the respective pole tips.

[7]When an armature conductor enters the interpolar region, it departs from the trailing edge and approaches the leading edge.

Net Flux in the Air Gap of a DC Machine

The net flux in the air gap of a DC machine (motor or generator) can be expressed in terms of the field mmf and an *equivalent demagnetizing mmf* caused by armature reaction. Expressed in equation form,

$$\mathcal{F}_{net} = \mathcal{F}_f - \mathcal{F}_d \tag{10–21}$$

$$\Phi_{gap} = \frac{\mathcal{F}_{net}}{\mathcal{R}} \tag{10–22}$$

where: $\mathcal{F}_{net}$ = net mmf (A-t/pole)
$\mathcal{F}_f$ = field mmf (A-t/pole)
$\mathcal{F}_d$ = equivalent demagnetizing mmf (A-t/pole)
$\mathcal{R}$ = reluctance (nonlinear)

Note: Although not exact, $\mathcal{F}_d$ is assumed proportional to the armature current. However, if a compensating winding (as described in Section 10.19) is used $\mathcal{F}_d = 0$.

10.18 BRUSH SHIFTING AS AN EMERGENCY MEASURE

The brush-neutral setting should not be changed to minimize sparking unless reasonable evidence indicates an incorrect setting.[8] The most likely causes of sparking at the brushes are an open coil in the armature, defective interpoles, incorrect brush-spring pressure, or wrong brush grade. An incorrect neutral setting may occur as a result of improper assembly after overhaul, or commutator wear. A commutator whose diameter was reduced by excessive wear and resurfacing may cause the brushes to contact the commutator at a slightly off-neutral position.

If the interpoles are defective, and repair or replacement requires an unacceptable downtime, shifting the brush rigging in the direction of rotation for a generator, and against the direction of rotation for a motor, will reduce sparking. Since the angular shift of flux caused by armature reaction is a function of the magnitude of the armature current, however, the brushes would have to be shifted to a different position for each value of load.[9]

Furthermore, shifting the brushes causes the armature mmf to have a component in a direction that weakens the field flux. This is illustrated in Figure 10.18 for a DC *generator*. The *normal position* of the brushes, shown in Figure 10.18(a), results in the armature mmf $\mathcal{F}_a$ to be normal to the field mmf $\mathcal{F}_f$. Shifting the brushes to the position shown in Figure 10.18(b) causes a component of armature mmf ($\mathcal{F}_d$) to be in opposition to the field mmf.

---

[8] See Reference **[1]** for more information on brush sparking and for test procedures to locate the correct brush-neutral setting.

[9] Before interpoles were invented, a brush-shifting lever was mounted on the brush rigging to provide for easy shifting of the brush axis as load came on and off.

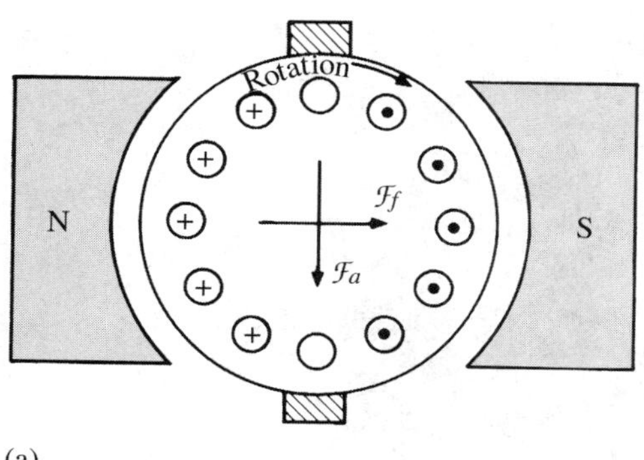

(a)

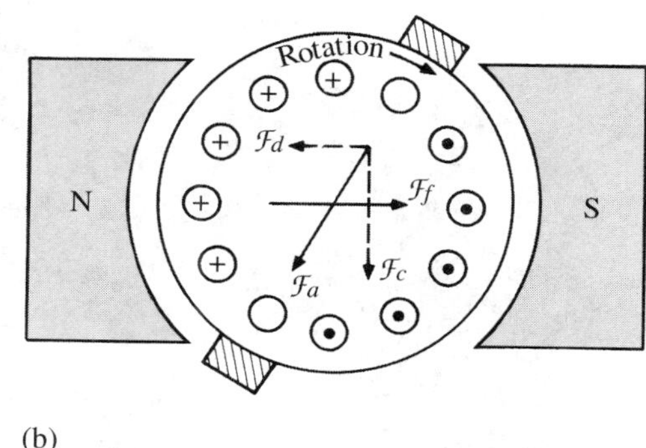

(b)

FIGURE 10.18
Effect of brush shift on the strength of the magnetic field of a DC generator: (a) brushes in magnetic-neutral plane; (b) brushes shifted to a new position in the direction of rotation.

Brush shifting should be done only as a last resort, and the brush rigging should be positioned to reduce sparking when operated at the average load. This is an extremely critical task that should be done by competent motor mechanics. If done incorrectly, injury to operating personnel and damage to equipment could occur.

10.19 COMPENSATING WINDINGS

The effect of armature reaction is particularly severe in applications where quick reversals and/or large and rapid load changes occur. Sudden large changes in armature flux induce high transient voltages in all armature coils by transformer action. If the transient voltages are high enough to cause arcing between commutator bars, a brush-to-brush flashover may occur. A *flashover* is a transient high-current arc accompanied by a loud explosive noise. Minor flashovers cause pitting of the commutator bars and burned spots on the edges of the brushes. A severe flashover may melt brush holders and commutator bars, as well as damage the armature winding and core. To prevent flashovers, DC machines designed for large and rapid load swings, such as in steel mills and ship propulsion, are equipped with *pole-face windings.* The pole-face winding, called a *compensating winding,* is shown in Figure 10.19. The compensating winding is connected in series with the armature, and essentially *eliminates armature reaction by setting up a mmf that is always equal and opposite to the armature mmf.*

Figure 10.20 illustrates the parallel arrangement of conductors and the relative directions of current in the armature and in the compensating winding (CW) for a one-coil elementary machine (generator and motor); the accompanying vector diagram shows the component mmf vectors. The compensating winding acts as a large coil surrounding the armature and connected in series with it.

Note that machines with compensating windings still require interpoles to counteract the emf of self-induction in the coils undergoing commutation.

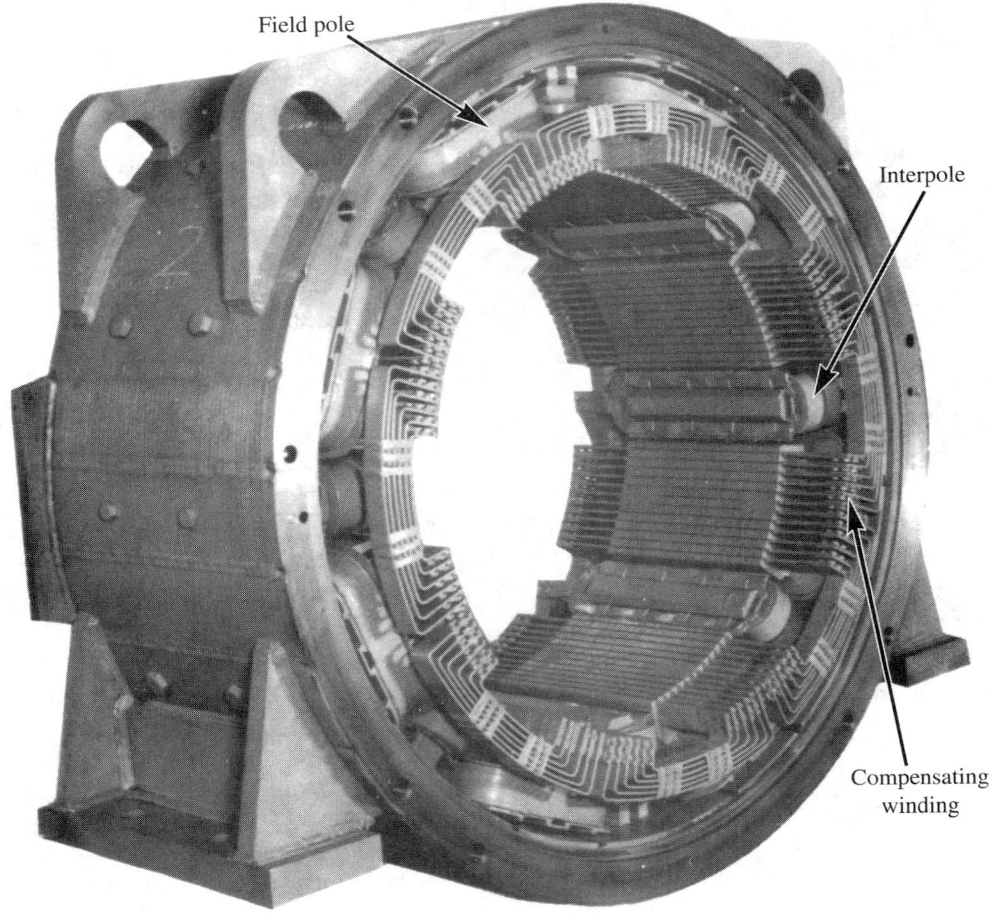

FIGURE 10.19
Direct-current machine with interpoles and compensating windings. (Courtesy Reliance Electric Company)

10.20 COMPLETE EQUIVALENT CIRCUIT OF A SEPARATELY EXCITED SHUNT GENERATOR

The complete equivalent-circuit diagram of a separately excited shunt generator that includes interpoles and compensating windings is shown in Figure 10.21. Applying Kirchhoff's voltage law to the armature circuit, and assuming steady-state conditions,

$$E_a = I_a R_{\text{acir}} + V_T \tag{10–23}$$

Defining:

$$R_{\text{acir}} = R_a + R_{\text{IP}} + R_{\text{CW}} \tag{10–24}$$

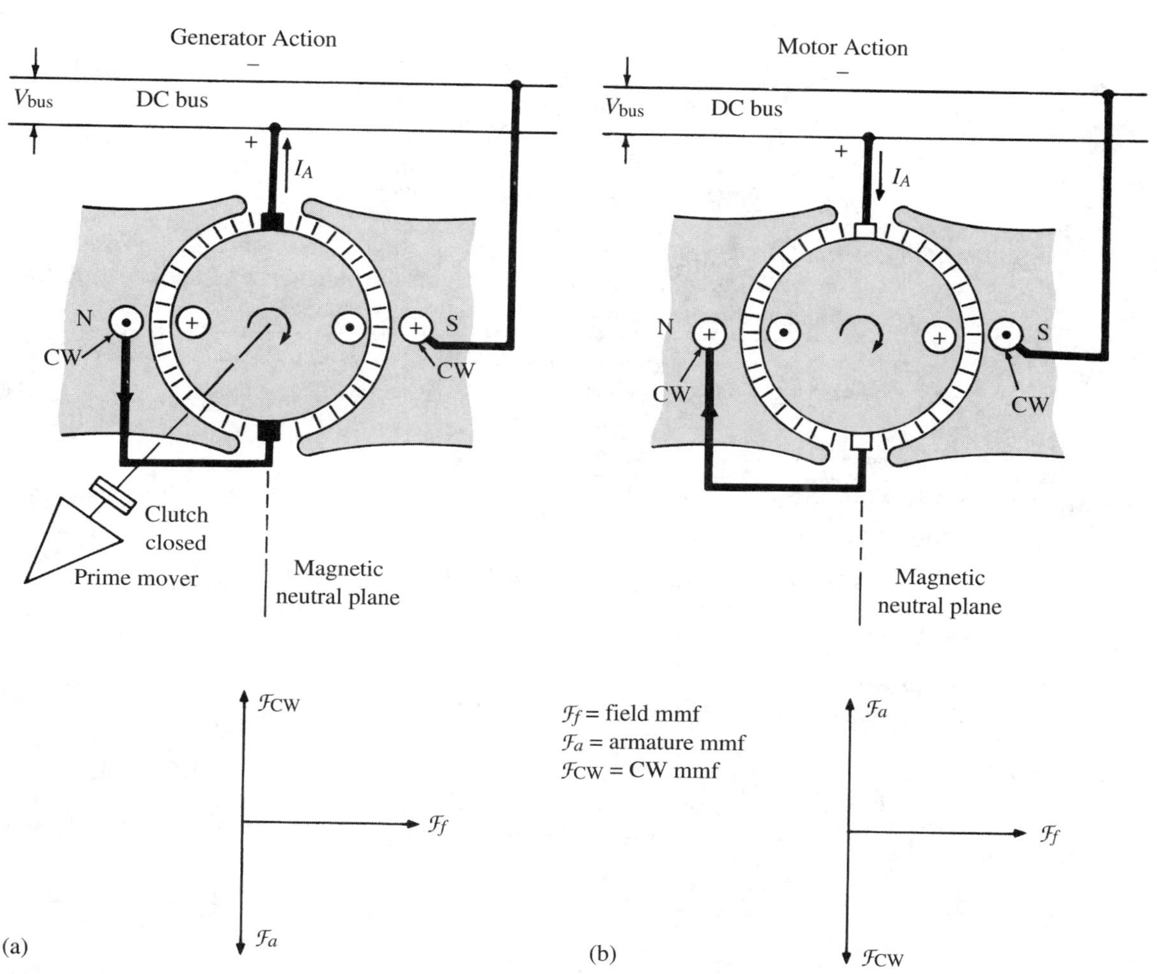

Generator Action
Vbus
DC bus
+
IA
N
S
CW
CW
Clutch
closed
Prime mover
Magnetic
neutral plane
Motor Action
Vbus
DC bus
+
IA
N
S
CW
CW
Magnetic
neutral plane
ℱCW
ℱf
ℱa
(a)
ℱf = field mmf
ℱa = armature mmf
ℱCW = CW mmf
ℱa
ℱf
ℱCW
(b)

FIGURE 10.20
Elementary diagram showing parallel arrangement of conductors and relative directions of current in armature and compensating winding for (a) generators; (b) motors.

FIGURE 10.21
Complete equivalent circuit of a separately excited shunt generator that includes interpoles and compensating windings.

Armature
Racir
+
Φp
Ea
Shunt
field
n
Prime
mover
Rf
IP + CW
VT
Load
To DC
source

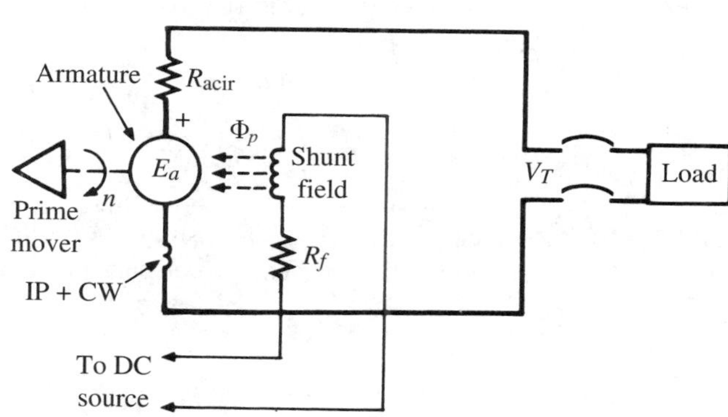

where: R_a = resistance of armature windings (Ω)
R_{IP} = resistance of interpole windings (Ω)
R_{CW} = resistance of compensating windings (Ω)
R_{acir} = resistance of armature circuit (Ω)

Note: Equation (10–23) ignores the relatively small voltage drop (called *brush drop*) across each carbon or graphite brush,[10] and ignores the inductance of the armature circuit. The inductance has no effect on steady-state performance.

[10] The brush drop cannot be ignored when determining losses. See Section 10.26. For more information on brush materials and brush drop, see References **[2]**,**[3]**.

EXAMPLE 10.6 A separately excited, compensated shunt generator, rated at 25 kW, 250 V, and 1450 r/min, has the following parameters;

$$R_a = 0.1053\ \Omega \qquad R_{IP} = 0.0306\ \Omega \qquad R_{CW} = 0.0141\ \Omega \qquad R_f = 96.3\ \Omega$$

Determine the induced emf.

Solution

The equivalent circuit is shown in Figure 10.21.

$$I_a = \frac{P}{V_T} = \frac{25{,}000}{250} = 100\text{ A}$$

$$R_{acir} = 0.1053 + 0.0306 + 0.0141 = 0.1500\ \Omega$$

$$E_a = V_T + I_a R_{acir} = 250 + 100 \times 0.150 = \underline{265\text{ V}}$$

10.21 COMPLETE EQUIVALENT CIRCUIT OF A SHUNT MOTOR

The complete equivalent-circuit diagram of a shunt motor that includes interpoles and compensating windings is shown in Figure 10.22.

Applying Kirchhoff's voltage law to the armature circuit, and assuming steady-state conditions,

$$V_T = E_a + I_a R_{acir}$$

$$E_a = V_T - I_a R_{acir} \qquad \textbf{(10–25)}$$

EXAMPLE 10.7 A 30-hp, 500-V, 850 r/min shunt motor draws a line current of 51.0 A when operating at rated conditions. The motor parameters are

$$R_a = 0.602\ \Omega \qquad (R_{IP} + R_{CW}) = 0.201\ \Omega \qquad R_f = 408.5\ \Omega$$

Determine the cemf.

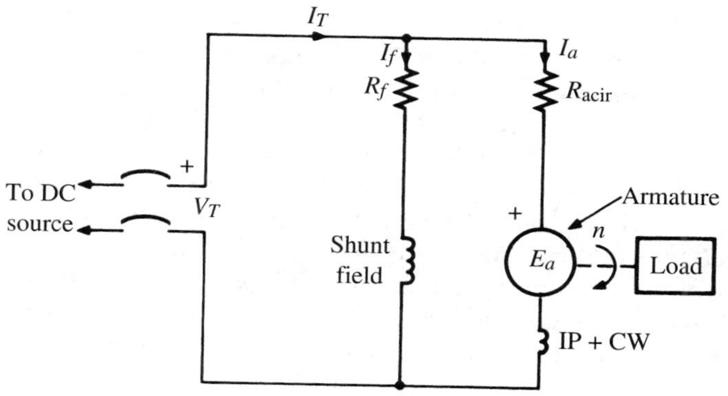

FIGURE 10.22
Complete equivalent circuit of a shunt motor that includes interpoles and compensating windings.

Solution

The equivalent circuit is similar to that shown in Figure 10.22.

$$I_f = \frac{V_T}{R_f} = \frac{500}{408.5} = 1.224 \text{ A}$$

$$I_a = I_T - I_f = 51.0 - 1.224 = 49.78 \text{ A}$$

$$R_{acir} = R_a + (R_{IP} + R_{CW}) = 0.602 + 0.201 = 0.803 \ \Omega$$

$$E_a = V_T - I_a R_{acir} = 500 - 49.78 \times 0.803 = \underline{460.0 \text{ V}}$$

10.22 GENERAL SPEED EQUATION FOR A DC MOTOR

The general speed equation for a DC motor is obtained by substituting Kirchhoff's voltage equation for the armature circuit into the basic speed equation. From Eq. (10–19)

$$n = \left.\frac{E_a}{\Phi_p k_G}\right|_{\Phi_p \neq 0}$$

Substituting Eq. (10–25) into Eq. (10–19),

$$n = \left.\frac{V_T - I_a R_{acir}}{\Phi_p k_G}\right|_{\Phi_p \neq 0} \quad \text{r/min} \qquad \textbf{(10–26)}$$

where:
R_{acir} = armature-circuit resistance (Ω)
I_a = armature current (A)
k_G = constant
n = speed (r/min)
Φ_p = flux produced by shunt field winding (Wb)

Motor speed Eq. (10–26) indicates the general trend of speed when the various parameters are changed, and can also be used to calculate the steady-state speed, *providing sufficient torque is developed to produce the necessary accleration. Note:* A break in the shunt-field circuit of a *lightly loaded* shunt motor (that is not protected with overcurrent or loss-of-field devices) may cause the machine to accelerate to dangerously high speeds. When the field circuit is opened, the flux does not instantaneously drop to zero; the hysteresis of the pole iron provides sufficient residual magnetism to permit acceleration to speeds that may result in destruction of the machine by excessive centrifugal forces. The reduction in cemf, caused by the reduction in flux, results in an increase in armature current. If the increase in armature current causes $B_p I_a$ to increase, the motor will accelerate [see Eq. (10–17)].

Base Speed

The base speed of a direct-current motor is its nameplate value, and is the speed obtained at rated voltage, rated shaft load, rated operating temperature, and with no added resistance in series with the armature and/or in series with the shunt field.

To accelerate a machine above its base speed, the developed torque must be made greater than the load torque on the shaft plus the windage and friction torque. Similarly, to decrease the speed, the developed torque must be made less than the combined shaft load, friction, and windage. As indicated in Eq. (10–17), changes in developed torque require changes in the flux density and/or armature current.

EXAMPLE 10.8 A 25-hp, 240-V shunt motor operating at 850 r/min draws a line current of 91 A when operating at rated conditions. A 2.14-Ω resistor inserted in series with the armature causes the speed to drop to 634 r/min. The respective armature-circuit resistance and field-circuit resistance are 0.221 Ω and 120 Ω. Determine the new armature current.

Solution

The circuit diagram is shown in Figure 10.23.

$$I_f = \frac{240}{120} = 2 \text{ A}$$

$$I_a = I_T - I_f = 91 - 2 = 89 \text{ A}$$

Using Eq. (10–26), and noting that the shunt field current was not changed,

$$\frac{n_1}{n_2} = \frac{[V_T - I_{a1}R_{\text{acir}}]}{[V_T - I_{a2}(R_{\text{acir}} + R_x)]}$$

$$I_{a2} = \frac{\left[V_T - \dfrac{n_2}{n_1} \times (V_T - I_{a1}R_{\text{acir}})\right]}{(R_{\text{acir}} + R_x)}$$

$$I_{a2} = \frac{\left[240 - \dfrac{634}{850} \times (240 - 89 \times 0.221)\right]}{(0.221 + 2.14)} = \underline{32.05 \text{ A}}$$

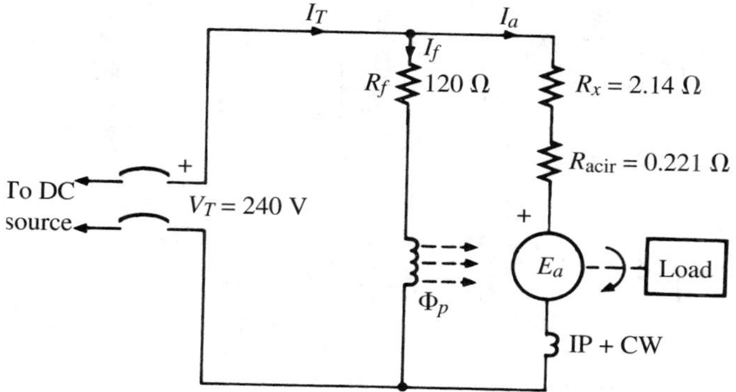

FIGURE 10.23
Circuit diagram for Example 10.8.

EXAMPLE 10.9

A shunt motor rated at 10 hp, 240 V, 2500 r/min, draws 37.5 A when operating at rated conditions. The motor parameters are $R_a = 0.213\ \Omega$, $R_{\text{CW}} = 0.065\ \Omega$, $R_{\text{IP}} = 0.092\ \Omega$, and $R_f = 160\ \Omega$. Determine (a) the steady-state armature current if a rheostat in the shunt field circuit reduces the flux in the air gap to 75.0 percent of its rated value, a 1.0-Ω resistor is placed in series with the armature, and the load torque on the shaft is reduced to 50 percent rated; (b) the steady-state speed for the conditions in (a).

Solution

The circuit diagram for rated conditions is shown in Figure 10.24(a).

(a) At rated conditions,

$$I_f = \frac{V_f}{R_f} = \frac{240}{160} = 1.50 \text{ A}$$

$$I_a = I_T - I_f = 37.5 - 1.5 = 36 \text{ A}$$

The circuit diagram for the new conditions is shown in Figure 10.24(b). From Eq. (10–17),

$$\frac{T_1}{T_2} = \frac{[\Phi_p I_a]_1}{[\Phi_p I_a]_2} \qquad \Rightarrow \qquad I_{a2} = I_{a1} \times \frac{T_2}{T_1} \times \frac{\Phi_{p1}}{\Phi_{p2}}$$

$$I_{a2} = 36 \times \frac{0.50\ T_1}{T_1} \times \frac{\Phi_{p1}}{0.75 \times \Phi_{p1}} = \underline{24.0 \text{ A}}$$

(b) $$R_{\text{acir}} = R_a + R_{\text{CW}} + R_{\text{IP}} = 0.213 + 0.065 + 0.092 = 0.370\ \Omega$$

From Eq. (10–26),

$$\frac{n_2}{n_1} = \left[\frac{V_T - I_a R_{\text{acir}}}{\Phi_p k_G}\right]_2 \times \left[\frac{\Phi_p k_G}{V_T = I_a R_{\text{acir}}}\right]_1$$

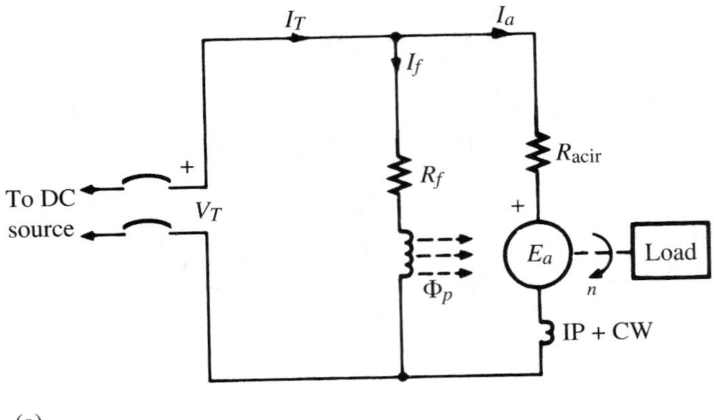

(a)

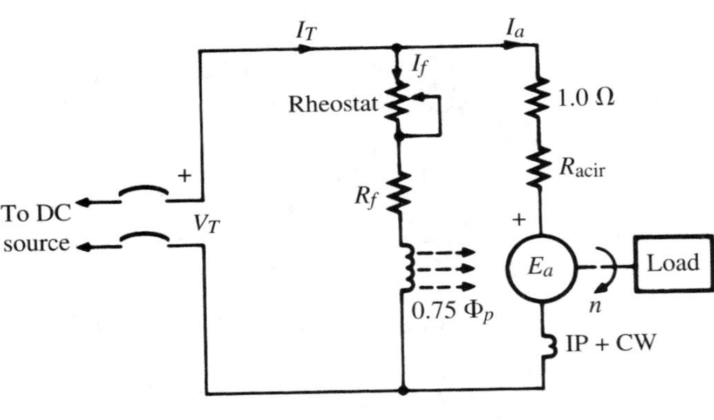

(b)

FIGURE 10.24
Circuit diagrams for Example 10.9: (a) rated conditions; (b) external resistor in series with the armature, and rheostat in series with the field.

$$n_2 = 2500 \times \left[\frac{240 - 24 \times (1 + 0.370)}{0.75\Phi_p k_G}\right] \times \left[\frac{\Phi_p k_G}{240 - 36 \times 0.370}\right]$$

$$n_2 = \underline{3046 \text{ r/min}}$$

10.23 DYNAMIC BEHAVIOR DURING SPEED ADJUSTMENT

Armature Control

To reduce the speed of a DC motor below its base speed, a rheostat or resistor must be inserted in series with the armature as shown in Figure 10.25(a). Figure 10.25(b) shows the dynamic behavior of armature current, developed torque, speed, and cemf

FIGURE 10.25
(a) Circuit for speed adjustment below base speed; (b) dynamic behavior of motor variables.

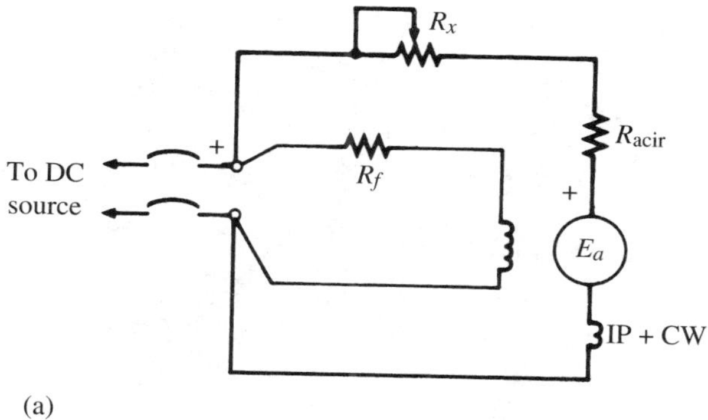

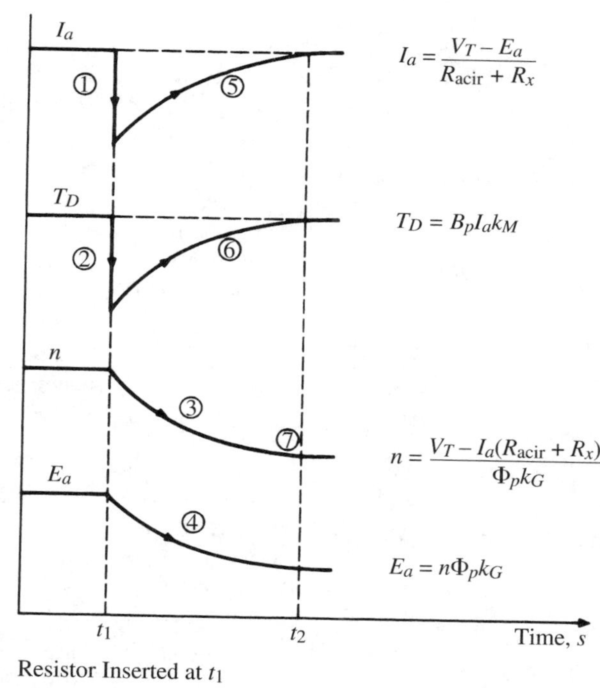

that occurs when a resistor is inserted in series with the armature of a shunt motor driving a constant-torque load. Although the sketches are not to scale, they do show the general behavior of the machine variables. The circled numbers indicate the sequence of events. Inserting a resistor in series with the armature causes a reduction in armature current that decreases the developed torque, and the machine slows down. As the machine decelerates, the cemf decreases, causing the newly reduced armature current to increase. When the armature current increases to a value that causes the developed torque to equal the load torque on the shaft plus windage and friction (time t_2), the machine will no longer decelerate, but will continue to run at the new lower steady-state speed. The reduction in windage and friction due to the lower speed is relatively small. Hence, the final steady-state armature current is essentially the same as before the resistor was inserted.

Shunt Field Control

To increase the speed of a DC motor above its base speed, a resistor or rheostat must be inserted in series with the shunt field circuit, as shown in Figure 10.26(a). Figure 10.26(b) shows (in simplified form) the dynamic behavior of the field current flux, cemf, armature current, developed torque, and speed that occurs when a resistor is inserted in series with the shunt field of a shunt motor driving a constant-torque load. The inductive time constant of the shunt field circuit prevents an instantaneous change in field current.

The sketches in Figure 10.26(b) are not to scale, but do indicate the general behavior of the machine variables. The circled numbers indicate the sequence of events. Inserting a resistor in series with the shunt field decreases the field current and thus reduces the flux. The reduction in flux decreases the cemf, which in turn allows the armature current to increase. If the percentage increase in armature current is greater than the percentage decrease in flux density, and the flux density is not zero, the torque will increase and the machine will accelerate. As the machine accelerates, its cemf increases proportionately with the increased speed, and the newly raised armature current decreases. When the armature current decreases to a value that causes the developed torque to equal the load torque on the shaft plus the friction and windage load (time t_2), the machine will no longer accelerate but will continue to run at the higher steady-state speed and draw a higher armature current. The machine can accelerate only while the internal torque developed is greater than the load torque.

10.24 PRECAUTIONS WHEN INCREASING SPEED THROUGH FIELD WEAKENING

Very high armature currents associated with large and rapid reductions in shunt field current may damage the commutator and brushes if the circuit breaker or other protective device fails to trip. Hence, speed changes through shunt field rheostat control should be made slowly. Large high-inertia machines are more susceptible to damage than are small machines. Small machines can accelerate rapidly, permitting a quick buildup of cemf, and thus a rapid reduction in armature current.

Centrifugal stresses and commutation problems limit the maximum allowable speeds that may be safely attained by shunt field control. Ratings of direct-current motors for industrial service that are specifically designed for speed adjustment through field control are listed in NEMA tables. Direct-current motors that are not specifically designed for operation above base speed should not use shunt field control. Such machines are constructed to withstand an *emergency overspeed* equal to 25 percent of base speed for only 1 minute **[4]**.

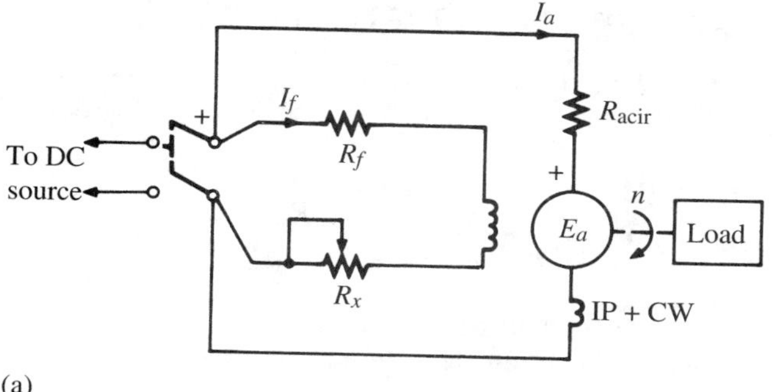

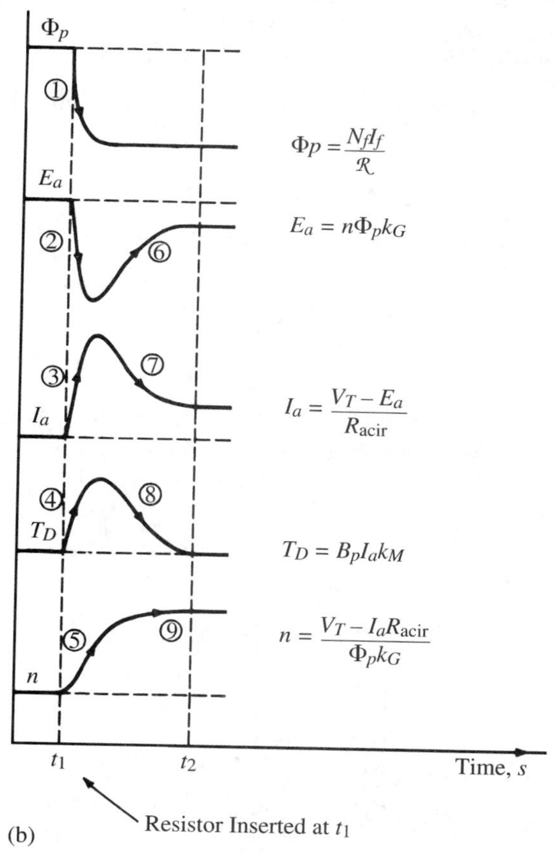

FIGURE 10.26
(a) Circuit for speed adjustment above base speed; (b) dynamic behavior of motor variables.

10.25 MECHANICAL POWER AND DEVELOPED TORQUE

The mechanical power developed by a DC motor is equal to the total power input to the armature circuit minus the copper losses in the armature circuit. Thus, referring to the shunt motor in Figure 10.27,

$$P_{mech} = V_T I_a - I_a^2 R_{acir} \quad \textbf{(10–27)}$$
$$R_{acir} = R_a + R_{IP} + R_{CW}$$

where:
P_{mech} = mechanical power developed (W)
V_T = applied voltage at terminals of motor (V)
I_a = armature current (A)
R_{acir} = resistance of armature circuit (Ω)
R_a = resistance of armature windings (Ω)
R_{IP} = resistance of interpole windings (Ω)
R_{CW} = resistance of compensating winding (Ω)

Applying Kirchoff's voltage law to the armature circuit in Figure 10.27, and neglecting the brush drop,

$$V_T = E_a + I_a R_{acir} \quad \textbf{(10–28)}$$

Multiplying both sides of Eq. (10–28) by I_a, substituting into Eq. (10–27), and simplifying,

$$V_T I_a = E_a I_a + I_a^2 R_{acir} \quad \textbf{(10–29)}$$

$$P_{mech} = E_a I_a \quad \textbf{(10–30)}$$

Examination of Eq. (10–30) indicates that mechanical work is done when the applied voltage causes current in the armature in a direction opposite to the cemf. This is somewhat analogous to an object moving at a uniform velocity, wherein the force of

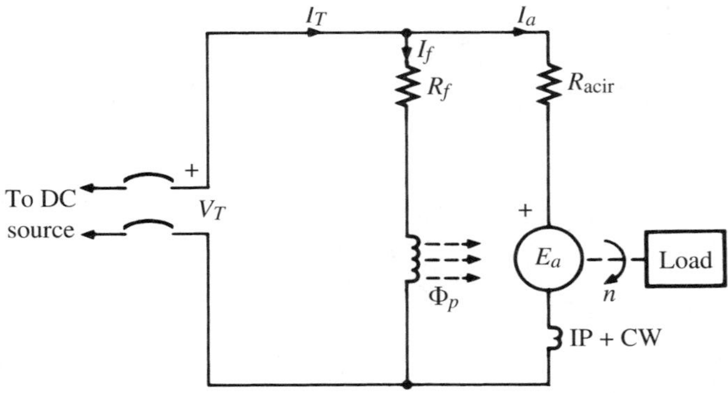

FIGURE 10.27
Circuit for derivation of mechanical power equation.

friction is equal and opposite to the applied force. Work is done by moving the object against the force of friction.

Motor Nameplates and NEMA

Motor nameplates and motor data supplied by manufacturers and the National Electrical Manufacturers Association (NEMA), are expressed in hp, r/min,[11] and lb-ft torque. Hence, these units will be used exclusively in all electric motor problems **[4]**.

The basic equation that relates horsepower to developed torque and rotor speed is

$$P_{\text{mech}} = \frac{T_D n}{5252} \quad \text{hp} \tag{10–31}$$

where: T_D = developed torque (lb-ft)
n = shaft speed (r/min)

Converting Eq. (10–30) to horsepower, substituting into Eq. (10–31), and solving for torque,

$$P_{\text{mech}} = \frac{E_a I_a}{746} \quad \text{hp} \tag{10–32}$$

$$\frac{E_a I_a}{746} = \frac{T_D n}{5252}$$

$$T_D = \frac{7.04 E_a I_a}{n} \quad \text{lb-ft} \tag{10–33}$$

Note: Equation (10–33) applies only to running machines. If the machine stalls or is prevented from starting (locked rotor), the speed is zero, the cemf is zero, and Eq. (10–33) reduces to

$$T_D = \frac{0}{0}$$

which is indeterminate. Torque at blocked rotor must be determined from Eq. (10–17), where,

$$T_D \propto B_p I_a$$

The presence of armature current and pole flux is positive proof that torque is developed, even if the speed is zero.

---

[11] Nameplates on machines used in the United States indicate r/min as RPM.

EXAMPLE 10.10 A 40-hp, 240-V, 2500 r/min shunt motor operating at rated conditions has a line current of 140 A. The armature-circuit resistance and field-circuit resistance are 0.0873 Ω and 95.3 Ω, respectively. Determine (a) the mechanical power developed; (b) torque developed; (c) shaft torque.

Solution

The circuit is shown in Figure 10.27.

(a)

$$I_f = \frac{V_T}{R_f} = \frac{240}{95.3} = 2.52 \text{ A}$$

$$I_a = I_T - I_f = 140 - 2.52 = 137.48 \text{ A}$$

From Kirchhoff's voltage law applied to the armature circuit,

$$E_a = V_T - I_a R_{acir} = 240 - 137.48 \times 0.0873 = 228.0 \text{ V}$$

$$P_{mech} = E_a I_a = 228.0 \times 137.48 = \underline{31{,}345 \text{ W}}$$

$$P_{mech} = \frac{E_a I_a}{746} = \frac{31{,}345}{746} = \underline{42.0 \text{ hp}}$$

(b)

$$T_D = \frac{7.04 E_a I_a}{n} = \frac{7.04 \times 31{,}345}{2500} = \underline{88.3 \text{ lb-ft}}$$

(c)

$$P = \frac{Tn}{5252} \quad \Rightarrow \quad T = \frac{P \times 5252}{n}$$

$$T_{shaft} = \frac{40 \times 5252}{2500} = \underline{84.0 \text{ lb-ft}}$$

10.26 LOSSES AND EFFICIENCY

Power-flow diagrams that illustrate the flow of power through a direct-current motor and a direct-current generator are shown in Figures 10.28(a) and (b), respectively. As indicated in the diagrams, the total power loss for the machine, whether operating as a generator or as a motor, is

$$P_{loss} = P_{acir} + P_b + P_{core} + P_{fcl} + P_{f,w} + P_{stray}$$

$$P_b = V_b I_a \qquad \textbf{(10–34)}$$

where:
P_{acir} = armature-circuit loss ($I_a^2 R_{acir}$), (W)
P_b = brush-contact loss (W)
P_{fcl} = field-circuit loss ($I_f^2 R_f$) (W)
P_{core} = core loss (W)
$P_{f,w}$ = friction and windage loss (W)
P_{stray} = stray-load loss (W)
V_b = brush-contact drop (V)

The core loss is the combined hysteresis and eddy-current losses in the armature and field iron. The stray-load loss includes losses in the coils undergoing commutation and losses due to eddy currents in the copper that are not accounted for in the calculated $I_a^2 R_a$ of the armature. Stray-load losses are approximately equal to 1 percent of the output power **[2]**.

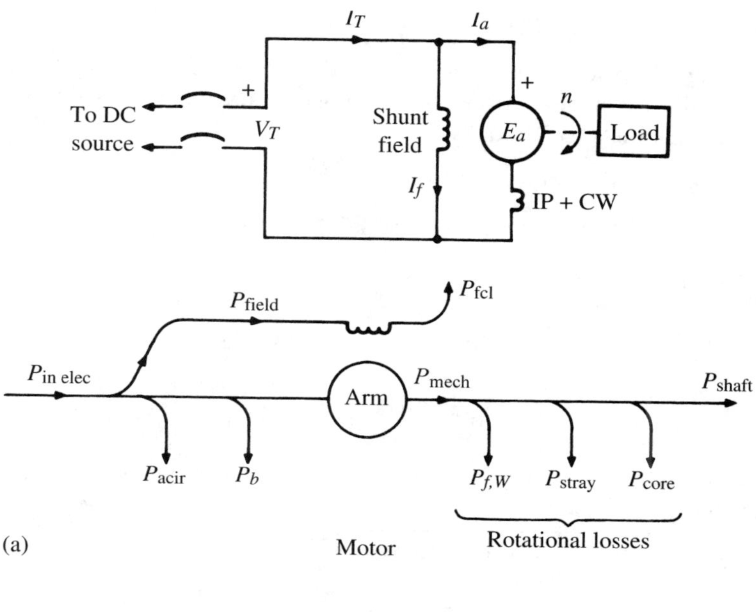

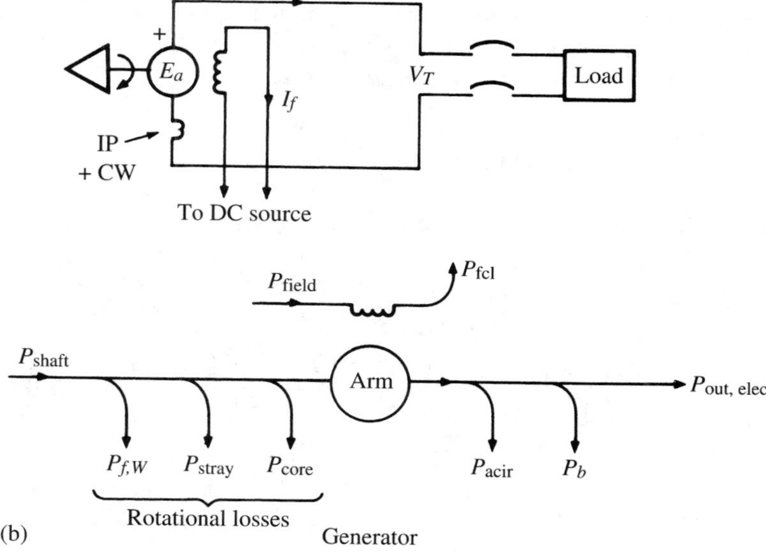

FIGURE 10.28
Power-flow diagrams: (a) shunt motor; (b) separately excited shunt generator.

Brush-Contact Drop

The voltage drop across the brushes, called *brush-contact drop* or *brush drop,* is relatively small. It includes brushes of both polarities and is assumed to have the following constant values for all loads[12]:

0.5 V for metal-graphite brushes

2.0 V for electrographitic and graphite brushes

Efficiency

The overall efficiency of a direct-current machine is given by

$$\eta = \frac{P_{\text{out}}}{P_{\text{in}}} \times 100 = \frac{P_{\text{out}}}{P_{\text{out}} + P_{\text{losses}}} \times 100 \qquad \textbf{(10–35)}$$

where η = efficiency in percent. If the efficiency is expressed in decimal form, it is called *per-unit* efficiency.

[12]For more information on brush materials and brush drop, see References **[2],[3]**.

EXAMPLE 10.11 A 150-hp, 240-V, 650 r/min shunt motor draws 420 A when operating at a reduced load of 124 hp. The brushes are graphite and the motor parameters are $R_a = 0.00872\ \Omega$, $R_{\text{IP}} + R_{\text{CW}} = 0.0038\ \Omega$, and $R_f = 32.0\ \Omega$. Determine (a) electrical losses; (b) rotational losses; (c) efficiency.

Solution

The circuit is shown in Figure 10.29.

(a)

$$I_f = \frac{V_T}{R_f} = \frac{240}{32.0} = 7.50 \text{ A}$$

$$I_a = I_T - I_f = 420 - 7.50 = 412.5 \text{ A}$$

$$P_f = I_f^2 R_f = (7.5)^2 \times 32.0 = 1800 \text{ W}$$

$$P_a + P_{\text{IP+CW}} = (412.5)^2(0.00872 + 0.0038) = 2130.36 \text{ W}$$

For graphite brushes, $V_b = 2.0$ V. Thus,

$$P_b = V_b I_a = 2.0 \times 412.5 = 825.0 \text{ W}$$

The total electrical losses are

$$P_f + P_{a+\text{CW}} + P_b = 1800.0 + 2130.36 + 825.0 = \underline{4755.4 \text{ W}}$$

(b)

$$V_T = E_a + I_a R_{\text{acir}} + V_b \quad \Rightarrow \quad 240 = E_a + 412.5 \times (0.00872 + 0.0038) + 2$$

$$E_a = 232.835 \text{ V}$$

$$P_{\text{mech}} = E_a I_a = 232.835 \times 412.5 = 96{,}044 \text{ W}$$

$$P_{\text{shaft}} = 124 \times 746 = 92{,}504 \text{ W}$$

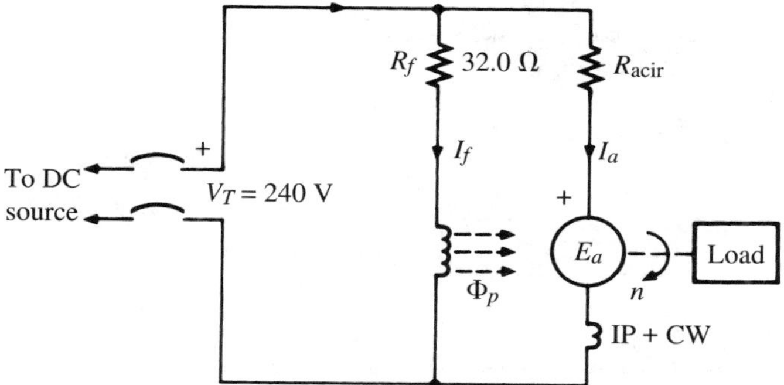

FIGURE 10.29
Circuit for Example 10.11.

From Figure 10.28(a),

$$P_{\text{rotational}} = P_{\text{mech}} - P_{\text{shaft}} = 96{,}044 - 92{,}504 = \underline{3540 \text{ W}}$$

(c)
$$\eta = \frac{P_{\text{out}}}{P_{\text{in}}} = \frac{92{,}504}{420 \times 240} \times 100 = \underline{91.8\%}$$

Alternatively,

$$\eta = \frac{P_{\text{out}}}{P_{\text{out}} + P_{\text{losses}}} = \frac{92{,}504}{92{,}504 + 3540 + 4755.4} \times 100 = \underline{91.8\%}$$

10.27 STARTING A DC MOTOR

When voltage is applied to a motor, the combined inertia of the motor and shaft load prevent instant rotation. The effect is the same as though the rotor were physically blocked. The values of current and developed torque that occur under these conditions are called the *blocked-rotor* or *locked-rotor* values.

The current drawn by the armature in Figure 10.30 is

$$I_a = \frac{V_T - E_a}{R_{\text{acir}}} \tag{10–36}$$

At locked-rotor conditions, however, the speed is zero. Hence, the cemf is zero and the current at locked rotor is

$$I_{a,\text{lr}} = \frac{V_T - 0}{R_{\text{acir}}} = \frac{V_T}{R_{\text{acir}}} \tag{10–37}$$

where $I_{a,\text{lr}}$ = armature current at locked rotor (A). To minimize copper losses, and hence increase efficiency, the armature is wound with conductors of relatively large cross section. Hence, to prevent excessively high armature currents when starting,

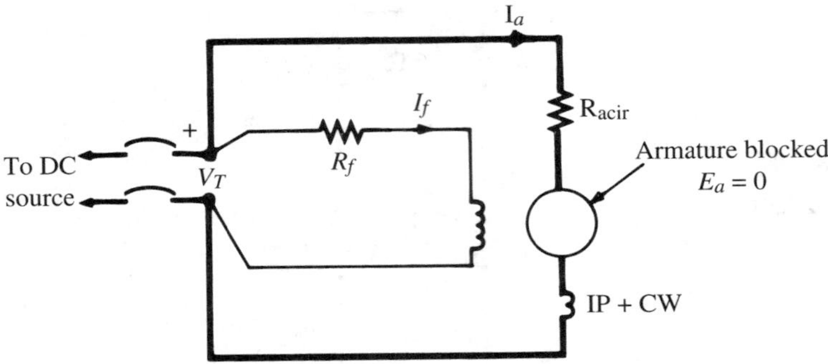

FIGURE 10.30
Circuit diagram for shunt motor with rotor blocked.

direct-current motors must have external resistors connected in series with the armature. Excessive armature current will cause destructive arcing and burning at the commutator–brush interface and the sudden impact to the driven equipment by the associated very high starting torque may cause mechanical damage to the driven equipment.

Manual Starter

With the exception of fractional horsepower motors, a DC motor requires the addition of external starting resistance to limit the current in the armature circuit when starting. This starting resistance must be left in the circuit until the cemf builds up sufficiently to reduce the armature current to a safe value. The value of starting resistance is generally selected to limit the current from 150 to 250 percent of rated value, depending on the starting torque required.

The shunt field is always connected across full-line voltage when starting, so that less armature current is needed to develop the required torque.

Figure 10.31 illustrates a rheostat type of manual starter connected to a shunt motor.[13] The circuit is closed when the lever is moved to the first contact, at which position all of the rheostat resistance is in series with the armature. As the machine accelerates, the lever should be moved slowly to the run position. The lever should not be left too long in an intermediate position or the starter will be damaged by overheating; the resistors are designed for intermittent duty, and should not be used for speed control.

The holding coil serves to hold the rheostat lever in the *run* position after all the resistance is cut out. It is connected in series with the shunt field of the motor, so that a break in the field circuit would de-energize the coil and permit the spring to pull the lever to the *off* position. The spring also returns the lever to the *off* position if there is a voltage failure or if the lever is left in some intermediate position.

---

[13]Automatic starters and motor controllers are discussed in Chapter 13.

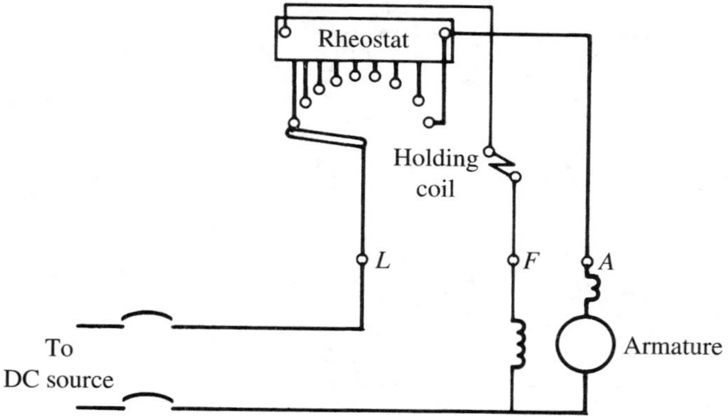

FIGURE 10.31
Connection diagram for a manually operated DC motor starter.

EXAMPLE 10.12 A 15-hp, 230-V, 1750 r/min shunt motor with a compensating winding draws 56.2 A when operating at rated conditions. The motor parameters are $R_{acir} = 0.280\ \Omega$ and $R_f = 137\ \Omega$. Determine (a) rated torque; (b) armature current at locked rotor if no starting resistance is used; (c) the external resistance required in the armature circuit that would limit the current and develop 200 percent rated torque when starting. (d) Assuming the system voltage drops to 215 V, determine the locked-rotor torque using the external resistor in (c).

Solution

(a) $$P = \frac{Tn}{5252} \quad \Rightarrow \quad T_{rated} = \frac{P \times 5252}{n} = \frac{15 \times 5252}{1750} = \underline{45.0 \text{ lb-ft}}$$

(b) The motor circuit is shown in Figure 10.32(a).

$$I_{a,lr} = \frac{V_T - E_a}{R_{acir}} = \frac{230 - 0}{0.280} = \underline{821.4 \text{ A}}$$

(c) The motor circuit with the starting resistor is shown in Figure 10.32(b).[14]

$$I_f = \frac{V_T}{R_f} = \frac{230}{137} = 1.68 \text{ A}$$

$$I_{a,rated} = I_{T,rated} - I_f = 56.2 - 1.68 = 54.52 \text{ A}$$

From Eq. (10–17), assuming constant flux density, the torque is proportional to the armature current. Hence, the armature current for 200 percent rated torque is

$$\frac{T_1}{T_2} = \frac{I_{a1}}{I_{a2}} \quad \Rightarrow \quad I_{a2} = I_{a1} \times \frac{T_2}{T_1} = 54.52 \times \frac{2T_1}{T_1} = \underline{109.0 \text{ A}}$$

---

[14]Practical motor controllers for large machines use two or more starting steps.

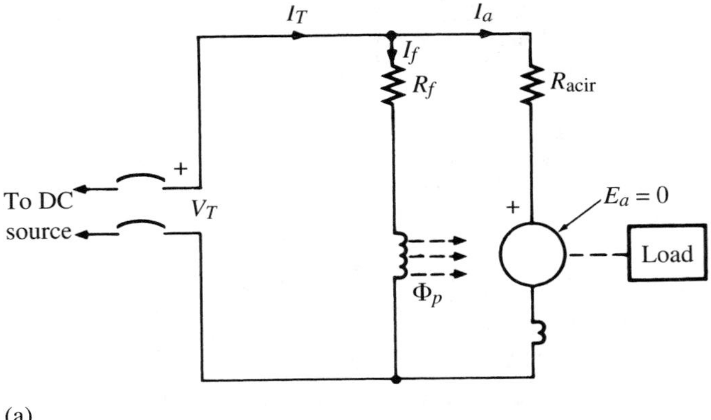

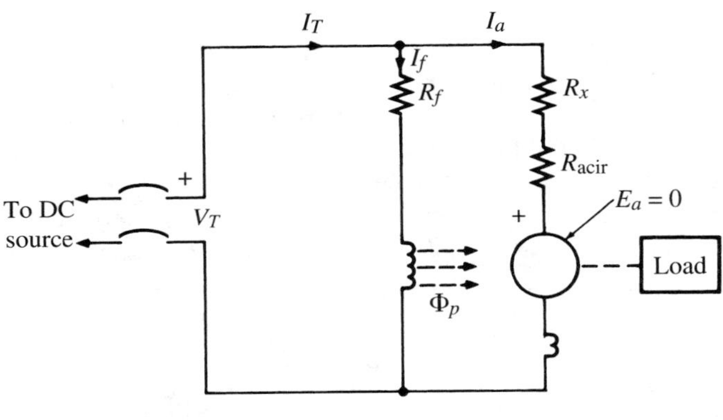

FIGURE 10.32
Circuits for Example 10.12: (a) without starting resistance; (b) with starting resistance.

Applying Kirchhoff's voltage law to the armature circuit with an external resistance in series with the armature, as shown in Figure 10.32(b),

$$V_T = E_a + I_a(R_{\text{acir}} + R_x) \qquad \Rightarrow \qquad R_x = \frac{V_T - E_a}{I_a} - R_{\text{acir}}$$

$$R_x = \frac{230 - 0}{109.0} - 0.280 = \underline{1.83\ \Omega}$$

(d) Since the magnetization curve for the specific machine is not available, it will be assumed that the flux density is proportional to the field current.

$$I_{f,215} = \frac{V_T}{R_f} = \frac{215}{137} = 1.57 \text{ A}$$

$$I_{a,215} = \frac{V_T - E_a}{R_{\text{acir}} + R_x} = \frac{215 - 0}{0.280 + 1.83} = 101.9 \text{ A}$$

Expressing Eq. (10–17) in terms of field current (reluctance assumed constant),

$$T_D \propto I_f I_a$$

Thus,

$$\frac{T_{D1}}{T_{D2}} = \frac{[I_f I_a]_1}{[I_f I_a]_2} \quad \Rightarrow \quad T_{D2} = T_{D1} \times \frac{[I_f I_a]_2}{[I_f I_a]_1}$$

Since the rotational losses such as windage and friction are not given, the rated torque will be taken as developed torque:

$$T_{D2} = 45.0 \times \frac{1.57 \times 101.9}{1.68 \times 54.52} = \underline{78.6 \text{ lb-ft}}$$

SUMMARY OF EQUATIONS FOR PROBLEM SOLVING

$$E_a = \frac{nPz_a\Phi_p}{60a} \quad \text{V} \tag{10–10}$$

$$E_a = n\Phi_p k_G \quad \text{V} \tag{10–12}$$

$$\Phi_p = \frac{N_f I_f}{\mathcal{R}} \quad \text{Wb} \tag{10–13}$$

$$E_a = \frac{nN_f I_f k_G}{\mathcal{R}} \quad \text{V} \tag{10–14}$$

$$I_f = \frac{E_{\text{bat}}}{R_f + R_{\text{rheo}}} \quad \text{A} \tag{10–15}$$

$$\text{VR} = \frac{V_{\text{nl}} - V_{\text{rated}}}{V_{\text{rated}}} \times 100 \quad \text{percent} \tag{10–16}$$

$$T_D = B_p I_a k_M \quad \text{lb-ft} \tag{10–17}$$

$$\text{SR} = \frac{n_{\text{nl}} - n_{\text{rated}}}{n_{\text{rated}}} \times 100 \quad \text{percent} \tag{10–20}$$

$$R_{\text{acir}} = R_a + R_{\text{IP}} + R_{\text{CW}} \quad \Omega \tag{10–24}$$

$$n = \left. \frac{V_T - I_a R_{\text{acir}}}{\Phi_p k_G} \right|_{\Phi_p \neq 0} \tag{10–26}$$

$$P_{\text{mech}} = E_a I_a \qquad \text{W} \tag{10–30}$$

$$P_{\text{mech}} = \frac{T_D n}{5252} \qquad \text{hp} \tag{10–31}$$

$$T_D = \frac{7.04 E_a I_a}{n} \qquad \text{lb-ft} \tag{10–33}$$

$$P_{\text{loss}} = P_{\text{acir}} + P_b + P_{\text{core}} + P_{\text{fcl}} + P_{f,w} + P_{\text{stray}} \qquad \text{W} \tag{10–34}$$

$$\eta = \frac{P_{\text{out}}}{P_{\text{in}}} \times 100 = \frac{P_{\text{out}}}{P_{\text{out}} + P_{\text{losses}}} \times 100 \tag{10–35}$$

SPECIFIC REFERENCES KEYED TO TEXT

1. Hubert, C. I. *Preventive Maintenance of Electrical Equipment,* Prentice-Hall, Upper-Saddle River, NJ, 2002.
2. Institute of Electrical and Electronic Engineers. *Test Procedures for Direct-Current Machines,* IEEE Standard 113-1985, IEEE, New York, 1985.
3. Kuhlmann, J. H. *Design of Electrical Apparatus.* Wiley, New York, 1940.
4. National Electrical Manufacturers Association. *Motors and Generators.* Publication No. MG-1-1998, NEMA, Rosslyn, VA, 1999.

GENERAL REFERENCES

Kloeffler, R. G., R. M. Kerchner, and J. L. Brenneman. *Direct-Current Machinery.* Macmillan, New York, 1949.

Langsdorf, A. S. *Principles of Direct-Current Machines.* McGraw-Hill, New York, 1959.

REVIEW QUESTIONS

1. What is the function of the commutator in (a) a DC generator; (b) a DC motor?
2. What is the neutral plane and where is it located?
3. Explain why the brushes of a DC machine must be located in the neutral plane.
4. What variables affect the torque developed in a DC machine?
5. What variables affect the voltage induced in a DC machine?
6. What apparatus is normally used to adjust the voltage of a shunt generator?
7. Is the induced emf in a motor or generator proportional to the field current? Explain.
8. Define voltage regulation.

9. Using appropriate sketches, explain in detail how torque is developed and voltage is generated in (a) motors; (b) generators.
10. What two methods can be used to reverse the direction of rotation of a shunt motor?
11. Using appropriate sketches, discuss the sequence of events that occur when loading a DC motor.
12. Using appropriate sketches, discuss the sequence of events that occur when removing load from a DC motor.
13. Define speed regulation.
14. How does the inductance of the armature affect the commutation process? Explain with the aid of sketches.
15. What are interpoles? Where are they located? How many are needed? What function do they serve?
16. What effect does saturation of the interpole iron have on the performance of DC generators when operating above rated load?
17. What effect does saturation of the interpole iron have on the performance of DC motors when operating above rated load?
18. Explain the effect of armature reaction on the performance of (a) DC generators; (b) DC motors.
19. What are compensating windings and what function do they serve? Explain how they work and why they are connected in series with the armature.
20. Do compensating windings eliminate the need for interpoles? Explain.
21. Under what conditions would it be necessary to shift the brush position of a modern DC motor? Explain.
22. (a) Sketch the complete equivalent circuit for a shunt generator and for a shunt motor, and label all parts. (b) Write Kirchhoff's voltage equation for the armature circuits in (a).
23. What is meant by the base speed of a DC motor?
24. Explain why inserting a resistor in series with the shunt field of a DC motor can cause an increase in speed.
25. Explain why a reduction in field flux will not always increase the speed of a DC motor.
26. Explain why inserting a resistor in series with the armature of a DC motor causes a decrease in speed.
27. Using appropriate sketches discuss the sequence of events that occurs when a resistor is placed in series with the armature of a running motor.
28. Using appropriate sketches discuss the sequence of events that occurs when a resistor is placed in series with the shunt field of a running motor.
29. What variables determine the torque at locked rotor?
30. Sketch the power-flow diagrams for a DC motor and for a DC generator. Label all losses.
31. What is brush-contact drop? What effect does it have on machine operation?
32. How may the efficiency of a DC motor be determined?
33. Explain why a resistor is needed in series with the armature when starting a DC motor, but is not needed when the motor is operating at its steady-state speed.

PROBLEMS

10–1/2 A 200-kW, eight-pole DC generator operating at 850 r/min has a generated emf of 200 V. Determine (a) the frequency of the induced emf; (b) the voltage and frequency if the speed is increased by 20 percent and the flux is decreased by 5 percent.

10–2/2 A 50-kW, 3500 r/min, 120-V, two-pole DC generator is operating at rated speed, rated air-gap flux, and no load. Determine the voltage and frequency of the generated voltage if the speed is increased by 15 percent and the flux is decreased by 6.2 percent.

10–3/7 The resistance of the shunt field winding and the resistance setting of the shunt field rheostat of a separately excited generator are 10.26 Ω and 14.23 Ω, respectively. The field circuit is connected to a 120-V DC source, and the magnetization curve for the machine is as shown in Figure 10.33. Sketch the equivalent circuit and determine (a) the induced emf; (b) the induced emf if a short in the rheostat caused the rheostat resistance to drop to 4.2 Ω.

10–4/7 A certain separately excited shunt generator has a shunt field winding resistance of 20.17 Ω and an armature resistance of 0.0014 Ω. The field winding is connected in series with a 40-Ω rheostat and supplied from a 240-V DC source. Assuming the magnetization curve for the machine is identical to that shown in Figure 10.33, determine (a) the induced emf if the rheostat is set to 0.0 Ω; (b) the setting of the rheostat (in ohms) required to reduce the generated voltage to 225 V.

10–5/8 A separately excited shunt generator rated at 1000 kW, 514 r/min, and 700 V has a no-load voltage of 725 V. Determine the voltage regulation.

10–6/8 A 100-kW, 1800 r/min shunt generator operating at rated load has a terminal voltage of 240 V. If the voltage regulation is 2.3 percent, determine the no-load voltage.

10–7/12 A 5-hp, 240-V, 1150 r/min shunt motor is operating at rated conditions. Determine the percent reduction in shunt field flux required to raise the speed to 1800 r/min. Assume the load on the shaft is adjusted to maintain rated armature current.

10–8/12 A 50-hp, 240-V, 650 r/min shunt motor operating at rated conditions draws a line current of 173 A. The motor has an armature resistance of 0.0705 Ω and a shunt field resistance of 81.63 Ω. If a reduction in load causes the armature current to drop to 70 percent of its rated value, determine (a) cemf; (b) speed; (c) line current.

10–9/12 A 5-hp, 120-V, 2500 r/min shunt motor draws 40 A when operating at rated conditions. Determine the new speed if the pole flux is reduced to 85 percent of its rated value and the load is adjusted so that the armature current is 28.6 A. $R_f = 63.8\ \Omega$ and $R_a = 0.252\ \Omega$.

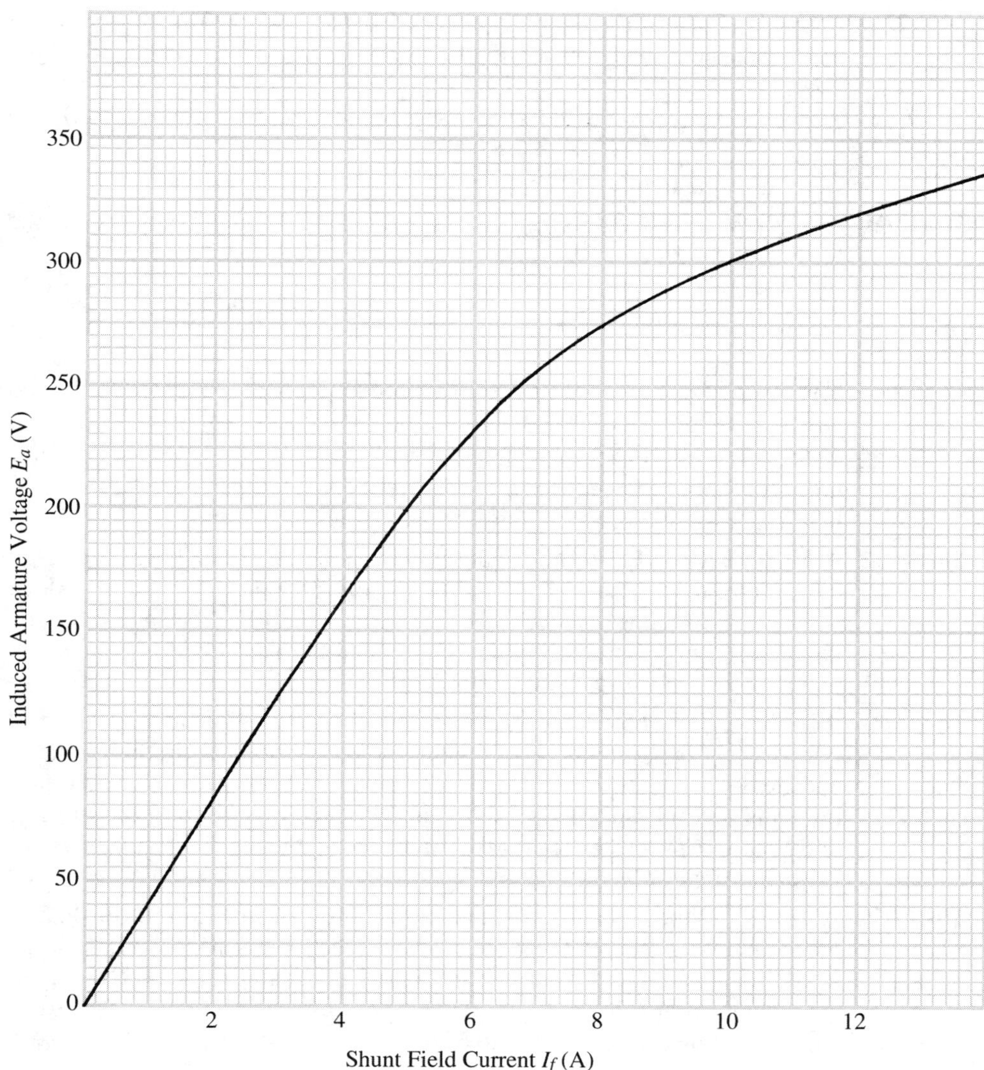

FIGURE 10.33
Magnetization curve for Problems 10–3/7 and 10–4/7.

10–10/12 A 15-hp, 240-V, 1150 r/min shunt motor draws 55 A when operating at rated conditions. The resistance of the shunt field windings is 109.1 Ω and the resistance of the armature is 0.364 Ω. Determine the no-load speed, assuming the total loss at no load is 970.6 W.

10–11/14 A 100-hp, 550-V, DC motor rated at 400 r/min has a speed regulation of 2.6 percent. Determine the speed if the load is removed.

10–12/20 A 50-kW, 1800 r/min, separately excited shunt generator operating at rated power has a terminal voltage of 248 V. The generator parameters are

$$R_a = 0.0487\ \Omega \qquad R_{\text{IP}} = 0.0111\ \Omega \qquad R_{\text{CW}} = 0.0125\ \Omega \qquad R_f = 75.6\ \Omega$$

Determine the induced emf.

10–13/20 A 125-kW, 250-V, 1450 r/min, separately excited shunt generator has the following parameters:

$$R_a = 0.0278\ \Omega \qquad R_{\text{IP}} = 0.0078\ \Omega \qquad R_f = 48.6\ \Omega$$

The machine is operating at part load with a terminal voltage of 260 V and an armature current of 300 A. Determine the induced emf.

10–14/22 A 240-V, 50-hp, 1150 r/min shunt motor is operating at rated speed. Determine (a) the new speed if the flux is reduced by 10 percent and the shaft load is adjusted to maintain rated armature current.

10–15/22 A 50-hp, 240-V, 1750 r/min shunt motor operating at rated conditions has a line current of 173 A. The armature-circuit resistance is 0.112 Ω and the shunt-field-circuit resistance is 70.2 Ω. If the motor has its shunt field flux reduced to 96 percent of its rated value and the shaft load is adjusted to maintain a constant speed, determine the new armature current.

10–16/22 A 60-hp, 240-V, 650 r/min compensated shunt motor has a line current of 206 A when operating at rated conditions. The armature-circuit resistance and field-circuit resistance are 0.084 Ω and 38.2 Ω, respectively. Determine the percent reduction in shunt field flux required to raise the speed to 1600 r/min. Assume the shaft load is reduced to a value that results in an armature current of 58.4 A.

10–17/22 A 120-V, 5-hp, 3500 r/min shunt motor draws a line current of 40.2 A when operating at rated conditions. The respective resistances of the armature circuit and shunt field circuit are 0.247 Ω and 66.4 Ω. Sketch the equivalent circuit and determine the speed if a reduction in shaft load causes the line current to drop to 32.1 A.

10–18/22 A 1-hp, 120-V, 1150 r/min shunt motor draws a line current of 9.5 A when operating at rated conditions. The respective resistances of the armature circuit and the shunt field circuit are 1.06 Ω and 252.6 Ω. Determine the resistance that must be connected in series with the armature to reduce the speed to 746 r/min, while developing the same torque.

10–19/22 A 240-V, 125-hp, 1150 r/min shunt motor operating at rated conditions, driving a constant-torque load, has a line current of 425 A. The armature-circuit resistance and field-circuit resistance are 0.0343 Ω and 47.1 Ω, respectively. Determine (a) the steady-state armature current if a 0.52-Ω resistor is connected in series with the armature, and a resistor inserted in series with the shunt field causes a 10 percent reduction in field flux; (b) the steady-state speed for the conditions in (a).

10–20/25 A DC shunt motor with a compensating winding draws 72.4 A from a 500-V bus when delivering 40 hp at 1740 r/min. The motor parameters are $R_a = 0.465\ \Omega$, $R_{\text{IP}} = 0.134\ \Omega$, $R_{\text{CW}} = 0.026\ \Omega$, and $R_f = 208.3\ \Omega$. Determine (a) cemf; (b) mechanical power developed; (c) developed torque; (d) shaft torque.

10–21/25 A 15-hp, 3500 r/min, 240-V shunt motor draws 54.4 A when operating at rated conditions. The machine parameters are $R_a = 0.112\ \Omega$, $R_{\text{IP}} = 0.036\ \Omega$, and $R_f = 177.2\ \Omega$. Determine (a) cemf; (b) mechanical power developed; (c) developed torque; (d) shaft torque.

10–22/25 A 30-hp, 1150 r/min, six-pole shunt motor has an efficiency of 88.5 percent when operating at rated load from a 240-V supply. The motor parameters are $R_a = 0.064\ \Omega$, $R_f = 93.6\ \Omega$, and $R_{\text{IP}} = 0.0323\ \Omega$. Determine (a) mechanical power developed; (b) developed torque; (c) shaft torque.

10–23/25 A 20-hp, 240-V, four-pole shunt motor operating at one-half rated load runs at 3042 r/min and has an efficiency of 86.0 percent. The motor parameters are $R_a = 0.0731\ \Omega$, $R_{\text{IP}} = 0.033\ \Omega$, and $R_f = 123\ \Omega$. Determine (a) input power to motor; (b) mechanical power developed; (c) shaft torque.

10–24/26 A 60-hp, 240-V, 600 r/min shunt motor draws 152 A when operating at 75 percent shaft power. The armature-circuit resistance and field resistance are $0.0482\ \Omega$ and $41.2\ \Omega$, respectively. The brushes are of electrographitic construction. Determine (a) electrical losses; (b) rotational losses; (c) efficiency.

10–25/26 A 150-kW, 250-V, 1750 r/min shunt generator is separately excited from a 240-V DC supply. The generator is operating at rated voltage and 50 percent load. The armature circuit has a resistance of $0.00728\ \Omega$ and the shunt field has a resistance of $64.0\ \Omega$. The machine uses metal–graphite brushes and has a shaft power input of 106 hp. Determine (a) electrical losses; (b) rotational losses; (c) efficiency.

10–26/27 A 300-hp, 500-V, 1750 r/min shunt motor, operating at rated conditions, has an efficiency of 92.0 percent. The motor parameters are $R_{\text{acir}} = 0.042\ \Omega$, $R_f = 86.2\ \Omega$. Determine (a) rated shaft torque; (b) mechanical power developed; (c) developed torque at rated load; (d) external resistance required in series with the armature circuit to limit the locked-rotor current to 225 percent rated armature current; (e) developed torque for the conditions in (d).

10–27/27 A 30-hp, 240-V, 1150 r/min shunt motor operating at rated conditions has an efficiency of 88.5 percent. The motor parameters are $R_a = 0.064\ \Omega$, $R_{\text{IP}} = 0.0323\ \Omega$, and $R_f = 93.6\ \Omega$. Determine (a) rated shaft torque; (b) mechanical power developed; (c) developed torque at rated load; (d) external resistance required in series with the armature circuit to limit the locked-rotor current to 175 percent rated armature current; (e) developed torque for the conditions in (d).

10–28/27 A 60-hp, 240-V, 850 r/min shunt motor has the following parameters: $R_a = 0.030\ \Omega$, $R_{IP} = 0.011\ \Omega$, and $R_f = 62.0\ \Omega$. The total losses when operating at rated conditions are 4.16 kW. Determine the external resistance required in series with the armature to limit the locked-rotor torque to 200 percent rated torque, if operating from a 220-V supply.

11

Direct-Current Motor Characteristics and Applications

11.1 INTRODUCTION

This chapter stresses the unique characteristics of shunt, series, and compound motors, and shows how the series field is used to shape the torque and speed characteristics of DC machines to meet the requirements of specific loads. Realistic calculations of motor speed using magnetization curves to account for magnetic saturation are compared with calculations using linear approximations. The adjustable-voltage method of speed control is introduced, along with applications for drive systems that require rapid reversing and speed adjustment over a wide range.

11.2 STRAIGHT SHUNT MOTORS

Straight shunt motors, previously discussed in Chapter 10, are essentially constant-speed machines. Slight weakening of the field due to armature-reaction effects, however, may cause a slight rise in speed with increasing load. Their principal fields of application include centrifugal pumps, fans, and other low starting-torque loads.

11.3 COMPOUND MOTORS

A compound motor has an additional winding, called a *series field* winding, wound with heavy copper conductor on top of the shunt field winding, as shown in Figure 11.1. The series field winding is connected in series with the armature as shown in Figure 11.2, so that its mmf will be proportional to the armature current, and in the *same direction* as the shunt field mmf.[1]

[1] Unless otherwise specified, compound motors are always assumed to be cumulative compound, that is, their shunt and series fields are additive.

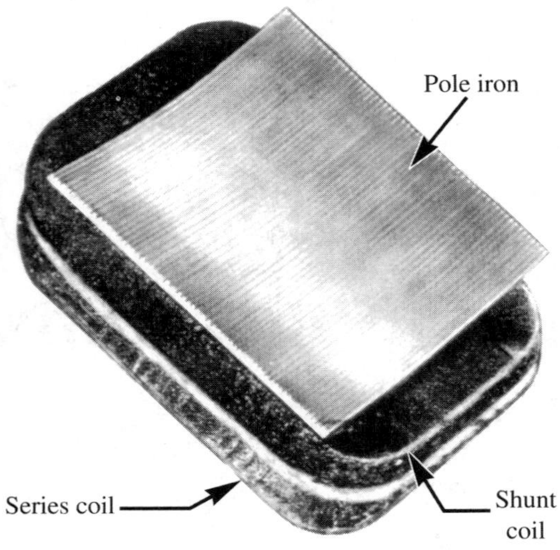

FIGURE 11.1
Pole iron with shunt and series coils. (Courtesy TECO Westinghouse)

Typical compound motors designed for industrial applications obtain approximately 50 percent of their mmf from the series field when operating at rated load.

Stabilized-Shunt Motor

Compound motors, whose series fields are designed to provide just enough mmf to nullify the equivalent demagnetizing mmf of armature reaction and provide a very slight speed droop, are called stabilized-shunt motors.[2] The series field windings of such machines generally have one-half to one and a half turns/pole, and depending on

[2] The series field does not eliminate armature reaction. Elimination of armature reaction requires a compensating winding. See Sections 10.17 and 10.19 of Chapter 10.

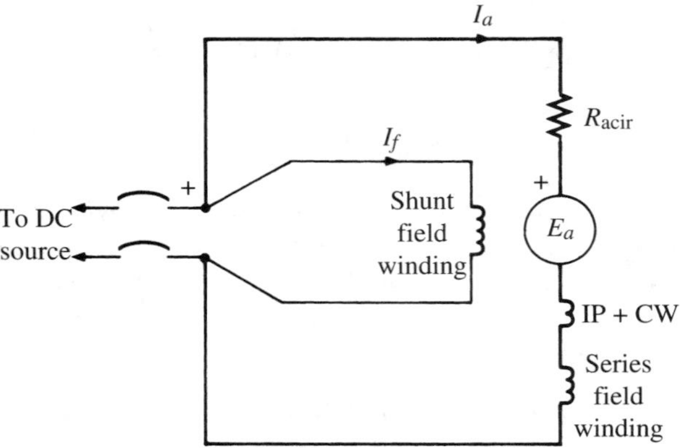

FIGURE 11.2
Circuit diagram for a compound motor.

the application, provide approximately 3 to 10 percent of the total field mmf at rated load. The speed of stabilized-shunt motors is fairly constant, with only a slight droop in speed with increasing load. Stabilized-shunt motors are used in applications that require a fairly constant speed and a moderate starting torque.

11.4 BEWARE THE DIFFERENTIAL CONNECTION

Connections to the series and shunt windings must be such as to cause their respective mmfs to be additive.[3] If the series field is accidentally reversed with respect to the shunt field (differential connection), its mmf will subtract from the shunt field mmf, causing the net flux to decrease with increasing load, resulting in excessive speed. Furthermore, because of the difference in inductive time constants, the series field current builds up faster than the shunt field current. Hence, the motor will start in the wrong direction, and then, depending on the load and the number of series field turns/pole, may (1) slow down and stop, causing the overcurrent device to trip the machine from the bus; (2) slow down, stop, reverse direction, and accelerate to dangerously high speeds; or (3) slow down, stop, reverse, slow down, stop, reverse, etc., cycling continuously until tripped off the bus.

11.5 REVERSING THE DIRECTION OF ROTATION OF COMPOUND MOTORS

Reversing the direction of rotation of compound or stabilized-shunt motors is accomplished by reversing the armature branch as shown in Figure 11.3, or reversing both the series field and the shunt field. Reversing only the series field or only the shunt

[3] See Section 11.12 to determine the correct connections for shunt and series windings.

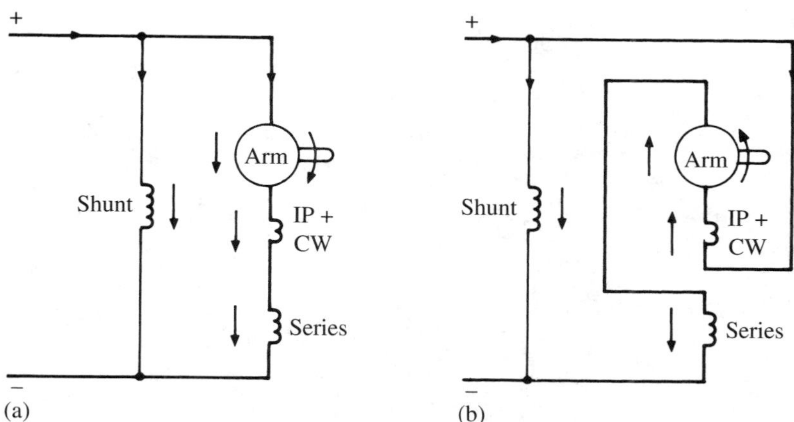

FIGURE 11.3
Reversing the direction of rotation of a compound motor by reversing the current in the armature, interpoles, and compensating winding.

field will result in a differential connection. The armature, interpoles, and compensating windings are treated as a single branch, and must be reversed as a unit. Reversing the armature without reversing the interpoles and compensating winding will cause severe arcing at the commutator–brush interface.

11.6 SERIES MOTOR

A series motor,[4] whose circuit is shown in Figure 11.4(a), is designed with a heavy series field, but has no shunt field to provide a base flux. With no shunt field to provide a base flux, a series motor will accelerate to damaging speeds if the shaft load is removed. Removing the load from a series motor causes $T_D > T_{\text{load}}$, resulting in an increase in motor speed. As the cemf starts to increase, it causes the armature current to decrease, which in turn causes the series field flux to decrease. The simultaneous opposing actions of increasing speed and decreasing flux, however, prevent the cemf from rising proportionately with speed. Thus, *I_a will decrease slowly with increasing speed.* Because the developed torque is proportional to $B_p I_a$, and because the armature current, and hence the flux density, do not decrease rapidly, loss of load will cause prolonged acceleration, resulting in destruction by centrifugal force.

Series motors must be connected directly to the load by solid couplings or gears, and the minimum load must be sufficient to limit the speed to a safe value. No belt drives are permitted.

Reversing the direction of rotation of a series motor is accomplished by reversing the current in the armature–interpole–compensating branch as shown in Figure 11.4(b), or reversing the current in the series field windings.

[4] See also universal motor in Section 7.9 of Chapter 7.

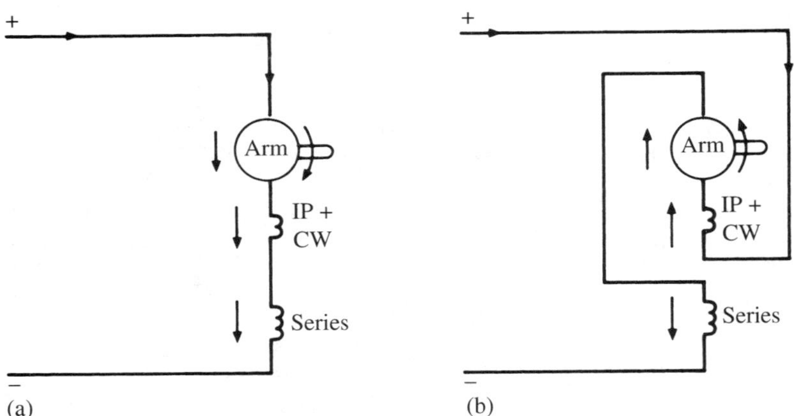

FIGURE 11.4
Reversing the direction of rotation of a series motor by reversing the current in the armature, interpoles, and compensating winding.

11.7 EFFECT OF MAGNETIC SATURATION ON DC MOTOR PERFORMANCE

Due to the effects of magnetic saturation, the pole flux is not directly proportional to the applied mmf. Hence, an accurate calculation of motor torque and motor speed for various operating conditions requires the use of a magnetization curve, such as that shown in Figure 11.5. Magnetization curves for specific machines are available from the manufacturer.

The net mmf that determines the pole flux includes components from the shunt field winding, series field winding, and armature reaction as applicable. For the general case, with all mmfs acting along the pole axis,

$$\mathscr{F}_{\text{net}} = \mathscr{F}_f + \mathscr{F}_s - \mathscr{F}_d \tag{11–1}$$

where: $\mathscr{F}_{\text{net}}$ = net mmf (A-t/pole)
$\mathscr{F}_f$ = shunt field mmf ($N_f I_f$) (A-t/pole)
$\mathscr{F}_s$ = series field mmf ($N_s I_a$) (A-t/pole)
$\mathscr{F}_d$ = equivalent demagnetizing mmf due to armature reaction (A-t/pole)

Note: Although not exact, $\mathscr{F}_d$ is assumed proportional to the armature current. If a compensating winding is used, however, $\mathscr{F}_d = 0$.

As previously shown in Chapter 10, the developed torque and speed of a DC motor can be determined from

$$T_D = B_p I_a k_M \tag{10–17}$$

$$n = \left. \frac{V_T - I_a R_{\text{acir}}}{\Phi_p k_G} \right|_{\Phi_p \neq 0} \tag{10–26}$$

$$R_{\text{acir}} = R_a + R_{\text{IP}} + R_{\text{CW}} + R_s$$

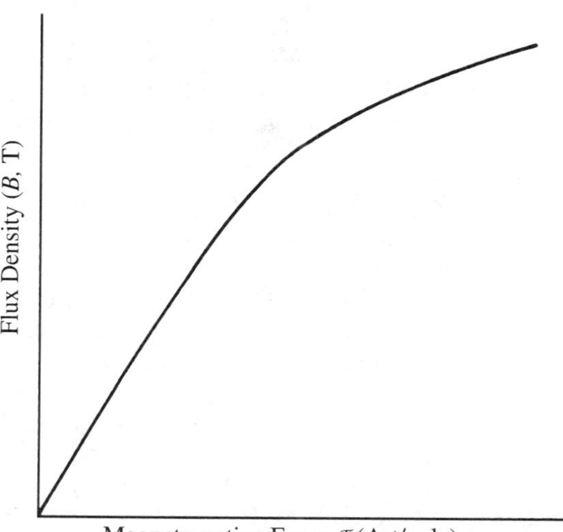

FIGURE 11.5
Typical magnetization curve for a DC motor.

where: R_{acir} = resistance of armature circuit (Ω)
R_a = resistance of armature windings (Ω)
R_{IP} = resistance of interpole windings (Ω)
R_{CW} = resistance of compensating windings (Ω)
R_s = resistance of series field winding (Ω)
B_p = air-gap flux density (T)
Φ_p = pole flux (Wb)

Constants k_G and k_M depend on the design of the machine, type of armature winding, and units used.

When solving problems to determine the effect of different values of pole flux and armature current on the developed torque and speed of the machine, it is convenient to use the following ratios based on Eqs. (10–17) and (10–26), respectively. Flux fringing is assumed to be negligible.

$$\frac{T_{D1}}{T_{D2}} = \frac{[B_p I_a]_1}{[B_p I_a]_2} \qquad \textbf{(11–2)}$$

$$\frac{n_1}{n_2} = \frac{\left[\dfrac{V_T - I_a R_{acir}}{\phi_p k_G}\right]_1}{\left[\dfrac{V_T - I_a R_{acir}}{\phi_p k_G}\right]_2}, \qquad \phi \neq 0 \qquad \textbf{(11–3)}$$

Subscripts 1 and 2 denote different sets of conditions.

Substituting $\phi_p = B_p \times A$ in Eq. (11–3), where A is the cross-sectional area of a pole, and simplifying,

$$\frac{n_1}{n_2} = \left[\frac{V_T - I_a R_{acir}}{B_p}\right]_1 \times \left[\frac{B_p}{V_T - I_a R_{acir}}\right]_2 \qquad \textbf{(11–4)}$$

EXAMPLE 11.1 A 240-V, 40-hp, 1150 r/min stabilized-shunt motor, operating at rated conditions, has an efficiency at rated load of 90.2 percent. The motor parameters are:

	Armature	Interpole	Series	Shunt
Resistance, Ω	0.0680	0.0198	0.00911	99.5
Turns/pole	—	—	$\frac{1}{2}$	1231

The circuit diagram and magnetization curve[5] for the motor are shown in Figure 11.6. Determine (a) the armature current when operating at rated conditions; (b) the resistance and power rating of an external resistance required in series with the shunt field circuit in order to operate at 125 percent rated speed. Assume the shaft load is adjusted to a value that limits the armature current to 115 percent of rated current.

---

[5] The motor parameters used in example problems and homework problems are, or closely approximate, the parameters of real motors. The magnetization curve shown in Figure 11.6 is hypothetical, but reasonable, and does indicate the general effect of magnetic saturation on motor performance.

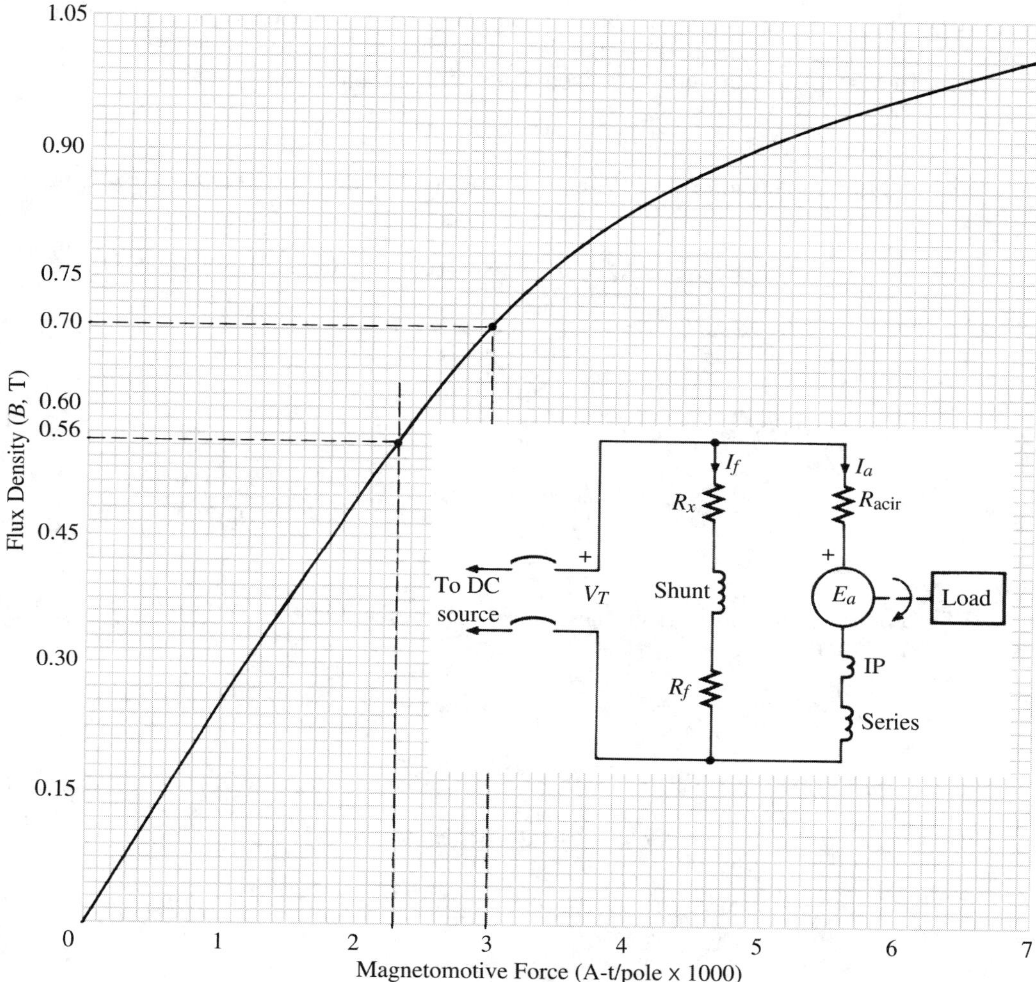

FIGURE 11.6
Magnetization curve and circuit diagram for Example 11.1.

Solution

(a) $P = V_T I_T \quad \Rightarrow \quad \dfrac{40 \times 746}{0.902} = 240 \times I_T \quad \Rightarrow \quad I_T = 137.84\ \text{A}$

$$I_f = \frac{V_T}{R_f} = \frac{240}{99.5} = 2.4121\ \text{A}$$

$$I_a = I_T - I_f = 137.84 - 2.41 = \underline{135.43\ \text{A}}$$

(b) The series field mmf of a stabilized-shunt motor is designed to be approximately equal and opposite to the equivalent demagnetizing mmf of armature reaction. Hence, the net flux in a stabilized-shunt motor is that due to the shunt field alone.

$$\mathscr{F}_{\text{net}} = \mathscr{F}_f = N_f I_f = 1231 \times 2.412 = 2969.2 \text{ A-t/pole}$$

From the magnetization curve in Figure 11.6, the flux density for a net mmf of 2969 A-t/pole is ≈0.70 T.

$$R_{\text{acir}} = R_a + R_{\text{IP}} + R_s = 0.0680 + 0.0198 + 0.00911 = 0.0969\ \Omega$$

$$\frac{n_1}{n_2} = \left[\frac{V_T - I_a R_{\text{acir}}}{B_p}\right]_1 \times \left[\frac{B_p}{V_T - I_a R_{\text{acir}}}\right]_2$$

$$B_{p2} = B_{p1} \times \frac{n_1}{n_2} \times \frac{[V_T - I_a R_{\text{acir}}]_2}{[V_T - I_a R_{\text{acir}}]_1}$$

$$B_{p2} = 0.70 \times \frac{1150}{1.25 \times 1150} \times \frac{240 - 135.43 \times 1.15 \times 0.0969}{240 - 135.43 \times 0.0969}$$

$$B_{p2} = 0.56 \text{ T}$$

The corresponding mmf from the magnetization curve in Figure 11.6 is $\mathscr{F}_f \approx 2.3 \times 1000 = 2300$ A-t/pole.

$$\mathscr{F}_f = N_f I_f \quad \Rightarrow \quad 2300 = 1231 \times I_f \quad \Rightarrow \quad I_f = 1.87 \text{ A}$$

$$I_f = \frac{V_T}{R_f + R_x} \quad \Rightarrow \quad R_x = \frac{V_T}{I_f} - R_f$$

$$R_x = \frac{240}{1.87} - 99.5 \approx \underline{28.8\ \Omega}$$

$$P_{R_x} = I_f^2 R_x = (1.87)^2 \times 28.8 \approx \underline{100.7 \text{ W}}$$

EXAMPLE 11.2 An 850 r/min, 125-hp, 240-V compound motor has the following parameters:

	Armature	**Interpole**	**Series**	**Shunt**
Resistance, Ω	0.0172	0.005	0.0023	49.2
Turns/pole	—	—	4.5	577

The circuit diagram and magnetization curve are shown in Figure 11.7. Approximately 10 percent of the series field mmf is used to offset the demagnetizing action of armature reaction. The efficiency at rated load is 85.4 percent, and the machine is driving a constant-torque load. Determine (a) shunt field current; (b) armature current; (c) developed torque; (d) the armature current if a resistor inserted in series with the shunt field caused the speed to increase to 900 r/min; (e) the external resistance required in series with the shunt field in order to obtain 900 r/min.

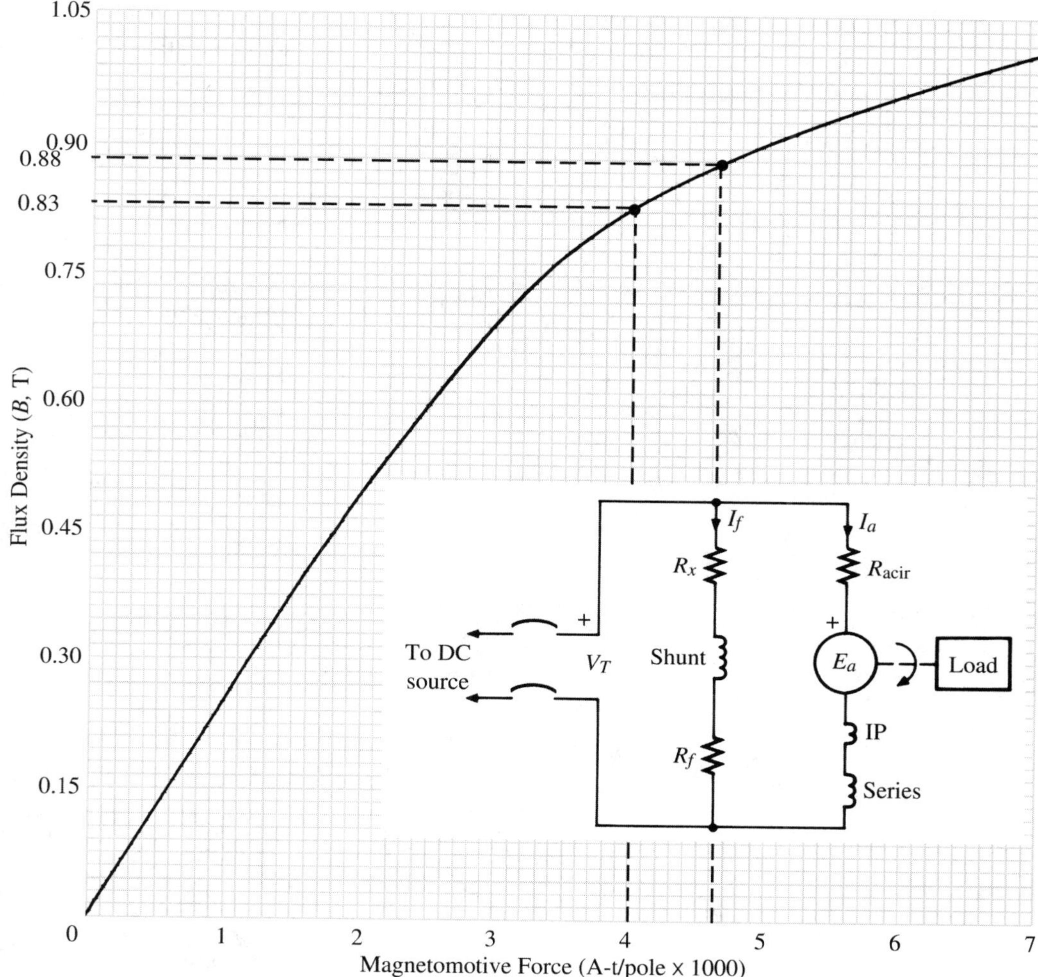

FIGURE 11.7
Magnetization curve and circuit diagram for Example 11.2.

Solution

(a)

$$I_f = \frac{V_T}{R_f} = \frac{240}{49.2} = \underline{4.88\text{ A}}$$

(b)

$$P_{in} = \frac{P_{out}}{\eta} \quad \Rightarrow \quad 240 \times I_T = \frac{125 \times 746}{0.854} \quad \Rightarrow \quad I_T = 454.97\text{ A}$$

$$I_a = I_T - I_f = 454.97 - 4.88 = \underline{450.09\text{ A}}$$

(c)

$$R_{\text{acir}} = R_a + R_{\text{IP}} + R_s = 0.0172 + 0.005 + 0.0023 = 0.0245\ \Omega$$

$$V_T = E_a + I_a R_{\text{acir}} \quad \Rightarrow \quad 240 = E_a + 450.09 \times 0.0245$$

$$E_a = 228.97\ \text{V}$$

$$P_{\text{mech}} = E_a I_a = 228.97 \times 450.09 = 103{,}057\ \text{W}$$

$$P = \frac{T \times n}{5252} \quad \Rightarrow \quad \frac{103{,}057}{746} = \frac{T \times 850}{5252}$$

$$T_D = \underline{853.6\ \text{lb-ft}}$$

(d)

$$\mathscr{F}_{\text{net}} = N_f I_f + N_s I_a - 0.10 N_s I_a$$

$$\mathscr{F}_{\text{net}} = 577 \times 4.88 + 4.5 \times 450.09(1 - 0.10) = 4638.6\ \text{A-t/pole}$$

From Figure 11.7, $B_{p1} \approx 0.88$ T.

$$\frac{T_{D1}}{T_{D2}} = \frac{[B_p I_a]_1}{[B_p I_a]_2} \quad \Rightarrow \quad 1 = \frac{0.88 \times 450.09}{B_{p2} \times I_{a2}}$$

$$I_{a2} = \frac{396.08}{B_{p2}} \tag{11–5}$$

Note: Equation (11–5) is specific to this problem.

$$\frac{n_1}{n_2} = \left[\frac{V_T - I_a R_{\text{acir}}}{B_p}\right]_1 \times \left[\frac{B_p}{V_T - I_a R_{\text{acir}}}\right]_2$$

$$B_{p2} = B_{p1} \times \frac{n_1}{n_2} \times \frac{[V_T - I_a R_{\text{acir}}]_2}{[V_T - I_a R_{\text{acir}}]_1}$$

$$B_{p2} = 0.88 \times \frac{850}{900} \times \frac{240 - 396.08 \times 0.0245/B_{p2}}{240 - 450.09 \times 0.0245}$$

$$B_{p2}^2 - 0.8711 B_{p2} + 0.0352 = 0$$

Using the quadratic formula,

$$B_{p2} = \frac{0.8711 \pm \sqrt{(-0.8711)^2 - 4 \times 0.0352}}{2}$$

$$B_{p2} \approx 0.83\ \text{T}, \qquad 0.043\ \text{T}$$

Substituting into Eq. (11–5)

$$I_{a2} = \frac{396.08}{0.83} = 477\ \text{A} \qquad I_{a2} = \frac{396.08}{0.043} = 9211\ \text{A}$$

Speaking "mathematically," the motor will operate at 900 r/min with a flux density of ≈0.83 T, or ≈0.043 T. A flux density of 0.043 T, however, will cause a prohibitively high armature current of 9211 A that is essentially equivalent to a short circuit; the motor will be severely damaged, if not totally destroyed, if protective devices do not "instantaneously" trip the machine from the line.

Given a choice between two values of flux density for a given load condition, the higher value of flux density will result in a lower armature current, and is the mandatory selection.

The value of B_{p2} may be approximated by using a simpler approach which assumes that the IR drop in the armature circuit will be very small for all load conditions up to rated load. Based on this assumption, the cemf will be essentially constant ($E_a = V_T - I_a R_{acir}$). From Eq. (10–13), Chapter 10,

$$I_a = \frac{nT_D}{7.04E_a}$$

Thus, with T_D constant, and E_a essentially constant, $I_a \propto n$.

$$\frac{n_2}{n_1} \approx \frac{I_{a2}}{I_{a1}} \quad \Rightarrow \quad I_{a2} = I_{a1} \cdot \frac{n_2}{n_1} = 450.09 \times \frac{900}{850} = 477 \text{ A}$$

Substituting into Eq. (11–5),

$$B_{p2} = \frac{396.08}{477} = 0.83 \text{ T}$$

(e) From the magnetization curve in Figure 11.7, the mmf ($\mathcal{F}_2$) corresponding to 0.83 T is ≈4000 A-t/pole.

$$\mathcal{F}_2 = N_f I_{f2} + N_s I_{a2} - 0.10 \times N_s I_{a2}$$

$$I_{f2} = \frac{\mathcal{F}_2 - 0.90 \times N_s I_{a2}}{N_f}$$

$$I_{f2} = \frac{4000 - 0.90 \times 4.5 \times 477}{577} \approx 3.58 \text{ A}$$

$$I_{f2} = \frac{V_T}{R_f + R_x} \quad \Rightarrow \quad R_x = \frac{V_T}{I_{f2}} - R_f$$

$$R_x = \frac{240}{3.58} - 49.2 \approx \underline{17.8\ \Omega}$$

EXAMPLE 11.3 A 100-hp, 650 r/min, 240-V series motor has an efficiency of 89.6 percent when operating at rated conditions. The series field has 14 turns/pole, and the equivalent demagnetizing mmf due to armature reaction is approximately 8.0 percent of the series field mmf. The motor parameters are:

	Armature	Interpole	Series
Resistance, Ω	0.0202	0.00588	0.00272

The circuit diagram and magnetization curve for the motor are shown in Figure 11.8. Determine the speed if the load is reduced to a value that causes the armature current to be 30 percent rated current.

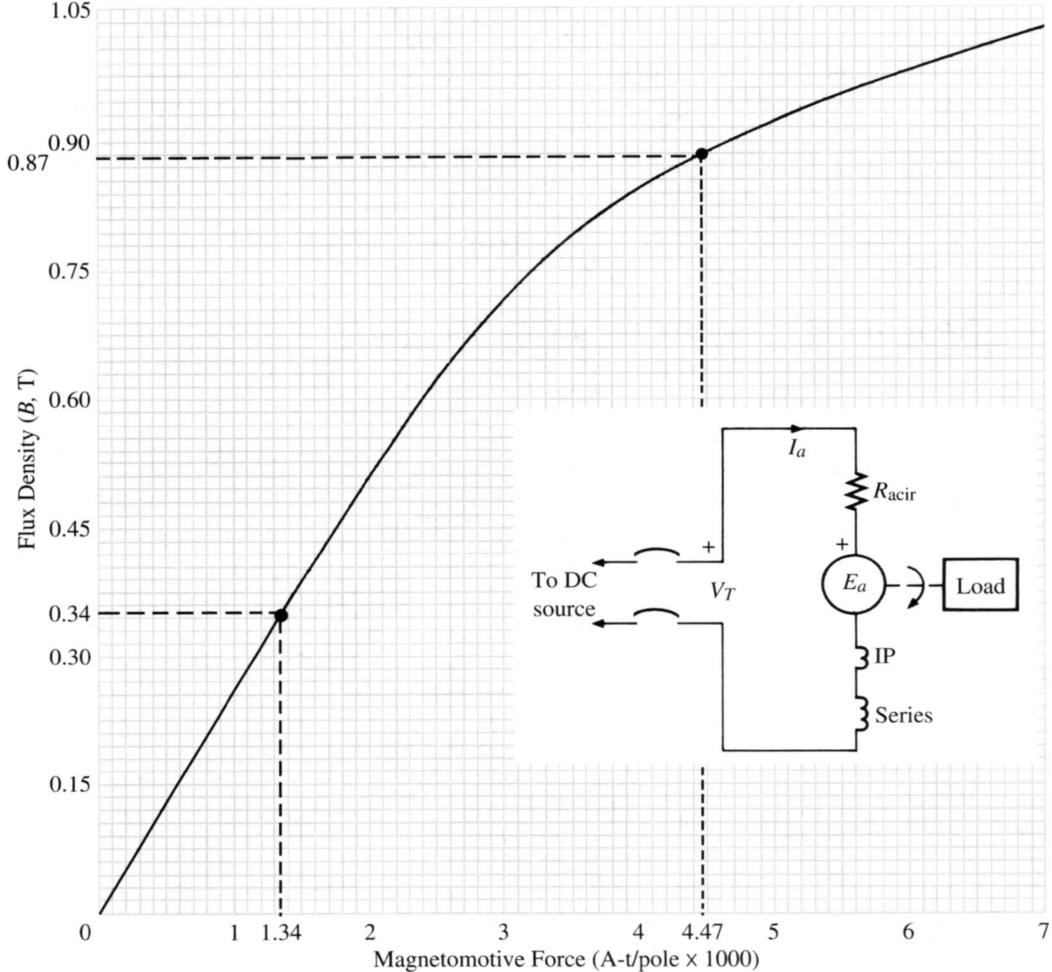

FIGURE 11.8
Magnetization curve and circuit diagram for Example 11.3.

Solution

$$P = V_T I_T = \frac{100 \times 746}{0.896} = 240 \times I_T \qquad \Rightarrow \qquad I_a = I_T = 346.91 \text{ A}$$

$$R_{\text{acir}} = 0.0202 + 0.00588 + 0.00272 = 0.0288\ \Omega$$

$$\mathcal{F}_{\text{net1}} = \mathcal{F}_s - \mathcal{F}_d = N_s I_a(1 - 0.080) = 14 \times 346.91 \times 0.92 = 4468.2 \text{ A-t/pole}$$

From the magnetization curve in Figure 11.8, $B_{p1} \approx 0.87$ T.

$$\mathcal{F}_{\text{net2}} = 0.30 \times \mathcal{F}_{\text{net1}} = 0.30 \times 4468.2 = 1340.5 \text{ A-t/pole}$$

From the magnetization curve, $B_{p2} \approx 0.34$ T.

$$\frac{n_1}{n_2} = \left[\frac{V_T - I_a R_{acir}}{B_p}\right]_1 \times \left[\frac{B_p}{V_T - I_a R_{acir}}\right]_2$$

$$\frac{650}{n_2} = \left[\frac{240 - 346.91 \times 0.0288}{0.87}\right] \times \left[\frac{0.34}{240 - 0.30 \times 346.91 \times 0.0288}\right]$$

$$n_2 = \underline{1714 \text{ r/min}}$$

11.8 LINEAR APPROXIMATIONS

If the specific magnetization curve is not available for the solution of problems involving changes in the magnetic field, a *rough approximation* may be obtained by assuming saturation effects to be negligible. It should be understood, however, that the series field of heavily compounded motors can drive the magnetic field into deep saturation, especially under locked-rotor and high-overload conditions. Thus, speed and torque calculations involving locked rotor and high overloads of compound motors will be substantially in error if saturation effects are ignored.

If the conditions of a problem indicate that the net mmf is to be reduced below its rated value, an approximation using a linear assumption will provide information as to the approximate value of required rheostat resistance.

A linear approximation of the torque and speed equations is obtained by expressing the flux and flux density in terms of the net mmf. Thus, assuming, saturation effects are negligible, Eqs. (11–2) and (11–4) can be expressed as

$$\frac{T_{D1}}{T_{D2}} = \frac{[\mathcal{F}_{net} I_a]_1}{[\mathcal{F}_{net} I_a]_2} \tag{11–6}$$

$$\frac{n_1}{n_2} = \left[\frac{V_T - I_a R_{acir}}{\mathcal{F}_{net}}\right]_1 \times \left[\frac{\mathcal{F}_{net}}{V_T - I_a R_{acir}}\right]_2, \qquad \phi \neq 0 \tag{11–7}$$

Series Motor

If the range of series-motor operation is in the unsaturated region, and armature reaction effects are negligible or compensated for, then Eq. (11–6) for the series motor becomes

$$\begin{aligned} T_{D,series} &\propto \mathcal{F}_{net} I_a = N_s I_a \cdot I_a \\ T_{D,series} &\propto I_a^2 \end{aligned} \tag{11–8}$$

As indicated in Eq. (11–8), for the conditions specified *the torque developed by a series motor is essentially proportional to the square of the armature current.*

EXAMPLE 11.4 Example 11.1 is re-solved using a linear approximation, and the solution is compared with the results obtained in Example 11.1.

Solution

Using the data and preliminary calculations in Example 11.1,

$$I_{f1} = \frac{V_T}{R_f} = \frac{240}{99.5} = 2.412 \text{ A}$$

$$I_{a1} = I_T - I_f = 137.84 - 2.412 = 135.43 \text{ A}$$

$$\mathcal{F}_{\text{net}} = \mathcal{F}_f = 1231 \times 2.412 = 2969.2 \text{ A-t/pole}$$

$$\frac{n_1}{n_2} = \left[\frac{V_T - I_a R_{\text{acir}}}{\mathcal{F}_{\text{net}}}\right]_1 \times \left[\frac{\mathcal{F}_{\text{net}}}{V_T - I_a R_{\text{acir}}}\right]_2$$

$$\mathcal{F}_{\text{net2}} = \mathcal{F}_{\text{net1}} \times \frac{n_1}{n_2} \times \frac{[V_T - I_a R_{\text{acir}}]_2}{[V_T - I_a R_{\text{acir}}]_1}$$

$$\mathcal{F}_{\text{net2}} = 2969.2 \times \frac{1150}{1.25 \times 1150} \times \frac{240 - 135.43 \times 1.15 \times 0.0969}{240 - 135.43 \times 0.0969}$$

$$\mathcal{F}_{\text{net2}} = 2354.8 \text{ A-t/pole}$$

$$I_f = \frac{\mathcal{F}_{\text{net}}}{N_f} = \frac{2354.8}{1231} = 1.91 \text{ A}$$

$$I_f = \frac{V_T}{R_f + R_x} \quad \Rightarrow \quad R_x = \frac{V_T}{I_f} - R_f$$

$$R_x = \frac{240}{1.91} - 99.5 = \underline{26.15 \ \Omega}$$

The rheostat resistance determined in Example 11.1 using the magnetization curve is 28.8 Ω. The error due to linear approximation is

$$\frac{28.8 - 26.15}{28.8} \times 100 \approx 9.2\%$$

The lower calculated value of rheostat resistance as determined through linear approximation would cause a slightly higher field current, and hence a slightly lower speed than the desired $1.25 \times 1150 = 1437.5$ r/min.

EXAMPLE 11.5 Example 11.2 is re-solved using a linear approximation to show the gross error that occurs when a linear assumption is applied to compound motors operating at overload conditions.

Solution

Using the data and preliminary calculations in Example 11.2,

$$\mathcal{F}_{\text{net,1}} = N_f I_f + N_s I_a - 0.10 \times N_s I_a$$

$$\mathcal{F}_{\text{net,1}} = 577 \times 4.88 + 4.5 \times 450.09 \times 0.90 = 4638.6 \text{ A-t/pole}$$

$$\frac{T_{D1}}{T_{D2}} = \frac{[\mathscr{F}_{\text{net}} I_a]_1}{[\mathscr{F}_{\text{net}} I_a]_2} \quad \Rightarrow \quad 1 = \frac{4638.6 \times 450.09}{\mathscr{F}_2 \times I_{a2}}$$

$$I_{a2} = \frac{2{,}087{,}787.5}{\mathscr{F}_2}$$

$$\frac{n_1}{n_2} = \left[\frac{V_T - I_a R_{\text{acir}}}{\mathscr{F}_{\text{net}}}\right]_1 \times \left[\frac{\mathscr{F}_{\text{net}}}{V_T - I_a R_{\text{acir}}}\right]_2$$

$$\mathscr{F}_2 = \mathscr{F}_1 \times \frac{n_1}{n_2} \times \frac{[V_T - I_a R_{\text{acir}}]_2}{[V_T - I_a R_{\text{acir}}]_1}$$

$$\mathscr{F}_2 = 4638.6 \times \frac{850}{900} \times \frac{240 - 2{,}087{,}787.5 \times 0.0245/\mathscr{F}_2}{240 - 450.09 \times 0.0245}$$

$$\mathscr{F}_2^2 - 4591.9\mathscr{F}_2 + 978{,}660 = 0$$

Using the quadratic formula,

$$\mathscr{F}_2 = \frac{4591.9 \pm \sqrt{(-4591.9)^2 - 4 \times 978{,}660}}{2}$$

$$\mathscr{F}_2 = 4367.8 \text{ A-t/pole}, \qquad 224.1 \text{ A-t/pole}$$

Taking the larger mmf, which results in a smaller armature current,

$$\mathscr{F}_{\text{net},2} = \underline{4367.8 \text{ A-t/pole}}$$

$$I_{a2} = \frac{2{,}087{,}787.5}{\mathscr{F}_2} = \frac{2{,}087{,}787.5}{4367.8} = 478.0 \text{ A}$$

$$\mathscr{F}_{\text{net},2} = N_f I_f + N_s I_a \times 0.90 \quad \Rightarrow \quad I_{f2} = \frac{\mathscr{F}_{\text{net}} - N_s I_a \times 0.90}{N_f}$$

$$I_{f2} = \frac{4367.8 - 4.5 \times 478.0 \times 0.90}{577} = 4.21 \text{ A}$$

$$I_{f2} = \frac{V_T}{R_f + R_x} \quad \Rightarrow \quad R_x = \frac{V_T}{I_{f2}} - R_f$$

$$R_x = \frac{240}{4.21} - 49.2 = \underline{7.74\ \Omega}$$

The rheostat resistance determined in Example 11.2 using the magnetization curve is 17.8 Ω. The error introduced by linear approximation is

$$\frac{17.8 - 7.74}{17.8} \times 100 \approx 56.5\%$$

11.9 COMPARISON OF STEADY-STATE OPERATING CHARACTERISTICS OF DC MOTORS

The steady-state operating characteristics of typical shunt, compound, and series motors, of the same torque and speed ratings, operating from a *constant-voltage* DC supply, are shown in Figure 11.9.

Shunt Motor

As indicated by its operating characteristics, the speed of a shunt motor is relatively constant from no load to rated load and is attributed to the essentially constant flux provided by the shunt field. The speed regulation of the shunt motor is approximately 5 percent. The internal torque developed by a shunt motor is dependent on the flux

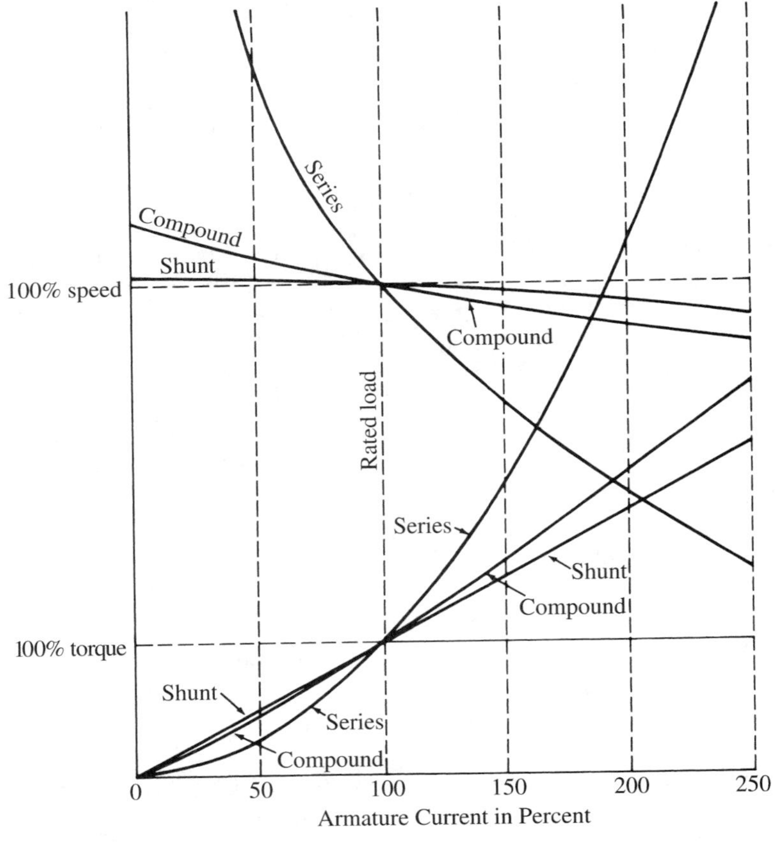

FIGURE 11.9
Steady-state speed and torque characteristics of typical shunt, series, and compound motors.

density and the armature current. Hence, the constant field of the shunt machine will cause the developed torque to vary almost linearly with the armature current. The shunt motor has its greatest field of application with loads that require essentially constant speed, and where high starting torques are not needed. Shunt motors are used to drive centrifugal pumps, fans, winding reels, conveyors, machine tools, and other loads of similar characteristics.

Compound Motor

The torque that a compound motor develops for values of armature current above the rated value is much higher than that of a shunt machine with the same rated power and base speed. The higher torque values of the compound machine, however, are accompanied by lower speeds. The speed regulation of compound motors is generally between 15 and 25 percent. The compound motor has its greatest application with loads that require high starting torques or have pulsating loads. They are used to drive electric shovels, metal-stamping machines, reciprocating pumps, hoists, compressors, and other loads of similar characteristics. Compound motors smooth out the energy demand required of a pulsating load. The motor slows down during the work stroke, giving up the energy stored in its moving parts, and at the same time developing a greater internal torque. During the recovery stroke, the motor accelerates and restores energy to the moving parts. Thus, the demand from the electrical system is less than if a shunt motor were used. A shunt motor would tend to maintain the same speed during the work stroke, causing excessive current demands from the electrical system.

Series Motor

The high ratio of light-load to full-load speed, coupled with its high starting torque, makes the series motor adaptable to driving hoists, electric locomotives, and other loads of similar characteristics. As indicated by its speed curve, completely removing the load from a series motor will cause it to "run away." Hence, series motors must be directly connected to the load by gears or a solid coupling and should not be applied to loads that can be removed or reduced sufficiently to cause overspeeding.

11.10 ADJUSTABLE-VOLTAGE DRIVE SYSTEMS

An adjustable-voltage drive system provides acceleration, deceleration, and speed control for *shunt* and *stabilized-shunt* motors by providing an adjustable voltage to the armature, while maintaining a constant voltage supply to the shunt field circuit.

An elementary circuit diagram of a simple adjustable-voltage drive system, called the *Ward–Leonard* system, is shown in Figure 11.10 for a small vessel. The prime mover driving the generator may be a diesel engine, turbine, or other machine that can provide an essentially constant speed. A small generator (not shown) supplies the excitation bus. The generator rheostat provides speed control from a creeping speed to the base speed of the motor, and also provides a means for reversing the rotation by reversing the polarity of the generator. The broken lines indicate the position

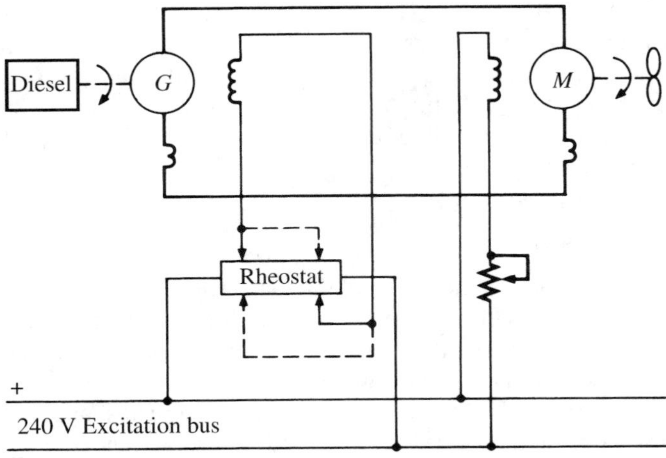

FIGURE 11.10
Elementary circuit diagram of a simple adjustable-voltage drive system.

of the sliding contacts for opposite rotation of the motor. The two sliding contacts are moved simultaneously in different directions by a single operating lever. The armature current is reduced to zero by centering the sliding contacts.

Although obsolete in terms of more modern methods of control, the Ward–Leonard system is still used in some of the older mine hoists, electric shovels, mills, etc. Drive systems for new plants and modernization of older plants use solid-state control.

EXAMPLE 11.6 The adjustable-voltage DC drive system shown in Figure 11.10 has an armature rated at 750 V, 1600 hp, 955 r/min. The armature current and field current at rated load are 1675 A and 5.20 A, respectively. The motor has a compensating winding, and the separately excited shunt field is fed through a rheostat from the 240-V excitation bus. The motor parameters are:

	Armature	**IP + CW**	**Shunt**
Resistance, Ω	0.00540	0.00420	14.70

The power required by the propeller load varies as the cube of the speed. Determine (a) torque developed when operating at rated speed; (b) developed torque required at half rated-speed; (c) armature voltage required for half rated-speed, assuming rated field current.

Solution

(a)

$$R_{acir} = R_a + R_{IP+CW} = 0.00540 + 0.00420 = 0.00960\ \Omega$$

$$V_T = E_a + I_a R_{acir} \quad \Rightarrow \quad 750 = E_a + 1675 \times 0.00960$$

$$E_a = 733.92\ \text{V}$$

$$P_{\text{mech}} = E_a I_a = 733.92 \times 1675 = 1{,}229{,}316\ \text{W}$$

$$P = \frac{T \times n}{5252} \quad \Rightarrow \quad \frac{1{,}229{,}316}{746} = \frac{T \times 955}{5252}$$

$$T_D = \underline{9062.5\ \text{lb-ft}}$$

(b)

$$P \propto n^3 \quad \Rightarrow \quad T \times n \propto n^3$$

$$\therefore T \propto n^2$$

$$\frac{T_1}{T_2} = \left(\frac{n_1}{n_2}\right)^2 \quad \Rightarrow \quad T_2 = T_1 \times \left(\frac{n_2}{n_1}\right)^2$$

$$T_2 = 9062.5 \times \left(\frac{0.5n_1}{n_1}\right)^2 = \underline{2265.6\ \text{lb-ft}}$$

(c)

$$\frac{T_1}{T_2} = \frac{[B_p I_a]_1}{[B_p I_a]_2} \quad \Rightarrow \quad \frac{9062.5}{2265.6} = \frac{B_p \times 1675}{B_p \times I_{a2}}$$

$$I_{a2} = 418.75\ \text{A}$$

$$\frac{n_1}{n_2} = \left[\frac{V_T - I_a R_{\text{acir}}}{\Phi}\right]_1 \times \left[\frac{\Phi}{V_T - I_a R_{\text{acir}}}\right]_2$$

$$V_2 = \frac{n_2}{n_1} \times \frac{\Phi_2}{\Phi_1} \times [V_T - I_a R_{\text{acir}}]_1 + I_{a2} R_{\text{acir}\,2}$$

$$V_2 = \frac{0.5n_1}{n_1} \times \frac{\Phi_1}{\Phi_1} \times [750 - 1675 \times 0.0096] + 418.75 \times 0.0096$$

$$V_2 = \underline{371.0\ \text{V}}$$

11.11 DYNAMIC BRAKING, PLUGGING, AND JOGGING

Direct-current motors may be decelerated quickly by converting the energy stored in the moving masses to electrical energy and dissipating it as heat through resistors. To do this, the motor armature is disconnected from the supply lines and connected across a suitable resistor while the shunt field is maintained at full strength. The motor behaves as a generator and feeds current to the resistor, dissipating heat at a rate equal to I^2R. The heat power rating of the resistor is determined by the material used and its physical dimensions. The value of R is generally selected to provide an armature current that will approximate 150 to 300 percent rated current. In accordance with Lenz's law, the armature current produced by dynamic braking will be in a direction to oppose the motion of the armature. It is this negative torque, or countertorque, that slows the machine. Dynamic braking is also very useful for limiting the speed of overhauling loads, such as would be produced by lowering heavy loads on elevators and winches or by electric trains going downgrade. Since the energy is expended in resistors, this type of braking is sometimes referred to as *resistive braking*.

A dynamic-braking circuit for a compound motor is shown in Figure 11.11. Figure 11.11(a) shows normal motor operations; contactors M1 and M2 are closed and

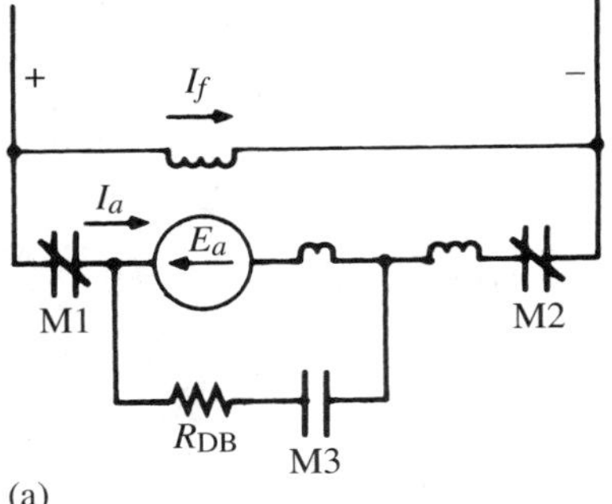

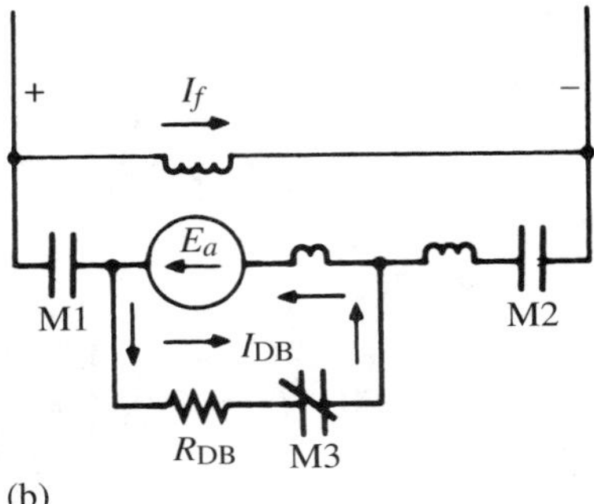

FIGURE 11.11
Compound motor with a dynamic-braking loop: (a) normal motor operation; (b) dynamic-braking loop.

contactor M3 is open. To apply dynamic braking, contactors M1 and M2 are opened and contactor M3 is closed. This is shown in Figure 11.11(b). Note that *the series field is not included in the dynamic-braking (DB) loop.* To include it would result in a differential connection; the current in the series field would set up an mmf in opposition to the shunt field mmf, causing a lower E_a, and hence a smaller braking effect.

When sufficiently slowed, the motor circuit is opened, and a mechanical brake is applied to bring the machine to a full stop and to prevent further rotation. A typical magnetically released spring-loaded holding brake is shown in Figure 11.12.

Regenerative Braking

Regenerative braking converts the energy of overhauling loads into electrical energy and pumps it back into the electrical system. An overhauling load will drive a DC motor faster than it would normally run with a given applied voltage. This will cause its cemf to become greater than the applied voltage and result in generator action. Regenerative braking is used extensively in electrified railroads and to some extent with elevators and winches. Regenerative braking cannot occur in the series motor unless the field is connected across the line with a current-limiting resistor in series, or by separately exciting the series field from a low-voltage DC generator.

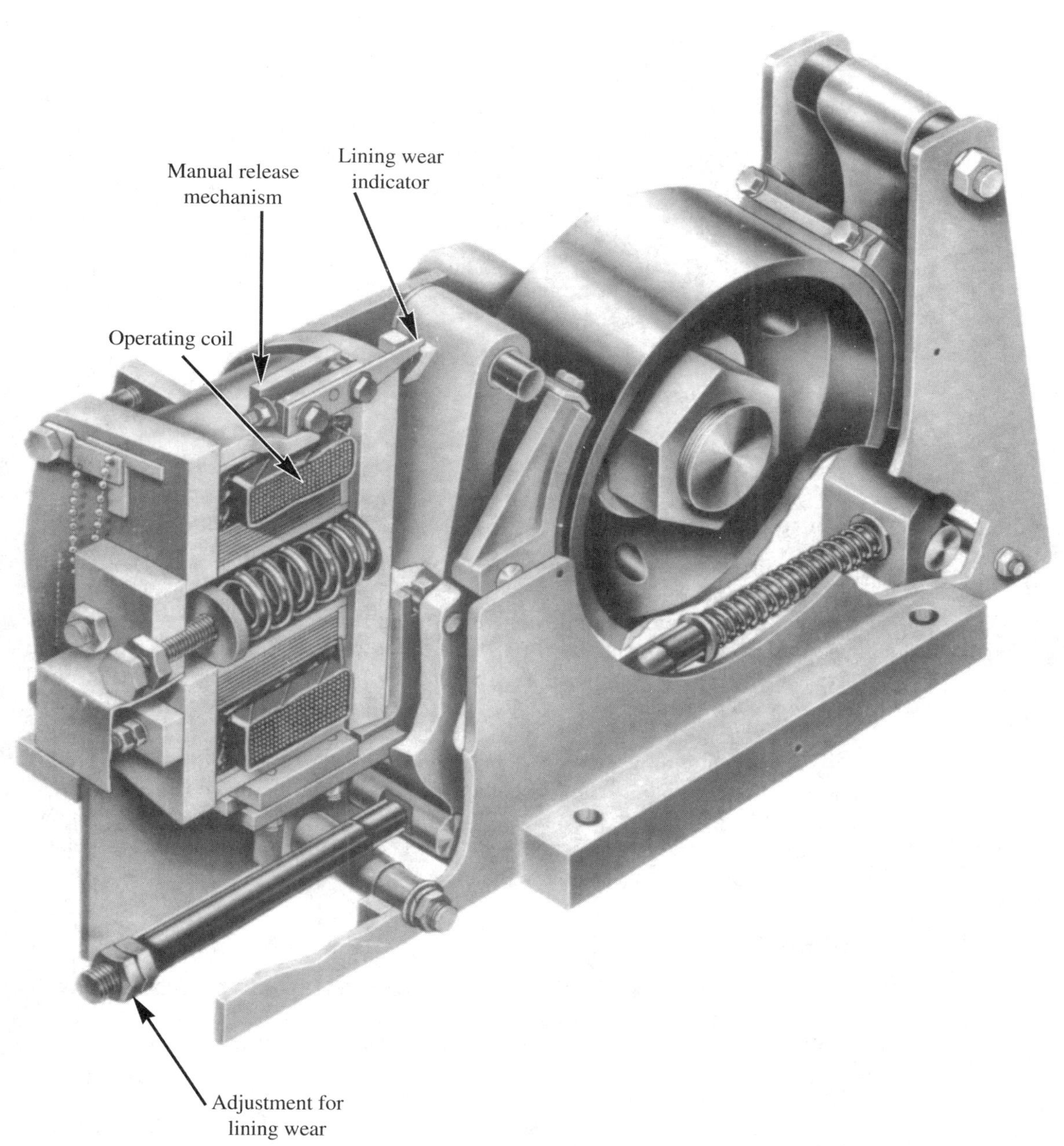

FIGURE 11.12
Magnetically released spring-loaded holding brake. (Courtesy GE Industrial Systems)

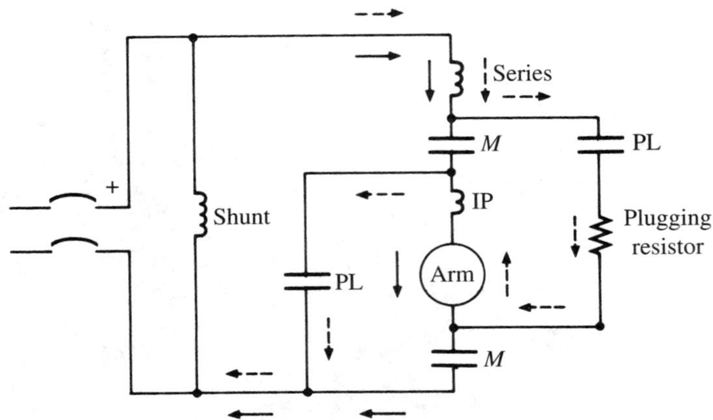

FIGURE 11.13
Simplified circuit for plugging operations.

Plugging

Plugging is the electrical reversal of a motor before it stops. It is used in those applications where a motor must be brought to an emergency stop, a quick reversal, or to eliminate the long coasting time inherent in high-inertia machines. The motor is switched to reverse by *reversing the voltage applied to the armature.* The current in the series and shunt fields must not be reversed. Sufficient resistance must be inserted in series with the armature circuit to prevent excessive current from tripping the breaker or damaging the commutator–brush interface.

A simplified circuit for plugging operations is shown in Figure 11.13. During normal motoring operation, the M contacts are closed and the PL contacts are open; the associated armature current path is shown with solid arrows. During plugging operation, the M contacts are opened and the PL contacts are closed; the associated armature current path is shown with broken arrows.

Jogging

Jogging is the *very brief* application of power to a motor, to cause a fraction of a revolution of the rotor. Jogging is used to position an elevator, to align shafts of different apparatus, etc. Resistance must be inserted in series with the armature to limit the armature current when jogging.

EXAMPLE 11.7 A 240-V, compensated shunt-motor driving a 910-lb-ft torque load is running at 1150 r/min. The efficiency of the motor at this load is 94.0 percent. The combined armature, compensating winding, and interpole resistance is 0.00707 Ω, and the resistance of the shunt field is 52.6 Ω. Determine the resistance of a dynamic-braking resistor that will be capable of developing 500 lb-ft of braking torque at a speed of 1000 r/min. Assume windage and friction at 1000 r/min are essentially the same as at 1150 r/min.

Solution

The dynamic braking circuit is similar to that in Figure 11.11(b).

$$P_{\text{shaft}} = \frac{T \times n}{5252} = \frac{910 \times 1150}{5252} = 199.257 \text{ hp}$$

$$P_{\text{in}} = \frac{P_{\text{shaft}}}{\eta} = \frac{199.257 \times 746}{0.940} = 158{,}134 \text{ W}$$

$$P_{\text{in}} = V_T I_T \quad \Rightarrow \quad 158{,}134 = 240 \times I_T \quad \Rightarrow \quad I_T = 658.89 \text{ A}$$

$$I_f = \frac{V_T}{R_f} = \frac{240}{52.6} = 4.56 \text{ A}$$

$$I_a = I_T - I_f = 658.89 - 4.56 = 654.33 \text{ A}$$

$$V_T = E_{a1} + I_{a1} R_{\text{acir}} \quad \Rightarrow \quad 240 = E_{a1} + 654.33 \times 0.00707$$

$$E_{a1} = 235.37 \text{ V}$$

Since the flux density remains constant,

$$\frac{T_1}{T_2} = \frac{[B_p I_a]_1}{[B_p I_a]_2} \quad \Rightarrow \quad \frac{910}{500} = \frac{654.33}{I_{a2}} \quad \Rightarrow \quad I_{a2} = 359.52 \text{ A}$$

$$\frac{E_{a1}}{E_{a2}} = \frac{[n\Phi_p k_G]_1}{[n\Phi_p k_G]_2} \quad \Rightarrow \quad \frac{235.37}{E_{a2}} = \frac{1150}{1000}$$

$$E_{a2} = 204.67 \text{ V}$$

Referring to Figure 11.11(b),

$$E_{a2} = I_{a2}(R_{\text{acir}} + R_{\text{DB}}) \quad \Rightarrow \quad 204.67 = 359.52(0.00707 + R_{\text{DB}})$$

$$R_{\text{DB}} = \underline{0.562\ \Omega}$$

11.12 STANDARD TERMINAL MARKINGS AND CONNECTIONS OF DC MOTORS

The standard terminal markings for DC motors and the proper connections for different directions of rotation are shown in Figure 11.14 **[2]**. The directions of rotation are as viewed from the end opposite the drive shaft. Differential connections are not shown. However, because of the hazards inherent in a differential connection, compound motors with unmarked terminals must be tested to determine the relative polarity of the shunt and series windings, and the leads must be appropriately marked, before connecting the motor to a starter. A test procedure to determine the relative polarity of the series and shunt fields is detailed in Reference **[1]**.

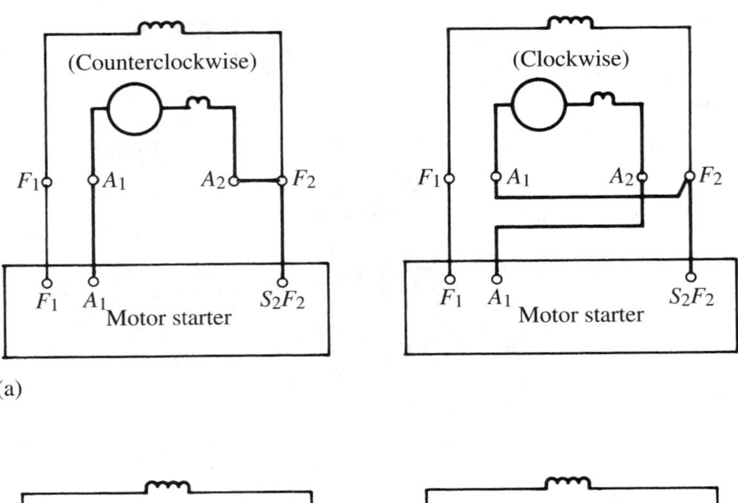

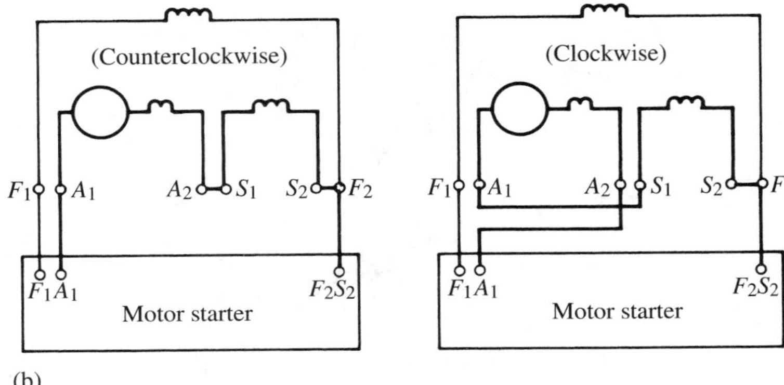

FIGURE 11.14
NEMA standard terminal markings for DC motors, and proper connections for different directions of rotation: (a) shunt motor; (b) compound motor.

SUMMARY OF EQUATIONS FOR PROBLEM SOLVING

$$\mathcal{F}_{net} = \mathcal{F}_f + \mathcal{F}_s - \mathcal{F}_d \quad \textbf{(11–1)}$$

$$T_D = B_p I_a k_M \quad \textbf{(10–17)}$$

$$n = \left. \frac{V_T - I_a R_{acir}}{\Phi_p k_G} \right|_{\Phi_p \neq 0} \quad \textbf{(10–26)}$$

$$\frac{T_{D1}}{T_{D2}} = \frac{[B_p I_a]_1}{[B_p I_a]_2} \quad \textbf{(11–2)}$$

$$\frac{n_1}{n_2} = \frac{\left[\dfrac{V_T - I_a R_{acir}}{\phi_p k_G}\right]_1}{\left[\dfrac{V_T - I_a R_{acir}}{\phi_p k_G}\right]_2}, \qquad \phi \neq 0 \tag{11–3}$$

$$\frac{n_1}{n_2} = \left[\frac{V_T - I_a R_{acir}}{B_p}\right]_1 \times \left[\frac{B_p}{V_T - I_a R_{acir}}\right]_2, \qquad \phi \neq 0 \tag{11–4}$$

Linear Assumption

$$\frac{T_{D1}}{T_{D2}} = \frac{[\mathscr{F}_{net} I_a]_1}{[\mathscr{F}_{net} I_a]_2} \tag{11–6}$$

$$\frac{n_1}{n_2} = \left[\frac{V_T - I_a R_{acir}}{\mathscr{F}_{net}}\right]_1 \times \left[\frac{\mathscr{F}_{net}}{V_T - I_a R_{acir}}\right]_2, \qquad \phi \neq 0 \tag{11–7}$$

$$T_{D,\text{series}} \propto \mathscr{F}_s I_a \cdot I_a \propto I_a^2 \tag{11–8}$$

SPECIFIC REFERENCES KEYED TO TEXT

1. Hubert, C. I. *Preventive Maintenance of Electrical Equipment.* Prentice Hall, Upper Saddle River, NJ, 2002.
2. National Electrical Manufacturers Association. *Motors and Generators.* NEMA Standards Publication No. MG-1-1998, NEMA, Rosslyn, VA, 1999.

GENERAL REFERENCES

Kloeffler, R. G., R. M. Kerchner, and J. L. Brenneman. *Direct-Current Machinery.* Macmillan, New York, 1949.

Langsdorf, A. S. *Principles of Direct-Current Machines.* McGraw-Hill, New York, 1959.

Smeaton, R. W. *Motor Application and Maintenance Handbook,* 2nd ed. McGraw-Hill, New York, 1987.

REVIEW QUESTIONS

1. Explain why some shunt motors require a stabilizing winding.
2. How does a compound motor differ in construction from a stabilized-shunt motor?
3. Sketch the circuit for a compound motor and indicate (with broken lines) on the diagram the correct connections to reverse the motor.

4. Approximately what percent of total mmf is supplied by the series field in (a) stabilized-shunt motors; (b) compound motors?
5. State an application that is specifically appropriate for (a) shunt motors; (b) series motors; (c) compound motors.
6. Explain why a series motor self-destructs due to overspeed when the load is removed.
7. Is there any difference between the rated torque developed by a compound motor and the rated torque developed by a shunt motor of equal power and speed ratings? Explain.
8. Explain how compound motors smooth the energy demand required by a pulsating load such as a flywheel punch press.
9. Explain why a differential connection of a compound motor should be avoided.
10. A 20-hp, 230-V compound motor is supplied with both an armature rheostat and a shunt field rheostat for speed control. The shunt field rheostat has a resistance range of zero to 300 Ω. The armature rheostat has a range of zero to 1.5 Ω. When the machine is started from standstill, what should the numerical value of resistance be in each rheostat? Justify your answer with a theoretical explanation.
11. Sketch the circuit diagram, and explain the operation of the Ward–Leonard system with respect to speed adjustment and reversing.
12. Differentiate between regenerative braking and resistive braking, and state an application for each.
13. Differentiate between plugging and jogging, and state an application for each.

PROBLEMS

11–1/7 A compensated shunt motor rated at 60 hp, 240 V, 400 r/min, draws 209.1 A when operating at rated conditions. The motor parameters are $R_{acir} = 0.0483\ \Omega$, $R_f = 39.1\ \Omega$, and $N_f = 1476$ turns/pole. The corresponding magnetization curve is shown in Figure 11.15. (a) Determine the resistance of an external resistor required in series with the shunt field to raise the speed to 600 r/min. Assume the new torque load is 40 percent rated torque. (b) Determine the new armature current; (c) power rating of the resistor.

11–2/7 A 75-hp, 240-V, 250-r/min compensated shunt motor has the following parameters: $R_{acir} = 0.0665\ \Omega$, $R_f = 25.6\ \Omega$, and $N_f = 835$ turns/pole. The corresponding magnetization curve is shown in Figure 11.15. The machine has an efficiency of 86.4 percent when operating at rated conditions. Determine the resistance and power rating of an external resistor required in series with the shunt field to raise the speed to 500 r/min. Assume a constant shaft-power load and no change in efficiency.

11–3/7 A 150-hp, 240-V, 650 r/min compensated shunt motor has the following parameters: $R_a = 0.00872\ \Omega$, $R_{IP} + R_{CW} = 0.0038\ \Omega$, $R_f = 32.0\ \Omega$, and $N_f = 1143$ turns/pole. The efficiency at rated load is 92.0 percent. The corresponding magnetization curve is shown in Figure 11.15. (a) Determine the external resistance required in series with the shunt field that will pro-

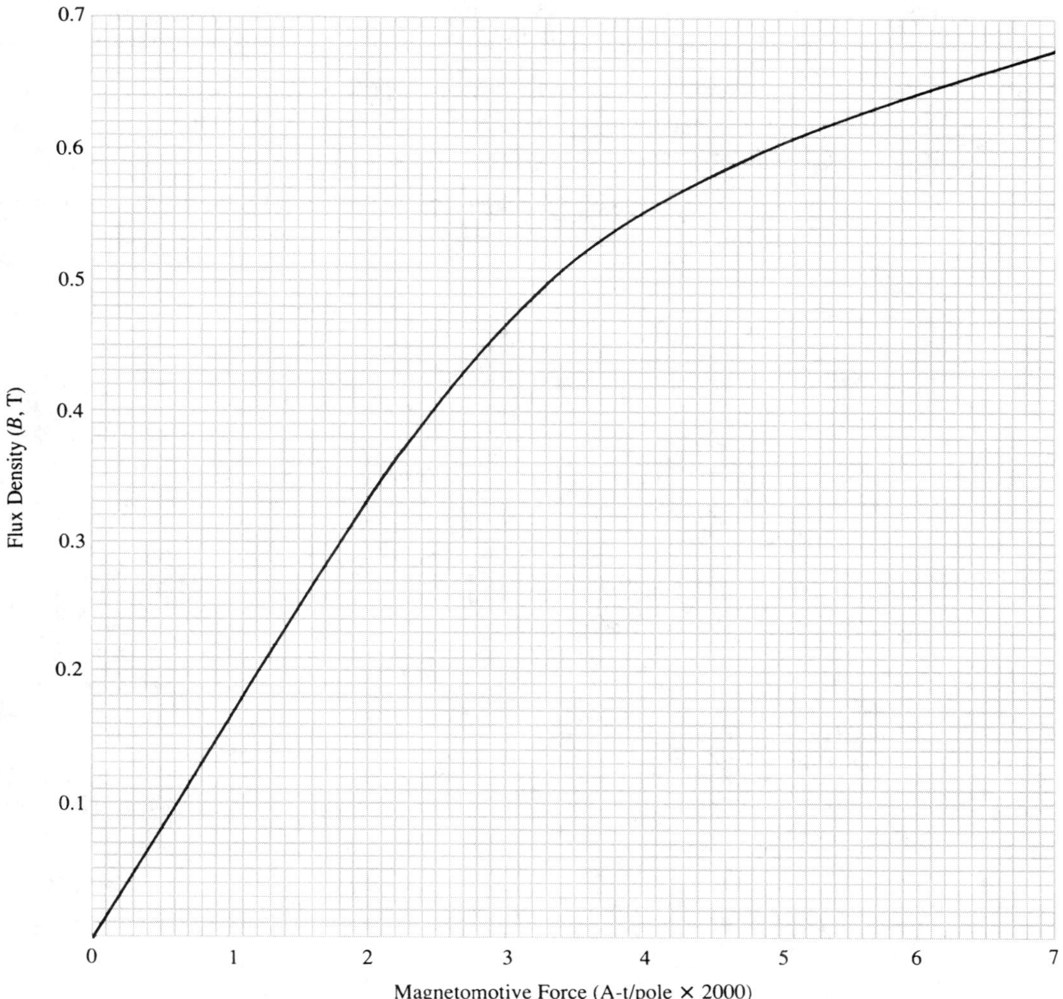

FIGURE 11.15
Magnetization curve for Problems 11–1/7 to 11–4/7.

vide a no-load speed of 2100 r/min. Assume the motor current at no-load is 18.4 A. (b) Determine the power rating of the resistor.

11–4/7 A 50-hp, 240-V, 1750 r/min compensated shunt motor operating at rated load has an efficiency of 88.7 percent. The machine parameters are $R_a = 0.0287\ \Omega$, $R_{\text{IP}} = 0.0138\ \Omega$, $R_f = 67.4\ \Omega$, and $N_f = 1850$ turns/pole. The corresponding magnetization curve is shown in Figure 11.15. (a) Determine the external resistance required in series with the armature in order to obtain a speed of 1024 r/min. Assume the machine is operating with 65 percent rated torque. (b) Determine the power rating of the resistor.

11–5/7 A 700-hp, 850 r/min, 500-V compound motor has an efficiency of 93.2 percent when operating at rated conditions. The motor parameters are:

	Armature	IP + CW	Series	Shunt
Resistance, Ω	0.00689	0.001374	0.000687	72.5
Turns/pole	—	—	6	995

The magnetization curve for the motor is shown in Figure 11.16. Determine (a) percent mmf supplied by the series field; (b) speed if an open occurs in

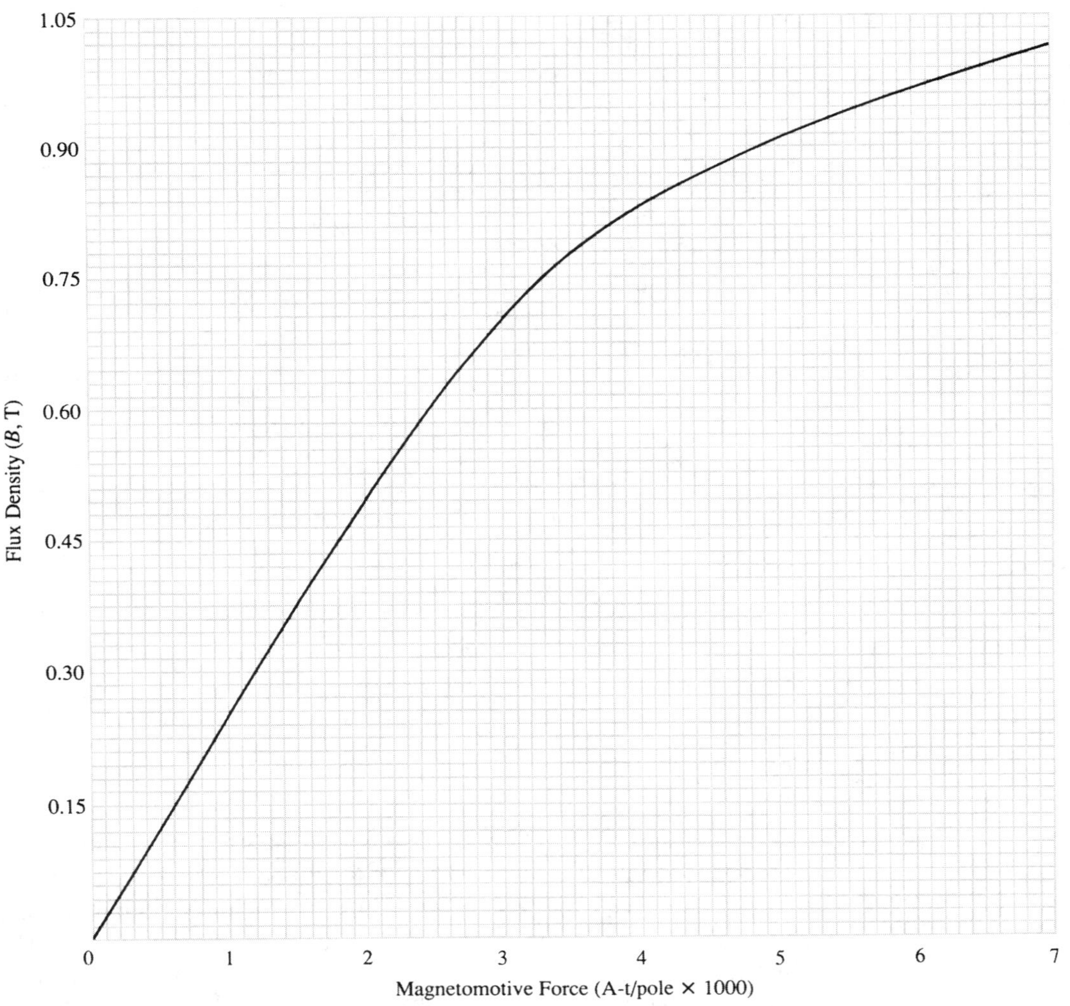

FIGURE 11.16
Magnetization curve for Problems 11–5/7 to 11–11/7.

the shunt field circuit. Assume the connected load is such that the armature current rises to 1540 A.

11–6/7 A 40-hp, 1150 r/min compensated shunt motor operating from a 240-V system is supplying a load whose torque varies directly as the speed. The efficiency of the motor at rated conditions is 90.2 percent, and the motor parameters are:

	Armature	IP + CW	Shunt
Resistance, Ω	0.0680	0.0289	99.5
Turns/pole	—	—	1231

The magnetization curve is shown in Figure 11.16. Determine the external resistance required in series with the shunt field circuit in order to operate at 125 percent rated speed.

11–7/7 A 400-hp, 250-V, 100 r/min compound motor with a compensating winding has an efficiency of 89.8 percent when operating at rated conditions. The motor parameters are:

	Armature	IP + CW	Series	Shunt
Resistance, Ω	0.002660	0.000774	0.000356	24.1
Turns/pole	—	—	2	400

The magnetization curve is shown in Figure 11.16. Determine (a) motor current; (b) armature current; (c) mechanical power developed. (d) If the torque required to start the machine is 115 percent rated torque, the machine is to be operated without its series field, and the armature current is limited to 200 percent rated armature current, will the motor start?

11–8/7 A 50-hp, 230-V, 500 r/min compound motor with a compensating winding has the following parameters:

	Armature	IP + CW	Series	Shunt
Resistance, Ω	0.071	0.0140	0.031	75
Turns/pole	—	—	6	1000

The magnetization curve is shown in Figure 11.16. The motor is operating at rated conditions and draws a line current of 190 A. Determine (a) cemf; (b) mechanical power developed; (c) rated shaft torque; (d) mechanical power required to overcome rotational losses. (e) Assuming a constant-torque load, determine the speed if the series field is not used.

11–9/7 A 60-hp, 468 r/min, 240-V stabilized-shunt motor has an efficiency of 89.0 percent when operating at rated conditions. The motor parameters are:

	Armature	Interpole	Series	Shunt
Resistance, Ω	0.0342	0.00924	0.00308	71.6
Turns/pole	—	—	1.5	1268

The magnetization curve is shown in Figure 11.16. Sketch the motor circuit and determine (a) rated motor current; (b) rated field current; (c) rated armature current; (d) rated shaft torque; (e) external resistance required in order to limit the locked-rotor armature current to 200 percent of its rated value; (f) additional series field turns of the same size conductor required to cause a 135 percent increase in locked-rotor torque, assuming armature current is limited to 200 percent rated armature current; (g) new steady-state speed, assuming additional series field turns, the motor is loaded to rated armature current, and no external resistance is inserted in series with the shunt or series field.

11–10/7 A 75-hp, 850 r/min, 500-V series motor has an efficiency of 90.2 percent when operating at rated conditions. The series field has 30 turns/pole, and the motor parameters are:

	Armature	IP + CW	Series
Resistance, Ω	0.1414	0.0412	0.0189

The motor has a maximum safe speed of 1700 r/min. Is it safe to operate the motor with a 10-hp shaft load? Use the magnetization curve in Figure 11.16, and show all work. Neglect changes in windage and friction.

11–11/7 A series motor rated at 150 hp, 400 r/min, and 240 V is used to drive a load that cycles between 120 percent rated power and 40 percent rated power. The equivalent demagnetizing mmf due to armature reaction is approximately equal to 7.5 percent of the series field mmf. The series field has 12 turns/pole, the efficiency at rated load is 93.2 percent, and the motor parameters are:

	Armature	Interpole	Series
Resistance, Ω	0.01346	0.00392	0.00181

Determine the upper and lower speeds. Use the magnetization curve in Figure 11.16.

11–12/8 A 240-V, 50-hp, 1150 r/min compound motor has the following parameters:

	Armature	IP + CW	Series	Shunt
Resistance, Ω	0.0673	0.0196	0.00902	85.6
Turns/pole	—	—	8	750

The efficiency at rated load is 88.7 percent. Determine the speed if a 20-Ω resistor is inserted in series with the shunt field, and the resultant speed causes an armature current of 200 A. Neglect saturation effects.

11–13/8 A 15-hp, 2500 r/min, 240-V compound motor draws 58.6 A when operating at rated conditions. The motor parameters are:

	Armature	IP + CW	Series	Shunt
Resistance, Ω	0.241	0.0700	0.322	138
Turns/pole	—	—	10	1360

If a 1.6-Ω resistor is connected in series with the armature, a 100-Ω resistor is connected in series with the shunt field, the load on the shaft is such as to cause the motor to draw 40 A from the line, and saturation effects are neglected, determine (a) speed; (b) mechanical power developed; (c) developed torque.

11–14/8 A 240-V, 30-hp, 650 r/min compound motor has an efficiency of 94.2 percent when operating at rated conditions. The motor parameters are:

	Armature	IP + CW	Series	Shunt
Resistance, Ω	0.1192	0.0347	0.0159	80
Turns/pole	—	—	7	513

Neglecting the effects of saturation, determine (a) speed at rated armature current if the series field is not used; (b) speed at rated armature current if the series field is differentially connected.

11–15/8 A 240-V, 20-hp, 300 r/min stabilized-shunt motor with an efficiency of 84.3 percent has the following parameters:

	Armature	Interpole	Series	Shunt
Resistance, Ω	0.168	0.0490	0.0226	100
Turns/pole	—	—	1.5	1096

Neglecting the effects of saturation, determine (a) speed at no load, assuming the no-load motor current is 15 A; (b) speed regulation in (a); (c) developed torque at 60 percent rated armature current.

11–16/8 A 100-hp, 3500 r/min, 240-V compound motor has an efficiency at rated load of 94.6 percent. The motor parameters are:

	Armature	IP + CW	Series	Shunt
Resistance, Ω	0.0358	0.0104	0.00480	52.3
Turns/pole	—	—	3	367

Neglecting the effects of saturation, determine (a) speed, if the load on the shaft is decreased to a value that results in a line current of 136 A; (b) developed torque for conditions in (a).

11–17/10 A coal conveyer is driven by a separately excited stabilized-shunt motor through an adjustable-voltage drive system. The coal conveyer is a constant-torque load on the motor. The motor armature is rated at 700 hp, 400 r/min, 250 V, 2230 A. The field is excited from a constant voltage 120-V DC supply. The motor parameters are:

	Armature	Interpole	Series	Shunt
Resistance, Ω	0.006294	0.001831	0.000843	10.8

Determine the applied armature voltage when operating at 100 r/min.

11–18/10 A DC adjustable-voltage drive system is used to drive a reel in a steel mill. As steel is wound on the reel, the reel diameter increases, increasing the torque load on the motor. To maintain constant power, the speed must vary inversely with the torque. In effect, the reel represents a constant-power load. The motor driving the reel is a stabilized-shunt motor rated at 1000 hp, 500 V, and draws an armature current of 1522 A at 400 r/min. The shunt field is separately excited through a rectifier from a 120-V, 60-Hz system. The motor parameters are:

	Armature	**Interpole**	**Series**	**Shunt**
Resistance, Ω	0.0115	0.00179	0.00156	15.76

Determine the applied armature voltage when operating at 700 r/min.

11–19/11 A separately excited shunt motor with a compensating winding, rated at 550 V, 400 hp, has a base speed of 1750 r/min. The shunt field is separately excited from a 120-V DC supply. The efficiency of the motor, less the shunt field, is 94.6 percent when operating at rated conditions. The motor parameters are:

	Armature	**IP + CW**	**Shunt**
Resistance, Ω	0.0192	0.00573	28

Assume the machine is operating at rated conditions when an overheated bearing starts to smoke. The armature is disconnected from the supply line and connected to a dynamic-braking resistor. The inertia of the load prevents an instantaneous slowing of the machine, and 200 percent rated torque is developed when the dynamic-braking circuit is closed. Determine the resistance of the dynamic-braking resistor. The circuit is shown in Figure 11.11(b).

11–20/11 A 150-hp, 1750 r/min, 240-V compensated shunt motor is operating at rated conditions with an efficiency of 92.3 percent. The motor parameters are:

	Armature	**IP + CW**	**Shunt**
Resistance, Ω	0.0233	0.0099	33.9

Determine the armature current at 500 r/min when dynamic braking through a 0.324-Ω resistor. The circuit is shown in Figure 11.11(b).

12

Direct-Current Generator Characteristics and Operation

12.1 INTRODUCTION

This chapter deals primarily with self-excited generators, their characteristics, operation, and problems, and builds on the theory developed in Chapter 10.

Although the expanded use of solid-state converters and control systems has reduced the demand for direct-current generators, they still have applications in steel mills, ice-breaking vessels, the electrochemical industry, and as a low-ripple DC source for some drive systems.

12.2 SELF-EXCITED SHUNT GENERATORS

A self-excited shunt generator whose equivalent circuit is shown in Figure 12.1(a) develops its initial emf from residual magnetism in the pole iron.[1] The switch shown in the field circuit is used only as an aid in the explanation of voltage buildup; field switches are rarely used with self-excited generators.

The magnetization curve and field resistance line for a self-excited generator are shown in Figure 12.1(b). The field-resistance line is an Ohm's law plot of applied voltage vs. current for the shunt field winding. One point on the line is the origin (0 V, 0 A). Another point (V_f, I_f) is obtained by applying approximately rated voltage to the shunt field winding and measuring the corresponding current.

The magnetization curve is obtained by running the machine at rated speed, *as a separately excited generator,* with no load connected to its output terminals, and plotting the induced emf vs. shunt field current as the field current is varied from zero to approximately 125 percent rated voltage. To avoid introducing hysteresis loops, the

[1]The magnetic retentivity of the pole iron in self-excited generators provides sufficient residual flux to generate approximately 5 percent rated voltage when operating at rated speed with the field switch open.

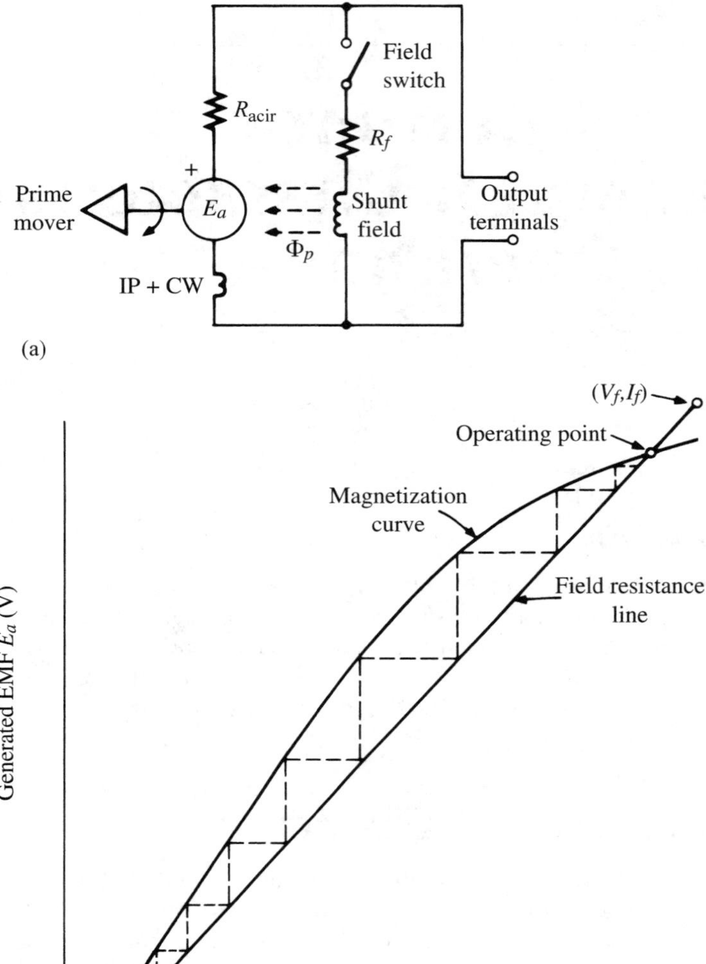

FIGURE 12.1
(a) Circuit diagram of a self-excited shunt generator; (b) illustration of voltage buildup in a self-excited shunt generator.

field current should not be decreased at any time during the test. If a planned test point is overshot, either start all over, or skip that point. The curve can be easily plotted on an *X-Y* plotter.

Voltage Buildup

Assuming the field switch in Figure 12.1(a) is open and the armature is rotating at rated speed, the residual magnetism will cause the voltage to build up to about 7 V. This is shown on the magnetization curve in Figure 12.1(b). Closing the field switch impresses 7 V across the field winding, causing a current of 0.6 A as shown on the field-resistance line. The 0.6 A of field current causes the voltage to increase to 13 V, as shown on the magnetization curve. The 13 V causes 1.1 A in the field winding, causing a further rise in voltage to 19 V, and so forth. The voltage continues to rise until a further increase in voltage no longer causes an increase in flux. This occurs at the intersection of the magnetization curve with the field-resistance line and is called the *operating point.* As the intersection is approached, the increment increase in emf gets less for a given increment increase in field current, and the buildup tapers off to a final value of generated emf and field current. Although described as a step-by-step process, the actual buildup is a smooth and fairly rapid rise to the operating point. The buildup of voltage is limited by magnetic saturation and the resistance of the field circuit.

Adjusting the No-Load Voltage

Adjusting the no-load voltage, also called *open-circuit voltage,* is accomplished by changing the slope of the field-resistance line.

Adding resistance in series with the shunt field increases the slope of the line, causing the point of intersection, and hence the operating point, to occur at a lower voltage. A family of field-resistance lines and the corresponding operating points (intersections) are shown in Figure 12.2 for different rheostat settings, where

$$R = R_f + R_{\text{rheo}}$$

The operating points in Figure 12.2 are the graphical solutions of two simultaneous equations: the magnetization curve, expressed as

$$E_a = n\Phi_p k_G = n\left(\frac{N_f I_f}{\mathcal{R}}\right)k_G \qquad \textbf{(12–1)}$$

and the field resistance line, expressed as

$$E_a = I_f(R_f + R_{\text{rheo}}) \qquad \textbf{(12–2)}$$

A graphical solution is used because the nonlinear nature of the reluctance $\mathcal{R}$ in Eq. (12–1) makes an algebraic solution very difficult.

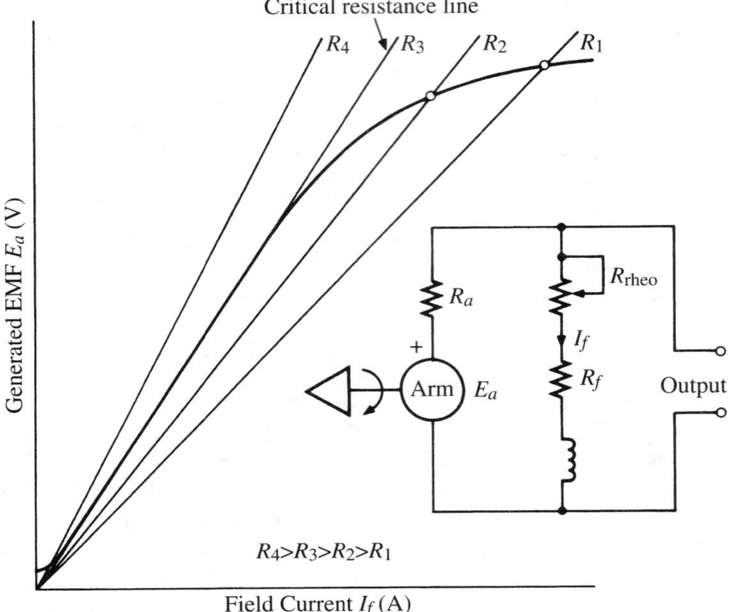

FIGURE 12.2
Family of field-resistance lines and corresponding operating points.

If the rheostat resistance is adjusted to a value that causes the field-resistance line to be to the left of the straight line portion of the magnetization curve, R_4 in Figure 12.2, the buildup of generator voltage will be only slightly higher than the value obtained by residual magnetism alone.

If the field-resistance line is coincident with the linear section of the magnetization curve, the voltage will be unstable and can assume any value along the coincident section. The value of field-circuit resistance that causes coincidence is called the *critical resistance.* To build up voltage, the field-circuit resistance must be less than the critical value. Thus an open in the field rheostat, too high a rheostat resistance, low brush pressure on the commutator, dirty brushes and/or commutator, or an open in the shunt field coils will prevent the buildup of voltage.

Basic Design

Self-excited generators are usually designed to obtain approximately 125 percent of rated voltage with the rheostat set for zero resistance and no load connected to the output terminals. The field rheostat is designed so that maximum rheostat resistance will limit the no-load voltage to approximately 50 percent of rated voltage.

12.3 EFFECT OF SPEED ON VOLTAGE BUILDUP OF A SELF-EXCITED GENERATOR

The critical resistance of a self-excited generator is closely related to the speed of the machine. A machine running at low speed will not build up to the same voltage as when running at a higher speed, using the same field-circuit resistance. Hence, the critical resistance will be different for different machine speeds. This is illustrated in Figure 12.3. For the field-resistance line shown, the intersection with the magnetization curve for high speed occurs at a relatively high voltage and the intersection with the magnetization curve for low speed occurs at a value of voltage slightly higher than that produced by the residual flux.

EXAMPLE 12.1 The magnetization curve for a certain self-excited, 125-V, 50-kW, 1750 r/min shunt generator is shown in Figure 12.4. The no-load voltage with the rheostat shorted is 156 V. Determine (a) field-circuit resistance; (b) field-rheostat setting that will provide a no-load voltage of 140 V; (c) armature voltage if the rheostat is set to 14.23 Ω; (d) field-rheostat setting that will cause critical resistance; (e) armature voltage at 80 percent rated speed, and the rheostat shorted; (f) rheostat setting required to obtain a no-load armature voltage of 140 V at 1750 r/min *if the shunt field is separately excited* from a 120-V DC source.

Solution

(a) From the magnetization curve in Figure 12.4, the shunt field current that corresponds to a no-load voltage of 156 V is ≈4.7 A. Thus,

$$R_f = \frac{E_a}{I_f} = \frac{156}{4.7} = 33.1915 \qquad \Rightarrow \qquad \underline{33.19\ \Omega}$$

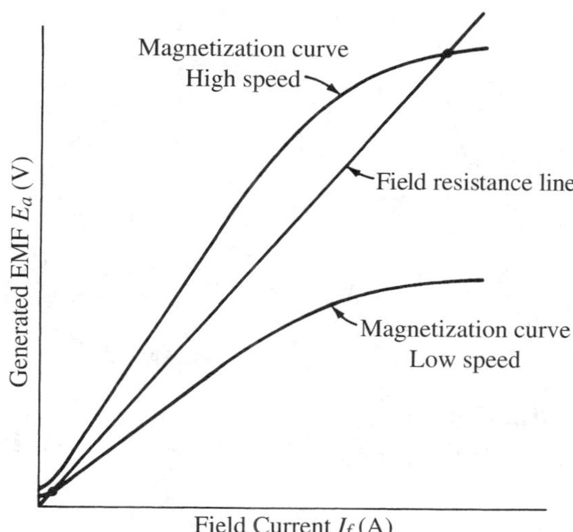

FIGURE 12.3 Effect of speed on voltage buildup in a self-excited generator.

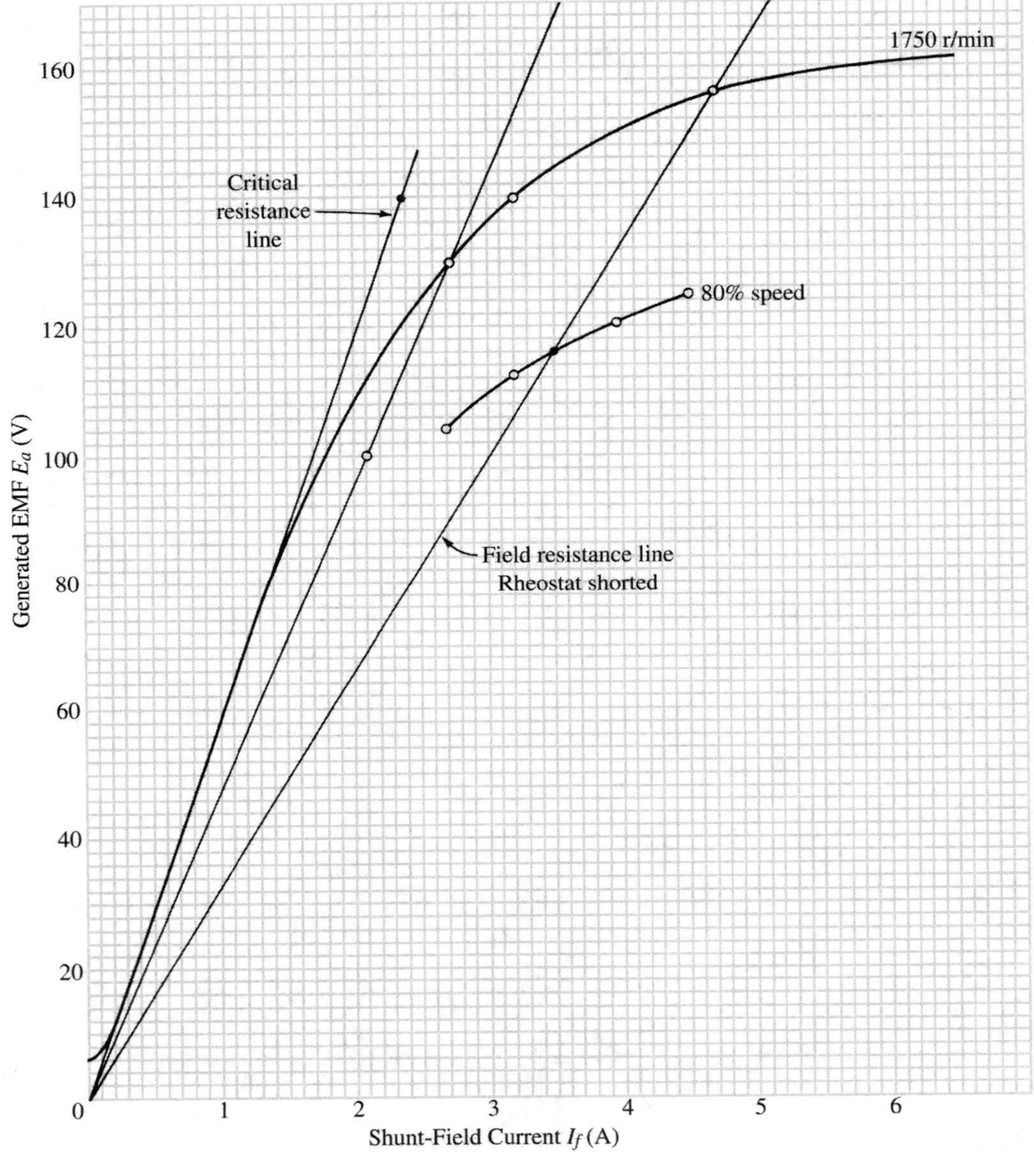

FIGURE 12.4
Magnetization curves and field-resistance lines for Example 12.1.

(b) From the magnetization curve, the field current that corresponds to a no-load voltage of 140 V is ≈3.2 A.

$$I_f = \frac{E_a}{R_f + R_{\text{rheo}}} \quad \Rightarrow \quad R_{\text{rheo}} = \frac{E_a}{I_f} - R_f$$

$$R_{\text{rheo}} = \frac{140}{3.2} - 33.19 = \underline{10.56\ \Omega}$$

(c)

$$R_f + R_{\text{rheo}} = 33.19 + 14.23 = 47.42\ \Omega$$

The operating voltage is determined from the intersection of the magnetization curve with the *new* field-resistance line, which is plotted with data obtained from Eq. (12–2). Thus, using two arbitrary voltages (0 V and 100 V), the corresponding field currents are

$$I_f = \frac{E_a}{R_f + R_{\text{rheo}}} = \frac{0}{47.42} = 0\ \text{A}$$

$$I_f = \frac{100}{47.42} = 2.1\ \text{A}$$

Drawing a straight line through coordinates (0 V, 0 A) and (100 V, 2.1 A) results in an intersection at <u>130 V</u>.

(d) The critical field-resistance line is sketched tangent to the linear section of the magnetization curve, and the critical resistance determined from any set of coordinates. Thus, selecting

$$E_a = 140\ \text{V} \qquad I_f = 2.35\ \text{A}$$

$$R_{\text{cr}} = \frac{E_a}{I_f} = \frac{140}{2.35} = \underline{59.6\ \Omega}$$

(e) The field-resistance line is the same as for part (a). A new magnetization curve, however, must be drawn for 80 percent speed. Since only a section of the curve is needed, voltage points are calculated for several values of shunt field current on either side of the field-resistance line. The required calculation is

$$E_a \text{ at } 80\% \text{ speed} = 0.80 \times E_a \text{ at rated speed}$$

Thus,

Field Current	Approximate Induced EMF (V)	
(A)	Rated Speed	80% Rated Speed
4.50	155	124
3.95	150	120
3.20	140	112
2.70	130	104

Plotting the tabulated 80 percent speed points results in the short-section low-speed magnetization curve shown in Figure 12.4. Intersection of the field-resistance line with the low-speed magnetization curve occurs at $\approx \underline{116\text{ V}}$.

(f) From the 1750 r/min magnetization curve, 140 V requires a field current of $\approx$3.2A.

$$I_f = \frac{V_f}{R_f + R_{\text{rheo}}} \quad \Rightarrow \quad R_{\text{rheo}} = \frac{V_f}{I_f} - R_f$$

$$R_{\text{rheo}} = \frac{120}{3.2} - 33.19 = \underline{4.31\ \Omega}$$

12.4 OTHER FACTORS AFFECTING VOLTAGE BUILDUP

Factors other than excessive field-circuit resistance or low speed that affect the buildup of voltage in a self-excited generator are reversed shunt field connections, reversed rotation, and reversed residual magnetism. These adverse effects can be visualized by studying the circuits in Figure 12.5 and using the right-hand rule to establish the direction of coil flux. For simplicity, only one field pole is shown. In each case:

$$\Phi_R = \text{flux due to residual magnetism}$$
$$\Phi_F = \text{flux due to field current}$$

Figure 12.5(a) represents normal operation; the prime-mover rotation is clockwise, and both the residual flux and the field-coil flux are directed to the left.

In Figure 12.5(b), reversed connections of the field circuit cause Φ_F to oppose Φ_R, and the voltage *builds down* from its original residual value.

In Figure 12.5(c), reversed rotation causes the armature voltage to reverse. This reverses the field current, causing Φ_F to oppose Φ_R, and the voltage builds down from its original residual value.

In Figure 12.5(d), reversed residual magnetism causes the armature voltage to reverse. This reverses the field current. In this case, both Φ_F and Φ_R are reversed. The result is voltage buildup in the reverse direction. The generator will operate at rated voltage with reversed polarity.

Correction of reversed polarity is accomplished by using an external DC source to remagnetize the iron in the correct direction. The procedure is called *field flashing* and is described in detail in Reference **[1]**.

12.5 EFFECT OF A SHORT CIRCUIT ON THE POLARITY OF A SELF-EXCITED SHUNT GENERATOR

Under severe fault conditions, such as may occur when a self-excited shunt generator is accidentally short circuited, the very high armature current causes complete saturation of the interpole iron, making the interpole ineffective (its field strength is no

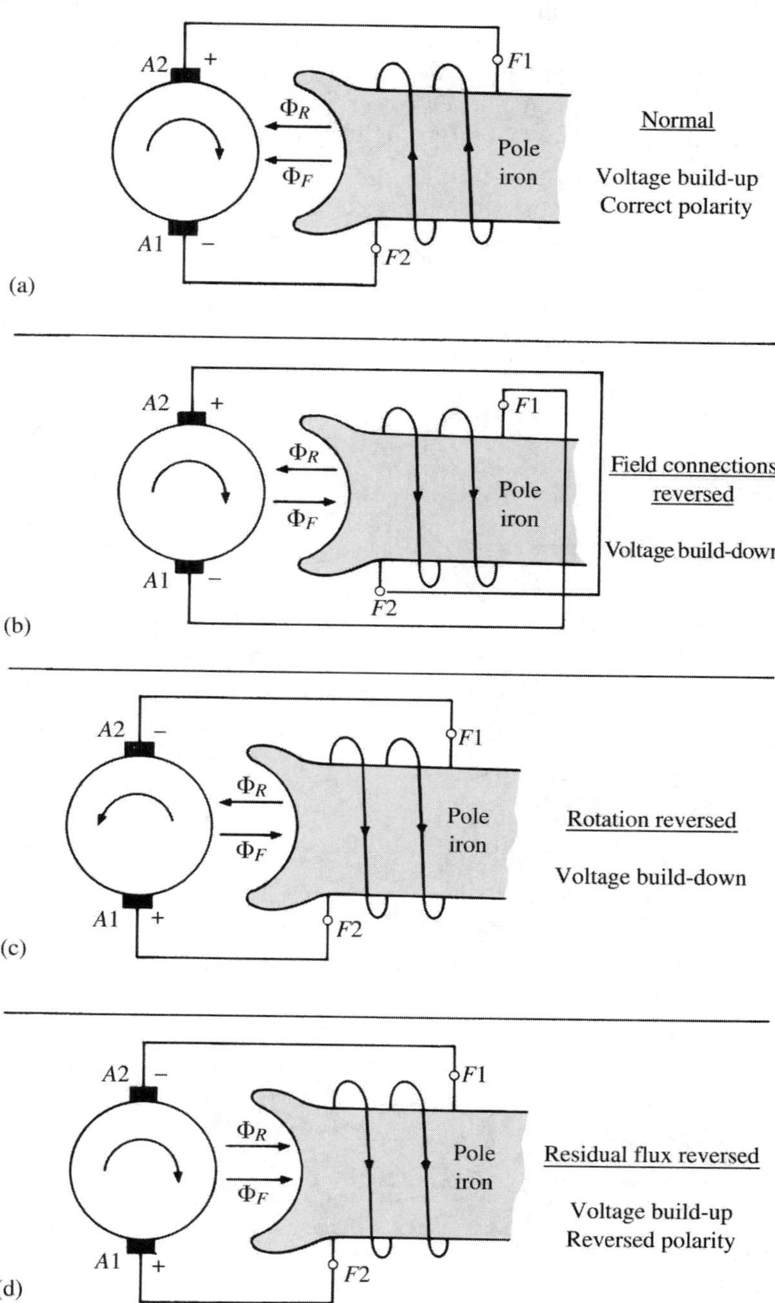

FIGURE 12.5
Factors affecting voltage buildup in a self-excited generator.

longer proportional to the armature current). The resultant high current in the commutated coil produces a strong demagnetizing mmf, shown as $\mathcal{F}_d$ in Figure 12.6(a). For simplicity, the shunt field windings and interpole windings are not shown. The direction of current in the commutated coil is determined by the direction of rotation and the polarity of the main field poles it left behind.

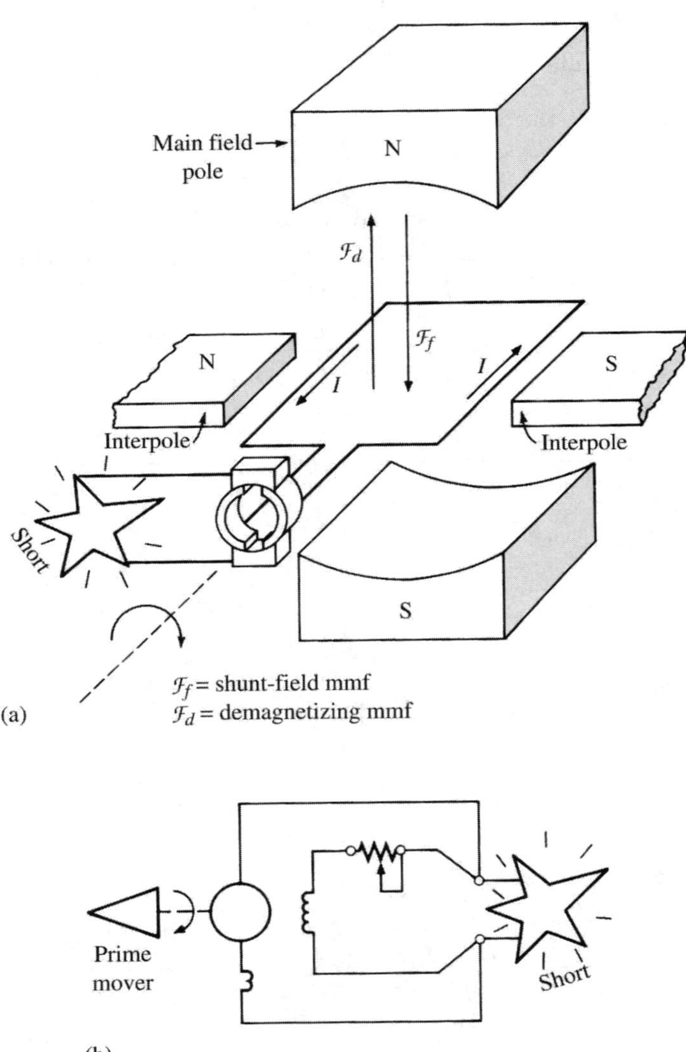

FIGURE 12.6
Effect of a short circuit on the polarity of a self-excited generator: (a) direction of component mmfs; (b) short circuiting the output terminals also short circuits the shunt field circuit.

Furthermore, referring to Figure 12.6(b), short circuiting the generator also short circuits the shunt field winding. This results in a very low shunt field current, and hence a very weak shunt field mmf. The combination of a weak shunt field mmf and a relatively high opposing mmf produced by the commutated coil reverses the residual magnetism in the main-pole iron. Thus, after the circuit breaker trips, the generator voltage will build up in the reverse direction.

12.6 LOAD-VOLTAGE CHARACTERISTICS OF SELF-EXCITED SHUNT GENERATORS

The behavior of self-excited shunt generators under various load conditions is more complex than that of separately excited machines. This is so because the field current of self-excited machines is dependent on the output voltage of the generator. The addition of load to a self-excited generator causes an *IR* drop in the armature circuit and a demagnetizing mmf due to armature reaction in uncompensated machines, both of which cause a lowering of the output voltage. The lowered output voltage reduces the shunt field current, causing a further reduction in voltage.

The dependency of field current on the output voltage, coupled with the nonlinearity of the magnetization curve, makes it necessary to use graphical techniques to determine the terminal voltage for a given load current. Such techniques are described and illustrated in References **[2]**,**[4]**.

Voltage Breakdown

The drop in output voltage due to a very heavy overload reduces the field current. This causes a further reduction in voltage and a rapid breakdown to almost zero volts, as shown in Figure 12.7. Should this occur, the generator circuit breaker must be tripped in order to rebuild voltage.

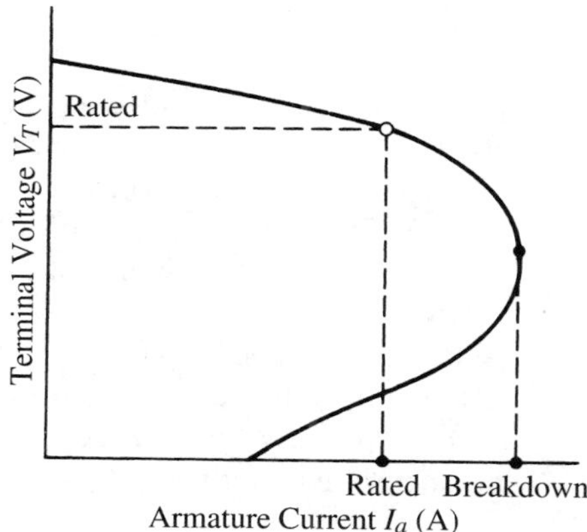

FIGURE 12.7
Load-voltage characteristic of a shunt generator that is loaded past the breakdown point.

12.7 GRAPHICAL APPROXIMATION OF THE NO-LOAD VOLTAGE

As previously discussed in Chapter 10, the voltage regulation of a DC generator is the percent change in terminal voltage from no load to rated load, with respect to rated voltage.

$$\text{VR} = \frac{V_{\text{nl}} - V_{\text{rated}}}{V_{\text{rated}}} \times 100 \qquad \text{percent} \tag{10–16}$$

where: VR = voltage regulation
V_{nl} = no-load (open-circuit) voltage (V)
V_{rated} = voltage indicated on nameplate of machine (V)

The no-load voltage may be obtained by operating the machine at its nameplate conditions, then removing the load and observing the no-load voltage. Although this is an exact method, it is not practical for very large machines.

An analytical method that provides a good approximation of the no-load voltage may be determined from the machine parameters and the associated magnetization curve. The magnetization curve must be used to account for magnetic saturation, the demagnetizing effect of armature reaction in noncompensated machines, and the effect of the series field mmf if the machine is a compound generator.

Separately Excited Shunt Generator With a Compensating Winding

Figure 12.8 shows a separately excited shunt generator, equipped with interpoles and a compensating winding. The machine is supplying rated load at the rated conditions specified on the generator nameplate.

Applying Kirchhoff's voltage law to the armature circuit,

$$E_a = V_T + I_a R_{\text{acir}} \tag{12–3}$$
$$R_{\text{acir}} = R_a + R_{\text{IP}} + R_{\text{CW}}$$

where: E_a = voltage induced in the armature (V)
V_T = voltage at output terminals, load switch closed (V)
R_a = resistance of armature winding (Ω)
R_{IP} = resistance of interpole winding (Ω)
R_{CW} = resistance of compensating winding (Ω)

Since the machine has compensating windings, armature reaction is eliminated and E_a is the voltage that appears across the output terminals when the load switch in Figure 12.8 is opened. Thus, for the special case of a separately excited shunt generator with compensating windings, $V_{\text{nl}} = E_a$.

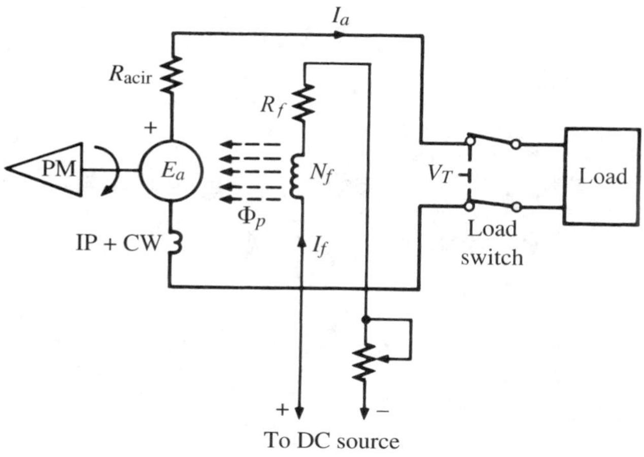

FIGURE 12.8
Separately excited shunt generator equipped with interpoles and a compensating winding.

Separately Excited Shunt Generator Without a Compensating Winding

If the generator in Figure 12.8 is designed without compensating windings, then

$$E_a = V_T + I_a(R_a + R_{\text{IP}}) \tag{12–4}$$

Equation (12–4) does not account for armature reaction nor the effect of magnetic saturation. The effect of armature reaction on the induced emf is shown on the magnetization curve in Figure 12.9. With the load switch open, the armature current is zero and voltage V_{nl} is due to the field mmf ($\mathscr{F}_f$) alone. With the load switch closed, the demagnetizing effect of armature reaction, shown as an equivalent demagnetizing mmf ($\mathscr{F}_d$) in Figure 12.9, causes a net reduction in flux:

$$\mathscr{F}_{\text{net}} = \mathscr{F}_f - \mathscr{F}_d$$

This results in a lowered induced emf E_a.

EXAMPLE 12.2 A 300-kW, 240-V, 900 r/min, separately excited, noncompensated shunt generator has the following parameters: $R_a = 0.00234\ \Omega$, $R_{\text{IP}} = 0.00080\ \Omega$, and $R_f = 18.1\ \Omega$. The shunt field has 1020 turns/pole, and is separately excited from a 120-V source through a rheostat. The circuit diagram and magnetization curve for the generator are shown in Figure 12.10. The equivalent demagnetizing ampere-turns caused by armature

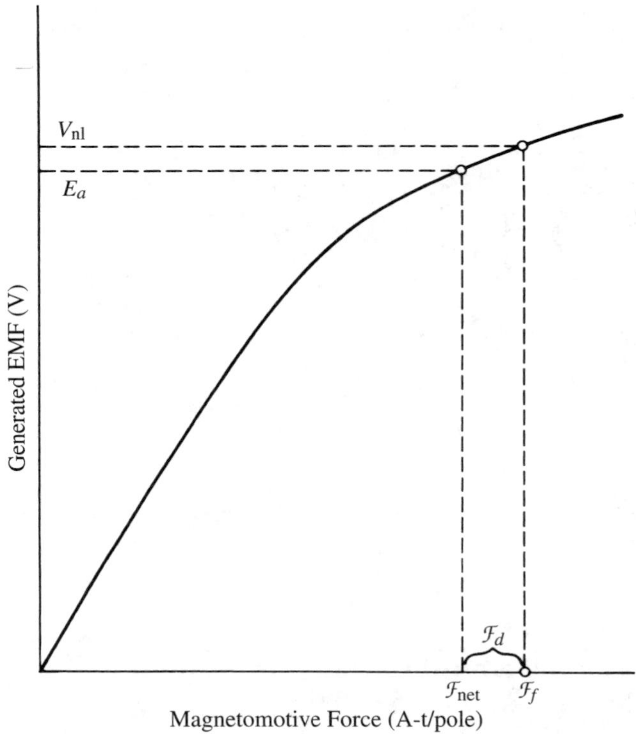

FIGURE 12.9
Magnetization curves showing the effect of armature reaction on the induced emf.

reaction, when supplying rated load at rated voltage and rated temperature, is assumed equal to 12.1 percent of the shunt field mmf. Determine (a) no-load voltage; (b) voltage regulation; (c) resistance setting of rheostat necessary to obtain rated voltage at rated conditions.

Solution

(a) $P = V_T I_a \quad \Rightarrow \quad 300{,}000 = 240 \times I_a \quad \Rightarrow \quad I_a = \underline{1250\text{ A}}$

$E_a = V_T + I_a(R_a + R_{\text{IP}}) = 240 + 1250(0.00234 + 0.00080)$

$E_a = 243.9\text{ V}$

The net mmf that results in $E_a = 243.9$ V is determined from the magnetization curve in Figure 12.10 to be ≈5100 A-t/pole.

$$\mathcal{F}_{\text{net}} = \mathcal{F}_f - \mathcal{F}_d = \mathcal{F}_f(1 - 0.121)$$

$$\mathcal{F}_f = \frac{5100}{1 - 0.121} = 5802\text{ A-t/pole}$$

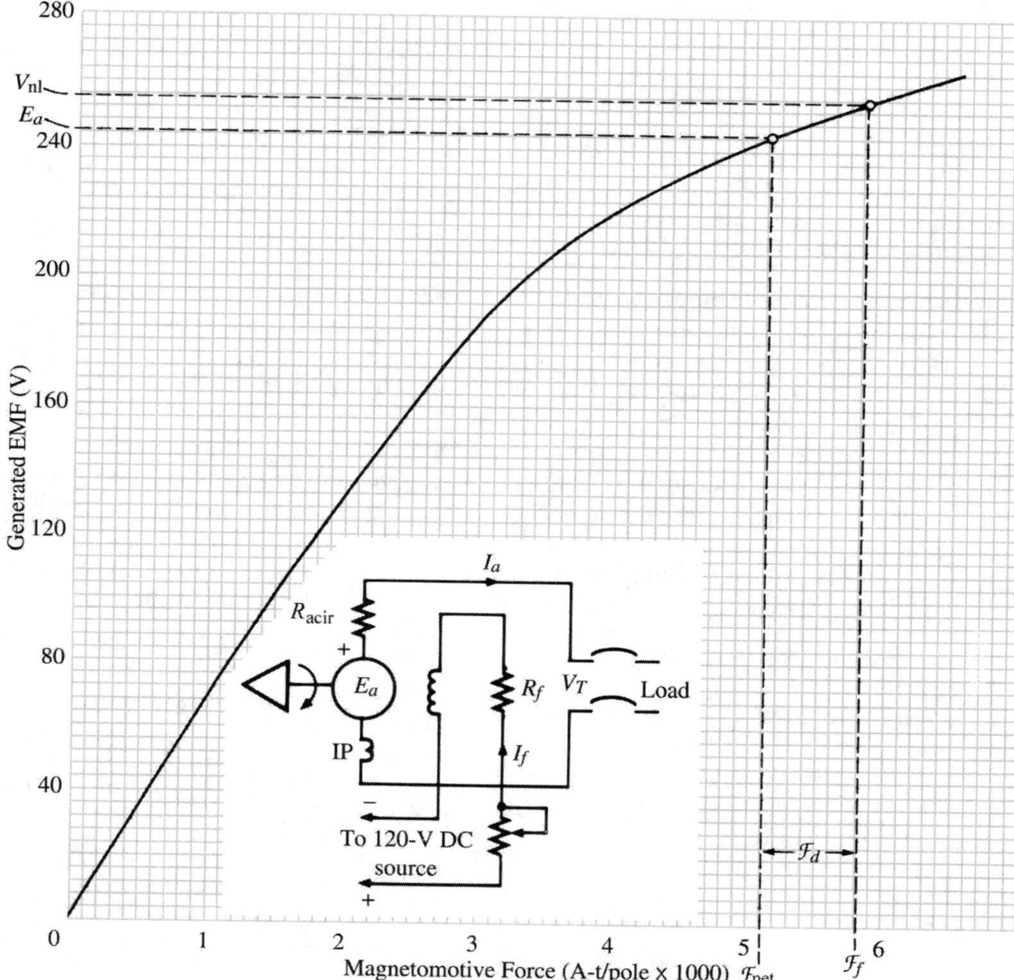

FIGURE 12.10
Magnetization curve and circuit diagram for Example 12.2.

The corresponding induced emf, as obtained from the magnetization curve in Figure 12.10, is the no-load voltage:

$$V_{nl} \approx \underline{255\text{ V}}$$

(b)

$$\text{VR} = \frac{V_{nl} - V_{rated}}{V_{rated}} \times 100 \quad \Rightarrow \quad \frac{255 - 240}{240} \times 100 = \underline{6.25\%}$$

(c) $\mathcal{F}_f = N_f I_f \quad \Rightarrow \quad 5802 = 1020 \times I_f \quad \Rightarrow \quad I_f = 5.69 \text{ A}$

$$I_f = \frac{V_f}{R_f + R_{\text{rheo}}} \quad \Rightarrow \quad R_{\text{rheo}} = \frac{V_f}{I_f} - R_f$$

$$R_{\text{rheo}} = \frac{120}{5.69} - 18.1 = \underline{3.0\ \Omega}$$

12.8 COMPOUND GENERATORS

The construction of compound generators is the same as for compound motors: Both have series and shunt fields. The shunt field of compound generators may be connected *short shunt,* as shown in Figure 12.11(a), or *long shunt,* as shown in Figure 12.11(b). The short-shunt connection avoids the small voltage drop across the series field and is generally preferred for generators. When solving problems involving self-excited generators, however, for ease of calculations, the long-shunt connection will be assumed. Numerical solutions assuming long-shunt or short-shunt connections are essentially the same, particularly so if graphical data from a magnetization curve is included.

The series field of compound generators may be connected cumulatively or differentially, depending on the application. Thus, expanding Eq. (12–1) to include the

FIGURE 12.11
Compound generator connections: (a) short shunt; (b) long shunt.

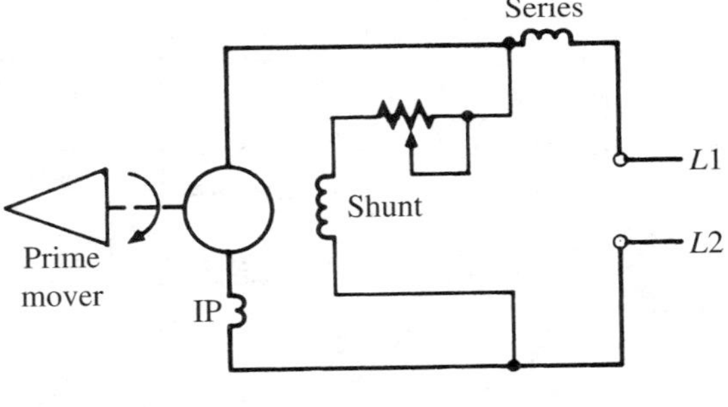

(a)

(b)

series field mmf (assuming cumulative compounding) and the equivalent demagnetizing mmf due to armature reaction,

$$E_a = n \cdot \left[\frac{\mathcal{F}_f + \mathcal{F}_s - \mathcal{F}_d}{\mathcal{R}} \right] k_G \tag{12–5}$$

where: $\mathcal{F}_f$ = shunt field mmf ($N_f I_f$) (A-t/pole)
$\mathcal{F}_s$ = series-field mmf ($N_s I_a$) (A-t/pole)
$\mathcal{F}_d$ = equivalent demagnetizing mmf due to armature reaction (A-t/pole)

Note: Although not exact, $\mathcal{F}_d$ is assumed proportional to the armature current. If a compensating winding is used, however, $\mathcal{F}_d = 0$. Thus, as indicated in Eq. (12–5), for a given armature current the increase in induced emf caused by the series field is dependent on the amount of compounding. More turns in the series field will produce higher emfs with the same load current. If enough turns are added to compensate for voltage drops caused by armature-circuit resistance and armature reaction, the terminal voltage will be essentially the same at no load as at full load, and the machine is said to be *flat compounded.* More than the minimum number of turns required for a flat-compound machine will cause the terminal voltage to rise with increasing load current; the machine is then said to be *overcompounded.* Less than the minimum number of turns will result in an *undercompound* machine. If the series field is differentially connected, its mmf will be negative, and an increase in the armature current will cause a decrease in flux and a corresponding decrease in generated emf. Figure 12.12 illustrates the variation in voltage from no load to full load for different types of compounding.

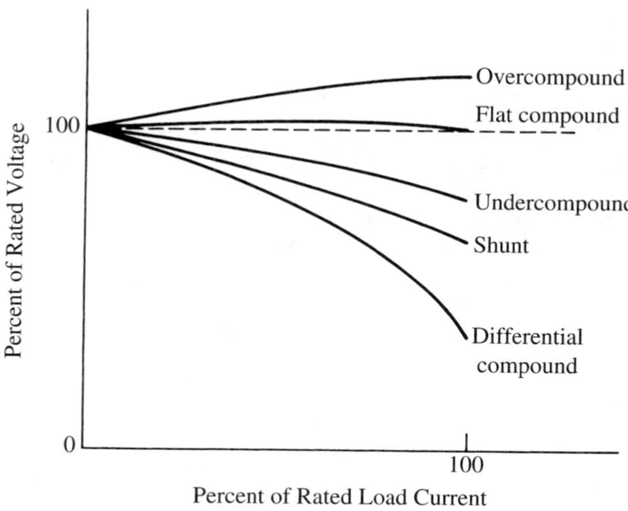

FIGURE 12.12
Comparison of load-voltage characteristics for different degrees of compounding.

Overcompound machines are used when DC power is to be transmitted over long distances; the rise in terminal voltage with load compensates for the voltage drop in the transmission line. A flat-compound machine is used when an essentially unvarying voltage is required and the distance of transmission is short. Shunt and undercompound generators are more stable for parallel operation than are flat and overcompound machines.

Differential compound generators are used when inherent overload protection is more desirable than a fixed voltage. Examples of its application are in generators supplying electric winches, and dredges, where an overload or short circuit would cause sufficient voltage drop to limit the current to a safe value.

EXAMPLE 12.3 A 250-V, 320-kW, 1150 r/min, self-excited, cumulative-compound, long-shunt generator is operating at rated conditions with the shunt field rheostat set for 7.70 Ω. The shunt field has 502 turns/pole, the series field has 1 turn/pole, and the generator parameters, expressed in ohms, are

Armature	IP + CW	Series	Shunt
0.00817	0.00238	0.00109	20.2

The circuit diagram and magnetization curve are shown in Figure 12.13. Determine (a) induced emf at rated load; (b) no-load voltage; (c) voltage regulation. (d) What is the type of compounding?

Solution

(a) $P_{\text{load}} = V_T I_{\text{load}} \quad \Rightarrow \quad 320{,}000 = 250 \times I_{\text{load}} \quad \Rightarrow \quad I_{\text{load}} = 1280.0 \text{ A}$

$$I_f = \frac{V_T}{R_f + R_{\text{rheo}}} = \frac{250}{20.2 + 7.70} = 8.96 \text{ A}$$

$$I_a = I_f + I_{\text{load}} = 8.96 + 1280.0 = 1288.96 \text{ A}$$

$$R_{\text{acir}} = 0.00817 + 0.00238 + 0.00109 = 0.01164 \ \Omega$$

$$E_a = V_T + I_a \times R_{\text{acir}} = 250 + 1288.96 \times 0.01164 = \underline{265.0 \text{ V}}$$

(b) At no load, the series-field mmf is zero and the induced emf is determined from the intersection of the field-resistance line and the magnetization curve. Using 250 V to determine the field-resistance line,

$$\mathcal{F}_f = N_f I_f = 502 \times 8.96 = 4498 \text{ A-t/pole}$$

Drawing a straight line through coordinates (0 V, 0 A-t) and (250 V, 4498 A-t) in Figure 12.13 results in an intersection at $\approx\underline{225 \text{ V}}$.

(c)
$$\text{VR} = \frac{V_{\text{nl}} - V_{\text{rated}}}{V_{\text{rated}}} \times 100 = \frac{225 - 250}{250} \times 100 = \underline{-10\%}$$

(d) The machine is overcompounded.

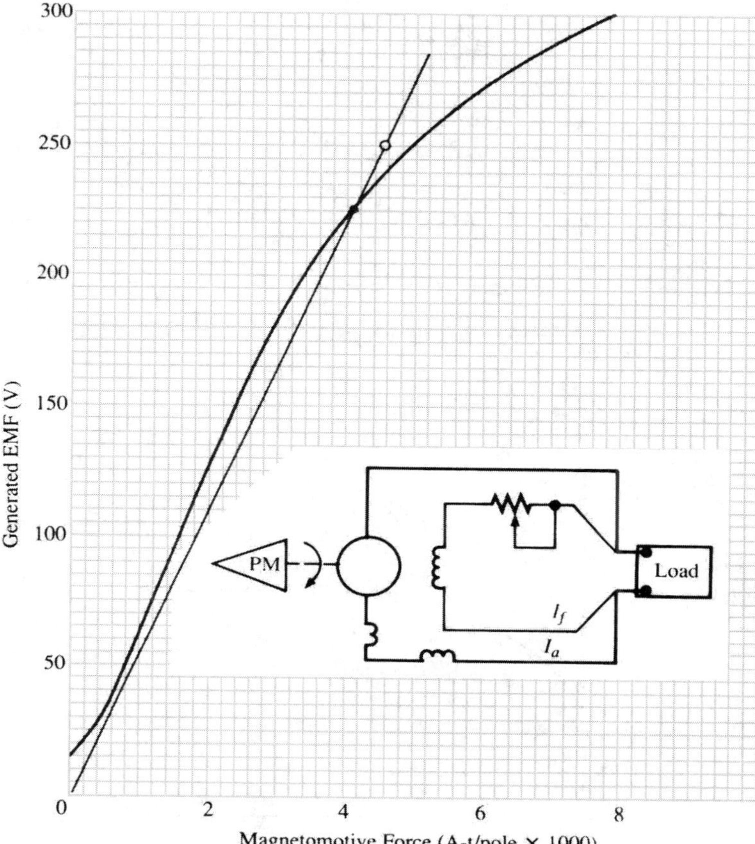

FIGURE 12.13
Magnetization curve and circuit diagram for Example 12.3.

12.9 SERIES-FIELD DIVERTER

The compounding of a generator may be reduced by diverting some of the armature current that would normally be in the series-field coils. This is illustrated in Figure 12.14(a), where a resistor, called a *diverter* or *series-field shunt,* is shown connected in parallel with the series field winding. The diverter is constructed of nichrome or other resistance material and mounted outside the generator housing.

The application of series-field diverters to generators that are subject to sudden heavy-load swings, however, requires that the diverter be made inductive by winding it on a laminated iron core. The reason for this is illustrated in Figure 12.14(b), assuming a noninductive diverter whose resistance is equal to the resistance of the series field. The noninductive diverter responds more rapidly to *changes in load current* than

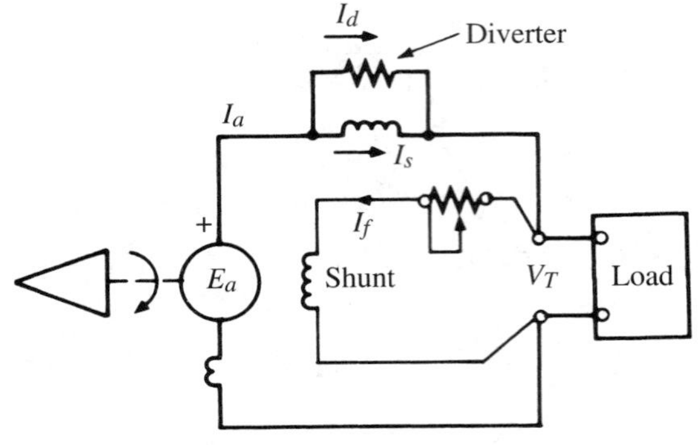

(a)

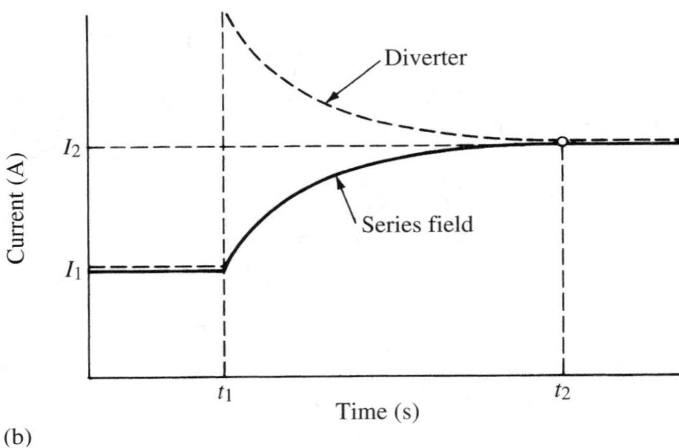

(b)

FIGURE 12.14
(a) Compound generator with a series-field diverter; (b) current transients in the series field and in the noninductive diverter for a suddenly applied load.

does the series field. Thus, as shown in Figure 12.14(b), the application of a sudden heavy load at time t_1 results in a delay in the buildup of current through the series field, and hence a delay in the buildup of generator emf. For optimum performance, diverters for compound generators with rapidly fluctuating loads should be designed to have the same inductive time constant as that of the series field.

EXAMPLE 12.4 The series field winding of a 170-kW, 250-V, 1450 r/min compound generator has a resistance of 0.00306 Ω. Determine (a) required resistance of a noninductive diverter that will bypass 27 percent of the total armature current; (b) power rating of the diverter.

Solution

The circuit is similar to that shown in Figure 12.14(a).

(a)

$$I_s = 0.73I_a \qquad I_d = 0.27I_a$$

Since the voltage drop across the diverter is the same as that across the series-field winding,

$$I_sR_s = I_dR_d \qquad \Rightarrow \qquad R_d = R_s \cdot \frac{I_s}{I_d} = 0.00306 \times \frac{0.73}{0.27} = \underline{0.00827\ \Omega}$$

(b)

$$I_a = \frac{P_{\text{load}}}{V_T} = \frac{170{,}000}{250} = 680\text{ A}$$

$$P_d = I_d^2R_d = (680 \times 0.27)^2 \times 0.00827 = \underline{279\text{ W}}$$

12.10 COMPOUNDING EFFECT OF SPEED

The speed of a generator has a pronounced effect on the degree of compounding. When operating at higher speeds, the flux contributed by the series field will be more effective in generating an emf. Although the shunt field rheostat can be adjusted to provide the same output voltage at different speeds, the series-field mmf is a function of the load current and cannot be adjusted by rheostat action. Hence, when operating above rated speed, a flat compound generator will behave like an overcompound machine, and when operating below rated speed it will behave like an undercompound machine.

12.11 PARALLELING DIRECT-CURRENT GENERATORS

When the load demanded from a system of generators exceeds the amount that can be safely supplied, additional generators must be switched onto the system to carry the additional load. Referring to Figure 12.15, GEN *B* is connected to the bus, and GEN *A* (the incoming machine) is about to be paralleled with GEN *B*. The disconnect switch permits isolation of the circuit breaker from the bus for purposes of test, maintenance, and repair.

Procedure for Paralleling

1. Close the generator disconnect switch of the incoming machine.
2. Start the prime mover and adjust it to rated speed.
3. Adjust the voltage of the incoming machine to be a few volts higher than the bus voltage.
4. Close the generator breaker.

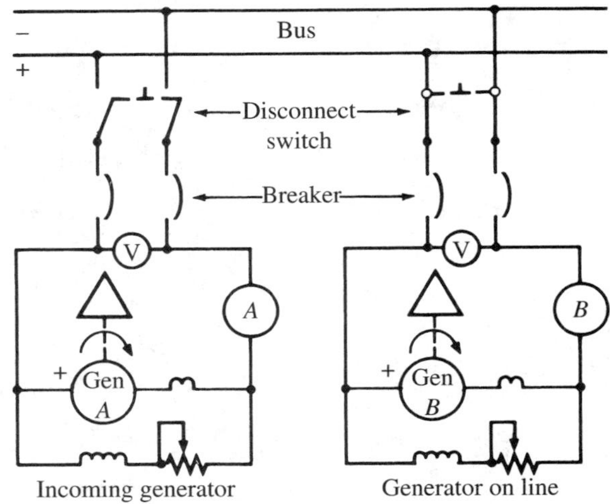

FIGURE 12.15
Circuit diagram showing connections for parallel operation of DC generators.

5. Turn the shunt field rheostat of the incoming machine in the "raise voltage" direction, and that of the other machine or machines on the bus in the "lower voltage" direction until the desired load distribution is attained. Load distribution is indicated on the respective ammeters; wattmeters are not used.

Before attempting to parallel a newly installed machine or one that had been disconnected for repairs, a polarity check should be made to ensure that the positive terminal of the incoming machine will connect to the positive terminal of the bus. A simple polarity check may be made with a DC voltmeter, testing across the top terminals of the circuit breaker and then across the bottom terminals; corresponding terminals must have the same polarity. Paralleling a machine with reversed polarity effectively short circuits the generator and results in damaged brushes, a damaged commutator, and a blacked-out plant. Generators that have been tripped off the bus because of a heavy *fault current* (short circuit) should be checked for reversed polarity before reclosing the breaker.

12.12 EFFECT OF FIELD-RHEOSTAT ADJUSTMENT ON THE LOAD-VOLTAGE CHARACTERISTICS OF DC GENERATORS

The circuit diagram and typical load-voltage characteristic of output voltage vs. armature current for a DC generator are shown in Figure 12.16. Although usually drawn as a straight line, the actual characteristic has a slight curve. The drooping characteristic provides inherent stability of operation when paralleled with other machines. Adjusting the shunt field rheostat raises or lowers the no-load voltage setting of the generator, but does not change the droop. Curves for different no-load voltage settings are shown with broken lines in Figure 12.16(b). For a given no-load voltage setting of the shunt field rheostat, a DC generator with a drooping characteristic can have only one value of load current for a given output voltage.

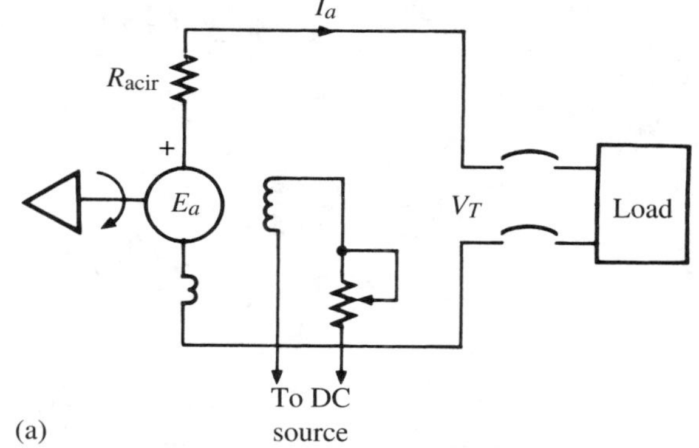

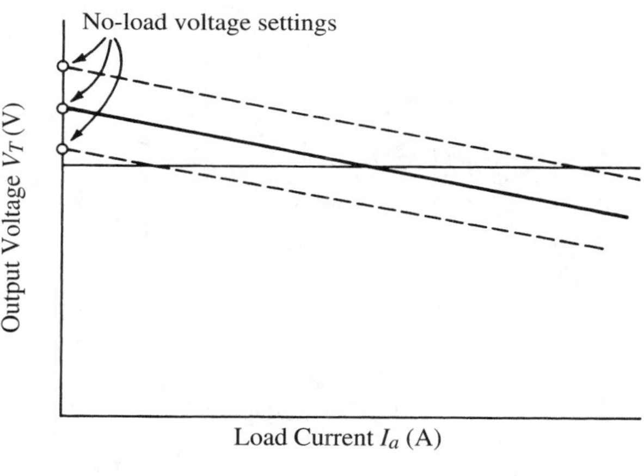

FIGURE 12.16
Circuit diagram and typical load-voltage characteristic of a DC generator, showing different no-load voltage settings.

12.13 DIVISION OF ONCOMING BUS-LOAD BETWEEN DC GENERATORS IN PARALLEL

The division of oncoming bus load between DC generators in parallel is determined by the voltage droop of the respective generators, which cannot be easily changed by the operator. This is shown in Figure 12.17(a), where two generators with different voltage droops are in parallel and supply equal shares of the total load current (subscript 1 on the curves).[2] The droop characteristics are exaggerated to illustrate the behavior of the machines as the bus load is increased or decreased. An increase in bus load causes an increase in the internal voltage drops, and thus a decrease in the bus voltage, shown with a broken line and subscript 2. Note that the generator with the least droop took a greater share of the *increase* in bus load.

[2]The one-line diagram in Figure 12.17(a) is an elementary representation of two generators connected to the bus.

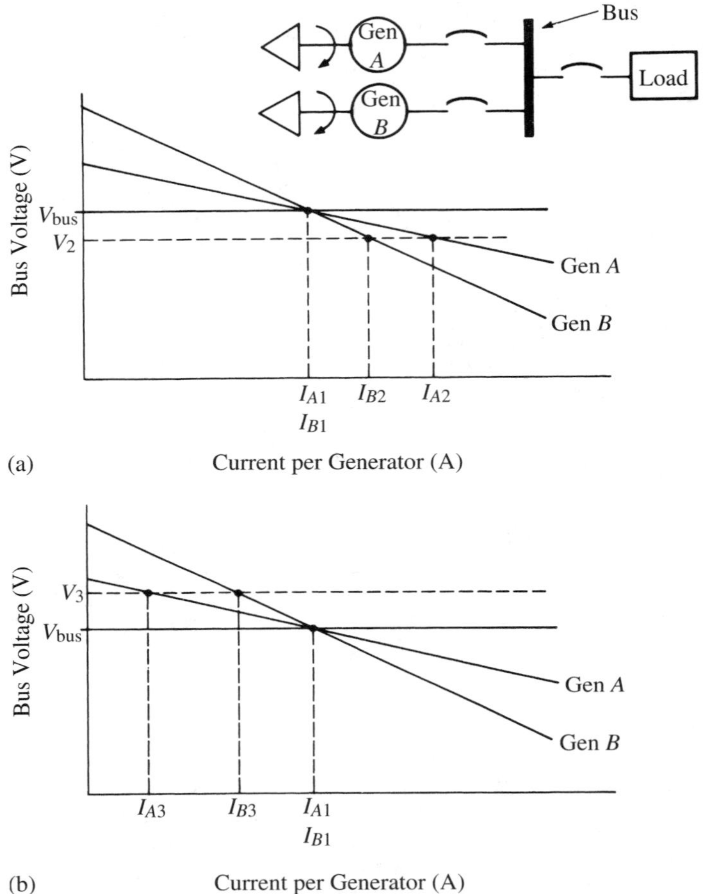

FIGURE 12.17
Load-voltage characteristics of two generators, with different voltage regulations, operating in parallel: (a) increasing bus load; (b) decreasing bus load.

Similarly, as shown in Figure 12.17(b), a decrease in bus load causes a decrease in the internal voltage drops and thus an increase in the bus voltage. As shown by subscript 3, the generator with the least droop assumed a greater share of the *change* in bus load. Whether the bus load is increasing or decreasing, *the machine with the least droop always assumes a greater portion of the change in bus load.*

For optimum performance, machines operating in parallel should have the same voltage regulation, and for stability of operation, the voltage regulation should be between 3 and 8 percent, obtained by governor adjustment and/or electrical design. *Paralleled machines with different power ratings but the same voltage regulation will divide any oncoming bus load in direct proportion to their respective power ratings.*

12.14 CHARACTERISTIC TRIANGLE AS A TOOL FOR SOLVING LOAD-DISTRIBUTION PROBLEMS BETWEEN PARALLELED DC GENERATORS

The solution of problems involving load distribution between DC generators in parallel may be accomplished in a straightforward manner by defining a *characteristic triangle* for each machine, and then using similar triangles to determine a solution.

Referring to Figure 12.18, the characteristic triangle is fixed for a given voltage droop and does not change with changes in the field-rheostat setting. The bus voltage is the terminal voltage (also called output voltage) of the generator.

An increase in bus load causes a decrease in bus voltage, as shown by the broken line in Figure 12.18. The intersection of the new voltage line with the load-voltage characteristic establishes a new triangle similar to the characteristic triangle. From the geometry of similar triangles,

$$\frac{\Delta V_{bus}}{\Delta I} = \frac{V_{nl} - V_{rated}}{I_{rated}} \tag{12–6}$$

From the voltage-regulation equation, Eq. (10–16),

$$V_{nl} - V_{rated} = \frac{VR}{100} \times V_{rated} \tag{12–7}$$

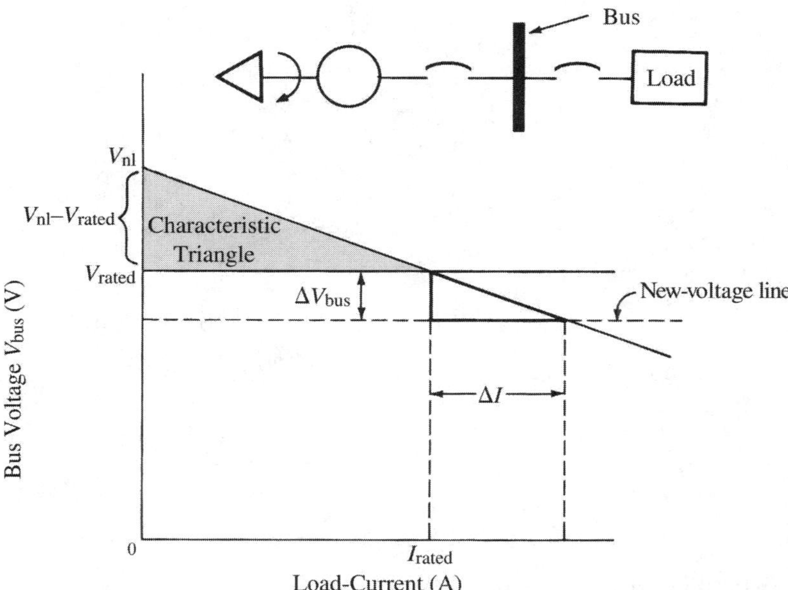

FIGURE 12.18
Characteristic triangle defines the droop characteristic of a generator.

Substituting Eq. (12–7) into Eq. (12–6),

$$\frac{\Delta V_{bus}}{\Delta I} = \frac{V_{rated} \times \dfrac{VR}{100}}{I_{rated}} \tag{12–8}$$

where: VR = voltage regulation
ΔI = change in generator load current (A)
ΔV_{bus} = change in bus voltage due to change in generator load current (V)

EXAMPLE 12.5 A 300-kW, 250-V DC generator *A* with a voltage regulation of 3.0 percent, and a 400-kW, 250-V DC generator *B*, with a voltage regulation of 5.0 percent, are operating in parallel and taking equal shares of a 350-kW, 250-V bus load. A one-line diagram of the paralleled machines is shown in Figure 12.19(a). If the bus load is increased to 2500 A, determine (a) new bus voltage; (b) current supplied by each generator.

Solution

(a) Neglecting the relatively small shunt field current,

$$I_{A,rated} = \frac{300{,}000}{250} = 1200 \text{ A} \qquad I_{B,rated} = \frac{400{,}000}{250} = 1600 \text{ A}$$

Substituting data from the characteristic triangles in Figure 12.19(b) into Eq. (12–8),

$$\frac{\Delta V_{bus}}{\Delta I} = \frac{V_{rated} \times \dfrac{VR}{100}}{I_{rated}}$$

$$\frac{\Delta V_{bus}}{\Delta I_A} = \frac{250 \times 0.03}{1200} \qquad \frac{\Delta V_{bus}}{\Delta I_B} = \frac{250 \times 0.05}{1600}$$

$$\Delta I_A = 160\Delta V_{bus} \qquad \Delta I_B = 128\Delta V_{bus}$$

The change in bus current is

$$\Delta I_A + \Delta I_B = \Delta V_{bus}(160 + 128) \tag{12–9}$$

$$\text{Original bus current} = \frac{350{,}000}{250} = 1400 \text{ A}$$

$$\text{New bus current} = 2500 \text{ A}$$

Thus,

$$\Delta I_A + \Delta I_B = 2500 - 1400 = 1100 \text{ A} \tag{12–10}$$

Substituting Eq. (12–10) into Eq. (12–9),

$$1100 = \Delta V_{bus} \times (160 + 128)$$

$$\Delta V_{bus} = 3.82 \text{ V}$$

Gen A
Gen B
Bus

(a)

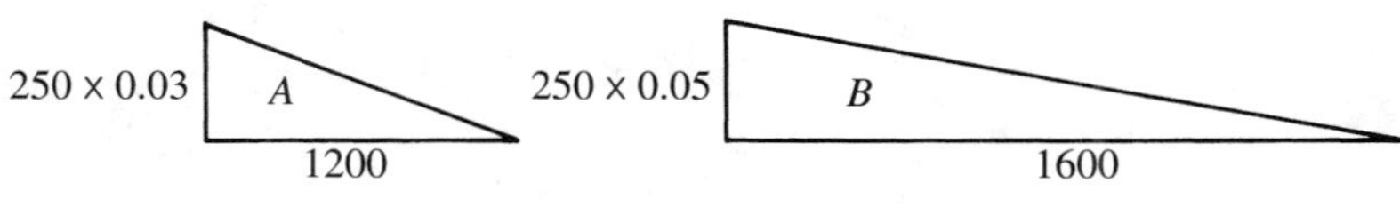

(b)

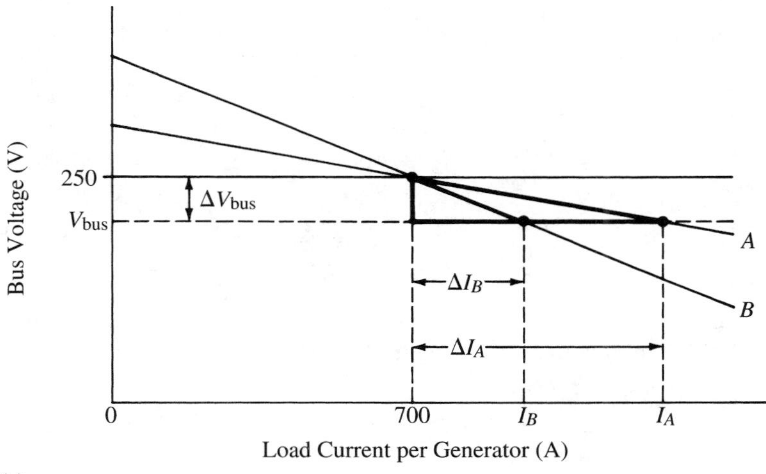

(c)

FIGURE 12.19
Illustrations for Example 12.5: (a) one-line diagram; (b) characteristic triangles; (c) load-voltage characteristics.

From Figure 12.19(c),

(b)

$$V_{bus} = 250 - \Delta V_{bus} = 250 - 3.82 = \underline{246.18 \text{ V}}$$

$$I_A = 700 + \Delta I_A = 700 + 160 \times 3.82 = \underline{1311 \text{ A}}$$

$$I_B = 700 + \Delta I_B = 700 + 128 \times 3.82 = \underline{1189 \text{ A}}$$

Note: Machine *A* is overloaded by

$$\frac{1311 - 1200}{1200} \times 100 = 9.2\%$$

EXAMPLE 12.6 A 100-kW, 250-V machine A is in parallel with a 300-kW, 250-V machine B. The voltage regulation of both machines is 4.0 percent. Machine A is carrying 200 A, and machine B is carrying 500 A. Determine (a) the increment increase in load on each machine if an additional 400-A load is connected to the bus; (b) current carried by each machine.

Solution

(a) Using the methods illustrated in Example 12.5,

$$I_{A,\text{rated}} = \frac{100{,}000}{250} = 400 \text{ A} \qquad I_{B,\text{rated}} = \frac{300{,}000}{250} = 1200 \text{ A}$$

$$\frac{\Delta V_{\text{bus}}}{\Delta I} = \frac{V_{\text{rated}} \times \dfrac{\text{VR}}{100}}{I_{\text{rated}}}$$

$$\frac{\Delta V_{\text{bus}}}{\Delta I_A} = \frac{250 \times 0.04}{400} \qquad \frac{\Delta V_{\text{bus}}}{\Delta I_B} = \frac{250 \times 0.04}{1200}$$

$$\Delta I_A = 40 \times \Delta V_{\text{bus}} \qquad \Delta I_B = 120 \times \Delta V_{\text{bus}}$$

The additional bus current = 400 A

$$\Delta I_A + \Delta I_B = 400$$

$$40\Delta V_{\text{bus}} + 120\Delta V_{\text{bus}} = 400 \quad \Rightarrow \quad \Delta V_{\text{bus}} = 2.5 \text{ V}$$

$$\Delta I_A = 40 \times \Delta V_{\text{bus}} = 40 \times 2.50 = \underline{100 \text{ A}}$$

$$\Delta I_B = 120 \times \Delta V_{\text{bus}} = 120 \times 2.50 = \underline{300 \text{ A}}$$

Note: Both generators have the same voltage regulation. Hence, any oncoming bus load will be divided in proportion to the power ratings of the respective machines.

$$\Delta I_A = \left(\frac{100}{100 + 300}\right) \times 400 = 100 \text{ A}$$

$$\Delta I_B = \left(\frac{300}{100 + 300}\right) \times 400 = 300 \text{ A}$$

A much easier solution!

(b)

$$I_A = 200 + 100 = \underline{300 \text{ A}}$$

$$I_B = 500 + 300 = \underline{800 \text{ A}}$$

12.15 THEORY OF LOAD TRANSFER BETWEEN DC GENERATORS IN PARALLEL

The transfer of bus load between generators is easily accomplished by using the respective field rheostats to change the no-load voltage settings of the respective machines. The machine that is to take some load should have its field rheostat turned in the "raise voltage" direction, and the machine that is to lose some load should have its field rheostat turned in the "lower voltage" direction. Observations of load are made on the generator ammeters. Wattmeters are not required and are rarely used. The transfer of load may be made by simultaneous adjustment of both generator rheostats or by separate adjustment on each machine until the desired transfer is accomplished.

Although the transfer of load between DC generators is accomplished indirectly through the medium of field excitation, the energy supplied by each generator is still dependent on the energy input to its respective prime mover; the machine that had its excitation increased generates a higher emf, causing an increase in its current output. Assuming the bus load is constant (e.g., a fixed lighting load), the increased current output of one machine will result in a decrease in current output of the other machine. The machine that accepted the additional load starts to slow down and its speed governor reacts to admit more energy to its prime mover. The machine that lost some load starts to speed up, and its speed governor reacts to decrease the energy input to its prime mover.

Figure 12.20 illustrates a two-step procedure for the transfer of load between shunt machines with identical load-voltage characteristics. Machine A carries a load of 200 A at 240 V, and machine B has just been paralleled and is carrying no load. The initial conditions are indicated by subscript 1 and the use of solid lines. To transfer

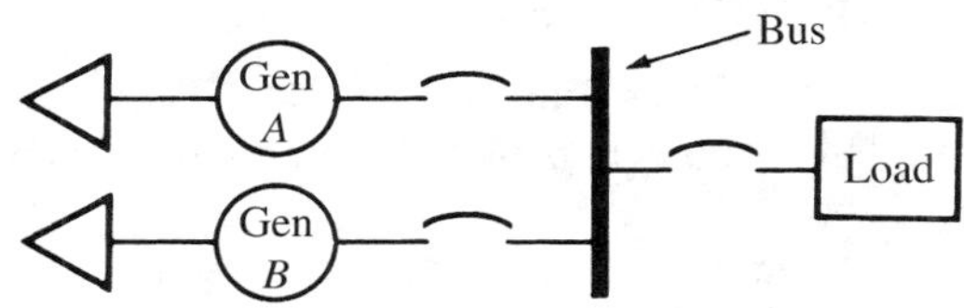

(a)

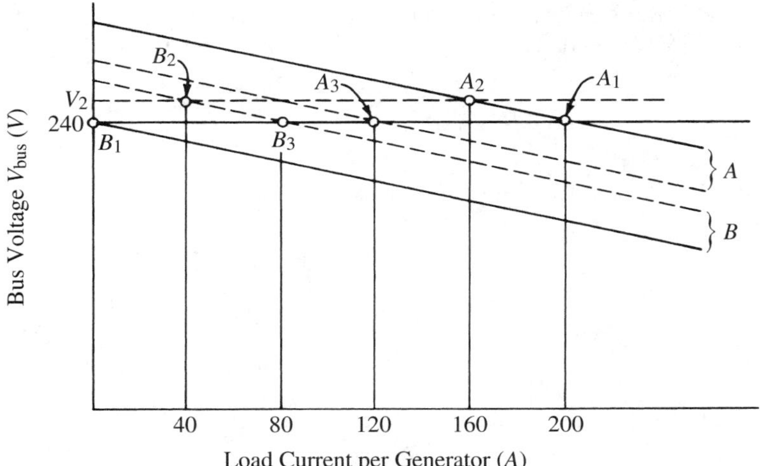

(b)

FIGURE 12.20
One-line diagram and load-voltage characteristics illustrating load transfer through shunt field rheostat control.

load from machine *A* to machine *B* and still maintain 240 V requires adjustment of the field rheostats of both machines.

Assume that 80 A are to be transferred from machine *A* to machine *B*. For the example shown in Figure 12.20, the field current of machine *B* is increased by adjustment of its field rheostat until one-half the amount of load that is to be transferred is shifted—namely, 40 A. This adjustment raises the no-load voltage setting, and hence the entire characteristic of machine *B*. The bus will then operate at some higher voltage V_2, and the characteristic of machine *A* will intersect the new voltage line at 160 A, as denoted by subscript 2. To complete the transfer of load, the field rheostat of machine *A* is adjusted to decrease its excitation until the remaining 40 A are transferred. The characteristic of machine *A* is thereby lowered, and the voltage of the system comes back to its original 240-V value. This is indicated by subscript 3, where machine *A* takes 120 A and machine *B* takes 80 A. The slope of the curves was not changed throughout the transition period. The entire characteristic of each machine was moved either up or down by adjustment of the field rheostats.

12.16 COMPOUND GENERATORS IN PARALLEL

For compound generators to operate successfully in parallel it is necessary to parallel their series fields. This is accomplished by an *equalizer connection,* such as is shown in Figure 12.21.

Consider two identical compound machines operating in parallel without an equalizer, and with the bus load carefully balanced between the two. Any attempt to transfer some of the load from one machine to the other by adjustment of the field rheostats would result in the machine with the greater excitation taking the entire load and driving the other as a motor; the machine that accepted some load would have its series field strengthened, whereas the machine that gave up some load would have its series field weakened. This would serve to increase the generated emf of the machine

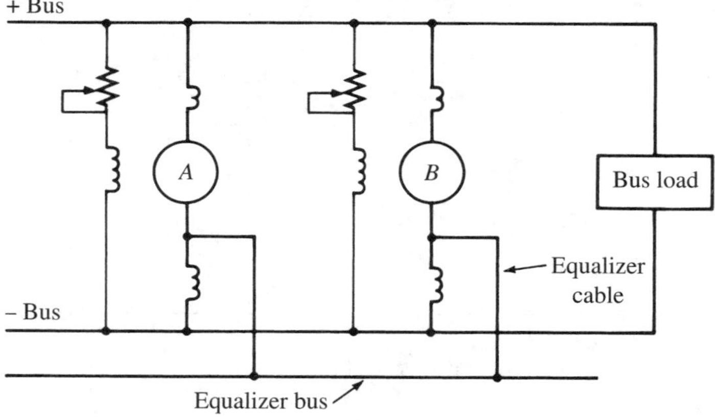

FIGURE 12.21
Compound generators in parallel, showing equalizer connections.

that accepted some load and to decrease the generated emf of the machine that gave up some load, resulting in an additional load transfer, etc.

Furthermore, motoring of a compound generator without an equalizer would result in a reversal of current through its series field, causing it to act as a differential compound motor. If the motoring generator is not immediately tripped off the bus, and the series field mmf is greater than the shunt field mmf, the polarity of the residual magnetism will be reversed; when the machine is restarted it will build up voltage in the reverse direction. When equalizers are used, however, accidental or deliberate motoring of a compound generator cannot cause reversed polarity.

The resistance of each equalizer cable (and its connections) should not exceed 20 percent of the resistance of the series field winding of the smallest of the paralleled machines **[3]**. This ensures that the series fields of paralleled generators will divide the total bus current in the approximate inverse ratio of their respective series-field resistances.

12.17 REVERSE-CURRENT TRIP

A reverse-current trip provides protection against sustained motoring. The layout and connections of one type of reverse-current trip are shown in Figure 12.22. The series coil is wound around an iron bar, called the armature, and connected in series with the

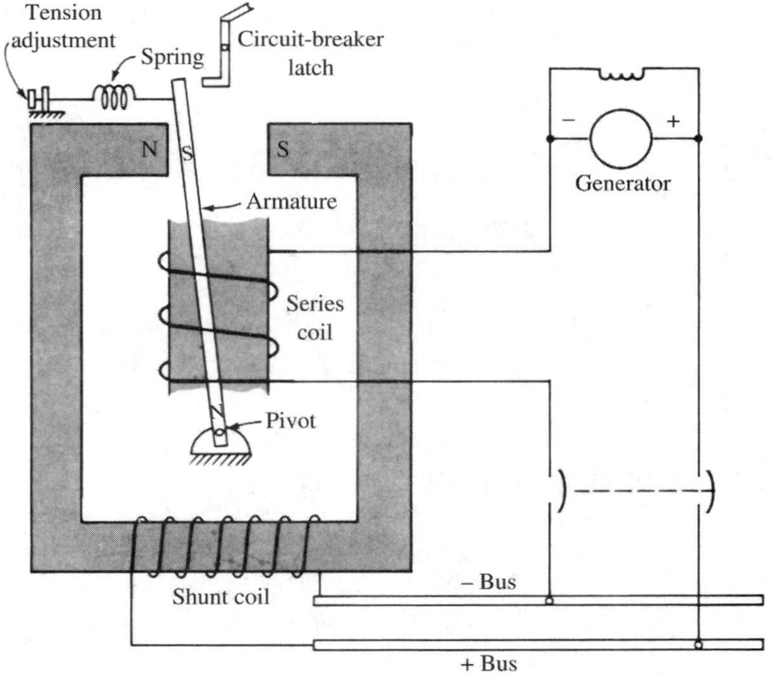

FIGURE 12.22
Reverse-current trip.

generator armature. The iron bar is pivoted at the bottom and free to move within the coil. Current in the series coil magnetizes the iron bar in the vertical direction. The shunt coil is connected across the bus and establishes the flux of the horseshoe magnet. The two coil connections are arranged so that as long as the generator current is in the proper direction the iron bar will be held in the nontripping position. If for some reason the generator voltage falls below that of the bus, however, the current in the generator will reverse its direction. This will reverse the polarity of the iron bar and cause it to be attracted by the opposite pole of the horseshoe magnet. Movement of the iron bar trips the circuit-breaker latch and the machine is tripped off the line. A spring with very light tension is used to hold the bar in the nontripping position. This is necessary to prevent vibration from accidentally tripping the circuit breaker when the generator is lightly loaded. The spring tension may be adjusted to vary the amount of reverse current required to operate the trip.

SUMMARY OF EQUATIONS FOR PROBLEM SOLVING

$$\text{VR} = \frac{V_{\text{nl}} - V_{\text{rated}}}{V_{\text{rated}}} \times 100 \text{ percent} \tag{10–16}$$

$$E_a = n\Phi_p k_G = n\left(\frac{N_f I_f}{\mathcal{R}}\right)k_G \tag{12–1}$$

$$E_a = I_f(R_f + R_{\text{rheo}}) \tag{12–2}$$

$$E_a = V_T + I_a R_{\text{acir}} \tag{12–3}$$

$$R_{\text{acir}} = R_a + R_{\text{IP}} + R_{\text{CW}}$$

$$E_a = n \cdot \left[\frac{\mathcal{F}_f + \mathcal{F}_s - \mathcal{F}_d}{\mathcal{R}}\right]k_G \tag{12–5}$$

$$\frac{\Delta V_{\text{bus}}}{\Delta I} = \frac{V_{\text{rated}} \times \dfrac{\text{VR}}{100}}{I_{\text{rated}}} \tag{12–8}$$

SPECIFIC REFERENCES KEYED TO TEXT

1. Hubert, C. I. *Preventive Maintenance of Electrical Equipment,* Prentice Hall, Upper Saddle River, NJ, 2002.
2. Langsdorf, A. S. *Principles of Direct Current Machines,* McGraw-Hill, New York, 1959.
3. National Electrical Manufacturers Association. *Motors and Generators.* Publication No. MG-1-1998, NEMA, Rosslyn VA, 1999.
4. Siskind, C. S. *Direct Current Machinery.* McGraw-Hill, New York, 1952.

GENERAL REFERENCE

Kloeffler, R. G., R. M. Kerchner, and J. L. Brenneman. *Direct-Current Machinery*. Macmillan, New York, 1949.

REVIEW QUESTIONS

1. (a) Sketch the magnetization curve and the field-resistance line for a self-excited generator. (b) Explain how the generator builds up its emf and what determines the maximum voltage that the generator may attain.
2. How may the voltage of a self-excited generator be adjusted?
3. What effect does the speed of a self-excited generator have on the buildup of voltage? Explain.
4. What basic faults can prevent a self-excited generator from building up voltage?
5. How can reversed polarity of a self-excited generator be detected and corrected?
6. (a) What is the function of the series field in a compound generator? (b) Explain why it is connected in series with the armature.
7. What is a diverter? How is it connected, and under what conditions would it be used?
8. How does a differential-compound generator differ from a cumulative-compound generator? State an application for the differential-compound machine.
9. Explain why changes in speed change the degree of compounding of a compound generator.
10. State in detail the procedure for paralleling one DC generator with another that is already on the bus. Include in your statement the instruments observed and the equipment operated.
11. How can the transfer of load be accomplished between two DC generators in parallel? Include in your statement the equipment operated and the instruments observed.
12. Explain why adjustment of the field excitation of a DC generator causes a shift in load between machines in parallel.
13. Two DC generators with identical drooping load-voltage characteristics are operating in parallel. Each machine carries an equal share of the bus load at 500 V. Sketch on one set of coordinate axes the approximate load-voltage characteristics and draw the 500-V line. With broken lines appropriately drawn and lettered to indicate the transition steps, show and explain how a portion of the load of one machine can be transferred to the other machine, and still have the same bus voltage when the transfer is completed.
14. Explain how the equalizer stabilizes the operation of compound generators in parallel.
15. If two cumulative-compound generators are connected in parallel without an equalizer, any attempt to transfer some load from one to the other may result in the motoring of one of the machines. Explain this behavior.
16. Explain the operation of the reverse-current trip as used in a generator breaker.

PROBLEMS

12–1/7 A 150-kW, 250-V, 1750 r/min, separately excited, noncompensated shunt generator has the following parameters: $R_{acir} = 0.0072\ \Omega$, $R_f = 18.6\ \Omega$, $N_f = 1491$ turns/pole. The field is excited from a 120-V source, and the magnetization curve is shown in Figure 12.23. The machine is operating at rated load, and the equivalent demagnetizing mmf caused by armature reaction is 15.2 percent of the shunt field mmf. Determine (a) no-load voltage;

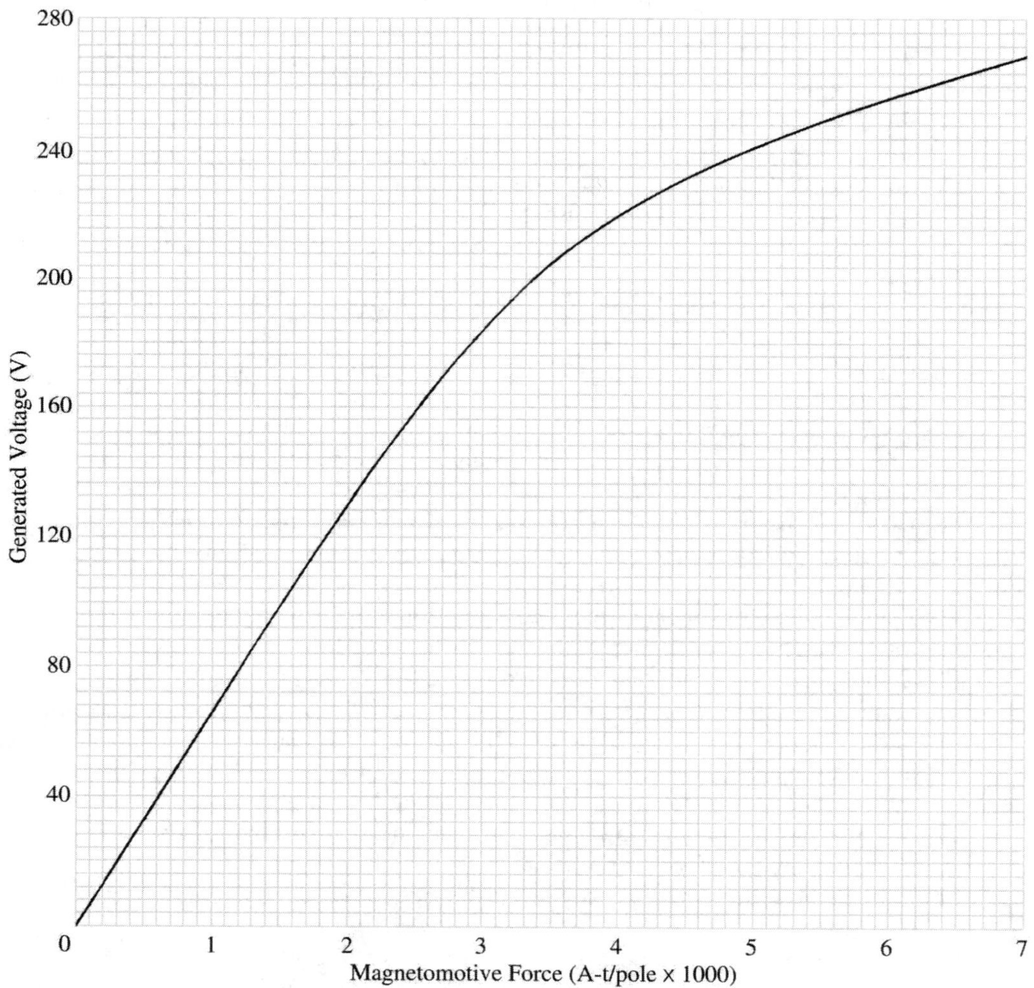

FIGURE 12.23
Magnetization curve for Problems 12–1/8 through 12–3/8.

(b) voltage regulation; (c) resistance of rheostat setting necessary to obtain rated voltage at rated conditions.

12–2/7 A 100-kW, 1750 r/min, 240-V, separately excited shunt generator without compensating windings has the following parameters: $R_{acir} = 0.026\ \Omega$, $R_f = 32\ \Omega$, $N_f = 1520$ turns/pole. The magnetization curve for the generator is shown in Figure 12.23. The machine is operating at rated load, is separately excited from a 220-V source, and the equivalent demagnetizing mmf caused by armature reaction is 10.4 percent of the shunt field mmf. Determine (a) no-load voltage; (b) voltage regulation; (c) resistance and power rating of rheostat setting necessary to obtain rated voltage at rated conditions.

12–3/7 A noncompensated, separately excited shunt generator rated at 250 V, 50 kW, 3450 r/min, has the following parameters: $R_a + R_{IP} = 0.023\ \Omega$, $R_f = 324\ \Omega$, $N_f = 1896$ turns/pole. The magnetization curve for the generator is shown in Figure 12.23. The machine is operating at rated load, and the equivalent demagnetizing mmf due to armature reaction is 8.65 percent of rated shunt field mmf. Determine the no-load voltage at 3000 r/min.

12–4/7 A 60-kW, 240-V, 1750 r/min, self-excited, compensated shunt generator has the following parameters, expressed in ohms:

Armature	**IP + CW**	**Shunt**
0.0415	0.0176	44.6

The shunt field has 1653 turns/pole, and the magnetization curve is given in Figure 12.24. The generator supplies rated power at rated voltage when operating at rated speed with the shunt field rheostat set for 20.0 Ω. Determine (a) load current; (b) shunt field current; (c) armature current; (d) induced emf when operating at rated conditions; (e) no-load voltage, assuming speed and field rheostat are not changed; (f) critical resistance; (g) no-load voltage if operating with the rheostat adjusted to 13.6 Ω and the machine operating at 60 percent speed.

12–5/7 A 170-kW, 850 r/min, 250-V, self-excited shunt generator has the following parameters expressed in ohms:

Armature	**Interpole**	**Shunt**
0.0154	0.0065	29.0

The machine delivers rated power at rated voltage when operating at rated speed and the field rheostat set for 17.0 Ω. The shunt field has 1121 turns/pole, and the magnetization curve in Figure 12.24 applies. Determine (a) induced emf at rated load; (b) no-load voltage; (c) voltage regulation; (d) critical field resistance; (e) no-load voltage if the rheostat is set for 9.0 Ω.

12–6/7 A six-pole, self-excited shunt generator has an output rating of 500 kW and 240 V at a speed of 1200 r/min. The combined resistance of the armature and interpole windings is 0.00475 Ω, and the brush drop is insignificant. The resistance of the shunt field and field rheostat are 15.2 Ω, and 7.80 Ω, respectively. The shunt field winding has 508 turns/pole, and the

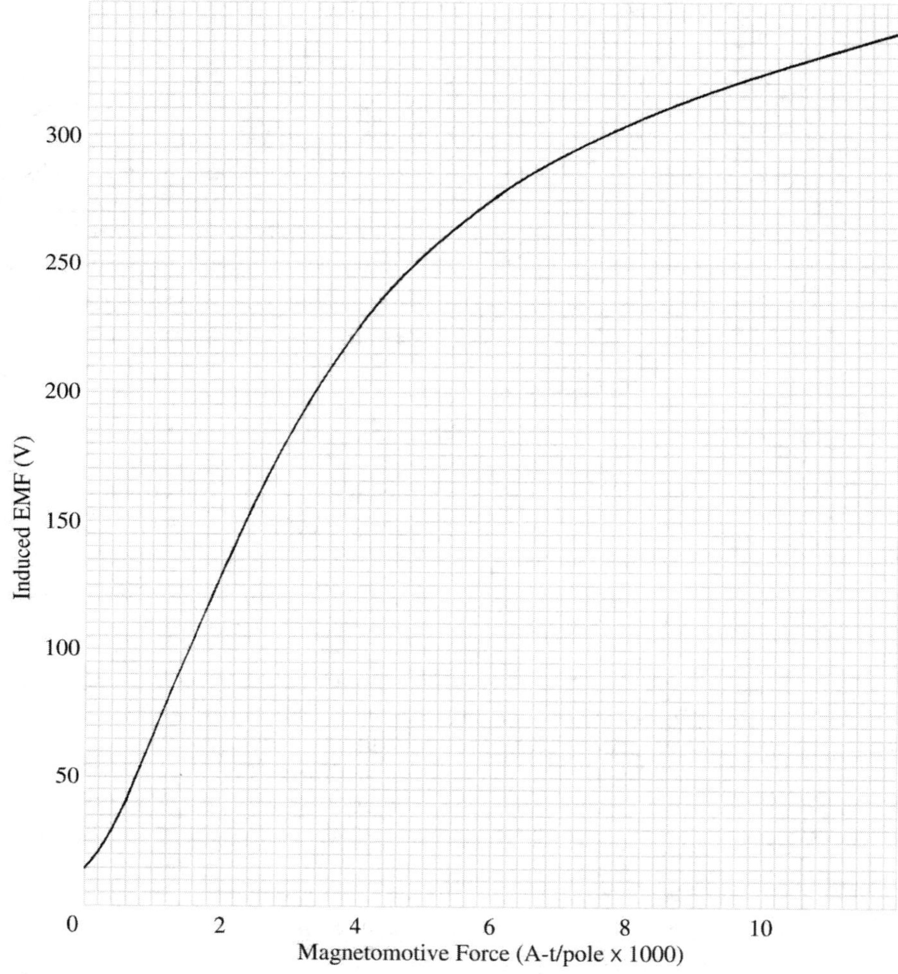

FIGURE 12.24
Magnetization curve for Problems 12–4/7 through 12–8/8.

magnetization curve in Figure 12.24 applies. Assuming the generator is delivering rated output, determine (a) load current; (b) shunt field current; (c) armature current; (d) induced emf; (e) no-load voltage; (f) voltage regulation; (g) frequency of the voltage generated in the armature; (h) no-load voltage if the rheostat is not changed, but the speed is cut in half; (i) critical resistance of the field circuit when operating at 1200 r/min; (j) losses in the armature circuit, shunt field winding, and rheostat when operating at rated conditions; (k) required driving torque of the prime mover if the generator is to deliver rated power at rated speed and rated voltage. Neglect windage and friction losses.

12–7/8 A 125-kW, 250-V, 1450 r/min, self-excited, cumulative-compound generator has the following parameters:

	Armature	Interpole	Series	Shunt
Resistance, Ω	0.02776	0.00808	0.00372	32.3
turns/pole			3	927

The rheostat is set for 13.2 Ω when operating at rated conditions. The magnetization curve in Figure 12.24 applies. Determine (a) rated load current; (b) shunt field current; (c) armature current; (d) induced emf at rated conditions; (e) no-load voltage; (f) voltage regulation; (g) type of compounding. (h) If the series field is disconnected because of damaged insulation, can the machine be operated as a shunt machine and obtain rated operating conditions through field-rheostat adjustment?

12–8/8 A 320-kW, 850 r/min, 250-V, self-excited, cumulative-compound generator is operating at rated conditions with the field rheostat set for 8.80 Ω. The magnetization curve in Figure 12.24 applies and the generator parameters are:

	Armature	IP + CW	Series	Shunt
Resistance, Ω	0.0131	0.00380	0.00175	27.2
turns/pole			2	630

Determine (a) load current; (b) shunt field current; (c) armature current; (d) no-load voltage; (e) voltage regulation; (f) critical resistance; (g) maximum no-load voltage attainable by rheostat adjustment; (h) total electrical losses at rated load.

12–9/9 A self-excited compound generator delivers its rated 400 kW at 500 V and 850 r/min, and the shunt field rheostat is set for 24 Ω. The generator parameters are:

	Armature	IP + CW	Series	Shunt
Resistance, Ω	0.01754	0.005105	0.002351	80

Determine (a) resistance of a diverter that will reduce the current in the series field to 365 A when operating at rated armature current; (b) power loss in the diverter.

12–10/9 A 480-V, 720 r/min, 500-V, self-excited compound generator has the following parameters:

	Armature	Interpole	Series	Shunt
Resistance, Ω	0.01432	0.00501	0.00216	82

The generator supplies rated load when operating at rated speed and the field rheostat is set for 18.2 Ω. If a series-field diverter of 0.00321 Ω is installed, and the shunt field rheostat is adjusted so that the generator is delivering its rated armature current, determine (a) voltage drop across the diverter; (b) current in the series field; (c) power loss in the diverter.

12–11/14 Two undercompound generators, *A* and *B,* are operating in parallel at 125 V and take equal shares of a 1000-A bus load. Each machine has a rating of 200 kW at 125 V. The voltage regulation of machine *A* is 6 percent, and that of machine *B* is 4 percent. If the bus load increases to 1800 A, determine (a) new bus voltage; (b) division of bus load between machines.

12–12/14 Two 500-kW, 250-V, undercompound generators *A* and *B* are operating in parallel. The voltage regulation of the machines are 2 and 3 percent, respectively. Both machines are operating at rated load and 250 V. Determine (a) new bus voltage; (b) division of load between machines when the total bus load drops to 2800 A.

12–13/14 Two DC generators are operating in parallel and take equal shares of a 900-kW bus load at 555 V. Generator *A* is rated at 500 kW, 600 V, and 3 percent regulation. Generator *B* is rated at 750 kW, 600 V, and 5 percent regulation. If the bus load drops to 1000 A, determine (a) new bus voltage; (b) current supplied by each machine.

12–14/14 Two identical 1000-kW, 600-V DC generators, whose regulation is 4 percent, are operating in parallel and take equal shares of the bus load. The bus load is 900 kW at 620 V. If one generator tripped off the line, determine (a) the voltage of the remaining machine; (b) voltage of the tripped machine.

12–15/14 Three 240-V DC generators are operating in parallel and taking equal shares of a 3000-A, 240-V bus load. The ratings of the generators are:

	A	***B***	***C***
Regulation (%)	2	4	6
Rated power (kW)	400	300	200

If the total bus load drops to 2000 A, determine (a) bus voltage; (b) current supplied by each machine.

13

Control of Electric Motors

13.1 INTRODUCTION

This chapter provides an introduction to magnetic, solid-state, and programmable controllers. Magnetic controllers are introduced first; they are the easiest to visualize, are universally used, and the associated ladder diagrams provide the background for future applications to programmable controllers. The sections on solid-state controllers cover only power circuits; firing circuits are beyond the scope of this text. The brief section on programmable controllers provides some insight into this expanding field.

13.2 CONTROLLER COMPONENTS

Magnetic controllers use magnetically operated relays and contractors to start and stop electric motors, and through the use of suitable control devices, limit current, limit torque, limit speed, change speed, change direction of rotation, provide uniform acceleration, apply dynamic braking, protect against damage due to overload, and so on. The control may be exercised automatically by a system of programming, using electric timing devices, sequence interlocks, and pilot switches. At the push of a button or the closing of a master switch the magnetic controller automatically starts the machine, limits its current, and provides proper acceleration; it takes the guesswork out of motor starting.

Magnetic Contactors

A magnetic contactor is a magnetically operated switch that serves to close or open an electric circuit. A cutaway view of a DC magnetic contactor is shown in Figure 13.1(a). When energized, the operating coil pulls on the armature, causing the contacts to close. The blow-out coil is connected in series with the stationary contact and provides a magnetic flux to "blow" the arc up the chute, where it is extinguished by elongation and cooling when the contacts are opened. Wear on the stationary and movable contacts is reduced by the arcing horn, which takes the brunt of the burning. The blow-out coil shifts the arc to the arcing horn and the upper curved part of the stationary

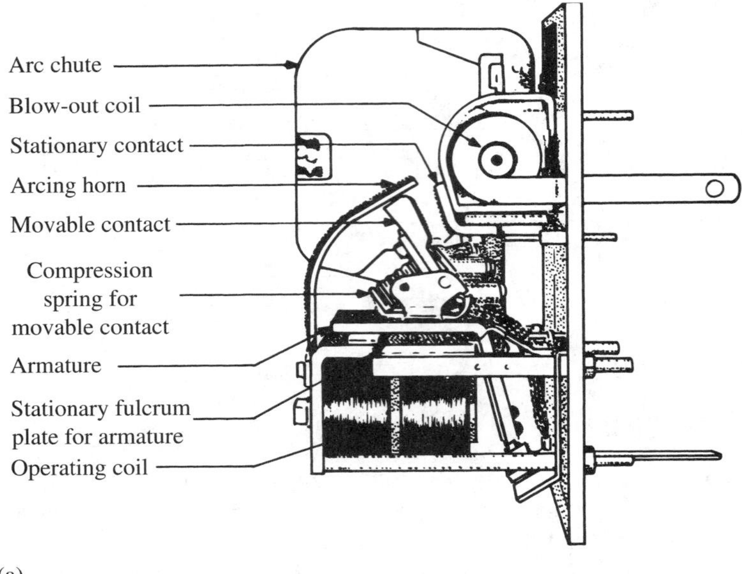

(a)

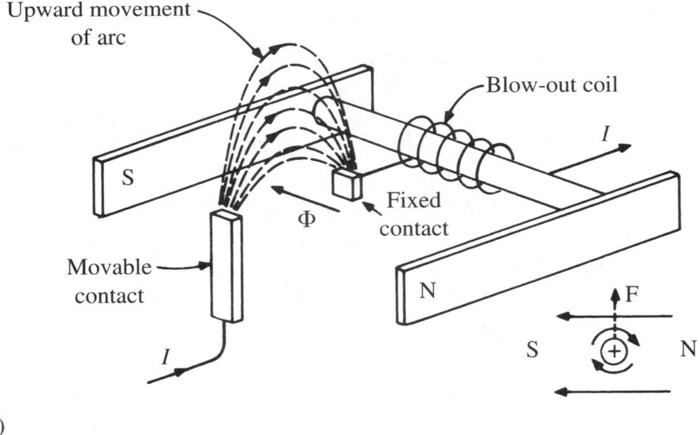

(b)

FIGURE 13.1
Magnetic contactor: (a) cutaway view of a DC contactor (Courtesy GE Industrial Systems); (b) operation of magnetic blow-out.

contact, where it is extinguished. Figure 13.1(b) illustrates the behavior of the magnetic blow-out. Current in the coil sets up a magnetic flux in the north and south direction, as shown. When the contacts separate, an arc is established in a direction perpendicular to the flux of the field. The resultant motor action pushes the arc upward, stretching it until it breaks. The upward movement of the arc may be verified by applying the flux bunching rule; see inset in Figure 13.1(b).

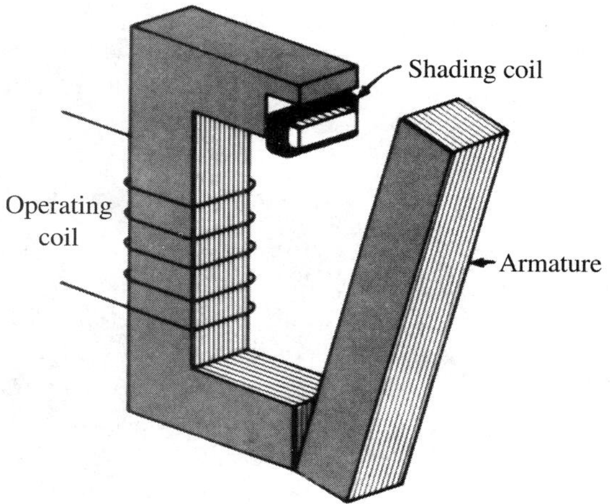

FIGURE 13.2
Magnetic circuit for an AC contactor, showing the shading coil.

Alternating-current contactors and relays also are equipped with arcing horns and blow-out coils. The magnetic core, however, is laminated to reduce eddy currents, and a pole shader is used to prevent the magnetic flux from dropping to zero each time the current in the coil goes through zero. The pole shader, shown in Figure 13.2, is a one-turn, short-circuited coil on the face of the core. The shading coil acts as the short-circuited secondary of a transformer. In accordance with Lenz's law, the shading coil causes the flux in the shaded part of the pole to lag behind the flux in the nonshaded part. This prevents the net flux from falling to zero, and thus reduces armature vibration (chatter).

Relays

A relay is a device used to control the operation of a magnetic contactor or other device. Relays operate as a function of current, voltage, heat, and pressure, and supply the "intelligence" that is necessary to provide automatic acceleration, protect against overload, undervoltage, excessive speeds, excessive torque, etc.

13.3 MOTOR-OVERLOAD PROTECTION

Excessive current drawn by an overloaded motor will, if allowed to continue for sufficient time, result in dangerous overheating of both motor and control. Correctly sized motor-overload relays with the proper time-current characteristics provide protection against sustained overloads, and yet permit the short-duration, high locked-rotor current necessary for motor starting. They also protect against overheating due to overcurrents caused by a stalled rotor, low-voltage, low-frequency, unbalanced voltages, and some other types of motor faults. Although such overcurrents are of relatively low magnitude compared with short circuits, they will, if sustained, shorten the life of the insulation.

Motor-overload relays do not protect against short circuits or grounds. Protection against short circuits and grounds must be provided by branch-circuit protection devices such as fuses or circuit breakers.[1]

Thermal-Overload Relay

The thermal-overload relay shown in Figure 13.3(a) consists of a bimetallic element, a heater element, normally closed contacts, normally open contacts, and a reset button.[2] The bimetallic element is formed from two strips of dissimilar metals with different coefficients of linear expansion that are bonded together to form a single element. The heater is connected in series with the motor and simulates the I^2R heating of the motor windings. Should an overload occur, the increase in heater temperature caused by the greater-than-normal motor current will cause the bimetallic element to deflect. This opens the normally closed contacts, which de-energizes the motor, and closes the normally open contacts, which energizes an alarm circuit. The reset arm is used to manually reset the relay contacts after the bimetallic element has cooled.

Figure 13.3(b) shows how the heater and associated normally closed contacts are connected into a controller circuit. Controllers for large motors use current transformers (CTs) to provide proportional but lower values of current to a proportionally lower rated heater; this is shown with broken lines in Figure 13.3(b). An overload of sufficient duration causes normally closed contact OL to open. This de-energizes coil M, causing the three normally open M contacts to fall to the open position.

The approximate current-time characteristics for tripping and resetting the bimetallic relay are shown in Figure 13.3(c). Note that the tripping time decreases with increasing current (inverse-time characteristic), but the reset time increases with increasing current. For the characteristics shown, an overload of 400 percent rated heater current causes tripping somewhere between 20 and 30 s. The resetting time for the specific family of relays, however, is roughly between 75 and 140 s, during which time the motor is at rest and cooling. The overcurrent relay shown in Figure 13.3(a) is provided with an adjustable setting to permit tripping at approximately 15 percent above or 15 percent below the heater rating.

The selection of an overcurrent protective device is determined by the type of overload the machine may encounter. Motors that drive fans, pumps, and most machine tools require protection against sustained overloads that may cause excessive heating of the insulation; such motors are generally protected by time-delay thermal-overload relays, whose tripping time decreases with increasing overload. In other applications, such as motor-driven elevators, winches, traction motors, and other loads that are subject to jamming, instantaneous magnetic-overload relays are used. Jamming of a motor will cause the current to rise to and hold at its locked-rotor value; if not immediately disconnected, the insulation may be destroyed.

---

[1]See Section 430 of the National Electrical Code (NEC) for proper selection of running-overload protection of motors, branch-circuit protection, and sizing of branch-circuit conductors **[4]**,**[5]**.

[2]See the references for other types of motor-overload relays, such as solder-pot and heater, solenoid and dashpot, and solid-state devices **[3]**.

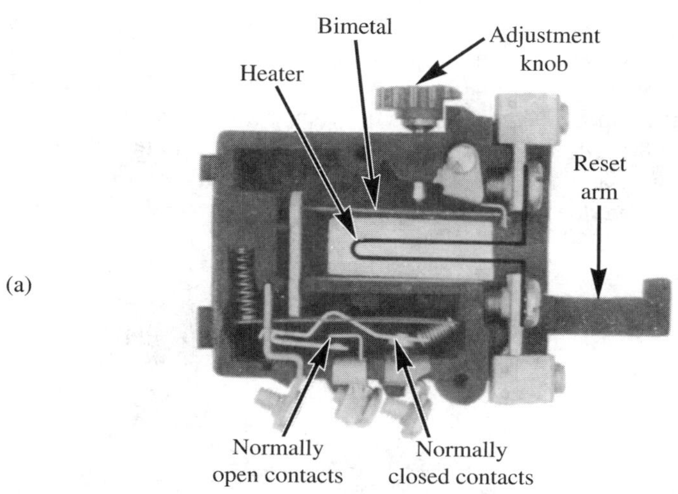

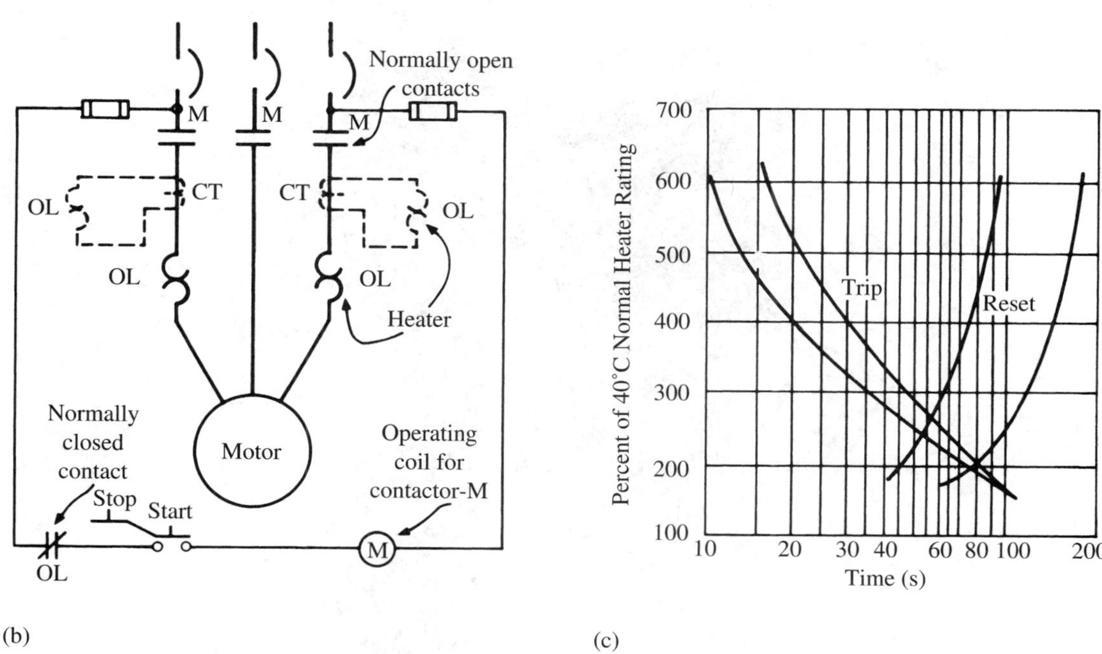

FIGURE 13.3
Thermal-overload relay (Courtesy GE Industrial Systems): (a) cutaway view; (b) circuit diagram; (c) current-time characteristics for tripping and resetting.

One type of magnetic-overload relay is shown in Figure 13.4. The operating coil is connected in series with the motor to be protected. When an overload occurs, the high value of flux attracts the armature to the pole face, causing the normally closed contact to open. The operating coil is supported on a heavy copper tube (called a copper jacket), which is in turn mounted on the magnetic core of the relay. The copper jacket acts as an inductive time-delay device to prevent the relay from tripping on transient overloads. Normal motor currents do not cause sufficient flux to trip the relay.

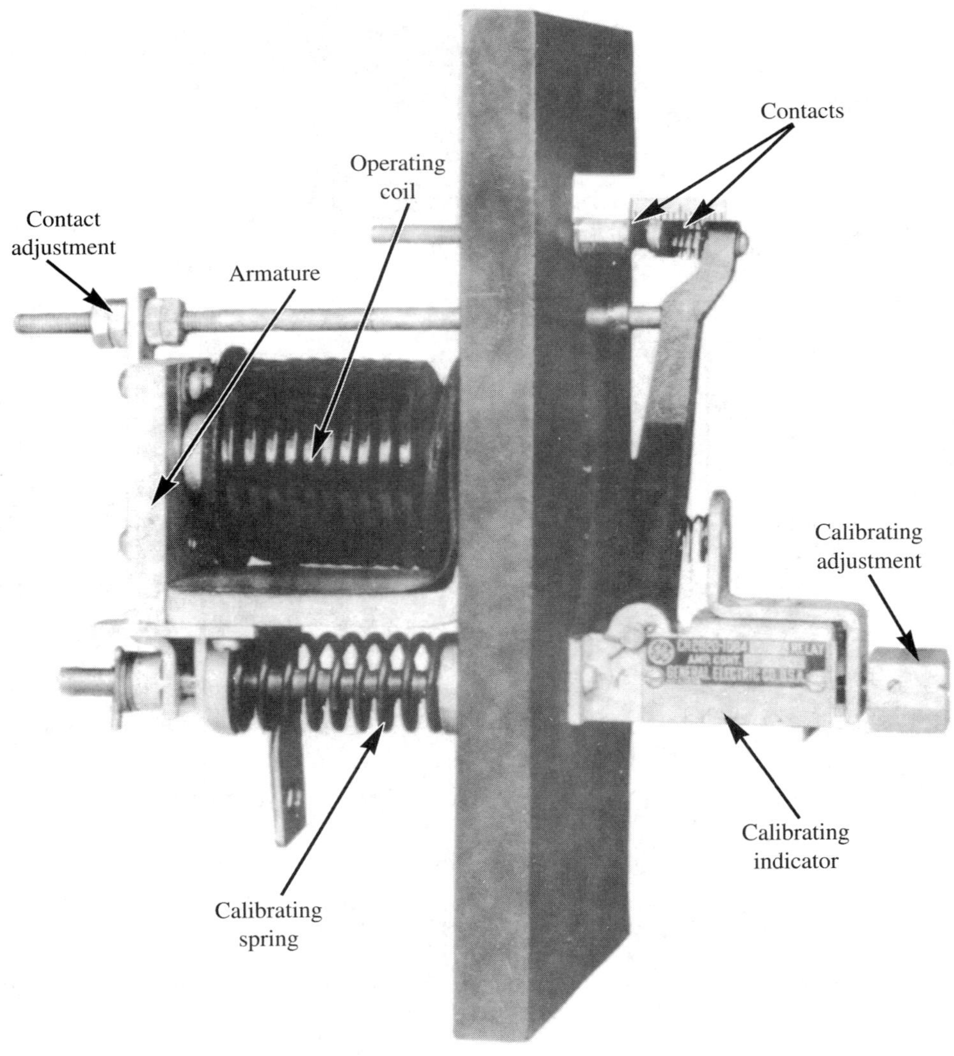

FIGURE 13.4
Instantaneous overcurrent relay. (Courtesy GE Industrial Systems)

13.4 CONTROLLER DIAGRAMS

The two general classifications of controller diagrams are the *connection diagram* and the *elementary diagram.*[3] The connection diagram, shown in Figure 13.5, shows the actual physical layout of the controller components and the location and routing of each wire. The connection diagram facilitates the location of the components for repair, replacement, or adjustment. Note that the power circuit, which carries the motor current, is drawn with heavy lines to indicate larger cross-section copper. The control circuit is drawn with light lines. The panel heater is used in damp environments to prevent condensation of moisture.

The elementary diagram, shown in Figure 13.6, is a simplified drawing that shows how the system works. The elementary diagram is often referred to as a *ladder diagram,* because the horizontal lines are like rungs and the two outside power lines are like the rails. The elementary diagram has a *power circuit* and a *control circuit.* The power circuit includes the motor components, starting resistors, and power contacts. The control circuit, also called the *logic circuit,* includes push buttons, operating coils, overload contacts, relay contacts, and limit switches that provide the logic and commands for operation of the motor. Note that the control components of a device shown in the connection diagram of Figure 13.5 are separated in the ladder diagram. For example, the three components of relay 2A in Figure 13.5 are shown separated in Figure 13.6. To avoid confusion, all components of a device are assigned the same letter designation in both the connection diagram and the ladder diagram. Furthermore, the numbering of the terminals in the ladder diagram corresponds to the numbering in the connection diagram. The numbering also serves as a guide for determining points of equal potential and is of considerable assistance when troubleshooting; terminals with the same numbering should have like potentials regardless of contactor position.[4]

The order of components in the ladder diagram is such that reading down the rungs, and from left to right, shows the sequence (or logic) of control operation. The ladder diagram is a simplified diagram that is easier to read than the connection diagram, and is easier to diagnose if malfunctioning occurs.

13.5 AUTOMATIC SHUTDOWN ON POWER FAILURE

All motor control circuits provide automatic shutdown on power failure, and either resequence the logic for automatic restarting when power is restored (undervoltage release) or remain in the off condition until manually restarted (undervoltage protection).

Undervoltage Release

The control circuit illustrated in Figure 13.7(a) provides for the automatic disconnection of the motor from the line in case of voltage loss and its automatic reconnection when voltage is restored. This type of control is generally applied to vital machinery,

[3]See Appendix D for graphic symbols used in controller diagrams.

[4]See Reference **[2]** for troubleshooting and maintenance of magnetic controllers.

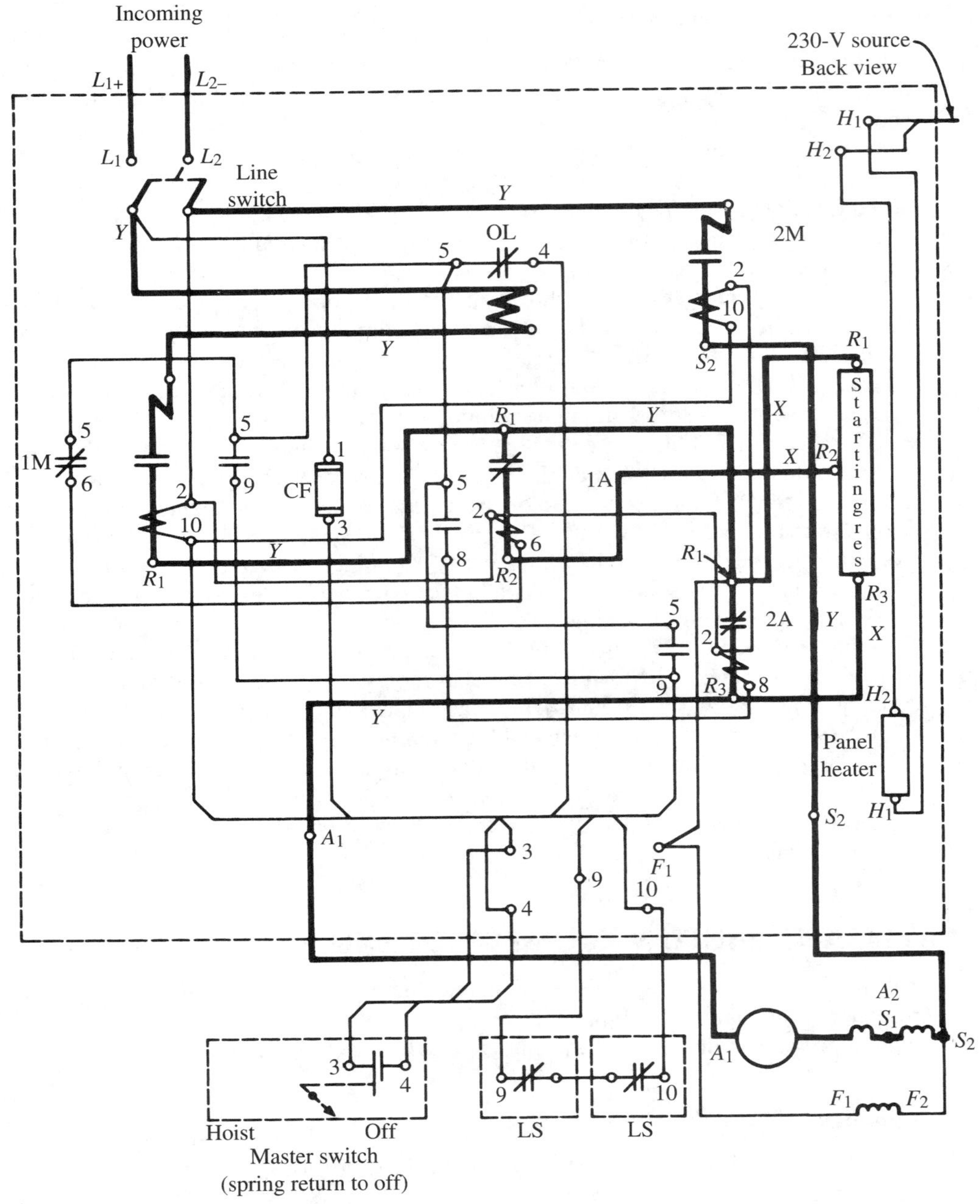

FIGURE 13.5
Typical connection diagram for a motor controller. (Courtesy GE Industrial Systems)

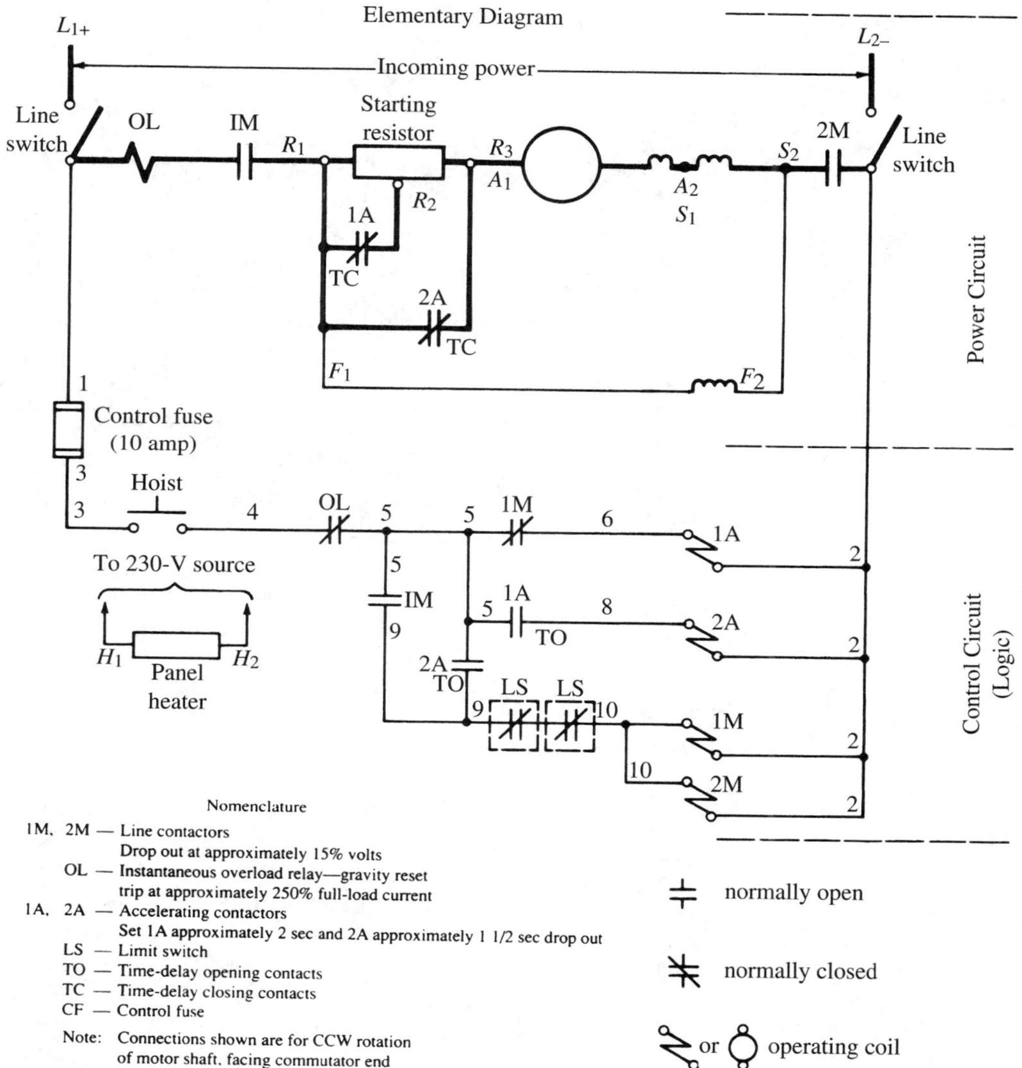

FIGURE 13.6
Elementary diagram for the controller in Figure 13.5. (Courtesy GE Industrial Systems)

in which it is imperative that the operation be automatic. Sump pumps and fire pumps are some of its applications. A non-spring-return type of push button is used in undervoltage release circuits. When the start button is pushed, it remains depressed and causes the control circuit to stay closed until the stop button is pushed. Hence, if voltage loss causes contactor M to open, the restoration of voltage will result in automatic reclosure.

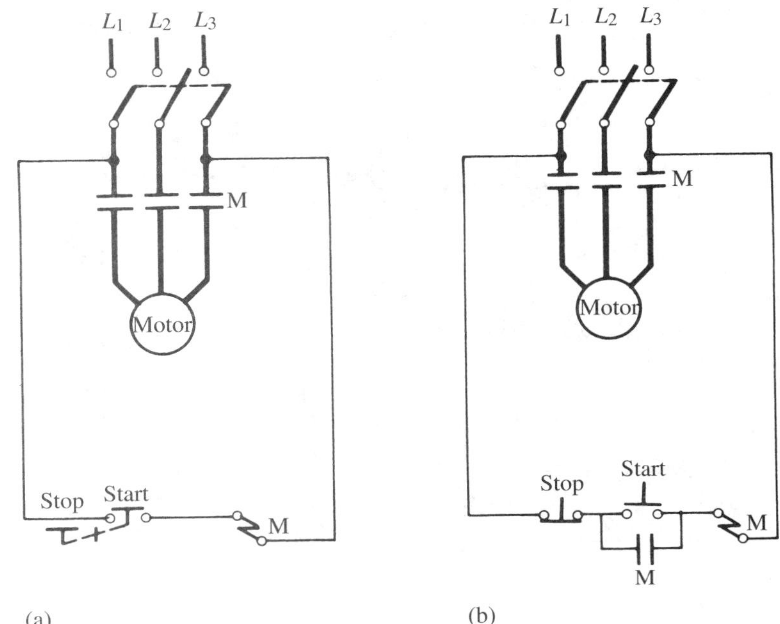

FIGURE 13.7
Circuits for automatic shutdown on power failure: (a) undervoltage release; (b) undervoltage protection.

Undervoltage Protection

Figure 13.7(b) illustrates an elementary control circuit that provides voltage loss protection. This type of control removes the machine from the line when voltage loss occurs and prevents automatic restarting when voltage is restored. The stop and start buttons are of the spring-return type and are not mechanically interlocked. When the start button is pushed, coil M becomes energized, closing the M contacts in the motor circuit and the auxiliary contact across the start button. The auxiliary contact, called a *sealing contact,* seals the starting circuit by short circuiting the start button. Hence, coil M remains energized after the start button is pushed and released. In the event of voltage loss, coil M is no longer energized, and the contacts fall to the normally open position. The motor will not start when voltage is restored unless the start button is pushed again. This type of control is generally applied to all nonvital machinery. It is particularly adaptable to those drives in which the unexpected starting of a saw, shaper, or other machine tool presents a serious hazard to operating personnel.

Note: Standard undervoltage protection and undervoltage release devices are not designed for dropout at any specific voltage reduction; their function is protection against voltage loss.

13.6 REVERSING STARTERS FOR AC MOTORS

A control circuit for an AC reversing across-the-line starter equipped with overload and voltage-loss protection is shown in Figure 13.8. The two forward buttons are mechanically interlocked, and the two reverse buttons are mechanically interlocked. Pushing the forward button closes its normally open contact and opens its normally closed contact. If the reverse button is pushed while the machine is running in the forward direction, the normally closed reverse button will open and de-energize the forward coil.

Coils F and R operate the respective directional contacts in the motor circuit and the respective sealing contacts in the control circuit.

13.7 TWO-SPEED STARTERS FOR AC MOTORS

The magnetic-starter circuit for a two-speed, two-winding, three-phase induction motor is shown in Figure 13.9. Each winding has its own overload protective device, an indication that the power and current ratings are different for the two speeds. An

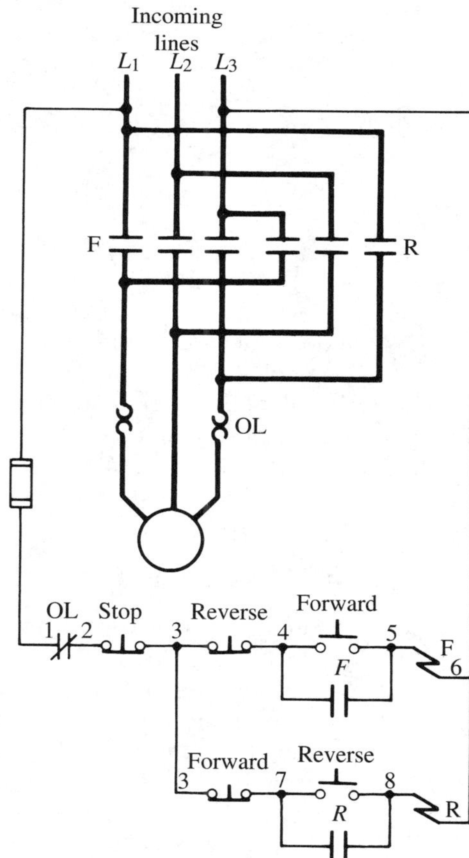

FIGURE 13.8
Reversing controller for a three-phase induction motor.

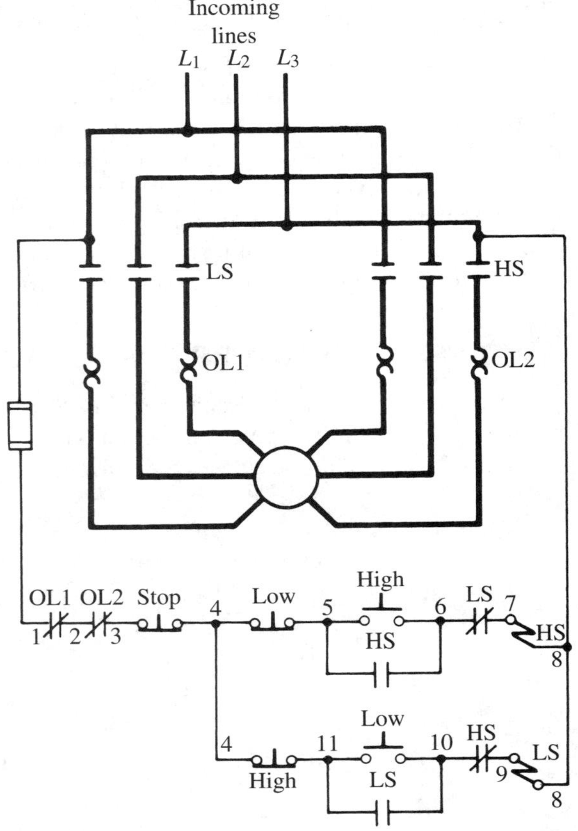

FIGURE 13.9
Controller for a three-phase, two-speed, two-winding induction motor.

additional electrical interlock (called a *sequence interlock*) on each magnetic contactor provides additional protection by requiring one contactor to drop out before the operating coil of the other can be energized. The sequence interlocks in Figure 13.9 are the normally closed contacts LS and HS. Pushing the high-speed button energizes the HS coil, which closes sealing contact HS, opens sequence interlock HS, and closes the three HS contacts in the power circuit.

13.8 REDUCED-VOLTAGE STARTERS FOR AC MOTORS

The circuit diagram for a typical three-phase, reduced-voltage starter is shown in Figure 13.10. The voltage taps are selected on the basis of the starting requirements, are permanently connected, and are not adjustable by the operator. Timing relay TR is used to permit starting at reduced voltage for a preset time, and then to connect the machine across full voltage. The timing relay has one normally open contact, two normally closed, time-opening contacts, one normally open, time-closing contact, a timing motor, and an operating solenoid.[5] When the start button is pushed, the

[5]Timing may be accomplished by dashpot, solid-state, mechanical (using an escapement mechanism), thermal, motor-drive, or pneumatic means.

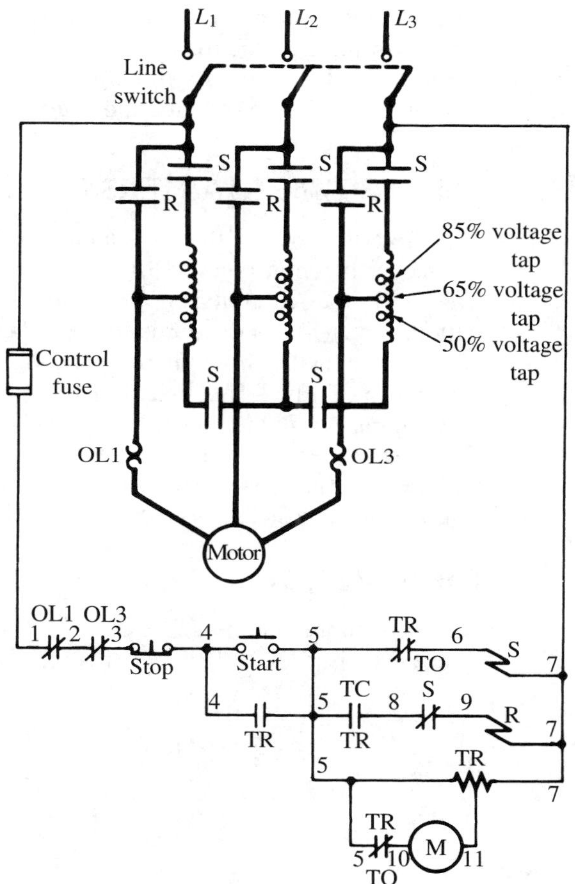

FIGURE 13.10
Controller for reduced-voltage starting of a three-phase induction motor.

current in the control circuit takes three parallel paths: through coil S to energize the autotransformer; through the timing motor to time the starting operation; and through relay coil TR to seal the starting circuit. Closing of contactor S starts the motor at reduced voltage. After a time delay of 1 to 10 s, adjustable, the timing motor causes the time-delay contacts to operate. Coil S is de-energized, the R coil becomes energized, and the circuit to the timing motor opens. The R coil closes the running contacts, connecting the motor across full voltage.

13.9 CONTROLLERS FOR DC MOTORS

Direct-current motors above 3/4 hp are started with reduced voltage. The starters may be divided into two general classifications: definite-time starters and load-sensitive starters. A definite-time starter will remove the starting resistance from the armature circuit after a preset time has elapsed. Thus, regardless of the load on the shaft or whether or not the motor is running, the armature will be connected across full voltage when the timing period runs out. A load-sensitive starter will adjust motor acceleration to the load. Heavy loads will be given longer starting time and lighter loads shorter starting time. If the motor does not start, the starting resistance will not be cut out.

13.10 DEFINITE-TIME STARTERS FOR DC MOTORS

A definite-time starter that uses an accelerating unit (AU) to provide the starting sequence is shown in Figure 13.11. Pushing the start button causes current in coil AU. This closes the AU contact in the motor circuit and the sealing contact across the start button. It also opens the normally closed contact across the resistor between points 4 and 5, reducing the current in the AU coil, making it more sensitive to reduced voltage. The closing of the AU contact in the motor circuit starts the machine with a current-limiting resistor in the armature circuit. The AU contacts connected to the starting resistor close after successive time delays, connecting the armature across the line. Note that the shunt field is connected directly across the line with no time delay when the start button is pushed. This assures full field current, and thus a high starting torque.

Flux-Decay, Time-Delay Contactor

A control circuit that uses flux-decay timing contactors is shown in Figure 13.12. The time delay is provided by a copper cylinder (called a copper jacket) surrounding

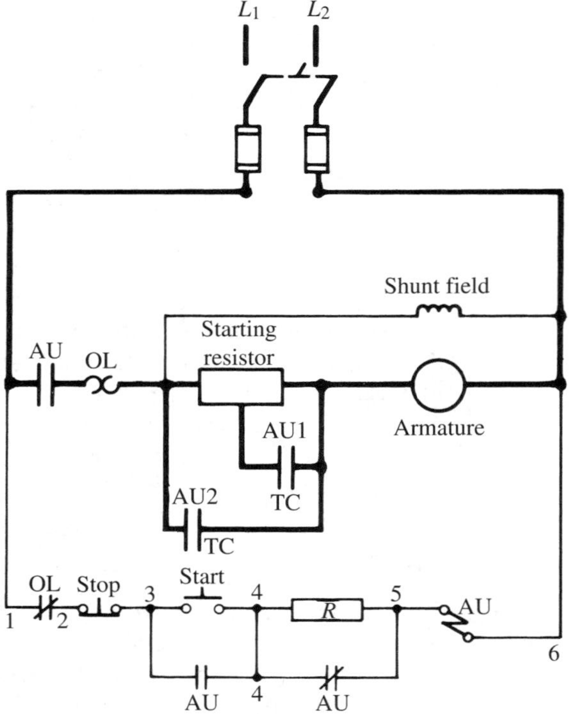

FIGURE 13.11
Definite-time starter that uses an accelerating unit to start a DC motor.

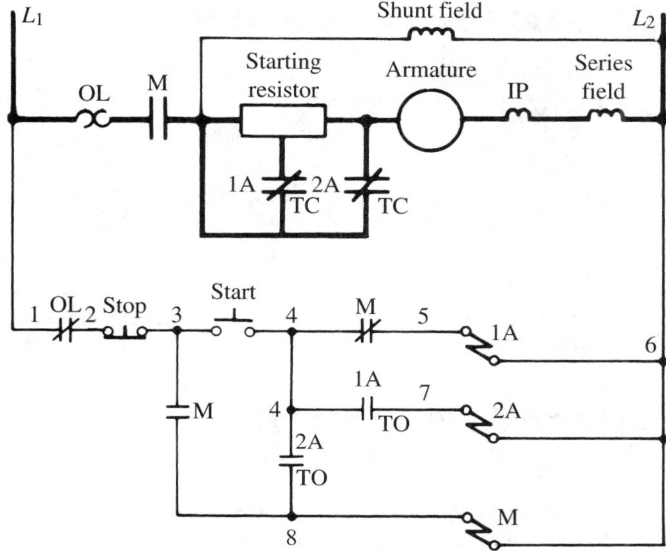

FIGURE 13.12
Definite-time starter that uses flux-decay, time-delay contactors to accelerate a DC motor.

the iron core, around which the operating coil is wound. The time-delay contactor is normally closed, and opens when the coil is energized. When the coil is de-energized, however, the copper jacket delays the decay of flux. Thus, the contact falls to the closed position after a time delay.

When the coil is de-energized, its magnetic field does not instantaneously fall to zero. As the field starts to collapse, it induces an emf in the copper jacket. In accordance with Lenz's law, the induced emf and resultant current in the copper jacket will be in a direction to oppose the decay of flux. The gradual reduction in flux will cause a time delay of several seconds before the contactor closes. The time delay may be adjusted by changing the reluctance of the magnetic circuit, changing the copper jacket, and/or adjusting the spring tension.

Referring to Figure 13.12, when the start button is pushed, current in coil 1A causes the simultaneous closing of auxiliary contact 1A and the opening of resistor-shorting contact 1A. Closing of auxiliary contact 1A energizes coil 2A, which simultaneously opens resistor-shorting contact 2A and closes auxiliary contact 2A. Closing of auxiliary contact 2A allows current in coil M. Coil M closes the motor line contactor M, which starts the motor with all of the starting resistance in series with the armature circuit. At the same time, auxiliary contact M between points 3 and 8 seals the starting circuit, and auxiliary contact M between points 4 and 5 opens. This de-energizes coil 1A, starting the timing cycle for motor acceleration. After a time delay of several seconds, resistor-shorting contact 1A closes, increasing the motor speed, and auxiliary contact 1A opens de-energizing coil 2A. After another time delay, resistor-shorting contact 2A closes, connecting the machine across full voltage.

13.11 COUNTER-EMF STARTER FOR DC MOTORS

The circuit for a cemf starter, shown in Figure 13.13, is an example of a load-sensitive starter. Coils 1A and 2A are connected across the armature so that their magnetic strength is proportional to the cemf. Pushing the start button energizes coil M, causing power contacts M to close, starting the motor. It also closes the auxiliary contact M in series with 1A coil and 2A coil. As the armature accelerates, its cemf increases, and when the voltage across coil 1A is high enough its contact closes, shorting out part of the starting resistor. Continued acceleration causes further increases in cemf. When high enough to actuate contact 2A, the armature is connected across the line. This type of starter will accelerate the motor faster when lightly loaded and slower when heavily loaded.

13.12 REVERSING STARTER WITH DYNAMIC BRAKING AND SHUNT FIELD CONTROL FOR DC MOTORS

The circuit diagram for a reversing definite-time starter with dynamic braking is shown in Figure 13.14. The timing periods are determined by copper-jacketed timing relays 1A and 2A, and reversing is accomplished by contactors 1R, 2R, 1F, and 2F in the armature circuit. The 1F *back coil* and the 1R *back coil* are magnetic interlocking coils connected across the armature. These holdback coils are excited by the armature cemf and prevent the application of reverse power until the motor slows down. The application of reverse power to a motor before it stops is called *plugging*. The physical layout of the operating coil and its corresponding back coil are shown in Figure 13.15.

Referring to Figure 13.14, if the motor is running in the forward direction and the reverse button is pushed, the forward contactors 1F and 2F will drop out, but the 1R operating coil will not be able to close its contact until released by the 1R back

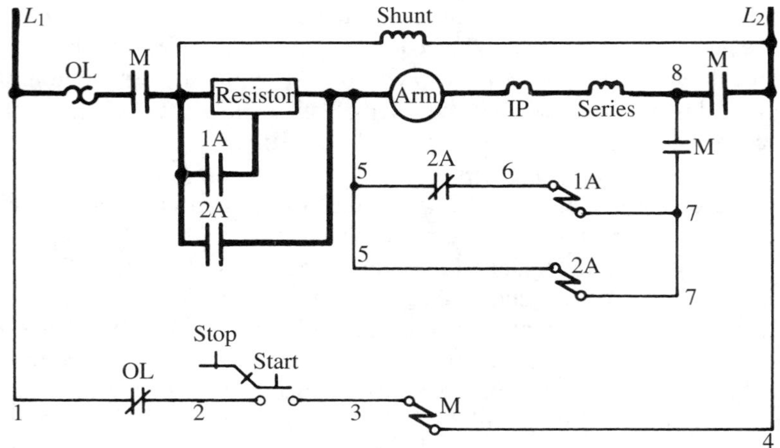

FIGURE 13.13
Counter-emf starter.

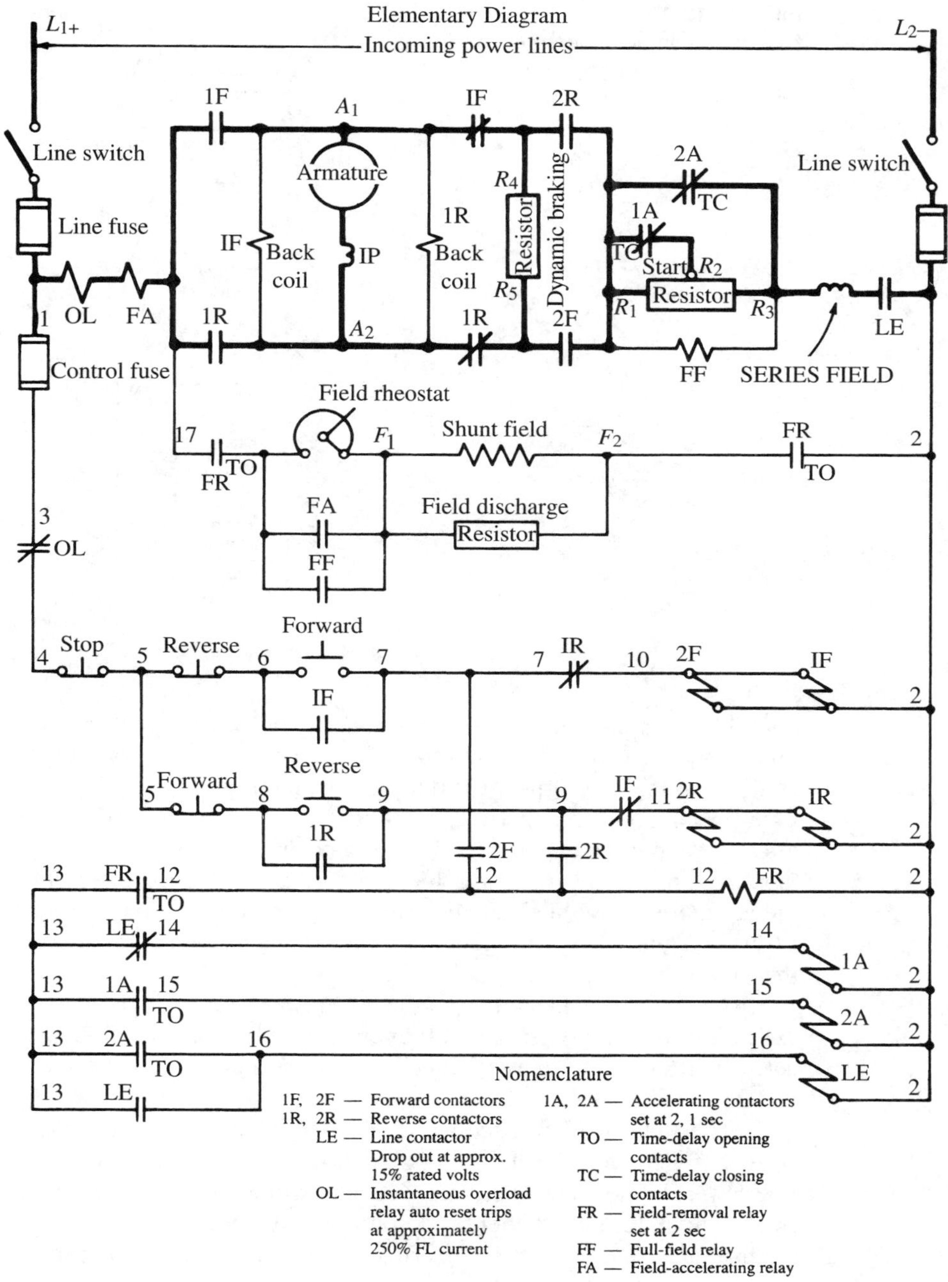

FIGURE 13.14
Reversing controller with dynamic braking. (Courtesy GE Industrial Systems)

FIGURE 13.15
Magnetic contactor with holdback coil and operating coil for the reversing controller in Figure 13.14.

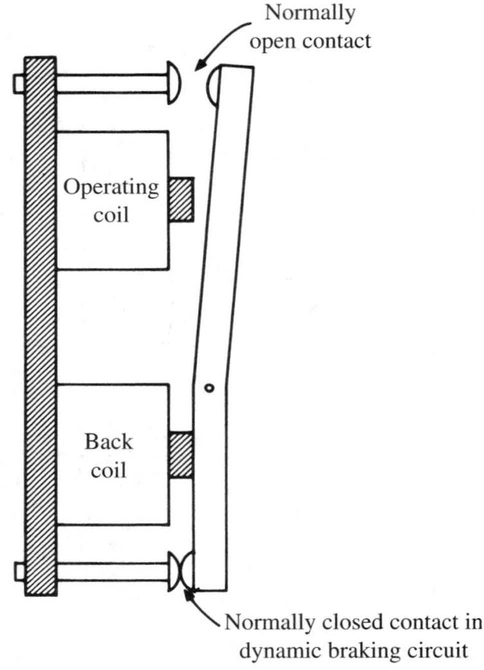

coil. A similar situation exists when running in the reverse direction and the forward button is pushed. The holdback coils also serve to hold the normally closed 1F and 1R contacts in the closed position during dynamic braking. Field-removal relay FR is a time-opening relay that permits the shunt field to remain energized for several seconds after the stop button is pushed. This permits dynamic braking to decelerate the motor quickly. The dynamic-braking loop, consisting of the dynamic-braking resistor and the armature, is established when the stop or reverse button is pushed. An "instantaneous" magnetic overload relay and undervoltage protection are also provided with this controller.

To ensure a high starting torque when starting, the shunt field rheostat is shorted out by full field relay FF. The FF operating coil obtains its voltage from the IR drop across part of the starting resistor. When all the starting resistance is shorted out, contact FF opens.

The shunt field rheostat is also shorted when the *field-accelerating relay* FA is sufficiently energized. The operating coil for this relay is connected in series with the line so that high values of armature current will automatically short out the field rheostat. A quick turn of the field rheostat in the direction to increase speed will weaken the flux and cause an increase in armature current sufficient to actuate relay FA. When the FA contact closes, the field increases in strength, decreasing the armature current, and thus allowing contact FA to open again. The FA contact flutters in and out until the high transient current (instigated by operation of the field rheostat) is sufficiently reduced, at which time contact FA remains open.

13.13 SOLID-STATE CONTROLLERS

Solid-state controllers use diodes, transistors, thyristors (SCRs), triacs, and other solid-state devices in different configurations to start, stop, reverse, brake, provide soft starting, and adjust the speed of electrical machinery. Since solid-state devices have no moving parts, they require considerably less maintenance than their magnetic counterparts. Furthermore, the absence of arcing and/or sparking makes them attractive for applications in explosive atmospheres. Solid-state devices may be used effectively and efficiently for the control of machinery from fractional horsepower units to tens of thousands of horsepower.

13.14 THYRISTOR CONTROL OF MOTORS

The thyristor, shown symbolically in Figure 13.16, is a silicon-controlled rectifier (SCR) that acts as a combined switch and diode. The SCR has two states, on and off. When in the off state, no current can pass in either direction. When in the on state, the SCR acts as a diode, permitting current only in the forward direction.

An SCR remains in the off state, called the *blocking state,* until its anode *and* gate are both positive with respect to the cathode. When both of these conditions are met the SCR "fires," and current conduction begins.

A basic SCR circuit is shown in Figure 13.17(a); the gate firing circuit, called a triggering circuit, contains the complex circuitry that determines when and for how long the gate is to be positive.

When the gate firing circuit triggers the SCR, it is switched to the on state, as shown in Figure 13.17(b); a pulse of only a few microseconds duration will switch on an SCR. Once fired, the SCR will remain in the on state, even after the gate circuit is opened; the load current keeps it in the on state. In effect, the SCR behaves like a latching contactor or latching relay. In order for the gate circuit to regain control, the anode current must cease. This may be accomplished by opening the switch in the anode circuit, by using an AC driving voltage in the anode circuit, or by forced commutation using a reverse-voltage impulse in the anode circuit. The process of turning off the thyristor is called *commutation.* Natural commutation occurs when the anode circuit is supplied by an AC driver. When using an AC driving voltage, the SCR switches to the off state each time the anode becomes negative, enabling the gate circuit to regain control.

FIGURE 13.16
Thyristor (SCR).

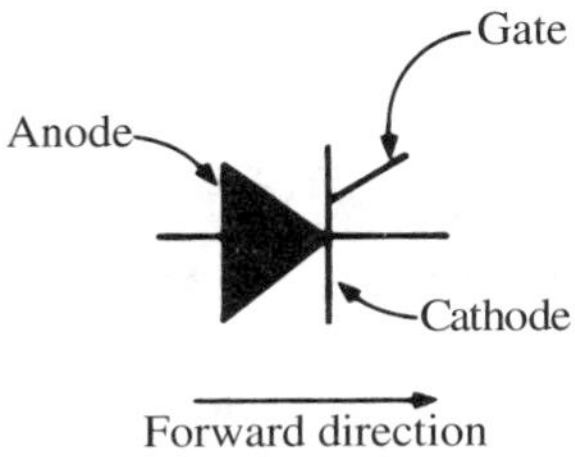

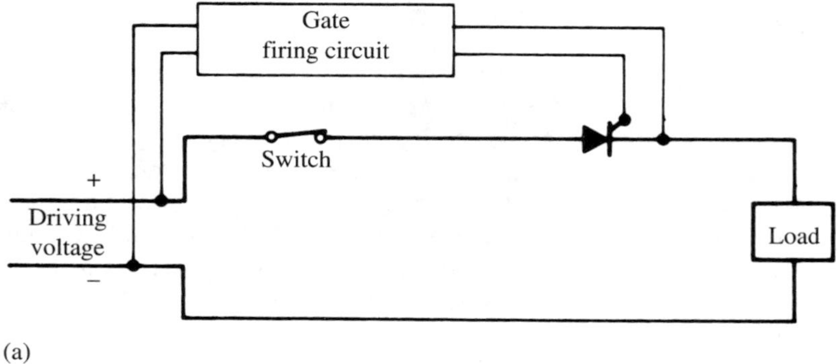

(a)

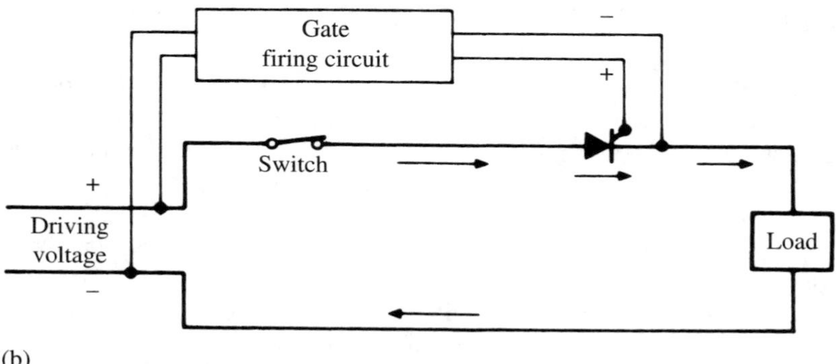

(b)

FIGURE 13.17
(a) Basic thyristor circuit; (b) thyristor in conducting state.

13.15 SOLID-STATE ADJUSTABLE-SPEED DRIVES

High-power, solid-state, adjustable-speed drives deliver precise speed control and high operating efficiency. The trigger circuits are controlled by microprocessors that switch the SCRs on and off in a programmed sequence that is easily varied by the operator.

The complexity of the firing circuit depends on the number of control features desired. Such features as soft start, speed control, reversing, torque limit, and current limit, all add to the complexity and to the cost.

Figure 13.18(a) shows the power circuit for an adjustable-speed, reversible DC motor drive system operating from a three-phase source. The field may be a permanent magnet, as shown, or separately excited through a rectifier. The firing circuits are not shown, and because of their complexity are generally not included in manufacturers' diagrams of solid-state motor controllers.[6] The firing circuits control the direction of

---

[6]For in-depth discussions of different types of solid-state drives and associated firing circuits see References **[6],[7]**.

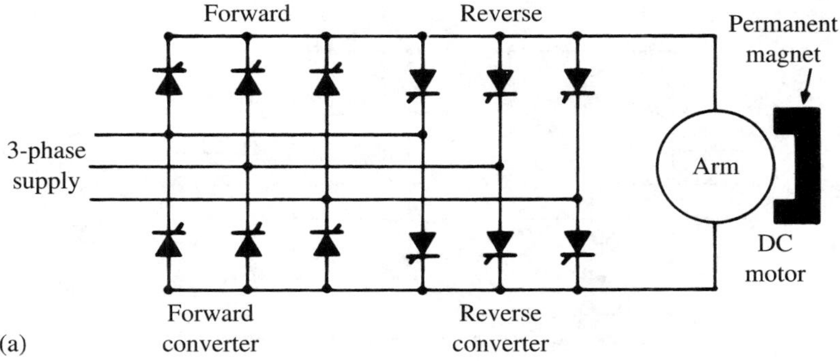

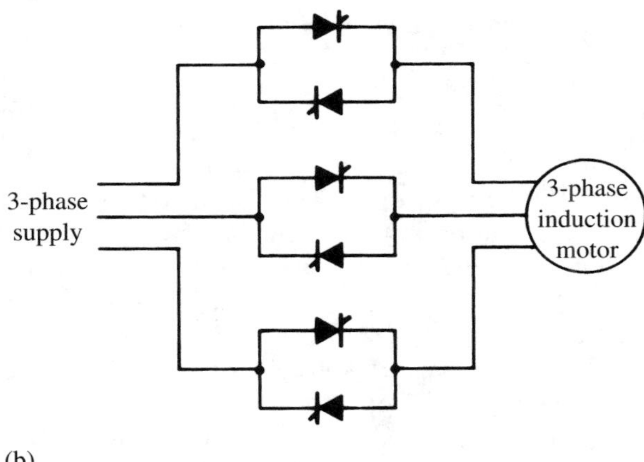

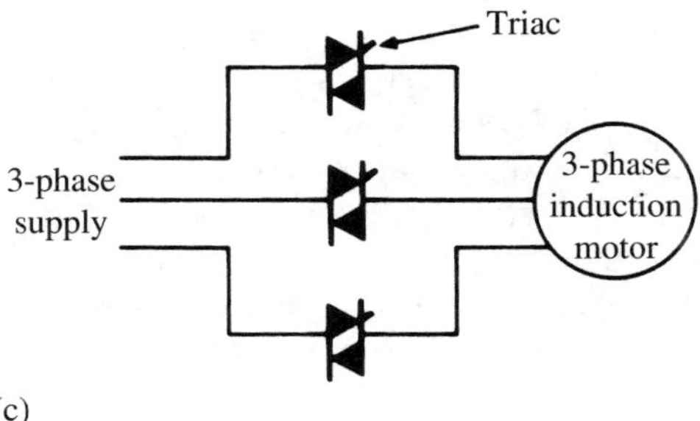

FIGURE 13.18
Thyristor power circuits: (a) adjustable-speed reversible drive system for a permanent-magnet DC motor; (b) soft-start motor controller with back-to-back SCRs for a three-phase induction motor; (c) soft-start motor controller with triacs for a three-phase induction motor.

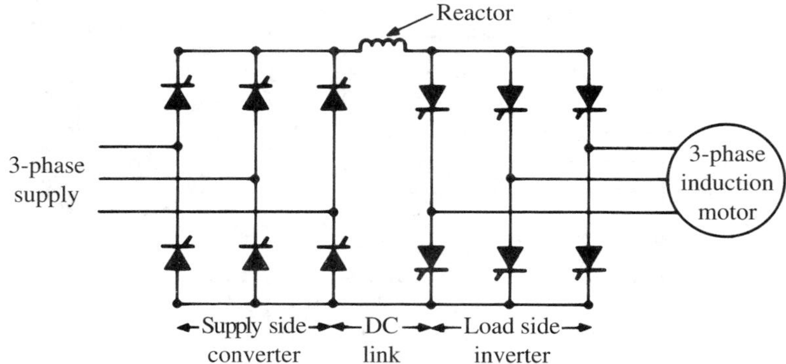

FIGURE 13.19
Adjustable-frequency drive system for a three-phase induction motor.

rotation by firing the forward converter or the reverse converter, and control the speed by controlling the voltage applied to the armature.

Figure 13.18(b) shows the power circuit for a soft start AC motor controller. Controlled firing of the back-to-back SCRs provides current limiting when starting and accelerating. A single device that provides characteristics similar to those provided by the back-to-back SCRs is called a *bidirectional thyristor* or *triac,* and is shown in Figure 13.18(c).

The circuit for one type of adjustable-frequency drive system for a three-phase motor is shown in Figure 13.19. The converter changes the fixed-frequency, fixed-voltage, three-phase input to DC, and the inverter changes the DC back to three-phase AC, whose frequency is controlled by the inverter firing circuits. The reactor in the DC link between the converter stage and the inverter stage serves to reduce harmonic currents caused by the rectification process. The firing circuit for the converter stage controls the voltage level of the DC by phase angle-control at the point of firing. The firing circuit for the inverter stage controls the frequency of the three-phase output by controlling the period of conduction of each SCR. The logic circuitry that controls the firing circuits maintains the voltage/hertz ratio constant, which is essential for proper operation of induction and synchronous motors (see Section 5.8). Turnoff of the inverter SCRs is accomplished by separate circuitry, not shown, that provides a negative-voltage impulse to the anodes; the very brief voltage pulse may be obtained by a programmed discharge of an *LC* circuit.

13.16 CYCLOCONVERTER DRIVES

A cycloconverter is a frequency converter that converts the incoming supply frequency to some other lower frequency. An elementary circuit for a cycloconverter drive is shown in Figure 13.20. It uses three dual converters to convert the three-phase input line frequency to an adjustable lower frequency of nearly sinusoidal waveform. The fixed-frequency, fixed-voltage AC input is converted to a variable-frequency,

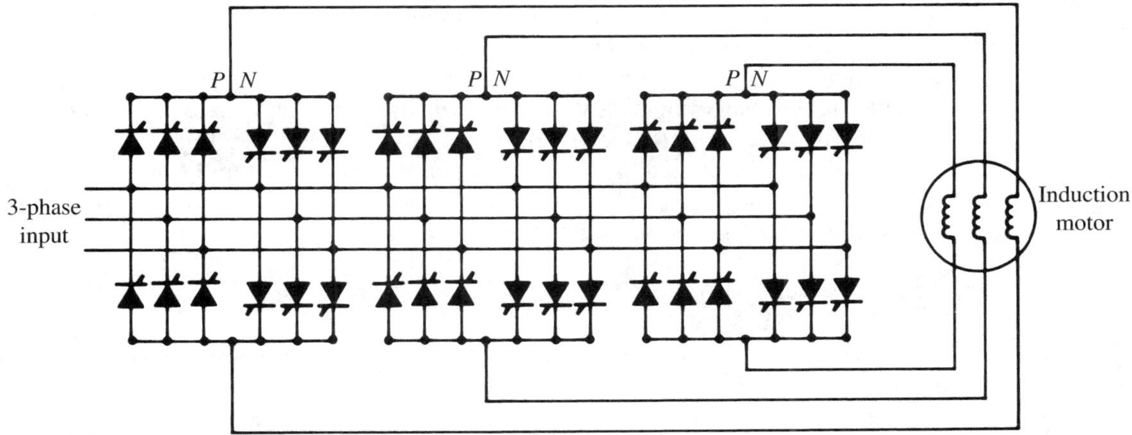

FIGURE 13.20
Cycloconverter supplying a three-phase induction motor.

variable-voltage output without having to go through a DC link. Each phase of the motor has its own dual converter, and obtains power from all three phases of the AC source. The positive half-cycle of current is supplied by the positive converters, and the negative half-cycle by the negative converters. These are indicated in Figure 13.20 as *P* and *N*, respectively.

The output frequency of the cycloconverter is smoothly adjustable from approximately half the input frequency with positive phase sequence, down to zero, and then back up to half the input frequency with negative phase sequence. Thus, given a 60-Hz input, the output frequency can be smoothly adjusted from 30 Hz to 0 Hz to −30 Hz. A three-phase induction motor or synchronous motor, driven by a cycloconverter, can be operated as a reversible adjustable-speed motor with regenerative braking. Controlling the firing angles of the individual converters determines the voltage and frequency applied to the motor phases.

The cycloconverter can be designed to handle loads of almost any size and any power factor. One example of the application of a high-power cycloconverter is in the drive system for the *QE 2* cruise ship; cycloconverters are used to control two 59,000-hp, three-phase, adjustable-speed, reversing, synchronous propulsion motors.

13.17 PROGRAMMABLE CONTROLLERS

A programmable controller, also called a programmable logic controller, PC or PLC, is defined by NEMA to be a "digitally operating electronic apparatus that uses a programmable memory for internal storage of instructions for implementing specific functions such as logic, sequencing, timing, counting and arithmetic, to control machines or processes through digital or analog input and output modules."

Programmable controllers were developed in the 1970s to replace hard-wired control panels in the automobile industry, and their success led to their gradual adoption in virtually all industries with large, complex control systems that must be modified periodically to meet changes in production. The principal advantage of PCs over hard-wired relay panels is the ability to program and reprogram relay logic through a computer terminal, instead of manually wiring, rewiring, or replacing hard-wired panels.

Programmable controllers are designed to withstand the high temperature, humidity, vibration, electrical noise, and power interruptions generally encountered in industrial environments, and are commercially available in small sizes ranging from 50–150 relays to 500–3000 relays in the larger sizes.

The Basic PC

The principal parts of a programmable controller are shown as components of a block diagram in Figure 13.21.

The central processing unit (CPU) contains logic memory, storage memory, power supply, and processor. The CPU receives data from input devices, performs logical decisions based on a stored program, and operates output devices to perform the desired functions.

The solid-state input/output (I/O) interface provides optical isolation to isolate the input and output voltages of field devices from the CPU, and also helps reduce electrical noise. Input field devices include start and stop buttons, limit switches, sensors, and other command devices. Output field devices include such items as motors, magnetic contactors, electric heaters, lights, and electrically operated valves.

The programming keyboard and monitor (TV screen) are used to enter the desired relay logic.

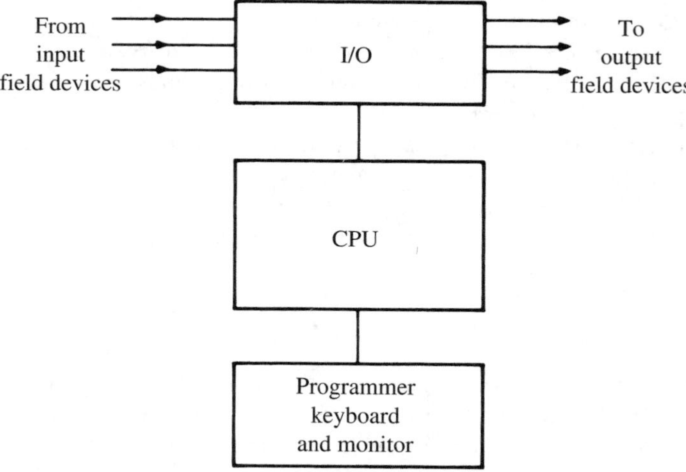

FIGURE 13.21
Basic components of a programmable controller.

Programming a PC

Programming a PC does not require any knowledge of computers, computer language, nor any programming skills. The required control system for a particular application is drawn as a ladder diagram by the design engineer, and the technician or plant engineer types the diagram into the CPU using a special keyboard.

The required number of relay contacts and their configuration, such as normally open, normally closed, series, or parallel, are typed into the computer memory. Then when changes are required, the computer logic can be easily reconfigured from the keyboard; contacts can be added, erased, reconnected, etc. The number of contacts associated with a specific relay coil is limited only by the available memory.

When in operation, a PC scans the programmed logic, starting from the left side of the top rung of the ladder diagram and ending at the right side of the bottom rung. When the last line is scanned, the process repeats, continuing from the time the PC is switched on to the time it is switched off.

For detailed descriptions, diagrams, and photographs of PCs, their mode of operation, and programming techniques, see References **[1],[8],[9]**.

SPECIFIC REFERENCES KEYED TO TEXT

1. Cox, R. A. *Technician's Guide to Programmable Controllers.* Delmar Publishers, Albany, NY, 1989.
2. Hubert, Charles I. *Preventive Maintenance of Electrical Equipment.* Prentice Hall, Upper Saddle River, NJ, 2002.
3. Kosow, I. L. *Control of Electric Machines.* Prentice Hall, Upper Saddle River, NJ, 1973.
4. Mc Partland, J. F. *National Electrical Code Handbook.* McGraw-Hill, New York, 1984.
5. National Fire Protection Association. *National Electrical Code,* 1999.
6. Pearman, R. C. *Power Electronics: Solid State Motor Control,* Reston Publishing, Reston, VA, 1980.
7. Pelly, B. R. *Thyristor Phase-Controlled Converters and Cycloconverters.* Wiley, New York, 1971.
8. Petruzella, F. D. *Programmable Logic Controllers.* McGraw-Hill, New York, 1989.
9. Webb, J. W. *Programmable Controllers, Principles and Applications.* Prentice Hall, Upper Saddle River, NJ, 1988.

GENERAL REFERENCES

Institute of Electrical and Electronic Engineers. *Graphic Symbols Used for Electrical and Electronic Diagrams.* IEEE Std. 315, IEEE, New York, 1993.

Institute of Electrical and Electronic Engineers. *Guide for AC Motor Protection.* IEEE Std. 588-1976, ANSI C37.96-1976, IEEE, New York, 1976.

Millermaster, R. A. *Harwood's Control of Electric Motors.* Wiley, New York, 1970.
National Electrical Manufacturers Association. *General Standards for Industrial Control and Systems.* Pub. No. ICS 1, NEMA, Rosslyn, VA, 1988.
Smeaton, R. W. *Switchgear and Control Handbook.* 2nd ed. McGraw-Hill, New York, 1987.

REVIEW QUESTIONS

1. Explain how a magnetic blow-out coil helps extinguish the arc when the contactor opens.
2. Explain how a shading coil in an AC contactor reduces chatter.
3. Explain the operation of a bimetallic type of motor-overload relay and state how it is connected in the control circuit.
4. What protection, other than motor-overload protection, does a thermal-overload relay provide?
5. Referring to the curves in Figure 13.3(c), if the motor load is such as to cause the current to be 200 percent rated heater current, determine (a) longest time that it will take for the relay to trip; (b) minimum time delay required for cooling before the relay can be reset.
6. What devices are used to protect motor-branch circuits from short circuits or grounds?
7. What type of motor-overload relay is used to protect motors that are subject to jamming? Explain how the relay works.
8. Describe the two general classifications of motor-controller diagrams and state the purpose of each.
9. Sketch a simple control circuit using undervoltage release. What is the characteristic behavior of this type of circuit and where would it be applied?
10. Sketch a simple control circuit using undervoltage protection. What is the characteristic behavior of this type of circuit and where would it be applied?
11. Sketch a full-voltage reversing starter for a three-phase motor. Assume undervoltage protection is required.
12. Explain how the flux-decay timing contactor obtains its time delay.
13. Describe the operation of a cemf starter. What advantage does it have over a definite-time starter?
14. What is the purpose of a field-accelerating relay? Explain how it works.
15. Explain the use of a field-removal relay in a dynamic-braking circuit.
16. What are the advantages of a solid-state starter over a magnetic starter?
17. What are the necessary conditions for switching a thyristor to the on state?
18. What two methods can be used to switch a conducting SCR to the off state?
19. Define commutation in terms of thyristor operation. What is natural commutation?
20. What is a cycloconverter? State an application and explain how it works.
21. What is a programmable controller and in what way does it differ significantly from a magnetic controller?
22. List the principal components of a programmable controller and state their function.

Balanced Three-Phase System

A.1 INTRODUCTION

This appendix is intended as a brief review of voltage, current, and power relationships in the three-phase system. The student is expected to have a working knowledge of phasors and complex numbers. A very detailed development of phasors, complex numbers, resonance, single-phase and three-phase, balanced and unbalanced three-phase circuits, and power measurement is available in power-oriented circuits texts such as those listed in the references.

A.2 LETTER DESIGNATIONS FOR VOLTAGES AND CURRENTS

Voltages and currents that are functions of time are expressed in terms of the following equations, where $\omega = 2\pi f$.

$$\begin{aligned} e &= E_{\max} \sin(\omega t + \theta_e) \\ v &= V_{\max} \sin(\omega t + \theta_v) \\ i &= I_{\max} \sin(\omega t + \theta_i) \end{aligned} \tag{A–1}$$

The corresponding root mean square values, also called rms or effective values, are expressed as

$$\begin{aligned} E &= \frac{E_{\max}}{\sqrt{2}} \\ V &= \frac{V_{\max}}{\sqrt{2}} \\ I &= \frac{I_{\max}}{\sqrt{2}} \end{aligned} \tag{A–2}$$

The complex number representations of phasors corresponding to the sinusoidal quantities in equation set (A–1) are expressed as

$$\begin{aligned} \mathbf{E} &= E\angle\theta_e \\ \mathbf{V} &= V\angle\theta_v \\ \mathbf{I} &= I\angle\theta_i \end{aligned} \tag{A–3}$$

The letters *e, E,* and **E** are generally used to represent voltage sources, and the letters *v, V,* and **V** are generally used to represent voltage drops or potential differences between two points.[1]

A.3 SERIES-CONNECTED CIRCUIT ELEMENTS

A circuit diagram, phasor diagram, and impedance diagram for the general case of series-connected circuit elements are shown in Figure A.1. The associated voltage, current, and impedance relationships are

$$\mathbf{Z}_s = R + jX_L - jX_C = Z_S\angle\theta_Z \tag{A–4}$$

where $X_L = 2\pi fL$ and $X_C = 1/(2\pi fC)$.

$$Z_S = \sqrt{R^2 + (X_L - X_C)^2} \tag{A–5}$$

$$\theta_Z = \tan^{-1}\left(\frac{X_L - X_C}{R}\right) \tag{A–6}$$

$$\mathbf{V}_T = \mathbf{V}_R + \mathbf{V}_L + \mathbf{V}_C$$

$$\mathbf{V}_T = \mathbf{I}_T\mathbf{Z}_S = \mathbf{I}_T(R + jX_L - jX_C)$$

$$\mathbf{I}_T = \frac{\mathbf{V}_T}{\mathbf{Z}_S} = I_T\angle\theta_i \tag{A–7}$$

The voltage drop across any one of two or more series-connected impedances may be determined by applying the *voltage-divider equation.* Referring to Figure A.2,

$$\mathbf{V}_k = \mathbf{V}_T \cdot \frac{\mathbf{Z}_k}{\mathbf{Z}_S} \tag{A–8}$$

where $\mathbf{Z}_S = \mathbf{Z}_1 + \mathbf{Z}_2 + \cdots + \mathbf{Z}_k + \cdots + \mathbf{Z}_n$.

---

[1] **Boldfaced type** is used to indicate complex quantities such as phasor current, phasor voltage, impedance, admittance, and complex power.

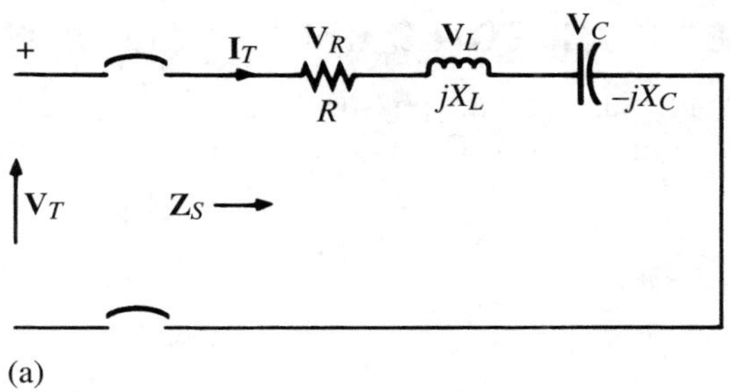

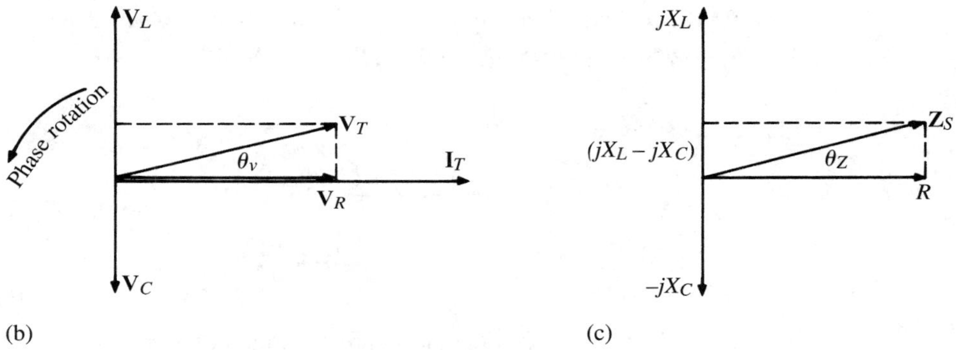

FIGURE A.1
(a) Series circuit; (b) phasor diagram; (c) impedance diagram.

FIGURE A.2
Circuit for voltage-divider equation.

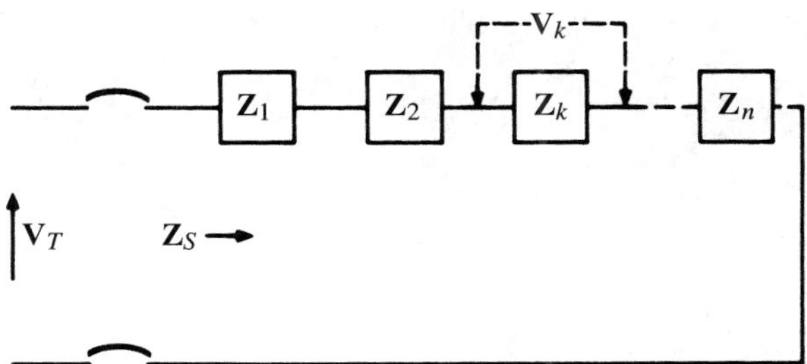

A.4 PARALLEL-CONNECTED CIRCUIT ELEMENTS

A circuit diagram and phasor diagram for parallel-connected circuit elements are shown in Figure A.3. The associated voltage, current, and impedance relationships are

$$\frac{1}{\mathbf{Z}_P} = \frac{1}{R} + \frac{1}{jX_L} + \frac{1}{-jX_C} \qquad \text{(A–9)}$$

$$\mathbf{I}_T = \mathbf{I}_R + \mathbf{I}_L + \mathbf{I}_C$$

$$\mathbf{I}_T = \frac{\mathbf{V}_T}{\mathbf{Z}_P}$$

Problems involving three or more parallel branches are generally solved using the *admittance method* as shown in Figure A.4, where

$$\mathbf{Y}_1 = \frac{1}{\mathbf{Z}_1} \qquad \mathbf{Y}_2 = \frac{1}{\mathbf{Z}_2} = \qquad \mathbf{Y}_n = \frac{1}{\mathbf{Z}_n} \qquad \text{(A–10)}$$

$$\mathbf{Y}_P = \mathbf{Y}_1 + \mathbf{Y}_2 + \cdots + \mathbf{Y}_n \qquad \text{(A–11)}$$

$$\mathbf{I}_T = \mathbf{V}_T \cdot \mathbf{Y}_P = \frac{\mathbf{V}_T}{\mathbf{Z}_P} \qquad \text{(A–12)}$$

Expressing the admittance in polar and rectangular components,

$$\mathbf{Y} = Y \angle\theta_y = G + jB \qquad \text{(A–13)}$$

where: Y = admittance in siemens (S)
G = conductance in siemens (S)
B = susceptance, in siemens (S)

The special case for two impedances in parallel reduces to the following much used and easily calculated formula:

$$\mathbf{Z}_P = \frac{\mathbf{Z}_1 \cdot \mathbf{Z}_2}{\mathbf{Z}_1 + \mathbf{Z}_2} \qquad \text{(A–14)}$$

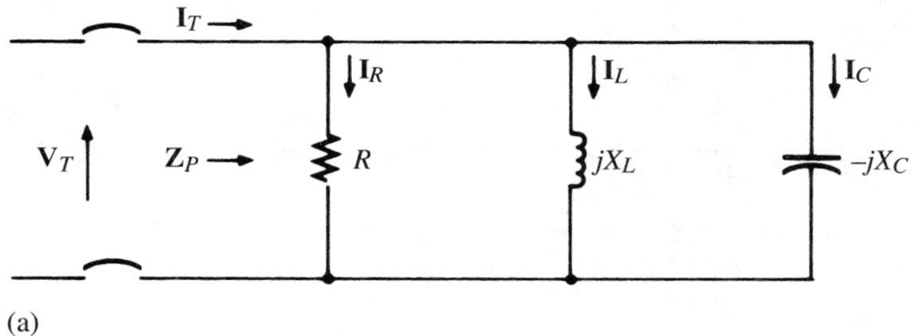

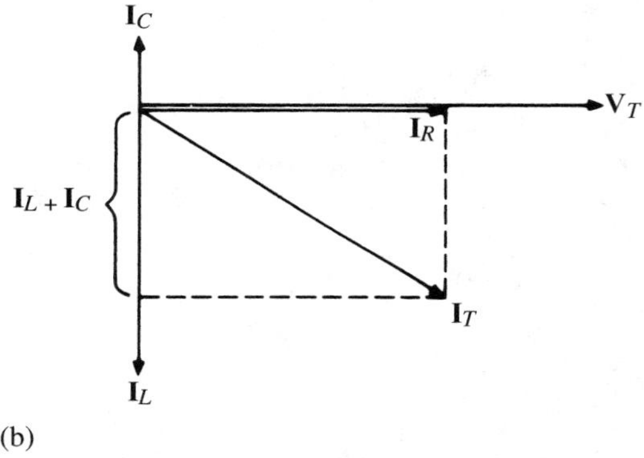

FIGURE A.3
(a) Parallel circuit; (b) phasor diagram.

FIGURE A.4
Impedance-admittance correspondence.

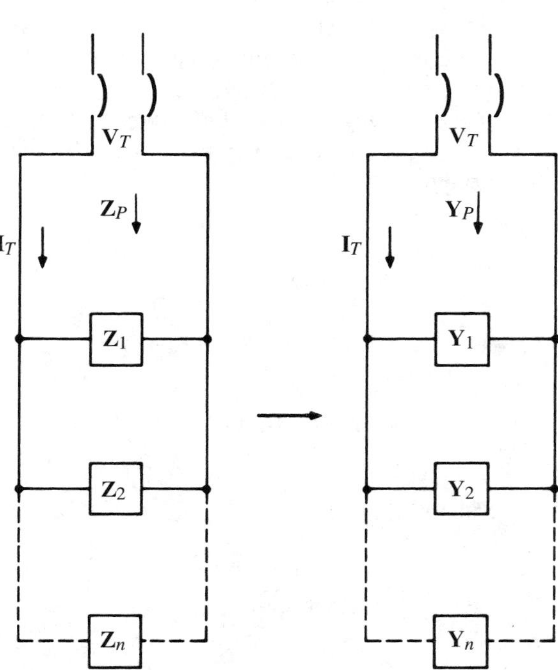

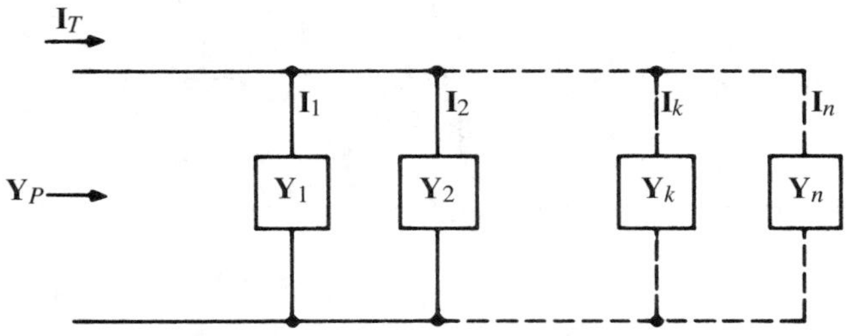

FIGURE A.5
Circuit for current-divider equation.

The current in any one of two or more parallel-connected admittances may be determined by the *current-divider equation.* Referring to Figure A.5,

$$\mathbf{I}_k = \mathbf{I}_T \cdot \frac{\mathbf{Y}_k}{\mathbf{Y}_P} \qquad \textbf{(A–15)}$$

where $\mathbf{Y}_P = \mathbf{Y}_1 + \mathbf{Y}_2 + \cdot\cdot\cdot + \mathbf{Y}_k + \cdot\cdot\cdot + \mathbf{Y}_n$.

A.5 POWER RELATIONSHIPS IN A SINGLE-PHASE SYSTEM

For the single-phase system shown in Figure A.6(a), the unknown circuit may have any combination of circuit elements in series, parallel, or series–parallel combinations. Regardless of the internal configuration, however, if the line voltage, line current, and corresponding phase angles are known, the active power P, reactive power Q, apparent power S, and power factor F_P can be determined from the product of the phasor voltage times the conjugate of the phasor current; this product is called *complex power* or *phasor power.* Thus, referring to Figure A.6(a), the complex power drawn by the circuit is

$$\mathbf{S}_T = \mathbf{V}_T \cdot \mathbf{I}_T^* \qquad \textbf{(A–16)}$$

From Figure A.6(b),

$$\mathbf{V}_T = V_T\underline{/\theta_v} \qquad \textbf{(A–17)}$$

$$\mathbf{I}_T = I_T\underline{/\theta_i} \qquad \textbf{(A–18)}$$

The conjugate of the current phasor is

$$\mathbf{I}_T^* = I_T\underline{/-\theta_i} \qquad \textbf{(A–19)}$$

Substituting Eqs. (A–17) and (A–19) into Eq. (A–16),

$$\mathbf{S}_T = V_T\underline{/\theta_v} \cdot I_T\underline{/-\theta_i} = V_T I_T\underline{/(\theta_v - \theta_i)}$$

Defining angle $\theta = (\theta_v - \theta_i)$ as the *power-factor angle,*

$$\mathbf{S}_T = V_T I_T\underline{/\theta} \qquad \textbf{(A–20)}$$

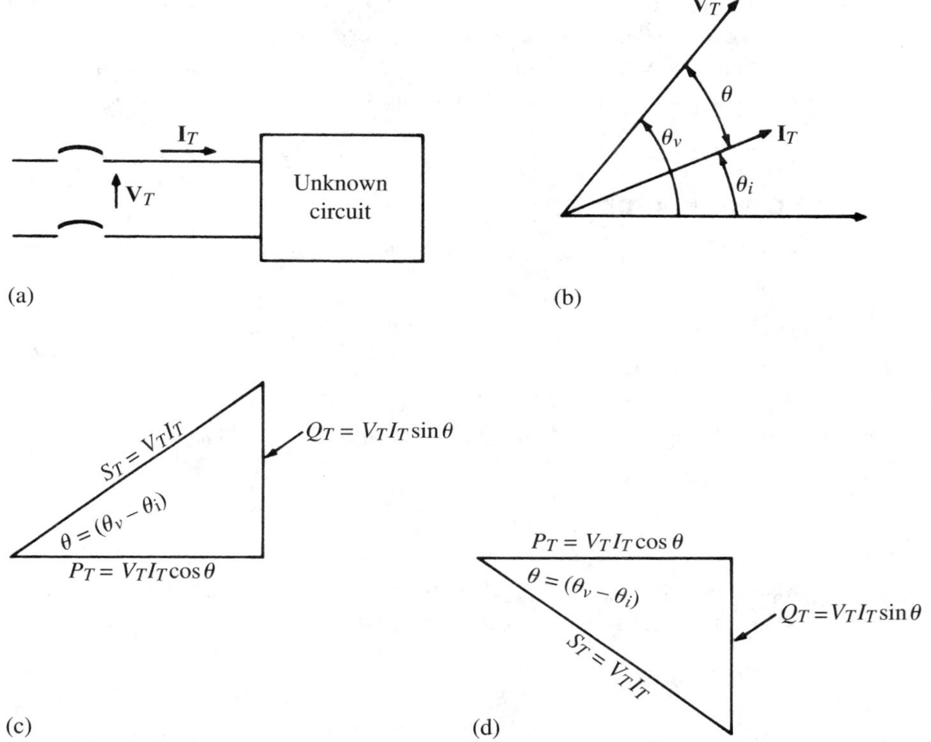

FIGURE A.6
(a) Unknown circuit; (b) phasor diagram; (c) power triangle for lagging power factor; (d) power triangle for leading power factor.

Expressed in rectangular form,

$$\mathbf{S}_T = V_T I_T \cos\theta + jV_T I_T \sin\theta$$

where:

$$\text{Active power (watts)} \quad P_T = V_T I_T \cos\theta \qquad \textbf{(A–21)}$$

$$\text{Reactive power (vars)} \quad Q_T = V_T I_T \sin\theta \qquad \textbf{(A–22)}$$

Power Triangle

Equations (A–21) and (A–22) represent two legs of the *power triangle* shown in Figure A.6(c). The hypotenuse $V_T I_T$ is the magnitude of the apparent power, and makes an angle θ from the zero-degree line. Thus, the apparent power may be conveniently expressed in terms of the magnitudes of its components:

$$S_T = \sqrt{P_T^2 + Q_T^2} \qquad \textbf{(A–23)}$$

If the unknown circuit is predominantly inductive, $\mathbf{I}_T$ lags $\mathbf{V}_T$, as shown in Figure A.6(b), angle θ is positive, and the power triangle will be as shown in Figure A.6(c). If the circuit is predominantly capacitive, however, $\mathbf{I}_T$ will lead $\mathbf{V}_T$, angle θ will be negative, and the power triangle will be as shown in Figure A.6(d).

Power Factor

The power factor of the circuit is defined as the ratio of active power to apparent power:

$$F_P = \frac{P}{S} \tag{A–24}$$

Substituting Eqs. (A–20) and (A–21) into Eq. (A–24),

$$F_P = \frac{V_T I_T \cos\theta}{V_T I_T} = \cos\theta \tag{A–25}$$

As indicated in Eq. (A–25), the power factor is numerically equal to the cosine of the phase angle between the voltage phasor and the current phasor.

EXAMPLE A.1 Assume the current and voltage supplied to a circuit are $125\angle 30^\circ$ A and $460\angle 20^\circ$ V, respectively. Determine (a) apparent power, active power, and reactive power; (b) whether the circuit is predominantly inductive or predominantly capacitive; (c) power factor of the load.

Solution

(a)

$$\mathbf{S} = \mathbf{V}\cdot\mathbf{I}^* = (460\angle 20^\circ)\cdot(125\angle 30^\circ)^* = 460\angle 20^\circ \cdot 125\angle -30^\circ$$

$$\mathbf{S} = 57{,}500\angle -10^\circ = 56,626.4 - j9984.8 \text{ VA}$$

Thus,

$$S = \underline{57.5 \text{ kVA}}$$
$$P = \underline{56.6 \text{ kW}}$$
$$Q = \underline{-9.98 \text{ kvar}}$$

(b) The negative reactive power indicates that the load is *predominantly capacitive.* This is also indicated by the given phase angles of current and voltage, which shows the current to be leading the voltage by 10°.

(c)

$$F_P = \cos(-10^\circ) = 0.985 \text{ or } \underline{98.5\% \text{ leading}}$$

A.6 DOUBLE-SUBSCRIPT NOTATION

Double-subscript notation is used in conjunction with assigned letter symbols for voltage in order to assist in circuit analysis and problem solving. The subscripts represent two nodes between which a voltage is measured, and the order of the subscripts indicates the direction of voltage measurement.

Thus, referring to Figure A.7, $\mathbf{V}_{bc}$ is the voltage at node b measured with respect to the voltage at node c. Voltage $\mathbf{V}_{bc}$ is considered a positive voltage if node b has a higher potential than node c, and will be considered a negative voltage if node b has a lower potential than node c.

Applying Ohm's law to impedance $\mathbf{Z}_2$, and noting the assumed direction of current,

$$\mathbf{I}_{bc} = \frac{\mathbf{V}_{bc}}{\mathbf{Z}_2}$$

Note: Voltage measurements from node c to node b are indicated as $\mathbf{V}_{cb}$. Since the direction of measurement is opposite to that of $\mathbf{V}_{bc}$,

$$\mathbf{V}_{cb} = -\mathbf{V}_{bc}$$

A.7 VOLTAGES IN A WYE-CONNECTED SOURCE

A three-phase, wye-connected system of voltages consists of three AC voltage sources, each equal in magnitude, but displaced from one another by 120 electrical degrees, and connected at a common point as shown in Figure A.8(a). The common point is called the *neutral connection.* The three voltage waves representing the three phase voltages are shown in Figure A.8(b), and the corresponding phasor diagram is shown in Figure A.8(c).

The voltage at the service entrance from terminal a to terminal b is determined by making a phasor summation of phase voltages while "walking" through the circuit from a to b. Thus,

$$\mathbf{E}_{a \text{ to } b} = \mathbf{E}_{a \text{ to } a'} + \mathbf{E}_{b' \text{ to } b}$$

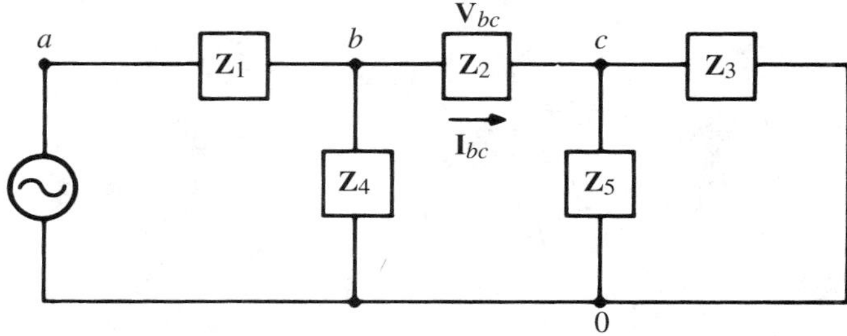

FIGURE A.7
Example of double-subscript notation.

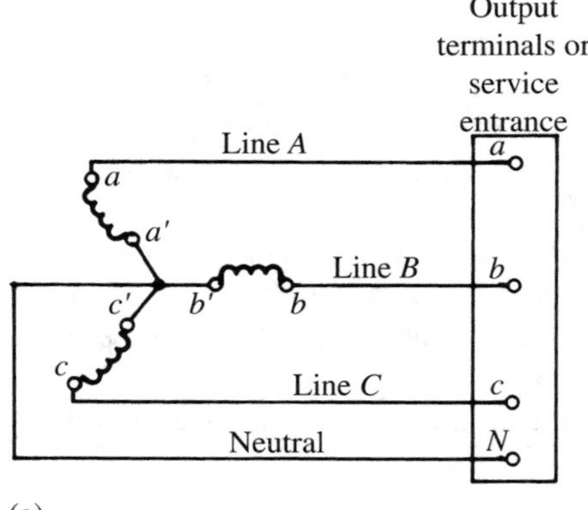

(a)

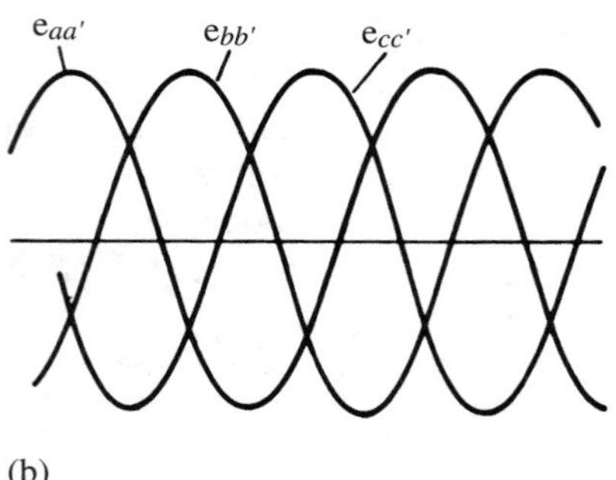

(b)

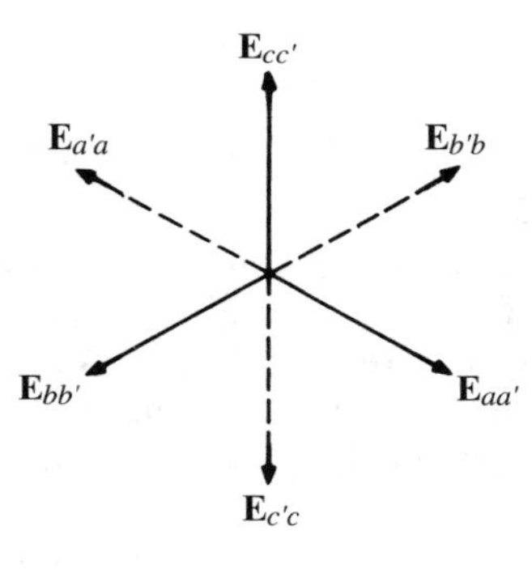

(c)

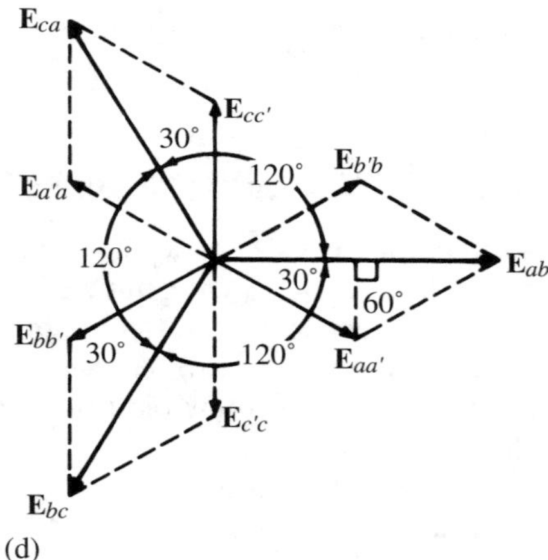

(d)

FIGURE A.8
(a) Wye-connected source; (b) voltage waves; (c) phasor diagram of component voltages; (d) graphical addition of voltages.

Or simply,

$$\mathbf{E}_{ab} = \mathbf{E}_{aa'} + \mathbf{E}_{b'b} \qquad \textbf{(A–26)}$$

Similarly,

$$\mathbf{E}_{bc} = \mathbf{E}_{bb'} + \mathbf{E}_{c'c} \qquad \textbf{(A–27)}$$

$$\mathbf{E}_{ca} = \mathbf{E}_{cc'} + \mathbf{E}_{a'a} \qquad \textbf{(A–28)}$$

The voltages between any two line terminals (a, b, *or* c) *are called line-to-line or line voltages, and the voltages between any line terminal and the neutral terminal are called branch voltages or phase voltages.*

A phasor diagram for the graphical addition of voltages in a wye-connected system is shown in Figure A.8(d). From the geometry of the phasor diagram,

$$E_{\text{line}} = \sqrt{3}\, E_{\text{phase}} \qquad \textbf{(A–29)}$$

A.8 VOLTAGES IN A DELTA-CONNECTED SOURCE

A three-phase, delta-connected system of voltages, shown in Figure A.9(a), consists of three AC voltage sources $\mathbf{E}_{aa'}$, $\mathbf{E}_{bb'}$, and $\mathbf{E}_{cc'}$, each equal in magnitude, but displaced from each other by 120 electrical degrees. The three voltages sources, called *phases,* are connected in series to form a closed loop, and lines from the three nodes connect the sources to the output terminals.

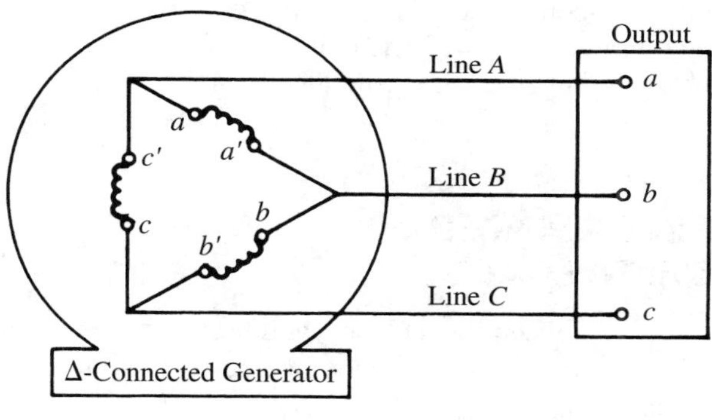

(a)

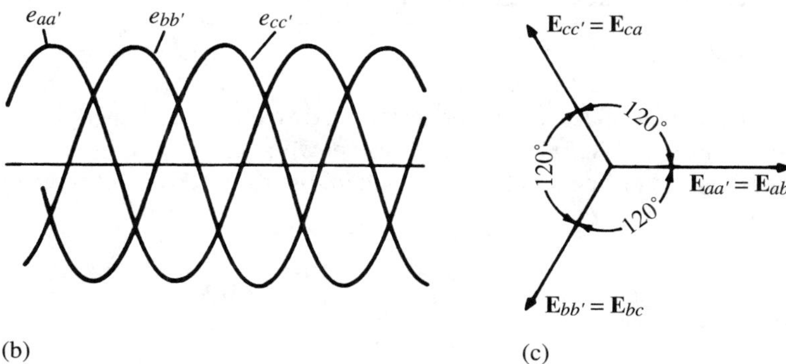

(b) (c)

FIGURE A.9
(a) Delta-connected source; (b) voltage waves; (c) phasor diagram.

The voltage waves for the three phase-voltages are shown in Figure A.9(b), and a phasor diagram for the corresponding phasors is shown in Figure A.9(c). *For standardization and convenience in problem solving, phasor* $\mathbf{E}_{ab}$ *is drawn at zero degrees for both wye and delta sources.*

The voltages measured between the output terminals of a generator, or between the service-entrance terminals at a factory, are called *line-to-line voltages,* or simply *line voltages.* Note that for a delta connection, the line voltage is equal to the corresponding phase voltage.

No current circulates in the closed circuit formed by the delta connection because the phasor sum of the three phase-voltages around the loop is equal to zero. This can be determined from the phasor diagram in Figure A.9(c), where the phasor summation is

$$\mathbf{E}_{aa'} + \mathbf{E}_{bb'} + \mathbf{E}_{cc'} = 0$$

Although the resultant voltage around the loop is at all times equal to zero and no current circulates around the closed delta, each of the three phases is still capable of supplying current to external loads. Interchanging the internal connections of any one phase of the delta-connected generator in Figure A.9(a), however, will result in a very high circulating current within the delta; the high current will cause rapid heating of the generator winding, damaging the insulation.

A.9 CURRENTS IN BALANCED WYE AND BALANCED DELTA LOADS

The relationship between line currents and branch currents for a balanced wye load is evidenced in Figure A.8(a), where the line current and the branch current are one and the same.[2] That is, $\mathbf{I}_{a'a} = \mathbf{I}_{\text{line } A}$; $\mathbf{I}_{b'b} = \mathbf{I}_{\text{line } B}$; and $\mathbf{I}_{c'c} = \mathbf{I}_{\text{line } C}$.

The relationship between line currents and branch currents for a balanced delta load cannot be determined by inspection, but will be demonstrated using an example.

[2]Each phase of a balanced three-phase load has the same impedance.

EXAMPLE A.2 A balanced three-phase, delta-connected load of $20.0\angle 40^\circ\ \Omega$ per phase is connected to a three-phase, 460-V, 60-Hz system, as shown in Figure A.10(a). The corresponding phasor diagram of line voltages is shown in Figure A.10(b). Determine the ratio of line current to phase current.

Solution

$$\mathbf{I}_A = \frac{\mathbf{E}_{ab}}{\mathbf{Z}} + \frac{\mathbf{E}_{ac}}{\mathbf{Z}} = \frac{460\angle 0^\circ}{20.0\angle 40^\circ} + \frac{-460\angle 120^\circ}{20.0\angle 40^\circ}$$

$$\mathbf{I}_A = 23\angle -40^\circ - 23\angle 80^\circ = 17.62 - j14.78 - (3.99 + j22.65)$$

$$\mathbf{I}_A = 13.63 - j37.43 = 39.83\angle -70.0^\circ \text{ A}$$

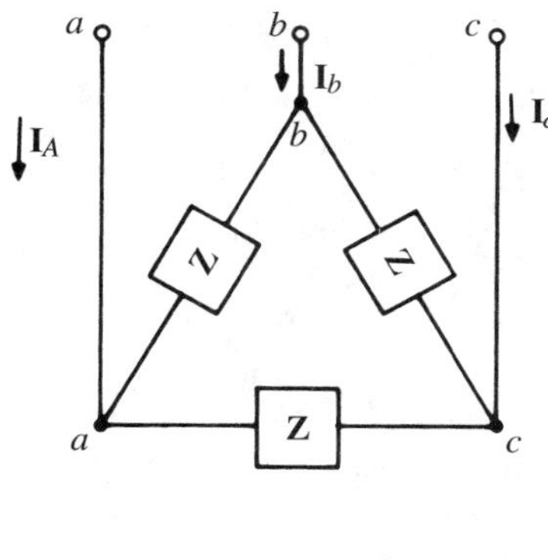

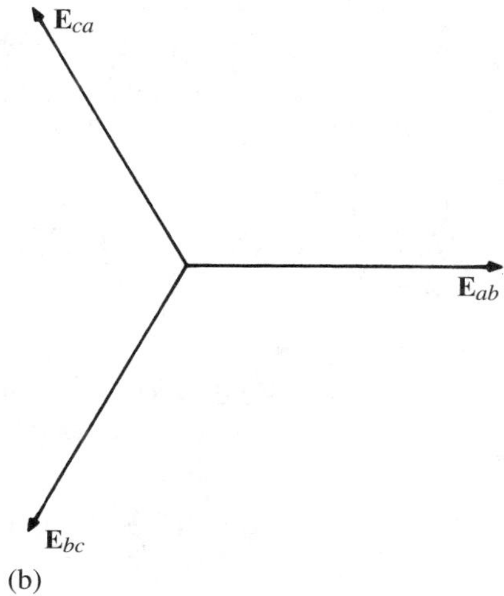

FIGURE A.10
Circuit and phasor diagram for Example A.2.

The ratio of the magnitude of the line current to the magnitude of the branch current (also called phase current) is

$$\frac{I_{\text{line}}}{I_{\text{phase}}} = \frac{39.83}{23} = 1.732 = \sqrt{3}$$

The same ratio holds true for all line currents in Figure A.10(a). Thus, for a balanced delta load,

$$I_{\text{line}} = \sqrt{3} \cdot \text{I}_{\text{phase}} \tag{A–30}$$

A.10 PHASE SEQUENCE

Phase sequence is the order or sequence in which the three line-voltages of a three-phase supply reach their maximum positive values.

Phase sequence may be determined from the voltage waves or from a corresponding phasor diagram, as shown in Figures A.11(a) and (b), respectively, where the indicated phase sequence is

$$\mathbf{E}_{ab}, \mathbf{E}_{bc}, \mathbf{E}_{ca}, \mathbf{E}_{ab}, \mathbf{E}_{bc}, \mathbf{E}_{ca}, \ldots$$

For brevity, however, the sequence is generally expressed in terms of only the first subscripts or only the second subscripts:

In terms of the first subscripts, the sequence is [*abc*]*abcabc* . . . , or simply *abc sequence.*

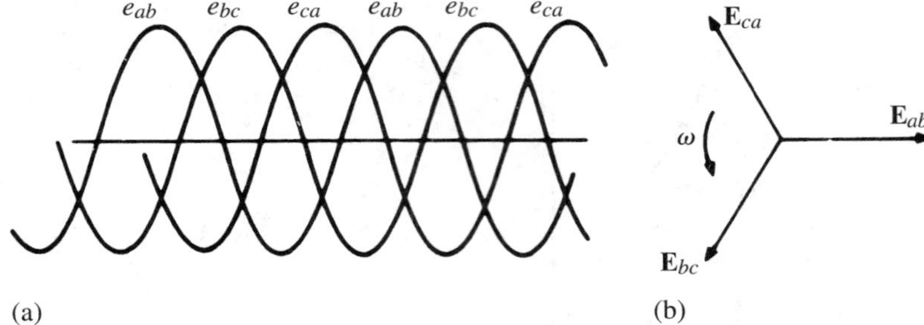

FIGURE A.11
Phase sequence as indicated by (a) voltage waves; (b) phasor diagram.

In terms of second subscripts, the sequence is *bc*[*abc*]*abca* . . . , which is also *abc sequence.*

Phase sequence at the load is indicated by reading the letter markings (or number markings) from top to bottom or from left to right as applicable **[1]**. Thus, referring to the circuit in Figure A.12(a), the phase sequence at the motor, reading from top to bottom, is *abc*. Interchanging any two of the three line leads reverses the phase sequence. This is shown in Figures A.12(b), (c), and (d), where interchanging any two line leads

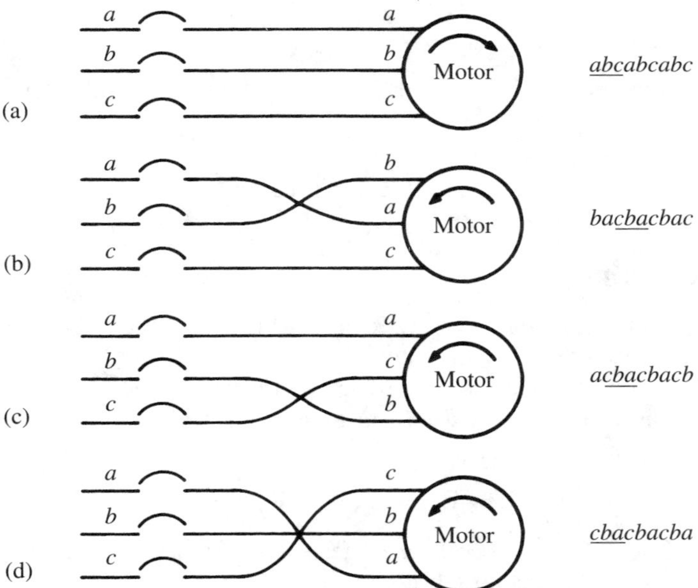

FIGURE A.12
Reversing the phase sequence by interchanging any two line leads.

changes the phase sequence from *abc* to *cba,* and reverses the direction of rotation of the motor. *Note:* There are only two possible phase sequences, *abc* and *cba.*

If the three-phase load has unbalanced impedances, reversing the phase sequence could cause major changes in the magnitudes and phase angles of the three line currents (see Section 21.7 in Reference [**2**]). If a three-phase generator is paralleled with another of opposite phase sequence, both machines may be severely damaged. It is therefore essential that phase sequence be taken into consideration when connecting three-phase loads or when paralleling three-phase generators (see Section 14.7 in Reference [**3**]).

A.11 CALCULATING LINE AND PHASE CURRENTS IN THREE-PHASE CIRCUITS

The procedure for calculating line and phase current in three-phase circuits is the same whether the circuit is wye or delta. Depending on the complexity of the circuit, the current may be determined by using Ohm's law, Kirchhoff's law, and/or loop and node analysis.

Guidelines for Solving Three-Phase Circuit Problems

1. Voltages and currents on the nameplates of electrical apparatus, and in technical literature concerning motors, generators, and other apparatus, are line-to-line rms voltages unless otherwise specified.
2. A universal phasor diagram and a table of load voltages, to be used when solving three-phase circuit problems, are shown in Figure A.13; line and phase voltages apply to wye loads, but only line voltages apply to delta loads. Note also that the voltage between any line and the wye junction (neutral) of a *balanced wye load* is the corresponding phase voltage, even though the junction is not connected to the source neutral.
3. If a wye load is *balanced* (identical impedances per leg) and the three-phase source has balanced voltages, there will be no current in a neutral line connecting the wye junction of the load to the source neutral. Hence, a neutral line connecting the source neutral to the wye junction of a *balanced* load is not necessary and is seldom used. Except for fault conditions (such as opens, shorts, and grounds), *three-phase motors are balanced loads,* and thus neutral lines are not required nor are they supplied for wye-connected motors.
4. Before starting the solution of problems involving multiple loads, an assumed direction of phasor current should be indicated on the diagram for each line and each phase being solved. For convenience and standardization, the direction of current in each line will be assumed to be *from the source to the load.* Once assigned, the assumed direction must not be changed during the solution process.

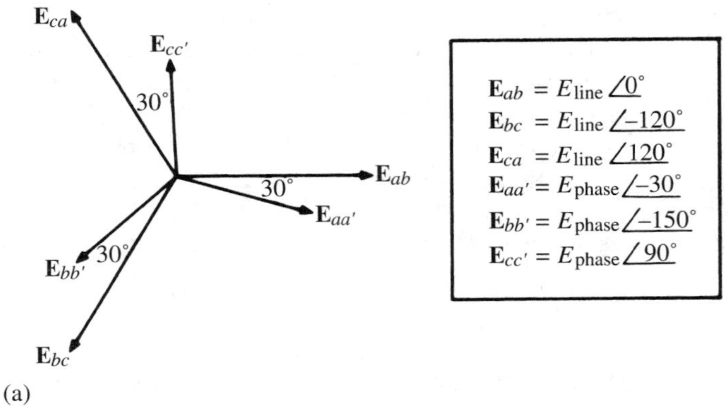

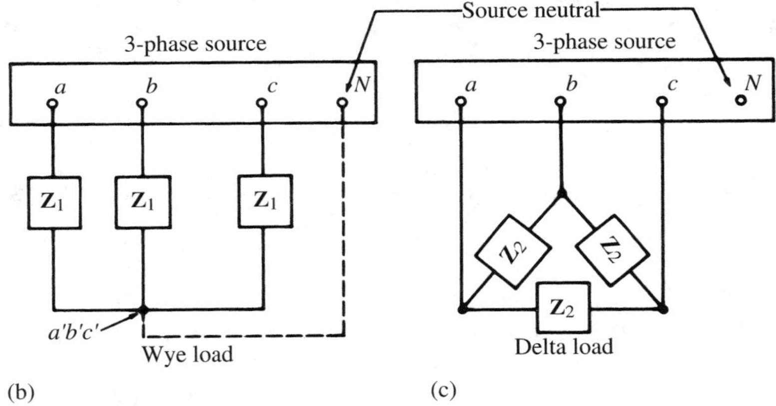

FIGURE A.13
(a) Universal phasor diagram and table of load voltages; (b) wye load; (c) delta load.

EXAMPLE A.3 For the circuit shown in Figure A.14,

$$\mathbf{Z}_1 = 10\underline{/30^\circ}\ \Omega \qquad \mathbf{Z}_2 = 15\underline{/10^\circ}\ \Omega \qquad \mathbf{Z}_3 = 20 + j20\ \Omega$$

Determine the ammeter reading.

Solution

Applying Ohm's law and Kirchhoff's current law to line A,

$$\mathbf{I}_A = \frac{\mathbf{E}_{aa'}}{\mathbf{Z}_1} + \frac{\mathbf{E}_{ab}}{\mathbf{Z}_2} + \frac{\mathbf{E}_{ac}}{\mathbf{Z}_2} + \frac{\mathbf{E}_{ac}}{\mathbf{Z}_3}$$

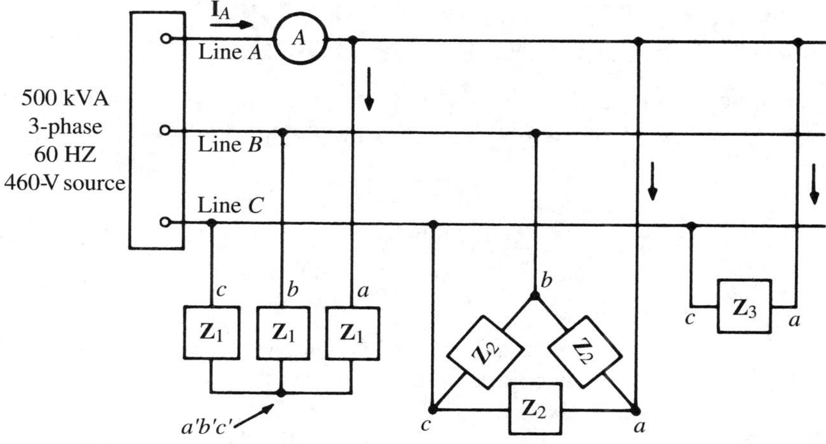

FIGURE A.14
Circuit for Example A.3.

Note that the wye-connected load is balanced. Hence, the wye junction is effectively at potential $a'b'c'$. From the source data, the magnitudes of line and phase voltages at the applicable loads are

$$E_{\text{line}} = 460 \text{ V} \qquad E_{\text{phase}} = \frac{460}{\sqrt{3}} = 265.6 \text{ V}$$

Using the table of voltages in Figure A.13(a) as a guide,

$$\mathbf{E}_{aa'} = 265.6\underline{/-30^\circ} \text{ V}$$

$$\mathbf{E}_{ab} = 460\underline{/0^\circ} \text{ V}$$

$$\mathbf{E}_{ac} = -\mathbf{E}_{ca} = -460\underline{/120^\circ} \text{ V}$$

Converting $\mathbf{Z}_3$ into polar form, and substituting the corresponding voltages and impedances,

$$\mathbf{Z}_3 = 20 + j20 = 28.28\underline{/45^\circ}$$

$$\mathbf{I}_A = \frac{265.6\underline{/-30^\circ}}{10\underline{/30^\circ}} + \frac{460\underline{/0^\circ}}{15\underline{/10^\circ}} + \frac{-460\underline{/120^\circ}}{15\underline{/10^\circ}} + \frac{-460\underline{/120^\circ}}{28.28\underline{/45^\circ}}$$

$$\mathbf{I}_A = 26.56\underline{/-60^\circ} + 30.67\underline{/-10^\circ} - 30.67\underline{/110^\circ} - 16.27\underline{/75^\circ}$$

$$\mathbf{I}_A = (13.28 - j23) + (30.20 - j5.33) + (10.49 - j28.82) + (-4.21 - j15.72)$$

$$\mathbf{I}_A = 49.76 - j72.87 = 88.23\underline{/-55.67^\circ}$$

The ammeter will read an rms value of <u>88.2 A.</u>

A.12 ACTIVE, REACTIVE, AND APPARENT POWER DRAWN BY BALANCED THREE-PHASE LOADS

The power supplied to a *balanced* three-phase load (wye or delta) is three times the power drawn by one branch. Expressed as complex power,

$$\mathbf{S}_{3\phi,\text{ bal}} = 3\mathbf{V}_{\text{br}} \cdot \mathbf{I}^*_{\text{br}} \tag{A–31}$$

where:

$\mathbf{V}_{\text{br}} = V_{\text{br}} \angle\theta_v$ = branch voltage phasor

$\mathbf{I}_{\text{br}} = I_{\text{br}} \angle\theta_i$ = branch current phasor

$\mathbf{I}^*_{\text{br}} = I_{\text{br}} \angle{-\theta_i}$

V_{br} = magnitude of branch voltage

I_{br} = magnitude of branch current

θ_v = phase angle of branch voltage

θ_i = phase angle of corresponding branch current

Expressing Eq. (A–31) in polar form

$$\mathbf{S}_{3\phi,\text{ bal}} = 3(V_{\text{br}}\angle\theta_v) \cdot (I_{\text{br}}\angle\theta_i)^* = 3V_{\text{br}} \cdot I_{\text{br}}\angle(\theta_v - \theta_i) \tag{A–32}$$

Defining $\theta = (\theta_v - \theta_i)$ as the *power-factor angle,* and expressing Eq. (A–32) in rectangular form,

$$\mathbf{S}_{3\phi,\text{ bal}} = 3V_{\text{br}} \cdot I_{\text{br}} \cos\theta + j3V_{\text{br}} \cdot I_{\text{br}} \sin\theta \tag{A–33}$$

where:

$$\text{Active power (watts)} = P_{3\phi,\text{ bal}} = 3V_{\text{br}} \cdot I_{\text{br}} \cos\theta \tag{A–34}$$

$$\text{Reactive power (vars)} = Q_{3\phi,\text{ bal}} = 3V_{\text{br}} \cdot I_{\text{br}} \sin\theta \tag{A–35}$$

$$\text{Apparent power} = S_{3\phi,\text{ bal}} = 3V_{\text{br}} \cdot I_{\text{br}} \tag{A–36}$$

Three-phase power may be expressed in terms of line voltage and line current by substituting the delta relationship or the wye relationship into Eqs. (A–34), (A–35), and (A–36). As previously shown,

If delta connected: $E_{\text{br}} = E_{\text{line}}$ and $I_{\text{br}} = \dfrac{I_{\text{line}}}{\sqrt{3}}$

If wye connected: $I_{\text{br}} = I_{\text{line}}$ and $E_{\text{br}} = \dfrac{E_{\text{line}}}{\sqrt{3}}$

Making the substitution and simplifying,

$$P_{3\phi,\text{bal}} = \sqrt{3}\, V_{\text{line}} I_{\text{line}} \cos\theta \tag{A–37}$$

$$Q_{3\phi,\text{bal}} = \sqrt{3}\, V_{\text{line}} I_{\text{line}} \sin\theta \tag{A–38}$$

$$S_{3\phi,\text{bal}} = \sqrt{3}\, V_{\text{line}} I_{\text{line}} \tag{A–39}$$

Power Triangle

As evidenced in Eqs. (A–37), (A–38), and (A–39), respectively, the active power and reactive power represent two legs of a right triangle whose hypotenuse is the apparent power. Thus,

$$S_{3\phi,\text{bal}} = \sqrt{P^2_{3\phi,\text{bal}} + Q^2_{3\phi,\text{bal}}} \qquad \textbf{(A–40)}$$

The power factor is

$$F_p = \frac{P_{3\phi,\text{bal}}}{S_{3\phi,\text{bal}}} = \cos\theta \qquad \textbf{(A–41)}$$

Note: Angle θ is the power-factor angle. It is the angle between the phase voltage and the phase current; it is not the angle between line voltage and line current! Note also that the power factor of a balanced three-phase load is the power factor of one phase. Substituting Eq. (A–41) into Eq. (A–37),

$$P_{3\phi,\text{bal}} = \sqrt{3}\, V_{\text{line}} I_{\text{line}} F_P \qquad \textbf{(A–42)}$$

Equation (A–42) is the expression generally used for calculating three-phase power.

EXAMPLE A.4 The phase voltage and phase current at one branch of a balanced delta load are determined to be $460\angle{-120^\circ}$ V and $10\angle{-160^\circ}$ A, respectively.

(a) Using the complex power equation, calculate the three-phase apparent power, active power, reactive power, and power factor.

(b) Is the load inductive or capacitive?

Solution

(a)

$$\mathbf{S}_{\text{br}} = \mathbf{V}_{\text{br}}\mathbf{I}^*_{\text{br}} = 460\angle{-120^\circ} \cdot 10\angle{160^\circ} = 4600\angle{40^\circ}$$

$$\mathbf{S}_{\text{br}} = 3523.80 + j2956.82$$

$$P_{3\phi} = 3P_{\text{br}} = 3 \times 3523.80 = 10{,}571.4 \text{ W} \quad \text{or} \quad \underline{10.6 \text{ kW}}$$

$$Q_{3\phi} = 3Q_{\text{br}} = 3 \times 2956.82 = 8870.46 \text{ var} \quad \text{or} \quad \underline{8.87 \text{ kvar}}$$

$$S_{3\phi} = 3S_{\text{br}} = 3 \times 4600 = 13{,}800 \text{ VA} \quad \text{or} \quad \underline{13.8 \text{ kVA}}$$

$$F_P = \frac{P_{3\phi,\text{bal}}}{S_{3\phi,\text{bal}}} = \frac{10{,}571.4}{13{,}800} = 0.766 \quad \text{or} \quad \underline{76.6\%}$$

(b) The reactive power is positive, indicating a lagging current caused by an inductive load.

A.13 POWER ANALYSIS AND POWER-FACTOR CORRECTION OF BALANCED THREE-PHASE LOADS IN PARALLEL

When data on balanced three-phase loads are expressed in kilovoltamperes, kilowatts, kilovars, power factor, horsepower, and η (efficiency), it is often more convenient to analyze the system on a power basis, as illustrated in the following example.

EXAMPLE A.5 A 440-V, 60-Hz, three-phase source supplies a distribution system containing the following three-phase loads:

Motor 1 Delta-connected induction motor rated at 60 hp and 1775 r/min operating at three-quarters rated load with an efficiency of 90 percent and a power factor of 94 percent.

Motor 2 Wye-connected induction motor rated at 75 hp and 890 r/min, operating at one-half rated load with an efficiency of 88 percent and a power factor of 74 percent.

Resistance Heater Delta connected resistor bank drawing 20 kW.

Determine (a) active power, reactive power, apparent power, and power factor of the system; (b) line current; (c) capacitance and voltage rating of each capacitor of a wye-connected capacitor bank required to correct the system power factor to 1.0 (unity power factor).

Solution

The problem will be solved by constructing a single power diagram that includes the individual power triangles of all loads. Furthermore, since all induction motors operate at a lagging power factor, the power factor angle $\theta = (\theta_v - \theta_i)$ is always positive for induction motors.

Motor 1

$$P_{\text{in}} = \frac{P_{\text{out}}}{\eta} = \frac{60 \times 3/4}{0.90} = 50 \text{ hp}$$

$$P_1 = 50 \times 746 = 37{,}300 \text{ W}$$

$$\theta_1 = \cos^{-1}0.94 = 19.95°$$

Motor 2

$$P_{\text{in}} = \frac{P_{\text{out}}}{\eta} = \frac{75 \times 1/2}{0.88} = 42.614 \text{ hp}$$

$$P_2 = 42.614 \times 746 = 31{,}790 \text{ W}$$

$$\theta_2 = \cos^{-1}0.74 = 42.27°$$

Resistance Heater

$$P_3 = 20{,}000 \text{ W} \qquad \theta_3 = 0°$$

The branch current and the branch voltage of a resistor are in phase, angle $\theta_r = (\theta_v - \theta_i) = 0°$, and the reactive power is zero. The individual power triangles are drawn in a common power diagram, as shown in Figure A.15, and the geometry of the individual triangles is used to determine the reactive power drawn by the respective three-phase load. Thus, from Figure A.15,

$$\tan 19.95° = \frac{Q_1}{37{,}300} \qquad \tan 42.27° = \frac{Q_2}{31{,}790}$$

$$Q_1 = 13{,}539 \text{ var} \qquad Q_2 = 28{,}895 \text{ var}$$

(a) The total active power, reactive power, and apparent power drawn by the system are

$$P_{\text{sys}} = P_1 + P_2 + P_3 = 37{,}300 + 31{,}790 + 20{,}000$$

$$= 89{,}090 \text{ W} \quad \Rightarrow \quad \underline{89.1 \text{ kW}}$$

$$Q_{\text{sys}} = Q_1 + Q_2 = 13{,}539 + 28{,}895 = 42{,}434 \text{ var} \quad \Rightarrow \quad \underline{42.4 \text{ kvar}}$$

$$S_{\text{sys}} = \sqrt{P_{\text{sys}}^2 + Q_{\text{sys}}^2} = \sqrt{89{,}090^2 + 42{,}434^2} = 98{,}679 \text{ VA} \quad \Rightarrow \quad \underline{98.7 \text{ kVA}}$$

$$F_P = \frac{P_{\text{sys}}}{S_{\text{sys}}} = \frac{89{,}090}{98{,}679} = \underline{0.903}$$

(b) $P_{\text{sys}} = \sqrt{3}\, V_{\text{line}} I_{\text{line}} F_P \quad \Rightarrow \quad 89{,}090 = \sqrt{3} \times 440 \times I_{\text{line}} \times 0.903$

$I_{\text{line}} = \underline{129.5 \text{ A}}$

FIGURE A.15
Power diagram for Example A.5.

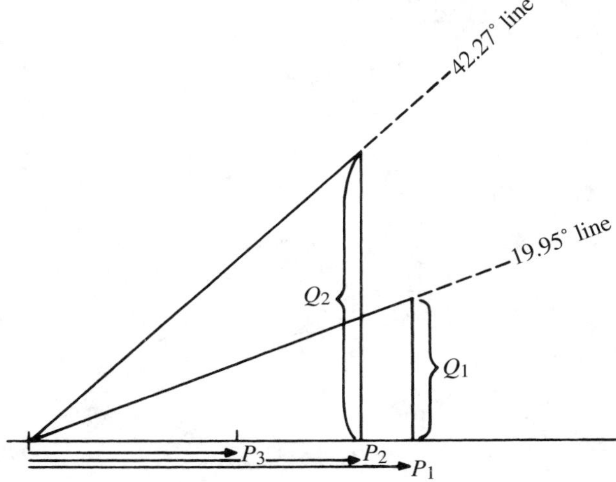

(c) To correct the system power factor to unity requires a three-phase capacitor bank with a var rating equal in magnitude to the lagging vars in the system. Thus, the required var rating of the capacitor bank is

$$Q_{3\varnothing} = 42{,}434 \text{ var}$$

$$Q_{\text{br}} = \frac{42{,}434}{3} = 14{,}145 \text{ var}$$

The voltage rating of each capacitor for a wye bank is $440/\sqrt{3} = \underline{254 \text{ V}}$.

$$Q_{\text{br}} = \frac{V_{\text{br}}^2}{X_C} \quad \Rightarrow \quad 14{,}145 = \frac{254^2}{X_C}$$

$$X_C = 4.56\ \Omega$$

$$X_C = \frac{1}{2\pi f C} = \quad \Rightarrow \quad 4.56 = \frac{1}{2\pi 60 C}$$

$$C = \underline{581\ \mu F}$$

SPECIFIC REFERENCES KEYED TO TEXT

1. American Society of Mechanical Engineers. *USA Standard Drafting Practices, Electrical and Electronic Diagrams.* USAS Y14.15-1966.
2. Hubert, C. I. *Electric Circuits AC/DC: An Integrated Approach.* McGraw-Hill, New York, 1982.
3. Hubert, C. I. *Preventive Maintenance of Electrical Equipment.* Prentice Hall, Upper Saddle River, NJ 2002.

B

Three-Phase Stator Windings

B.1 TWO-POLE WINDING

Three-phase motors have three separate but identical stator windings, each producing its own set of north and south poles. Figure B.1(a) shows the coil layout for a representative two-pole three-phase induction motor. The three sets of north and south poles (A and A'), (B and B'), and (C and C') are displaced from each other by 120 electrical degrees. The relationship between space degrees and electrical degrees is[1]

$$\text{Space deg.} = \frac{2 \times \text{elec. deg.}}{P}$$

Thus, for a two-pole motor, the number of space degrees is equal to the number of electrical degrees; for a four-pole motor the number of space degrees is equal to one-half the number of electrical degrees, etc.

The stator winding for the two-pole motor in Figure B.1(a) has one coil per pole per phase. Coils A, B, and C contribute, in turn, to one magnetic polarity, whereas coils A′, B′, and C′ contribute, in turn, to the opposite magnetic polarity. All coils are wound in the same direction, have one or more turns, and span one-half of the circumference (full pitch). Full pitch is 180 electrical degrees, which is equal to 180 space degrees for a two-pole stator.

The connection diagram for the two-pole stator winding is shown in Figure B.1(b). All coils for a particular phase are connected in series in a manner that will result in alternate north and south poles. Terminal markings b and e indicate the respective beginning and end of each coil. Thus, when the direction of current through coil A is from b to e, the direction of current through coil A′ will be from e to b, causing opposite polarity in the A′ coil.

[1]For more information on the relationship between electrical degrees and space degrees, see Section 1.15, Chapter 1.

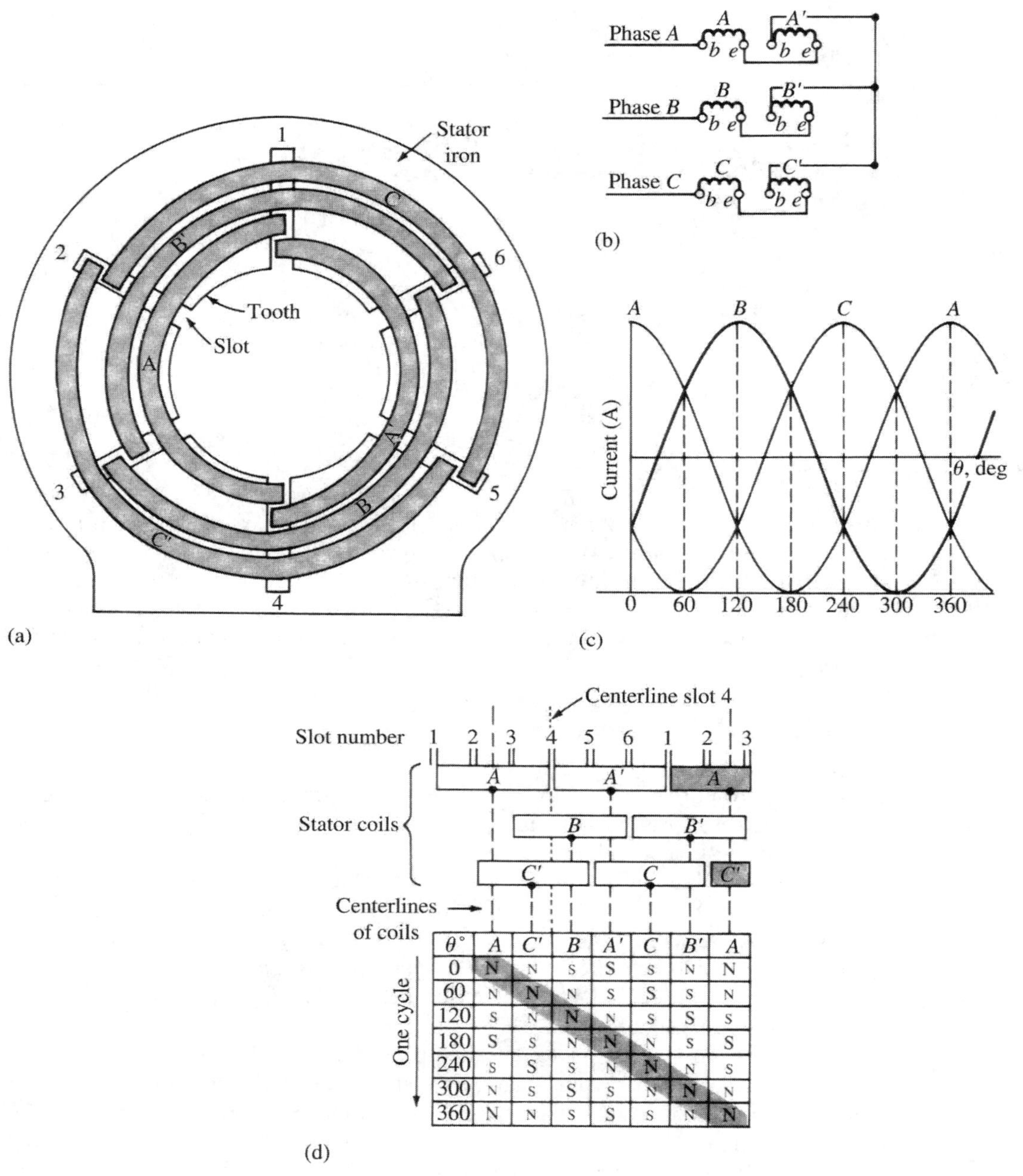

θ°	A	C′	B	A′	C	B′	A
0	N	N	S	S	S	N	N
60	N	N	N	S	S	S	N
120	S	N	N	N	S	S	S
180	S	S	N	N	N	S	S
240	S	S	S	N	N	N	S
300	N	S	S	S	N	N	N
360	N	N	S	S	S	N	N

FIGURE B.1

(a) Symmetrically spaced stator coils of a two-pole, three-phase induction motor; (b) connection diagram; (c) current waves; (d) developed view of coils in (a).

For the coils shown in Figure B.1(a), a north pole is formed in the unprimed coils when its respective current is positive, and a south pole is formed in the unprimed coils when its respective current is negative. The primed coils have opposite polarity with respect to the corresponding unprimed coils. Thus, when the current in phase A is positive, coil A will be north and coil A′ will be south, etc.

The coil span (pitch) for this stator is three slots: slots 1 and 4 for coil A, slots 3 and 6 for coil B, etc. The top of slot 4 contains one side of coil A and the bottom of slot 4 contains one side of coil A′, the top of slot 5 contains one side of coil C′ and the bottom of slot 5 contains one side of coil C, etc. Note that each coil has two sides, and each slot contains two coil sides. Hence, in effect there are the same number of coils as slots.

The two-pole, three-phase stator shown in Figure B.1(a) is redrawn as a developed view in Figure B.1(d). The developed view was obtained by imagining the coils in Figure B.1(a) to have been removed from the stator iron, while maintaining their relative positions and slot numbers, and then spread flat using slot number 4 as a "hinge."[2] *Note:* There are only two coils per phase; the shaded sections on the extreme right are the same coils repeated. The table appended to Figure B.1(d) shows the relative magnitudes and directions of the flux developed by each coil for the specific phase angles of the three current waves shown in Figure B.1(c); the relative magnitudes of the flux are indicated by the size of the letter (N, S, N, S). The broken lines in Figure B.1(d) connect the centerlines of each coil to the corresponding columns. Reading vertically down, each column shows the respective changes in magnetic polarity that occur within each coil as the three-phase current goes through one cycle, from 0 to 360 electrical degrees.

The shifting of the magnetic poles, with the changing current, is shown as a shaded area in the table. Note that the north pole moves from its "starting" position midway between slots 2 and 3 (at zero electrical degrees), returning to its initial position at 360 electrical degrees. Thus, for the two-pole motor in Figure B.1(a), the rotating flux makes one revolution per cycle (360 electrical degrees) of applied three-phase stator voltage.

B.2 FOUR-POLE WINDING

Figure B.2(a) shows the coil layout for a representative four-pole three-phase induction motor. The three sets of north and south poles are displaced from each other by 120 electrical degrees; in terms of space degrees,

$$\text{Space deg.} = \frac{2 \times \text{elec. deg.}}{P} = \frac{2 \times 180}{4} = 90^\circ$$

Coil sets (A_1 and A_2), (B_1 and B_2), and (C_1 and C_2) contribute in turn to one magnetic polarity, and coil sets (A_1' and A_2'), (B_1' and B_2'), and (C_1' and C_2') contribute in turn

[2] This flat layout of stator coils represents a linear motor. (See Section 7.8, Chapter 7.)

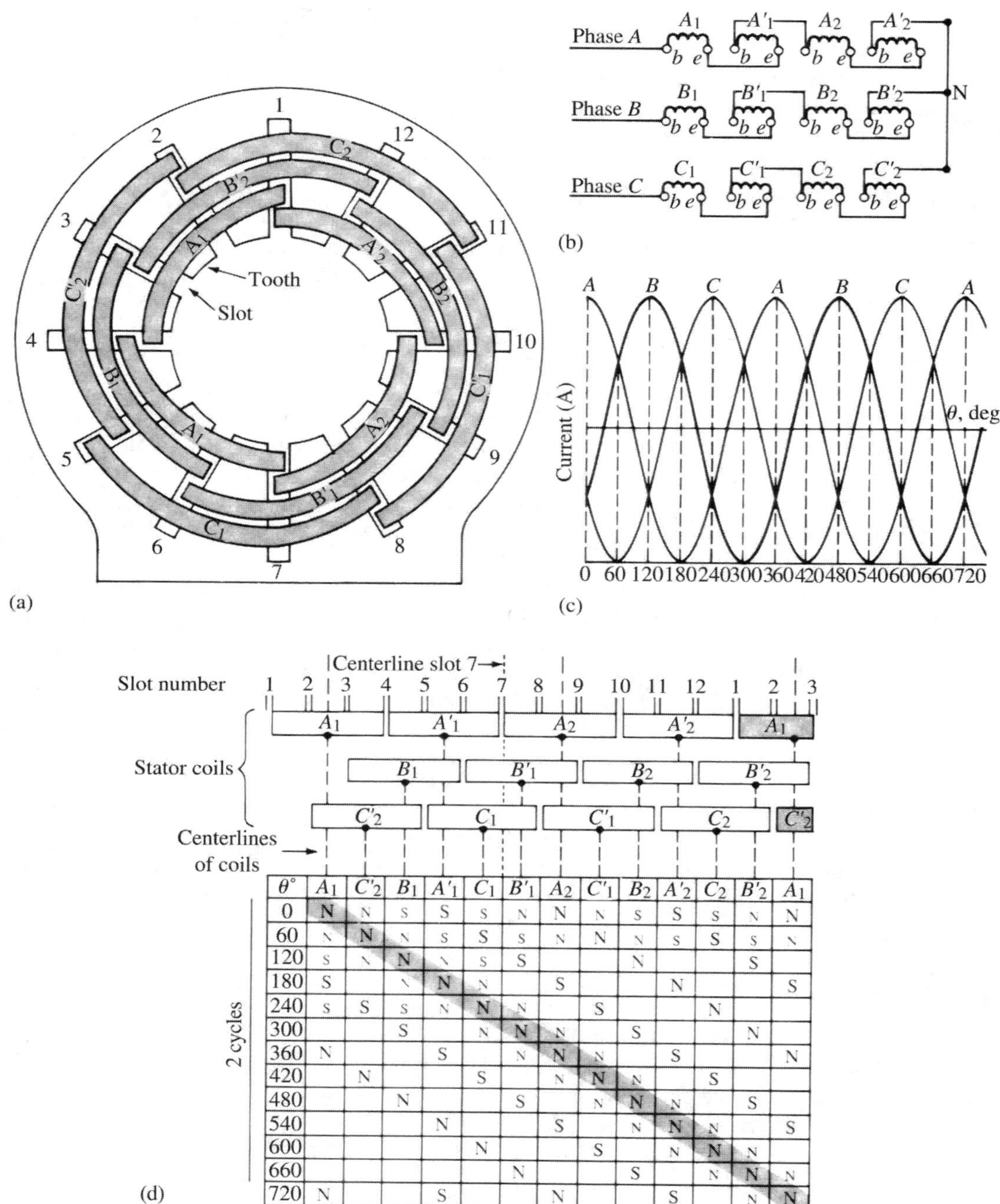

FIGURE B.2
(a) Symmetrically spaced stator coils of a four-pole, three-phase induction motor; (b) connection diagram; (c) current waves; (d) developed view of coils in (a).

to the opposite magnetic polarity. Each coil of the four-pole stator spans one-quarter of the circumference (90 space degrees), equivalent to 180 electrical degrees.

To obtain alternate north and south poles, the coils for each phase are connected in series, end to end, and beginning to beginning, as shown in Figure B.2(b). As with the previous example for the two-pole motor, a north pole is formed in an unprimed coil when its respective current is positive, and a south pole is formed when its respective current is negative. The primed coils have opposite polarity with respect to the corresponding unprimed coils. Thus, when the current in phase A is positive, coils A_1 and A_2 will be north and coils A_1', and A_2' will be south, etc.

The four-pole three-phase stator shown in Figure B.2(a) is redrawn as a developed view in Figure B.2(d). The developed view was obtained in a manner similar to that used for the two-pole winding, with slot number 7 as the "hinge." *Note:* There are only four coils per phase; the shaded sections on the extreme right are the same coils repeated. The table appended to Figure B.2(d) shows the relative magnitudes and directions of the flux developed by each coil for the specific phase angles of the three current waves in Figure B.2(c); the relative magnitudes are indicated by the size of the letter (N, S, N, S). The broken lines in Figure B.2(d) connect the centerlines of each coil to the corresponding columns. Reading vertically down, each column shows the changes in magnetic polarity that occur within each coil as the three-phase current goes through two cycles, from 0 to 720 electrical degrees. The blank spaces are left as an exercise to be filled in by the student.

The shifting of the magnetic poles, with the changing current, is shown as a shaded area in the table. Note that the north pole moves from its "starting" position midway between slots 2 and 3 at zero electrical degrees to a position midway between slots 8 and 9 at 360 electrical degrees, making only one-half a revolution per cycle of applied voltage. It takes 720 electrical degrees for the rotating flux to make one complete revolution, returning to its starting position midway between slots 2 and 3. This is twice the number of electrical degrees and, hence, twice the time for the rotating flux to make one revolution. Thus, contrasted with the previously discussed two-pole motor, the rotating flux of a four-pole motor revolves at half the speed of the rotating flux of a two-pole motor.

B.3 FULL-PITCH WINDING

The coil span for a full-pitch winding is

$$\delta = \frac{S}{P} \tag{B–1}$$

where: S = number of stator slots
P = number of poles
δ = full-pitch slots/pole

The coil slot locations for full pitch coils are 1 and $1 + \delta$, 2 and $2 + \delta$, etc.

EXAMPLE B.1 Determine (a) the coil span for the stator of a six pole, three-phase, 54-slot induction motor; (b) list slot locations for several coils.

Solution

(a)

$$\delta = \frac{S}{P} = \frac{54}{6} = \underline{9}$$

(b) Coil slots locations: 1 and 10, 2 and 11, etc.

B.4 FRACTIONAL-PITCH WINDINGS

If the stator coils have a span less than full pitch (<180 electrical degrees), the winding is called fractional pitch, short pitch, or chorded **[1]**. The difference between a full-pitch winding and a fractional-pitch winding is shown in Figure B.3. Assuming the stator has 24 slots and is to be wound for three phases with four poles, the coil span for full pitch will be

$$\delta = \frac{S}{P} = \frac{24}{4} = 6 \text{ slots}$$

Thus, the coil sides for a representative full-pitch coil will be in slots 1 and 7.

The shorter end turns, also called end connections, of a fractional-pitch winding result in a savings in copper, lower resistance, less heat loss in the windings, and higher efficiency than for an equivalent full-pitch winding. Furthermore, the lower leakage reactance of a fractional pitch winding increases the maximum torque that the machine can develop and provides a general overall improvement in machine operation.

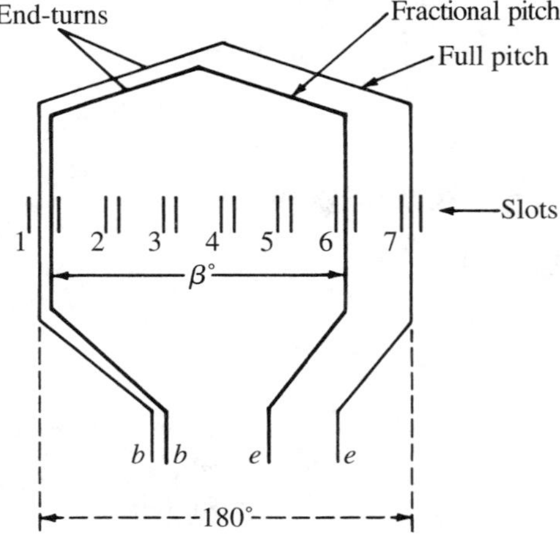

FIGURE B.3
Comparison of coils for a full-pitch winding and a fractional-pitch winding.

B.5 DISTRIBUTED WINDINGS

The windings illustrated in Figures B.1 and B.2 are called concentrated windings in that all of the turns per pole per phase are concentrated in one coil. A distributed winding **[1]** distributes the turns into two or more series connected coils in adjacent slots, as shown in Figure B.4. These are called *pole phase groups.* To accommodate the additional coils, the stator core for a distributed winding has two or more times the amount of slots that would be required if it were housing a concentrated winding. The number of slots per pole phase group may be determined from:

$$S' = \frac{S}{P \times \text{phases}} \tag{B–2}$$

where: S' = number of slots per pole phase group
S = number of slots
P = number of poles

Although the many narrow and shallow slots of the distributed winding cause more flux pulsations per revolution than does the concentrated winding, the pulses are of much lower amplitude and result in a smoother flux distribution. This results in smoother torque, lower amplitudes of vibration, and a better distribution of heat losses in the iron. The voltage produced by a distributed winding is slightly less than that produced by a concentrated winding with the same number of turns. The voltage ratio

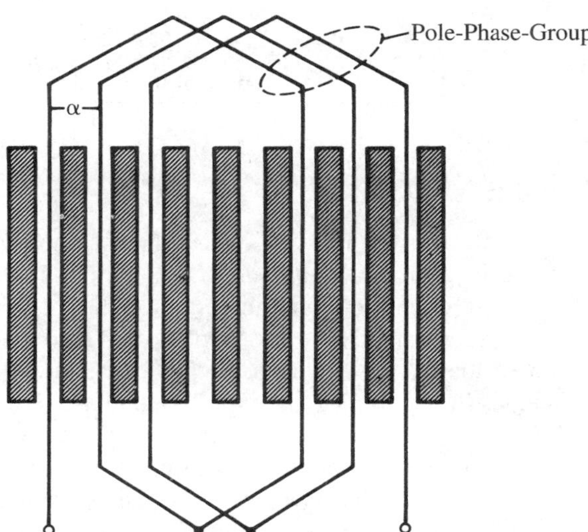

FIGURE B.4
Distributed winding.

of V-distributed to V-concentrated, called the distribution factor, spread factor, breadth factor, or belt factor, can be calculated from:

$$k_d = \frac{\sin(S' \times (\alpha/2)}{S' \times \sin(\alpha/2)} \qquad \textbf{(B–3)}$$

$$\alpha = \frac{P \times 180}{S}$$

where: k_d = distribution factor
α = number of electrical degrees between the centers of adjacent slots
S' = number of slots per pole phase group

EXAMPLE B.2 Given a four-pole, three-phase, 48-slot, full-pitch stator for a 100-hp 460-V motor, determine (a) the coil pitch and a representative span for one coil; (b) the number of slots per pole phase group; (c) the electrical degrees between centers of adjacent slots; (d) the distribution factor.

Solution

(a)

$$\delta = \frac{S}{P} = \frac{48}{4} = \underline{12 \text{ slots}}$$

Span is from slot 1 to slot 13.

(b)

$$S' = \frac{S}{P \times \text{phases}} = \frac{48}{4 \times 3} = \underline{4}$$

(c)

$$\alpha = \frac{P \times 180}{S} = \frac{4 \times 180}{48} = \underline{15^\circ}$$

(d)

$$k_d = \frac{\sin(S' \times \alpha/2)}{S' \times \sin(\alpha/2)} = \frac{\sin(4 \times 15/2)}{4 \times \sin(15/2)} = \underline{0.958}$$

B.6 CONSEQUENT-POLE MOTORS

A two-speed motor with a speed ratio of 2:1 can be obtained from a single winding that is specifically designed for *consequent-pole* operation. The coil pitch of 90 electrical degrees for a consequent pole winding is one-half that for a standard machine. High and low speeds are obtained by disconnecting and reconnecting the windings or by using a two-pole, double-throw selector switch, as shown in Figure B.5(a). For simplicity, all sketches in Figure B.5 show only one phase of a three-phase winding. When the switch is in the up position, coils 2 and 4 will have opposite polarity with respect to coils 1 and 3, and the stator will exhibit four poles, as shown in Figure B.5(b). When the switch is in the down position, all four coils will have the same polarity, and the stator will exhibit eight poles, as shown in Figure B.5(c). Making all four coils north poles forces south poles to form between them; the four south

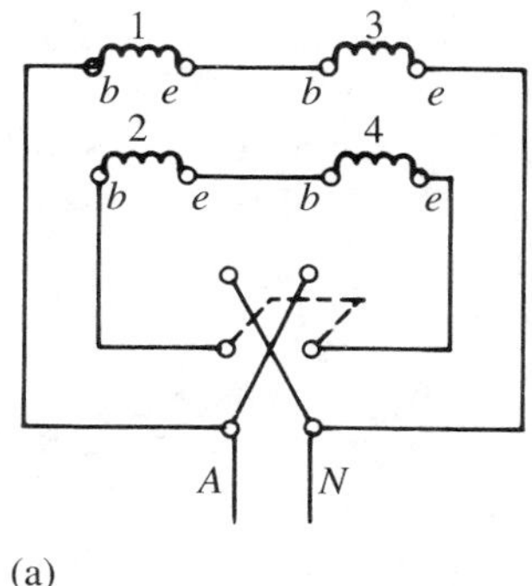

(a)

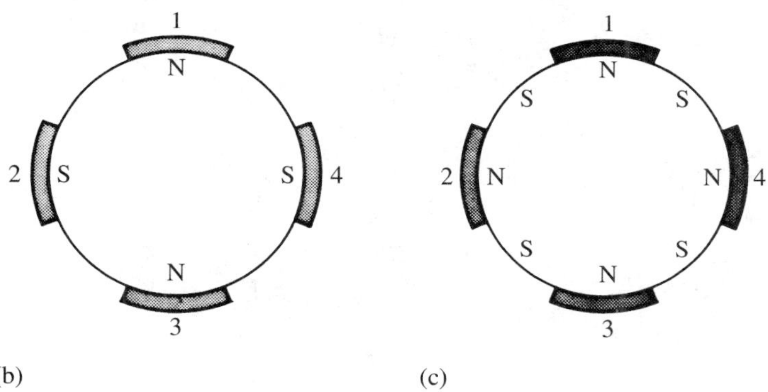

(b) (c)

FIGURE B.5
Consequent-pole windings; (a) winding connections to selector switch; (b) coil polarities for four-pole operation; (c) coil polarities for eight-pole operation.

poles in Figure B.5(c) are called consequent poles because they were formed as a consequence of this connection. Note that with respect to the eight-pole connection in Figure B.5(c), the coils have full pitch; the same coils are half-pitch for the four-pole connection.

A more generalized concept of the consequent pole machine is the PAM (pole amplitude modulation) motor **[1],[2]**. The PAM motor is a squirrel-cage induction motor whose *single winding stator* can be connected to provide speed ratios of other than 2:1. Selective switching of coil polarities using a switching arrangement similar to that used for consequent-pole machines are used to obtain the desired number of poles. Representative pole arrangements are shown in Figure B.6. Reversing coils 2, 3, and 4 of the six-pole winding in Figure B.6(a) results in a four-pole winding; reversing coils 2, 3, 4, and 5 in the 8-pole winding in Figure B.6(b) results in a 6-pole winding; reversing coils 2, 5, 6, 8, and 9 of the 10-pole winding in Figure B.6(c) results in a 4-pole winding.

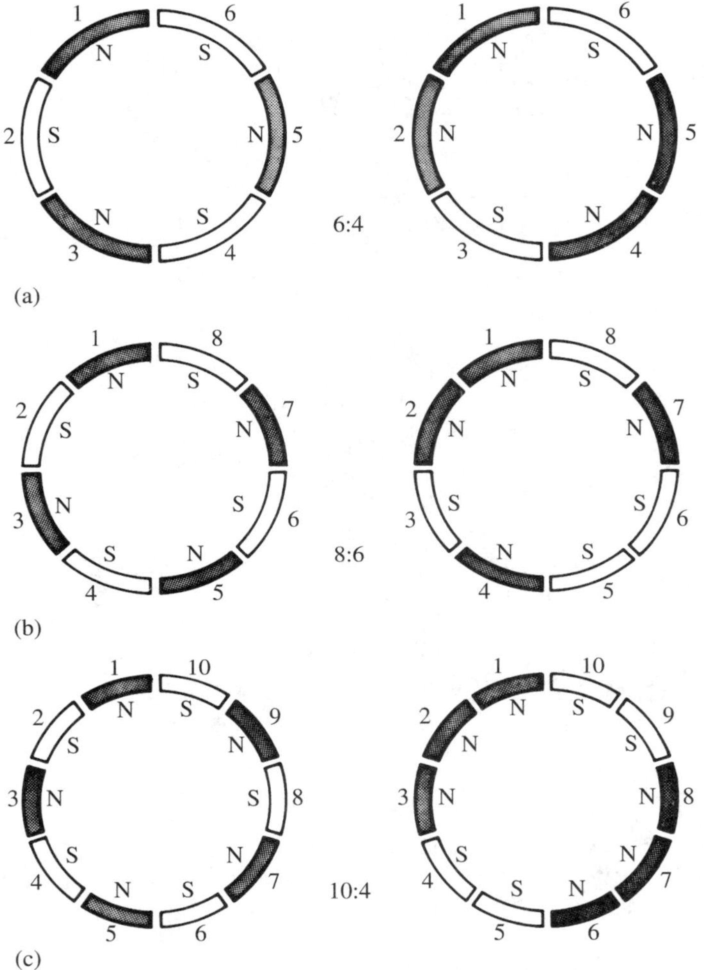

FIGURE B.6
Representative pole arrangements for three representative PAM motors: (a) 6 and 4 poles; (b) 8 and 6 poles; (c) 10 and 4 poles.

Unlike a conventional stator, the PAM stator uses irregular coil groupings that produce space harmonics in the rotating field. Since space harmonics can cause low starting torque, excessive noise, and sharp dips in torque during acceleration, more consideration must be given to the selection of an appropriate winding arrangement. PAM motors, however, have a slightly higher efficiency than a comparable two-speed, two-winding motor; and are smaller, lighter in weight, and generally less expensive than conventional two-speed two-winding motors.

SPECIFIC REFERENCES KEYED TO TEXT

1. McPherson, G. *An Introduction to Electrical Machines and Transformers.* Wiley, New York, 1981.
2. Ratcliffe, R. The change-speed PAM motor and its application in the rubber and plastics industries. *IEEE Trans. Industry General Applications,* Vol. IGA-6, No. 2, Mar./Apr. 1970.

C

Constant-Horsepower, Constant-Torque, and Variable-Torque Induction Motors

There are three general groups of multispeed squirrel-cage induction motors, each group designed for a specific type of application. They are constant-horsepower, constant-torque, and variable-torque induction motors **[1],[2]**.

C.1 CONSTANT-HORSEPOWER MOTOR

A constant-horsepower, multispeed motor is designed to deliver approximately the same rated horsepower with every synchronous speed connection. Hence, the rated torque for the different speed connections must vary inversely with the synchronous speed. The mathematical relationship involved is the basic power equation $P = Tn/5252$. Thus, for a constant-horsepower, multispeed induction motor,

$$\frac{T_{LO}n_{LO}}{5252} = \frac{T_{HI}n_{HI}}{5252} \quad \Rightarrow \quad \frac{T_{LO}}{T_{HI}} = \frac{n_{HI}}{n_{LO}} \qquad \textbf{(C–1)}$$

For example, the torque that can be delivered at a 900 r/min connection would be twice that at an 1800 r/min connection. Multispeed motors of this type are used for lathes and other machine tools that often require a constant rate of doing work. *It is important to note, however, that the motor will not deliver the same horsepower at all speed connections unless the load demands it.*

C.2 CONSTANT-TORQUE MOTOR

A constant-torque multispeed motor is designed to deliver approximately the same torque with every synchronous speed connection. Hence, the rated horsepower for the different speed connections must vary directly with the synchronous speed. From the basic power equation, $T = 5252P/n$. Hence,

$$\frac{5252P_{\text{LO}}}{n_{\text{LO}}} = \frac{5252P_{\text{HI}}}{n_{\text{HI}}} \quad \Rightarrow \quad \frac{P_{\text{LO}}}{P_{\text{HI}}} = \frac{n_{\text{LO}}}{n_{\text{HI}}} \tag{C–2}$$

For example, the horsepower rating for a 900 r/min connection would be half the horsepower rating at the 1800 r/min connection. Multispeed motors of this type are used for conveyers, compressors, reciprocating pumps, printing presses, and similar loads. *It should be noted, however, that a constant-torque motor will not deliver constant torque unless the load demands it.*

C.3 VARIABLE-TORQUE MOTOR

A variable-torque multispeed motor is designed to have its rated torque vary in direct proportion to the synchronous speed for every speed connection. Hence, its horsepower rating for the different speed connections will be in proportion to the square of the synchronous speed. Expressed mathematically,

$$\frac{T_{\text{LO}}}{T_{\text{HI}}} = \frac{n_{\text{LO}}}{n_{\text{HI}}} \tag{C–3}$$

$$\frac{P_{\text{LO}}}{P_{\text{HI}}} = \frac{n_{\text{LO}}^2}{n_{\text{HI}}^2} \tag{C–4}$$

For example, the torque that can be developed at a 900 r/min connection would be half that at the 1800 r/min connection. Hence, the 900 r/min connection would have one-quarter the horsepower rating of the 1800 r/min connection. Multispeed motors of this type are used for fans, centrifugal pumps, or other loads with similar characteristics. The power requirements for fans and blowers is directly proportional to the cube of the speed. Hence, a lower speed connection requires significantly less power from the motor.

EXAMPLE C.1 A 20-hp, 460-V, 60-Hz, variable-torque induction motor has speeds rated at 1750 and 1150 r/min. What is its horsepower rating at each speed?

Solution

The horsepower rating for the higher speed connection is always the nameplate value. Thus, in this example the horsepower rating for the 1750 r/min connection is 20 hp. The respective synchronous speeds are 1800 r/min and 1200 r/min. The horsepower rating for the 1150 r/min connection is

$$\frac{P_{\text{LO}}}{20} = \frac{1200^2}{1800^2}$$

$$P_{\text{LO}} = \underline{8.89 \text{ hp}}$$

SPECIFIC REFERENCES KEYED TO TEXT

1. Heredos, F. P. Selection and application of multi-speed motors. *IEEE Trans. Industry Applications,* Vol. IA-23, No. 2, Mar./Apr. 1987.
2. National Electrical Manufacturers Association, *Motors and Generators.* Standards Publication No. MG-1-1998, NEMA, Rosslyn, VA, 1999.

D

Selected Graphic Symbols Used in Controller Diagrams

Symbol	Device
	Ground connection
	Fuse
	Resistor
	Rheostat
	Indicating lamp
	Capacitor
	Diode
	Silicon controlled rectifier (SCR)
or	Overload heater
	Blowout coil
or	Operating coil
	Contact normally open
	Contact normally closed
	Spring-return push button normally open
	Spring-return push button normally closed
	Sustaining-type push button
	Plug-type contact

Symbol	Device
	Mechanical interlock
	Mechanical interlock with fulcrum
	Crossing conductors not connected
	Connected conductors
	Transformer
	Current transformer
	Three-pole circuit breaker
	Three-pole power breaker for AC circuits rated in excess of 1500 V
	Switch
	Reactor or field winding

Bell Buzzer Horn or siren

Limit Switches

Symbol	Device
	Normally open contact
	Normally open contact held closed
	Normally closed contact
	Normally closed contact held open

Full-Load Current in Amperes, Direct-Current Motors

The following values of full-load currents* are for motors running at base speed.

hp	Armature Voltage Rating*					
	90 V	120 V	180 V	240 V	500 V	550 V
$\frac{1}{4}$	4.0	3.1	2.0	1.6		
$\frac{1}{3}$	5.2	4.1	2.6	2.0		
$\frac{1}{2}$	6.8	5.4	3.4	2.7		
$\frac{3}{4}$	9.6	7.6	4.8	3.8		
1	12.2	9.5	6.1	4.7		
$1\frac{1}{2}$		13.2	8.3	6.6		
2		17	10.8	8.5		
3		25	16	12.2		
5		40	27	20		
$7\frac{1}{2}$		58		29	13.6	12.2
10		76		38	18	16
15				55	27	24
20				72	34	31
25				89	43	38
30				106	51	46
40				140	67	61
50				173	83	75
60				206	99	90
75				255	123	111
100				341	164	148
125				425	205	185
150				506	246	222
200				675	330	294

*These are average direct-current quantities.

Full-Load Current in Amperes, Single-Phase Alternating-Current Motors

The following values of full-load currents are for motors running at usual speeds and motors with normal torque characteristics. Motors built for especially low speeds or high torques may have higher full-load currents, and multispeed motors will have full-load current varying with speed, in which case the nameplate current ratings shall be used.

To obtain full-load currents of 208- and 200-V motors, increase corresponding 230-V motor full-load currents by 10 and 15 percent, respectively.

The voltages listed are rated motor voltages. The currents listed shall be permitted for system voltage ranges of 110 to 120 and 220 to 240.

hp	115 V	230 V
$\frac{1}{6}$	4.4	2.2
$\frac{1}{4}$	5.8	2.9
$\frac{1}{3}$	7.2	3.6
$\frac{1}{2}$	9.8	4.9
$\frac{3}{4}$	13.8	6.9
1	16	8
$1\frac{1}{2}$	20	10
2	24	12
3	34	17
5	56	28
$7\frac{1}{2}$	80	40
10	100	50

G

Full-Load Current, Two-Phase Alternating-Current Motors (Four-Wire)

The following values of full-load current are for motors running at speeds usual for belted motors and motors with normal torque characteristics. Motors built for especially low speeds or high torques may require more running current, and multispeed motors will have full-load current varying with speed, in which case the nameplate current rating shall be used. Current in the common conductor of a two-phase, three-wire system will be 1.41 times the value given.

The voltages listed are rated motor voltages. The currents listed shall be permitted for system voltage ranges of 110 to 120, 220 to 240, 440 to 480, and 550 to 600 V.

	Induction-Type Squirrel-Cage and Wound-Rotor (A)				
hp	115 V	230 V	460 V	575 V	2300 V
$\frac{1}{2}$	4	2	1	0.8	
$\frac{3}{4}$	4.8	2.4	1.2	1.0	
1	6.4	3.2	1.6	1.3	
$1\frac{1}{2}$	9	4.5	2.3	1.8	
2	11.8	5.9	3	2.4	
3		8.3	4.2	3.3	
5		13.2	6.6	5.3	
$7\frac{1}{2}$		19	9	8	
10		24	12	10	
15		36	18	14	
20		47	23	19	
25		59	29	24	
30		69	35	28	
40		90	45	36	
50		113	56	45	
60		133	67	53	14
75		166	83	66	18
100		218	109	87	23
125		270	135	108	28
150		312	156	125	32
200		416	208	167	43

Full-Load Current, Three-Phase Alternating-Current Motors

For full-load currents[1] of 208- and 200-V motors, increase the corresponding 230-V motor full-load current by 10 to 15 percent, respectively.

The voltages listed are rated motor voltages. The currents listed shall be permitted for system voltage ranges of 110–120, 220–240, 440–480, and 550–660 V.

[1]These values of full-load current are for motors running at speeds usual for belted motors and motors with normal torque characteristics. Motors built for especially low speeds or high torques may require more running current, and multispeed motors will have full-load current varying with speed, in which case the nameplate current rating shall be used.

	Induction-Type Squirrel-Cage and Wound-Rotor (A)					Synchronous-Type Unity Power Factor* (A)			
hp	115 V	230 V	460 V	575 V	2300 V	230 V	460 V	575 V	2300 V
$\frac{1}{2}$	4	2	1	0.8					
$\frac{3}{4}$	5.6	2.8	1.4	1.1					
1	7.2	3.6	1.8	1.4					
$1\frac{1}{2}$	10.4	5.2	2.6	2.1					
2	13.6	6.8	3.4	2.7					
3		9.6	4.8	3.9					
5		15.2	7.6	6.1					
$7\frac{1}{2}$		22	11	9					
10		28	14	11					
15		42	21	17					
20		54	27	22					
25		68	34	27		53	26	21	
30		80	40	32		63	32	26	
40		104	52	41		83	41	33	
50		130	65	52		104	52	42	
60		154	77	62	16	123	61	49	12
75		192	96	77	20	155	78	62	15
100		248	124	99	26	202	101	81	20
125		312	156	125	31	253	126	101	25
150		360	180	144	37	302	151	121	30
200		480	240	192	49	400	201	161	40

*For 90 and 80 percent power factor the preceding figures shall be multiplied by 1.1 and 1.25, respectively.

I

Representative Transformer Impedances for Single-Phase 60-Hz Transformers

Rating (kVA)	Voltage 2400		7200	
	% R	% X	% R	% X
10	1.51	1.78	1.60	1.62
50	1.30	2.25	1.29	2.10
100	1.20	2.31	1.20	3.53
250	1.01	4.70	1.00	5.16
500	1.00	4.75	1.00	5.24

J

Unit Conversion Factors

Force: lb $\times$ 4.448	= N
Length: ft $\times$ 0.3048	= m
Magnetics: Oersteds $\times$ 79.577	= A-t/m
Lines $\times 10^{-8}$	= Wb
Lines/in.^{2} $\times 1.55 \times 10^{-5}$	= T
Gausses $\times 10^4$	= T
Power: hp $\times$ 746	= W
Rotational speed: r/min $\times$ 0.1047	= rad/s
Torque: lb-ft $\times$ 0.7376	= N-m

Answers To *Odd-Numbered* Problems

Chapter 1

1. (a) 0.40 Wb, (b) 8 Ω
3. 64.18 V
5. 1.013 T
7. (a) 1499.1 A-t/m, (b) 1.25 T, 0.10 Wb, (c) 663.5, (d) 22486 A-t/Wb
9. (a) 1499.1 A-t/m, (b) 0.48 T, 0.0384 Wb, (c) 254.8, (d) 58557.2 A-t/Wb
11. −47.1%
13. (a) 7.21 V, (b) 7.96 V
15. 65.89 m/s
17. 24 Hz, 89.5 V
19. (a) 474.87 V, (b) 672 cos(28t) V
21. 48.98 W

Chapter 2

1. (a) 121 t, (b) 0.0862 Wb
3. (a) 126 t, 630 t, (b) 3.0 A
5. (a) 2.60 A, (b) 0.460 A, (c) 2.56 A, (d) 220.8 W
7. (a) 15.87 A, (b) 0.173, (c) 1875 var
9. (a) 6600 V, (b) 45.83 A, (c) 1375 A, (d) 0.160 $\angle 46^\circ$ Ω, (e) 210.1 kW, 217.6 kvar, 302.5 kVA
11. (a) 124.8 V, (b) 624 V, (c) 3.12 A, (d) 1651.05 W, 1031.69 var, 1946.88 VA
13. (a) 8.97 $\angle 61.63^\circ$ Ω, (b) 0.56 $\angle 61.63^\circ$ Ω
15. (a) 13.78 $\angle 63.21^\circ$ Ω, (b) 531.11 $\angle 41.95^\circ$ Ω, (c) 13.89 A, (d) 7377 V, (e) 0.427 A, (f) 17276.4 $\angle -75.5^\circ$ Ω
17. (a) 0.0435 $\angle 83.5^\circ$ Ω, (b) 495.27 V, (c) 3.18 percent
19. (a) 2387.8 V, (b) −0.51 percent, (c) 51.68 $\angle -15.21^\circ$ Ω

21. (a) 243 V, (b) 5.68 percent, (c) 1341.1 $\angle 48.82^\circ$ Ω, (d) 42321 $\angle 75.99^\circ$ Ω
23. (a) 0.194 Ω, (b) 0.012 Ω
25. (a) 94,696 A, (b) 3.47 percent
27. 2.84 percent
29. 2.61 percent
31. (a) 3.26 percent, (b) 237.5 V, (c) 464.7 V
33. Plot
35. (a) 345 W, (b) 97.05 percent
37. Plot
39. (a) $R_{eq} = 1.743\ \Omega$, $X_{eq} = 3.233\ \Omega$, $R_{fe} = 11901\ \Omega$, $X_M = 2961.9\ \Omega$, (b) 2.51 percent, (c) 97.4 percent
41. (a) 6572.6 Ω, (b) $R_{PU} = 0.0121$, $X_{PU} = 0.0384$, (c) 98.8 percent, (d) 2.44 percent, (e) 235.6 V, (f) 4712 V

Chapter 3

1. (a) 225 A, (b) $I_{HS} = 44.03$ A, $I_{tr} = 181.1$ A
3. (a) 800 V, (b) 6 A, (c) 2 A, (d) 1600 VA, (e) 3200 VA, (f) 0.015 Wb
5. (a) 1.176, (b) 1.200, (c) 122.4 V
7. (a) 1.100, (b) sketch, (c) 750 A, 681.8 A
9. 10.07
11. (a) 416.67 $\angle -43.95^\circ$ A, (b) $I_A = 234.1 \angle -44.14^\circ$ A, $I_B = 182.5 \angle -43.70^\circ$ A
13. $I_A = 38.17\%$, $I_B = 34.99\%$, $I_C = 26.94\%$
15. No, B will overheat.
17. $I_{LS,phase} = I_{LS,line} = 479.2$ A, $I_{HS,phase} = 69.45$ A, $I_{HS,line} = 120.3$ A
19. 69.28 kVA
21. 10,249 A

Chapter 4

1. (a) 1800 r/min, (b) 50 r/min, (c) 0.278
3. (a) 1800 r/min, (b) 0.01388, (c) 25 r/min, (d) 0.833 Hz
5. (a) 20 r/min, (b) 1.0 Hz, 3.6 V
7. (a) 50 r/min, (b) 2.78% (c) 180 r/min
9. (a) 500 r/min, (b) 0.040, (c) 63.10 A, (d) 218.83 lb-ft, (e) 2.0 Hz
11. (a) 22030 W, (b) 26.58 hp, (c) 2203 W, (d) 586 r/min, (e) 238.2 lb-ft, (f) 506 W
13. (a) 706.9 r/min, (b) 945.4 lb-ft, (c) 928.7 lb-ft, (d) 0.805, (e) 1683.8 W
15. (a) 1746.2 r/min, (b) 18.8 lb-ft, (c) 1.23 lb-ft
17. (a) 22.0 hp, 3150.5 r/min, 84.3%

Chapter 5

1. (a) 47.89 lb-ft, (b) 73.68 lb-ft, (c) 36.84 lb-ft
3. (a) 37.7 lb-ft, (b) 29.57 lb-ft, (c) 26.61 lb-ft

5. (a) 11.96 $\angle 37.58^\circ$ Ω, (b) 27.77 $\angle -37.58^\circ$ A, (c) 21916.6 W, 16866.9 var, 27655.6 VA, 0.793 lagging, (d) 22.8 A, (e) 861.2 W, (f) 608.7 W, (g) 764.2 W, (h) 20291.2 W, (i) 19682.5 W, (j) 119.04 lb-ft, (k) 25.92 hp, (l) 116.95 lb-ft, (m) 88.23%, (n) sketch.
7. (a) 7.89 $\angle 40.31^\circ$ Ω, (b) 33.68 $\angle -40.31^\circ$ A, (c) 20461.9 W, 17356.4 var, 26831.6 VA, 76.26% lagging, (d) 27.11 A, (e) 643.38 W, (f) 421.38 W, (g) 865.3 W, (h) 18963.8 W, (i) 18542.8 W, (j) 148.47 lb-ft, (k) 2280.1 W, (l) 24.37 hp, (m) 88.66%, (n) sketch, (o) LR = 360.54 lb-ft, BD = 342.52 lb-ft, PU = 252.38 lb-ft
9. (a) 810.5 r/min, (b) 392.2 lb-ft
11. No
13. (a) 0.958%, (b) 1782.8 r/min, (c) 215.7 A, (d) 191.9 hp
15. (a) 1169 r/min, (b) 20.1 A, (c) 17.2 hp
17. 26.8 hp
19. (a) 479.2 V, (b) 104.2 hp, (c) 750 r/min, (d) 734.25, (e) 745 lb-ft
21. 0.135 Ω/phase
23. (a) 71 A, (b) 267.3 A $\leq I_{lr} <$ 301.2 A
25. (a) 3.17%, (b) 108°C, (c) 5.7 years, (d) 52.8 hp
27. $R_1 = 0.2539\ \Omega$, $X_1 = 0.6836\ \Omega$, $R_2 = 0.1872\ \Omega$, $X_2 = 1.0282\ \Omega$, $X_M = 21.42\ \Omega$
29. PUR_1 = 0.0333, PUX_1 = 0.0562, PUR_2 = 0.030, PUX_2 = 0.084, PUX_M = 1.227, PUR_{fe} = 42.38
31. $R_1 = 0.1915\ \Omega$, $R_2 = 0.1895\ \Omega$, $X_1 = 0.5745\ \Omega$, $X_2 = 1.3404\ \Omega$, $X_M = 14.92\ \Omega$, fwcor = 405.1 W/phase
33. 28.8 kW
35. 186.53 kW
37. (a) 135.78 A, (b) 441.34 lb-ft, (c) 168 V, (d) 76.38%, (e) 553.7 A, (f) 422.9 A
39. (a) 248.2 A, (b) 595.46 lb-ft, (c) 318.57 A, (d) 1.444
41. (a) 2.358 Ω, (b) 154.4 V, (c) 45.7 lb-ft

Chapter 6

1. (a) 12.34 Ω, (b) Auxiliary 5.53 $\angle -16.83^\circ$ A, Main 20.84 $\angle -46.26^\circ$ A, (c) 25.78 $\angle -40.67^\circ$ A
3. (a) 222.7 μF, (b) 25.6 $\angle -19.7^\circ$ A
5. (a) 11500 μF, (b) 1325 μF, (c) 33.33 hp
7. (a) I_{line} = 22.41 A, I_{phase} = 12.94 A, (b) I_{line} = 38.81 A, I_A = 12.94 A, I_B = 25.87 A

Chapter 7

1. 30.54% increase
3. (a) 170%, (b) 170%
5. (a) 200, (b) 30 r/s, (c) 26

7. (a) 1000 steps/rev, (b) 210
9. 0.20

Chapter 8

1. 150 r/min
3. 80 poles
5. (a) 5835.6 lb-ft, (b) 161 A, (c) 13669.5 V, (d) −29.2°, (e) 12389.6 lb-ft
7. (a) 496.0 V, (b) −27.8°, 1332.5 lb-ft, (c) 666.2 lb-ft
9. Phasor diagrams
11. (a) plot, (b) sketch, (c) rated: 7280 V/phase, 75% rated: 5500 V/phase, 50% rated: 3680 V/phase
13. (a)−18°, (b) 189,666 W, −26,1053 var, (c) 58.8% leading, (d) 742 lb-ft
15. (a) 0.4931 leading, (b) 542.3 V, (c) −15.57°
17. (a) 78.1% lagging, (b) 0.895 lagging
19. (a) 4084.9 lb-ft, (b) 3063.7 lb-ft, (c) 20,424.4 lb-ft

Chapter 9

1. 1500 r/min
3. (a) 314.46 V, (b) 64.08 Hz
5. (a) 670 hp, (b) 859.1 V, (c) 580.8 A, (d) 499843 W, 338348 var, (e) 82.8% lagging
7. 61.29 Hz, $P_1 = 75.79$ kW, $P_2 = 74.21$ kW
9. (a) 60.15 Hz, (b) $P_A = 313$ kW, $P_B = 373$ kW
11. (a) 24.77 Hz, (b) $P_A = 471.68$ kW, $P_B = 601.87$ kW, $P_C = 626.45$ kW
13. (a) Phasor diagram, (b) $F_P = 76.6\%$ leading, $\delta = -17°$
15. (a) Sketch, (b) 35°, (c) 27°
17. $P_A = 347.76$ kW, $P_B = 115.92$ kW, $Q_A = 143.92$ kvar, $Q_B = 170.09$ kvar
19. (a) Sketch, (b) 59.89 Hz, $P_A = 690.90$ kW, $P_B = 595.45$ kW, $P_C = 563.63$ kW, (c) 73.5% leading
21. 1.449
23. (a) 271.8 V, (b) 65.7°, (c) 274 V, (d) 1.1%, (e) 208 V
25. 7.9%
27. 483 V
29. (a) 1.32 Ω, (b) 1.07
31. (a) 0.5047 Ω, (b) 1.52 Ω, (c) 1.60 Ω

Chapter 10

1. (a) 56.67 Hz, (b) 68 Hz, 228 V
3. (a) 200 V, (b) 280 V
5. 3.57%
7. 36.1%

9. 3005.1 r/min
11. 410.4 r/min
13. 270.7 V
15. 248.5 A
17. 3563.4 r/min
19. (a) 466.6 A, (b) 1268.7 r/min
21. (a) 232.1 V, (b) 12314.5 W, (c) 24.8 lb-ft, (d) 22.5 lb-ft
23. (a) 8674.4 W, (b) 8082.1 W, (c) 17.27 lb-ft
25. (a) 1767.2 W, (b) 3208.8 W, (c) 93.8%
27. (a) 137.0 lb-ft, (b) 23655 W, (c) 144.8 lb-ft, (d) 1.24 Ω, (e) 253.4 lb-ft

Chapter 11

1. (a) 33.2 Ω, (b) 119.7 A, (c) 365.9 W
3. (a) 97.4 Ω, (b) 335.0 W
5. (a) 49.3%, (b) 969.4 r/min
7. (a) 1329.2 A, (b) 1318.8 A, (c) 323109 W, (d) yes
9. (a) 209.6 A, (b) 3.35 A, (c) 206.2 A, (d) 673.3 lb-ft, (e) 0.5347 Ω, (f) 6 turns, (g) 423.3 r/min
11. 700 r/min, 377 r/min
13. (a) 3121 r/min, (b) 5963 W, (c) 13.4 lb-ft
15. (a) 319 r/min, (b) 6.3%, (c) 224 lb-ft
17. 77.5 V
19. 0.442 Ω

Chapter 12

1. (a) 268 V, (b) 7.2%, (c) 7.3 Ω
3. 227.8 V
5. (a) 265 V, (b) 290 V, (c) 16.0%, (d) 73.1 Ω, (e) 315 V
7. (a) 500 A, (b) 5.49 A, (c) 505.49 A, (d) 270 V, (e) 257 V, (f) 2.8%, (g) under, (h) yes
9. (a) 0.001951 Ω, (b) 377.4 W
11. (a) 123.5 V, (b) I_A = 820 A, I_B = 980 A
13. (a) 562.1 V, (b) I_A = 483.6 A, I_B = 516.4 A
15. (a) 241.9 V, (b) I_A = 351.35 A, I_B = 756.76 A, I_C = 891.89 A

Index

Acyclic machine, 24n
Additive polarity, 93
Admittance, 542
Airgap
  fringing at, 9, 7
  induction motor, 137
Airgap power, 148
Alternator. *See* Synchronous generator
Amortisseur winding, 306
Arcing horn, 514
Armature
  DC machine, 389
  synchronous machine, 305
Armature coil, 354, 389
Armature reaction
  and compensating windings, 415
  and interpoles, 413
  in a DC generator, 413, 488
  in a DC motor, 413
  in a synchronous machine, 312
Askarels, 40
Autotransformers, 95
  for motor starting, 230

B-H curve, 6
Balanced three phase system, 539
Base impedance, 65
Base speed, 420
Base voltage and current, 65
Basic impulse level (BIL), 95
Bimetal element, 516
Blow-out coil, 513
BLV rule, 22
Braking
  dynamic, 461, 528
  mechanical, 462
  regenerative, 462
  resistive, 461
Branch voltage, 549
Brush contact drop, 418, 430
Brushes, 394
Buck-boost transformer, 101
Burden, instrument transformer, 125

Central processing unit, 536
Characteristic triangle
  DC generator, 496
  synchronous generator, 357
Chorded winding, 527
Circle diagram of induction motor rotor, 148, 149
Circulating currents in
  paralleled transformers, 107, 119
  transformer iron, 28
Code letter, 205, 206
Coercive force, 16
Cogeneration, 220
Coil face, 28
Coil pitch, 137
Coil side, 20
Coil window, 20, 25
Commutating poles, 410
Commutating zone, 408
Commutation, 408, 410
Commutation period, 408
Commutator, 394, 395
Compensating winding, 415
Compensator, 230
Complex numbers, 540
Complex power, 544
Conductance, 542
Conjugate phasor, 544
Consequent-pole motor, 141, 568
Constant-hp motor, 573
Constant-torque motor, 573
Control of electric motors, 513
Controllers. *See* Starters, Drives
Copperjacket timing relay, 518
Core loss
  DC machine, 428
  induction motor, 158
  synchronous machine, 377
  transformer, 71
Core material, transformers, 38
Counter-emf (cemf), 27
  in a DC motor, 403
  in a synchronous motor, 312, 313
  in a transformer, 41

Counter-force, 22
Counter-torque, 22, 221
Couple, 21
Crawling, 137, 156,
Current divider equation, 549
Current transformer (CT), 125
  burden, 125
Cycloconverter, 534

Damper winding, 306
DC generator. *See also* DC machines
  applications, 492
  characteristic triangle, 496
  compound, 490
    cumulative, long and short shunt, 490
    differential, 491
    flat, over, under, 491
  compounding effect of speed, 495
  critical resistance, 478
  demagnetizing mmf, 413, 487
  diverter resistor, 403
  equalizer connection, 504
  equivalent circuit, 416
  field flashing, 482
  field resistance line, 475, 480
  interpole saturation, effect of, 482, 483
  load-voltage characteristic, 485
  long shunt, 490
  magnetization curve, 480, 489
  motoring of, 504
  parallel operation of, 495
    theory of load transfer, 502
  polarity
    reversed, 482–485, 496, 505
    short-circuit, effect on, 482
  self-excited, 475
    basic design, 478
  separately excited, 486, 487
  short shunt, 452
  shunt generator, 416
  voltage breakdown, 485
  voltage buildup, 477

DC generator (*continued*)
voltage buildup, factors affecting
armature reaction, 487
field circuit resistance, 478
reversed field connections, 482
reversed residual magnetism, 482
reversed rotation, 482
speed, 479
voltage regulation, 400, 486
DC machines, general. *See also* DC motor, DC generator
armature inductance and commutation, 408
armature reaction, 412, 413
armature winding, 396
basic DC generator, 398
brush contact drop, 418, 430
brush position, 398
brush shifting, an emergency measure, 414
commutating zone, 408
commutation, 394, 408
commutation period, 408
commutator, 394, 395
compensating windings, 415
construction, 394
efficiency, 450
end-connections, 393
end-turns, 393
Faraday-Lenz relationship, 390
flashover, 415
flux in air gap, 389, 412–414
interpoles, 410
and armature reaction, 413
and armature inductance, 408
effect of incorrect connections, 411
effect of magnetic saturation, 411
lap winding, 396
leading edge of a pole, 412
losses and efficiency, 428
magnetization curve, 400, 489
neutral plane, 390, 395, 409
pole-face winding, 415
power flow diagram, 429
series and shunt field coils, 394, 444
sparking and arcing at brushes, 409
trailing edge of pole, 412
voltage regulation, 400
DC motor
applications, 459
basic DC motor, 403
base speed, 420
braking, 461
compound, 443, 459
cumulative, 443n
differential, 445, 465
counter-emf (cemf), 403
current
locked rotor, 431
rated (table), 579
dynamic behavior
during speed adjustment, 422–425
when loading and unloading, 406, 407
equivalent circuit, 405, 418
efficiency, 430
flux density in air gap, 403
field
break (open) in field circuit, 420
function of, 404
generator-to-motor transition and vice versa, 401
jogging, 464
magnetic saturation, effect of, 446
magnetization curve, 447, 449, 451, 454
manual starter, 432
mechanical power, 426
nameplates and NEMA, 427
overhauling load, 461
plugging, 461, 464, 528
regenerative braking, 462
reversing rotation, 403, 445
series, 446, 455, 459
shunt, 303, 458
speed adjustment
emergency overspeed, 424
precautions in, 424
through armature control, 422
through shunt-field control, 424
speed equation, 419
linear approximation of, 455
speed regulation, 406
stabilized shunt, 444
starting, 431
steady-state characteristics, 458
steady-state speed, 458
straight shunt, 398, 443
terminal markings, 465
torque, 403, 426
linear approximation of, 455
Delta connection, 549
Derating curve, insulation, 211
Design letter, induction motor, 169, 204, 216
Distributed winding, 567
Distribution factor, 568
Diverter, series field, 493
Double-subscript notation, 547
Drives. *See also* Starters
adjustable voltage, 459, 532
cycloconverter, 534
solid-state, 532
Ward-Leonard, 459
Droop rate, 351
Dynamic braking
DC motors, 461, 528, 530
induction motors, 227
synchronous motors, 331

Eddy current, 28
Eddy current loss, 28
in transformers, 71
Eddy voltages, 28
Efficiency
DC machines, 428
induction motors, 157, 159
nominal, 203
synchronous generators, 377
synchronous motors, 323
transformers, 71, 74
Electric/magnetic analogy, 12
Electrical degrees, 29
Electromagnetism, 1
Electron spins, 1
End-connections, 20, 393
End-turns, 20, 393
Energy conversion in rotating machines, 27
Equalizer connection, 504
Equivalent magnetic circuit, 12
Error angle, 347, 348
Exciter, 308
Exciting current
induction motor, 156, 179
transformer, 43, 112

Faraday's law, 21
Faraday-Lenz relationship, 390
Ferromagnetic materials, 5
B-H curves, 6, 8
Field flashing, 482
Flashover of a DC machine, 415
Fluorocarbon gas C2F6, 38
Flux
bunching, 17, 135

cutting, 22, 24
fringing at airgaps, 7, 9
Forces
on adjacent conductors, 17
short-circuit, 18
Fourier series expansion, 112
Fractional pitch winding, 137, 566
Frequencies, standard, 27
Friction losses, 157

Gap power, 148
Generator action, elementary, 21, 25
Governor characteristic, 350
Governor droop, 351
Graphic symbols for control diagrams, 577

Harmonic currents
in power lines, 112
in single-phase transformers, 110
in three-phase transformers, 121
Harmonic suppression, 123
Harmonics, space, 156, 570
Horsepower equation, 427
Homopolar machine, 24n
Hunting of synchronous motors, 306
Hysteresis, 14
Hysteresis loop, 15, 111
in magnetization curves, 475
Hysteresis loss, 15, 16
in induction motors, 157
in transformers, 71
Hysteresis motor, 282
unique features of, 286
Hysteresis torque, 284
Hysteresis-reluctance motor, 286

Ideal transformer, 49, 178
Impedance angle, induction motor, 191
Impedance diagram, general case, 505
Impedance matching transformer, 51
Impedance multiplier, 51
Impedance voltage, per unit, 64
Induction generator, 209, 219, 367
capacitance line, 225
counter-torque, 221
critical capacitance, 227
equivalent circuit, 224
emergency overspeed (table), 223
failure to build up voltage, 225n
isolated operation, 224
loss of residual magnetism, 225n
motor to generator transition, 220
power, torque, current, 222
pushover torque, 221
self-excited, 225
disadvantages of, 227
single-generator operation, 224
starting, 221
voltage buildup, 227
Induction motors, single phase
locked-rotor torque, 256
quadrature field theory, 254
NEMA standard ratings, 270
phase splitting, 256
locked rotor torque, 256
rated current (table), 581
reversing, 269
shaded pole, 269
split-phase, 256
capacitor-start, 262
permanent-split capacitor, 265
resistance-start, 256–261
two-value capacitor, 265
standard power ratings, 270
Induction motor, three phase, 133
acceleration of, 155
airgap, 137
airgap power, 148
applications, 170
behavior during loading and breakdown, 155
blocked rotor test, 214
braking
dynamic, 227
with DC injection, 228
with capacitors, 228
branch circuits, 238
breakdown, 150
bumps and dips in characteristic, 156
circle diagram of rotor, 148, 149
classification and performance characteristics, 168
code letter, 205, 206
consequent pole, 140, 568
constant hp, 573
constant torque, 573
construction, 136
core loss, 157
crawling of, 137, 156
DC test, 213
derating curve, 211
design letter, NEMA, 169, 204, 216, 181
upgrading problem, 176
distributed winding, 567
efficiency, 157, 159
guaranteed, 204
nominal, 203
emergency overspeed, 223
energy policy act (EPACT), 168n
equivalent circuit, 143, 178
approximate, 183
rotor, 143
exciting current
magnitude of, 158, 219
frame number, 204
friction and windage, 157
frequency
constraints (NEMA), 189
effect on locked-rotor current, 192
effect on locked-rotor torque, 192
effect on running torque, 190
off-rated, 189
60 Hz motor on 50 Hz, 194
volts/hertz ratio, 194
harmonic torques, 156
high inertia loads, 208
hysteresis loss, 157
impedance, input, 179
angle at locked rotor, 191, 237n
inrush current, 206
insulation class, 204
insulation life, 210
and number of starts, 208
and temperature, 210
and unbalanced line voltages, 209
locked rotor, 141
inrush current, 206
phase angle of, 237n
locus of rotor current, 146
losses, 157
in percent of total loss, 158
magnetizing reactance
determination of, 178
mechanical power, 150
multi-speed pole-changing, 141
constant horsepower, 573
constant torque, 573
variable torque, 574
nameplate data, 202
NEMA-design, 168, 170
no-load conditions, 156
no-load current in percent of rated, 219
no-load test, 216

Induction motor, three phase, (*continued*)
  normal running conditions
    squirrel cage, 186
    wound rotor, 137, 195–202
  open-wye motor, 271
  operation from a single-phase line, 270
  overload conditions, 186
  PAM winding, 569
  parameter determination, 213
  parasitic torques, 156
  per-unit parameters, 212
    determination of, 213
  plugging, 196, 464
  power factor, 159
  power, torque, and speed calculations, 179
  power-flow diagram, 158, 182
  reactive power, 148
  reclosing out-of-phase, 209
  reversal of rotation, 135
  rotating field, 135
  rotor impedance diagram, 144
  rotor leakage reactance, 145
  service factor, 204
  shaft-power out, 159
  shaping the torque-speed curve, 182
  single-phasing (fault), 272
  slip, 141
    at maximum torque, 183
    effect on rotor frequency, 142
    effect on rotor voltage, 143
  space harmonics, 156, 570
  speed
    constraints, 221, 223
    subsynchronous, 137, 156
    synchronous, 137
  squirrel cage rotor, 137
  starting, 229
    autotransformer, 230
    full voltage, 229
    part winding, 238
    reclosing out of phase, 209
    series impedance, 234
    solid state, 238
    wye-delta, 232
  stator windings, 561
  Steinmetz equivalent circuit, 151
  stray power losses, 158
  temperature rise, 205
  torque
    breakdown (maximum), 150, 155, 156, 183, 184
    breakdown, minimum (NEMA tables), 173, 174
    developed, 151
    harmonic, 156
    locked-rotor, 153
    locked-rotor, minimum (NEMA tables), 171, 172
    parasitic, 156
    pull-up, 156, 157,
    pull-up, minimum (NEMA tables), 175, 176
  torque-speed characteristic, 153
  turns ratio, stator/rotor, 178, 197
  upgrading problem, 176
  variable torque, 574
  voltage
    constraints (NEMA), 189
    off-rated, effect of, 189
    unbalanced, effect of, 209
  wound rotor, 137, 195
    applications, 198
    behavior during rheostat adjustment, 198
    normal running and overload conditions, 200
    rheostat, 137, 198
Infinite bus, 305
Input impedance
  of an induction motor, 179, 191
  of a transformer, 50, 55
Inrush current
  induction motor, 183
  transformer, 101
Instrument transformer, 125
  accuracy, 126
  phase angle error, 126
  polarity, 126
  polarity test, 126
Insulating liquid, 40
Insulation
  class, 204
  derating curve, 211
  life, 210
  relative life, 210
Interaction of magnetic fields, 17
Interpole, 410, 413
  saturation, effect of, 411, 482
Isochronous machine, 353
  paralleling with, 355

Jogging, 464

Ladder diagram, 519
Laminated cores, 28
Lap winding, 396
Leading and lagging edge of a pole, 413
Leakage flux of a transformer, 48
Leakage reactance
  induction motor, 145, 178
  transformer, 51
Lenz's law, 22
  and commutation, 408
  and copper-jacket time-delay relays, 527
  and DC machines, 390, 402
  and dynamic braking, 461
  and generators, 25, 401
  and induction motor action, 135
  and shaded pole motors, 269
  and shading coils, 515
  and single-phase motors, 254
  and transformer action, 41
Line voltage
  delta, 549
  wye, 547
Linear induction motor (LIM), 295
  applications of, 299
Load angle, 311
Locked rotor current, 192
Locked rotor torque, 153, 155
Logic circuits, 519

Magnet torque, 284, 311
Magnetic
  circuit, 2, 4
  domains, 1, 16
  drop, 4
  field, 1
    intensity, 3
    flux density, 4
  flux lines, 2
  hysteresis, 15
  materials, 5
  mmf and mmf gradient, 3
  permeability, 5
  potential difference, 4
  reluctance, 4
  saturation, 5
Magnetic/electric analogy, 12
Magnetization curve, 5, 8
  induction generator, 225
  knee, linear, and saturation regions, 5, 7
  self-excited DC generator, 475
  separately excited DC generator, 400
  synchronous generator, 370
Magnetomotive force, 3
Mechanical degrees, 29

Mechanical force on a conductor, 19
Moment arm, 20
Moment of inertia, 330
Montsinger, A. M., 210
Motor action, 18
Motor control
  arcing horn, 513
  blow-out coil, 513
  diagram
    connection, 520
    elementary, 521
    ladder, 519
    logic, 519
  magnetic contactors, 513
  manual starter, 432
  operating coil, 513
  overload protection, 515
  pole shader, 515
  relays, 515
  shading coil, 515
  undervoltage protection, 522
  undervoltage release, 519
Motor controller. *See* Starters
Motor types. *See* Consequent pole motor, Constant hp motor, Constant-torque motor, DC motor, Hysteresis motor, Induction motor, Linear-induction motor, Multi-speed motor, PAM motor, Reluctance motor, Shaded pole motor, Stepper motor, Synchronous motor, Universal motor, Variable-torque motor
Multi-polar machines, 29
Multi-speed induction motor, 141, 568, 573
Mutual flux, 41

Nameplate
  DC motors, 427
  induction motors, 202
  transformers, 94
National Electrical Code (NEC), 239
  branch circuit protection, 238
  full load motor current
    DC, 579
    single phase, 581
    three phase, 585
    two phase, 583
National Electrical Manufacturers Association (NEMA), 151, 167
NEMA standards
  for DC motors
    emergency overspeed, 424
    nameplates, 427
    terminal markings, 465
  for single phase motors, 270
  for three phase induction motors
    constraints on unbalanced voltage, 211
    constraints on voltage and frequency, 189
    designs, 168–177
    efficiency, 168n, 203
    emergency overspeed, 223
    insulation class, 204
    locked rotor kVA/hp, 205
    nameplate interpretation, 202
    torque, 168–177
  for synchronous motors, 330
  for transformers, 92, 94
Neutral connection, 547

Oersted, 6
One-line diagram, 345
Overhauling load, 461, 462
Overload protection of motors, 515

PAM motor, 569
Parallel circuit relationships, 542
Parallel operation
  of DC generators, 495
  of single phase transformers, 104
  of synchronous generators, 345
  of three phase transformers, 119
    effect of 30° phase shift, 119
Parasitic torques, 156
PCBs, 40
Per-unit
  efficiency, 74, 159
  impedance, transformers, 64
  impedance voltage, 64
  parameters, induction motor, 212, 213
  regulation, 67
Permeability, 5
  of free space, 5
  relative, 5
Phase sequence, 348, 551
Phase splitting circuit, 256
Phase voltage, 549
Phasor power, 544
Pigtail, 395
Plugging, 196, 464, 528
Polarity
  additive and subtractive, 93
  reversed, 482–485, 496, 505
  test, 126
Pole-amplitude modulation, 569
Pole face winding, 415
Pole phase group, 528
Pole pitch, induction motor, 137, 561
  linear induction motor, 298
Pole shader, 515
Pole slipping, 309, 329
Polychlorinated biphenyls, 40
Potential transformer, (PT), 125
Power angle, 311
Power factor, 546
  angle, 544
  improvement with capacitors, 558
  improvement with synchronous motors, 324
  induction motor, 159
Power
  active, reactive, apparent, 544
  complex, 544
  diagram, 504, 522
  phasor, 504
  single phase, 544
  three phase, 556
  triangle, 545, 557
Power-directional relay, 356
Power-flow diagram
  DC machine, 429
  induction motor, 158, 182
  synchronous generator, 378
  synchronous motor, 323
Prime mover, 25, 344
  characteristics of, 350
Pull-out power, 343
Pumped storage, 337
Pythagorean theorem, 45, 67

Quadratic formula, 195, 452
Quadrature field theory, 253

Reclosing out-of-phase, 209
Reflected impedance, 56
Relative permeability, 5
Relays, 515
  accelerating, 530
  full-field, 530
  overload, 515
  timing, 528
Reluctance, 4
  in parallel, 12
  in series, 12
  transformer core, 46

Reluctance motor, 279
Reluctance torque, 279, 281
Reluctance-synchronous motor, 142, 327
Residual magnetism, 15
in DC generators, 477
in induction generators, 225
in transformers, 110
Residual voltage, 209
Resonance
effect on transformers, 64, 123,
Reverse-current trip, 505
Rheostat, three phase, 137, 139
Right hand rule, 1, 25
Rotating flux, 135, 563
Rotor core, 29

Salient-pole generator, 343
Saturation curves. *See* Magnetization curve
Series circuit relationship, 540
Series field, 444
Series motor
AC, 299
DC, 299, 446
Service factor, 204
Shaded-pole motor, 269
Shading coil, 515
in a motor, 269
Short-circuit
forces, 18
ratio (SCR), 376
Shunt generator, 363
Siemens, 542
Single phase circuit
parallel, 542
power, 544
power factor, 546
series, 540
Single-phasing (a fault condition), 272
Sinusoidal emfs, 25
Skewed rotor slots, 137
Skewing angle, 19
Skin effect, 73, 215
Slip, 141
negative, 221
Slip rings, 172
Slip speed, 141
Space degrees, 31
Space harmonics, 156, 570
Speed regulation, governor, 350
Speed voltage, 22
Spider, 307
Split-phase motor, 256
Squirrel cage rotor, 133, 138
NEMA design, 168
Stability limit, 321
Starters. *See also* Drives
AC, reduced voltage, 524
AC, reversing, 523
AC, two-speed, 523
diagrams
connection, 520
elementary, 521
components, 513–515
DC, cemf, 528
DC, definite time, 526
DC, flux decay time delay, 526
DC, manual, 432
DC, reversing, 528
overload protection, 515
programmable, 535
Static torque, 153
Stator, 18
Steinmetz equivalent circuit, 151
Steinmetz exponent, 16n
Stepper motors, 286
drive circuits, 292
half-step operation, 289
holding torque, 290
microstepping, 289
permanent-magnet stepper, 291, 293
resolution, 287
static-torque curve, 292
step accuracy, 291
step angle, 287
stepping frequency, 287
variable reluctance stepper, 287
Stray-power losses, 158
Subsynchronous speed, 137, 156
Subtractive and additive polarity, 93
Sulfurhexafloride gas SF6, 38
Superconducting generators, 378
Susceptance, 542
Symbols for controller diagrams, 578
Synchronizing AC generators, 346
Synchronizing lamps, 348
Synchronous condenser, 305, 324
Synchronous generator,
armature reaction, 312
construction, 337
cooling, 378
counter torque, 344
determination of parameters, 373
load, power factor and the prime mover, 344
losses and efficiency, 377
magnetization curve, 370
motor-to-generator transition, 338, 344
motoring, 356
parallel operation, 345
characteristic triangle, 357
division of active power, 351
division of reactive power, 363
error angle when paralleling, 347, 348
governor characteristics, 350
droop, 316
speed regulation, 350
isochronous, 353, 355
loss of field excitation, 367
procedure for paralleling, 346
procedure for safe shut-down, 356
synchronizing, 346
per-unit parameters, 367
phasor diagram, 339, 340
power
angle, 339
equation, 342
power-flow diagram, 378
pull-out power, 343
salient pole, 343
short circuit ratio (SCR), 376
superconducting, 378
voltage regulation, 368
Synchronous motor, 305
airgap flux, 312
amortisseur winding, 306
armature reaction, 312
flux, 312
reactance, 315
voltage, 314
braking, 331
brushless excitation, 309, 310
construction, 305
counter-emf, 312, 313
cylindrical rotor, 306
damper winding, 306
efficiency, 323
equivalent circuit, 315
excitation
effect on current, pf and power angle, 320
normal, under, and over, 321
excitation voltage, 313
from graph, 322
exciter, 307
field winding, 306

hunting, 306
loading, effect on current, pf and load angle, 311, 318
losses, 323
phasor diagram, 315
pole face winding, 306
pole slipping, 309, 329
power
angle, 311
magnet, 311, 328
reluctance, 311, 327
power equation, 316, 327
power factor improvement with, 324
power-flow diagram, 323
reactance, direct and quadrature, 327
reversing, 311
round rotor, 306
salient pole, 307, 326
normal operation, 327
speed control, 330
speed voltage, 313
stability limit, 321
starting, 309
synchronism, loss of, 311, 319
synchronizing, 309
lamps, 348
synchronous impedance, 316
synchronous reactance, 316
torque
angle, 311
blocked rotor, 330
magnet, 311, 328
pull-in and moment of inertia, 329
pull-out, 317, 319, 328
reluctance, 311, 327
V-curves, 321
Synchronous speed, 135, 137
Synchroscope, 346

Temperature rise, 205
Ten-degree rule, 210
Tertiary coil, 123
Three-phase system
active, reactive, and apparent power, 544
balanced wye and delta loads, 550
calculating line and phase currents in three-phase loads, 553
delta connection, 549
phase sequence, 551
power factor correction, 558, 560
power triangle, 545, 557, 559
universal phasor diagram, 554
wye connection, 547
Thyristor, 531
Time degrees, 31
Torque
angle, 311
blocked rotor, 153, 330
breakdown (maximum), 155, 183, 184
developed, 20, 150
eddy current, 306n
hysteresis, 284, 306n
locked rotor, 153, 330
magnet, 311, 328
reluctance, 279, 311
pull-in and moment of inertia, 329
pull-out, 317, 319, 328
pull-up, 157, 175, 176
pushover, 221
Trailing and leading edge of a pole, 412
Transformer
action, 21, 40
auto, 95
bank, circulating current in, 104, 119
single phase, 104
three phase, 119, 113
BIL, basic impulse level, 95
buck-boost, 101
available ratios, 103
connections, three phase
delta-delta, 115
open-delta (V-V), 115
wye-wye-delta, 123
construction, 37
core material, 37
core type, 37, 118
shell type, 37, 118
efficiency, 71, 74
from per-unit values, 74
equivalent circuit models, 43, 51, 52, 54, 55, 56, 57, 58, 59
equivalent core-loss, resistance, 43
equivalent impedance, 55
equivalent magnetizing reactance, 44
exciting current, 43
core-loss component of, 43
harmonics in, 110
magnetizing component of, 44
phase angle of, 44
fault current, 49, 66
gas-filled dry type, 38
harmonics, 110, 121
triplen, 123
high-side low-side, 57
ideal, 49, 178
impedance matching, 51
impedance
input, 50, 60
multiplier, 51
per-unit, 64
phase angle, values of, 66n
reflected, 56
table of, 587
inrush current, 109
instrument, 125
leading pf, effect on, 48
leakage flux, effect on, 48
leakage reactance, 51
leakage reactance drop, 54
liquid immersed, 39
loading and unloading, 46
losses and efficiency, 71
magnetizing
ampere-turns, 46
current, 43, 111
reactance, 44, 55
mutual flux, 41
nameplate, 94
nameplate ratio, 50
no-load conditions, 43, 46
mmf and components, 45
parameter determination, 75
parallel operation of
single phase, 104
three phase, 119
per-unit and percent impedance of, 64
phase angle of exciting current, 43, 44
polarity, 92
additive and subtractive, 93
instrument, 126
power factor, 44
principles, 47
reflected (referred) impedance, 56
rerating for open delta connections, 117
resonance, 64, 123
shell type, 38, 118
sinusoidal voltage, 41
specialty, 91
terminal markings, 92
tertiary winding, 123

Transformer, (*continued*)
  three-phase connections, 113
    harmonics in, 121
    paralleling, 119
    phase shift in, 119
    transient behavior, 46
  turns-ratio, 42, 49
  ventilated dry type, 38
  voltage ratings, 94
  voltage-ratio, 49
  wye-wye-delta, 123
  voltage regulation, 62
    at other than rated load, 70
    from per unit values, 67
Triac, 534
Triplen harmonics, 123
Turns ratio, 42, 49

Unbalanced voltages, 129
Unipolar machine, 24n
Unit conversion factors, 589
Universal motor, 299
Utilization voltage, 101n

V-curves, 321
Variable-torque motor, 574
Varistor, 309
Velocity
  angular, 27
  rotational, 27
Voltage divider, 540
Voltage regulation
  AC generator, 368
  DC generator, 400
  transformers, 62, 67
Voltage unbalanced, 129

Ward-Leonard system, 459
Windage losses, 157
Windings
  distributed, 567
  four-pole, 563
  two-pole, 561
  pitch, 137
    fractional, 137, 566
    full, 137, 565
Window of a coil, 20, 25
Wk2, 330
Wound-rotor motor. *See* Induction motor, three-phase
Wye connection, 547

Yoke, 396

Zero-sequence currents, 113